Odors and Air Pollutants Conference 2018

Portland, Oregon, USA
25 - 28 March 2018

ISBN: 978-1-5108-6743-7

Printed from e-media with permission by:

Curran Associates, Inc.
57 Morehouse Lane
Red Hook, NY 12571

Some format issues inherent in the e-media version may also appear in this print version.

Printed by Curran Associates, Inc. (2018)

Water Environment Federation (WEF)
601 Wythe Street
Alexandria, Virginia 22314
USA

Phone: 1-800-666-0206
Fax: 1-703-684-1545

csc@wef.org

Additional copies of this publication are available from:

Curran Associates, Inc.
57 Morehouse Lane
Red Hook, NY 12571 USA
Phone: 845-758-0400
Fax: 845-758-2633
Email: curran@proceedings.com
Web: www.proceedings.com

TABLE OF CONTENTS

SESSION 5

SESSION 6

SESSION 7

SESSION 8

SESSION 9

SESSION 10

SESSION 11

SESSION 12

SESSION 13

NEW CONCEPT AND NEW RESULTS FOR INSTRUMENTAL MEASUREMENT OF ODOROUS COMPOUNDS

Jean-Michel GUILLOT and Sofia LAFHAL-SAKIOU

IMT Mines Alès, University of Montpellier, Environmental Engineering Laboratory, 6 avenue de Clavieres, 30319 Alès cedex, FRANCE

jean-michel.guillot@mines-ales.fr ; sofia.sakiou@mines-ales.fr

Key words- E-nose, Odor, SPR, Wellmess, Detection, Monitoring

Abstract

Each year, new technologies or approaches are proposed to measure air contaminants such as odorous compounds. From lost cost sensors to high techniques like PTR-MS, it exist a market for different applications. As an example of such new technologies, this study deals with results obtained with an innovative type of instrument to measure odorous volatile compounds. The device developed by a French company (Aryballe Technologies, Grenoble – France), in based on a combination of Nanotechnology, Biotechnology, Data processing… The core technology is based on Surface Plasmon Resonance (SPR) imaging. Such a technology is commonly used for liquid samples but is quite new for gas samples and gaseous compound detection. The device has been developed to propose an alternative to technologies generally used for sensor arrays and e-noses.
Our laboratory is a participant of a large project called WISE (**W**ellness & medical d**I**agnostics olfactory **SE**nsors). The project is to continue the development of the device and applications.

Different potential applications are actually listed:

- For people suffering from anosmia (loss of smelling), the device is able to recognize more than 50 different day to day odors;
- For odor panels that can refer to value of the device as a reference to protect original sensorial recipes;
- For pollution monitoring and also disagreement assessment caused by malodorous compounds;
- For gustatory and safety quality of high added value food with intelligent kitchen equipment including cooking assistance;
- For some disease detection in the case this disease can be characterized by a low specific odor and when trained dogs are generally used for this kind of diagnosis;
- …

This non-exhaustive list shows how odor measurement can be useful in different domains and sometimes can be essential as for safety applications (gas leak detection with continuous monitoring).
The project, with this new approach, has the objective to propose monitoring for situations that cannot be considered with actual automatic analysis. With actual development, better applications could be in the field of flavors and fragrances, home automation… The device needs more development for environmental analysis.

1. Introduction

During two decades, developments about sensors for odorous measurements have been carried out. Different technologies were used such as metal oxide sensors, polymer sensors, surface wave sensors… Theses sensors were the basis of electronic nose development (called e-noses) with the use of sensor array, data treatment to analyze all signals and training the system to link a combination of signals (fingerprint) to an odor. All these techniques present advantages and drawbacks and globally, efficiencies are typically linked to the application. Actually, if the reference is human nose, the different techniques can present lack of sensitivity, specificity, repeatability…and so all equipment can be considered as odorant and non-odorant monitoring devices but not typically as "noses". Actually, a lot of new technologies appear and in the next future, we can imagine a tendency to converge in direction of human nose efficiency (including data treatment as brain combines signals from sensorial receptors). New technologies can be new micro-sensors, biological sensors or typically new

approaches. Because it's difficult to imagine which technology will give the better result (and for which duration), this domain of measurement is a real challenge for the future. The system must be applicable for large situations independently to variations of humidity, temperature that interfere strongly on the signal. It must be applicable to a large range of concentration and for lots of applications.
Because the need of environmental monitoring is real and especially odor monitoring, the European committee for standardization started to work on "instrumental Odor monitoring".
Independently to the system response at the presence of odorant or non-odorant compounds, and without fixing a specific mathematical treatment of collected signals, the standard has the objective to be applied indoor air, ambient air, odorous gas emissions and odor removal efficiency of end-of-pipe. The output of devices should be an odor metric linked to the presence and attributes of odor including a method of validating the sensor signal based metrics to the presence and attributes of odor. As every standard, some aspects must be considered because a device cannot cover all the properties of odors. That's why typical human feelings are excluded such as hedonic tone or assessment of annoyance. And of course because odor concentration (expressed in ou_E/m^3) is measured according to EN 13725, the output of a device cannot be expressed as a result obtained with the human nose.

Assembly of sensors for odorous volatile compounds is the base of electronic noses and in most cases, sensors are not very specific.
The specific approach is typically the reality in biology and bio-analytical field with lock/key-recognition-model. Surface Plasmon Resonance Imaging (SPRi) of biological interactions is commonly used in liquid phase and the development of such a technique in gas phase opens applications on volatile organic compounds (VOCs) and odorous compounds. The interactions of odorous compounds with receptors (SPR) are converted into an image that gives a visual representation of an odorous atmosphere. The device was designed as hand-portable analyzer called Neose (with an interface as on a smartphone). Because this device is quite new on the market, all actual developments must help to select application fields.

2. Description of the device

The e-noses are devices designed to monitor odors. Odors are a biological perception, while the instrumental odor monitoring systems provide data only of odorants. In a qualitative point of view, strong algorithms for data treatment can perform recognition/distinction of odors. This aspect needs system training to create a database of different odors. Actually, human brain is a very powerful treatment process to analyze signals sent by nasal receptors. One limitation of e-noses is the number of sensors comparatively to human olfactory epithelium, but SPRi define a new way to increase number of sensors.
As shown on Figure 1, the key element of the gas-imaging device is a SPR prism. This prism is chemically functionalized with a specific set of several tenths of biosensors which can react with a wide range of molecules and then can allow the identification of odorous mixtures. The measurement device is completed by microfluidic system, image acquisition and processing software. The concept is illustrated by Figure 1 and Figure 2 illustrates the global device with all key equipment.

Figure 1: From left to right - Binding of VOCs on the chemical ligands arrayed on prism covered by a gold layer; interrogating the surface by SPRi; image analysis and data transfer.

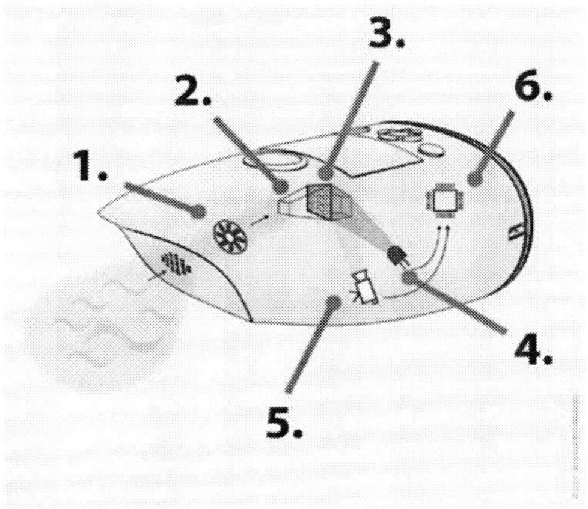

Figure 2: Global scheme of measurement device: 1- Pumping of air and odorous molecules; 2- Transport of compounds to the chamber; 3- Interaction of compounds with chemical sensors arrayed on a prism; 4- Lighting the prism with LED lamp; 5- Recording optical signal transduction with a camera; 6- Storage (database) and analysis of results.

When volatile organic compounds bind on the sensors, dynamic images are measured. This record corresponds to association/dissociation phases of the interaction/recognition process between compounds and sensors. With training of known odors (real mixtures of compounds), new measurements and new images are compared with stored images in the database. If the mixture is known, identification and quantitation can then be carried out. Results are given as a normalized radar chart allowing for a straightforward recognition process or can be used in another format. Of course, the sample is not modified by a measurement cycle because the interaction is not a chemical reaction and the measurement is carried out at ambient temperature.

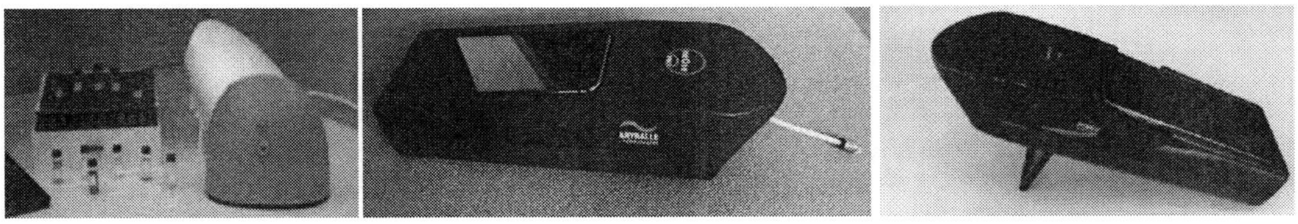

I	II	III
2016	2017	2018

Figure 3: Three generations of Neose instruments - On the left: the device (until end 2016); middle and on the right, recent devices (2017 and 2018) that includes a smartphone as interface.

This device based on SPR imaging detects a wide panel of compounds with threshold detection around ppmv level. If currently, around 50 sensors are present on the prism, the number can be extended to 100 allowing a precision improvement of the measurement. It's typically a great increase of sensor numbers comparatively to e-noses used during last decades. It's by such an increase that precision can be improved because human receptors are around 400 as actives. The device is simple to use as illustrated in Figure 3. Device is hand portable with simple interface to use.

As all sensor arrays, plots with the response of all individual sensors are obtained, and these data can also illustrated as a radar plot (very common output as a fingerprint of an odor detected by an e-nose). In the case of Neose, odors can be also registered as a bar-code to create a library of known odors. This type of registration is shown on Figure 4. A bar code is a way to transcript a registered odor and then a way store data of registered odor.

Figure 4: Possible output: Bar code as a transformation of odor to visual data.

3. Results

In the project, each partner is in charge of development of a part of the equipment (prism, data treatment, test of efficiency…). At IMT Mines Alès, experiments are carried out on odorous compounds. Efficiencies are tested with compounds place in flasks and vapor collection on headspace is carried out by the internal pump of the monitoring device. Because the system is based on biosensors, it cannot be used actually in hard conditions. A blank with air (and humidity) is registered previously to analysis and the sample must be collected with constant flow to warranty the same interaction equilibrium between compounds and biosensors.

Results as radar plot illustrates that this new measurement device gives a more informative fingerprint than classical sensor array (with 4, 8, 12 sensors). This strongest information offers a warranty to distinguish odors. Some results with two aldehydes and an alcohol are show on Figure 5.

Octanal n-Butanol 3- methylbutyraldéhyde

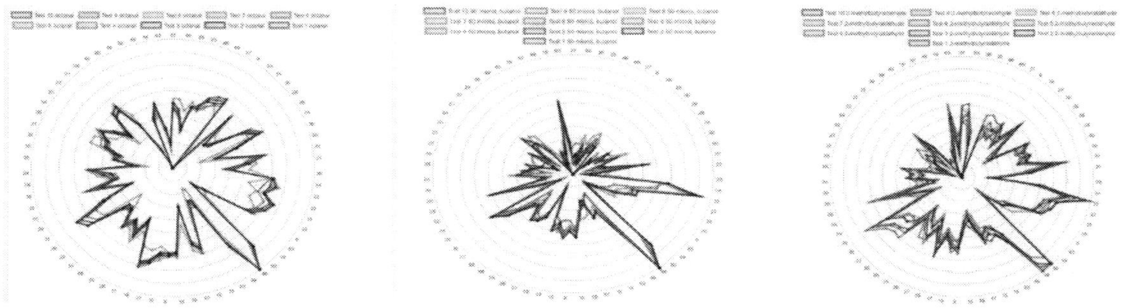

Figure 5: Example of results as radar plots (repetitions of 10 analyses for each compound)- results obtained with an equipment of second generation

As shown with Figure 5, the number of biosensors helps the recognition because each point of the radar plot corresponds to the response of a biosensor on the functionalized prism of the SPRi cell. We can see that the two aldehydes present a different fingerprint on the radar plot even if both compounds have the same chemical function (aldehyde R-CHO) and only differ on the carbon chain number. But in order to verify the capacity to discriminate compounds, a mathematical treatment by Principal Component Analysis (PCA) was tested. Four compounds were analyzed: n-butanol, octane, 1-octen-3-ol and 3,7-dimethyl-3-octanol. Each compound was analyzed 10 times and with PCA as shown in Figure 6, It is clear that these compounds are clearly separated by this kind of analysis. It shows that different alcohols are distinguished and that compounds with C-8 to C-10 are also separated because chemical functions (unsaturation or substitution) are sufficient enough to produce a different signal. Such results confirm that the number of biosensors is a key point to go on the way of recognition and differentiation efficiencies.

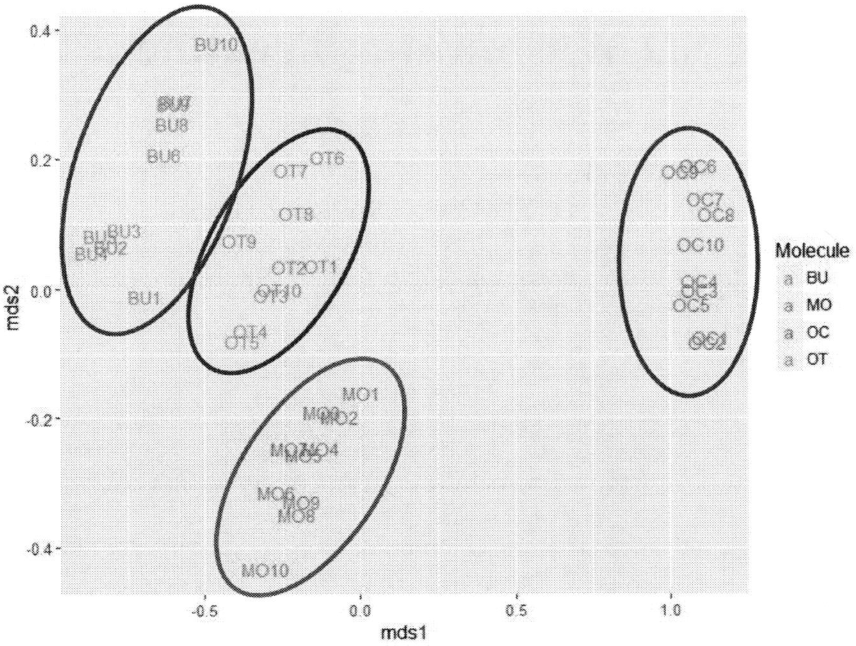

Figure 6: Example of PCA results (repetitions of 10 analyses for each compound)- results obtained with an equipment of second generation

4. Conclusion

This new measurement device offers a new technology for odorous atmosphere analysis and for new applications. As all instrumental device based on sensors, it's a complement of other techniques. Identification of odorous compounds is always obtained with GC/MS or CG/MS-O and odor concentration measurement by olfactometry. The objective of sensors is multiple but the first advantage is to propose continuous measurement. Other advantages are known odors identification (comparison of mixture fingerprints); detection of abnormal situations (in everyday life at home or for an industrial process)…

With the Neose device, a new measurement approach is to change a complex chemical mixture (characteristic of an odor) into an image (by SPRi) that can be analyzed by a classical image recognition process. So this device can be considered as a tool to visualize odors and then to represent one sensation (olfaction) by another one (view).

This example demonstrates that new technologies appear and that actual measurement devices could be replaced by new equipment based on new technologies.

5. Acknowledgement

The authors thank all national and regional supports that financially support the WISE Project and so developments of this new Technology. This project is based on a wide consortium with 8 French partners: Aryballe Technologies (leader), CEA-INAC, GIPSA-Lab, GIP Mind, ARMINES / Centre LGEI of IMT Mines Alès, Prestodiag, Cemag Consulting, SEB.

How Reduced Sulfur Compounds in Trunkline Foul Air Vary Over Diurnal Fluctuations at OCSD Wastewater Treatment Plant

Tadeo G. Vitko

Orange County Sanitation District
10844 Ellis Avenue
Fountain Valley, CA. 92708, USA

Abstract

Many odorants are generated in the treatment process of domestic wastewaters and all can contribute to public nuisance. The most common group of odorants are those called reduced sulfur compounds. Among the reduced sulfur compounds, hydrogen sulfide (H_2S) is the odorant that can most easily be determined and continuously monitored. As a result, H_2S is used as an indicator of odors. However, not all reduced sulfur compounds behave in the same manner when analyzed over the diurnal flow and odor fluctuation. This paper shows that when odors are analyzed at the head of the treatment plant, from the foul air taken from the incoming trunklines, when H_2S increases 2.2 times, methyl mercaptan only increases by 1.5, dimethyl disulfide increases by 1.1, and dimethyl sulfide decreases by 0.8 times. This finding will affect results when daily and seasonal corrective factors are used based on H_2S trends.

Introduction

Wastewater Treatment Plants (WWTPs) invariably produce odors (Dague, 1972; Van Langenhove et al., 1985; Hwang et al., 1995). The chemical compounds that cause odors are called odorants. Odorants found in sewer foul air can be classified as those emitted by direct discharge, such as solvents or detergents, and those generated by biochemical processes, such as the reduced sulfur compounds (especially hydrogen sulfide, methyl mercaptan, dimethyl sulfide, and dimethyl disulfide). Odors generated by biochemical processes can be thought as being proportional to the amount of biodegradable organic matter in the sewage entering the treatment plant. Higher temperatures increase biological activity and decreases oxygen solubility in water, both of which increase odor production in sewers (Metcalf & Eddy, 1991). Diurnal flows, the trend between human activity and collection system flows over time, change throughout the day. The arrival time of peak daily flows at WWTPs depend on factors, such as population, length and slope of trunklines, type of industrial or commercial activity in the service area, etc.

Incoming flows to WWTPs normally have two daily peaks; one around noon and another before midnight (Metcalf & Eddy, 1991). Figure 1 shows the variation of BOD, suspended solids, and

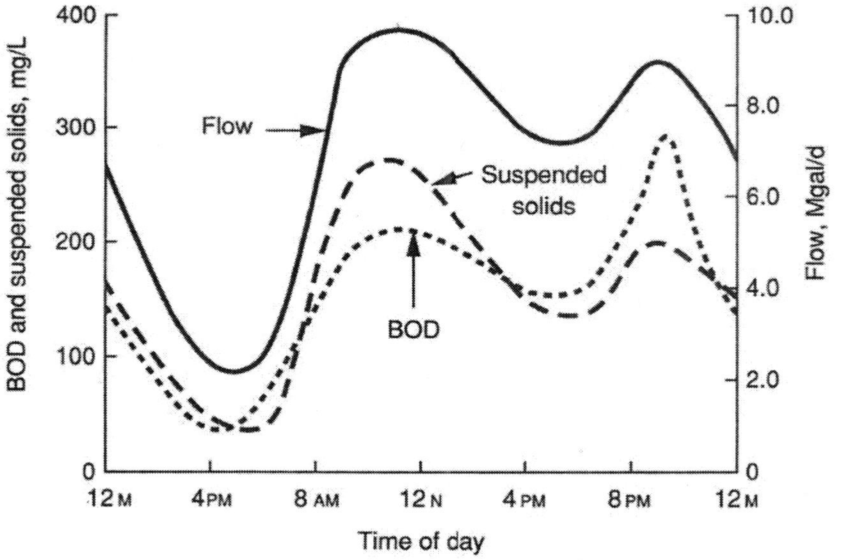

Figure 1 – Typical Variation of Biochemical Oxygen Demand (BOD), Total Suspended Solids, and Flow (Metcalf & Eddy, 1991)

flow at a typical WWTP. Thus, the mass loading of organic matter into WWTPs vary with time of day. As the flow and mass of organic matter increases, so does the odors. Odors also depict a two-daily peak pattern. However, there is normally a lag between the peak daily flows and the peak odors, as there is an inherent retention time of the foul air in the collection system. This retention time is unique for each WWTP. Because of this, odor emissions at WWTPs are constant but not uniform, especially at the head of the treatment process. Odors and corrosion at the trunk lines are the by-product of anaerobic bacteria breaking down organic matter (sulfate and nutrients) in sewage, as shown in Figure 2. As the sewerage enters the treatment plant, odors can become a nuisance to the public living near WWTPs if the odorous plant processes are not covered, if the foul air is not collected, and if the collected foul air is not adequately treated.

Figure 2 – Mechanism for the Generation of Odors in the Collection System

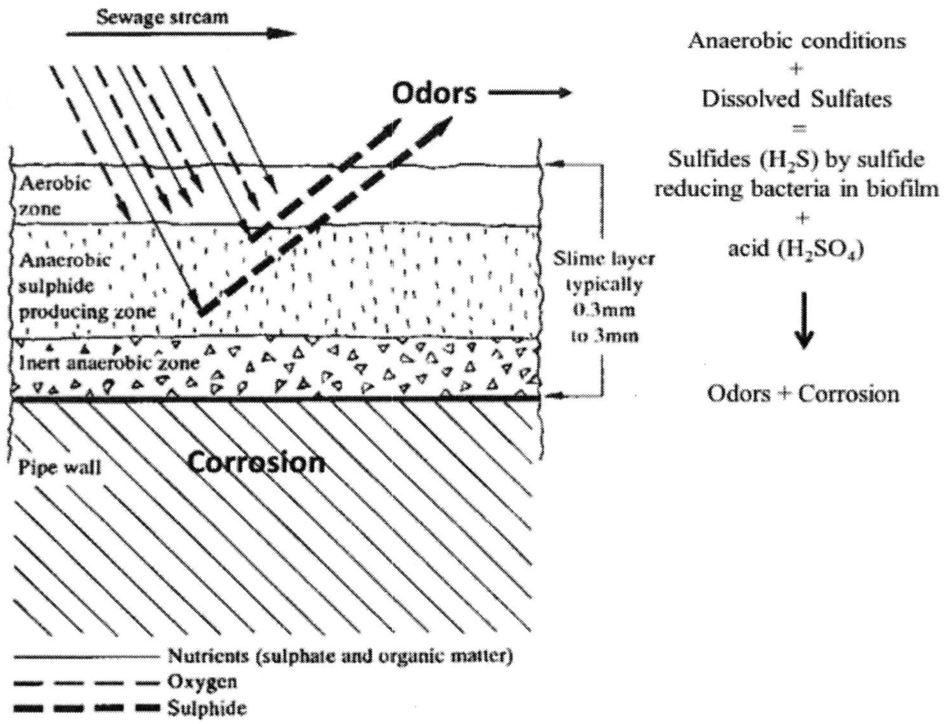

When designing odor abatement systems for WWTPs it is important to consider the existing worse-case scenarios. This not only includes the use of worse-case odorant emission concentrations at a particular location, but worse-case dispersion situation for that emission at or beyond the WWTPs' fence line. The gravity of the resulting odor nuisance, and consequent public complaint, depends on how long the odors linger beyond the fence lines.

When attempting to characterize odors, there are sample holding time limitations that hinder sampling the foul air at exactly the peak odor conditions. In those situations, one relies on hydrogen sulfide continuous monitors to provide information about specific daily and seasonal fluctuations, so that the sampled data points can then be adjusted by correction factors. In the past, studies that have attempted to apply correction factors, have done so by assuming all reduced sulfur compounds behave like hydrogen sulfide and assigned the same corrective factor to all reduced sulfur odorants (OCSD, 2017). However, laboratory tests performed to Trunkline foul air during a portion of the diurnal fluctuation show that not all reduced sulfur compounds concentrations increase at the same rate as hydrogen sulfide. Therefore, each reduced sulfur odorant has a particular correction factor.

Methods

Foul air samples were obtained from Orange County Sanitation District (OCSD) Plant 2 (Huntington Beach) Trunkline prior to odor treatment. Samples were collected in Tedlar bags at locations of positive pressure (after a fan) in order to avoid possible dilutions with atmospheric air. Tedlar bags were rinsed twice with the targeted foul air before collecting the sample. Because the bags were collected over a period of several hours, they were temporarily stored in a cool place and protected from direct sunlight. Once all samples were collected, they were taken to the OCSD lab and analyzed by EPA Method 637.

Results and Discussion

Raw sewage typically arrives at WWTPs through several trunklines. Each trunkline has an air space between the surface of the water and the top of the pipe. This air space is odorous because, under the anaerobic conditions created on the submerged portion of the pipe surface, the sulfide reducing bacteria break down the dissolved sulfates in sewage producing hydrogen sulfide (H_2S) and other odorants. To minimize odors at the head of the plant, WWTPs typically vacuum this foul air and conveys them to trunkline scrubbers. The samples obtained for this study were taken in the trunkline conveyance system before treatment.

Table 1 and 2 show the results of two separate sampling runs. It compares the field readings conducted with a Jerome[1] with the results of the laboratory chemical analyses. For those laboratory results reported as non-detect (ND), the tables show in green a concentration based on the method reporting limit (MRL) for that compound. In both sampling runs, the results show a clear ramp-up for H_2S with time. Table 1 shows a gradual increase throughout the sample period because it was done late morning through early afternoon. Table 2 shows the early morning dip followed by the late morning through early afternoon ramp-up because the sampling started early in the day.

The factors were calculated as the ratio between the largest and the lowest concentration for each of the reduced sulfur constituents during the test. Comparing the factors between the two tests show that this variation is relatively constant when comparing the two separate tests done two months apart. While the H_2S field tests (Jerome) experience a change by a factor of 1.5-1.6 over the portion of daily cycle, the laboratory H_2S changes by a factor of 2.1-2.2. However, methyl

[1] The portable Jerome® 631-X Hydrogen Sulfide Analyzer with a range of 0.003-50 ppm H_2S

mercaptan only changes by a factor of 1.3-1.5 over that same daily cycle. Dimethyl sulfide, in turn, only changes (decreases) by a factor of 0.6-0.8 and Dimethyl disulfide does not change. Figure 1 and 2 show these results graphically.

Table 1 – Reduced Sulfur Results for OCSD Trunklines in Late Morning and Early Afternoon

Parameters	Sample	Sample	Sample	Sample	Factor
Date	7/28/2016	7/28/2016	7/28/2016	7/28/2016	
Time	11:00 AM	12:00 PM	1:00 PM	2:00 PM	
Jerome H_2S	9	11	12	14	1.5
Hydrogen Sulfide	6.7	10.2	11.5	13.9	2.1
Methyl Mercaptan	0.511	0.555	0.504	0.678	1.3
Dimethyl Sulfide	0.057	0.037	0.037	0.035	0.6
Dimethyl Disulfide	0.03	0.03	0.03	0.03	1.0

Table 2 – Reduced Sulfur Results for OCSD Trunklines in Early Morning and Early Afternoon

Parameters	Sample	Sample	Sample	Sample	Sample	Sample	Factor
Date	9/28/2016	9/28/2016	9/28/2016	9/28/2016	9/28/2016	9/28/2016	
Time	7:30 AM	9:30 AM	10:30 AM	11:30 AM	12:30 PM	1:30 PM	
Jerome H_2S	12	8.6	10.3	11	12	14	1.6
Hydrogen Sulfide	10.4	6.24	8.48	12.3	13.6	13.9	2.2
Methyl Mercaptan	0.585	0.567	0.641	0.703	0.783	0.829	1.5
Dimethyl Sulfide	0.027	0.057	0.054	0.049	0.050	0.048	0.8
Dimethyl Disulfide	0.09	0.09	0.10	0.08	0.09	0.10	1.1

Conclusions

Incoming sewage is an important source of odors at WWTPs. This source of odors is continuous, but becomes more intense as the organic mass loading increases and the air space in the pipes decreases with higher incoming flows. This variation is seen daily as two peaks, but is also apparent seasonally, peaking during the hot summer days. In an attempt to correct for these variations, odor results can be adjusted with corrective factors obtained from continuous H_2S monitors. Since most of the odors are reduced sulfur compounds, there is a tendency to apply these corrective factors indiscriminately to all odorants. This paper shows that doing so would cause an over estimation of worse-case concentrations for methyl mercaptan by 32%, dimethyl sulfide by 64%, and dimethyl disulfide by 50%.

Future research should collect additional data (field and laboratory results) over an extended period of time to confirm the observations presented here. Also, it would be beneficial to test in the same manner the foul air from other plant process areas to determine if similar daily corrective factors are obtained.

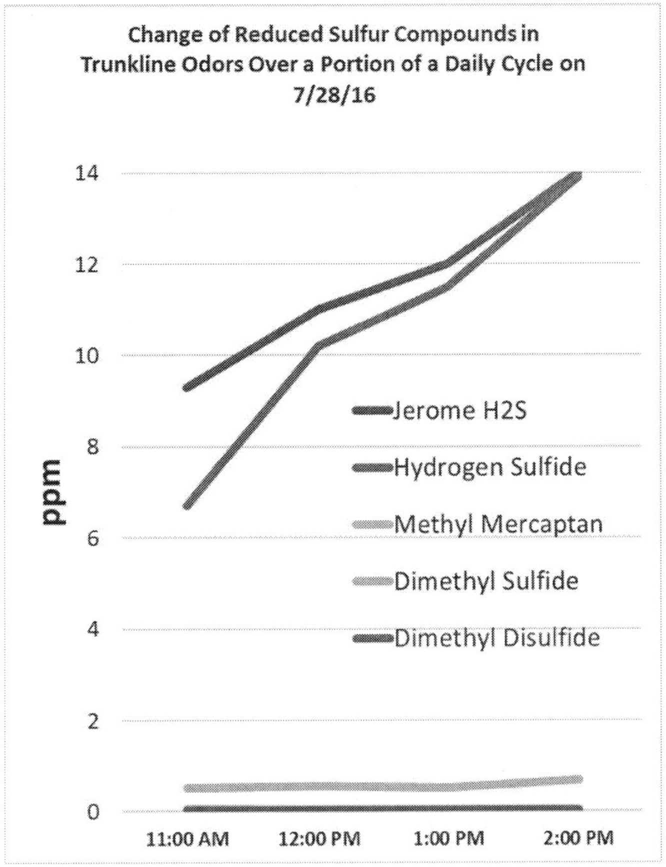

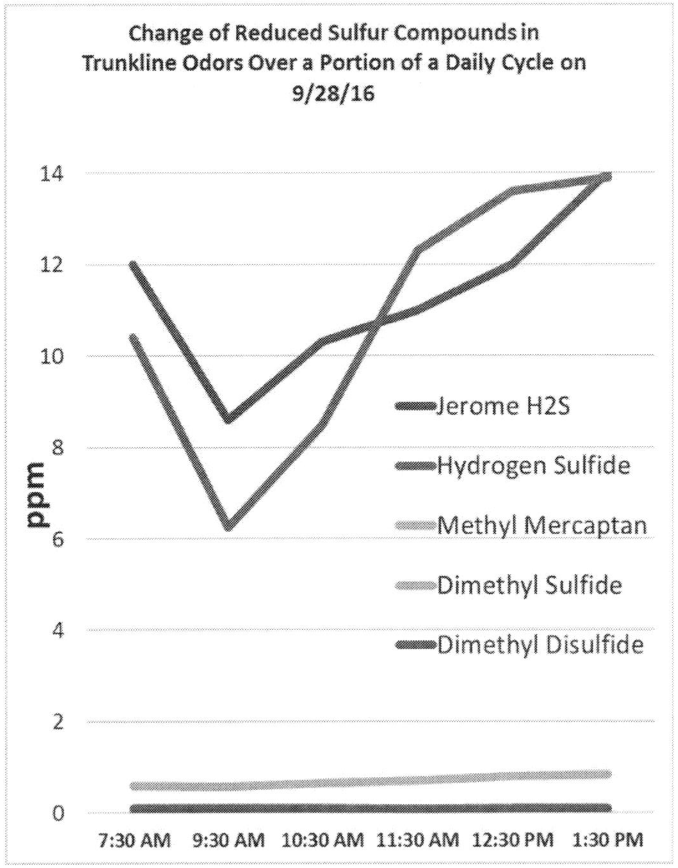

Figure 1 and 2 – Graphical Results of Field and Reduced Sulfur Laboratory Results for OCSD Trunklines

References

Yongwoo Hwang, Tomonori Matsuo, and Keisuke Hanaki. Identification and quantification of sulfur and nitrogen containing odorous compounds in wastewater. Water Research. Volume 29, Issue 2, February 1995, Pages 711-718.

H.Van Langenhove, K.Roelstraete, N.Schamp, and J.Houtmeyers. GC-MS identification of odorous volatiles in wastewater. Water Research. Volume 19, Issue 5, 1985, Pages 597-603

Richard R. Dague. Fundamentals of Odor Control. Journal (Water Pollution Control Federation) Vol. 44, No. 4 (Apr., 1972), pp. 583-594. Published by: Water Environment Federation

Metcalf & Eddy. Wastewater Engineering: Treatment, Disposal, Reuse. Third Edition, 1991. Ed. Tchobanoglous, G. and Burton, F. L.

OCSD, 2017. Odor Control Master Plan. Orange County Sanitation District, Fountain Valley, CA. CH2M-Hill Corp.

Odor Threshold Determination – A review of the effect of number of assessors

Michael McGinley, P.E., Charles McGinley, P.E., Kirsten Weiss
St. Croix Sensory, Inc.
1150 Stillwater Blvd N
Stillwater, MN 55082

ABSTRACT

Over 10 years ago the European standard test method for determination of odor concentrations (EN13725) made significant advancements in the field of olfactometry. Steps were taken to standardize the olfactometer and the assessors. Testing with standard odorant n-butanol allowed laboratories to select assessors and continually monitor their performance with a universally agreed upon metric. Standardized assessors led to the agreement that environmental odor threshold tests could be determined with a sample of four assessors completing two testing rounds (eight responses). ASTM International Standard E679 was revised in 2004 to acknowledge these advancements.

Since the adoption of these standards, the question is often raised if four assessors provide sufficient precision for determination of an odor threshold value for environmental odor samples. This topic has been part of discussions at recent CEN Working Group meetings for the review and revision of EN13725.

A new study conducted by St. Croix Sensory has reviewed multiple detection threshold calculations determined with triangular forced choice testing utilizing n=1 to n=10 assessors. Statistical analysis was completed for all possible combinations of assessor responses for each sample. This paper will discuss detailed results of the analysis and provide recommendations for number of assessors needed for specific odor study objectives, e.g. source comparisons, routine testing, performance testing, etc.

KEYWORDS: EN13725, olfactometry, n-butanol, assessors, threshold, odor unit

INTRODUCTION

Olfactometry Standard EN13725:2003 provides a framework for QA/QC programs in olfactometry laboratories (CEN, 2003). This standard includes requirements for important elements such as physical design of the laboratory, equipment operation, and equipment performance testing. The other principle component of this standard involves selection of the human assessors used to complete the odor observations.

Historically, ASTM E679 and other olfactometry standards recommended the use of assessors with olfactory sensitivity representative of the general population (ASTM, 2004). The expectation was that these responses obtained from **eight to twelve assessors** assigned to a test session would be similar to the population average.

During the 1980's and 1990's, research in Europe was being conducted as part of development of a universal olfactometry standard. The inter-laboratory studies conducted at that time showed it was impossible to represent the population while meeting agreed upon repeatability criteria with the small sample size of assessors typically convened on odor panels (Hermans, 1989; Heeres, 1990).

Van Harresveld presented a clear conclusion resulting from this study in his 1999 publication in the Journal of the Air & Waste Management Association, "A Review of 20 Years of Standardization of Odor Concentration Measurement by Dynamic Olfactometry in Europe." During the development of the new olfactometry standard, "the notion that the panel should be representative of the general population was explicitly abandoned…" (van Harresveld, 1999). It was determined that the sensor involved in the odor testing, i.e. the human noses, must be standardized.

Further work defined the use of n-butanol as a standard reference odorant for selection of odor panel assessors. Research conducted in Europe during the 1990's, including several inter-laboratory studies, arrived at an agreed upon criteria for n-butanol. It was determined that the n-butanol reference odor threshold value was 40-ppb. An assessor must have a certain sensitivity and repeatability in their responses to tests of this reference odorant. The assessor is required to have an average sensitivity to n-butanol in the range of 20 to 80-ppb, with the standard deviation of the log threshold values less than 2.3.

This selection of assessors with n-butanol creates a standardized sensor; with the intent that a small sample of these, essentially interchangeable assessors, create a panel that perform as an instrument and provide results within a necessary range of accuracy and precision. EN13725 requires a minimum of four assessors, who evaluate each sample twice to provide eight total responses. Since its publication, practitioners have often questioned if four assessors are sufficient to provide quality results.

The subject of this paper is to expand the investigation of assessor responses by testing multiple environmental odorous air samples with a panel of ten assessors and considering the results of all combinations of panels of 1-10 total assessors.

METHODOLOGY

St. Croix Sensory, a commercial sensory testing laboratory (Stillwater, Minnesota), conducts odor evaluations for various consulting firms, sanitation districts, industries, universities, and government agencies throughout the U.S. and Canada.

At St. Croix Sensory, odor threshold values are determined using triangular forced-choice testing procedure on an AC'SCENT® International Olfactometer operating with a presentation flow rate of 20-lpm, following dynamic dilution olfactometry standards CEN EN13725:2003, *Air Quality – Determination of Odour Concentration by Dynamic Olfactometry*, and ASTM International E679-04, *Standard Practice for Determination of Odor and Taste Threshold by a Forced-Choice Ascending Concentration Series Method of Limits* (Appendix X.3). The AC'SCENT Olfactometer has 14 dilution levels with an operating range of 60,000 down to 8 dilutions.

Multiple environmental odorous air samples were evaluated with a panel of 10 assessors, trained by St. Croix Sensory, qualified to EN13725 selection criteria, and experienced in threshold testing evaluations. Each sample was evaluated in two rounds by all assessors for a total of 20 individual threshold estimate (ITE) responses.
For each sample tested, all combinations of assessor responses were investigated to evaluate all possible results for 1 to 10 assessors (2-20 total responses). Figure 1 graphically displays the total number of possible combinations of individual assessors for each value of n, number of assessors. For example, from a group of 10 assessors, there are 210 possible combinations of four (n=4) specific assessors.

Figure 1. Number of possible combinations of assessors for each value of n, number of assessors, from a group of 10 individual assessors.

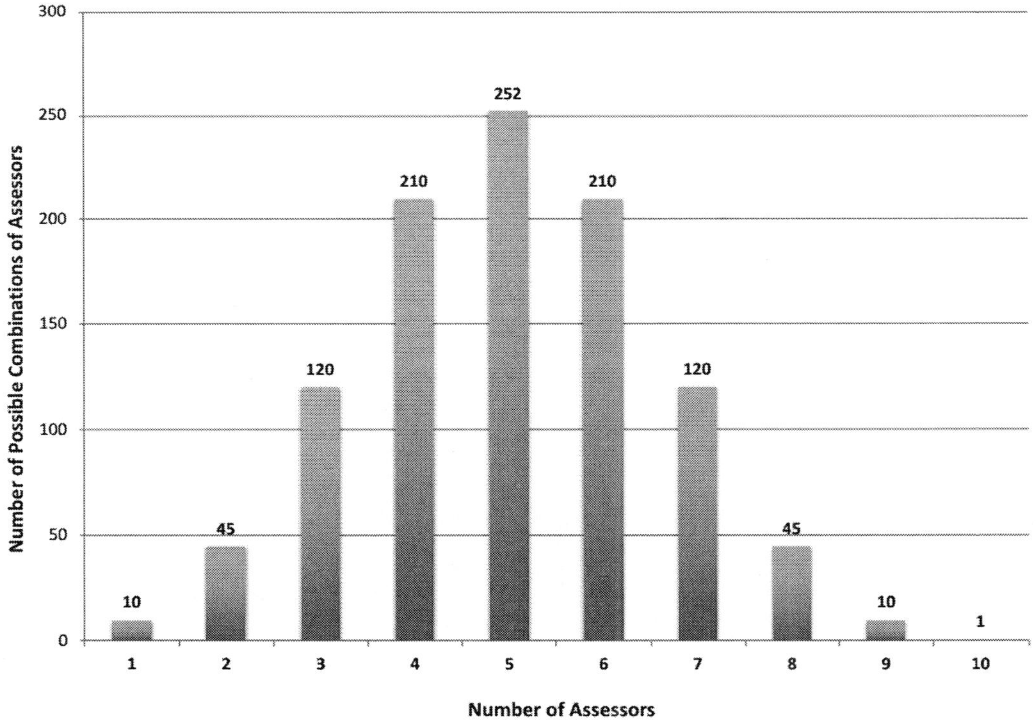

RESULTS

Nine odorous air samples with average detection threshold (DT) values of 55 to 37,500 ou_E/m^3 were investigated on five different testing days. Additionally, on two of those testing days, standard odorant n-butanol test samples were also investigated with all ten assessors. Figure 2 is a box plot diagram of the results for one of the n-butanol tests, where the detection threshold was determined to be near 800 ou_E/m^3. This corresponds to a butanol threshold concentration of 50ppb, which is well within the EN13725 requirement of 20-80ppb.

The box plot provides a visual representation of the variability in the panel results. The bottom of each box represents the 1st quartile, where only 25% of all values lie below the box. The top of the box represents the 3rd quartile, where only 25% of values lie above the box. The median is represented by the line bisecting the box. The red cross represents the mean value, which is the same for all combinations of the represented number of assessors. In Figure 2, the results for the standard odorant, n-butanol, show decreasing variability with more assessors. As the number of assessors increase, the maximum and minimum values become closer to the mean value. The box showing the middle 50% of scores becomes narrower at 4 assessors and again at 6 assessors, where it then continues to get smaller with additional assessors.

Figure 2. Box plot of odor detection threshold values determined by odor panels for standard odorant, n-butanol (40ppm). The results show a detection threshold of 800 ou_E/m^3, which corresponds to a threshold concentration of 50 ppb.

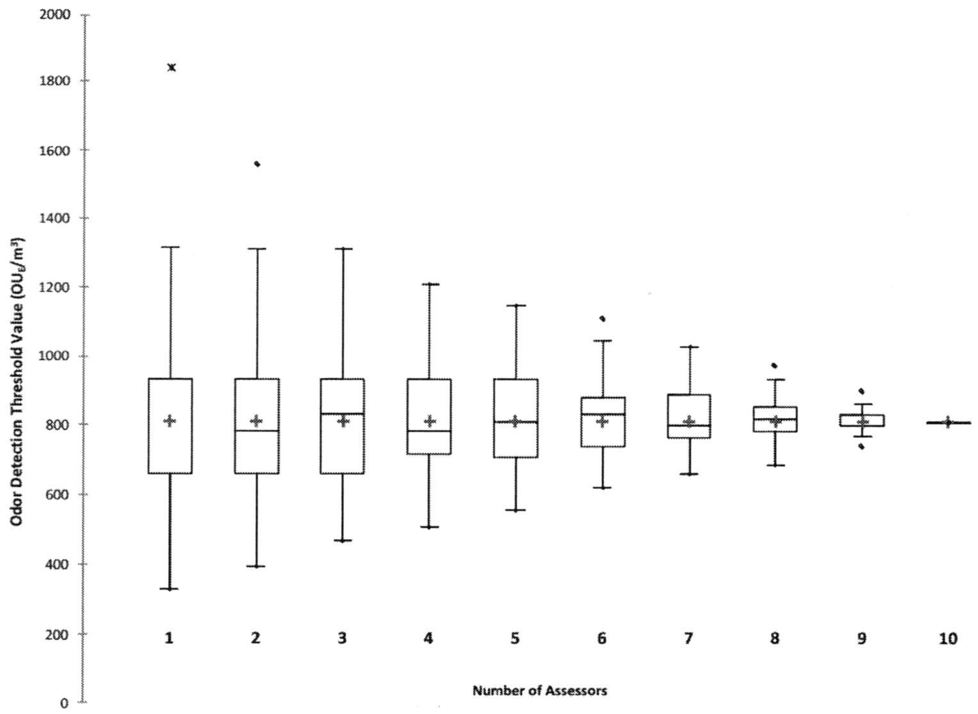

Figures 3 and 4 provide two more example box plot diagrams. Figure 3 is a box plot for a test sample with a detection threshold value of 600 ou_E/m^3. As with the standard butanol sample, it is visually seen there is an initial reduction of variability at 4 assessors, and again at 6 assessors, with a continual decrease in variability with more assessors.

Figure 4 provides a similar picture for a detection threshold value near 7,000 ou_E/m^3. Note that the variability looks greater than in Figure 3, due to the logarithmic nature of odor results; however, there is still a decrease in variability at 5 assessors and again at 8 assessors.

Figure 3. Box plots for sample with Detection Threshold of 600 ou_E/m^3 for increasing number of assessors.

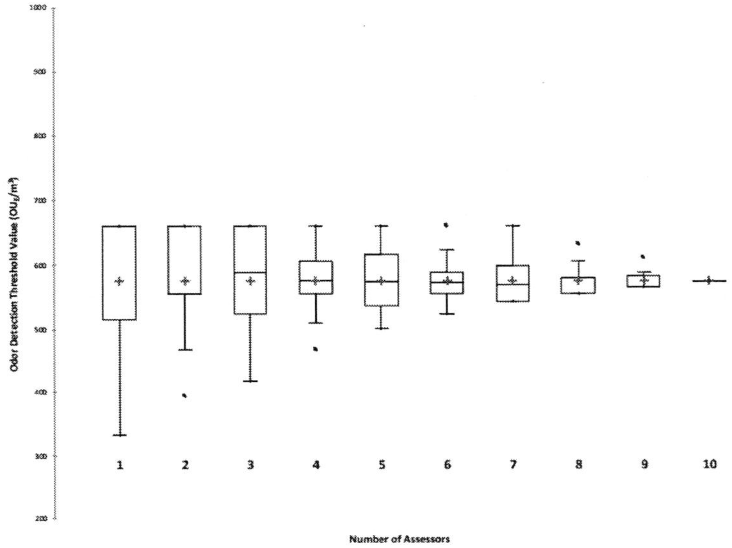

Figure 4. Box plots for sample with Detection Threshold of 7,000 ou_E/m^3 for increasing number of assessors.

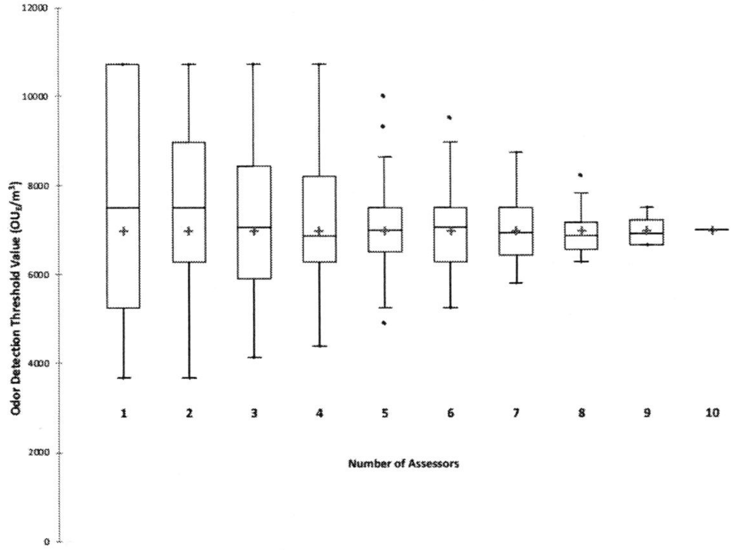

Figure 5 provides an additional way to investigate the variation in the results by reviewing the ratio of the maximum panel detection threshold to the minimum panel detection threshold based on all combinations of results from the specified number of assessors. Choinièr, et. al. showed in their study the average ratio was 4.13 for 4 assessors, 2.94 for 5 assessors, 2.04 for 6 assessors, and 1.44 for 7 assessors (Choinièr, 2014). Our results in Figure 5 show these ratios also decrease with increasing number of assessors. For 4 assessors, the ratios were mostly between 2 and 3, with an average of 2.48. For 5 and 6 assessors, the ratios were closer to 2, with averages of 2.1 and 1.8, respectively, and then below 2.0 for 7 and more assessors, with values of 1.6, 1.3, and 1.2 for 7, 8 and 9 assessors.

For the standard odorant n-butanol, EN13725 specifies that the panel threshold should be between 20 – 80ppb. This is a factor of 2 above and below the agreed upon threshold value of 40ppb. This range of acceptable values gives a maximum to minimum ratio of 4.0, thus providing a measure of what has already been agreed to as acceptable, shown as a red line in Figure 5. Experience at St. Croix Sensory has shown that panel variability within a laboratory should be better than this EN13725 "acceptable" range. Ideally, this ratio should be closer to 2-3, as is seen here for 4 and more assessors.

Figure 5. Ratios of the maximum calculated panel detection threshold value and the minimum calculated panel detection threshold value for each number of assessors in a panel.

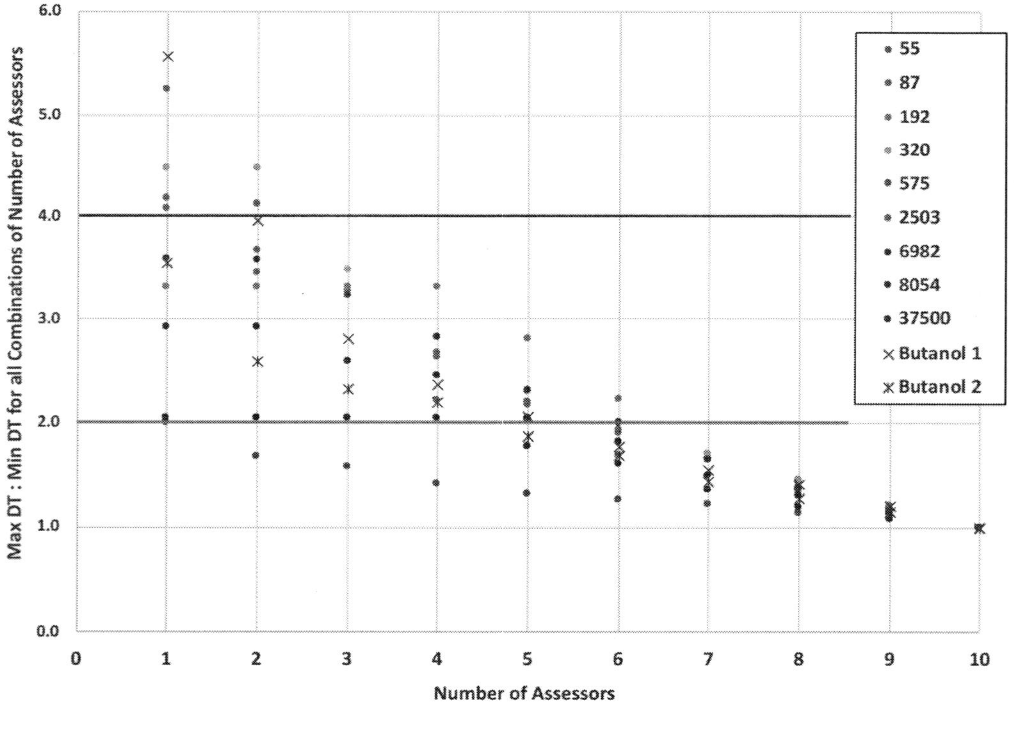

These box plots and ratio values both visually present the variability of the results; however as seen in the two box plot examples, the logarithmic nature of olfactometry and odor perception prevents direct comparison of the real magnitude of the variance in samples with different detection threshold values. It is better to utilize uncertainty measures provided in sections of EN13725 addressing quality assurance / quality control, which, like the ratio value in Figure 5, are independent of the magnitude of the detection threshold.

Figure 6 presents standard deviation values for the computed threshold values from the multiple combinations of assessors. Previous papers have discussed the use of logarithmic transformations for analysis of detection threshold results (McGinley, 2006). As outlined in EN13725, n-butanol is utilized to test multiple panels and develop a repeated measure. This calculated repeatability is then transferable to other odorous air samples. One EN13725 criteria for variability of repeated samples is to have a standard deviation of the logarithmic detection threshold values less than 0.172 (red line in Figure 6). At St. Croix Sensory, these within laboratory standard deviations for n-butanol have been found to be 0.10 (green line) and historically confirmed through testing of multiple environmental odor samples (McGinley, 2006). The within panel standard deviations (same panel repeating the same sample) has shown to provide standard deviations closer to 0.05.

Figure 6 shows this variability is very high for testing with only 1 or 2 assessors. With combinations of different panels of 4 or more assessors, the standard deviation drops to a maximum of 0.10. By 7 assessors, this standard deviation maximum has decreased to 0.05, values more consistent with repeated measures of one panel of the same 4 or 5 assessors.

Figure 6. Standard deviations of the logarithmic detection threshold values for all panel combinations for each number of assessors. EN13725 laboratory QA/QC criteria provides a limit of 0.172 (red line) for the standard deviation of multiple repeated measures. The value of 0.10 (green line) provides a target goal for a quality laboratory.

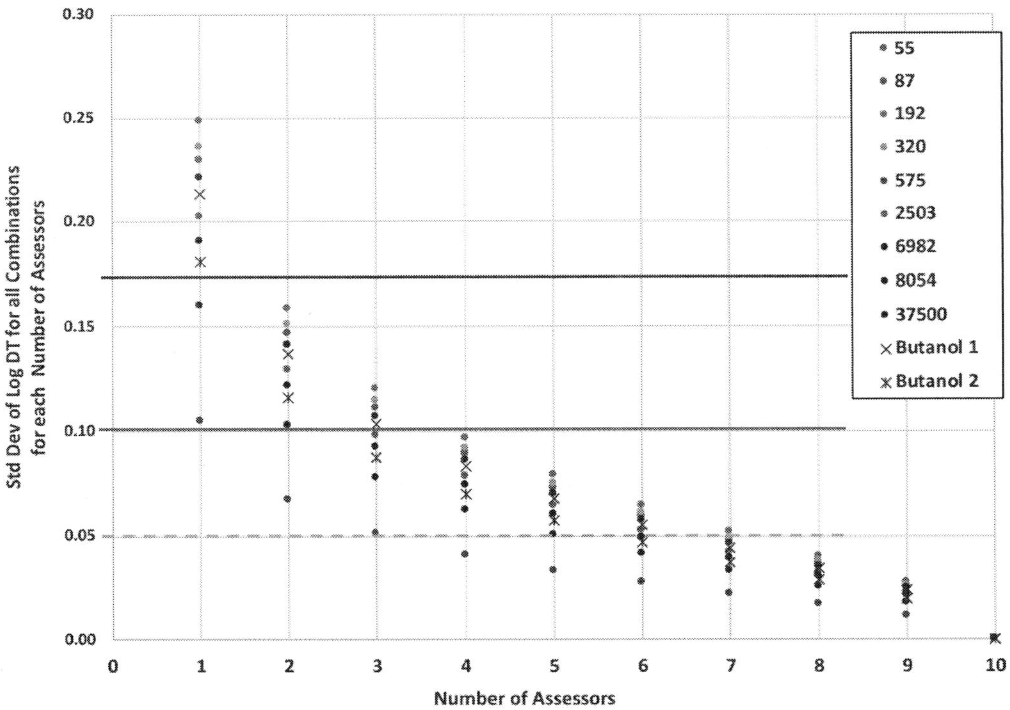

EN13725 outlines specific calculations for a measure of repeatability, which is a function of this standard deviation of multiple repeated panel threshold tests. The EN13725 criteria for the repeatability calculation is a value of equal to or less than 0.477. To place more meaning to this logarithmic value, a further review is conducted by calculating $10^{0.477} = 3.0$. The meaning of this value is that any two repeated measures will be within a factor of 3 in 95% of the cases.

In practice, a quality laboratory will have repeatability results much better than this criterion. Figure 7 shows the repeatability results computed from all combinations of the panel results with each quantity of assessors. The EN13725 criteria of 0.477 is shown with a red line in Figure 7. Most results are below this criterion with only two assessors. Consistent with previous testing at St. Croix Sensory, with 4 or more assessors, all results are below a target maximum repeatability value of 0.3, which refers to a repeatability with a 95% confidence interval bound by only a factor of 2. This value decreases with additional assessors, with very consistent results by 8 assessors within a factor of 1.3 in 95% of measures.

Figure 7. Values for the EN13725 Repeatability criteria. EN13725 laboratory QA/QC criteria provides a limit of 0.477 (red line) for the repeatability measure. The value of 0.30 (green line) provides a target goal for a quality laboratory, which refers to results within a factor of 2 of the mean in 95% of measures.

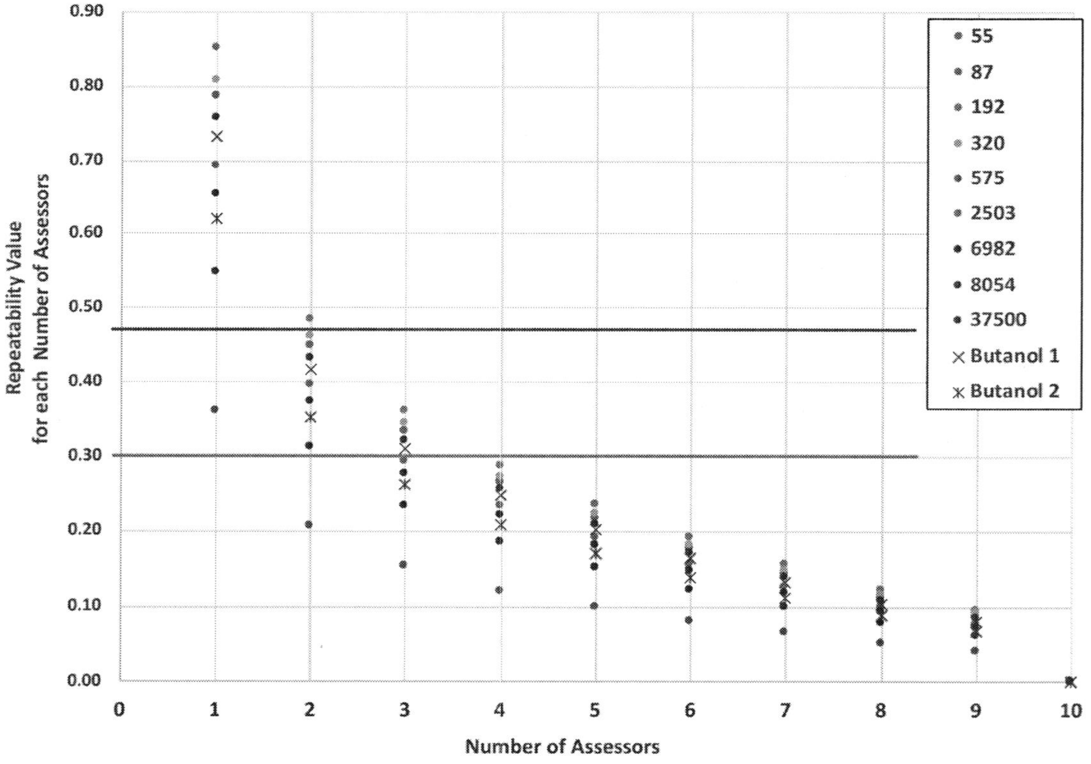

DISCUSSION

The repeatability values determined by a laboratory, through multiple tests of n-butanol, and further confirmed with periodic repeating of test samples within a panel and also within a laboratory with multiple panels, can be utilized to estimate confidence intervals of test sample odor detection threshold values.

The results from this study provide estimates of the logarithmic standard deviations of detection threshold values for 1-10 assessors at St. Croix Sensory. Table 1 summarizes the average logarithmic standard deviations identified for multiple samples tested with combinations of 1-10 assessors. Recall EN13725 requires a laboratory have results with a standard lower than 0.172.

Table 1. Logarithmic standard deviation for all thresholds determined by multiple combinations of the number of assessors.

Number of Assessors	Logarithmic Standard Deviation
1	0.20
2	0.13
3	0.10
4	0.08
5	0.06
6	0.05
7	0.04
8	0.03
9	0.02
10	n/a

The confidence interval of any odor detection threshold result can be computed form multiple repeated measures of the same sample; however, commonly only one test is run per sample. An estimate of the 95% confidence interval can be made based on these determined standard deviation values. Figure 8 shows these estimated 95% confidence intervals for a test sample with a detection threshold of 500 ou_E/m^3 as determined by panels of 4-8 assessors. For this detection threshold determined by a panel of 4 assessors, the estimated confidence interval around 500 ou_E/m^3 is 350-720. If this samples result was determined with 8 assessors, the confidence interval is 440-570.

Figure 9 provides a second example with an odor detection threshold of 20,000 ou_E/m^3. Here the estimated 95% confidence interval is 14,000-29,000 for 4 assessors and 17,000-23,000 for 8 assessors.

Figure 8. The estimated 95% confidence interval upper (orange) and lower (green) bounds for one odor sample with an odor threshold value of 500 ou$_E$/m^3 tested with 4 to 8 assessors. The red lines show the upper and lower bounds of the EN13725 laboratory quality criterion.

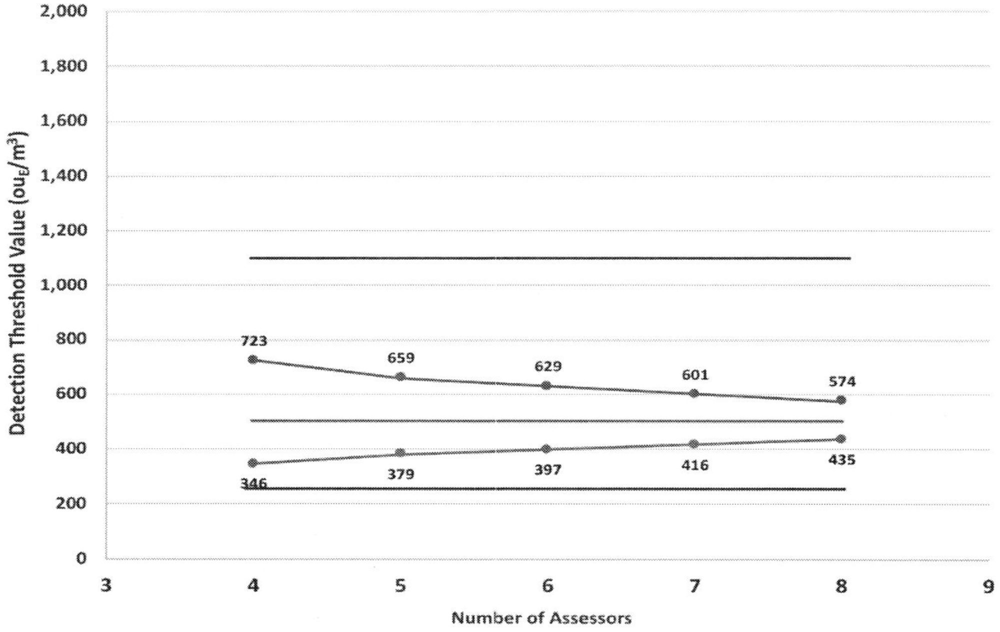

Figure 9. The estimated 95% confidence interval upper (orange) and lower (green) bounds for one odor sample with an odor threshold value of 20,000 ou$_E$/m^3 tested with 4 to 8 assessors. The red lines show the upper and lower bounds of the EN13725 laboratory quality criterion.

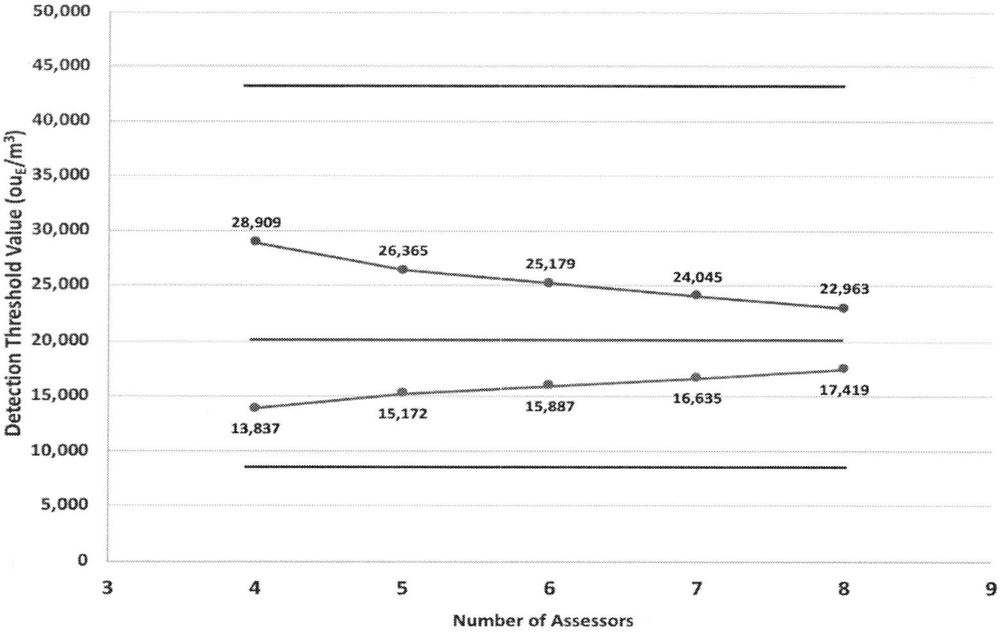

The logarithmic nature of odor threshold values makes the variability appear greater for higher values; however, the values can be normalized through log base 10 transformation and the confidence intervals can be further expressed as deviations from the mean. Figure 10 displays the estimated 95% confidence interval with this normalization. The odor detection thresholds value is represented at 0.0 and the confidence interval is shown as the bounds deviating from this mean.

Figure 10 shows the upper and lower bound criteria for EN13725, computed from the standard deviation limit of 0.172. This variance leads to a logarithmic limit of ±0.344, which is a factor of ±2.2. Additionally, the dashed line represents the upper and lower bound for a target repeatability for a good laboratory quality program, which is based on a standard deviation of 0.10. This is a a logarithmic limit of ±0.30, or a factor of ±2.0. The estimated 95% confidence intervals, shown here for the results of this specific study, show the decrease in variability with the use of more assessors. While Figure 10 shows results directly applicable only to the one odor laboratory from this study; it is expected other odor laboratories would show the same trend in reduced variability with increasing number of assessors.

Figure 10. The estimated 95% confidence interval upper (orange) and lower (green) bounds for one odor sample. The graph is normalized to show deviation from the threshold results and allows for computing values for any threshold result (ou_E/m^3). Data represents result of the laboratory used for this testing.

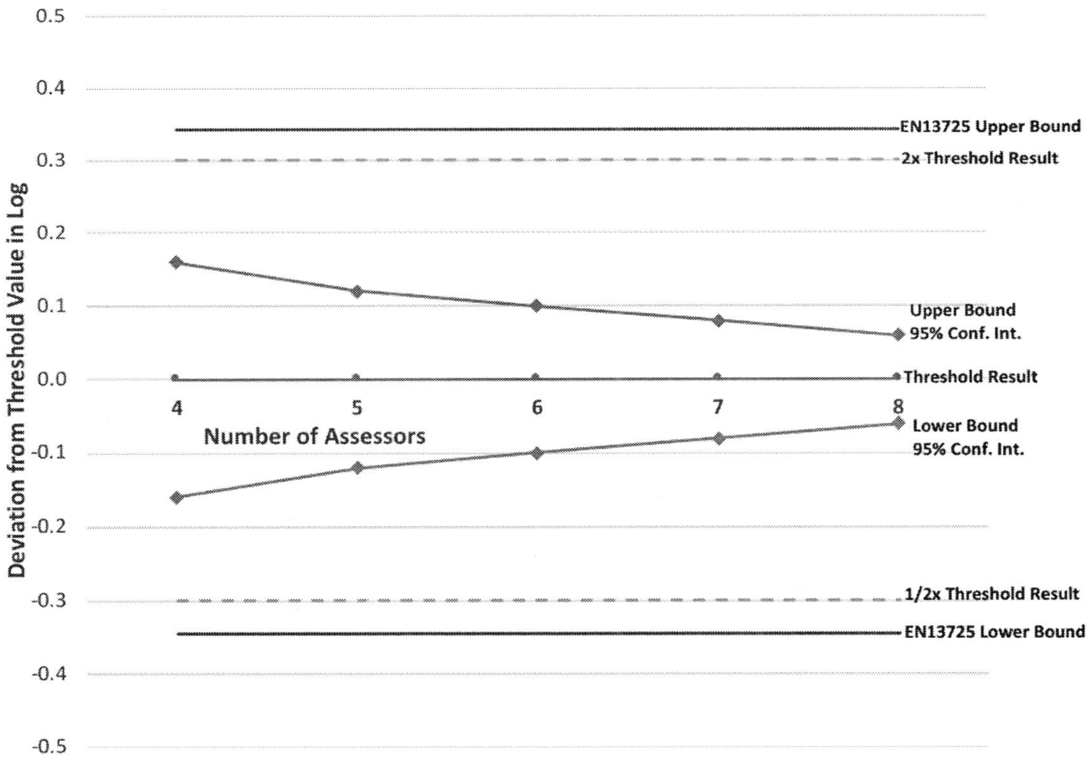

Conclusions

The results of this study do confirm that additional assessors will decrease variability of odor detection threshold testing results. An odor laboratory can provide estimates of the variability based on repeated testing with standard odorant n-butanol and with selected real environmental odor samples.

However, these result comparisons are made without consideration of all other elements that affect variability of odor source evaluation, most importantly sample collection and process dynamics. These other important elements are expected to lead to much higher levels of variability in results than the number of well-trained, experienced assessors utilized.

The use of the standard 4-5 trained assessors as prescribed in EN13725 will be sufficient in most cases for completing a site odor inventory. With similar assessor variability across all odor samples, ranking and prioritizing sources would likely be unchanged with the use of more assessors. Likewise, routine, periodic site source monitoring will also be sufficient with this lower number of assessors. Source monitoring over time would look for much larger changes in detection threshold values than would be seen from variability due to the number assessors.

Additional assessors improve *precision* (repeatability), however, additional assessors do not improve *accuracy*. Accuracy implies closeness to a "true value", and a true value is not known for individual environmental odor measurements.

Therefore, practitioners involved in collecting odor threshold data may want to consider the use of a higher number of assessors in situations where *perceived* precision is important, such as:

1. Determination of odor threshold for a source being tested for regulatory or permitted limit.
2. Determination of odor threshold for a source being tested for a contract performance standard.
3. Determination of odor threshold for a source being tested for inputs to odor dispersion modeling.

In these cases, it is also important to consider when collection of replicate samples is warranted to provide information about the odor source variability.

REFERENCES

ASTM International. 2004. E679-04: *Standard Practice for Determination of Odor and Taste Threshold by a Forced-Choice Ascending Concentration Series Method of Limits*. Philadelphia, PA, USA.

CEN - Committee for European Normalization. 2003. *Air Quality – Determination of odour concentration by dynamic olfactometry*; EN13725:2003; Brussels, Belgium.

Choinière, D.; Giard, D.; Kazemi, H.; Barrington, S. Determination of the Impact of Panel Size and Presentation Mode on the Variability of Odour Detection Threshold Concentration. *Proceedings of The Water Environment Federation Specialty Conference: Odors and Air Pollutants*; Miami, FL; 31 May 2014 – 3 June 2014.

Heeres P.; Harssema H. 1990. Progress of the Standardization of Olfactometers in the Netherlands. *Staub Reinh. Der Luft.* Vol. **50**, 185-187.

Hermans L. 1989. *Ringonderzoek Olfactometers. Publikatiereeks Lucht No. 80.* Ministry of the Environment, Leidschendam, The Netherlands.

McGinley, M.A.; McGinley, C.M. 2006. Precision of Olfactometry and Odor Testing Results. *Proceedings of The Water Environment Federation / Air & Waste Management Association Specialty Conference: Odors and Air Emissions*; Hartford, CT; 9-12 April 2006.

Van Harreveld, A.P. 1999. A Review of 20 years of Standardization of Odor Concentration Measurement by Dynamic Olfactometry in Europe; *Journal of Air & Waste Management Association*, **49**, 705-715.

ODOR CONTROL – A JOURNEY OF 14 YEARS

By John Bennett, WEF Fellow and Manager of Trinity River Authority's Denton Creek Regional Wastewater System

Abstract: Odor Control - A journey of 14 years.

This presentation is both an informative and humorous look at lessons learned over the past fourteen years while developing an odor control program at an advanced wastewater treatment facility in one of the fastest growing communities in Texas. This presentation will demonstrate that a successful odor program is not a one-size-fits-all endeavor and requires commitment from every level of staff and the surrounding community.

The presenter will demonstrate a timeline of growth both at the facility and within the community served combined with an increase in odor-related complaints. He will identify what technology has been embraced while providing examples of both successes and failures. Additionally, he will speak to creating a community outreach program that has proven to be successful in limiting the number of issues associated with odors common to a treatment facility.

Keywords: Odor Control, Wastewater Treatment, Hydrogen Sulfide, Odor Reports, Magnesium Hydroxide, Vapex Hydroxyl Ion

Introduction: This presentation will focus on the challenges a wastewater system has in regards to odors in a robust and growing community, tracking these challenges and responses by the facility manager and staff as they transitioned from a very small, rural facility to an advanced treatment program with neighbors literally sharing a border with the facility.

Background: The Trinity River Authority's (TRA) Denton Creek Regional Wastewater System (DCRWS) became operational in 1990, constructed as a .75 million gallon per day (MGD) treatment facility in what was then a rural part of the Dallas/Fort Worth area of North Texas. The DCRWS facility treats all or part of wastewater flows from 11 customer cities with the majority of the flows being characterized as domestic. The author became the facility's manager in 2003 while the facility was expanding from 2.5 MGD to 5.0 MGD. The expansion was in response to the growth in the surrounding communities; however, the area around the facility was still rural.

In 2004, DCRWS received the first of what was soon to become numerous odor reports. The reports were coming from a community approximately one and a half miles from the facility. This community had built on and around a 30-inch interceptor pipeline. The interceptor transitioned from force main to gravity at a drop manhole that was located in the middle of the development. In addition to the velocities created by the drop manhole configuration, the lift station only cycled 3-4 times per day, creating detention times in the pipe as long as 24 hours. The anaerobic conditions in the pipeline combined with the gas-generating conditions in the manhole resulted in measured sulfide concentrations in excess of 700 parts per million. Given that the odor threshold of sulfides are 3 – 5 ppm and levels above 100 ppm are considered to be immediately dangerous to life and health, we knew that we needed to take immediate action.

Community odor reports resulted in the involvement of the City Council and concurrent investigations by local investigative reporters from two news agencies and the Texas Commission of Environmental Quality. Although it was determined that the odors being detected in the neighborhood were not being generated by the facility, it was still our desire to minimize these reports and ensure the safety of staff working in the collection system as well as the people living next to it.

Plant operations staff uses magnesium hydroxide, supplied by Premier Magnesia located in Waynesville, North Carolina, at the facility to assist with alkalinity deprivation, noted to be the result of strong ammonia loadings related to Texas Motor Speedway events. While in use, this product not only provided alkalinity to the facility but kept hydrogen sulfide odors to a minimum by limiting the release of the sulfide from its liquid state. In response to these odor reports and in an attempt to be good neighbors, we moved the magnesium hydroxide feed station from the facility to the upstream lift station. After internally investigating, we also began using Vapex Environmental Technologies, located in Cocoa, Florida to mist a radical hydroxyl ion into the drop manhole, chemically restructuring the hydrogen sulfide and minimizing the odors. The two technologies combined worked extremely well, reducing measurable sulfides to less than 2ppm and as long as both technologies were functioning, we received no further odor reports from that area.

In 2009, the facility was to undergo another expansion, this time from 5.0 MGD to 11.5 MGD. Additionally, the facility was to convert from a traditional advanced nitrification system to a full biological nutrient removal facility. During the planning of the expansion, we learned that the land to the immediate south and west of the facility had been sold and would be developed into a high end residential community with plans to build both homes and an elementary school on the border of the facility. With this in mind, numerous modifications to the design were developed in order to minimize odor emissions as well as light and noise pollution. Low cost alternatives such as the planting of cedar trees, covering process locations that have turbulent water, utilizing LED directional lighting and soundproofing the blower building were initiated, fulfilling our desire to be good neighbors.

The expansion was completed in late 2010, and we began utilizing the new processes. However; on Labor Day weekend of 2011, a planned change in the process resulted in a substantial odor outbreak. The odors were so strong and consistent that many in the surrounding neighborhood tried to educate themselves on what caused the odors. Once they determined it was hydrogen sulfide, there was an outcry from the community over health concerns. Additionally, the City Council became involved once again, and the lack of institutional memory proved to be a challenge. The community was angry and concerned.

Methods:

We continued to use magnesium hydroxide and the radical hydroxyl ion in the collection system to mitigate odors at that location. In response to these new complaints we installed eight odor loggers to track H2S concentration, both within the facility and the yards of our new neighbors. The data was tracked and logged in order for us to ensure sulfide concentrations in the neighborhood were not to a level that could be hazardous. We also used the data collected to assist us with verification of odor reports and in turn were able to associate certain reports with particular process within the facility; providing us with a better understanding of how our operations could in fact produce or reduce odors. Additionally, we contracted with a consulting engineer to perform an odor survey of both the treatment facility and the neighborhood. Based on the results of this survey a five-year capital improvements program was designed. Year one would be the study, with the following years used to construct odor mitigation equipment in the areas deemed to be most critical. The following year we would conduct another survey. Upon its conclusion we would install additional odor mitigation equipment. After four years of this process we have effectively reduced the amount of odors that once routinely were detected to beyond our borders to odors that are now rarely detectable beyond our borders.

In addition, TRA decided that the facility manager would begin a community outreach program to educate the new neighbors on the basics of wastewater treatment, current odor control technologies and our future plans to continue to mitigate odors through the afore mentioned capital improvement program. Many hours were invested providing after hours' tours of the facility. We conducted four town hall type meetings at the local school as well as at the treatment facility. These meetings focused on the basics of wastewater treatment and served to inform the public on exactly what we do as well as our capital planning and odor control efforts. Penning articles in the Homeowner Association's newsletter and meeting with City Council. Additionally, we undertook the creation of an odor brochure and an online odor reporting site.

The odor brochure was distributed to the home owners during the town hall meetings, tours and displayed at City Hall. Additionally, an electronic version was created that realtors could provide to prospective homeowners as well as potential investors in the area. The odor reporting site provided the technical data such as time, date, strength of odor, description of odor, accurate location of the person reporting, contact name of the person reporting, wind direction, temperature and barometric pressure. This data was all captured and logged so comparisons in reports, times and locations could help to corroborate that the facility either was or was not responsible for the odor. It also allowed for a timely and appropriate response from the facility manager, who ensured that all reports of odors was acted upon with phone calls, emails and often the manager or staff would arrive at house of the person reporting the odor to smell exactly what they were reporting.

Results: All of our efforts have paid off. The neighbors surrounding our facility now all understand that we are a wastewater treatment facility and as such, there are going to be occasions that they will detect odors. They also understand that there is a substantial cost to mitigate odors and the staff at the DCRWS facility along with TRA management is committed to fist treat wastewater that exceeds our DMR requirements and secondly are committed to being good neighbors while minimizing the effects of odors on their neighborhood while being fiscally responsible to all eleven of our customer cities. Our efforts in providing excellent customer service, modifying the operation of our facility, aggressively advancing our capital improvement program and evaluating various odor reducing technologies has paid off. We went from having more odor reports than we could count in 2011 to only four reports in 2016, of these four reports, only two could be verified to be associated with the facility itself.

Conclusion: It is important to the facility staff for the community to see them as protectors of the environment and not as just creators of odors. The dissemination of the message and the earnest desire to be a good neighbor drove the efforts required for the community outreach to succeed. Along the way, we learned numerous important steps and lessons on odor control and dealing with the general public. Most notable are: you must find technology that fits your facility, and there are no "one size fits all" technologies. As an operator, it is important that we understand what makes up our flows, what creates the odors and to discover the best way to treat them. Equally important are lessons learned when dealing with the community. Operators should learn to use terms like "Hello Neighbor" while being honest, upfront, factual, knowledgeable and compassionate. But perhaps most important, take ownership and make the problem yours!

Orange County Sanitation District's Proactive Strategies for Preventing Odor Complaints

Reza Sobhani*, Brian Reed, MarcoPolo Velasco, Christina Thomas, Umesh Murthy
Orange County Sanitation District
10844 Ellis Ave, Fountain Valley, CA 92708
*Email: rsobhani@ocsd.com

ABSTRACT

Orange County Sanitation District (OCSD) is a public agency that provides wastewater collection, treatment, and disposal services for approximately 2.6 million people in 479-square-mile dense populated area in Southern California. Treatment of wastewater is inherently an odorous process, and to combat the persistent odors, OCSD has implemented several operational and process improvements using a combination of innovative and well-established odor control strategies to reduce public odor complaints. These efforts have resulted in a considerable decrease of odor complaints attributed to OCSD's treatment facilities and collection system. This paper illustrates the comprehensive approach implemented to reduce odor complaints and presents a summary of proactive tasks and steps that were taken to meet the Agency's goal. Furthermore, it sheds light on the challenges that the odor control team faced during this attempt.

KEYWORDS
Odor control, prevention, proactive strategy, OCSD

INTRODUCTION

Orange County Sanitation District (OCSD) is a special agency governed by a 25-member Board of Directors comprised of 20 cities, four special districts, and one representative from the Orange County Board of Supervisors. OCSD provides wastewater collection, treatment, and recycling for approximately 2.6 million people living within a 479-square-mile area of central and northwestern Orange County, California (Fig. 1). Its facilities include 396 miles of sewer pipes located throughout the county. Each day approximately 185 million gallons of wastewater are treated from residential, commercial, and industrial clients located in the second most densely populated county in the state and is sent to two treatment plants: Plant No. in Fountain Valley and Plant No. 2 Huntington Beach, CA. OCSD is considered the third largest wastewater agency west of the Mississippi River.

The conveyance and treatment of wastewater is naturally an odorous process. Today, there is an increase in the public vocalizing their concerns and their intolerance of odors from wastewater facilities. In response, OCSD has instituted a proactive and public program for controlling and mitigating odor emissions to be a "good neighbor." Twelve major sewer trunk lines collect wastewater and transfer collected solids and wastewater to the treatment plants.

As a commitment to our neighbors and our public, the 5-Year Strategic Plan was developed in November 2013 with our agency's levels of service outlined. Our promise to the community was to minimize the odor-related nuisances from both plants and the collection system. The plan adopted an organizational culture adhering to our set of core values while making efficient and effective use of all available resources. The Strategic Plan centers efforts in several categories, including providing exceptional customer service.

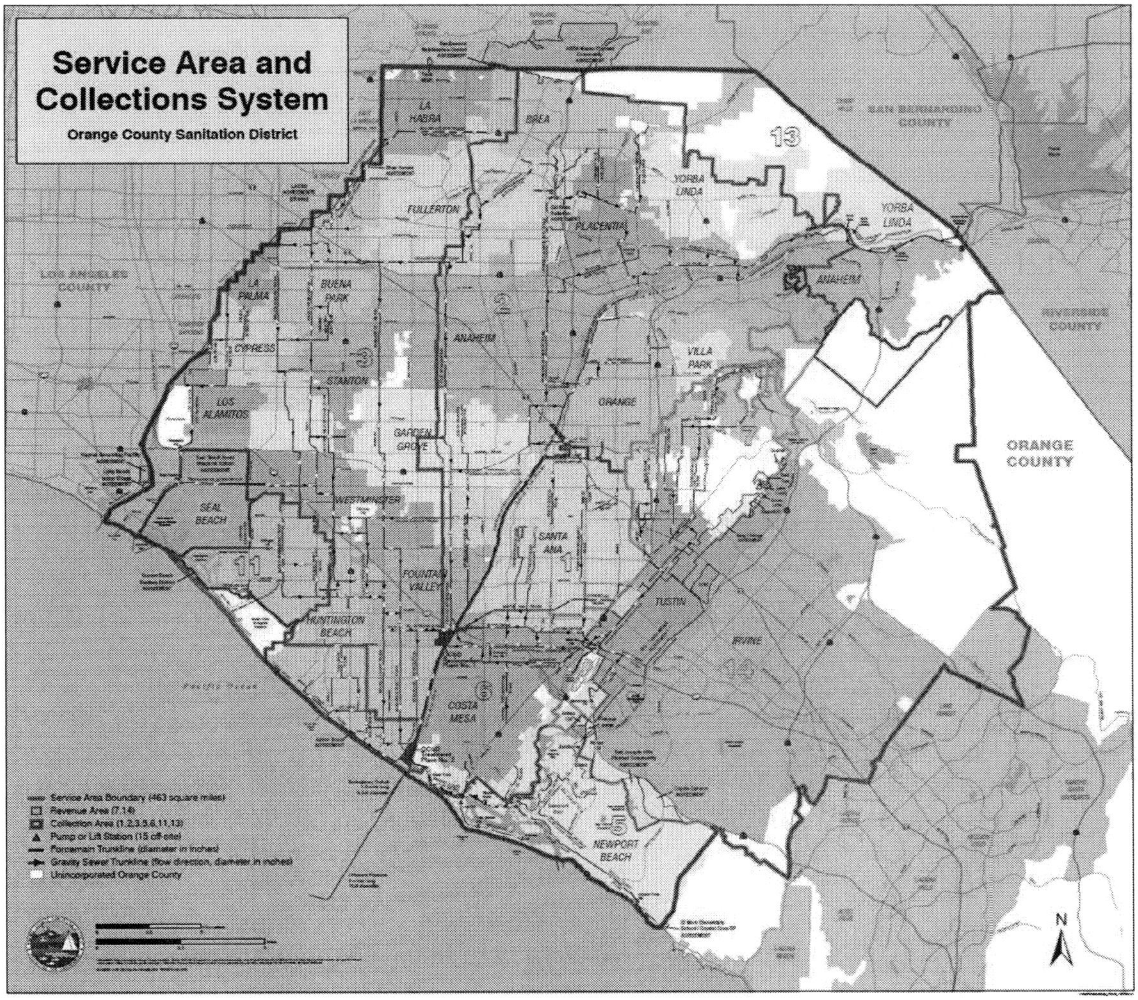

Figure 1. OCSD Service Area that covers 2.6 million people in 20 cities

During fiscal years (FY) 2015-2017 the number of odor complaints received was higher than the established goal of 12 or fewer complaints a year. The developed proactive strategies to prevent odor complaints started during the first quarter of 2017, and have made great improvements in the number of odor complaints. The aim of this project was to reduce the number of odor complaints to meet OCSD's Strategic Goals and its Level of Service to the community. Table 1 illustrates the Agency's strategic goals and the levels of service.

Table 1 – OCSD strategic goals and the levels of service

Agency Strategic Goals (Levels of Service)	
Providing Exceptional Customer Service	**LOS Target**
1.Odor Control – Complete the Odor Control Master Plan.	**100%**
Treatment plants odor complaint response within 1 hour.	**100%**
Collection system odor complaint response within 1 working day.	**100%**
Number of odor complaints: (Under normal operating conditions) • Reclamation Plant No. 1 • Treatment Plant No. 2 • Collection System	 **0** **0** **<12**

METHODOLOGY

The Odor Prevention Plan was created to design and implement a comprehensive strategy to prevent fugitive odor emissions, reduce attributable odor complaints, and achieve OCSD's established goals. It encompasses a broad range of activities to prevent odor formation, minimize air release, and remove odorous compounds. The following is a summary of the taken steps:

Real-Time Monitoring and Control

To have a robust odor monitoring and complaint prevention program, it is vital to collect and monitor real-time H_2S concentration, which is the main source of odor complaints in the collection system. OCSD's odor team utilizes the H_2S data loggers that continuously measure H_2S concentrations (in gaseous phase) in OCSD's trunk lines, and transmit the logged data through in-built modems to a dedicated FTP server accessible to OCSD odor team. By monitoring the H_2S concentration trends, odor team makes appropriate actions to control the H_2S level at the collection system. Most of the deployed data loggers are permanent, while some are temporarily used for specific studies. In addition to the real time H_2S monitoring in gaseous phase, the H_2S concentrations are measured periodically in liquid phase as well.

Enhanced Up-Stream H_2S Removal

Generally, the treatment strategies for liquid and solids begin at the treatment plants. However, for OCSD, the odor treatment begins in the collection system, where foul air emissions from 400 miles of sewer pipes and 15 pump stations could cause nuisance odor complaints. OCSD odor team mitigates odors and decreases H_2S loads to the treatment plants with chemical additions of ferrous chloride, magnesium hydroxide, calcium nitrate, and/or sodium hydroxide up-stream in the collection system.

These chemicals are added continuously to the wastewater trunk-lines by designated chemical dosing stations, or periodically through chemical batches provided by chemical tankers. Below are more explanations regarding these two approaches:

A. Optimization of Continuous Chemical Dosing

Optimization of chemical dosing is achieved by close monitoring of chemical additions in the collection system. The odor team and associated contractors remotely monitor the tank levels in the collection system, adjust customized dose profiles, perform system diagnostics, and respond to alarms that warn staff if the system is not running properly.

Currently the odor team operates 11 dosing stations at 9 locations to minimize the odor and fugitive emissions at four OCSD trunk lines. The dosing sites utilize nitrate addition, iron salt, and pH adjustment at the chemical dosing sites presented in Figure 2. The effective chemical dosing strategies mitigates the odor formation in collection system, decreases the attributable odor complaints in surrounding areas, and maintains overall trunk-line health. The nitrate addition (i.e., calcium nitrate) prevents H_2S formation in trunk-lines by replacing sulfate with nitrate, and removing pre-existing dissolved sulfide in collection system. The iron salt (ferrous chloride) chemically combines with dissolved H_2S to form relatively insoluble precipitates. The pH adjustment (i.e., magnesium hydroxide) holds dissolved sulfide in solution phase and mitigates vapor-phase odor in collection system.

The odor team also facilitates the operations of the new permanent dosing stations. These new sites will improve the odor control among the trunk lines that currently utilizes chemical dosing stations and will reduce odor release to the ambient air among the trunk lines that previously had no permanent dosing station.

B. Chemical Batch (Chemical Shock)

The odor team implements chemical shocks periodically by utilizing sodium hydroxide (caustic) to control and mitigate the H_2S formation in our trunk lines. Each shock can effectively impact the trunk lines for 1 to 2 weeks. In this approach, increasing the pH level destroys the slime layer and causes it to slough. The odor team has a regular schedule (i.e., every 2 weeks) for caustic shock at pre-determined points (i.e. manholes). During the hot weather, the intervals between the shocks might be reduced as needed, based on monitoring data. Further dosing points might be utilized for chemical shocks based on observed H_2S concentration in the trunk-lines.

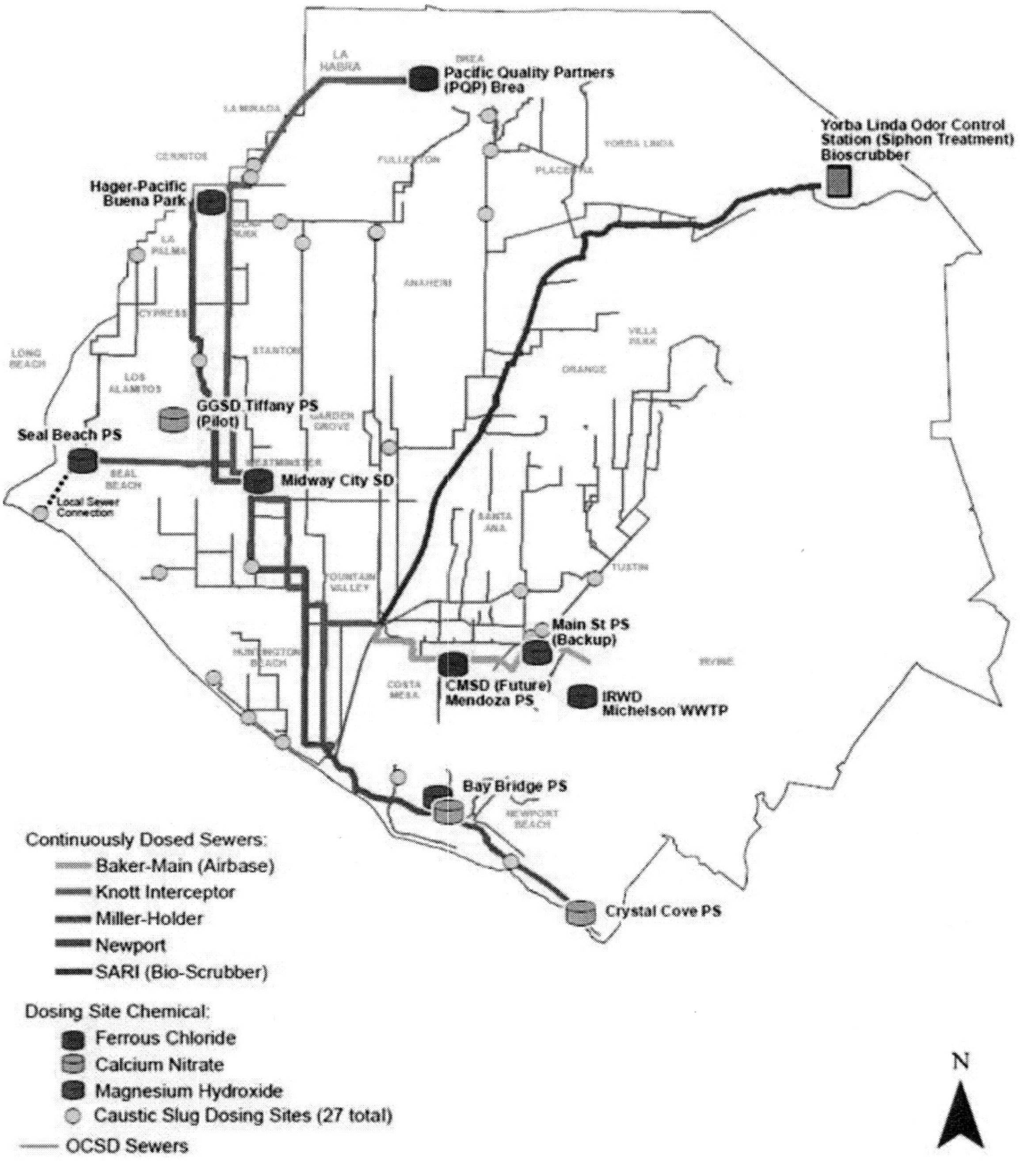

Figure 2. Enhanced up-stream H_2S removal by continuous chemical dosing and periodic chemical shocks

Utilization of best business practices

During this approach, the odor team utilized the best business practices across the organization and implemented a unified approach for odor prevention at both plants, within the collection system, and during maintenance and construction projects. The following assets and tools are the examples of best practices that were deployed for achieving odor complaint reduction (Fig. 3):

A. Odor Boss: a mobile platform that generates a vapor to neutralize and treat the air with an iodine solution (Fig. 3);
B. Odor control chemical atomizer, which reduces odor employing an iodine solution at the source with a low-pressure chemical/water feed and minimal chemical and energy consumption; Chemical and biological scrubbers at both plants;
C. Smoke testing to see exactly where the odor originates;
D. Infrared optical gas imaging camera; and
E. Nasal Ranger which is a portable olfactometer.

Figure 3. Examples of best business practices at OCSD

Visualizing odor complaints in a GIS environment

The OCSD odor team developed a Geographic Information System (GIS) Map to effectively visualize the odor complaints in a spatial context (Fig. 4.A). The Odor Complaint Map was integrated with other layers of OCSD GIS data, including the information of sewer trunk lines and manholes, permitted manholes for industrial discharge, dosing stations and monitoring locations, pump stations, and wind direction. The odor team used the developed map to facilitate:

A. Identification of the hot spots and critical points in collection system and plants' surrounding areas (Fig. 4.B);
B. A more effective H_2S concentration monitoring and control;
C. Provision of a reliable chemical supply for each trunk line; and,
D. A more effective odor complaint prevention.

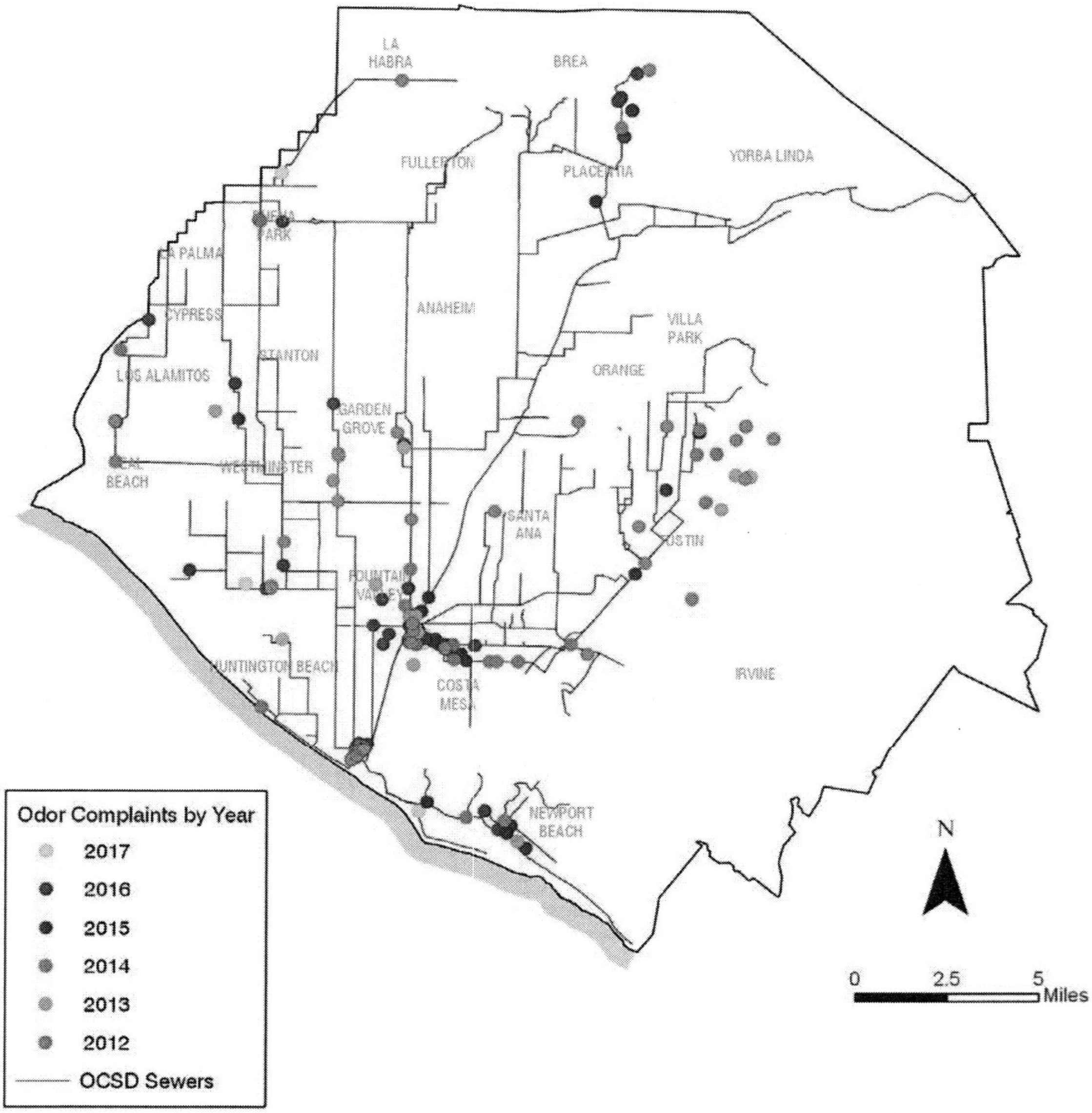

Figure 4.A. Developed GIS maps to visualize odor complaints in a spatial environment

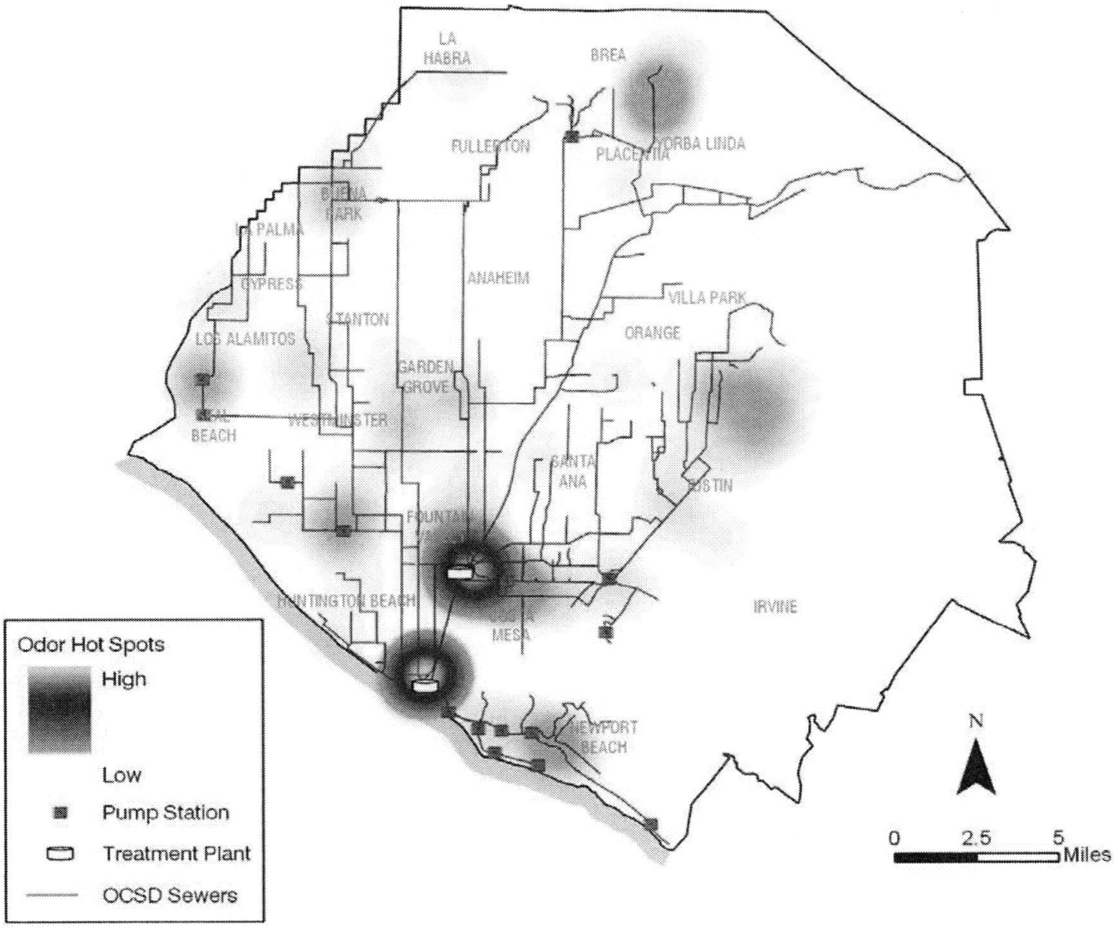

Figure 4.B. Utilizing a GIS map to identify critical points of attributable odor complaints

Effective community outreach

Effective community outreach with our neighbors includes facility tours, one-on-one meetings, collateral materials, and advocacy in our community. In this approach, residents and business owners can call the 24-hour OCSD hotline to report an odor complaint. OCSD takes odor complaints seriously and responds to complaints within one hour. Once the smell has been identified, OCSD will have staff get in contact with the resident or business owner to ensure the problem is resolved. Collateral information and community open houses are provided at both treatment plants, where the neighbors are invited, and the odor related projects in OCSD Capital Improvement Program (CIP) are discussed.

RESULTS AND DISCUSSION

OCSD's multi-pronged approach in reducing the number of attributable odor complaints reflected a strong leadership and the technical skills, with a high level of dedication to our Agency's Mission and Levels of Service. By implementing the discussed proactive strategies, OCSD consistently reduced the number of attributable odor complaints both in collection system and the plants' surrounding areas during the last few years (Fig. 4).

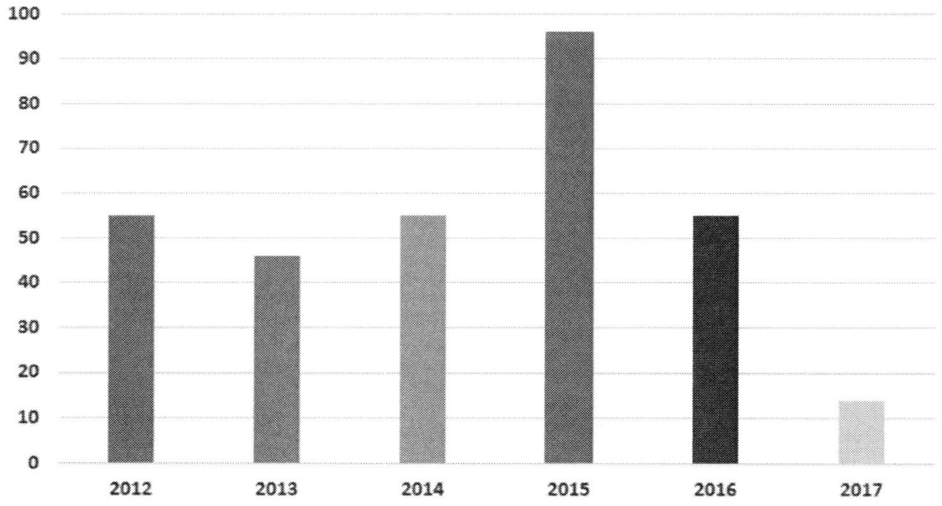

Figure 4 – Total attributable odor complaints received between 2012 to 2017 (Calendar Year)

Also, comparing the results between FY 2015-16, FY 2016-17, and the first 2 Quarters of FY 2017-18 indicates that the total odor complaints decreased from 107 complaints in FY 2015-16 to 46 complaints in FY 16-17, and reduced to 6 complaints during the first 2 Quarters of FY 2017-18 (Table 2). The successful implementations of proactive strategies were the result of successful team work and input from several departments. This effort also improved OCSD operational effectiveness by working together across the organization to pilot test new technology and document field findings.

During this project, OCSD combined best management practices, and integrated strategies used in the collection system with practices and strategies used in wastewater treatment plants. These original and innovative strategies include the following:

A. *Integrating odor treatment from the collection system to the treatment plants*: OCSD utilized a comprehensive approach to treat wastewater odors from the point of origination through the final treatments, by establishing and implementing the proactive Odor Prevention Plan.

B. *Prevention*: In efforts to prevent any odor complaints, OCSD monitors real-time data trends, analyzes critical points, and identifies and treats hot spots.

C. *Working across the organization*: The results achieved in FY17-18 were attained through effective collaboration among several OCSD teams.

Table 2 – Number of attributable odor complaints received during FYs 2015-16 to 2017-18

Agency Strategic Goals (Levels of Service)

Providing Exceptional Customer Service	**FY 15-16 Year-end Results**	**LOS Target**
1.Odor Control – Providing reliable, responsive and affordable services in line with customer needs and expectations.		
Treatment plants odor complaint response within 1 hour.	100%	**100%**
Collection system odor complaint response within 1 working day.	100%	**100%**
Number of odor complaints: (Under normal operating conditions)		
• Reclamation Plant No. 1	63	**0**
• Treatment Plant No. 2	20	**0**
• Collection System	24	**12**

Agency Strategic Goals (Levels of Service)

Providing Exceptional Customer Service	**FY 16-17 Year-End Results**	**LOS Target**
1.Odor Control — Providing reliable, responsive and affordable services in line with customer needs and expectations.		
Treatment plants odor complaint response within 1 hour.	100%	**100%**
Collection system odor complaint response within 1 working day.	100%	**100%**
Number of odor complaints: (Under normal operating conditions)		
• Reclamation Plant No. 1	10	**0**
• Treatment Plant No. 2	11	**0**
• Collection System	25	**12**

Agency Strategic Goals (Levels of Service)

Providing Exceptional Customer Service	FY 17-18 2nd Quarter Results	LOS Target
1. Odor Control – – Providing reliable, responsive and affordable services in line with customer needs and expectations		
Treatment plants odor complaint response within 1 hour.	100%	**100%**
Collection system odor complaint response within 1 working day.	100%	**100%**
Number of odor complaints: (Under normal operating conditions)		
• Reclamation Plant No. 1	1	0
• Treatment Plant No. 2	1	0
• Collection System	4	12

Complexity

During the implementation of developed proactive strategies, the odor team faced complex challenges mainly due to the following issues:

1- OCSD is located in one of the most population dense areas in the United States;
2- Agency's collection system includes 396 miles of trunk-lines, and 15 pump stations that continuously encounters foul air movement and fugitive emissions;
3- Southern California faced long drought and subsequent water conservation during the last decade that caused a decrease influent flow with an increase in suspended solids;
4- Agency had many construction activities within the wastewater treatment facility that require complex flow diversions;

Contribution to Social or Economic Advancement

There is generally an increase in public concern and intolerance of odors and other air contaminates from all aspects of the process, including the treatment plants, pump stations, and the collection system. Implementation of the Odor Prevention Plan has decreased the odor nuisance in the service areas which is amongst one of the most populated areas of the state. This allows the residents to utilize the region's natural resources and enjoy the events without any negative impacts associated with fugitive sewer odors. Furthermore, the utilization of an integrated strategic plan allowed OCSD to advance in odor prevention without impacting operational costs due to the effectiveness and efficiency of the system's operation.

CONCLUSIONS

OCSD is a public agency that provides wastewater collection, treatment, and water reclamation services for approximately 2.6 million people in 479-square-mile dense populated area. OCSD's odor team developed and implemented a proactive Odor Prevention Plan to reduce public odor complaints from the collection system and Plants' surrounding areas.

The proactive plan includes:

a. Real-time H_2S monitoring and control,
b. Enhanced up-stream H_2S removal,
c. Optimization of continuous chemical dosing,
d. Deployment of chemical shocks,
e. Utilization of best business practices in odor control,
f. Development of GIS maps to visualize odor complaints in a spatial environment, and integrate these map with other Agency's information, and
g. Effective community outreach program through Agency's neighborhood.

These efforts significantly decreased odor complaints attributed to OCSD's treatment facilities and collection system.

In the World of Odor Control – The Community Rules

Richard J. Pope, P.E., BCEE Hazen and Sawyer
rpope@hazenandsawyer.com

Abstract

The anticipated response by an owner to a community alarmed about nuisance odors from the collection system or water reclamation facility nearby is to investigate where the odors are originating and mitigate them with the implementation of a cost-effective odor control technology that will return the owner back to the ranks of being a "Good Neighbor". The owners are finally getting it, the community will not go away quietly, and that effective odor control is required to address the prime driving force – the community. But just as important as installing the right odor control technology may be in appeasing the community, so is communicating with the residents of the community to:

- listen to their voices
- understand their concerns
- ask for their assistance
- find out where, when, how strong, and how long the odors last
- let them know the steps that are being taken to mitigate the odors
- let them know of the time it will take to implement controls.

Simply put, what you don't tell the community, they won't know! And under that scenario the rumors about doing nothing will proliferate and the owner will likely suffer from the loss of credibility.

It is up to the owner then to take the steps necessary to ensure that the loss of credibility is not permanent. Accordingly, creating a **dialogue** with the community is a critical first step that will go a long way to have them starting to think that you are listening, you do care and that you are doing something to mitigate the bothersome nuisance odors in their community. The steps taken will determine how the community responds, but the necessary steps can vary from simple baby steps to more sophisticated and involved giant steps. Credibility can be regained! Nevertheless, do not expect the community to jump on your band wagon. Understand that the community will always tend to be skeptical of what you do, will want you to do more than is required by regulation, and will demand that you avoid any backsliding. However, after you create that dialogue they will be more likely to stop outwardly fighting the owner at every turn and perhaps even support your efforts. Be prepared, the road may be long and difficult but the benefits will be worth the journey.

This paper will discuss the varied approaches that different owners have taken to open a dialogue with affected residents through a community outreach program.

Keywords

Community outreach, nuisance odors, odor dialogue, working with community, odor control

Background

The one fundamental premise that we all recognize is that wastewater by its very nature is odorous. After all we flush it down our drains and into the wastewater collection system in order to get it out of our homes, quickly. This effective removal network may have led to the advancement in sanitation that was critical to our growth as a human race as we were able to control water/wastewater borne diseases like cholera, but it also does a good job with odor control within our homes. What do we then expect to happen when the wastewater from hundreds, thousands and/or millions of homes combines and reaches the resource recovery facility after traveling in a closed conduit for extended lengths of time? Nuisance odors at wastewater plants are inevitable, like death and taxes, only the degree of community impact will be different based on many variables linked to site specific conditions.

Understanding the potential of a plant to release nuisance odors and controlling them before they can reach the community is what being a good neighbor is all about. Nevertheless, there are times when things get out of control despite the best efforts of the utility. It is worth reminding the utility management and staff that they must be careful about making statements or promises, or supporting their elected officials when they say that their facility will be odor free. It is somewhat reckless to assert that a facility that treats wastewater can ever be odor free 100% of the time. This is especially true when you consider that at least three things are in play at all times:

- The nature of the wastewater is that it is odorous!
- Mechanical systems are used to treat the wastewater.
- Humans operate those mechanical systems, and the plant.

It is impossible to control all three of these elements all the time. At any given moment, one of these can go wrong: a motor can shut down, electricity can be interrupted, a gear can get caught up, an operator can press the wrong button, fall asleep at the controls, forget to initiate start-up, etc. Regarding the plant operation and odor control, back-ups and fail safes are installed to prevent problems, and monitoring is conducted along with preventative maintenance to keep systems functioning effectively and continuously. With these provisions in operation the likelihood that odor control may be interrupted is minimized and the duration of any excursion is reduced to a limited duration. However, announcing that the facility will be odor free all the time cannot be supported.

Community outreach is an important aspect of being a "Good Neighbor", and yet why are many utilities struggling with implementing the simplest elements of community outreach? Some typical answers received from the utilities regarding the lack of enacting a community outreach program include, but are not limited to:

- We are fixing the odor problems already
- Our nuisance odor complaints are down
- We are working as fast as we can to provide relief
- The community will witness the benefits of the odor control soon enough

- The residents have retained a law firm and are ready to sue
- The community is not the problem, the regulators are

While these answers appear capable of convincing a utility to avoid implementing a community outreach program, they are denying the relevance of the community and their impact on the utility's facility. Consider that without communities there would be no nuisance odor complaints. Without nuisance odor complaints, there would likely be no odor control mitigation measures taken. In reality, the communities are there, and they do complain. The prime driving force for mitigating nuisance odors, therefore, is the community.

Nevertheless, in the past utilities have tried to be inconspicuous by creating buffer zones around the wastewater facilities in an attempt to be out of sight and out of mind. According to years of experience at a few hundred wastewater treatment plants, it seems inevitable that wastewater facilities will ultimately lose the buffer zones that at one point used to surround their plants and keep them to some degree insulated from the surrounding community. Historically, wastewater treatment plants have been placed at what was then considered to be safe distances from the neighbors in undeveloped land or in locations where adjacent industrial operations themselves were considered to be a nuisance. While these planning decisions were considered to be safe at the time, no one wanted to predict the inevitable expansion of the urban and suburban landscape that ultimately spread out and consumed this available open space. Two examples from diverse areas of the country that come to mind based on firsthand experience are: the City of Las Vegas, NV treatment plant, which when first built was miles away from downtown and now is surrounded by residents and a nearby golf course; and the Atlantic Plant, part of the Hampton Roads Sanitation District, which used to be surrounded by farm land before a community with schools joined the neighborhood. Attribute some of this buffer land grab to zealous developers who were looking for inexpensive land to build on without concern for the wary and uninformed consumer; as well as local planning boards who were looking for more assets to add to their tax legers as they were asked to do more with less from the federal and their own state government's; and consumers looking for a good deal. Regardless of the reason why, buffer zones have become a thing of the past for many utility owners, and the reality of dealing with the community regarding their operations and how they may potentially impact the residents is key to being a "Good Neighbor".

Nuisance Odor Complaint System

The utilities must develop a way to communicate with the key nuisance odor control driving force, the community, as the way to answer their concerns about the impact of the wastewater treatment facilities on their way of life. The community's approach to communicating with the utility is through nuisance odor complaints, especially when they have no other means of dialogue with the utility. Nuisance odor complaints are the communities' outreach to the utility, as an expression that a problem exists. Accordingly, one of the first steps in a community outreach program should be to focus on establishing a formal nuisance odor complaint program whereby the utility treats the complaint with the importance that it deserves, as a record of conditions perceived by the community regarding the facility that are not meeting expectations. The resident nuisance odor complaint represents the community's report card on the effectiveness of the facility to control nuisance odors. At the same time, if a resident considers the conditions are bad enough to take the time to call and complain, the utility should feel obligated to treat the call and information seriously and take the time to record the information the complainant is willing to report and ultimately respond with follow-up information to the complainant about the complaint. This is how dialog with the community starts, and it is up to the utility to keep it going. Yet it is true that all nuisance odor complaints are not valid, and that there are residents who may have a separate agenda independent of nuisance odors, but that is all the more reason to establish a formal nuisance odor complaint system to identify and verify the true odor complaints from the others.

There are many basic steps needed to establish a nuisance odor complaint program, including the following:

- Develop a nuisance odor complaint log form
- Assign a telephone number or e-mail address
- Select operators and provide training
- Establish a review and response procedure
- Create an internal reporting hierarchy
- Determine follow-up practices

Nuisance Odor Complaint Log Form

When the community is willing to call in a complaint, the utility should be equally as willing to collect the most relevant information that would be valuable to them in investigating the veracity of the complaint and/or the conditions that may be causing the complaint, whether it be from the utility facility or another source that is in the neighborhood. Here are some of the key basic questions that a utility may want to include as part of their nuisance odor complaint log form:

- What is the name and address of the name of the person complaining
- What are the weather conditions at the time the odor was observed, which may be different from the weather conditions at the time the person complaining is calling? The delay in calling may be attributed to not having access to a phone or the time to make a call (caring for their children, making dinner, working in the yard, etc.) at the time of the nuisance odor observance.
- How fast/strong and which direction is the wind blowing? Make a distinction between which direction the wind is blowing, to or from. Set a convention, for example, which way the wind is blowing **FROM**, and stick with it. This way there is less confusion over interpreting the data later.
- How strong is the odor? Provide them with a scale such as: faint; mild; moderate; strong; very strong.
- What does the odor smell like? Provide them with some typical wastewater odor characters such as: rotten egg; fishy, fecal, greasy, putrid, rotten cabbage, etc. Consider

including the odor character of other potential odor sources in the area such as landfills and commercial and/or industrial facilities.
- When did you smell the odor and how long did it last? Make certain that the complaining person indicates whether the odor is still on-going or when it happened in the past. This could dictate the timing of the field response to the odor complaint call. For example, if the call is regarding a nuisance odor complaint that the resident experienced last week it is not necessary to go out to the field to verify the complaint. The duration of the odor event is important to understand, especially when attempting to track down the potential cause of the odor complaint. It is also valuable to ask and understand if this is a frequent odor or a one time observation. Residents may not always call in an odor complaint every time they detect a nuisance odor. Their annoyance factor is contingent on frequency and duration along with the unpleasantness of the odor detected.
- Where were they when they smelled the odor? Do not assume they the resident complaining was at home and the address given at the start of the log form is where they were when they detected the odor. Ask them to describe explicitly where they were when making the nuisance odor observation (At the park, driving to school, at work, etc.)
- Leave some additional space for any additional comments that the resident complaining wants to make.

The section of the nuisance odor complaint log form completed by the resident complaining should be simple and straightforward with check boxes and short responses. Minimize the amount of writing the person filling out the form has to do.

Assign a telephone number or e-mail address
Select a way that the residents of the surrounding community and others can contact the utility and initiate a nuisance odor complaint. Assigning a separate or using the standard utility call in number is the traditional way to provide the community with access to the odor complaint log form through an operator that is trained to complete the form. The utility may also want to make the form available through their website which can be accessed by computer, tablets or phones. This approach eliminates the operator who would be trained to ask key questions and to ensure the form is more completely filled out, however, depending on the age of the resident complaining may be the path more frequently used.

Select operators and provide training
Operators answering the odor complaint calls need to be trained on the questions on the log form and follow-up questions that they should be asking. This mining of information will help in responding to the complaint and perhaps better understanding the underlining root cause(s).

Establish a review and response procedure
Once the information is received the utility needs to assign a staff member with the responsibility to review the complaint, verify the information, assess the potential plant based odor source(s), respond to the location of the complaint, meet with the person complaining, collect any additional information that may be useful to the complaint, assess whether the nuisance odor is still present, obtain relevant plant-based or nearby meteorological data (particularly wind speed

and direction), and report out whether the plant should take any action in the event that a source or sources at the plant can be attributed with causing the odor complaint. Some utilities establish a timeframe for responding to the location of the complaint and complaintant. The staff member reviewing the complaint and conducting the investigation will file the data collected, determine whether there should be a follow-up course of action to mitigate off-site plant-based odors, and make a report to management staff based on the conclusion of the investigation.

Create an internal reporting hierarchy
Nuisance odor complaints are investigated and the staff member assigned this task must also evaluate whether the complaint is valid. Depending on the findings discovered during the follow-up and investigation, a course of action may be recommended that includes mitigating plant-based nuisance odors that were found to be impacting the surrounding community. A reporting hierarchy should be set up for the investigating staff member to inform management of important nuisance odor based findings. Most complaints can be filed without management notification. However, substantiated odor complaints should be noted and reported the first level of management. Nuisance odor complaints that cause the plant to take mitigation measures, and/or multiple verified odor complaints supported by an action at the plant (ex. Shutting down a process tank, or emptying an anaerobic digester) should be reported to the first and upper level of management where a defined course of action can be crafted and subsequently implemented, and statements can be prepared to send to or respond to media inquiries. The reporting hierarchy presented above, first and upper levels of management, would be selected based on the utility's organization.

Develop follow-up procedures
Set up procedures for following-up with the individual who complained. The procedures should establish how (in person, by e-mail, letter, phone) and when (within a proscribed amount of time, day, two days, week) to conduct the follow-up.

Mining Odor Complaint Data
Although nuisance odor complaints may not be the best barometer for how the facility is performing from a nuisance odor perspective, the information provided can be an important way to understand facility trends. Typical ways to manage the data collected is to use it to develop graphs and figures that evaluate:

- Number of complaints
- Frequency
- Location
- Character
- Meteorology
- Timing

Graphical illustrations are an excellent way to reduce the data and evaluate the data for trends that can help understand the odor impacts, migration of odors from the facility, topographical and meteorological significance, timing of the day/week/month that odor complaints dominate, etc. The following graphs, charts and aerial illustrate how the odor complaint data can be reduced, reviewed and used.

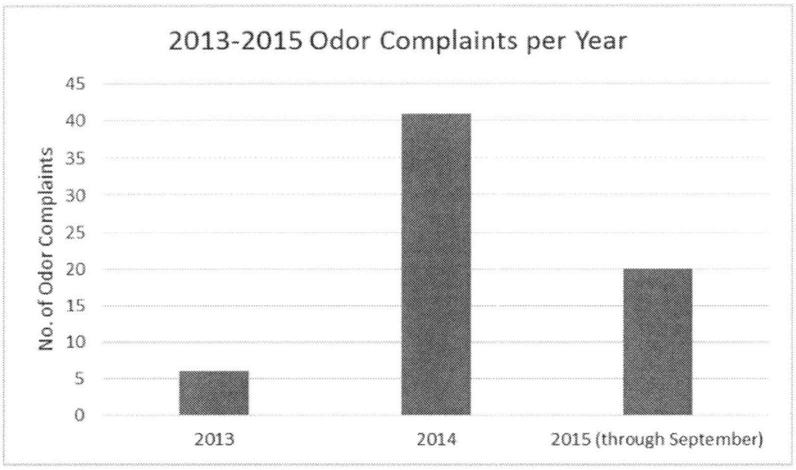

Figure 1 – Nuisance Odor Complaints by year.

Figure 1 illustrates how the odor complaints vary by year and indicates that some event(s) in 2014 triggered an increase in complaints that dropped to some degree in 2015 but are still not at the desired level of 2013. Figure 2 takes the same data and graphs it by the month of the year for the same three years used in Figure 2. May is by far the biggest month for nuisance odor complaints. Typically, the warmer summer months, July and August, are anticipated to be the highest nuisance odor complaint months. What could the facility be doing in May that might cause the complaint numbers to rise to this level? Could it be the changeover from winter to summer conditions, or a facility activity?

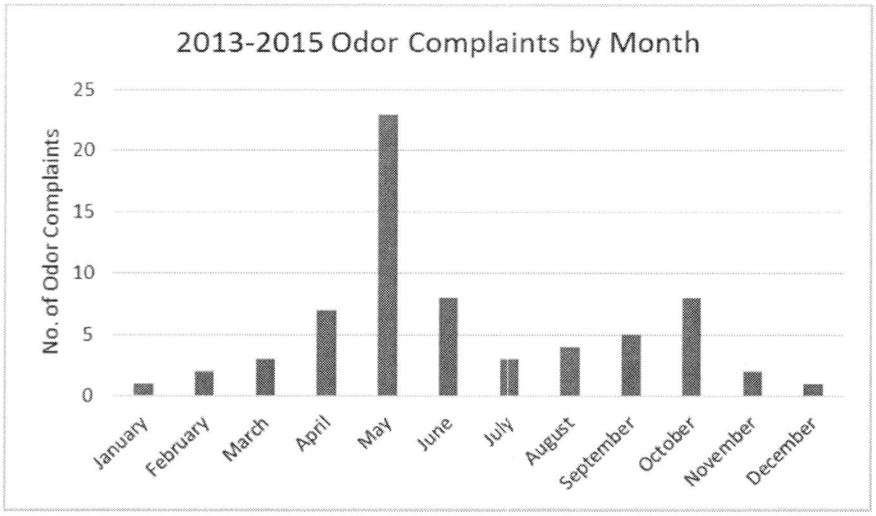

Figure 2 – Nuisance Odor Complaints by Month

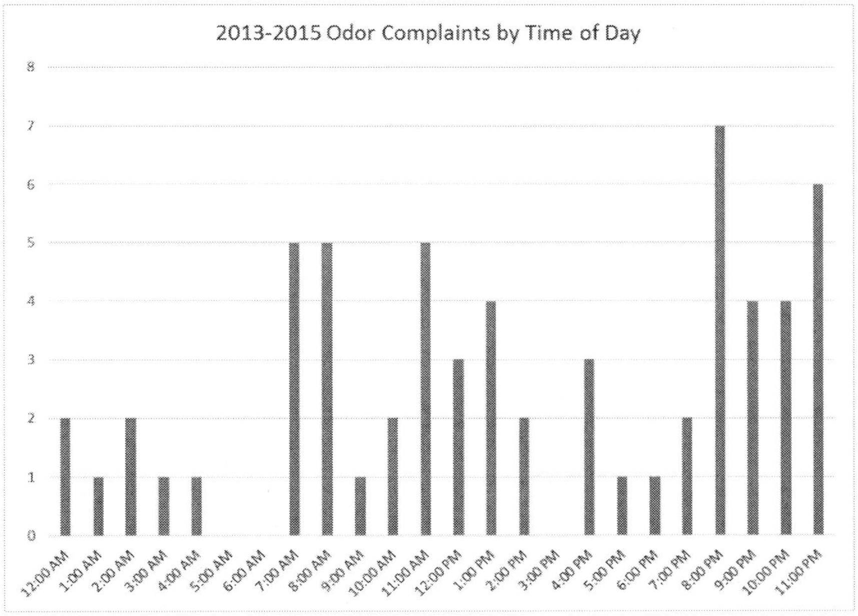

Figure 3 – Nuisance Odor Complaints by Time of Day

The complaints illustrated by the time of day in Figure 3 show a good spread of complaints over the day with concentrations that appear to be consistent with the diurnal wastewater flow to the facility. Early morning, mid-day, and evening are common high wastewater flow and correspondingly high odor release times of the day and are fairly consistent with other facilities. The complaints are not inconsistent with common facility odor release time frames.

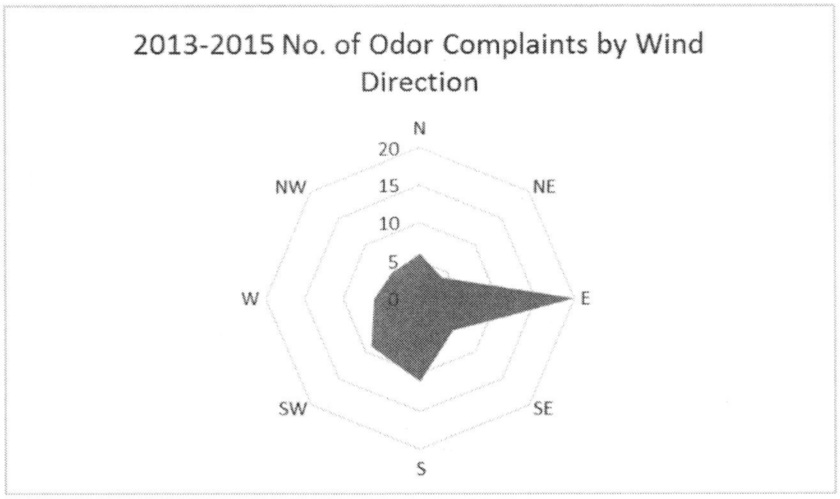

Figure 4 – Nuisance Odor Complaints by Wind Direction.

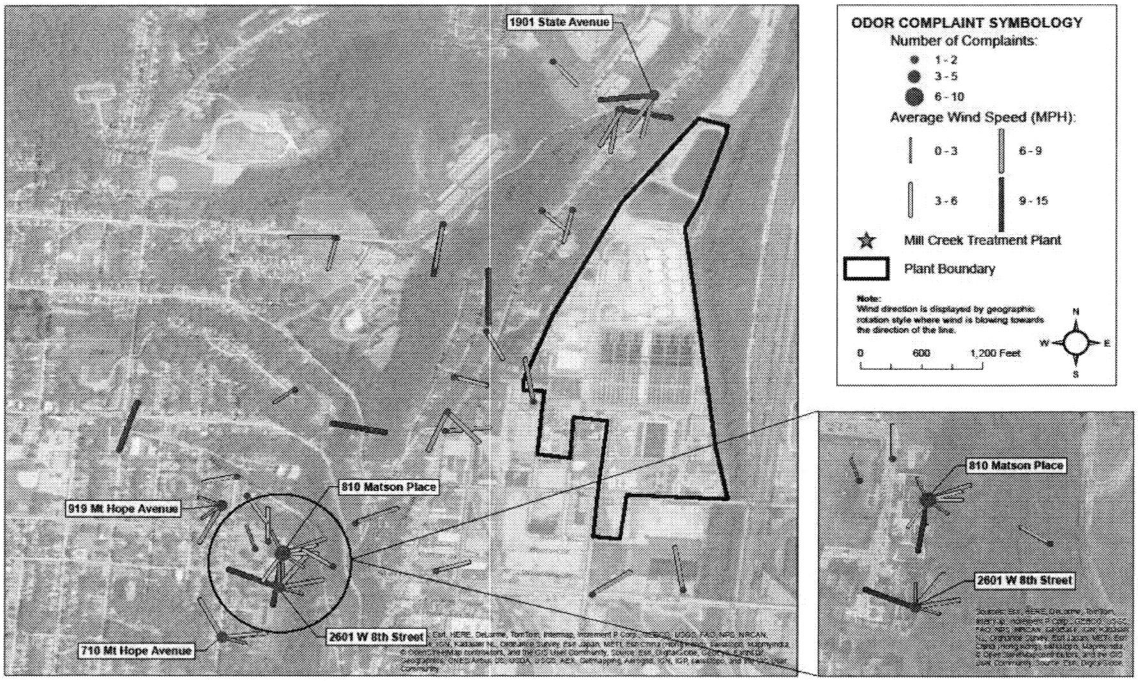

Figure 5 – Nuisance Odor Complaints by Location, Frequency, and Wind Speed.

Figure 4 illustrates that the worst off-site odor complaints occur when there is an east wind. The facility may consider monitoring the wind direction when they plan on conducting any scheduled process operation that has the potential to release nuisance odors. Figure 5 is an illustration of how to combine base elements of the data mined from the odor complaint log form: aerial of the community; location and frequency of complaints, and wind speed and direction. This combined information is helpful when trying to assess impacts from the plant.

Reviewing the nuisance odor complaint trends provide valuable information to the utility about the actual and/or perceived impact that the facility is attributed to having on the community. It provides an understanding of where the facility has a greater or lesser impact on the community and perhaps supports the fact that the facility can be detected in the community.

Community Outreach

Even if nuisance odor complaints stop being received, there are no assurances that the residents stopped detecting the facility. There are times when the community will throw up its hands and stop complaining, especially when they do not hear back from the utility. They can reach a point where their complaints are perceived as falling on deaf ears. The utility may perceive this as a positive, no nuisance odor complaints means that they must be doing a good job at being a "good neighbor". And so the dynamic intensifies, where the community gets angrier at the utility and loses faith in their ability to address their needs and to respect their quality of life; and the utility assumes that they are doing a good job and therefore does not want to do anything that they perceive would damage existing conditions.

Nuisance odor complaints do not go away, especially when they are not addressed; they tend to galvanize a community and attract media attention and political involvement (particularly around election time). As a result, you ignore complaints at your own peril. These actions are definitely not to the advantage of the utility who are themselves culpable for much of the outcome. Think about it, what the utility does not tell the community through open dialogue…they will not know. And so, the community will begin to speculate about what is being done; rumors will begin to rise and spread like wildfire with nothing being done to put out the flames; doubt will creep in about the ability of the utility to mitigate odors; and along with doubt and fear will be a loss of credibility and distrust in the utility.

The answer for the utility to gain back part of what has been lost is to begin reaching out to the community as a way to:

- educate
- inform
- respond
- engage
- allay fears

And the community outreach building blocks include achieving these through:

- open dialogue
- community involvement
- fostering partnerships
- improving credibility

with the understanding that this is an on-going challenge that needs commitment and determination to keep it going.

The community outreach road map is not the same for every utility. Each utility must evaluate the various approaches that are available and select the elements that they can implement. This will generally require the utility to step outside its comfort zone, but this is what will be necessary to succeed.

Below are some examples of successful approaches used by other utilities. They range from simple and straightforward to more complicated and engaging. In some instances, combinations of these approaches have been employed. A word of advice to utilities that have not conducted community outreached programs or have not found consistent and effective approaches is to make it a utility commitment, start off slow and steady, and try and make it part of the utility culture that is viewed by the community as emanating from management.

Brochure

Develop a simple brochure or flyer that contains information that the utility believes is important for the community to know about nuisance odors in the community. The brochure could:

- Educate and inform the community about conditions the utility recognizes may be causing problems.
- Inform about upcoming studies that are being conducted to understand where the odors are, how strong they are, how they could be investigate, and how much it may cost.
- Highlight odor control systems that are being installed/constructed
- Present a schedule for immediate and/or long term odor control measures

The content possibilities are endless, but the idea is to relay information to the community in a somewhat informed manner. The text should be brief, not overly technical, and accompanied with graphics (tables, graphs, photos) to make the brochure easy on the eyes and comprehensible on many levels.

Traditionally the brochure was handed out at meetings/functions, included in mailings, placed as door hangers, left on counters at utility structures and other business and commercial establishments, etc. Today, the brochure can be viewed and distributed through websites, phones, tablets, and social media sites. The message would remain the same for these newer distribution methods, however the length should be modified to address the nature of these methods and the shorter attention span associated with this audience. Whereas the brochure informs and educates the community, it tends to be a one-way communication avenue. Therefore, as an initial step it is a good start, but the objective is to initiate a two-way dialogue.

Stakeholder Meeting

The stakeholder meeting satisfies the objective of creating a two-way dialogue between the community and the utility in a controlled environment. The utility sets out to identify key community individuals that represent the community at large or segments of the community. They are known as stakeholders. Stakeholders are usually:

- leaders in the community
- respected & responsible individuals
- religious officials
- political representatives

Stakeholders are viewed as level-headed individuals that have a calming effect on the community.

A stakeholder meeting is intended to educate and inform the stakeholders about the actions being taken by the utility. These meetings will provide a depth of understanding about the actions of the utility so that the stakeholders can related what is important to their constituents. To overcome the credibility gap between the community and the utility, the stakeholders are asked to share the information heard at the meetings with the community that they typically interact with on a routine basis. In this way, the positive odor mitigation actions to be taken by the utility will reach a considerably broader audience and from a source that they trust and who will be given all the information necessary to come across as knowledgeable.

The stakeholder meetings are arranged by invitation only in order to manage and educate the selected individuals. These meetings provide two-way dialogue in a controlled and focused group. The stakeholders are used strategically to control the nuisance odor narrative that is disseminated to the public. Having that dialogue with the community is invaluable.

Public Forum

Public forum meetings are generally open meeting where either presentations are made and/or Town Hall style discussions are conducted. An agenda is developed and the meeting is advertised in the press, local TV, utility website, and social media (Facebook, Twitter, etc.). An overview of the actions by the utility are presented and discussed with limited detail. This is the utilities chance to communicate directly with the community about the issues that are very important to them. It is intended that the stakeholders have briefed the residents before attending this meeting so that there is some prior knowledge and understanding of the utility actions.

Regardless of any prior understanding by the attendees, the utility can expect this meeting to cover any topic that has bothered the community recently or in the past, and that individuals will want to vent their pent-up frustrations. It is understandable that the individuals will want to vent, and they should be allowed to do so, but the utility should avoid letting the meeting get subverted to mob psychology. A word of advice, stick to the agenda, keep focused, remain calm, cool, and collected. Keep the meeting informal, keep it moving, answer questions, and avoid any appearances of hiding information. The community can get irrational about odors – remember that the human sense of smell is the only one of the five senses that is directly hardwired to the brain and to the area of the brain that is the center for memory and emotions – and so be prepared to deal with them at that raw emotional level by listening, being calm, being supportive and being open. The community is looking for transparency. Remind the attendees that the past is over and the utility is moving forward to a new future where the control of nuisance odors is a key element. Also understand that the proof to the statements made at these meetings is action, demonstrating the commitment of the utility to follow through. The community is looking for actions and results and not mere words.

Plant Tours

Plant tours are a tremendous way to enable the community to witness firsthand the size and complexity of the wastewater treatment facility. Most residents do not know or understand how the treatment plant functions, how much depends on the everchanging wastewater, characteristics that are received from the community, and the extent of the odors associated with each unit operation, starting with the raw wastewater as it exits the collection system and is exposed to the atmosphere for the first time inside the facility.

This plant tour is an excellent opportunity to demonstrate the size of the facility, nature of the wastewater and processes that treat it, and the environmental good that is achieved by discharging clean effluent, generating methane gas to develop a more sustainable operating facility, removing and/or mining nutrients, creating a usable biosolids product, etc. It is also a great opportunity to explain and demonstrate how various unit operations around the facility generate/release nuisance odors, and the steps being taken by the utility to cover, contain, convey, and treat them to levels that are non-detectable in the community.

The plant tours make the facility and nuisance odors real to the community as they recognize that the task of treating the wastewater combined with nuisance odor control is more difficult than the realized or understood. Plant tours also provide an excellent opportunity for the utility to guide the narrative of the tour, showcase the key aspects of plant operations and odor control, and open a two-way dialogue between the utility and the community.

Plant tours provide an opportunity for the citizens to ask questions and witness a show and tell of the facility's unit operations and odor control equipment and practices.

As with any tours of a wastewater facility, the utility must stress and the attendees must be aware of safety issues. The utility should also take the time before the tour to hand out appropriate safety protective wear such as hard hats, as necessary, and discuss safety practices where walking around moving equipment and uncovered tanks, particularly if there is construction at the site. In the event of an emergency at the facility, the attendees should also be warned of how they need to evacuate, exits to be taken, and places to marshall.

With proper planning, plant tours can be safe and informative, as well as an effective community outreach mechanism.

Community Action Committee

A community action committee is a group of average citizens that are selected by the community to represent them and be their voice while they monitor the progress of nuisance odor control steps being implement by the utility. When the utility develops an odor control program (the program can be multi-phased, consisting of investigations, planning, design, construction and start-up, or single phase) the community action committee, or CAC, is crated to perform the following functions:

- Meet with the utility and their team (engineers, contractors, etc.)
- Get briefed by the utility team regarding the odor control program and schedule
- Routinely meet with the utility team to get updated on the progress of the odor control program elements
    - Ask questions on the progress
- Relay the updated progress to the community

- Elicit comments from the community
- Bring community comments back to routine meetings with the utility team

The CAC acts as the liaison between the utility team and the community.

The CAC is another effective tool used to open the dialogue with the community using residents to reach more people than the utility might typically be able to do, and to engage them in open conversation, and allow them to provide comments/questions. Although this may not be a direct communication between the utility and the community, getting the utility message about their odor control program out to as many residents in the community is key.

Community Training

During the course of a community outreach program it may become evident to the utility that the community is looking to take a more proactive role with the utility to help understand more basic details about the perception of odors in the community. One way that the utility can get them more involved is to ask if resident volunteers would be willing to become community noses. Community noses are individuals who are asked to fill out odor incident forms every time they detect odors in the community. The odors are not solely related to the utility facility, but any odor in general that they encounter that they would consider to be at a level that they would complain about. The individuals would initially be screened and then trained to rank odors based on odor intensity (light, mild, moderate, strong, very strong), to characterize odors (what does is smell like, using descriptors), to judge the pleasantness/unpleasantness of an odor, and to differentiate between odors. The screening will determine whether the candidate has an average sense of smell, does not smoke, is not pregnant and, therefore, is acceptable to be a community nose. The training cannot make them smell any better, but it can help them to pay attention to their olfactory sense and be more cognizant of what they smell so they can understand differences in what they are smelling. This has helped in identifying what key odor sources at the facility are causing a problem off-site (St. Paul, MN), and in identifying an outside source that was operating only at night under the shades of darkness (New York City, NY).

Each person is trained and provided incident odor log forms to fill out every time an odor is detected. These are returned to the utility on a routine basis and reviewed for pertinent information, just as odor complaint forms are mined for information. The forms can also be placed on-line so that when they are completed they can be emailed back to the utility.

The community training is another tool that can be effective when used in the right community. It opens up the interaction between the utility and the community and has the community helping to understand and solve nuisance odor occurrences in the neighborhoods surrounding the facility, and becoming part of the solution rather than continuing to be vocal adversaries. This approach is not for every utility, but it can be an extremely useful tool when applied wisely.

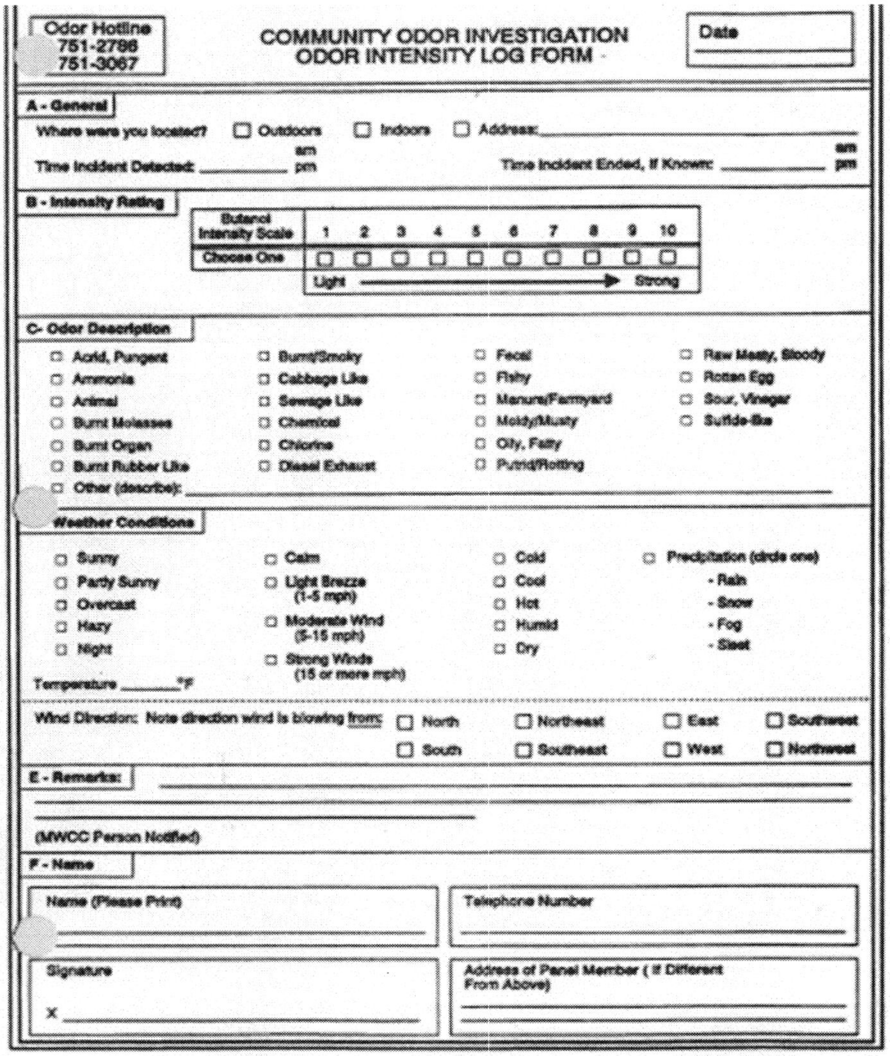

Odor Hotline
751-2786
751-3067

COMMUNITY ODOR INVESTIGATION
ODOR INTENSITY LOG FORM

Date

A - General

Where were you located? ☐ Outdoors ☐ Indoors ☐ Address:

Time Incident Detected: ______ am pm Time Incident Ended, If Known: ______ am pm

B - Intensity Rating

Butanol Intensity Scale	1	2	3	4	5	6	7	8	9	10
Choose One	☐	☐	☐	☐	☐	☐	☐	☐	☐	☐
	Light → Strong									

C- Odor Description

- ☐ Acrid, Pungent
- ☐ Ammonia
- ☐ Animal
- ☐ Burnt Molasses
- ☐ Burnt Organ
- ☐ Burnt Rubber Like
- ☐ Burnt/Smoky
- ☐ Cabbage Like
- ☐ Sewage Like
- ☐ Chemical
- ☐ Chlorine
- ☐ Diesel Exhaust
- ☐ Fecal
- ☐ Fishy
- ☐ Manure/Farmyard
- ☐ Moldy/Musty
- ☐ Oily, Fatty
- ☐ Putrid/Rotting
- ☐ Raw Meaty, Bloody
- ☐ Rotten Egg
- ☐ Sour, Vinegar
- ☐ Sulfide-like
- ☐ Other (describe):

Weather Conditions

- ☐ Sunny
- ☐ Partly Sunny
- ☐ Overcast
- ☐ Hazy
- ☐ Night
- ☐ Calm
- ☐ Light Breeze (1-5 mph)
- ☐ Moderate Wind (5-15 mph)
- ☐ Strong Winds (15 or more mph)
- ☐ Cold
- ☐ Cool
- ☐ Hot
- ☐ Humid
- ☐ Dry
- ☐ Precipitation (circle one) - Rain - Snow - Fog - Sleet

Temperature ______°F

Wind Direction: Note direction wind is blowing from: ☐ North ☐ South ☐ Northeast ☐ Southeast ☐ East ☐ West ☐ Southwest ☐ Northwest

E - Remarks:

(MWCC Person Notified)

F - Name

Name (Please Print)

Telephone Number

Signature

X

Address of Panel Member (If Different From Above)

Figure 6 - Odor Investigation log form

Example

The New York City Department of Environmental Protection (DEP) was upgrading their largest wastewater treatment plant, the Newtown Creek Wastewater Treatment Plant (Newtown Creek) to comply with regulatory treatment standards. The plant, treating an average dry weather flow of 310 mgd, was required to continue to comply with effluent discharge limits in their SPDES permit throughout their upgrade construction activity. This along with a constrained site prolonged the time to complete the project to beyond 10 years and cost the city over $4 billion. The site has no buffer and there are 9,100 residents living within two blocks of the plant. The community became aware of the project and started their campaign to voice their concerns before construction began. Because of the extended construction timeframe, the closeness of the community, their vocal campaign, etc., the city recognized that they needed a substantial community outreach program that started before and went through and beyond the end of construction. The DEP understood they had to work very closely with the community on a regular basis to keep them informed about the progress of the work, warn them of upcoming work that could impact their community, and to address issues as they arose (And they did!).

The community was concerned with more than just nuisance odors, they were worried about quality of life issues during construction including: Noise; traffic; odor; vibration; parking; etc. With regards to nuisance odor, an Environmental Impact Statement comment received from a resident stated:

- "With all the money being spent on this project, the final outcome must be an odorless plant."

The DEP was placed on notice that the community's expectations went beyond any regulatory requirements.

While planning the plant upgrade, the DEP's main objectives were:

- Satisfying community concerns
- Upgrading the plant while operating
- Addressing odor concerns Today, Tomorrow, and the Future
- Being a "Good Neighbor"

With this understanding, the DEP initiated a comprehensive outreach program that began before the first shovel went in the ground and lasted till the construction ended. The core of the outreach program centered on the formation of the Newtown Creek Monitoring Committee (NCMC). The NCMC consisted of key representatives from the community, a selected individual from the community that had a desk in the resident engineer's trailer, and two independent engineering consultants that acted as the community's advocate. The NCMC met with the plant and resident engineering staff at least monthly and at times during heavy construction periods more frequently to ask questions, hold the DEP and contractor accountable for maintaining schedules, and understand progress being made or not. The individual who was situated in the resident's trailer could be found at her desk on a daily basis, checking on the status of construction, raising issues with quality of life excursions like nuisance odors in the community, following up on change requests, etc. This person was supported by the DEP. The two independent engineering consultants were also supported by the DEP to be the community's advocate. They reviewed all documents, drawings and specifications, change requests, basis of design reports, monthly odor reports, etc. and provided comments, and discussed their comments and responses with the DEP and responsible parties. They represented the community on technical issues that were too much for the average citizen to comprehend. And they pushed the DEP, the designers and contractors not only to comply with regulatory requirements, but to do even better than that. These two groups were tireless community advocates and would often meet with the DEP to discuss the technical issues and merits before presenting the information to the rest of the community at their regular monthly meetings. This step was a critical part of the community outreach program, and kept the community well informed even though the information being transmitted was not always good news. The plant superintendent did an incredible job of always making himself available to the community, helping to explain complex technical engineering and/or construction concepts in regular terms that they could understand. He was always the calm voice during contentious discussions who would tell it like it was, for good or bad. He became someone the community trusted to give them the truth without any sugar coating. He was one of the most important pieces of the community outreach program.

Figure 7 -Newtown Creek WWTP Nature Walk

In addition to the community representatives, the DEP also committed to providing the community with the following:

- State-of-the-Art odor control systems that contained all the odor sources at the plant and treated the collected foul odor in activated carbon vessels before being discharged to the atmosphere.
- Nature walk along the edge of the waterway which was the boundary of the plant.
- Visitor's center at the entrance to the plant that the community could also use for local environmental education and community gatherings.
- Planted hundreds of trees in the neighborhood.

Figure 8 - Newtown Creek WWTP Visitor's Center

In the end, the Newtown Creek community outreach program was effective in working with the community, educating them, keeping them informed of activities that were going on, gaining the community trust, and ultimately achieving the DEP objective of being a "Good Neighbor".

Summary

The community is the driving force when it comes to implementing nuisance odor control at wastewater facilities. They communicate with the utility in different ways, but none as important as the nuisance odor complaint mechanism, which represents the perception of the facility by the public in their neighborhood. This report card of sorts is the community's attempt to reach out to the utility and highlight that things are not going to their liking and the resultant nuisance odors are impacting their quality of life. The utility needs to seize the opportunity to review the data provided by the complaints and observe trends that they can use to improve on the plant's ability to capture, contain and treat the nuisance odors that are released from the facility. In addition, the utility needs to implement a community outreach program of its own to address the residents' perception and inform and educate on the steps they are taking to address the community's concerns with nuisance odors. The utility also has to remember that what they don't tell the community, they community won't know. So, seize the opportunity to reach out to the community, especially when there is positive news, such as the implementation of nuisance odor control measures.

With the help of technology the ability to open a dialogue, particularly with the younger generation, has vastly improved the owners resources to communicate with the community, including avenues to reach out and to receive comment back from the community through the internet using: e-mail, Facebook, Twitter, Instagram, etc. The ability to communicate with the community has expanded and also become easier. But as always, controlling the message and reporting the relevant information is still the key aspect of communicating through any media. Careful and planned messages are important.

There are many community outreach approaches that are available, from simple to complex. The utility must select the approach that fits best and move forward with its implementation. It is through an effective community outreach program that a utility is able to achieve and maintain that "Good Neighbor" status.

INFLUENCE OF BIOSOLIDS PROCESSING ON THE PRODUCTION OF ODOROUS EMISSIONS AT WASTEWATER TREATMENT PLANTS

Ruth Fisher[1], James Hayes[1], Juan Pablo Alvarez-Gaitan[1], Robert Aurisch[2], Dammika Vitanage[2] and Richard Stuetz[1*]

1. UNSW Water Research Centre, School of Civil and Environmental Engineering, UNSW Sydney, NSW, 2052 Australia
2. Sydney Water, Sydney, NSW 2150 Australia

*Corresponding author: r.stuetz@unsw.edu.au

HIGHLIGHTS

- Malodours from biosolids are a barrier to the beneficial reuse of biosolids
- Odours throughout biosolids processing at 8 wastewater treatment plants analysed
- Relationships between odorants, odours, composition and process operation studied
- Biosolids Processing Odour Wheels produced based on odour and operational data

Extended Abstract

Odours from wastewater treatment plants (WWTPs), in particular the biosolids processing area, are frequently perceived as a nuisance by the surrounding community (Hayes et al., 2014). With the increasing need for the beneficial reuse of biosolids and increasing community expectations for quality of life, potential nuisance impacts associated with the odours from the production and re-use of biosolids must be minimised (Hayes et al., 2017a).

Odorants emissions throughout biosolids processing at 8 WWTPs were analysed to identify the full spectra of odorants and olfactory properties. The WWTPs varied in scale, sewer catchment characteristics, stabilisation methods (anaerobic or aerobic digestion), dewatering equipment and onsite storage duration (Figure 1, Table 1). Additionally, the compositions of the sludge and biosolids samples were measured and the operational conditions at the sites reported. The effect of biosolids storage, temporal variations and dewatering on emissions from the biosolids were also studied in a case study (Table 1).

Key odorants

Volatile sulfur compounds (VSCs) were the dominant contributors to nuisance odours (Fisher et al., 2018). The primary sludge and biosolids after dewatering were key locations producing nuisance odours. Other non-sulfur odorants such as *p*-cresol, trimethylamine, indole, hexanal and a range of volatile fatty acids butanoic acid were also identified based on their frequency of detection, odour detection thresholds and/or olfactory analysis methods.

Relationships with process operation

Intra and intersite variations in emissions were attributed to different unit process configurations (specifically thickening, digestion and dewatering), sewer catchments characteristics, chemical dosing and onsite storage durations.

Emissions of non-sulfur compounds throughout biosolids processing were linked to their origin and properties such as solubility or how they sorb to organic matter (Fisher et al., 2017). Greater numbers of tradewaste emitting industries in the sewer catchment appeared to be related to higher concentrations and ranges of aromatics, alkanes and halogenated compounds from the dewatered biosolids. Their emission behaviour during biosolids storage depended on wether they were produced during storage or present from upstream sources.

The olfactory and analytical measurements from the WWTPs were used to identify links between odour characters, key odorants and biosolids processing performance (Hayes et al., 2017b). These were then used to construct the Biosolids Processing Odour Wheel (Figure 2), an easy to use tool to aid onsite odour management and community engagement.

This research identified the range of odorants and sensorial properties of emissions throughout biosolids processing. By understanding why the different odorants classes are present, onsite odour management can be more targeted and specific.

References

Fisher R., Le-Minh N., Alvarez-Gaitan J.P., Moore S. and Stuetz R.M. (2018) Emissions of volatile sulfur compounds (VSC) through wastewater biosolids processing. *Science of the Total Environment* **616-617**, 622-631 (DOI: 10.1016/j.scitotenv.2017.10.282)

Fisher R., Sivret E.C., Le-Minh N., Gaitan J.P., Moore S. and Stuetz R.M. (2017) Distribution and Sensorially relevance of volatile organic compounds emitted through biosolids processing. *Science of the Total Environment* **599-600**, 663-670 (DOI: 10.1016/j.scitotenv.2017.04.129)

Hayes J.E., Stevenson R.J and Stuetz R.M. (2014) The impact of malodour on communities: a review of assessment techniques. *Science of the Total Environment*, **500-501**, 395-407 (DOI: 10.1016/jscitoenv.2014.09.003)

Hayes J.E., Stevenson R.J. and Stuetz R.M. (2017a) Survey of the effect of odour impact on communities. *Journal of Environmental Management* **204** (1), 349-354 (DOI: 10.1016/j.jenvman.2017.09.016)

Hayes J.E., Fisher R.M., Stevenson R.J. Mannebeck C. and Stuetz R.M. (2017b) Unrepresented community odour impact: Improving engagement strategies. *Science of the Total Environment* **609**, 1650-1658 (DOI: 10.1016/j.scitotenv.2017.08.013)

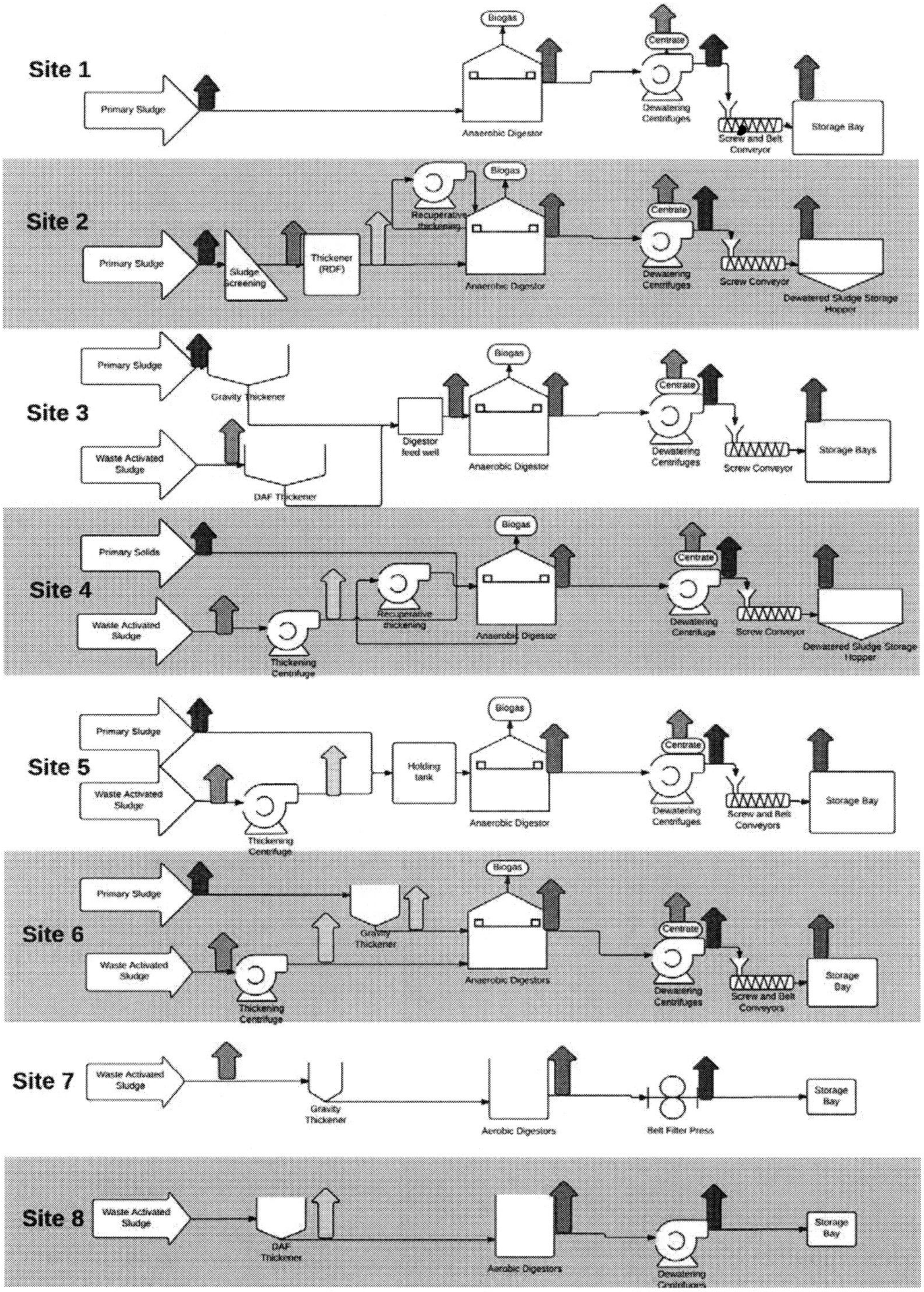

Figure 1. Sludge and biosolids sampling locations throughout the 8 WWTPs analysed.

Table 1. Characteristics of the 8 WWTPs analysed in the overall project.

	Site 1	Site 2	Site 3	Site 4	Site 5	Site 6	Site 3 R **	Site 7	Site 8
PE	1 500 000	1 300 000	240 000	70 000	80 000	200 000	240 000	35,000	115,000
No. of sampling locations	5	7	7	7	7	8	1	3	3
Sewer catchments	Municipal and industrial	Municipal and industrial	Municipal and industrial	Municipal and industrial	Municipal (no registered industrial)	Municipal and industrial	Municipal and industrial	Municipal	Municipal and industrial
Inlet works (PSTs)	Screens, grit removal, PST	Rotary drum screens, grit tanks, high rate PST	Step screens+ aerated grit + PST	Step screens, aerated/vortex grit chambers, PST	Step screens, aerated/vortex grit chambers, PST	Step screens, grit tank, PST	Step screens+ aerated grit + PST	Step screens and vortex grit traps	Step screens and vortex grit traps
Secondary treatment	NA	NA	CAS	CAS	CAS	BNR	CAS	CAS	CAS
Primary thickening		Rotary drum (+ sludge press for screening)	Gravity			Gravity	Gravity	NA	NA
WAS thickening			DAF with polymer	Centrifuge with polymer	Centrifuge	Centrifuges	DAF with polymer	Gravity drainage deck	DAF
Digestion type	Anaerobic	Anaerobic (+recuperative)	Anaerobic	Anaerobic (+recuperative)	Anaerobic	Anaerobic	Anaerobic	Aerobic	Aerobic
Digester SRT (days)	10-12	19	12	20	32	19	21	17 – 23 *	20 *
Centrifuge type	High G	High G	Low G	High G	Low G	High G	High G	Belt filter press	speed unknown

Dissolved air flotation - DAF; *NA* is used where criteria is not applicable; primary sedimentation tank – PST; Population equivalents – PE
*** Averages from 2014;**
**** Site 3 R was used for the case study into the effects of temporal variations and dewatering on odour emissions**

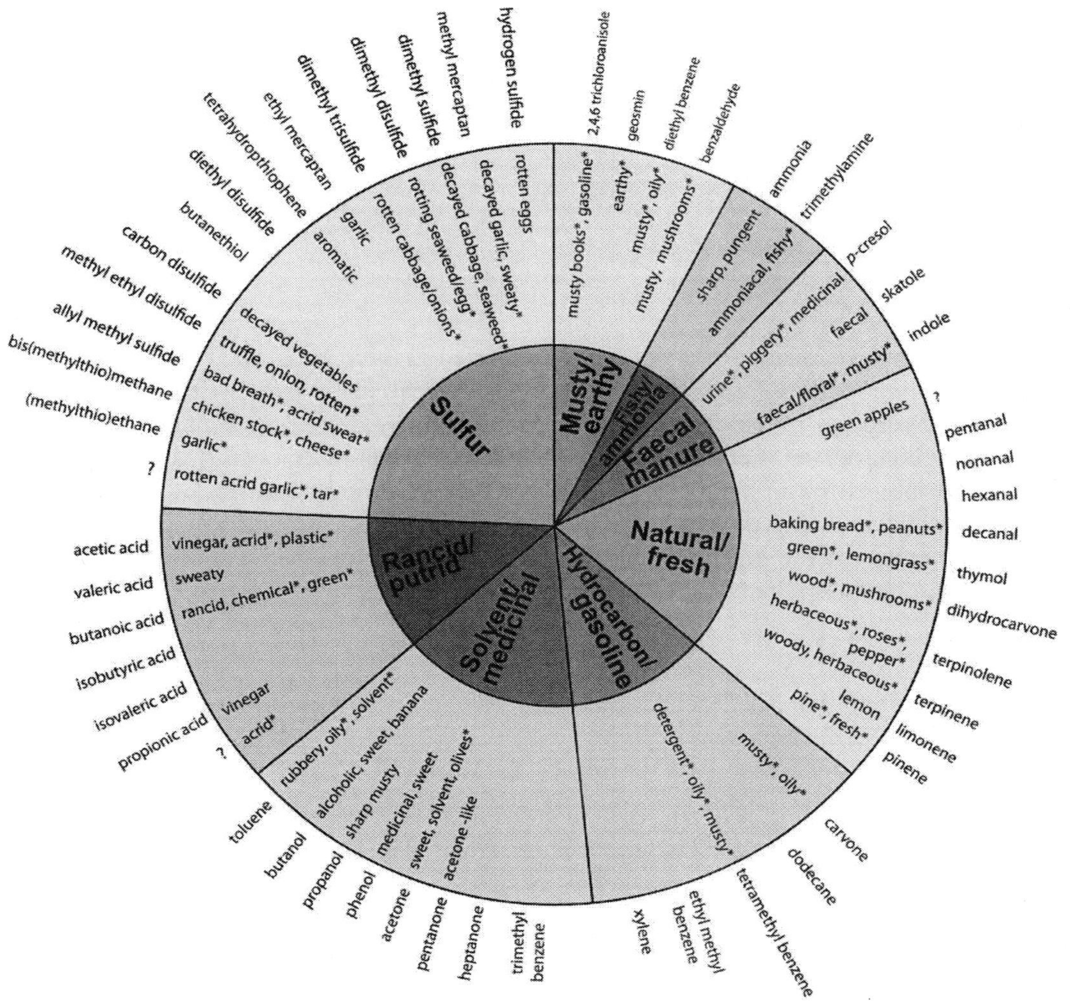

Figure 2. Biosolids processing odour wheel, based on data from various locations throughout the biosolids processing at 8 WWTPs.

Optimizing Hydrogen Peroxide Dosing for Odor Control Originating from the Septage Hauler Dumping Area at the City of Kenosha, Wisconsin's Wastewater Treatment Plant

Michael L. Nelson[1], Brandon Sotiroff[1], Melissa Arnot, P.E.[2], Katrina Karow[2].

[1] USP Technologies, Atlanta, GA.

[2] Kenosha Water Utility, Kenosha, WI.

ABSTRACT

This paper presents the results of a field study initiated in June of 2017 at the City of Kenosha, Wisconsin, Public Utilities Department, Wastewater Treatment Plant, with USP Technologies implementing hydrogen peroxide (H_2O_2) as odor control for a septage hauler dumping area. Furthermore, a dual dosing method consisting of hydrogen peroxide dosing at the septage hauler dump site and the grit basin simultaneously was tested in order to control odors in the primary clarifiers. It was found that dosing into both the hauler dump site and grit basin provided the lowest H2S averages at the grit basin and primary clarifier effluent, while using either the same or less hydrogen peroxide than dosing at either location exclusively.

KEYWORDS

Odor control, septage, landfill leachate, hauler, hydrogen peroxide, hydrogen sulfide, Kenosha, grit, primary clarifier, dosing optimization

BACKGROUND

The City of Kenosha Wastewater Treatment Plant, situated on Lake Michigan between Chicago, Illinois and Milwaukee, Wisconsin, regularly accepts landfill leachate and other septage hauling trucks. The septage hauler dumping area, which is situated outdoors, is large enough to accommodate up to three trucks unloading simultaneously. The septage hauler trucks are allowed to dump between 6 a.m. and 4 p.m. on Monday through Friday and between 6 a.m. and 12:00 p.m. on Saturday. The septage is unloaded by gravity into an underground vault where it mixes with flush water from the final clarifiers. From there it flows by gravity to the headworks (~268 meters or 880 feet). From the headworks building the flow enters an outdoor, uncovered grit basin, after which it flows through a into uncovered primary clarifiers (Figure 1).

The dumping of septage hauler trucks contributes a variable loading to the plant, averaging 20 to 36 trucks/day. This equates to an average of 643 m^3 (170,000 gallons/day). From January through September 2017 the average total daily influent volume to the plant was 100,000 m^3 (27 million gallons), average biological oxygen demand (BOD_5) was 160 mg/L, and average total suspended solids (TSS) was 160 mg/L. The facility is located adjacent to a popular park, beach, nature preserve, and single family housing, making odor control a priority. Kenosha has applied odor control agents in the past, including potassium permanganate.

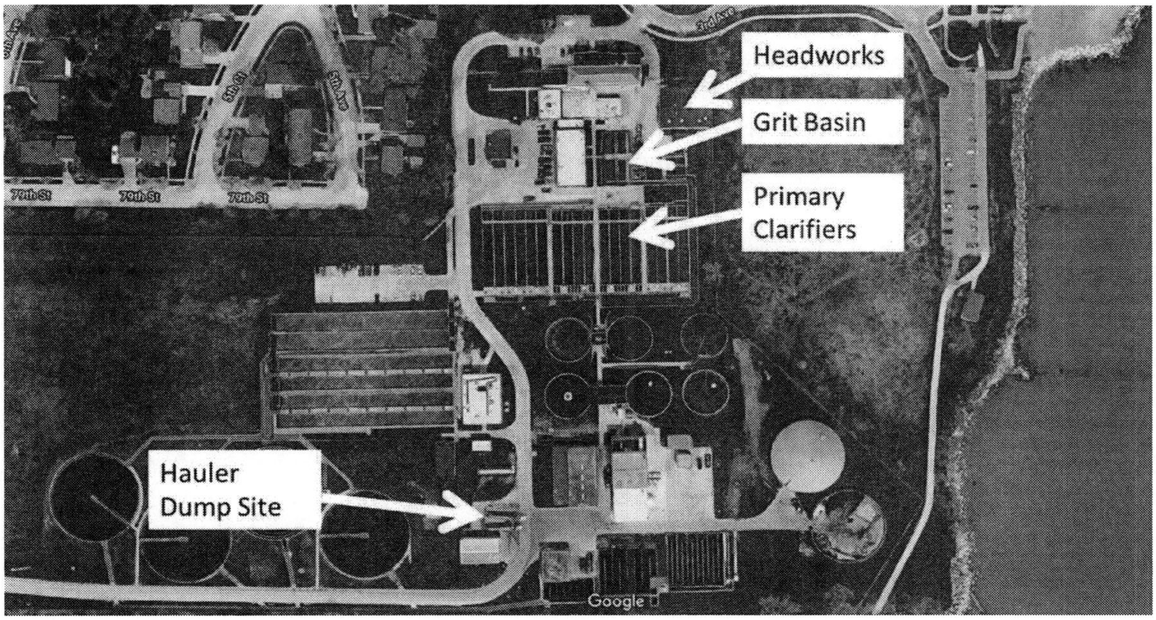

Figure 1. Overview of Kenosha, WI's WWTP with odor control locations labelled. Image courtesy of Google Maps, accessed January 19, 2018.

APPROACH

Within the grounds of the WWTP and upstream of the aeration basins, odorants present could be volatilized from the uncovered grit chambers, flow conveyance channels, and primary clarifiers. The three main sources of potential odors that could volatilize in this section of the plant include: odorants from the septage hauler dumping stream, the influent to the plant (odorants generated in the collection system), and odors generated within the plant (most significantly within the primary clarifiers). The odor control solution developed aimed to address all three of the potential odorant sources at the plant. USP began collaborating with Kenosha in 2012 on measuring dry weather odor conditions, the summary of which is presented below in Table 1.

Table 1. Summary of odor survey conducted by USP, September – October 2012.

Site	Average of Total Sulfide (mg/L)	Maximum H_2S Observed (PPM)
14th Ave/67th St 84"	1.1	58
48" Line (within grounds)	0.2	N/A
72" Line + 84" Line (near entrance)	1.3	N/A
84" Line (outside grounds)	1.3	N/A
Grit Structure	0.4	58
Primary Basin	0.5	11
Aeration basins (influent gates)	0.5	6

In order to provide seasonal and on-demand odor control for a large, uncovered section of the plant, liquid phase treatment was best suited. While ferric chloride is already dosed into the aeration basins for chemical phosphorus removal, concerns about corrosion of the headworks infrastructure ruled out the most accessible existing dosing point into the headworks. It likely warrants further evaluation to determine if dosing of an iron salt further upstream could improve odor control while also meeting the phosphorus removal targets.

Hydrogen peroxide was selected as the odor control agent based on the presence of reduced sulfur compounds (primarily hydrogen sulfide, or H_2S), the detention times between potential injection points and control points, and its relatively low cost. Hydrogen peroxide reacts with hydrogen sulfide under acidic to neutral conditions according to the following reaction:

$$H_2O_2 + H_2S \rightarrow S + 2H_2O$$

The primary objective of this study was to determine the most efficient way to reduce odors in the septage hauler area and the primary clarifier using hydrogen peroxide. Effectiveness was gauged by continuous measurement of gaseous hydrogen sulfide compared against the amount of hydrogen peroxide dosed. The hydrogen peroxide injection points and H_2S monitoring locations are shown below in Figure 2.

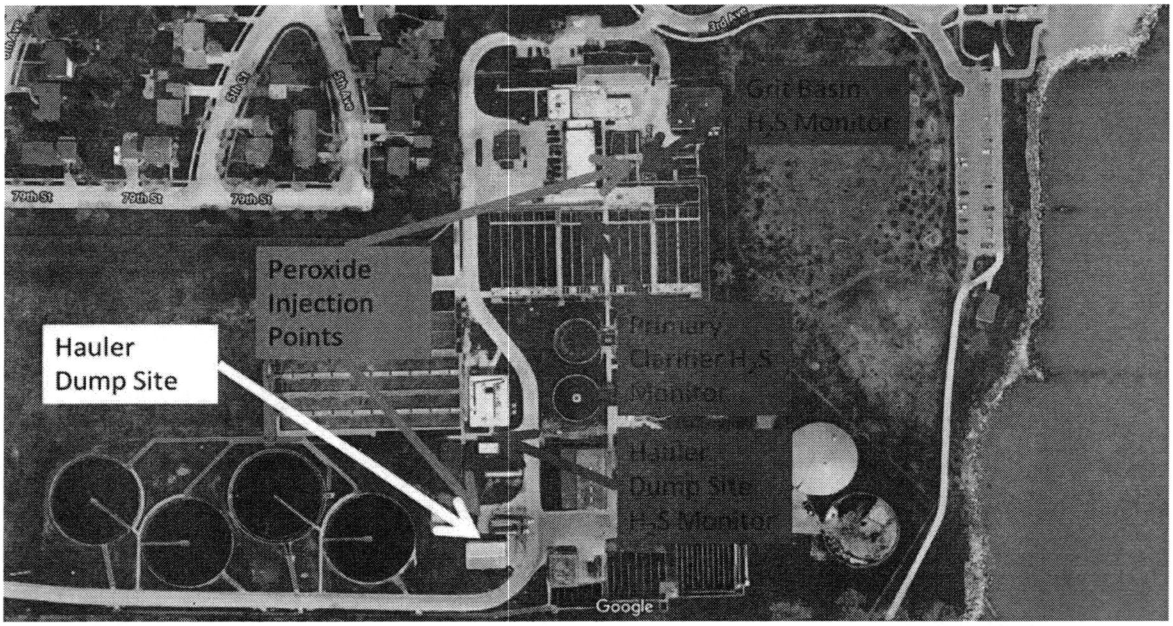

Figure 2. Overview of Kenosha, WI's WWTP with H_2O_2 injection points and H_2S monitoring locations labelled. Image courtesy of Google Maps, accessed January 19, 2018.

The septage hauler trucks arrive at the plant between 6 a.m. and 4 p.m. on Monday through Friday and between 6 a.m. and 12:00 p.m. on Saturday, with none allowed on Sunday. The testing included trying various dosing rates of H_2O_2 solution between these times when haulers would be dumping, with the daily masses of H_2O_2 used (100% basis) ranging from 49.4 kg/day to 56.7 kg/day.

A double-walled minibulk tank containing a 50% H_2O_2 solution was used at the Hauler Dump site injection point and totes of 27% H_2O_2 were used at the Grit Basin injection point. Odor was measured using H_2S meters (Odalogs™) downstream of the Hauler Dump area, at the Grit Basins, and at the weirs at the midpoint of the north end of the primary clarifiers. Gaseous H_2S levels are used throughout this paper as a way to quantify the level of odor since it is highly prevalent in septic waste water, toxic in high concentrations, and easily measured.

RESULTS

The comparable baseline average H_2S levels throughout the summer of 2017, without hydrogen peroxide dosing were: 25.0 ppm at the Hauler Dump site, 1.1 ppm at the Grit Basin, and 1.9 ppm at the Primary Clarifier Effluent (Table 2). The H_2S meter for the Hauler Dump site was suspended within a vault 47 meters (100 feet) downstream along the gravity line through which the septage and leachate flow to the headworks. The Grit Basin and Primary Clarifier H_2S meters were deployed approximately 60 centimeters (2 feet) above the average water level of those locations, which are only partially protected from winds. They were suspended as close to the water level as possible while also avoiding incidental splashing.

The first test condition comprised dosing hydrogen peroxide at 5.67 kg/hour continuously for ten hours/day when the haulers were dumping. This resulted in a daily usage of hydrogen peroxide Monday through Saturday of 56.7 kg and lowered the Hauler Dump Site H2S average to 0.70 ppm and the Primary Clarifier Effluent H2S average to 0.051 ppm.

After this initial phase demonstrated the effectiveness of using H_2O_2 in controlling odor, the focus moved to optimizing chemical usage to meet overall odor control goals. In the next phase of the trial, 3.40 kg/hour H_2O_2 was dosed at a steady rate into the Hauler Dump site vault in addition to 1.70 kg/hour H_2O_2 into the Grit Basin influent splitter box, with a daily total of 49.9 kg. By using 12% less hydrogen peroxide per day, the H_2S average at the Hauler Dump site increased to 2.8 ppm, the H_2S averaged 0.004 ppm at the Grit Basin, and the H_2S average decreased to 0.028 ppm at the Primary Clarifier Effluent.

Table 2. Untreated baseline and treated dosing trial data. Note H_2O_2 quantities are denoted based on mass of actual H_2O_2 used, not the weight of the aqueous solutions.

Condition	Kg/day H_2O_2 (100% basis)	Hauler Dump Site Avg. H_2S (ppm)	Grit Basin Avg. H_2S (ppm)	Primary Clarifier Effluent Avg. H_2S (ppm)
Untreated	0	25.0	1.1	1.9
5.67 kg/h at Hauler Dump Site During Hauler Dumping	56.7	0.70	N/A	0.051
3.40 kg/h at Hauler Dump Site & 1.70 kg/h at Grit Basin During Hauler Dumping	49.9	2.8	0.004	0.028
4.94 kg/h at Grit Basin During Hauler Dumping	49.4	N/A	0.14	0.21

In the final phase of the trial, 4.94 kg/hour H_2O_2 was dosed at the Grit Basin dosing point, while nothing was dosed at the Hauler Dump site. The H_2S average at the Grit Basin was 0.14 ppm while the H_2S average at the Primary Clarifier was 0.21 ppm.

DISCUSSION

The results, shown in Table 2, suggest that the most effective dosing strategy is to dose at both the Hauler Dump Site and Grit Basin or just at the Hauler Dump Site. Unfortunately data was not available from the Grit Basin monitoring location during the Hauler Dump site-only dosing test. When a comparable volume of hydrogen peroxide was dosed exclusively at the Grit Basin H_2S levels at the Grit Basin and Primary Clarifier monitoring locations were higher, we can rule out dosing exclusively at the Grit Basin as the most effective approach.

H_2S was maintained at a 55% lower concentration at the Primary Clarifier monitoring location with 12% less overall hydrogen peroxide dosing when dosing was done at both the Hauler Dump site and Grit Basin as opposed to exclusively at the Hauler Dump site. This appears to be the most cost-effective strategy.

Injecting hydrogen peroxide in two locations allows the flexibility to address all three potential sources of odors in this section of the plant: odorants originating from the collection system entering at the headworks, odorants originating from the hauled in septage and landfill leachate, and odorants generated on-site within the primary clarifiers. Dosing hydrogen peroxide where the haulers dump allows a headstart on oxidizing odorants before they can volatilize at the grit basins. When the haulers are not dumping, which is frequent because they show up rather unpredictably during their allowed times, the hydrogen peroxide would be dosed into low-strength return flows so that a significant residual should arrive at the headworks to oxidize odorants present there entering from the collection system. Finally, dosing hydrogen peroxide at the grit basins allows for treatment of odorants entering the grit basins and provides a way to chemically boost the oxidation-reduction potential (ORP) incrementally within the primary clarifiers to control odorant generation within.

Rainfall was noted to be a significant factor worth considering for two reasons. First, when there are heavy rains around Kenosha plant influent volumes increase significantly. This has been routinely observed to lower the odorant concentrations from the collection system through to the clarifiers. During those time periods hydrogen peroxide dosing could conceivably be reduced. Second, a large percentage (85%) of the hauled-in waste is landfill leachate. The volume of this increases after large rain events.

CONCLUSIONS

Dosing hydrogen peroxide at the Hauler Dump site and at the Grit Basins, either together or each on their own, was able to significantly reduce measured H_2S concentrations at the downstream control points. In order to reduce H_2S concentrations within the conveyance channels, at the grit basins, and through the primary clarifiers it seems that it is most economical to dose hydrogen peroxide concurrently at both the Hauler Dump site and Grit Basin injection points. By doing so odorants levels entering with the septage and landfill leachate haulers, from the collection system, and generated within the primary clarifiers can be controlled. This trial was conducted in the summertime, through wet and dry flow conditions, offering a fairly wide selection of potential odorant loading conditions. These conditions have historically had a higher relative frequency of odor complaints, but odor monitoring may be worthwhile throughout the year as well.

Further investigation may be warranted to determine the appropriate H_2S target concentrations to minimize public nuisance at the lowest cost. Since the odorant concentrations entering the plant and being generated within may vary in an unpredictable manner, online measurement and automation may offer further improvements in efficiencies. Finally, shifting some or all of the ferric chloride dosing for phosphorus removal from the aeration basins upstream into the primary clarifiers may benefit odor control and reduce the demand for hydrogen peroxide.

REFERENCES

Davidson, S., Muller, T., Berrelez, M., & Hinch, K. (2009). ENHANCED CHEMICAL TREATMENT CALMS PERFECT STORM OF ODOR CONDITIONS. *Proceedings of the Water Environment Federation*, *2009*(11), 4619-4633.

Ferrero, T. (2004). PRETREATMENT OF SEPTAGE AND GREASE TRAP WASTES. *Proceedings of the Water Environment Federation*, *2004*(9), 465-472.

Hentz, L. H., Southard, T., Balchunas, B. M., Ferguson, C., & Mody, S. S. (2004). ODOROUS EMISSIONS FROM BIOSOLIDS AND SEPTAGE STORAGE AND DEWATERING FACILITIES. *Proceedings of the Water Environment Federation*, *2004*(1), 554-575.

Reidy, J. G., Moulton, P. W., & McAuley, J. B. (2002). HYDROGEN PEROXIDE PRETREATMENT: IT'S NOT JUST FOR BLONDES ANYMORE (THE USE OF HYDROGEN PEROXIDE TO REDUCE HYDROGEN SULFIDE LEVELS AND TO ACHIEVE OPERATIONAL SAVINGS). *Proceedings of the Water Environment Federation*, *2002*(14), 329-329.

Stallings, R. B., & Alosso, J. (2002). CONTROLLING SEPTAGE HANDLING ODORS AT THE EDGARTOWN, MASSACHUSETTS WASTEWATER TREATMENT FACILITY. *Proceedings of the Water Environment Federation*, *2002*(5), 82-97.

Szczucki, C., Gotham, B., & Young, C. (2014). ODOR CONTROL OF WASTEWATER CONTAINING LEACHATE IN A 14 MILE SEWER. *Proceedings of the Water Environment Federation*, *2014*(3), 1-13.

Zhang, L., De Schryver, P., De Gusseme, B., De Muynck, W., Boon, N., & Verstraete, W. (2008). CHEMICAL AND BIOLOGICAL TECHNOLOGIES FOR HYDROGEN SULFIDE EMISSION CONTROL IN SEWER SYSTEMS: A REVIEW. *Water research*, *42*(1), 1-12.

Food Waste Impact Study: Sewer Process Modelling for Future Options Assessment

N.J.R. Kraakman[1], B. Johnson[2], C. Ransom[3], D. Bassett[4], G. Kedan[5]
[1] CH2M/Jacobs, Portland, OR, USA
[2] City of Los Angeles, Los Angeles, CA, USA
[3] CH2M/Jacobs, San Francisco, CA, USA
[4] Stantec, Los Angeles, CA, USA
[5] CH2M/Jacobs, Los Angeles, CA, USA

ABSTRACT

The City of Los Angeles Bureau of Sanitation (LASAN) has conducted a study to assess the impacts associated with conveying food waste using an existing wastewater collection system. A proposed Organics Processing Facility (OPF) is located 29 kilometer (17 miles) upstream of the Hyperion Wastewater Reclamation Plant (HWRP). The OPF will separate food waste and other digestible organics from non-processable and non-digestible materials to create an organic food waste slurry. The combined waste stream (existing municipal wastewater and food waste slurry) will be treated at HWRP with a portion of the waste converted into renewable energy via anaerobic digestion. This study examines (1) the food waste characteristics and (2) predicts impacts to the conveyance system and the influent wastewater quality to HWRP.

To determine the impacts, the volumes and constituents, including the physical, chemical, and biochemical characteristics of the food waste slurry relevant to process analysis, are defined. Physical-chemical characteristics of the food waste slurry, as well as its biochemical characteristics are detailed based on available technical literature. Sewer process modeling using the Interceptor-powered-by-WATS model is undertaken to predict the impact of the food waste on the HWRP influent water quality and the wastewater conveyance, including sewer ventilation odor treatment systems. The sewer model covers microbial and chemical transformation processes in the conveyance system network related to relevant process parameters such as organic matter, pH and sulfurous compounds under aerobic, anoxic, and anaerobic conditions.

The impacts to the HWRP influent key constituent characteristics are estimated. This includes an approximation of changes to the properties of the food waste slurry within the conveyance system prior to reaching the HWRP. The impact of the additional food waste to the wastewater conveyance system on the existing sewer ventilation odor treatment systems are also estimated. Limitations relative to prediction of waste characteristics and the potential impacts of the additional food waste are noted, with some recommendations provided for further verification and/or analyses.

KEYWORDS: Source Separated Organics, Food Waste Characterization, Sewer Process Modelling, Interceptor-powered-by-WATS, Wastewater Conveyance System, Renewable Energy.

INTRODUCTION

Food waste (FW) refers to disposed mainly unsold food, food preparation leftovers and uneaten food from households, restaurants and large producers such as collective caterers and supermarkets. The carbon footprint of FW is estimated, if FW was a country, as the third top greenhouse gases emitter, after China and USA (Jan et al., 2013). FW is more typical of lifestyles observed in high-income countries (Jan et al., 2013). Nowadays, more FW reaches landfills than any other single material, with over 20% of the municipal solid waste in USA (EPA, 2011). Once in landfills, FW follows natural biodegradation, principally in an anaerobic way, releasing environmental harmful elements as leachates and gases.

The City of Los Angeles (the City) is in the process of reducing the City's landfill disposal and greenhouse gas emissions and is developing facilities capable of food waste recycling. LASAN's Program Management Support Services (PMSS) Team was directed to review the feasibility of food waste processing using existing wastewater infrastructure. The currently proposed Organics Processing Facility (OPF) will result in the generation of several waste streams, including a concentrated food waste slurry. The current approach to managing the food waste slurry is to dilute it to a concentration that allows it to flow in a pipe, and discharge the diluted slurry to the sewer collection system tributary to the Hyperion Water Reclamation Plant (HWRP) for further processing. The proposed OPF will process 300 tons per day (TPD) of source-separated organics (SSO) and food waste materials received from commercial source-separated food waste collection programs. The OPF will separate food waste and other digestible organics from non-processable and non-digestible materials to produce a wet organic slurry that can be anaerobically digested to generate renewable energy at the HWRP or a dedicated waste conversion digester facility.

The purpose of the OPF project is to produce renewable energy and reduce the City's landfill disposal and greenhouse gas emissions as part of its Zero Waste Plan. The OPF will also be an important asset in the City's efforts to provide solutions to local businesses for compliance with the California State Assembly Bill 1826 (State of California, 2014), which requires processing of organic wastes for many organic and food waste generators. It requires businesses that generate a specific amount of organic waste per week to arrange for recycling services for that waste according to a tiered implementation schedule. The law, also known as the Mandatory Commercial Organics Recycling (MORe) program, requires the state's commercial sector, including restaurants, supermarkets, large venues and food processors, to separate their food scraps and arrange for an organics recycling service.

The proposed OPF is to be located at 2201 E. Washington Boulevard in the City of Los Angeles within the existing CLARTS facility. This existing solid waste Transfer Station facility is permitted to receive and process solid waste. The primary goal of the CLARTS

OPF is to provide receiving and processing of source-separated organic (SSO) wastes, expected to be generated and collected in compliance with the Zero Waste LA Exclusive Franchise System. Zero Waste LA is a public-private partnership designed to address the 3-million tons of waste disposed annually by businesses, consumers and residents. The system establishes a waste and recycling collection program for all commercial, industrial and large multifamily customers in the City of Los Angeles. A schematic of the proposed processes at CLARTS OPF to produce the food waste slurry is shown in **Figure 1**.

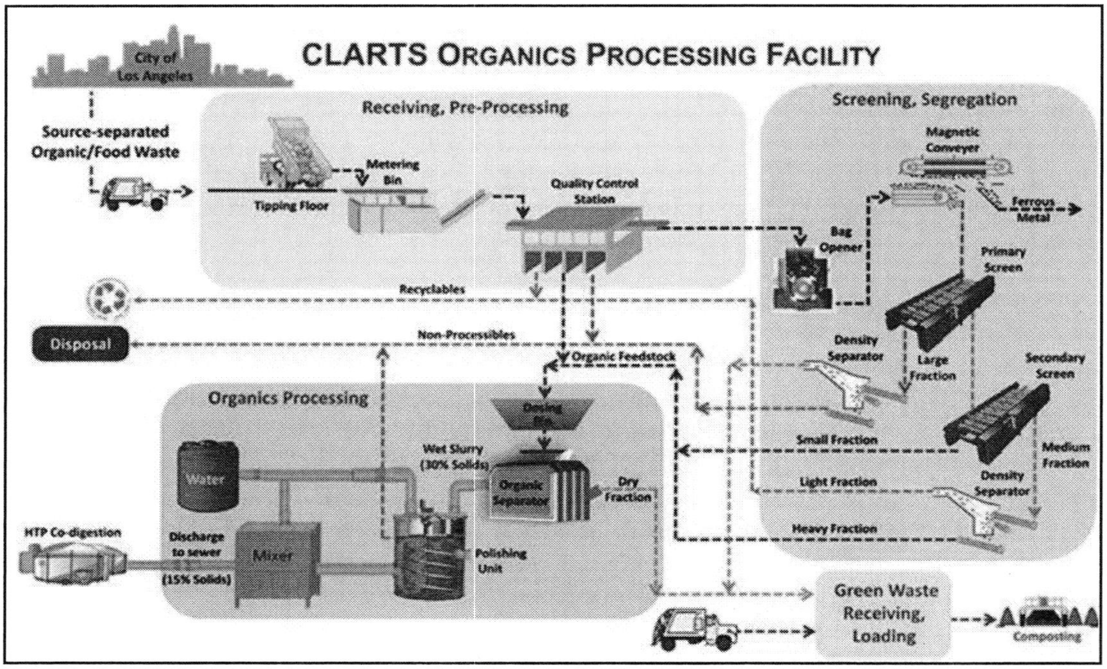

Figure 1: Schematic processing overview of the CLARTS Organic Processing Facility for the option of processed organic slurry conveyance by sewer to HWRP (figure courtesy of JR Miller Associates).

One of the operating options for CLARTS OPF involves the diversion of a food waste slurry from the new recycling facility to the East Central Interceptor Sewer (ECIS)/ North Outfall Replacement Sewer (NORS) sewer interceptors of the HWRP. The routing of the food waste from CLARTS OPF via the ECIS and the NORS Interceptors to the HWRP is shown on **Figure 2**. The total length of the conveyance interceptor system is approximately 29 kilometers (17 miles). Further physical details of the conveyance system are discussed in the section Sewer Process Model Setup Methodology.

At the HWRP, the combined waste stream (existing municipal wastewater and food waste) will be treated with a portion of the waste converted into renewable energy via anaerobic digestion.

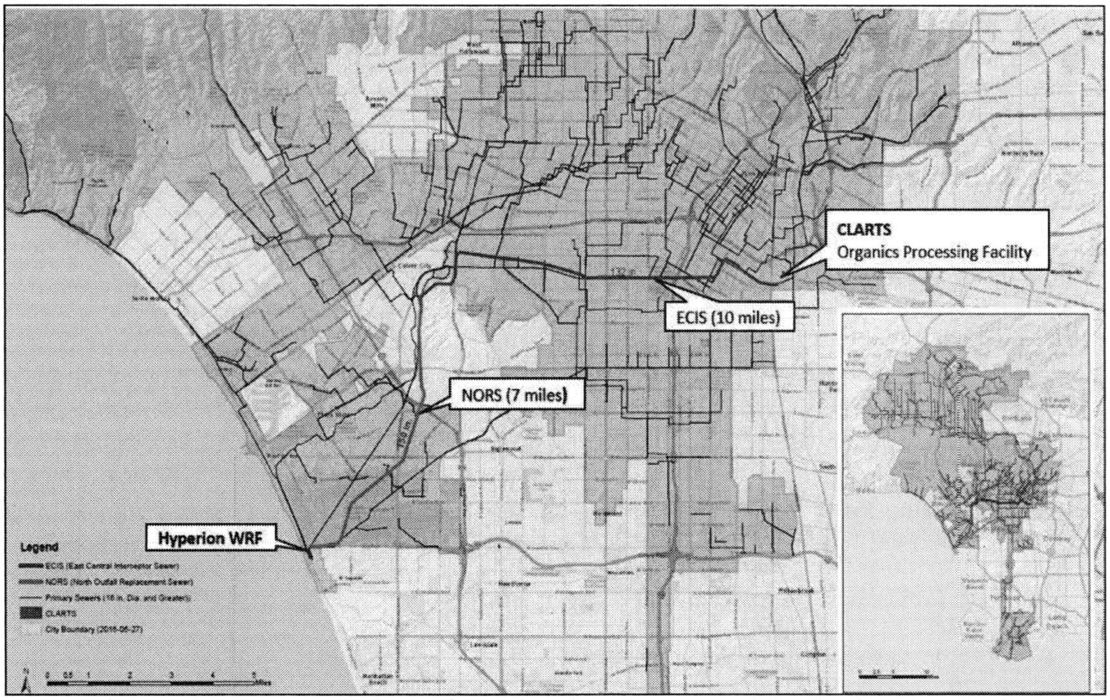

Figure 2: Catchment area of the HWRP showing the route of the food waste via the ECIS and the NORS from the CLARTS OPF to the HWRP.

METHODOLOGY

Food Waste Characterization

The CLARTS OPF represents a new approach to FW management for the City. As a result, there is limited existing information available related to the City's FW characteristics, and as such an evaluation of this approach requires reliance on the information from similar systems in other locations as cited in the literature. Physical-chemical characteristics of the FW, as well as its biochemical characteristics are predicted based on peer reviewed literature. The most relevant FW characteristics are drawn from analyses of FW collected in the City of San Francisco (Zhang et al., 2007), which is also a large city in the state California with similar conditions for collection from similar FW sources. The FW characteristics should therefore be relatively similar to the FW collected and processed at the CLARTS OPF. These FW characteristics were compared and augmented with several other peer reviewed literature sources. Kim et al. (2015), for example, studied a FW that was prepared according to the typical US diet based on information collected by the United States Department of Agriculture. Fisgative et al. (2016) reviewed the FW characteristics on seventy technical papers and give an overview of the general mean values of FW characteristics. This study also highlights the origins of the major variations of FW and discusses the anaerobic digestion behavior against these variations.

Sewer Process Modeling

The model selected for this assessment is the INTERCEPTOR-powered-by-WATS model. INTERCEPTOR is Jacobs's software used to characterize specific in-sewer wastewater transformations within sanitary sewers. To take advantage of recent advances in the field of sewer process modeling, the INTERCEPTOR model was adapted to incorporate the WATS model as an additional engine for predicting sewer system behavior. The WATS model was developed at the Aalborg University of Aalborg in Denmark through pioneering research over the past several decades to improve the understanding of biological, chemical, and physical processes within wastewater conveyance facilities, the understanding of which is key to solving problems related to in-sewer wastewater transformations, such as odor formations and microbial induced corrosion (MIC). The WATS model is continuously extended and improved as knowledge of in-sewer processes increases. In brief, sewer process modeling typically aims at understanding one or more of the following in-sewer problems to develop related asset management strategies:

- Concrete corrosion caused by hydrogen sulfide
- Odor emissions and odor control methods
- Hydrogen sulfide impacts on human health
- Off-gassing "hot spots" along sewer alignments due to in-sewer pressurization
- Analysis of wastewater quality at inflows to wastewater treatment plants

With appropriate physical inputs and field sampling data, the model can be calibrated and then used to simulate future conditions, identify prospective intervention strategies, optimize those strategies, estimate costs for the best asset management options, and develop design guidance to inform capital improvement projects.

A description of the current state of the WATS model can be found in the book Microbial and Chemical Process Engineering of Sewer Networks (Hvitved-Jacobsen et al., 2013) and related journal publications. The book provides detailed descriptions and mathematical formulations of the key processes included in the model.

Sewer Process Model Setup Methodology

A systematic approach was taken to setting up the model using the best available data. Two model scenarios were developed, the *Baseline Without Food Waste* model is the sewer system without the food waste input. The *Baseline With Food Waste* model is identical to the baseline model, but has the food waste discharge point to the sewer with the food waste having its own defined characteristics. The model setup methodology are detailed below, including the conveyance system physical geometry, wastewater flow, and baseline wastewater quality.

Conveyance System Physical Geometry

Pipe physical geometry was obtained from the City's MIKE URBAN hydraulic model. The piping configuration among the list of 56 subject pipes was checked and includes two siphons which were modelled as being force mains. One section of the ECIS involves a short 100 meter (300 feet) underpass under the road, consisting of three parallel siphons, which have been consolidated into a single pipe with a cross-sectional area equivalent to the sum of the three pipes. Because the lengths between nodes were maintained, pipe consolidation will have minimum effect on the process model. A second siphon, approximately a mile-long underneath Interstate 405, was identified and also modeled as a force main (see **Figure 2**).

Two large odorous air treatment facilities are located along the conveyance system between the CLARTS OPF and HWRP. The Odor Control Facilities ECIS/LCSFVRS ATF #809 (located on the corner of Jefferson & La Cienega) and NORS/ECIS Carbon Scrubber #811 (located on 10100 Jefferson Pl Baldwin Hills, Culver City). They extract odorous air from the interceptor at a design rate of respectively 20,000 cubic feet per minute (cfm) and 10,000 cfm (see **Figure 2**). Data gathered on ATF #809 indicate a flowrate from the ECIS interceptor of approximately 13,500 cfm.

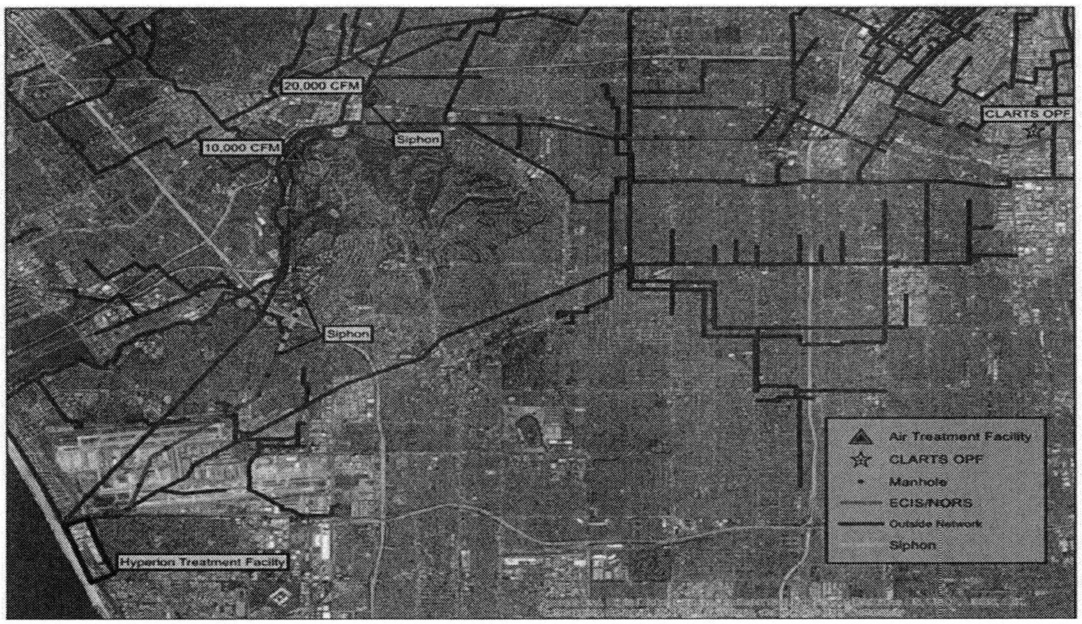

Figure 2: A map of the conveyance system (ECIS/NORS) from the CLARTS OPF to the HWRP showing the locations of the odor control facilities and siphons.

Drop structure configuration influences re-aeration and H_2S mass transfer (stripping) at drops. In general drop height was modeled as the difference in invert elevations at nodes minus the tail water depth. The tallest drop in the system is approximately 1.4 meter (4.5 feet). No vortex drops, down-pipe drops, or other special engineered drop configurations are present in the sewer modeled domain.

Figure 3 shows the modeled conveyance system profile of the sewer and in relation to the ground level between the food waste dosing point (CLARTS OPF) and the discharge point (HWRP).

Pipe material (that is the inner pipe surface material) has an important influence on the headspace H_2S concentration with different materials acting as a H_2S 'sink' to varying degrees. Pipe material is all reinforced concrete with a corrosion protective liner.

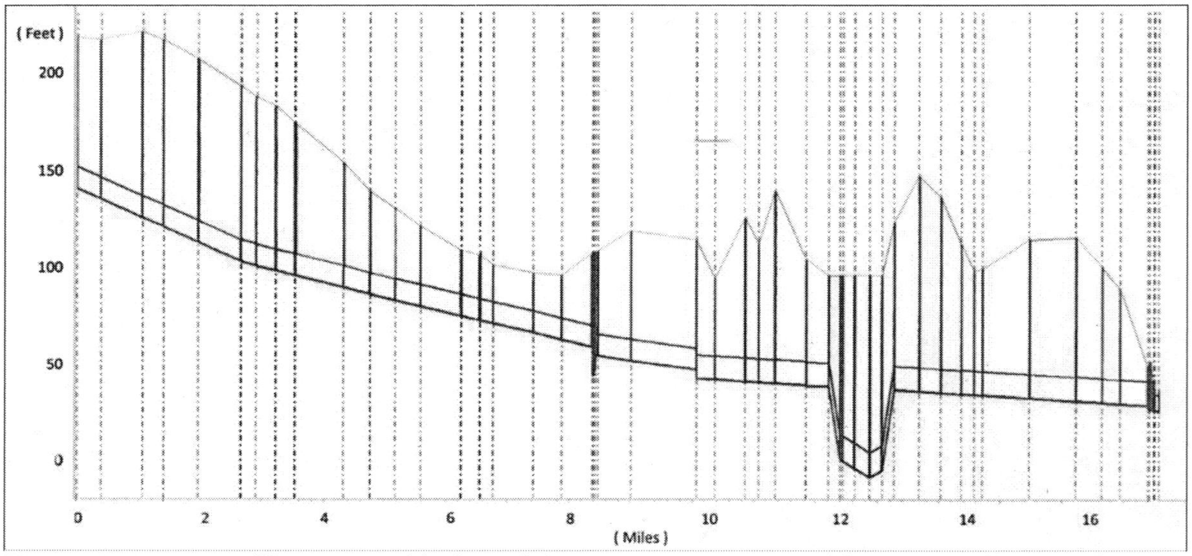

Figure 3: Sewer profile measured in feet of the conveyance system (blue colored lines) and in relation to the ground level (green colored line) of the sewer modeled domain.

Wastewater Flow Input

Wastewater flow is a key input parameter to the sewer process model. Wastewater flow influences detention time and the relative volumes of water and air in a gravity pipe. It also influences ventilation, septicity, and fermentation, all which mechanisms are included in the sewer process model. Flow inputs were taken from the hydraulic model and the average dry weather scenario was implemented – average conditions being germane to biogenic corrosion, average odor loadings to the relevant existing odor control facilities and average wastewater transformations in the sewer. The hydraulic model flow inputs were checked against plant influent records for dry weather conditions. The sum of inputs for the average dry weather hydraulic model scenario agrees with the average 2016 dry weather influent to the HWRP to within approximately 15 percent. The individual water flows into the interceptor from all side streams is shown in the schematic diagram below on **Figure 4**.

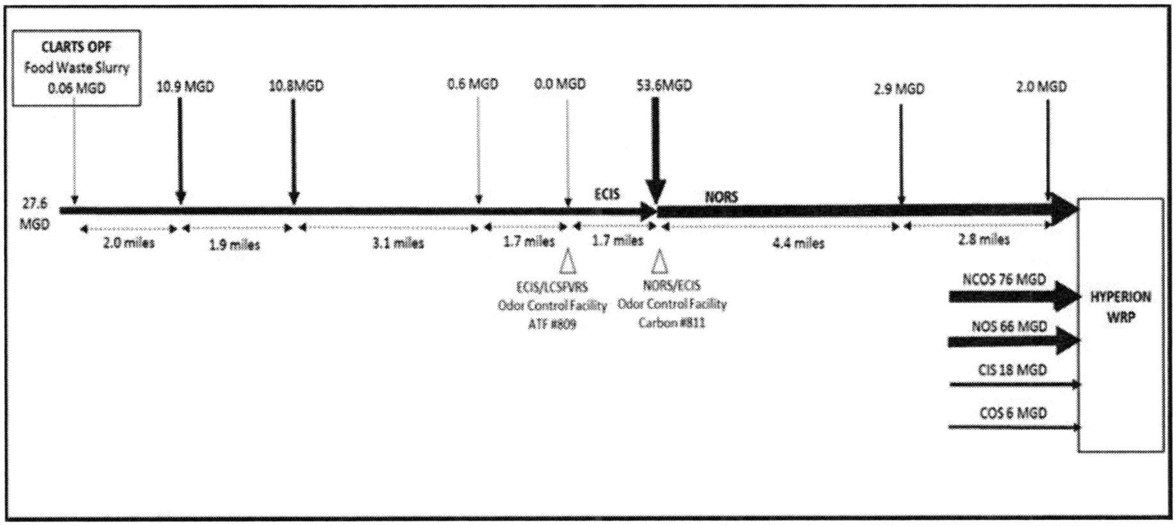

Figure 4. Individual water flows and the relative distances between the side streams of the interceptor between CLARTS OPF and HWRP.

The City operates several major flow diversion structures used to modulate flow between the different main interceptors. Since they are not used routinely, the process model does not incorporate any flow diversions.

Wastewater, Sewer Headspace and Food Waste Quality Input

Default model (baseline) parameters for defining influent wastewater characteristics are based on the City of San Francisco sewer network, and were used in this model except for pH, Chemical Oxygen Demand (COD), and temperature. The pH value used in the model was 7.71, calculated from the HWRP 2016 average dry season pH influent values. This pH is relatively high for typical wastewater, due to the continuous chemical dosing of magnesium hydroxide slurry as a non-hazardous means to regulate the pH of its wastewater to control odors. The interceptor COD concentration (977 mg COD per liter [COD/L]) was calculated from the 2016 HWRP average dry season influent BOD concentrations. An assumed BOD : COD ratio of 1 : 2.3 was used to convert BOD concentration to COD concentration. In addition, the average wastewater temperature was set at 27 °C (80 °F), which was obtained from data collected at a nearby sewer network.

Sewer headspace H_2S data at three locations was used to calibrate the model. H_2S that was measured at manhole 53509022 near the NORS/ECIS Junction close to the inlet of the existing Odor Control Facility (Carbon #811). H_2S that was measured at manhole 55905006 along NORS before the syphon under the highway at intersection Bristal Pkwy and Hannum. The average H_2S concentrations were taken from the City of Los Angeles LA

Sanitation Collection System Odor Control Master Plan 2017 report (City of Los Angeles, 2017). In addition, H_2S was measured at the inlet of the existing Odor Control Facility (ATF #809) during the initial start-up. The average H_2S concentration was taken of the data collected over several months.

Food Waste processing is assumed to occur during the day 7 days per week, creating a steady food waste slurry output of 140 tons/day into the conveyance system. The FW quality is detailed showing physical-chemical characteristics as well as its biochemical characteristics of the food waste slurry. The three main classes of organic solids in the wastewater are carbohydrates, proteins, and lipids. Other, more inert classes of organic compounds such as hemicellulose, cellulose and lignin are also present in both wastewater and food waste. Under aerobic conditions, some proteins and carbohydrates can decompose within hours under the right conditions, but lipids and the more inert organic compounds remain largely unchanged.

Traditionally, total suspended solids (TSS) in wastewater is defined based on a filtration procedure. The classification of wastewater in terms of size distribution is normally done from a practical point of view, where a distinction is made between soluble, colloidal and suspended. However, when dealing with microbiological processes in sewers, an extension of the definition "solids" is required. From a sewer process modeling perspective, it is more important to distinguish organic compounds by rate of decomposition rather than by organic class or filtration procedure. It is important to distinguish between soluble substrates that can penetrate the cell wall of the micro-organisms responsible for decomposition, and particulate substrate that must undergo hydrolysis before utilization. Typically, organic compounds are either readily biodegradable substrates (S_S) or hydrolysable substrates (X_S). **Figure 5** illustrates these size classifications of wastewater, which are also applicable to food waste slurry.

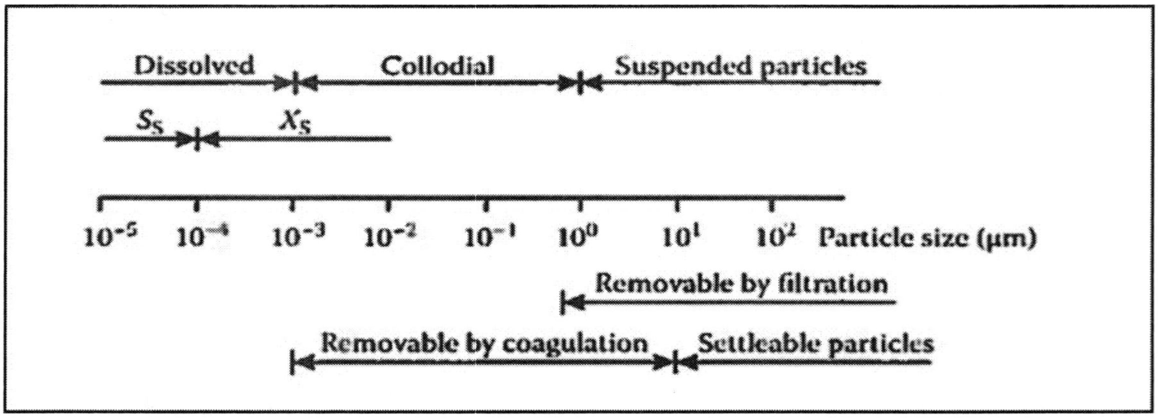

Figure 5: Size classification of wastewater components (Hvitved-Jacobsen et al., 2013)

Readily biodegradable substrates (S_S) can be divided up into readily (fermentable) biodegradable substrate (SF) and volatile acids/fermentation products (S_A). The hydrolysable substrates (X_S) can be divided up into fast, medium, and slow hydrolysable.

Heterotrophic organisms are the organisms predominantly responsible for wastewater degradation, and are expressed here as heterotrophic biomass in water (X_{HW}). All the above fractionations are typically expressed in units of mg COD/L. The input values for the sewer process modeling are shown in **Table 1**.

Table 1. COD Fractionation Input Values for the Sewer Process Modeling

COD fraction	Wastewater (% of total COD)	Food Waste Slurry (% of total COD)
Readily (fermentable) Biodegradable Substrate (S_F)	6	10
Volatile acids/Fermentation Products, mostly VFA (S_A)	0	3
Fast Hydrolysable Substrates (X_{Sf})	25	20
Medium Hydrolysable Substrates (X_{Sm})	30	22
Slow Hydrolysable Substrates (X_{Ss})	36	45
Heterotrophic Biomass in Water (X_{HW})	3	0

Sensitivity Analyses

A sensitivity analysis was performed to understand the sensitivity towards important parameters and determine the impact on the wastewater quality entering the HWRP and the impact on the conveyance system.

Sensitivity analyses were performed on *the Baseline Without Food Waste Model* for COD, dissolved oxygen, flow, temperature, and dissolved H_2S. *The Baseline Without Food Waste Model* sensitivity analysis explores variations in the wastewater entering HWRP from the NORS Interceptor as a result of changes in the existing sewer wastewater parameters.

Seasonal changes resulting from production changes at large food producers can generate significant increases during the peak seasons. Sensitivity analyses were therefore also performed on the food waste itself in *the Baseline With Food Waste Model* for pH, total COD, and COD fractionation. *The Baseline With Food Waste Model* sensitivity analysis explores variations in the wastewater entering HWRP from the NORS Interceptor as a result of changes to the assumptions of the food waste parameters.

RESULTS

PART 1: Food Waste Characterization

The OPF will process organics and waste materials received from SSO waste collection programs. Source-separated FW collection programs are basic to LASAN's vision and strategy to enable commercial businesses within the City to comply with State regulations for organics collection and processing. SSO wastes are expected to be generated and collected in compliance with the Zero Waste LA Exclusive Franchise System. These SSO wastes will come primarily from high food waste generators such as food service establishments, food markets and distribution and businesses with internal food scrap separation programs. The FW slurry will be mainly composed of unsold food, food preparation leftovers, and uneaten food from households, restaurants, and large producers such as collective caterers and supermarkets. The CLARTS OPF is therefore based on a source-separated food waste collection program. The OPF will separate food waste and other digestible organics from non-processable and nondigestible materials to create an organic slurry. The characteristics of the FW is detailed below.

Chemical Composition

For the characterization of the FW, the basic formula $C_aH_bO_cN_dS_e$, is used, where:

Coefficient	Value	Description	Reference
a	21.53	Stoichiometric coefficient for carbon	Diggleman and Ham, 2003
b	34.21	Stoichiometric coefficient for hydrogen	Diggleman and Ham, 2003
c	12.66	Stoichiometric coefficient for oxygen	Diggleman and Ham, 2003
d	1	Stoichiometric coefficient for nitrogen	Diggleman and Ham, 2003
e	0	Stoichiometric coefficient for sulfur	Diggleman and Ham, 2003

The value for sulfur is not zero, but close to zero as can be seen from the elementary composition reported by Fisgativa et al. (2016) when expressed as percent on dry weight basis as shown:

Element	Value		Description	Reference
C	45.5	%	percent on dry weight basis	Fisgativa et al., 2016
H	7.0	%	percent on dry weight basis	Fisgativa et al., 2016
O	33.2	%	percent on dry weight basis	Fisgativa et al., 2016
N	2.8	%	percent on dry weight basis	Fisgativa et al., 2016
S	0.4	%	percent on dry weight basis	Fisgativa et al., 2016

Solids Content

The solids content of the food waste (FW) can be expressed by a variety of parameters, each providing specific information on the waste characteristics.

Parameter	Value	Unit	Description	Reference

FW_{TS}	30.9	%	Total solids of food waste (percent on wet weight basis)	Zhang et al., 2007
FW_{VS}	26.35	%	Volatile solids, or organic matter (percent on wet weight basis)	Zhang et al., 2007
$FW_{TSS:TS}$	62	%	Total suspended solids to total solids ratio of food waste	Li et al., 2002
$FW_{density}$	31	pounds (lb) per cubic foot	Density of food waste	Tchobanoglous et al., 1993
FW_C	46.8	%	Total carbon of food waste (percent on dry weight basis)	Zhang et al., 2007
FW_N	3.2	%	Total nitrogen of food waste (percent on dry weight basis)	Zhang et al., 2007
FW_P	0.5	%	Total phosphorus of food waste (percent on dry weight basis)	Zhang et al., 2007

Volatile Solids to Total Solids Ratio
This ratio provides a relative measure of the organic fraction of the FW.

- $FW_{VS:TS} = FW_{VS} / FW_{TS} = 0.85$

Molecular Weight
The molecular weight is expressed in terms of the principal components.

- FW_{MW} $12.01\ C_a + 1.01\ H_b + 16\ O_c + 14.01\ N_d + 32.07\ S_e = 510$ g/ mol

Carbon : Nitrogen (C:N) Ratio
The ratio of carbon to nitrogen in the waste is an important parameter relative to biological treatment efficiency and nutrient removal capabilities. Using the City of San Francisco's measured ratio (Zhang et al., 2007), the projected ratio for the City's food waste is:

- C:N ratio = 15:1 Carbon to nitrogen ratio (g/g)

The C:N ratio for FW can vary significantly and typical ranges from 12:1 to 31:1 (Fisgativa et al., 2016).

Micro Elements
The content of various micro elements is based on wet weight basis as determined by Zhang et al. (2007) and converted to content on dry weight basis (FW with a total solid content of 30.9 percent).

Parameter	Value (ppm on wet weight basis)	Value (ppm on dry weight basis)[1]
S (total)	2508	8117
Al (total)	1202	3890
Fe (total)	766	2479
B (total)	12	39
Zn (total)	76	246
Mn (total)	60	194
Cu (total)	31	100
Cd (total)	< 1	< 3
Cr (total)	3	10
Pb (total)	4	13
Ni (total)	2	6
[1] Dry weight basis of 30.9 percent.		

FW has relatively low presence of heavy metals as also reported by Fisgativa et al. (2016) with the following averages and standard deviations on dry weight basis: Cu = 23 parts per million (ppm) (37), Cd = 0.3 ppm (0.4), Cr = 28 ppm (71), Pb = 18 ppm (46), Ni = 10 ppm (25), Hg = 0.3 ppm (0.4).

pH

The pH of food waste can change significantly depending upon its "age." When the food waste is "fresh," generally the pH will be slightly acidic, around a pH of 5 (Kim et al., 2016; Fisgativa et al., 2016). This pH can vary and will gradually decrease over time during the storage, collection, and processing of the food waste to a slurry. The increase in acidity is the result of fermentation processes generating volatile acids such as acetic acid, lactic acid, propionic acid, and butyric acid. The food waste slurry to be produced at CLARTS is expected to have a pH as low as approximately 4.0 when introduced into the conveyance system.

Chemical Oxygen Demand

Chemical Oxygen Demand (COD) is one way to measure the relative organic strength of a waste. For complete conversion of organic matter, aerobic decomposition occurs via the following formula:

- $C_aH_bO_cN_dS_e$ +(2e+(4a+b-2c+3d)/4) O_2 --> aCO_2 + ((b-3d)/2) H_2O + dNH_3+eSO_4^{2-}

On this basis, the calculated COD for the food waste is:

- FW_{COD} = 1.54 lb/lb COD of food waste (lb COD per dry lb food waste)

This may represent a conservative value since it is calculated using the aforementioned chemical formula of food waste based on typical formulae for proteins, carbohydrates, lipids, fats, etc. A lower experimentally-measured COD is often observed, possibly because a large part of the organics in food waste such as carbohydrates has a lower COD value than average wastewater and that some may not be completely oxidized in the COD test as discussed by Kim et al. (2015). Their research showed an experimentally-measured FW_{COD} of 1.21 lb/lb using food waste that was prepared according to the typical U.S. diet collected by the U.S. Department of Agriculture.

Biochemical Characteristics of the Food Waste

The projected biochemical characteristics of the food wastes are summarized below.

Parameter	Value	Unit	Description	Reference
Carbohydrates	36	%	Sugars/carbohydrates[1] of the total volatile solids	Fisgativa et al., 2016
Proteins	21	%	Proteins of the total volatile solids	Fisgativa et al., 2016
Fats	15	%	Fats of the total volatile solids	Fisgativa et al., 2016
Hemicellulose	9	%	Hemicellulose of the total volatile solids	Fisgativa et al., 2016
Cellulose	9	%	Cellulose of the total volatile solids	Fisgativa et al., 2016
Lignin	7	%	Lignin of the total volatile solids	Fisgativa et al., 2016
[1] Excluding structural carbohydrates such as cellulose, hemicellulose, and lignin.				

Biochemical Oxygen Demand

The biochemical oxygen demand (BOD) represents one means to measure the biodegradability and relative organic strength of a waste. FW_{BOD} = 82 percent of food waste VS that is biodegradable (assume VS is reasonable approximation for BOD) (Tchobanoglous et al., 1993)

- FW_{BOD} = 1.26 lb/lb BOD of food waste (lb BOD per dry lb food waste)

Biomethane Potential

The biomethane potential (BMP) provides information on biodegradability of the waste and potential for methane production under anaerobic conditions.

- BMP = 435 milliliter (mL) of CH_4 per kilogram (kg) of total volatile solids digested (Zhang et al., 2007)

The BMP of food waste is typically higher than that of most organic waste (for example, yard waste, cattle manure, and sewage sludge). The above value was measured from FW collected in the City of San Francisco, California, after a period of 28 days of digestion with approximately 80 percent of the methane obtained after the first 10 days of digestion. This methane yield is lower than values reported by others including 472 mL/kg VS by Cho and Park (1995), which investigated Korean food waste, and 460 mL/kg VS by Fisgativa et al. (2016) who averaged results from 70 technical papers on analyses on food waste from countries in Asia, Europe, and North America.

Suspended and Soluble Fractionation of the Food Waste

Grinding food wastes impacts the particle size and hence solubilization of organics, nitrogen, and phosphorus. Wastewater treatment plants like HWRP consist of liquid and solid treatment trains; therefore, understanding the fractionation of the food waste organics, nitrogen, and phosphorus in both particulate and liquid phases is important. The values of the soluble fraction are obtained assuming fresh food waste based on the typical U.S. diet collected by the U.S. Department of Agriculture, but they may be slightly underestimated because they do not assume storage, collection, and processing of food waste to a slurry, which would initiate hydrolyses and fermentation processes. The food waste collection and processing time is expected to be one to one and a half weeks for the CLARTS OPF.

The projected suspended and soluble fractionation values for food waste are summarized below.

Parameter	Value	Unit	Description	Reference
$FW_{S\text{-}COD}$	40-45	%	Percent soluble COD of the total COD	Kim et al., 2015
$FW_{S\text{-}BOD}$	30-35	%	Percent soluble BOD of the total BOD	Kim et al., 2015
$FW_{S\text{-}N}$	25-30	%	Percent soluble nitrogen of the total nitrogen	Kim et al., 2015
$FW_{S\text{-}inorganic\text{-}N}$	2.5	%	Percent soluble inorganic nitrogen of the total nitrogen	Kim et al., 2015
$FW_{S\text{-}P}$	< 5	%	Percent soluble phosphorus of total phosphorus	Kim et al., 2015

This means 90 percent of the soluble nitrogen is organic nitrogen and 10 percent inorganic nitrogen. All the inorganic nitrogen can be assumed to be all total ammonia nitrogen (TAN), which is NH_3 and NH_4^+. This assumption assumes negligible amounts of nitrites and nitrates, although some studies have shown evidence of small amounts of nitrites and nitrates in food waste (Fisgativa et al., 2016).

PART 2: Sewer Process Modeling

Impact on Influent HWRP Wastewater Quality

The impact of food waste on the HWRP influent wastewater quality from the NORS Interceptor is summarized in **Table 3**.

Table 3. Impact of Food Waste on the HWRP Influent Wastewater Quality.

Parameter	Increase	Units
Flow[a]	0.062	MGD
pH[a]	-0.35	
Dissolved sulfide	170	kg-S/day
Sulfate	-260	kg-S/day
COD	59,800	kg-COD/day
Total Nitrogen	1,240	kg-N/day
Ammonia (+ ammonium)	31	kg-N/day
Total Phosphorus	204	kg-P/day
Total Dissolved Solids	24,327	kg/day
Total Suspended Solids	14,910	kg/day

a. Flow and pH only impact the NORS interceptor portion (about 30%) of the total HWRP influent. The effect on pH entering the plant will be less due to dilution with the wastewater flows from other interceptors (see Figure 4).

kg/day = kilogram(s) per day
kg-N/day = kilogram(s) of nitrogen per day
kg-S/day = kilogram(s) of sulfide per day

Figure 6 shows how COD changes through the ECIS/NORS conveyance system from the food waste injection point for the *Baseline Without Food Waste Model* and *the Baseline With Food Waste Model*. This figure illustrates that the concentration of COD in *the Baseline Without Food Model* gradually decreases from CLARTS to the HWRP through the ECIS/NORS Interceptor. When Food Waste is added to the Interceptor at CLARTS, the COD concentration in the sewer immediately downstream of the injection point is much higher and a dilution effect can be seen at each location where another sewer line intersects the main interceptor. The addition of food waste increases the COD concentration in the ECIS interceptor just downstream of the CLARTS OPF from 976 mg/L to 1550 mg/L. The addition of food waste increases the COD concentration at the end of the NORS interceptor (entering HWRP before mixing with other interceptors) from 960 mg/l to 1100 mg/L. The impact of the food waste on the individual wastewater COD fractions as it enters the HWRP is shown in **Table 4.**

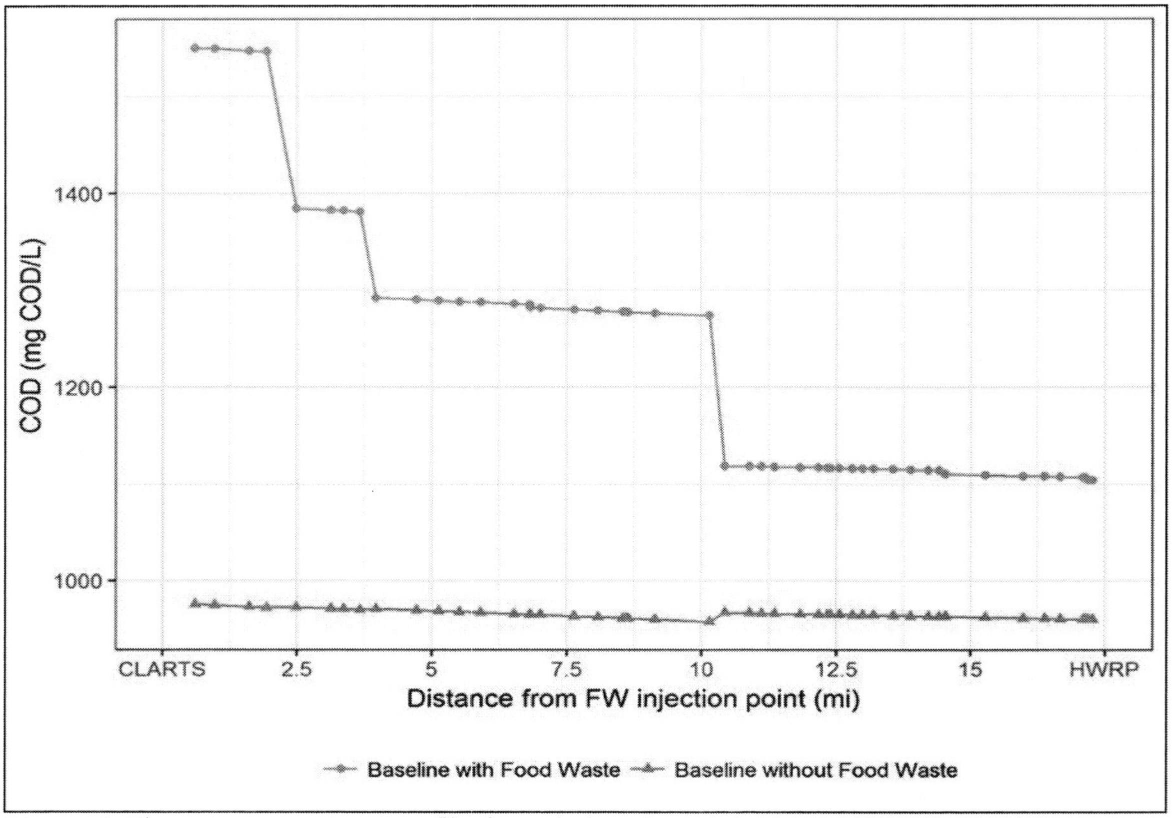

Figure 6: The wastewater COD profile through the conveyance system from the food waste injection point (CLARTS) for *the Baseline Without Food Waste Model* and *Baseline With Food Waste Model* to the treatment facility (HWRP).

Table 4. Impact of Food Waste on the HWRP Influent Wastewater COD Quality.

Parameter	**Increase (kg-COD/day)**
Fermentable Substrates (S_F)	6,010
Fermentation Products (S_A)	826
Heterotrophic Biomass (X_{HW})	48
Sulfide Forming Biomass (X_{SW})	538
Fast Hydrolysable Substrates (X_{Sf})	11,930
Medium Hydrolysable Substrates (X_{Sm})	13,300
Slow Hydrolysable Substrates (X_{Ss})	27,170
Total COD (sum of COD fractions)	**59,800**

Impact on Sewer Network

Sewer headspace H_2S data from several locations is used to calibrate the model. H_2S that was measured at manhole 53509022 near the NORS/ECIS Junction close to the inlet of the existing Odor Control Facility (Carbon #811) was on average 22 ppm. H_2S that was measured at manhole 55905006 along NORS before the syphon under the highway at intersection Bristal Pkwy and Hannum was on average 28 ppm. In addition, H_2S that was measured at the inlet of the existing Odor Control Facility (ATF #809) during the initial start-up. The average H_2S concentration was taken of the data collected over several months and was 35 ppm. Recent H_2S data were also obtained from NORS/ECIS Carbon Scrubber #811, but there is currently construction in the area and the sewer flow has been diverted, so that recent H_2S data are likely to be lower than under standard operations without the flow diverted. Given the recent construction and diversion, it was determined that the H_2S data received from NORS/ECIS Carbon Scrubber #811 inlet were not representative and therefore not used to calibrate the model.

Figure 7 shows the modelled H_2S concentrations in the headspace of the ECIS/NORS interceptor for *the Baseline Without Food Waste Model* and *the Baseline With Food Waste Model*. **Figure 7** also shows the odor control facility extraction points. This figure illustrates that there will be an increase in H_2S levels throughout the sewer with the addition of food waste. **Table 5** shows the increase in H_2S and methyl mercaptans (MM) at the two odor control facilities as a result of the food waste addition. The predicted H_2S increase at the carbon scrubber is anticipated to be higher than the H_2S increase at the biotrickling filter based on the calibrated model with limited measured data.

Table 5. H_2S and Methyl Mercaptan Increase Predicted at the Odor Control Facilities as a Results of the Food Waste Addition.

Location	H_2S increase (ppm)	MM increase (ppm)
ATF #809 (Biotrickling Filtration with Activated Carbon polishing)	40	0.01
CS #811 (Activated Carbon Scrubber)	65	0.03

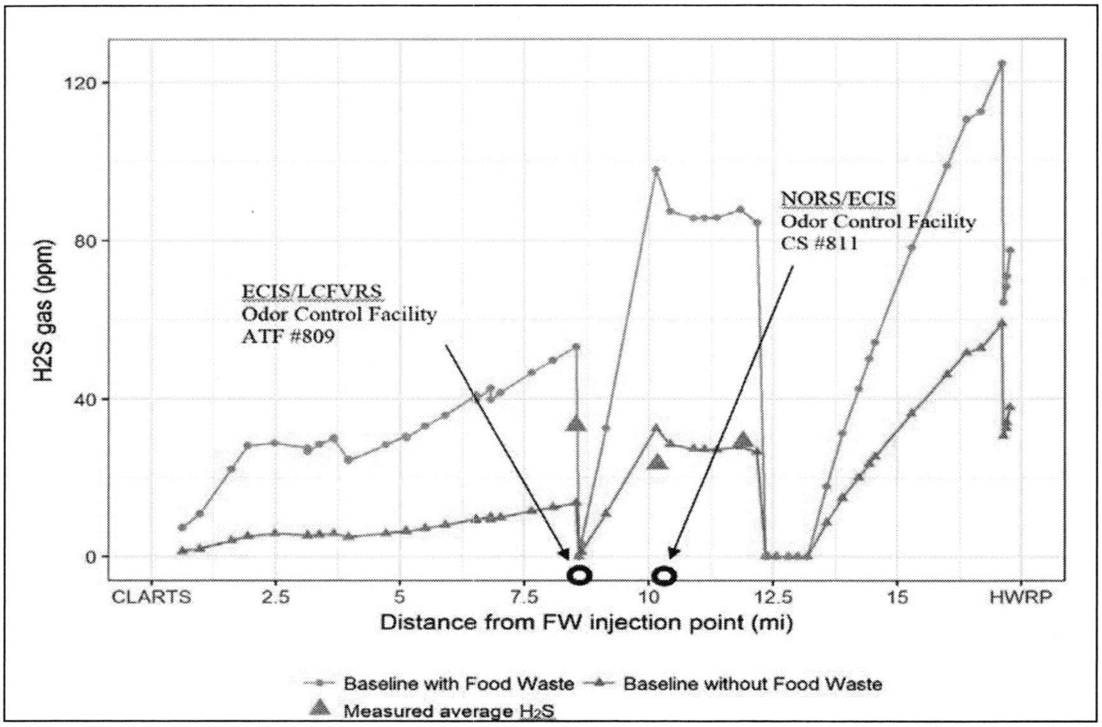

Figure 7. The sewer headspace H_2S profile through the conveyance system from the food waste injection point (CLARTS) for *the Baseline Without Food Waste Model* and *Baseline With Food Waste Model* to the treatment facility (HWRP).

The inner pipe surface material of the ECIS/NORS interceptor is all reinforced concrete. Both the ECIS and the NORS interceptor sewer are relatively young, 10-12 years and 30-40 years, respectively, and each have been built with liners in them to minimize the impact of corrosion. The FW will therefore not increase any corrosion rates along the interceptor.

The risk of solid deposition is minimum in the ECIS/NORS interceptor due to the relatively high water velocities. Nevertheless, in the second siphon, the approximately a mile-long underneath Interstate 405, the water velocities drop significantly. The resulting bottom shear stress (a risk measure for sewer sediment build-up) indicates an increased risk of permanent sediment formation during the low flow periods of the day. The bottom shear stress is the average shear at the wet pipe perimeter, which can be calculated from the average flow, the specific density, the hydraulic perimeter and the slope. The risk exists only certain periods of the day due to the diurnal wastewater flow patron through the interceptor.

PART 3: Sensitivity Analyses

Sensitivity analyses were performed on *the Baseline Without Food Waste Model* for COD, dissolved oxygen, flow, temperature, and dissolved H_2S. *The Baseline Without Food Waste Model* sensitivity analysis explores variations in the wastewater entering HWRP from the NORS Interceptor as a result of changes in the existing sewer wastewater parameters.

Sensitivity analyses were also performed on the food waste itself in *the Baseline With Food Waste Model* for pH, total COD, and COD fractionation. *The Baseline With Food Waste Model* sensitivity analysis explores variations in the wastewater entering HWRP from the NORS Interceptor as a result of changes to the assumptions of the food waste parameters. *The Baseline With Food Waste Model* sensitivity analysis does not look at variations in the existing sewer wastewater quality.

Baseline Without Food Waste Model Sensitivity

Sensitivity analyses on *the Baseline Without Food Waste Model* were conducted for variations of the following wastewater quality parameters:

- Total COD, which had a default value of 977 milligrams per liter [mg/L] and was increased by 25% and reduced by 25%, with the default COD fractionation retained.

  Note: The COD range used covers the 20th to 85th percentiles of the 2016 dry season variability into the HWRP plant based on BOD data measured by the City.

- Dissolved Oxygen (DO), which had a default value of 1.0 mg/L and was reduced to 0.5 mg/L, and

- Dissolved Sulfide, which had a default concentration of 0.5 mg-S/L, was increased to 1.0 mg-S/L.

The results from these sensitivity analyses are shown in **Table 6**. The second column in **Table 6** shows the impacts of the FW as previous discussed for comparison (i.e. the difference between *the Baseline Without Food Waste Model* and *the Baseline With Food Waste Model*).

Variations of the wastewater COD concentrations have the greatest effect on the model outputs. The effect of a 25% increase in wastewater COD includes a corresponding increase in all predicted parameters except for sulfate. The decrease in sulfate reflects the increased anaerobic conditions due to increased COD concentration in the sewage. The effect of a 25% decrease in wastewater COD is a corresponding decrease in all predicted parameters except for sulfate. The increase in sulfate reflects the reduced anaerobic conditions due to reduced COD concentration in the sewage.

The results show that the 25 percent variability in wastewater COD (representing about 20th to 85th ‰ of the wastewater COD variability) has a greater impact on the HWRP influent

COD than the addition of food waste (99,100 kg-COD/day versus 59,800 kg-COD/day). The results also show that while increasing COD in the sewer reduces sulfate loading to the plant, it increases the dissolved sulfides which are the precursor for H_2S, and increases dissolved methyl mercaptans which are an important volatile odorous compound.

The reduction in DO and increase in dissolved sulfides did not significantly impact the wastewater quality of the NORS Interceptor entering HWRP. There is a slight increase in methyl mercaptans and H_2S in the sewer headspace, with minimal impact on the odor control facilities. Addition of food waste into the sewer system will not increase corrosion rates as the interceptor path have been built with liners in them to minimize corrosion.

Sensitivity analyses were also performed on the following physical parameters of the wastewater:

- Wastewater Flow from each branch into the modeled domain (+/- 10%), and
- Temperature with +/- 3°C (5.4 °F) from an average of 27 °C (81 °F).

The results from these sensitivity analyses on the HWRP influent are shown in **Table 7**. The results are shown along with the impacts of the food waste (second column in Table 7) as previously discussed for comparison.

Varying the wastewater flow contributed by each interceptor branch by 10 percent had a significant impact on the COD loading into HWRP, but not as much as adding the food waste. Decreased wastewater flow increases hydraulic retention time, while increase wastewater flow increased COD and turbulence, which all potentially can affect dissolved sulfide and hydrogen sulfide.

The effect of 3 degrees Celsius wastewater temperature increase, slightly increases the dissolved sulfide, fermentable products, and H_2S and MM headspace concentrations, while sulfate concentration, total COD, fermentable substrates and hydrolysable substrates decrease. This change in these constituents reflects increased overall microbial activity resulting in mostly increased hydrolyses reflected by the significant reduction in hydrolysable substrates.

Table 6. Sensitivity Analysis Results of Wastewater COD, Dissolved Sulfides, and Dissolved Oxygen Variation on the Influent Wastewater to HWRP and the Odor Control Facilities ATF#809 and CS#811.

Parameter	*Impact of FW*	**Baseline COD + 25%**	**Baseline COD – 25%**	**Baseline dissolved oxygen – 0.5 mg-O/L**	**Baseline dissolved sulfides + 0.5 mg-S/L**	**Units**
Impact on the Influent Wastewater to HWRP						
Dissolved sulfides	*166*	485	-600	99	172	kg-S/day
Sulfate	*-267*	-509	620	-107	21	kg-S/day
Dissolved MM	*2.1*	23	-23	1	0	kg-MM/day
Total COD	*59,800*	99,100	-98,800	41	-82	kg-COD/day
Fermentable substrates (S_F)	*6,000*	4,350	-3,940	-205	-246	kg-COD/day
Fermentation products (S_A)	*703*	1,440	-1,270	-82	-82	kg-COD/day
Heterotrophic biomass (X_{HW})	*48*	2,420	-2,340	-205	-82	kg-COD/day
Sulfide forming biomass (X_{SW})	*620*	123	-1,270	123	164	kg-COD/day
Fast hydrolysable substrates (X_{Sf})	*11,900*	23,800	-23,900	328	164	kg-COD/day
Medium hydrolysable substrates (X_{Sm})	*13,300*	29,900	-29,900	41	41	kg-COD/day
Slow hydrolysable substrates (X_{Ss})	*27,200*	36,200	-36,300	0	0	kg-COD/day
H_2S - sewer gas at HWRP	*39*	13	-14	1.5	3.7	ppm
MM - sewer gas at HWRP	*0.08*	0.91	-0.92	0.04	0.01	ppm
Impact on the Odor Control Facilities ATF#809 and CS#811						
H_2S - sewer gas at ATF #809	*40*	7.1	-6.2	0.85	2.3	ppm
MM - sewer gas at ATF #809	*0.00*	0.52	-0.55	0.10	0.09	ppm
H_2S - sewer gas at CS #811	*65*	15	-14	1.3	3.5	ppm
MM - sewer gas at CS #811	*0.03*	0.65	-0.73	0.05	0.07	ppm

Table 7. Sensitivity Analysis Results of Wastewater Flow and Temperature Variation on the Influent Wastewater to HWRP and the Odor Control Facilities ATF#809 and CS#811.

Parameter	***Impact of FW***	**Baseline flow -10%**	**Baseline flow +10%**	**Baseline temperature + 3°C**	**Baseline temperature - 3°C**	**Units**
Impact on the Influent Wastewater to HWRP						
Flow	*0.06*	4,277	5,278			MGD
Dissolved sulfides	*166*	-58	55	390	-423	kg-S/day
Sulfate	*-267*	-792	789	-415	439	kg-S/day
Dissolved MM	*2.1*	-5	5	26	-22	kg-MM/day
Total COD	*59,800*	-39,900	40,000	-1,070	1,070	kg-COD/day
Fermentable substrates (S_F)	*6,000*	-2,280	2,300	-1,400	1,400	kg-COD/day
Fermentation products (S_A)	*703*	-357	382	1,800	-1,360	kg-COD/day
Heterotrophic biomass (X_{HW})	*48*	-1,060	1,060	-123	164	kg-COD/day
Sulfide forming biomass (X_{SW})	*620*	-144	136	944	-944	kg-COD/day
Fast hydrolysable substrates (X_{Sf})	*11,900*	-9,640	9,650	-1,990	1,520	kg-COD/day
Medium hydrolysable substrates (X_{Sm})	*13,300*	-12,000	12,000	-287	246	kg-COD/day
Slow hydrolysable substrates (X_{Ss})	*27,200*	-14,500	14,500	-164	82	kg-COD/day
H_2S - sewer gas at HWRP	*39*	3.5	-2.7	8.9	-8.8	ppm
MM - sewer gas at HWRP	*0.08*	0.31	-0.26	1.6	-1.2	ppm
Impact on the Odor Control Facilities ATF#809 and CS#811						
H_2S - sewer gas at ATF #809	*40*	0.84	-0.70	3.2	-3.0	ppm
MM - sewer gas at ATF #809	*0.00*	0.09	-0.07	1.1	-0.76	ppm
H_2S - sewer gas at CS #811	*65*	9.3	-8.4	10	-8.7	ppm
MM - sewer gas at CS #811	*0.03*	0.56	-0.55	1.3	-1.0	ppm

Food Waste Sensitivity

Sensitivity analyses on *the Baseline With Food Waste Model* were conducted for variations of the following food waste parameters:

- the COD of the food waste has a default value of 1.54 lb COD/dry lb food waste and was increased by 25% and decreased by 25%, with the default COD fractionation of the food waste held constant,

- the pH of the food waste slurry, which had a default value of 5.0, was increased by one pH unit (to pH 6.0) and decreased by one pH unit (pH 4.0).

The results from these sensitivity analyses are shown in **Table 8**. The results are shown along with the impacts of the food waste using the default values (second column in Table 8) for COD and pH for comparison.

The effect of a 25% increase in food waste COD include a corresponding increase in all predicted parameters except for sulfate. The decrease in sulfate reflects the increased anaerobic conditions due to increased COD concentration in the sewage. The effect of a 25% decrease in food waste COD include a corresponding decrease in all predicted parameters except for sulfate. The increase in sulfate reflects the reduced anaerobic conditions due to reduced COD concentration in the sewage. These results show also that 25 percent variability in food waste COD has an impact on the potential HWRP influent COD. The impact is smaller than the impact of 25 percent variability in wastewater COD (representing about 20th to 85th ‰ of the wastewater COD variability) as discussed before.

Lowering and raising the pH by one unit does not have a significant impact on the influent wastewater.

Sensitivity analyses on the fractionation of the COD in the food waste was also conducted. For these analyses, total COD of the food waste remained constant, but the fermentation substrate, fermentation products, and fast hydrolysable substrate fraction were changed. For each of these scenarios, the concentration of the parameter being tested was increased (or decreased) by 50 percent and the equivalent mass was removed from (or added to) the slowly hydrolysable substrate. **Table 9** shows the results from these analyses. Changes in fractionation of the COD in the food waste does not cause a significant change in predicted parameters except for the parameter itself.

Table 8. Sensitivity Analysis Results of the Food Waste COD and pH Variation on the Influent Wastewater to HWRP and the Odor Control Facilities ATF#809 and CS#811.

Parameter	*Impact of FW*	**Food Waste COD - 25%**	**Food Waste COD + 25%**	**Food Waste pH - 1**	**Food Waste pH + 1**	**Units**
Impact on the Influent Wastewater to HWRP						
Dissolved sulfides	*166*	-90	99	-8	16	kg-S/day
Sulfate	*-267*	95	-111	4	-16	kg-S/day
Dissolved MM	*2.1*	-0.8	0.4	0	0	kg-MM/day
Total COD	*59,800*	-14,900	14,900	41	0	kg-COD/day
Fermentable substrates (S_F)	*6,000*	-1,320	1,320	41	-41	kg-COD/day
Fermentation products (S_A)	*703*	-205	247	0	0	kg-COD/day
Heterotrophic biomass (X_{HW})	*48*	0	-41	0	-41	kg-COD/day
Sulfide forming biomass (X_{SW})	*620*	-247	205	-41	0	kg-COD/day
Fast hydrolysable substrates (X_{Sf})	*11,900*	-3,040	3,000	-41	0	kg-COD/day
Medium hydrolysable substrates (X_{Sm})	*13,300*	-3,330	3,330	0	0	kg-COD/day
Slow hydrolysable substrates (X_{Ss})	*27,200*	-6,780	6,820	0	0	kg-COD/day
H_2S - sewer gas at HWRP	*39*	-3.5	3.9	0.44	-1.3	ppm
MM - sewer gas at HWRP	*0.08*	-0.04	0.03	0.00	0.00	ppm
Impact on the Odor Control Facilities ATF#809 and CS#811						
H_2S - sewer gas at ATF #809	*40*	-3.8	3.8	0.58	-1.7	ppm
MM - sewer gas at ATF #809	*0.00*	-0.04	0.03	-0.01	0.02	ppm
H_2S - sewer gas at CS #811	*65*	-7.7	8.0	0.71	-1.8	ppm
MM - sewer gas at CS #811	*0.03*	-0.05	0.04	-0.01	0.02	ppm

Table 9. Sensitivity Analysis Results of the Food Waste COD Fractionation Variation on the Influent Wastewater to HWRP and the Odor Control Facilities ATF#809 and CS#811.

Parameter	***Impact of FW***	**Ferment. Products (S_A) -50%**	**Ferment. Products (S_A) +50%**	**Ferment. Substrates (S_F) -50%**	**Ferment. Substrates (S_F) +50%**	**Fast Hydrol. Substrates (X_{Sf}) -50%**	**Fast Hydrol. Substrates (X_{Sf}) +50%**	**Units**
Impact on the Influent Wastewater to HWRP								
Dissolved sulfides	*166*	-16	12	-58	58	-107	131	kg-S/day
Sulfate	*-267*	16	-16	62	-66	115	-144	kg-S/day
Dissolved MM	*2.1*	0	0	-2	1	0	0	kg-MM/day
Total COD	*59,800*	41	0	164	-123	288	164	kg-COD/d
Fermentable substrates (S_F)	*6,000*	-164	82	-2,710	2,750	247	-247	kg-COD/day
Fermentation products (S_A)	*703*	-616	781	0	0	247	-247	kg-COD/day
Heterotrophic biomass (X_{HW})	*48*	0	0	0	-41	0	-41	kg-COD/day
Sulfide forming biomass (X_{SW})	*620*	-41	41	-164	123	-288	247	kg-COD/day
Fast hydrolysable substrates (X_{Sf})	*11,900*	-41	0	-41	0	-6,000	6,450	kg-COD/day
Medium hydrolysable substrates (X_{Sm})	*13,300*	0	0	0	0	0	0	kg-COD/day
Slow hydrolysable substrates (X_{Ss})	*27,200*	904	-904	3,040	-3,000	6,040	-6,040	kg-COD/day
H_2S - sewer gas at HWRP	*39*	-0.63	0.63	-2.3	2.2	-4.1	4.9	ppm
MM - sewer gas at HWRP	*0.08*	0.00	0.00	-0.06	0.04	-0.01	0.00	ppm
Impact on the Odor Control Facilities ATF#809 and CS#811								
H_2S - sewer gas at ATF #809	*40*	-0.81	0.84	-2.4	2.4	-4.4	4.8	ppm
MM - sewer gas at ATF #809	*0.00*	0.00	0.00	-0.04	0.03	-0.03	0.03	ppm
H_2S - sewer gas at CS #811	*65*	-1.4	1.6	-4.9	5.0	-8.9	10	ppm
MM - sewer gas at CS #811	*0.03*	-0.01	0.01	-0.06	0.05	-0.03	0.03	ppm

DISCUSSION

Variation in the quantities and characteristics of the waste generated by the CLARTS OPF can be expected. The variation may relate to the time of day and the day of week due to the schedule of food waste collection and processing at the CLARTS OPF. Variations may also be related to the season, because large food producers may ramp up production during harvest or holiday seasons. A sensitivity analysis was therefore performed on both *the Baseline Without Food Waste Model* and *the Baseline With Food Waste Model* to understand the sensitivity towards important parameters and determine the impact on wastewater quality entering the HWRP and the impact on the conveyance system.

Wastewater quality data such as dissolved sulfide and headspace H_2S from the sewer network are typically used to calibrate the model. No dissolved sulfide water quality was available and therefore were not used, therefore the model was calibrated with limited measured data. It is recommended that additional data be collected on dissolved sulfide and H_2S concentrations to better predict the impact of the addition of food waste on headspace H_2S concentrations in the sewer system. The accuracy of the model could be further refined if additional data were collected from the food waste and the existing sewage.

The wastewater flow velocities through the conveyance system are generally higher than 1 meter per second (3 fps) which is typical for sewer conveyance systems without too much risk of permanent settled solids. On the other hand, reduced water velocities will occur in the two siphons present in the conveyance system from the CLARTS OPF to the HWRP. One section of the ECIS involves a short (100 metre) underpass under the road and the second siphon at the NORS is approximately a kilometer long and drops about 10 meter (30 feet) as can be seen in the profile illustrated on **Figure 3**. The settling properties of the food waste slurry are unknown and could be different than typical wastewater. Increased settled solids in the sewer might therefore occur, especially in the second siphon with the relatively low average water velocity flows of only 0.4 meter per second (1.3 fps) at dry weather flow conditions. The settling properties are also important for determining how well the food waste slurry will separate in a primary clarifier at the HWRP and what COD fraction will be removed. The settling properties of the food waste slurry and the associated risk to settled solids in the siphons as well as the mitigation mechanisms should be investigated further.

Siphons can act as settling tanks which could accumulate particulate matter from wastewater including the food waste slurry. The deeper layer of the accumulated matter below the interface with the free-flowing wastewater is likely to be populated by anaerobic bacteria for which dissolved BOD, oxygen, nitrate, and sulfate may not be available, since those constituents would be consumed by the layer of microbes closest to the free-flowing wastewater. Therefore, anaerobic digestion of volatile settled solids could occur deeper in the sludge layer, allowing large amounts of sulfate reducers and methanogens to form. This could potentially create increased H_2S-issues and also potentially generate increased amounts of methane. Over-sized sewers relative to flow have resulted in a tendency toward anaerobic digestion of settled solids and formation of methane, H_2S, nitrogen, and carbon dioxide gases within the force main (Easter et al., 2012).

Mitigating measures of preventing solids deposits or methane generation might have to be implemented. Dosing the Food Waste Slurry only during the higher flow periods of the day might be an option to consider which would significantly reduce this risk. Alternatively, deactivation of methanogenic activity could be implemented using occasional caustic shock dosing. An inactivation of anaerobic biofilm can be obtained when pH is increased above 8.6. Methanogenic activity can also be prevented in the presence of nitrate (nitrate dosing).

CONCLUSIONS

The City of Los Angeles Bureau of Sanitation (LASAN) has conducted a study to assess the impacts associated with conveying food waste (FW) using an existing wastewater collection system. A proposed Organic Processing Facility (OPF) will separate FW and other digestible organics from non-processable and non-digestible materials to create an organic FW slurry. The FW slurry will be treated at the Hyperion Water Recovery Facility (HWRP) with a portion of the waste converted into renewable energy via anaerobic digestion.

The impacts to the HWRP influent key constituent characteristics are predicted. This includes an approximation of changes to the properties of the FW slurry within the conveyance system prior to reaching the HWRP. The impact of the additional FW to the wastewater conveyance system on the existing sewer ventilation odor treatment systems are projected. In conclusion, this study has shown the following:

The addition of 280 ton/day FW slurry (15% dry weight) from CLARTS OPF is minimal in terms of flow, but organically concentrated and is predicted to increase loadings at HWRP as follows:

- COD by 59,800 kg-COD/day,
- total nitrogen by 1,240 kg-N/day,
- total phosphorus by 204 kg-P/day,
- ammonium by 31 kg-N/day,
- total dissolved solids (TDS) by 24,327 kg/day,
- total suspended solids (TSS) by 14,910 kg/day and
- slightly decrease the pH.

Addition of FW into the sewer system will increase H_2S levels in the sewer headspace that will impact the total H_2S load into the existing odor control systems.

Addition of FW into the sewer system will not increase corrosion rates as the interceptor path have been built with liners in them to minimize the impact of corrosion.

Sensitivity analyses were performed for COD (+/-25%), dissolved oxygen (-0.5mg/l), flow (+/-10%), temperature (+/-3°C), and dissolved H_2S (+0.5mg/l) of the wastewater. Sensitivity analyses were also performed on the FW itself for pH (+/- one pH unit), total COD (+/-25%), and COD fractionations (+/-50%). Of these tested parameters, total COD variability in the wastewater and the total COD variability of the food waste has the most significant impact on the HWRP influent wastewater quality and have the greatest impact on H_2S levels in the sewer, followed by wastewater flow variability.

Acknowledgements:
We thank LASAN staff for their participation and contributions connected with this effort that included meeting attendance, document review, response to questions, and providing input and guidance. In addition, we would like to thank Alex Demith, Matthew Ward, Jeff Friesen and Cameron Clark from Jacobs, as well as professor Jes Vollertsen of the University of Aalborg for the discussions and their valuable input.

REFERENCES

Cho, J.K., and S.C. Park. 1995. "Biochemical Methane Potential and Solid State Anaerobic Digestion of Korean Food Wastes." Bioresource Technology, Vol. 52, No. 3: 245-253.

City of Los Angeles (2017) LA Sanitation Collection System Odor Control Master Plan 2017 report.

Diggleman, C., and R.K. Ham. 2003. "Household Food Waste to Wastewater or to Solid Waste? That Is the Question." Waste Management and Research, Vol. 21, pages 501-514.

Easter, C., M. Ward, C. Bott, J. Sparks. 2014 Use of Force Main Modeling to Evaluate Gas Emissions from Long Force Main Systems: Case Studies of Methane and Nitrogen Gas Control. WEF Odors and Air Pollutants Conference, May 31 – June 3, 2014. Miami, Florida

Fisgativa, H., A. Tremier, and P. Dabert. 2016. "Characterizing the Variability of Food Waste Quality: A Need for Efficient Valorization Through Anaerobic Digestion." *Waste Management*, 50: 264-274.

Hvitved-Jacobsen, T., J. Vollertsen, A.H. Nielsen. 2013. *Microbial and Chemical Process Engineering of Sewer Networks.* Second Edition, CRC Press, ISBN 978-1-43-988177-4.

Jan, O., Tostivint, C., Turbé, A., O'Connor, C., Lavelle, P., 2013. Food Wastage Footprint Impacts on Natural Resources Technical Report. FAO Natural Resources and Management Department

Kim, M., M.M.I. Chowdhuray, G. Nakhla, and M. Keleman. 2015. "Characterization of Typical Household Food Wastes from Disposers: Fractionation of Constituents and

Implications for Resource Recovery at Wastewater Treatment." *Bioresource Technology*, Vol. 183: 61-69.

Li, Y.Y., H. Sasaki, K. Yamashita, K. Seki, and I. Kamigochi. 2002. High-Rate Methane Fermentation of Lipid-Rich Food Wastes by a High-Solids Co-Digestion Process." Water Science and Technology, Vol. 45, No. 12, pages 143-150.EPA, U.S., 2011. Municipal Solid Waste in the United States, 2011 Facts and Figures.

State of California. 2014, *Assembly Bill No. 1826*. [https://leginfo.legislature.ca.gov/faces/billNavClient.xhtml?bill_id=201320140AB1826] accessed on 27 February 2018.

Tchobanoglous, G., H. Thiesen, and S. Vigil. 1993. *Integrated Solid Waste Management*, Chapter 4. McGraw-Hill, Inc., New York.

Zhang, R., H.M. El-Mashad, K. Hartman, F. Wang, G. Liu, C. Choate, and P. Gamble. 2007. "Characterization of Food Waste as Feedstock for Anaerobic Digestion." *Bioresource Technology*, Vol. 98, pages 929-935.

The Balance Between Lean-Burn Anaerobic Digestion Power Production, Formaldehyde Formation, and Public Health Concerns

Michael Lannan*, Tech Environmental, Inc., mlannan@techenv.com
Katherine Mears, Tech Environmental, Inc.

ABSTRACT

Digesting farm wastes and food wastes and land applying the digestate reduces the waste loading on landfills, and provides a valuable natural fertilizer. Unfortunately, what may be considered cost-effective for air pollution control on a commercial-sized anaerobic digester project, may not be for a farm-based system. Farm based systems are typically limited in capacity by manure production rates. With smaller systems, "top-case BACT" may not be economically feasible for nitrogen oxide control. In those cases, lean burn may be still be cost effective, even if a CatOx is added to address the increase in carbon monoxide from lean burn. Although carbon monoxide is created in all engines, formaldehyde formation, at measurable levels, is limited to lean burn. Any CatOx system will also reduce the formaldehyde, but it cannot eliminate it. This paper discusses the balance between potential lean burn operations, formaldehyde emissions, and the overall environmental benefit of the project.

KEYWORDS: Anaerobic digestion, waste-to-energy, lean-burn, formaldehyde, top-case BACT

INTRODUCTION

Aerobic and anaerobic digestion has been used on farms for years to take care of animal and associated animal feed. In most cases, the digestion process is very basic. Manure is often kept in lagoons, pits or tanks until it can be applied to the fields. When it is, there is that telltale odor of "that farm smell" from decaying manure. In the spring and fall, the partially stabilized material that has been traditionally land applied becomes fully stabilized as it sits in the field.

Manure, although somewhat low in energy value still has some residual energy available, so it is an excellent item for farmers to exploit for energy to increase revenue. An article in the Wall Street Journal from February 26, 2018 (Bunge and Newman, 2018) discussed how farms are looking for other revenue streams to balance the high cost of farming and the reduction in agricultural income. In the article titled, "To Stay on the Land, Farmers Add Extra Jobs", it discusses how small farmers today receive 82% of farm household income from off-farm work.[1]

Tech Environmental (Tech) has been working with a number of fourth and fifth generational dairy farmers in the northeast to help them develop energy from their manure. Tech has worked with farmers that are completing the digestion process themselves; farmers that are having someone build a digestion and power plant with the intent to operate it initially with an agreement to turn over the entire process to the farmers in 10 years; and environmental venture capitalist partnerships with farmers. In this last case, the farmers provide manure and ultimately use the digestate, but all digestion operations are run by a separate operations firm. All of these methods have helped ensure that these farms have a new revenue stream to help them pass the farm on to the next generation. This process has some excellent other environmental benefits as well. It provides a better digestate for the farmers to add to their feed fields, and a combined heat and power component that can be used at the farm. In some cases, the dairy farmers use a dewatering system to press the digestate to remove any residual hay which can then be recycled back to the farm as bedding material for the maternity barn. The farms that have done this have seen a great return for their investment.

Traditionally, manure digestion has simply been the practice of covering a lagoon or tank with a tarp, collecting the gas, and flaring it to minimize odor potential. The primary goals here for digestion were stabilization and odor potential reduction. The problem was that it was not cost-effective to build a true digester with a manure only feedstream. There just was not enough gas generated to justify any additional digestion or process equipment. That changed recently with diversion of food waste from landfills. Food waste that was traditionally landfilled as a waste is now considered a valuable energy source. When mixed with manure, the total heat value increases so that a "reasonably in-depth" digestion and power system can be installed. The "reasonably in-depth" definition is an important point. This paper focuses on a side-stream of power production from a farm digester project, formaldehyde. Formaldehyde is formed, at measurable levels, when a system is run in a lean-burn configuration to reduce the formation of nitrogen oxides (NO_x). In a larger commercial digester, separate NO_x removal systems are traditionally required and can be cost-effective. Unfortunately, for a farm-based system, a stand-alone catalyst and urea injection NO_x removal system is cost-prohibitive. Although this system is cost-prohibitive, NO_x can still be reduced by operating the engine in lean-burn configuration. A lean-burn engine has additional products of incomplete combustion, such as carbon monoxide (CO) and the formation of formaldehyde, which are reduced and essentially non-existent in a rich burn system, respectively. These can, and are often reduced via a catalytic oxidizer (CatOx), but there is still residual formaldehyde remaining. This paper discusses the unique situation of addressing potential concerns from residual formaldehyde that remains after the CatOx system.

PROBLEM STATEMENT

Anaerobic digestion facilities create a biogas primarily comprised of methane and carbon dioxide. When combusted in an engine/generator set, the facility produces electrical energy and heat that can be used on site. Excess electricity can be sold back to the electrical grid. Like most fuel-combusting facilities, during the permitting process anaerobic digester facilities are required to complete a Best Available Control Technology (BACT) analysis to determine the maximum degree of nitrogen oxide (NO_x) reduction that most benefits the environment, while limiting energy consumption and cost. A BACT analysis for small anaerobic digestion farm facilities most often indicates that it is not cost-effective for a facility to employ "top-case" BACT, such as Selective Catalytic Reduction (SCR) technology. Although "top-case" BACT may not be cost-effective, there are combustion controls that can be applied to reduce nitrogen oxide formation at a lower cost, specifically configuring an engine to draw in excess air during combustion (lean-burn).

Although lean-burn engines drastically reduce NO_x emissions, it also increases products of incomplete combustion such as carbon monoxide (CO) and formaldehyde. CO is regulated as a criteria pollutant, while formaldehyde is considered a potential carcinogen. Unlike CO, which is present in digester gas prior to combustion, formaldehyde is not. Formaldehyde is formed in the "fringes" of combustion when excess oxygen is minimized to reduce NO_x. Although there is significantly more CO than formaldehyde formed, formaldehyde has much lower health thresholds.

A BACT analysis was originally designed to consider economic costs, along with energy and environmental "costs". What is often missing, is the balance of the potential "go/no go" of a project and all of its environmental, economic, or energy benefits with a cost-prohibitive air pollution control technology. In some states, presumptive BACT, defined as the "top-case BACT" technology with the highest level of control that was economically feasible for a number of projects previously, is the only option in many cases. Unfortunately, what works for commercial facilities are simply not cost-effective for small farm-based digester-to-energy projects. It is important to realize that these are farms first, and digester projects second. In other words, the limiting ingredient is typically manure. Enough food waste is imported to properly balance the manure, so that digester gas can be created. Food waste is not to be imported until there is sufficient gas created to support a commercial grade NO_x control system. If it were, then the farm with a digester would become a power plant with a farm attached. This would create an entirely different set of issues, since most farms are permitted to operate differently than a commercial facility. In conclusion, Tech has been able to show that even though farm digester projects do not have the "top-case" BACT technology installed, they can lower NO_x emissions via lean-burn configuration all while diverting food waste from landfills. Unfortunately, formaldehyde emissions are increased from lean-burn operations, so a discussion on formaldehyde formation, regulation, and health concerns is necessary.

ANAEROBIC DIGESTER AIR POLLUTION CONTROL

Nitrogen oxides (NO_x) are a group of gases that are emitted into the atmosphere primarily through the combustion of fuel. Atmospheric air is comprised of 78% nitrogen. During combustion, the engine draws in air and the nitrogen creates nitrogen oxides through a range of chemical and thermal reactions. In digester gas, this happens primarily through thermal reactions between nitrogen and oxygen. After formation, NO_x emissions can react with atmospheric gases and form nitrogen dioxide (NO_2) and ground-level ozone, which is a health concern. NO_x pollution creates haze and contributes to nutrient pollution in waters by causing algal blooms, which depletes the water's oxygen supply; disrupting ecosystems. These concerns lead the EPA to regulate NO_x emissions through the National Ambient Air Quality Standards (NAAQS).

NAAQS were developed by the EPA to set standards for certain pollutants that are considered harmful to the public and environment. Table 1 shows the current NAAQS pollutants and each pollutant's corresponding limits.

Table 1. NAAQS Criteria Pollutants and Limits

Pollutant	Averaging Period	NAAQS ($\mu g/m^3$)
NO_2	1-hour Annual	188 100
Sulfur Dioxide (SO_2)	3-Hour 24-Hour Annual	1,300 365 80
CO	1-Hour 8-Hour	40,000 10,000
2.5 µm Particulate Matter ($PM_{2.5}$)	24-Hour Annual	35 15
10 µm Particulate Matter (PM_{10})	24-Hour	150
Lead (Pb)	Rolling 3-month average	0.15

EPA has established both short-term (1-hour) and long-term (annual) NAAQS for NO_2. In the short-term, the 98th percentile of 1-hour NO_2 concentrations in a year is not to exceed 188 $\mu g/m^3$ or 100 ppb, when averaged over 3 years. In the long-term, the maximum arithmetic mean in a year is not to exceed 100 $\mu g/m^3$ or 53 ppb.

Meeting these standards is often times only the first step towards demonstrating compliance. In addition to meeting the NAAQS, most fuel-combusting facilities are required to follow EPA's BACT guidance to determine whether one has applied the best air pollution control technology that is feasible for the project. BACT was implemented as a method for complying with the federal Clean Air Act (CAA). BACT does not define what technology is "top-case", leaving interpretation up to state and local regulating agencies. Many state or local permitting agencies have their own designation of what technology is considered "top-case BACT" for each pollutant.

In most cases, selective catalytic reduction (SCR) is considered "top-case BACT". SCR systems use a catalyst to convert NO_x in the exhaust stream to nitrogen and water vapor. This is done by injecting a reducing agent, such as aqueous ammonia or urea, into the exhaust gas stream upstream of a catalyst bed. The mixture then comes into contact with a catalyst, where some of the NO_x is converted and reduced to nitrogen gas and water vapor.

However, SCR can be problematic when used to treat non-combusted digester gas in the exhaust stream. Digester gas inherently contains a broad range of compounds due to the varying quality of waste that is fed to the system, and these compounds can damage the sensitive catalyst. Over time, the SCR catalyst activity decreases, and requires catalyst replacement. The rate at which the catalyst activity decreases is highly dependent on the quantity and quality of the digester gas. The nature of the waste stream that comprises the digester gas makes it impossible to maintain a consistent high quality gas.

Having to frequently replace the catalysts adds an unpredictable additional cost. Even without frequent catalyst replacement, this technology is very expensive since a large volume of catalyst is required to convert an adequate amount of NO_x. In EPA's Air Pollution Control Technology Fact Sheet for SCR, it is estimated that the cost of catalyst is approximately \$10,000 per cubic meter of catalyst (EPA-452/F-03-010).[2] Additionally, for sufficient NO_x conversion with SCR technology, the temperature of the exhaust stream needs to be in the range of 1200° to 2000° F. For gas engines where exhaust temperatures do not meet this range, the exhaust stream would need to be reheated prior to the SCR, such as with additional fuel combustion, which is an additional expense.

Although "top-case BACT" may not be cost-effective for small digestion farm facilities, there are combustion controls that can be applied to reduce NO_x formation at a lower cost, specifically configuring an engine to draw in excess air (lean-burn).

APPLICABLE FORMALDEHYDE REGULATIONS AND GUIDELINES

Formaldehyde (CH_2O) is a colorless, flammable, odorous chemical. It is used in building materials and as a preservative in mortuaries and medical laboratories. Formaldehyde also occurs naturally in the environment. It is produced in small amounts by most living organisms as part of normal metabolic processes and it is present in large quantities in forest fires. In gas engines, formaldehyde is a byproduct of methane oxidation. Methane has a lower minimum temperature for oxidation than formaldehyde. Therefore, when methane has been completely combusted, it is possible that there is formaldehyde remaining in the edges of the combustion chamber where temperatures are less stable.

Airborne formaldehyde exposure limits are regulated by federal agencies and in some cases additional guidelines are established by state agencies. In the case of occupational exposure to formaldehyde, the Occupational Safety and Health Administration's (OSHA) Permissible Exposure Limits (PEL) may not be exceeded. OSHA has established a PEL of 0.75 parts of formaldehyde per million parts of air (ppm) for an 8-hour Time-Weighted Average (TWA). Similarly, a Short-Term Exposure Limit (STEL) has been established to address concerns related to short-term formaldehyde concentration spikes. OSHA's STEL limits the permissible 15-minute maximum workplace concentration to 2 ppm (Code of Federal Regulations, Title 29).[3]

The National Institute for Occupational Safety and Health (NIOSH) has Recommended Exposure Limits (REL) of 0.016 ppm for an 8-hour TWA and 0.1 for a 15-minute maximum workplace concentration (C.[4] Another occupation safety organization, Association Advancing Occupational and Environmental Health (ACGIH), has recommended a 15-minute maximum workplace Threshold Limit Value (TLV) of 0.3 ppm and an 8-hour TWA TLV of 0.1 ppm.[5] Although OSHA is the only federally-mandated formaldehyde regulating agency, other regulators have established stricter exposure trigger limits. Although these limits are only recommendations for further evaluation, they are often used in the process of determining state-level permitting requirements.

Oregon's Department of Environmental Quality (ODEC) has Ambient Benchmark Concentrations (ABC) that address 52 air toxics of concern. The ABCs are annual thresholds based on levels that when exposed for a lifetime, would result in a cancer risk of one-in-a-million. The ABC for formaldehyde is 0.0024 ppm. While not a regulatory standard, it is intended to provide a benchmark for which air toxic problems can be identified, and further evaluated.[6]

Additionally, Oregon's Portland State University (PSU) has implemented a "Formaldehyde Hazard Communication Program". The program applies to all University employees who are responsible for, or assigned to, work areas where there is a likelihood of exposure to formaldehyde in excess of 0.1 ppm. This is an additional safety measure put into place at PSU, beyond that of Oregon's OSHA regulation, to reduce health concerns among faculty and students.[7]

California's Office of Environmental Health Hazard Assessment (OEHHA) has chronic, 8-Hour and acute REL guidelines of 0.007 ppm, 0.007 ppm, and 0.044 ppm, respectively. The acute REL is specifically related to reducing eye irritation, while their 8-hour and chronic guidelines are specifically to reduce effects of inhaled formaldehyde on the respiratory system.[8]

Massachusetts Department of Environmental Protection (MassDEP) has a more stringent formaldehyde ambient air guideline. In Massachusetts, the Ambient Air Limit (AAL) on an annual average is $6.0E^{-5}$ ppm. Massachusetts also has a 24-hour average Threshold Effect Exposure Limit (TEL) of 0.002 ppm. The 24-hour guideline is even more stringent than NIOSH's 8-hour REL of 0.016 ppm.[9]

Commission on Environmental Quality (TCEQ) has chronic and acute Reference Value (ReV) guidelines. The two chronic ReVs are based on noncarcinogenic and carcinogenic effects. The lesser threshold of 0.009 ppm is a guideline intended to prevent irritation to the eyes, nose and airway. The higher threshold of 0.015 ppm is intended to prevent possible carcinogenic effects, as observed in rats at similar concentrations. The two acute ReVs of 0.041 ppm are based only on irritation to the eyes and nose on a 1-hour and 24-hour average.[10]

Many states simply follow the federal Hazardous Air Pollutant (HAP) guideline if an emission threshold is exceeded. For example, Texas Ohio EPA's Division of Air Pollution Control (DAPC) requires that any permit applicants for a new or modified combustion source that produces more than 1 ton per year of formaldehyde should perform an air dispersion modeling analysis to determine impacts at neighboring residences. The results of these impacts are compared to ACGIH's TLVs. If the applicant is not successful using the TLV-based approach, they may instead compare to DAPC-developed annual average concentration of $6.5E^{-4}$ ppm and maximum 1-hour concentration of 0.039 ppm, but both conditions must be met.[11]

New Hampshire's Department of Environmental Services (NHDES) has developed Annual and 24-Hour AALs of $7.2E^{-4}$ ppm and 0.001 ppm, respectively.[12]

Table 2 shows the regulations or guidelines for each agency and some states. It includes the averaging period, and the corresponding limit for each condition.

Table 2. Applicable Formaldehyde Regulations and Guidelines

State or Agency	Averaging Period	Limit (ppm)
OSHA	8-Hour PEL	0.75
	15-Minute STEL	2
NIOSH	8-Hour REL	0.016
	15-Minute Ceiling	0.1
ACGIH	8-Hour TWA	0.1
	15-Minute Ceiling	0.3
Oregon	Annual Ambient Benchmark Concentration (ABC)	0.0024
Massachusetts	Annual AAL	6.0E-5
	24-Hour TEL	0.002
California	Chronic REL	0.007
	8-Hour REL	0.007
	Acute REL	0.044
Texas	Chronic ReV	0.009
	Chronic ReV	0.015
	Acute ReV (1-Hour)	0.041
	Acute ReV (24-Hour)	0.041
Ohio	Annual Average	6.5E-4
	1-Hour Maximum	0.039
New Hampshire	Annual AAL	7.2E-4 ppm
	24-Hour AAL	0.001 ppm

"TRUE" HEALTH CONCERNS

What are the "true" health concerns from formaldehyde today? Formaldehyde gained most of its negative notoriety when it was included in insulation, commonly referred to as UFFI (Urea Formaldehyde Foam Insulation). In the late 1970s and early 1980s, UFFI was identified as a continuous emitter of formaldehyde in homes. This issue became more of a concern right after the energy crisis, when there was a significant push to tighten up homes with better windows, insulations, sealing cracks, etc. These two items together meant that formaldehyde was emitted within homes and then not allowed to escape. Many states started to listed formaldehyde as a major indoor air concern.

Table 3. Average Formaldehyde Levels in Air[13]

	Canadian Study	EPA Study	CPSC Study
UFFI Homes	0.054 ppm	0.073 ppm*	0.07 ppm
Non-UFFI Homes	0.036 ppm	0.03 ppm**	0.027 ppm
Outdoor Air	less than 0.01 ppm	0.005 ppm (urban)	0.01 ppm
		0.0004 ppm (rural)	

* 10.3 percent of all UFFI homes registered levels of formaldehyde exceeding 0.1 ppm (age of UFFI installation unknown)
** 2.85 percent of all non-UFFI homes register levels of formaldehyde exceeding

By the 1990s, UFFI had been banned for some time and existing sources had either been mitigated, or the formaldehyde had been emitting and the insulation was no longer off-gassing.

Why is this relevant today? Most formaldehyde exposure potential today is much more intermittent. Individuals that smoke are exposed to formaldehyde while smoking, but not in-between, and individuals exposed to large combustion sources typically must meet complete combustion requirements for CO and other pollutants. Emissions from mobile combustion sources, such as vehicles, also see intermittent exposure even with catalytic convertors. For example, in the state of Washington all but a small single digit percent comes from anything but wild fires and vehicle emissions.

The point here is that when considering exposure from a small farm digester for formaldehyde, it should also be okay to consider things such as (1) is it even possible for someone to be exposed where the maximum modeling results may exceed a state trigger threshold, and (2) if so, is it realistic that someone can truly be exposed at this location for the time periods of concern?

CONCLUSIONS

Tech has successfully demonstrated that the potential for formaldehyde exposure from a farm based digester is very low, especially in areas of complex terrain, or rural areas with farms. While it is possible that these areas may change their land use over time, the digester projects have a limited life cycle also, so the potential for exposure should be considered on a life-cycle basis only, when compared to the benefits of saving farms today for future generations by providing an additional income stream, closing the loop on manure reuse, and providing renewable energy.

As Chris Marrow, a farmer in northwestern Missouri mentioned for the Wall Street Journal article discussed previously, with other sources of revenue he can make his 350 acres work, but he would need 1,000 to 1,500 acres to make it on farming alone, and that quantity of land is just not available today.

REFERENCES

[1] Bunge, J.; Newman, J. To Stay on the Land, American Farmers Add Extra Jobs. *The Wall Street Journal*, Feb 25, 2018.

[2] *Air Pollution Control Technology Fact Sheet*; U.S. Environmental Protection Agency, EPA-452/F-03-010.

[3] Occupational Safety and Health Administration, Department of Labor, *Code of Federal Regulations*, Part 1910, Title 29.

[4] *NIOSH Recommendations for Occupational Safety and Health, Compendium of Policy Documents and Statements*; U.S. Department of Health and Human Services; January, 1992.

[5] United States Department of Labor, OSHA. https://www.osha.gov/dts/chemicalsampling/data/CH_242600.html (accessed Feb 25, 2018).

[6] Oregon State Air Toxics Program; *Ambient Benchmarks for Air Toxics*; 340-246-0090; Department of Environmental Quality.

[7] Sumner, R.; Cooper, C. Portland State University, Portland, OR. Formaldehyde Hazard Communication Program; Jun 6, 2010.

[8] Dodge, D. D.; Harris, G. Appendix L: OEHHA/ARB Health Values for Use in Hot Spot Facility Risk Assessments. *Air Toxics Hot Spots Program*. Office of Environmental Health Hazard Assessment. 2015.

[9] Massachusetts Department of Environmental Protection. https://www.mass.gov/service-details/massdep-ambient-air-toxics-guidelines (accessed Feb 25, 2018).

[10] Haney, J. T.; Curry, A. 24-Hour Ambient Air Monitoring Comparison Value for Formaldehyde. *Development Support Document*; June 16, 2014.

[11] *Engineering Guide #70 – Air Toxic Analysis*; Ohio Environmental Protection Agency; May 9, 2017.

[12] New Hampshire Department of Environmental Services; *New Hampshire Code of Administrative Rules*, Part Env-A 1450 Table of All Regulated Toxic Air Pollutants.

[13] *UFFI Information Sheet*; The Commonwealth of Massachusetts Executive Office of Health and Human Services Department of Public Health.

Using WATS Sewer Process Model for Project Pre-Design

Guo Ji Chiu[1], Nohemy Revilla[1], Jes Vollertsen[2]
[1]City and County of San Francisco, Public Utilities Commission, Wastewater Enterprise
[2] Aalborg University, Department of Civil Engineering, Thomas Manns Vej 23, 9220 Aalborg, Denmark

Abstract

The City and County of San Francisco, SFPUC, Wastewater Enterprise (WWE) operates and maintains a combined sewer system which transports both dry weather (DW) and wet weather (WW) flows. This collection system consists of approximately 1000 miles of main line sewers and over 2000 miles of side sewers and catch basin culverts. The main line sewers include large transport/storage (T/S) boxes and large circular sewers to convey sanitary flows and to capture and convey storm runoff. WWE has decided to follow such an approach and apply a novel sewer processes model (the WATS model) to analyze the problems and develop solutions for odor management. In the past several years, multiple odor complaints were received by the Bay Area Air Quality Management District and the City around 3rd Street and Cargo Way all the way to the Custer Avenue. Recently, the WATS model was used for Cargo Way Odor Reduction Project during pre-design phase to determine the feasibilities of some of the alternatives in terms of liquid sulfide reduction, shear stress (also called tractive force, an indicator of solid settling), corrosion control, and chemical dosage. The purpose of the Cargo Way Odor Reduction project is to alleviate odors from the collection system along Cargo Way from Jennings Street to 3rd Street and on 3rd Street from Cargo Way to Custer Avenue.

Five (5) initial alternatives were developed at a brainstorming session held with Operations, Engineering and Maintenance (OEM) representatives and the project team. They are as follows: 1. Manhole inserts; 2. Regulatory compliance and manhole inserts; 3. Hunters Point Dry-Weather flow diversion to Jennings St into Cargo Way and manhole inserts; 4. Flushing line; and 5. Dedicated Dry-Weather sewer pipe for Amador Street force main and manhole inserts. Sampling campaign was conducted to calibrate the WATS model for the project area before the WATS model was used for this project. Based on the modeling results for alternatives 3, 4, and 5, and careful consideration, only alternative 4 remained into the conceptual engineering report. Both alternative 3 and 5 will not be able to meet our project targets of liquid sulfide less than 0.5 mg/L (Target #1) and shear stress (tractive force) greater than 0.5 Pa (Target #2) after the completion of this project.

After the alternative was chosen, WATS model was used to determine the project area and minimum flush rate to meet both project targets. WATS results have shown that upstream of Mendell St, liquid sulfide is below 0.5 mg/L, which indicates that the flushing line is not needed to start at Jennings St. Average flushing rate was determined based on the WATS simulation results with the target of liquid sulfide below 0.5 mg/L at the project area, and the minimum flushing rate was simulated based on the shear stress rate above 0.5 Pa.

Introduction

The City and County of San Francisco, SFPUC, Wastewater Enterprise (WWE) operates and maintains a combined sewer system which transports both dry weather (DW) and wet weather (WW) flows. This collection system consists of approximately 1000 miles of main line sewers and over 2000 miles of side sewers and catch basin culverts. The main line sewers include large transport/storage (T/S) boxes and large circular sewers to convey sanitary flows and to capture and convey storm runoff. Odors and asset corrosion is an issue for cities around the world. San Francisco is one of these, and its Wastewater Enterprise is constantly working towards minimizing odors in sensitive areas and ensuring that asset corrosion stays within acceptable bounds. WWE has decided to follow such an approach and apply a novel sewer processes model (the WATS model) to analyze the problems and develop solutions for odor management. In the past several years, multiple odor complaints were received by the Bay Area Air Quality Management District and the City around 3rd Street and Cargo Way all the way to the Custer Avenue. Recently, the WATS model was used for Cargo Way Odor Reduction Project during pre-design phase to determine the feasibilities of some of the alternatives in terms of liquid sulfide reduction, shear stress (indicator of solid settling), corrosion control, and chemical dosage. The purpose of the Cargo Way Odor Reduction project is to alleviate odors from the collection system along Cargo Way from Mendell Street to 3rd Street and on 3rd Street from Cargo Way to Custer Avenue.

Sewer System Background

San Francisco sewer system is divided into the Westside and Bayside drainages. The Westside basin covers approximately 36 percent of San Francisco's land area, drains to the Pacific Ocean, and generates 25 percent of annual combined sewer flows. The Bayside basin covers approximately 64 percent of San Francisco's land area, drains to Central San Francisco Bay and generates 75 percent of annual combined sewer flows. See Figure 1 for the System Map.

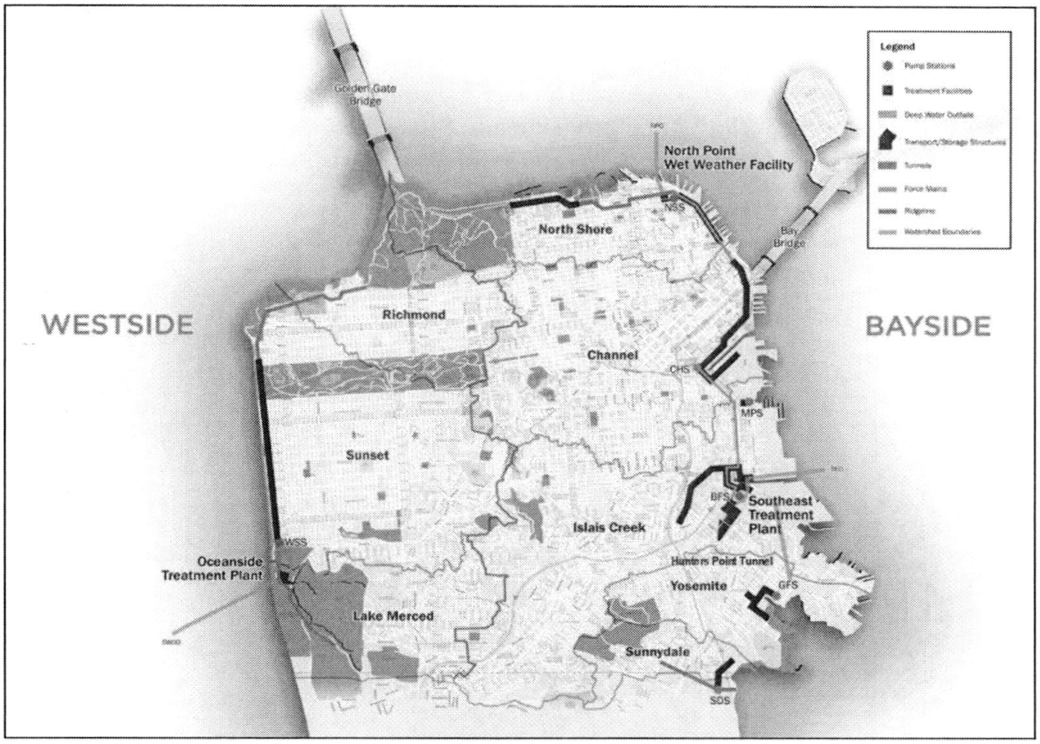

Figure 1 – System Map

Cargo Way is located at the southeast drainage area of Islais Creek. Under wet weather operation or during a dry weather shutdown of Southeast Treatment plant (SEP) inlet pump structure (SEP 011) lift pumps (such as flow equalization), the upper tier sewer system (indicated in green in Figure 2) fills up to 10 feet deep, equivalent to elevation -9 feet before overtopping internal weirs into the Islais Creek Transport/Storage (T/S) Box. The T/S box is considered part of the lower tier sewer system (indicated in red in Figure 2) and is much deeper than the upper tier system (T/S box invert elevation is at -47.0 feet at the connection to the Bruce Flynn Pump Station (BFS)). In these conditions, BFS activates to provide additional wet weather capacity to supplement SEP 011 sump pumps, lifting flow into the SEP from the Islais Creek T/S box (ICT). The ICT has a flushing system that allows it to be cleaned after storm events; however, the upper tier collection system which is connected to SEP 011 lift pumps does not. Furthermore, flow equalization occurs during dry weather conditions on a daily basis in the upper tier system. When the backup of sanitary flow caused by flow equalization extends from SEP 011 to Cargo Way, it becomes a significant source of odor. Any solids that overtop the internal weir at 3rd Street and Cargo Way into the ICT can deposit there until the next wet weather event or flush. See Figure 3 for the extents of flow equalization in the upper tier system. In addition, the rendering plant, Darling International, located on Amador Street, is the major odor source in the area. It collects and recycles inedible animal byproducts and used cooking oil or grease from restaurants and food eateries in San Francisco and converts it into commercial products. Its operation also involves mincing and boiling animal carcasses in large vats at high temperatures. The animal fat is then stripped converted into tallow, fats and oils that are eventually used as ingredients in the production of such products as pet food, soaps, cosmetic, and biofuels. The rendering plant's wastewater is the major contributor to Amador Pump Station's force main flow, which discharges into the Cargo Way sewer near Amador Street. After extensive field investigations, the force main was determined to be the source of an atypical sewer odor described as "burned bone" odor by the project team. Since the rendering plant's wastewater is the major contributor to the force main, the rendering plant is assumed to be the source of "burned bone" odor.

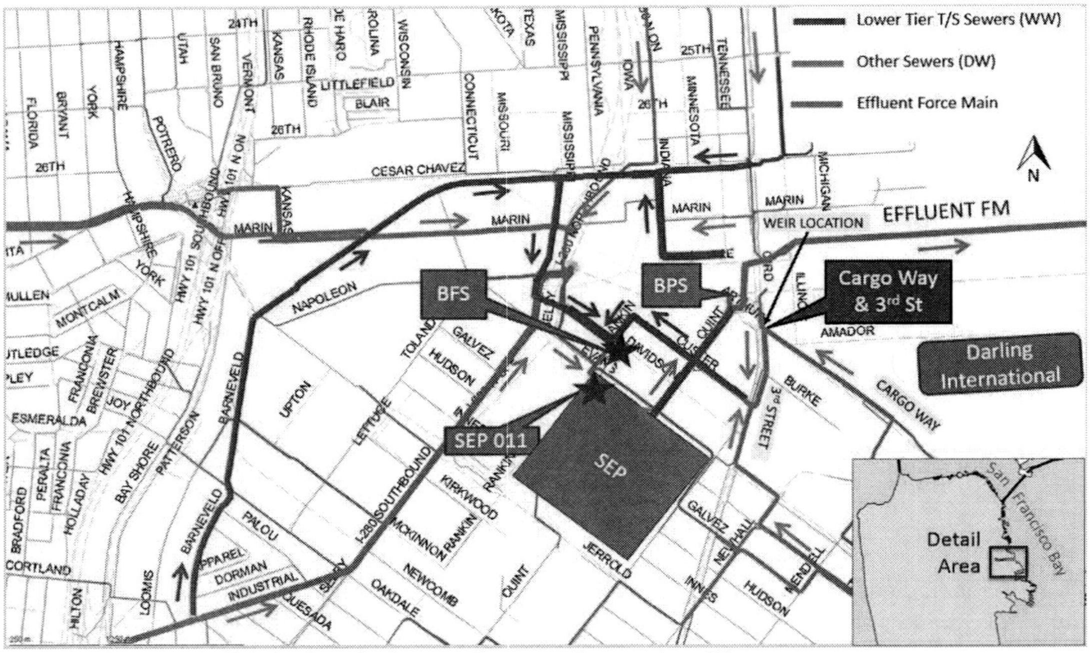

Figure 1 - Islais Creek Sewer System Operation Plan (BPS: Booster Pump Station)

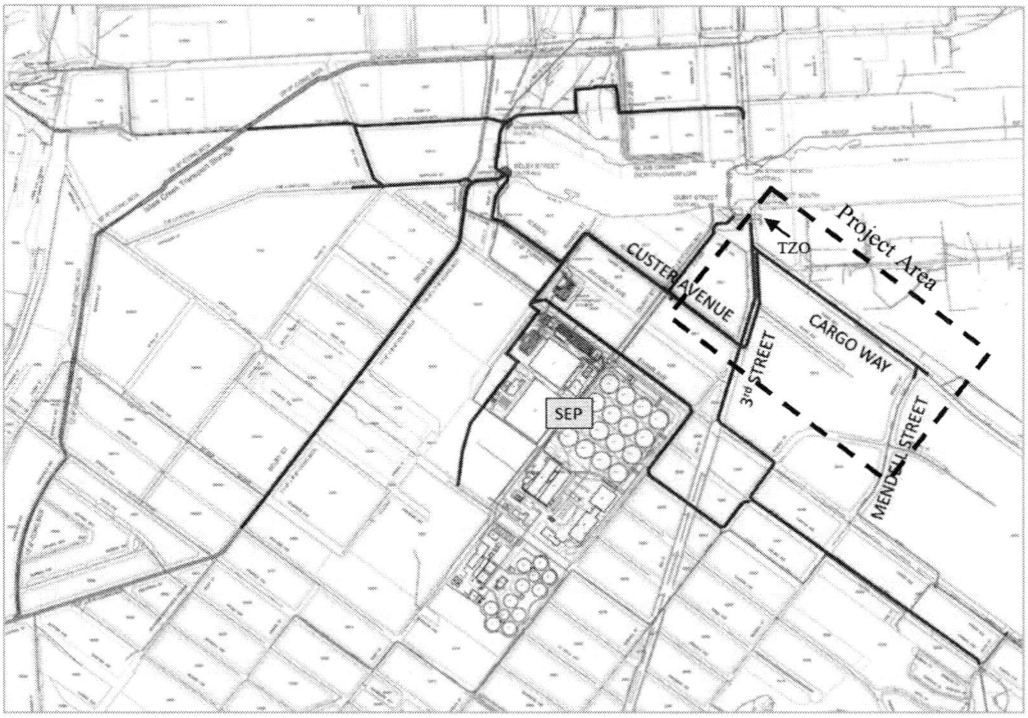

Figure 2 - Extents of Flow Equalization in Upper Tier System (TZO: Third Street Outfall)

Five (5) initial alternatives were developed at a brainstorming session held with Operations, Engineering and Maintenance (OEM) representatives and the project team. They are as follows: 1. Manhole inserts; 2. Regulatory compliance and manhole inserts; 3. Hunters Point Dry-Weather flow division to Jennings St into Cargo Way and manhole inserts; 4. Flushing line; and 5. Dedicated Dry-Weather sewer pipe for Amador Street force main and manhole inserts.

The following measurable metrics were identified as screening criteria when comparing alternatives;

1. Aqueous sulfide concentration of 0.5 mg/L is the threshold before moderate sulfide problems occurs per Sewer Processes guidelines.
2. A shear stress of 0.5 Pascal (Pa) is assumed sufficient to maintain the sewer's self-cleaning ability per Tractive Force Design for Sanitary Sewer Self Cleansing. It will be used as a measure of the flow's ability to suspend and transport particles downstream.

Sampling campaign was conducted to calibrate the WATS model for the project area before the WATS model being used for this project.

The purpose of this report is to summarize the WATS model calibration results of the project area and the use of WATS modeling for pre-design alternative selection between alternatives 3, 4 and 5, and design criteria for the selected alternative.

The WATS Concept

WATS stands for Wastewater Aerobic/anaerobic Transformations in Sewers. The WATS concept describes the most important biological, chemical, and physicochemical processes occurring in a sewer environment. The WATS concept can be viewed as a matrix of non-linear differential equations. The WATS model discretizes these equations and links the WATS concept to other processes important for the overall processes description in sewer networks: the wastewater flow, its pH, the mass transfer between liquid and air, and the movement of air in the network.

WATS Model Calibration

The WATS model has been calibrated several times since 2013 and was calibrated particularly targeting the project area during planning phase of the project. Sampling campaign was conducted on 8/29 through 9/16/16 within the project area as shown in Figure 4. Results were used to calibrate the WATS model. Table 1 summarizes the sampling results in average and the number of samples per sampling location. Table 2 has the calibrated modeling results showing that the simulated values are correlated well with the sampling results.

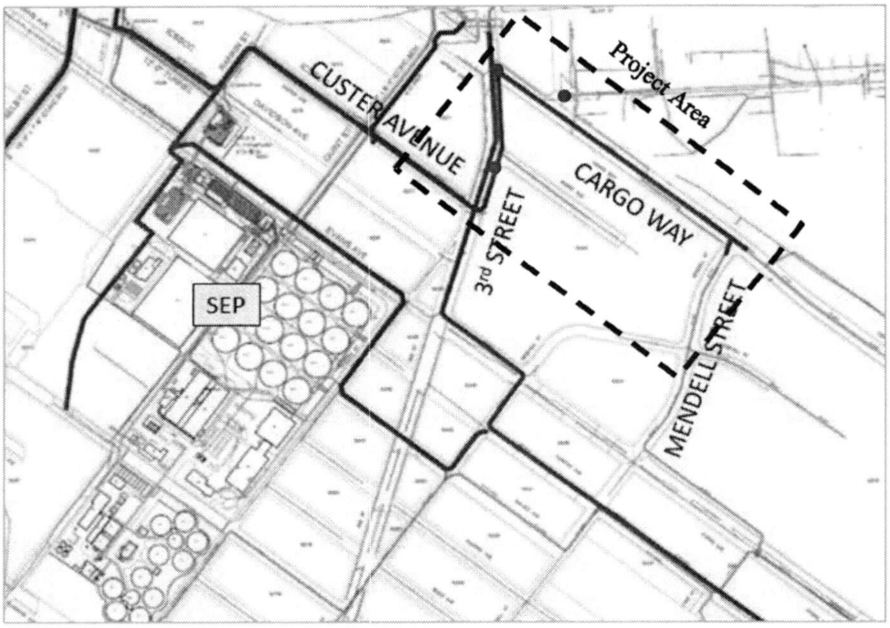

Figure 4 - Sampling Campaign Locations

Table 1 - Sampling average vales and number of samples per site per parameter

Constituents	Average Value [Number of Samples]		
	Illinois St & Amador St	**3rd St & Cargo Way**	**3rd St Near Custer Ave**
COD, mg/L	6118 [24]	3988 [28]	4559 [16]
pH	7.2 [24]	7.2 [28]	9.0 [16]
Liquid Sulfide, mg/L	2.8 [28]	3.0 [27]	5.2 [17]
DO, mg/L	4.3 [16]	3.5 [20]	2.7 [16]
Temperature, °C	23 [23]	23 [27]	22 [16]
H_2S Gas, ppm	46.53 [677]	1.15 [2781]	0.68 [1245]

Table 2 – WATS Simulated Baseline Results after Calibration

Baseline						
Asset ID	Location Description	Flow in the pipe [MGD]	Water velocity in the pipe [ft/s]	Bottom shear stress of gravity pipes [Pa] (> 0.5 Pa)	H_2S Liquid, average concentration in pipe [mg/L] (<0.5 mg/L)	Sampling Results, mg/L
25550-34549	Cargo Way and Jennings St	0.01	0.66	0.3	0.13	-
34549-25549	Cargo Way toward Mendell	0.02	0.68	0.31	0.23	-
25549-23924	Cargo Way toward Mendell	0.02	0.69	0.32	0.27	-
23924-34550	Cargo Way toward Mendell	0.02	0.69	0.32	0.32	-
34550-34551	Cargo Way toward Mendell	0.02	0.72	0.35	0.31	-
34551-34552	Cargo Way toward Mendell	0.03	0.74	0.35	0.38	-
34552-34553	Cargo Way toward Mendell	0.03	0.74	0.35	0.51	-
34553-23915	Cargo Way toward Mendell	0.03	0.70	0.31	0.65	-
23915-34554	Cargo Way toward Amador	0.06	0.90	0.47	0.61	-
34554-23862	Cargo Way toward Amador	0.06	0.76	0.33	1.23	-
23862-23853	Cargo Way toward Amador	0.06	0.69	0.26	2.55	-
23853-282659	Cargo Way toward Amador	0.06	0.47	0.11	3.48	-
282660-282659	Amador to Cargo Way	0.06	0.46	0.38	**2.30**	**2.8**
282659-23852	Cargo Way from Amador to 3rd St	0.12	0.59	0.16	**2.54**	**3.0**
23852-180202	3rd St South of Cargo Way toward Custer Ave	0.12	1.38	0.91	3.78	-
180202-85745	3rd St South toward Custer Ave	0.13	1.39	0.9	**5.02**	**5.2**
85745-23884	3rd St South and Custer Ave	0.13	1.36	0.87	5.97	-

WATS Modeling Results for the Alternatives

Alternative 3

The WATS model was used to simulate alternatives 3 to 5. Based on the simulation results shown in Table 3, alternative 3 - Hunters Point Dry-Weather flow diversion to Jennings St into Cargo Way and manhole inserts, will not be able to meet the project targets of liquid sulfide less than 0.5 mg/L (Target #1) and shear stress greater than 0.5 Pa (Target #2) after the completion of this project, even though with an assumption that there was no liquid sulfide from the Hunter's Point stream. Therefore, alternative 3 is not feasible.

Table 3 – WAST modeling results assuming flow from Hunter's Point containing zero liquid sulfide

Pipe ID	Location Description	Flow in the pipe [MGD]	Water velocity in the pipe [ft/s]	Bottom shear stress of gravity pipes [Pa]	H_2Swater, average concentration in pipe [mg/L]	H2Sgas average concentration in pipe [ppm]
23862-23853	Cargo Way Before Amador	0.86	0.48	0.89	0.40	0.02
23853-282659	Cargo Way South of Amador	0.86	0.33	0.39	0.54	0.02
282659-23852	Cargo Way from Amador to 3rd St	0.90	0.33	0.40	0.69	1.61
23852-180202	3rd St South of Cargo Way toward Custer Ave	0.90	0.78	2.28	1.03	0.03
180202-85745	3rd St South toward Custer Ave	0.91	0.77	2.19	1.43	0.04
85745-23884	3rd St South and Custer Ave	0.91	0.76	2.17	1.66	0.07

Alternative 4

Alternative 4 is connecting the Islais Creek Booster Station (SEP 550) manifold near the 3rd Street and Cargo Way intersection with a force main to continuously supply SEP effluent water to flush the Cargo Way and 3rd Street Sewer. The original proposed discharge point reached to Jennings Street as shown in Figure 5. Based on the baseline WATs modeling results in Table 3, most of the sections between Jennings St. and Mendell St. are below 0.5 mg/L. With moving the last discharge point to Mendell St. will save about half of the pipe length as a result of capital saving. Table 4 shows that with a flushing rate of 0.65, and the liquid sulfide in the project area will reduce to less than 0.5 mg/L. However, two sections of Cargo Way sewer lines (up stream of Third Street) were less than 0.5 Pa. In order to exceed 0.5 Pa for theses sections, a flushing rate of 1.5 MGD will be needed. Modeling results are detailed in Table 5. Per *Tractive Force Design for Sanitary Sewer Self Cleansing*, minimum of 1 hour per day of 0.5 Pa of shear stress is required for the sewer pipe to maintain self-cleaning ability.

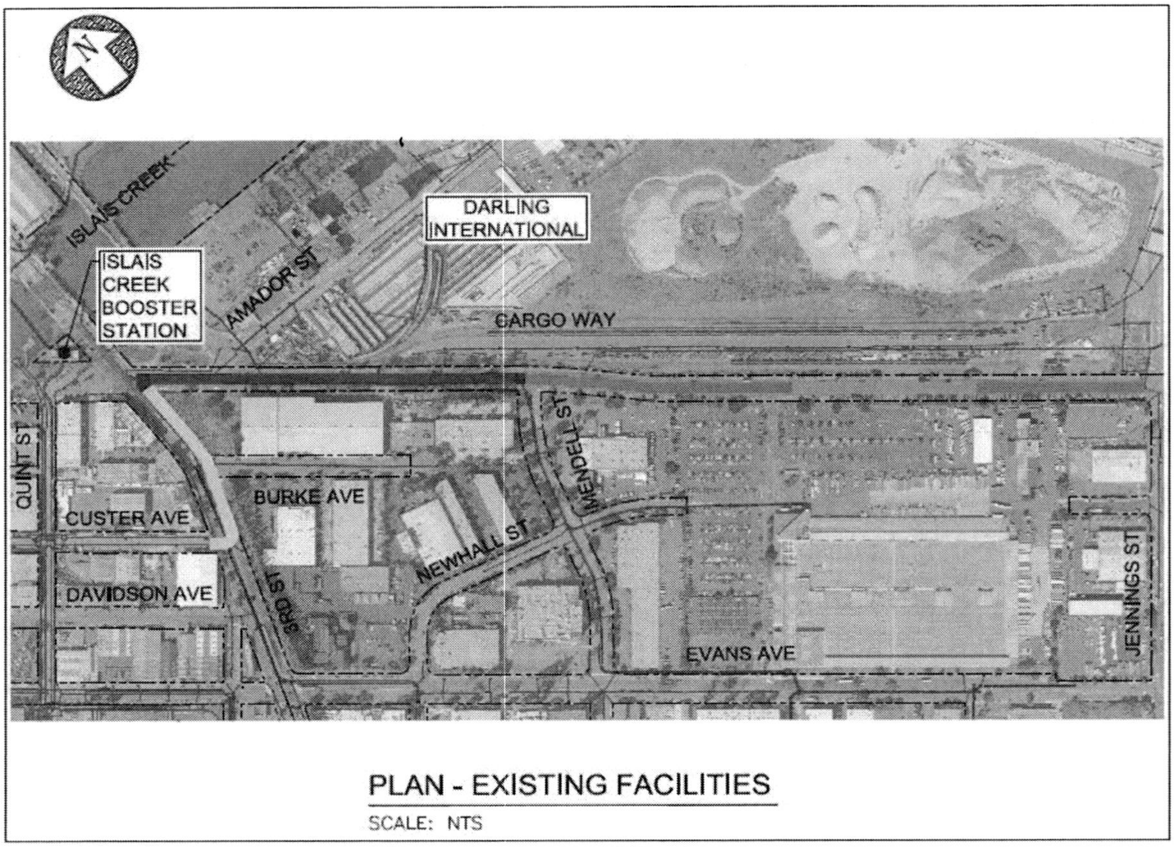

Figure 5 – Proposed Project Area

Table 4 - WATS modeling results for Alternative 4 with a flow rate of 0.65 MGD

Flushline – Flow Rate of 0.65 MGD							
Asset ID	Location Description	Flow in the pipe [MGD]	Water velocity in the pipe [ft/s]	Bottom shear stress of gravity pipes [Pa] (> 0.5 Pa)	Dissolved sulfide in the wastewater at downstream node [mg-S/L]	H2S gas in the sewer air at downstream node [ppm]	Average corrosion rate in the pipe [mm/year]
34553-23915	Cargo Way toward Mendell	0.68	1.89	1.32	0.00	0.00	0.005
23915-34554	Cargo Way toward Amador	0.71	1.97	1.50	0.00	0.00	0.000
34554-23862	Cargo Way toward Amador	0.71	1.67	1.04	0.00	0.00	0.000
23862-23853	Cargo Way toward Amador	0.71	1.49	0.82	0.02	0.00	0.001
23853-282659	Cargo Way toward Amador	0.71	1.01	0.36	0.02	0.00	0.002
282659-23852	Cargo Way from Amador to 3rd St	0.77	1.04	0.37	0.19	0.00	0.099
23852-180202	3rd St South of Cargo Way toward Custer Ave	0.77	2.44	2.12	0.35	0.01	0.026
180202-85745	3rd St South toward Custer Ave	0.78	2.41	2.05	0.43	0.01	0.031
85745-23884	3rd St South and Custer Ave	0.78	2.38	2.02	0.45	0.01	0.042

Table 5 - WATS modeling results for Alternative 4 with a flow rate of 1.50 MGD

Flushline - Flow Rate of 1.50 MGD							
Asset ID	Location Description	Flow in the pipe [MGD]	Water velocity in the pipe [ft/s]	Bottom shear stress of gravity pipes [Pa] (> 0.5 Pa)	Dissolved sulfide in the wastewater at downstream node [mg-S/L]	H2S gas in the sewer air at downstream node [ppm]	Average corrosion rate in the pipe [mm/year]
34553-23915	Cargo Way toward Mendell	1.53	2.41	1.92	0.00	0.00	0.004
23915-34554	Cargo Way toward Amador	1.56	2.51	2.16	0.00	0.00	0.000
34554-23862	Cargo Way toward Amador	1.56	2.11	1.49	0.00	0.00	0.000
23862-23853	Cargo Way toward Amador	1.56	1.89	1.18	0.00	0.00	0.000
23853-282659	Cargo Way toward Amador	1.56	1.28	0.51	0.00	0.00	0.000
282659-23852	Cargo Way from Amador to 3rd St	1.62	1.30	0.52	0.04	0.00	0.078
23852-180202	3rd St South of Cargo Way toward Custer Ave	1.62	2.44	2.12	0.00	0.00	0.004
180202-85745	3rd St South toward Custer Ave	1.63	2.99	2.84	0.00	0.00	0.000
85745-23884	3rd St South and Custer Ave	1.63	2.00	1.56	0.00	0.00	0.001

Alternative 5

Alternative 5 is to dedicate Dry-Weather sewer pipe for Amador Street force main and manhole inserts. This means that the discharge from the rendering plant will be discharged further downstream and will be mixed with a high flow coming from another sewer line as shown in Figure 6. Even through redirecting the discharge from the rendering will reduce the sulfide level downstream of Cargo Way and Amador Street, alternative 5 also will not be able to meet our project targets of liquid sulfide less than 0.5 mg/L (Target #1) and shear stress greater than 0.5 Pa (Target #2) after the completion of this project due to high liquid sulfide coming upstream of Cargo Way from Mendell to Amador Street and low flowrate in results of low sheer stress. WATS modeling results are detailed in Table 6.

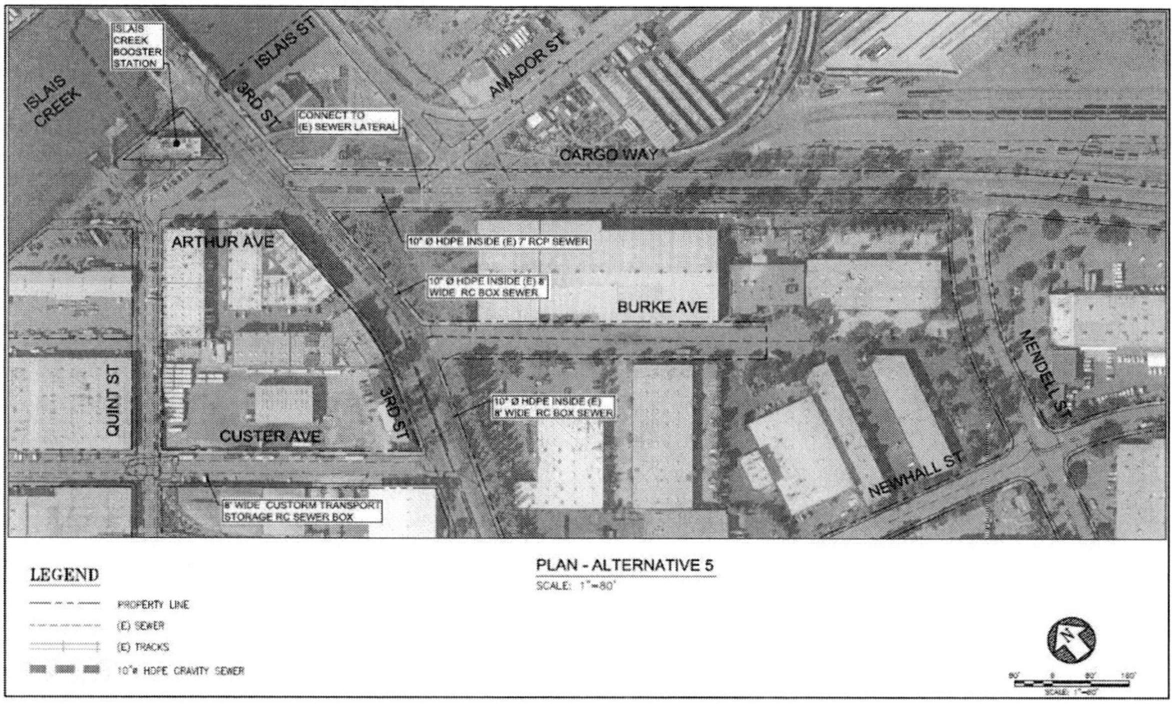

Figure 6 - Alternative 5 Schematic

Table 6 – WATS modeling results for alternative 5

Alternative 5 - Dedicated Dry Weather Sewer Pipe							
Pipe ID	Location Description	Flow in the pipe [MGD]	Water velocity in the pipe [ft/s]	Bottom shear stress of gravity pipes [Pa] (> 0.5 Pa)	H_2S Liquid, average concentration in pipe [mg/L] (<0.5 mg/L)	H_2S gas average concentration in pipe [ppm]	H_2S gas average concentration in pipe [ppm]
34553-23915	Cargo Way toward Mendell	0.03	0.70	0.31	0.55	0.07	0.06
23915-34554	Cargo Way toward Amador	0.06	0.90	0.47	0.49	0.28	0.07
34554-23862	Cargo Way toward Amador	0.06	0.76	0.33	0.84	0.07	0.10
23862-23853	Cargo Way toward Amador	0.06	0.69	0.26	1.58	0.17	0.20
23853-282659	Cargo Way toward Amador	0.06	0.47	0.11	3.42	0.57	0.19
282659-23852	Cargo Way from Amador to 3rd St	0.06	0.47	0.11	3.60	0.50	0.22
23852-180202	3rd St South of Cargo Way toward Custer Ave	0.06	1.11	0.66	4.54	0.28	0.24
180202-85745	3rd St South toward Custer Ave	0.07	1.15	0.67	4.48	0.38	0.20
85745-23884	3rd St South and Custer Ave	0.07	1.09	0.62	4.72	0.31	0.24

Based on the modeling results, alternative 4 was chosen over alternatives 3 and 5 because alternative 4 will meet the project target of liquid sulfide below 0.5 mg/L and shear stress rate above 0.5 Pa. WATS modeling results were able to provide a guideline to determine the project area and minimum flush rate to meet both project targets. Average flushing rate was determined based on the WATS simulation results with the target of liquid sulfide below 0.5

mg/L at the project area, and the minimum flushing rate was simulated based on the shear stress rate above 0.5 Pa.

Conclusion

A calibrated WATS model can be a very powerful tool to manage collection system odor related issues and provide guidelines for odor control improvement projects. As described above, based on WATS model simulated results, the best alternative was selected and design flow criteria were developed to meet the project targets for the Cargo Way odor control project.

Reference

1. LaVere B. Merritt (2015). Tractive Force Design for Sanitary Sewer Self-Cleansing. ASCE
2. Hvitved-Jacobsen T, Vollertsen J, Nielsen AH (2013). Sewer Processes: Microbial and Chemical Process Engineering of Sewer Networks. Second Edition, pp 408, CRC Press, ISBN 978-1-43-988177-4

Hydrogen Sulfide Control in Wastewater Ponds Using Strategic Dosing of Hydrogen Peroxide at Southern Minnesota Beet Sugar Cooperative

Michael L. Nelson[1] & Sagar Sunkavalli[2].

[1] USP Technologies, Atlanta, GA.

[2] Southern Minnesota Beet Sugar Cooperative (SMBSC), Renville, MN.

ABSTRACT

This paper details an effective strategy by which Southern Minnesota Beet Sugar Cooperative (SMBSC), a preeminent sugar producer in rural Minnesota, in partnership with USP Technologies, controls hydrogen sulfide (H_2S) emissions from their ponds in order to comply with a standard of 30 ppb measured at the property line.

SMBSC regularly has a surplus volume of water retained due to seasonal discharge of treated water. Since discharge to the receiving stream is only allowed from September through March, the net balance of water received or generated can be greater than the volume allowed for discharge. Water composition and quality can vary greatly from pond to pond as well as throughout the season, with conditions sometimes favorable for sulfate reducing bacteria (SRB's) proliferation possible. This can lead to H_2S emissions that would need to be controlled for compliance with the Minnesota Pollution Control Agency (MPCA) seasonal limits.

Many potentially useful chemical additives are contraindicated for use due to US FDA Generally Recognized as Safe (GRAS), Food Grade, and/or Kosher for Passover requirements, since water from some of these ponds can be returned to the factory. USP Technologies developed a method by which an oxidant (hydrogen peroxide) diluted with pond water is distributed rapidly across the surface of the ponds. The goal of this treatment method is to only treat the top few inches of the pond surface to provide rapid control of H_2S release through oxidation. With diligent monitoring of water quality parameters and other influencing meteorological conditions, chemical usage can be effectively managed when needed and curtailed while H_2S emissions are controlled.

KEYWORDS

Hydrogen peroxide, hydrogen sulfide, Southern Minnesota Beet Sugar Cooperative, beet sugar, industrial water, facultative lagoon, aeration lagoon, dosing optimization, high COD industrial water, property line monitoring

THE CHALLENGE

SMBSC must comply with stringent H_2S limit of 30 ppb set by the Minnesota Pollution Control Agency (MPCA) along their facility boundaries from April through October. H_2S concentrations are monitored by in-situ monitoring network set-up across the boundaries comprising three to four monitors.

In addition, restricted seasonal discharge of treated water creates imbalances in 'net volume in and volume out' as there can be more water generated from beet processing than the facility's ability to discharge. Great effort has been made by SMBSC to reduce the total amount of water used and generated, but this imbalance remains despite the reuse of a majority of the water it generates in addition to treatment via land application. Land application is practiced through irrigation of adjacent land on which grass is grown and harvested as hay seasonally.

SMBSC maintains seven storage ponds with capacities ranging from 170,000 – 246,000 m^3 (45 – 65 million gallons) each. Chemical oxygen demand (COD) in these ponds typically ranges from 100 – 35,000 mg/L, depending on the time of the year. The pH profile in these ponds also can vary greatly with the lower strength ponds typically in the neutral range whereas high strength ponds are at levels ranging from 4.5 – 6.5. Lime is used in the production of beet sugar so basic pH water can also need to be stored in these ponds. SMBSC also operates an adjacent treatment plant.

Figure 1. Overview of SMBSC's water storage ponds. Image courtesy of Google Maps, accessed January 22, 2018.

With a wide range of potential water quality parameters possible, anaerobic conditions favorable to sulfate reducing bacteria (SRB) proliferation can occur in many of the storage ponds. Additionally, water entering one end of a pond may be from a different source and thus have very different properties than the water already in the pond, making detailed monitoring of each pond essential.

Ponds 1, 2, 3, and 4 are horseshoe shaped, with flow generally entering the southwest corner and leaving at the northeast corners (shown in Figure 1). Each of Ponds 1 through 4 is approximately 760 meters (2,500 feet) long, 120 meters (400 feet) wide, and up to 4 meters (12 feet) deep. Ponds 1 and 2 typically store cleaner water, such as from the condenser water loops. Ponds 3 and

4 are usually used to store high strength water. The higher strength ponds are typically dredged as-needed, which is generally every two years. The water from Ponds 1 through 4 may be reused, while that in Ponds 5 and 6 will only be used for irrigation. Pond 7 stores the cleanest water and is not a potential source of H_2S emissions.

The property line borders these storage ponds in several areas, which requires that tight control of H_2S emissions be maintained to comply with the limit. Since water from several of these ponds could be returned to the factory, any water additives should be deemed Generally Recognized As Safe (GRAS) by the U.S. Food and Drug Administration, and furthermore if any additives leave residuals they must also be certified as Kosher for Passover.

THE SOLUTION

Throughout the H_2S control season (April through October) it is possible for sulfide generation to occur. The conditions that have been most conducive to H_2S generation are: when water temperatures are warmest, when medium to high COD ponds (5,000 – 35,000 mg/L) develop septic conditions, when pH rises above 6, and when large COD loadings enter the ponds.

SMBSC has focused on preventing as much loss of sugar to the ponds as possible through a focused employee education campaign. This strategy reduces wasted sugar that would otherwise be sold while also potentially lowering the costs of water treatment. However, it is not feasible to completely prevent some level of sugar loss to the ponds, such as from the beet wash water, on-site storm water, factory clean out events, and when upsets throughout the process require large discharges to the ponds. While some of these events can be predicted, there is always a significant degree of unpredictability that requires diligent monitoring and process management.

In the past when these types of events occurred, sections of the ponds would generate H_2S. With great effort SMBSC was able to control these H_2S emissions by gravity feeding hydrogen peroxide onto the surface of the ponds from chemical totes. Hydrogen peroxide directly oxidizes the sulfide present, in a near 1:1 ratio, according to the below reaction in acidic to slightly alkaline conditions:

$$H_2S + H_2O_2 \rightarrow S_0 + 2\ H_2O$$

In the presence of soluble iron and/or catalase, an enzyme produced by microorganisms exposed to aerobic conditions, hydrogen peroxide will decompose into dissolved oxygen in a 2:1 ratio according to the following reaction:

$$2\ H_2O_2 \rightarrow O_2 + 2\ H_2O$$

This decomposition reaction is a liquid to liquid transition, which can temporarily supersaturate the water with dissolved oxygen levels above the normal solubility limit.

Furthermore, in the high strength ponds, soluble iron concentrations range from 2 – 4 mg/L. Above pH 6.5 ferrous iron will bind with sulfide, forming ferrous sulfide, that will prevent volatilization as H_2S. Hydrogen peroxide dosing in excess of that needed to oxidize soluble sulfide can oxidize the sulfide in the FeS compound, leaving the ferrous iron "regenerated" and able to bind to another sulfide. Hydrogen peroxide also can oxidize the ferrous iron (Fe^{2+}) to the ferric (Fe^{3+}) state, which is then capable of oxidizing sulfide. Through these interactions with iron, hydrogen peroxide dosed above the demand immediately needed for sulfide oxidation can provide long-acting sulfide control.

When substantially higher concentrations of hydrogen peroxide were required peracetic acid also was dosed. Peracetic acid, an equilibrium mixture of hydrogen peroxide and acetic acid with antimicrobial effects, effectively slows the oxygen uptake rates within the water column (Figure 3) and rapidly minimizes H_2S release.

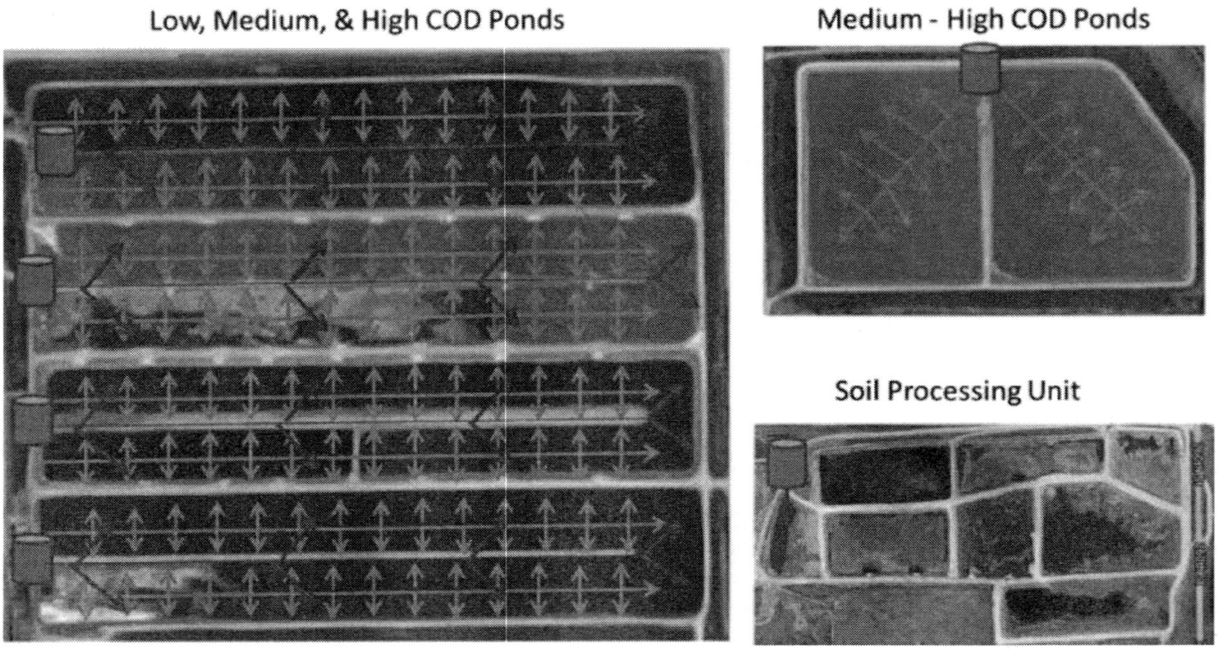

Figure 2. Hydrogen peroxide dosing schematic, SMBSC water storage ponds, 2017 season. Blue indicates pond water & H2O2 dosing line. Red indicates H2O2-only injection points.

USP Technologies developed a dosing method to distribute a diluted solution of hydrogen peroxide or peracetic acid gently across the surface of large ponds, allowing for rapid and targeted treatment of the surface layer when needed (simplified schematic shown above in Figure 2).

The hydrogen peroxide or peracetic acid can be dosed from 0.38 – 30 liters/minute (0.1 – 8 gallons/minute) along with 1100 liters/minute (300 gallons/minute) of pond water continuously circulating through the dosing lines. The floating dosing lines consist of a trunk line with smaller offshoots that terminate in small nozzles, the distribution and orifice sizes engineered based on pond dimensions and dosing site constraints.

In practice this allows for lower overall volumes of chemical control agent to be dosed, since dosing is targeted at the surface layer and not the entire volume of the pond. H_2S, which is typically generated most heavily in the anaerobic layer at the bottom of the pond, must pass through the surface layer in which more oxidative conditions can be maintained through continuous application of H_2O_2 and/or peracetic acid (Figures 2 and 4).

There are often very dissimilar dosing needs in different sections of the same pond, such as when a slug load of high COD water enters and gradually makes its way through the pond. In those instances, additional chemical dosing lines are used to target treatment more heavily in sections where H_2S generation is the highest or is anticipated (Figure 4). Overall, strict operational and management practices developed by SMBSC minimize loading in the ponds. However, there are often challenges and upsets that are beyond human control which will necessitate the continued reliance on treatment systems.

In order to comply with the H_2S limit at the lowest cost possible, SMBSC vigilantly monitors incoming and outgoing pond water volumes and compositions, parameters within many sections of the ponds (COD, pH, DO, ORP, temperature), H_2S emissions throughout the facility, and atmospheric conditions. When rapid control of H_2S emissions is needed, H_2O_2 and peracetic acid can be used to ensure compliance. USP Technologies' systems employed 4G communication capabilities so that chemical dosing rates and application sites could be adjusted 24/7.

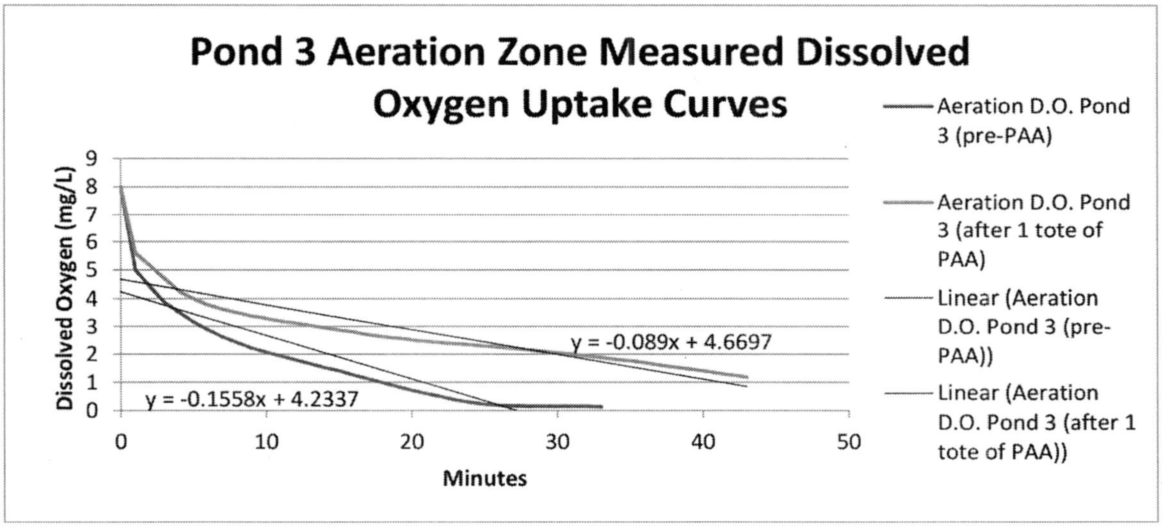

Figure 3. Hydrogen peroxide dosing. Diluted mixture in background, concentrated dose in foreground.

Further evaluation may be warranted into additional online measurements to help better target and time H_2S treatments. The treatment of the lower strength ponds with hydrogen peroxide when needed is cost-effective as COD levels follow the most predictable pattern. The demand for hydrogen peroxide and peracetic acid is highest in the high-strength ponds, which also have the most unpredictable H_2S generation rates. The high strength ponds can potentially receive loadings from more sources, making the overall control of H_2S emissions there the greatest challenge.

The combination of hydrogen peroxide and peracetic acid, used mostly when large slugs of high COD water entered ponds and depleted surface dissolved oxygen concentrations, appears to have a longer-lasting impact than either agent alone, so further investigation into the mechanisms and ideal relative rates of each solution and application method may undercover opportunities for further efficiency improvements. Finally, being limited to US FDA GRAS, Food Grade, and/or Kosher for Passover control agents makes it critical to apply the limited chemical options in the most economical way.

Figure 4. Hydrogen peroxide dosing. Diluted mixture in background, concentrated dose in foreground.

REFERENCES

Bicudo, J. R., Westerman, P. W., & Oleszkiewicz, J. A. (2000). Anaerobic and aerobic treatment of animal manure–a review. *Proceedings of the Water Environment Federation*, *2000*(16), 495-530.

Bowker, R. P., McGinley, M. A., & Schubert, J. (2004). Analysis of ambient odor data from an industrial area with multiple odor sources. *Proceedings of the Water Environment Federation*, *2004*(3), 374-393.

Cesca, J., & Cunnington, M. (2008). Case Study: Odour Risk Management at the Western Treatment Plant, one of Australia's Largest and Most Unique Wastewater Treatment Plants. *Proceedings of the Water Environment Federation*, *2008*(4), 1-15.

Dobson, R., Townsend, D., & Boxley, B. (2002). Permeable covers for odor and emissions control from swine manure lagoons. *Proceedings of the Water Environment Federation*, *2002*(5), 860-866.

Schafer, P. (2009). Lagoon Digestion-Old Dog, New Tricks. *Proceedings of the Water Environment Federation*, *2009*(14), 3085-3093.

Weber, M. D. (2000). Effective Odor Control & Responsive Community Relations In the Sonoran Desert. *Proceedings of the Water Environment Federation*, *2000*(3), 101-108.

City of Los Angeles Pilot Test on a Suspended Carbon Vessel Design

Cyrous Gilani, P.E., Project Manager
Hamed Haddad-Zadegan, Ph.D., P.E. Assistant Project Manager
Sheri Symons, P.E. HWRTP
Ali Poosti, P.E., Division Manager
Adel Hagekhalil, P.E., B.C.E.E., Assistant Director

Wastewater Engineering Services Division
City of Los Angeles, LA Sanitation
2714 Media Center Drive, Los Angeles, CA 90065

ABSTRACT

LA Sanitation owns and maintains 10,800 kilometers (6,700 miles) of sewer and operates three Air Treatment Facilities (ATF) and ten Carbon Odor adsorbers located along the sewer system for removing odor and other related constituents from the collection system. The ATFs utilize Bio-trickling Filters (BTFs) technology followed by carbon adsorbers as polishing units while the Carbon scrubber sites utilize carbon adsorbers as the only treatment for foul air.

In order to improve carbon life and reduce the foot-print required for carbon vessels, the City started conducting a series of tests to evaluate alternate carbon odor adsorbers with suspended media for all-around air flow.

The purpose of this test was to evaluate the performance of the new carbon vessel design with suspended media chamber (SMC) and all-around air flow which provides larger media surface area, smaller foot print, and capability to separate moisture from the inlet air stream. This testing project conducted a side by side comparison between a deep bed traditional vertical upward air flow and the new all-around air flow design.

Keywords

Carbon scrubber, odor control, hydrogen sulfide, radial design, carbon lifetime, adsorbers

Introduction

Moisture in the air stream entering the carbon bed is one of the main challenges which reduces the carbon life cycle in City's odor control facilities. Although in all City's odor control facilities some type of de-humidification or demister filter is used at inlet air stream to the carbon vessels, water particles still pass to the carbon bed. This phenomenon occurs by condensation taking place after demister filter. In order to improve the ATFs and current carbon scrubber performance at City's odor control facilities, a pilot test started in September of 2015 to evaluate and improve the performance of the new carbon vessel design in comparison with traditional carbon vertical flow.

Methodology

The carbon units test includes a new 850 m^3/hr (500 cfm) carbon vessel with all-around flow system manufactured by AfrasAir, and a more traditional 500 cfm typical upward/vertical flow carbon vessel. Each unit treats foul air from the primary battery tanks channel. 55o lbs of coconut shell virgin carbon was used in side each unit.

The traditional vertical flow vessel was three feet in diameter and contains three feet of activated carbon media. Foul air enters the open space at the bottom of the vessel and travels uniformly upward through the carbon.

The AfrasAir vessel is an all-around flow design. The foul air enters the vessel from the side of the housing, fills the open annular space between the vessel and the media chamber. Foul air travels through the carbon bed from all sides of the media chamber before discharging through an open exhaust system in the center of the vessel. A deflector system is installed inside the vessel at a right angle to the point where the inlet air flow enters the vessel and is designed to separate the moisture from the inlet air stream by impaction and then pipe the water collected through a drain system at the bottom of the vessel. Both vessels used coconut shell, virgin carbon for the test.

The concentrations of H_2S and Total Non-Methane Hydro-Carbon (TNMHC) were measured on the inlet and outlet of each carbon unit once a day during the testing period. The amount of water captured by each carbon vessel is also measured during the testing period.

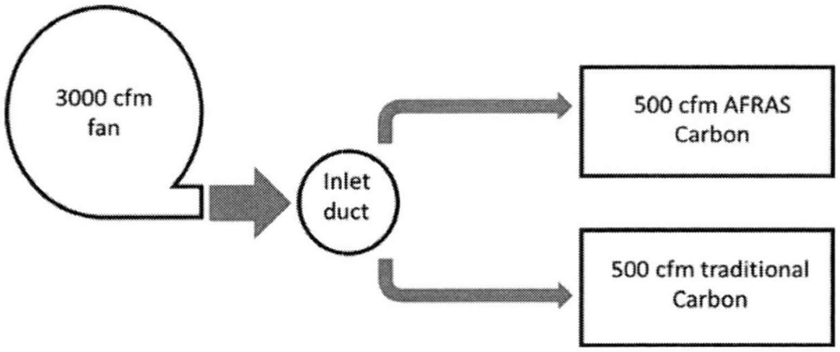

Figure 1: Carbon Pilot Test Units Arrangement

Figure 2: Inlet of Carbon Test Units: AfrasAir (Right), Traditional Carbon Filter (Left)

Figure 3: Outlet of Carbon Test Units: AfrasAir (Left), Traditional Carbon Filter (Right)

Test Results

Both carbon test units were operating continuously since October 14, 2015. In addition to daily collected data for H_2S and TNMHC, the inlet and outlet of each unit were also tested two times during this test period for speciated VOCs and Odor evaluation. Daily data collected from inlet and outlet of the traditional carbon vessel with vertical up-ward flow indicated that this unit started the initial sign of break-through on November 11. During this test period only a few drops of water was collected on daily basis from the drain of this unit. This is a clear indication that high level of inlet humidity was passed through the carbon bed. The level of H_2S on the outlet of this unit reached above the normal AQMD permit level (1.0 ppm) in few days thereafter, and finally on November 19th this unit was shut down and taken out of the testing. The H_2S and TNMHC removal graphs for this unit end at November 18, 2015 as shown on Figures 4 and 5.

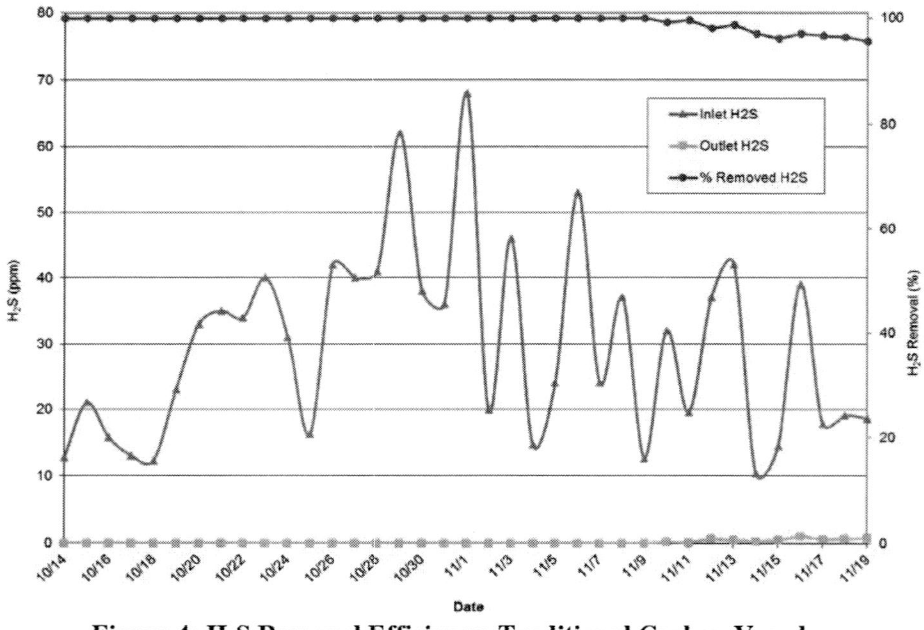

Figure 4: H_2S Removal Efficiency, Traditional Carbon Vessel

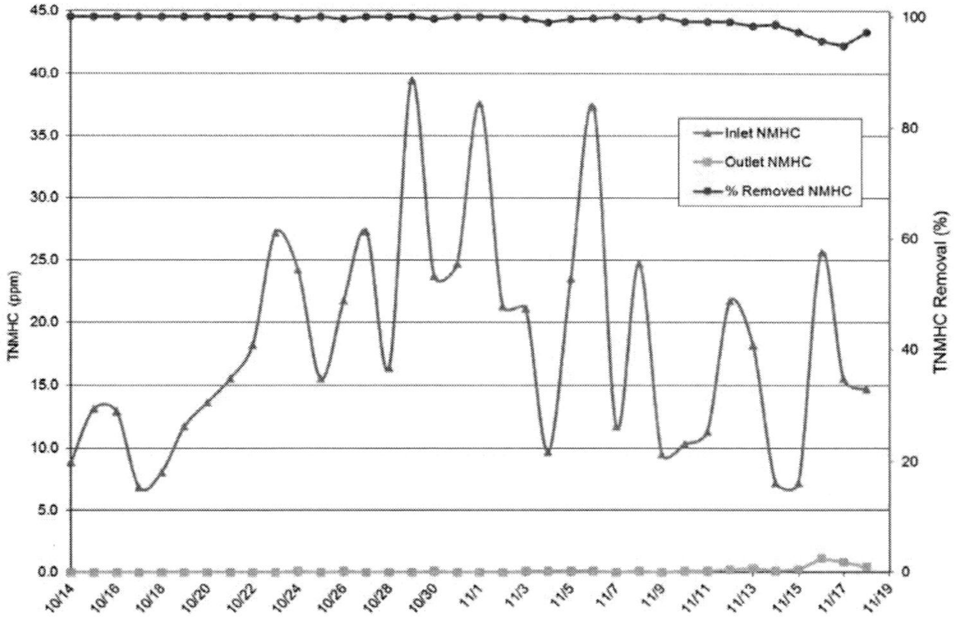

Figure 5: TNMHC Removal Efficiency, Vertical Flow Carbon Vessel

During the 4 week life span of this test, the average inlet H_2S and TNMHC concentrations for the vertical flow carbon bed were 29.6 ppm and 18.3 ppm, respectively.

However, the AfrasAir unit operated with 850 m^3/hr (500 cfm) air flow during the month of November with no sign of breakthrough.

The AfrasAir was operating until December 7th, before initial sign of break-through was detected. The carbon in AfrasAir system lasted about 54 days as shown in Figures 6 and 7. Per the result of this test, the carbon in AfrasAir vessel lasted about 75% longer in life compared to the carbon inside the traditional vertical flow vessel.

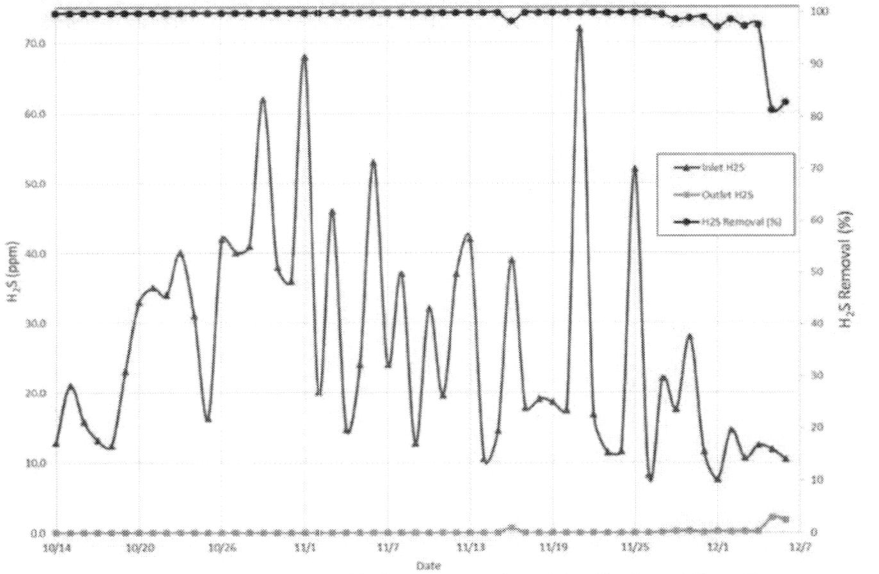

Figure 6: H_2S Removal Efficiency, AfrasAir Carbon Vessel

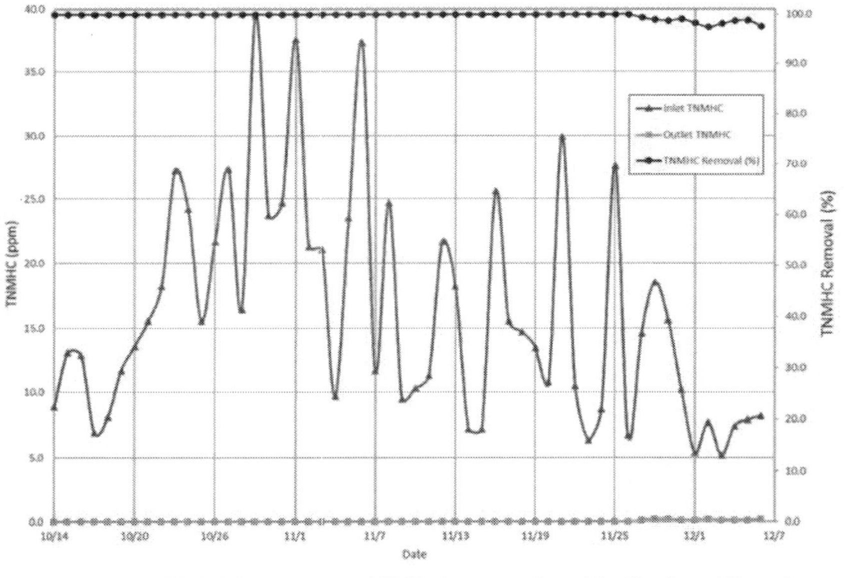

Figure 7: TNMHC Removal Efficiency, AfrasAir Carbon Vessel

The average H_2S and TNMHC for AfrasAir vessel were 26.5 ppm and 16.1 ppm respectively during the 7 week life cycle.

The daily average for H_2S and TNMHC for both test units were calculated based on data collected once a day. The actual daily average could be different as concentrations vary during the day.

Treatment capacity of the carbon units is determined by breakthrough with respect to the contaminant. As the carbon media has the tendency to fill the inner pore structure with moisture that comes from the odor source air flow, the performance from a total capacity perspective can be impacted by elevated inlet relative humidity/moisture levels. Therefore, the lower the inlet moisture levels, the greater the ability of the activated carbon to adsorb compounds since less of the activated carbon sites where adsorption takes place will be taken up by water molecules. The AFRAS unit was designed to condense and capture the inlet moisture to extend the life of the media. To assess the impact of water on each unit, a comparison was conducted between the AFRAS unit which removes the condensable water by installing an impact plate at the vessel entry point and the conventional deep bed which takes no steps to address the inlet moisture. Under normal condition at City's Air Treatment Facilities and carbon scrubber sites, A demister filter is used before the air enter the carbon vessels. However due to condensation some moisture forms and reaches the carbon bed. For this test no demister was used and both carbon vessel test units were expose to the same type of foul air pulling out of Hyperion primary channel. This comparison result with H_2S and VOC removals are depicted in Figures 8 and 9.

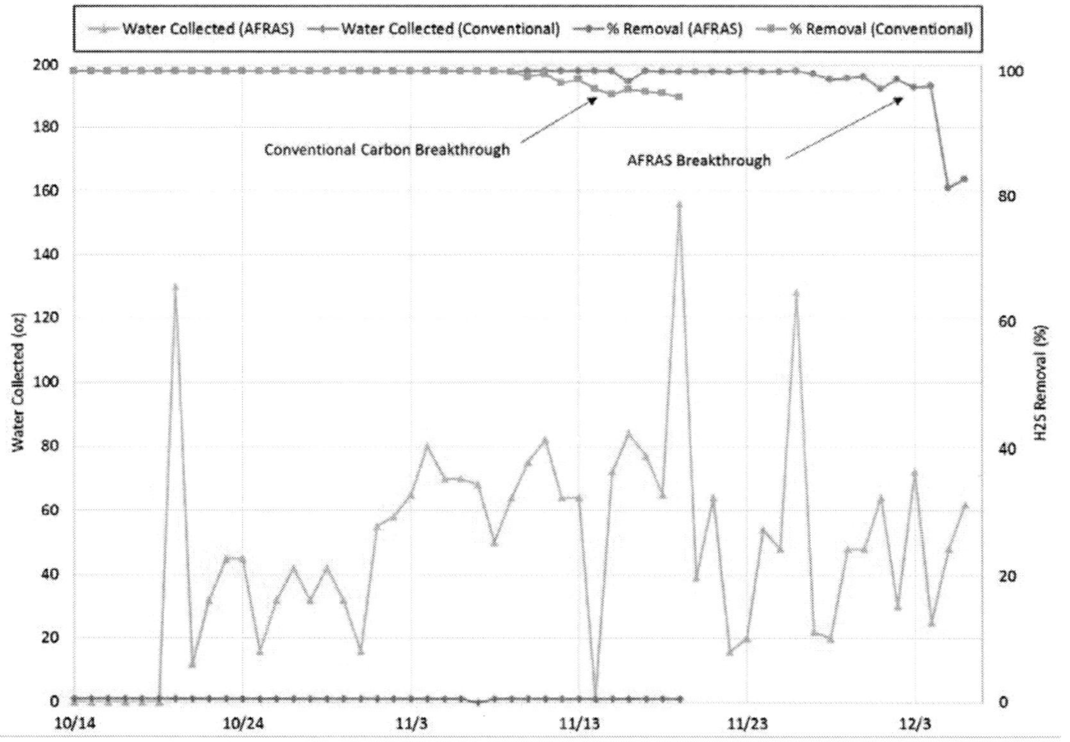

Figure 8: H_2S Removal vs. Water Collected

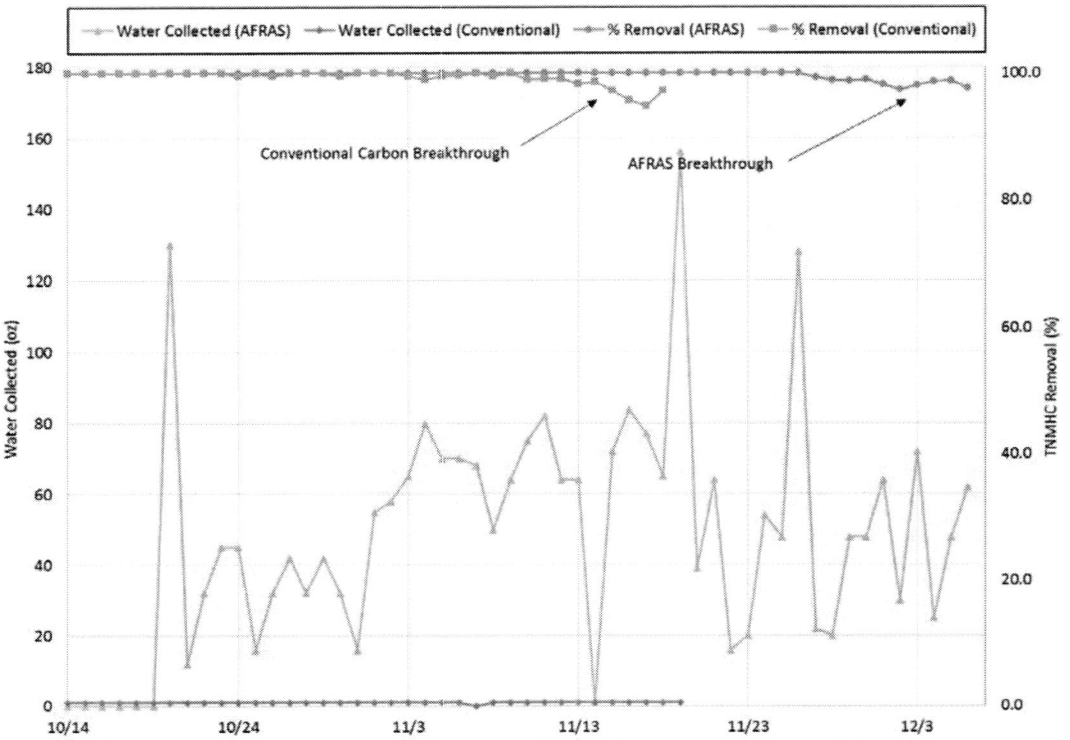

Figure 9: TNMHC Removal vs. Water Collected

Discussion

The water collected for the AfrasAir system far exceeded the amount collected for the conventional carbon unit, with daily measured volumes up to 4.61 liters (156 fluid ounces) (AfrasAir) compared to the 0.03 L (1 fluid oz) measured daily by the conventional unit. The moisture not removed by the conventional carbon vessel was either adsorbed by the media or passed through the carbon and into the ambient atmosphere through the discharge stack. Based on the shorter time it took for the conventional unit to reach capacity, it can be surmised that some of that moisture that entered the activated carbon vessel was taken up by the media, thereby reducing the number of sites available for compound removal and shortening its cycle time. The average daily volume of water from the inlet moisture collected for the AfrasAir unit was approximately 1.6 L (54 fluid oz). By removing this volume of the moisture from the incoming odor source, the AfrasAir unit was able to better utilize the carbon media as it did not become saturated with moisture. As a result, the AfrasAir unit effectively treated H_2S, TNMHC, and VOCs for nearly three weeks longer than the conventional canister. H_2S and VOC breakthrough occurred in the conventional carbon unit within approximately four weeks, whereas breakthrough occurred in the AfrasAir unit in about seven weeks. The extended capacity demonstrated by the AfrasAir unit is a positive result that is worth exploring further in a full scale setting at one of the interceptor collection system sites.

Larger carbon bed surface area (in smaller foot-print) was other factor for extended life cycle for AfrasAir which provided larger area on carbon's Mass Transfer Zone to adsorb more odorous compounds entering this vessel. The weight of carbon was the same for both test units, however the surface area of carbon in direct contact with foul air was larger in AfrasAir unit.

The pH of drain water captured in the AfrasAir was around 2.5 while the pH of water collected from the more traditionally designed carbon vessel was higher, at around 5.

Two samples were collected from the activated carbon pilot comparison and sent to an independent odor laboratory for evaluation. The first sample was collected in the middle of the pilot test while the second sample was collected at the time when the conventional unit was at the point of reaching its capacity and the removals reported earlier showed the concentrations of H_2S and VOC began slipping. As a result, the second set of odor detection/recognition threshold results for the AfrasAir and conventional units cannot be directly compared (Figures 10 and 11).

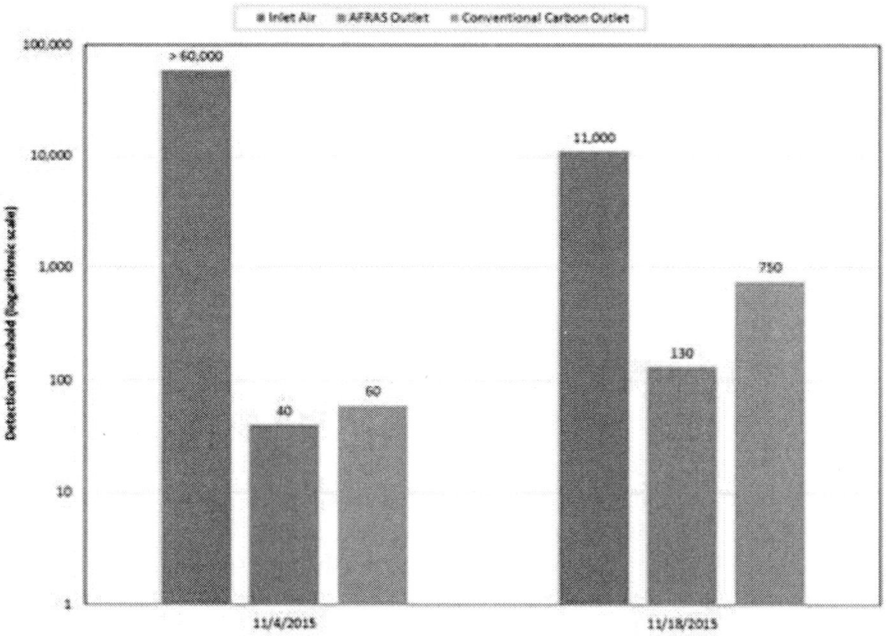

Figure 10: Detection Threshold Graph

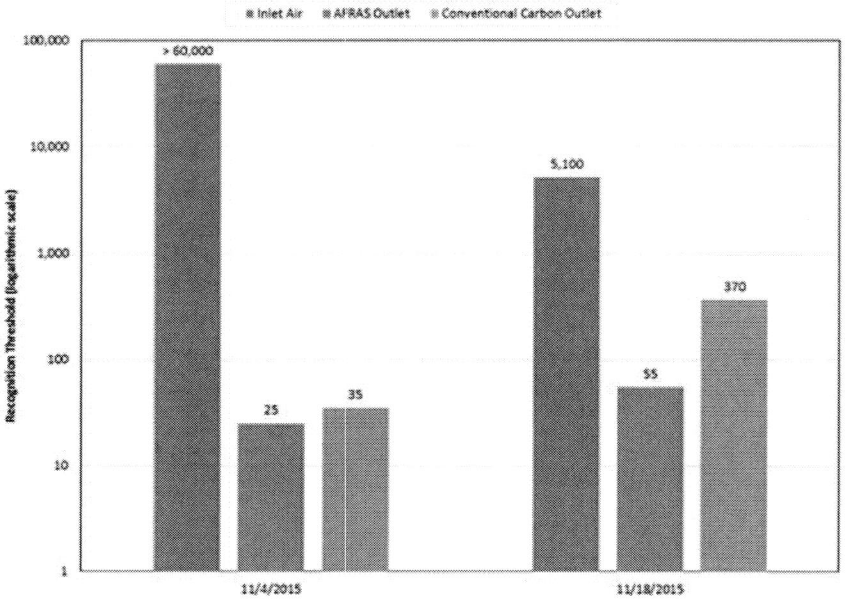

Figure 11: Recognition Threshold Graph

Based on the results from the first odor samples, the results in the discharge from both the AfrasAir and the conventional unit for detection and recognition threshold were the same. Although the AfrasAir unit had lower values, the difference between 40 and 60 (detection threshold for AfrasAir and conventional, respectively) and 25 and 35 (recognition threshold for AfrasAir and conventional, respectively) is nearly indistinguishable to the average community resident. Both levels demonstrate excellent performance by the activated carbon systems and represent discharge levels that would be diluted down to imperceptible levels in the ambient atmosphere before they reach the nearby community.

The hedonic tone data depicted in Figure 12 shows that both activated carbon units, AfrasAir and conventional, continue to be negative or unpleasant. The data also indicates that AfrasAir unit was less unpleasant and offensive than the conventional unit based on the panelist's subjective observations. The odor characterization has chemical as the key principle odor descriptor for the AfrasAir discharge while the conventional descriptor is still offensive, which is a positive for the AfrasAir system. This is more encouraging data for the AfrasAir unit that suggests the moisture in the conventional unit may have more of an odor perception impact than just taking up sites on the activated carbon. Perhaps the moisture carried through the conventional unit exacerbates the offensiveness of the odor that is evaluated by the panelists. This concept, however, cannot be confirmed with the information collected from this pilot study.

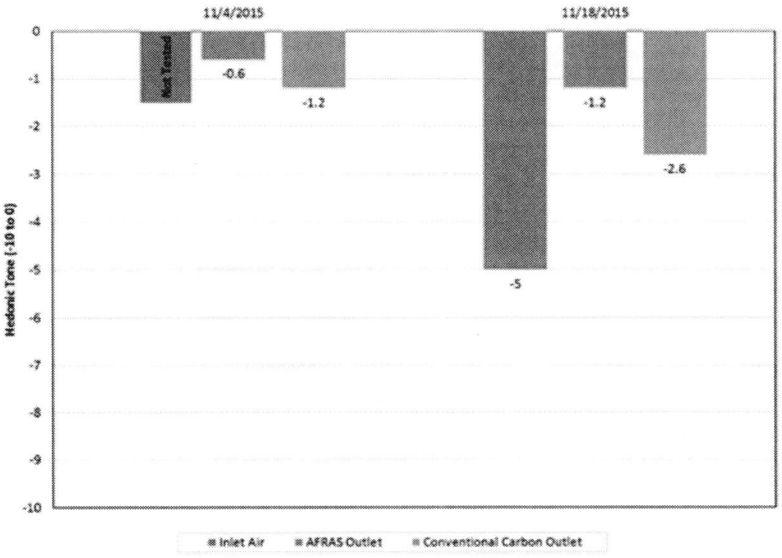

Figure 12: Hedonic Tone Graph

Results and Conclusions

The results show that activated carbon will remove H_2S and VOCs to very low discharge levels that comply with SCAQMD air permit limits, regardless of the moisture that is removed from the air inlet stream, until they reach saturation. The moisture that is allowed to enter the activated carbon bed acts to reduce the capacity of the bed to remove compounds, but does not interfere with the removals up to the point of saturation. The fact that the AfrasAir unit can reduce the inlet moisture that enters the activated carbon bed and thereby increase the available sites for H_2S and VOC removal and extending the capacity of the bed beyond the conventional unit operation is an encouraging conclusion that should be examined further on a full scale basis.

At City's Air Treatment Facilities and carbon scrubber sites a demister filter is used upstream of the carbon vessels, as previously stated. However moisture forms after the demister filter due to condensation and reaches the carbon bed.

Future Work

To verify the performance of the full scale system, City has decided to install a 16,990 m^3/h (10,000 cfm) AfrasAir carbon scrubber system at one of the City carbon scrubber site. This unit is scheduled to be installed in early 2018. The performance of this full-scale vessel will be closely monitored to confirm the longer carbon life cycle. After performance confirmation at the full-scale unit, City will replace the carbon scrubber vessels at 7 City's carbon scrubber sites with new AfrasAir carbon vessels.

This scrubber system may replace the polishing carbon scrubbers at City's ATFs as the next phase of rehabilitation at Collection System and Treatment Plants.

The Pressure is On – Enforceable Odor Limits

Anders Wistrom, CentriairUSA
Jack Delin, Scandinavian Centriair

ABSTRACT

The design, implementation and validation of the odor control installation servicing the wastewater treatment plant at a Nestle Purina pet food manufacturing facility is discussed. Strict performance criteria had been established by the facility owner and included a Not-to-Exceed odor concentration threshold at the exhaust stack and a required uptime during which the threshold odor concentrations could not be exceeded. Together, both criteria govern not only the nominal efficiency of odor reduction during normal operation but also the odor removal requirements during shutdowns for planned and emergency maintenance activities.

KEYWORDS

Odor control, AERMOD, EN13725, wastewater treatment, UV oxidation, performance guarantee, validation.

INTRODUCTION

In many jurisdictions, odor impacts are regulated under the nuisance provisions of a local or municipal code. However, the explicit conditions that establish whether a nuisance condition exists are not easily defined. Due to this shortcoming, local jurisdictions are recognizing the need to introduce objectivity when setting allowable odor limits. One approach is the introduction of an objective odor standard that can help wastewater treatment facilities, engineers, and regulators to develop a consistent basis for odor testing. To that end, the European olfactometry standard EN13725 is used to guide the quantification of odors when setting odor control performance criteria and when determining system performance during the warranty period.

In this paper, we present the case of Nestle Purina, a pet food manufacturer located adjacent to sensitive receptors and a potential source of odor complaints from neighbors as well as from the community at large. The management of this potential odor source was a major concern for both Nestle Purina and the local authorities who must control odor emissions. Strict odor emissions criteria combined with a specified uptime were included in the procurement documents, which guided the subsequent technology selection, design, and implementation of the odor control facility.

ODOR CONTROL FACILITY DESIGN

Before setting performance requirements on the odor control facility the facility owner's consultant first analyzed the site surroundings to identify the most sensitive odor receptors. Here, the closest receptor was determined to be at the shortest distance from the odor source, here exhaust stack, to the "fence line" and was classified as being "very sensitive". In a second step, the relevant not-to-exceed threshold odor concentration was determined at this location. The third step was to assess the duration in time or *uptime* when the threshold odor concentrations cannot be exceeded. And lastly, the allowable odor concentration at the odor control facility exhaust "stack" was calculated.

The threshold concentration at the stack was estimated by a simple multiplication of the threshold odor concentration at the fence line with the dispersion factor, where the dispersion factor was determined using atmospheric modeling software, e.g. AERMOD developed by U.S. EPA. An example is shown in Figure 1 where the extent of equal odor concentrations have been plotted.

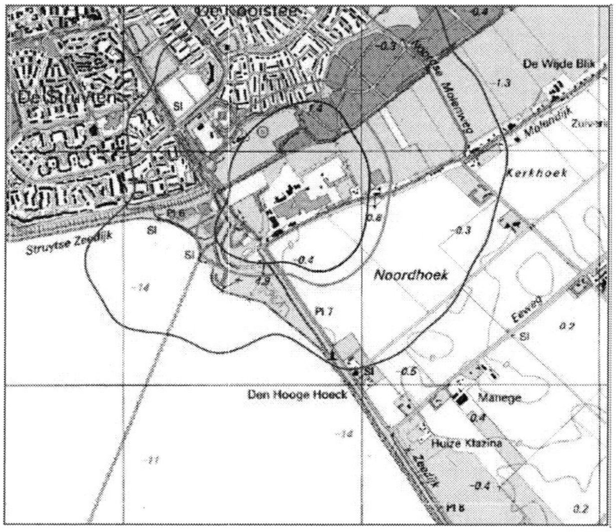

FIGURE 1. Modeling Threshold Odor Concentrations at Recipient Locations

Measuring the Odor Profile

A starting point for all design is a review of the "odor library" whereby past project data and literature values are used to provide a general assessment of site conditions. When a "new" odor source is encountered additional information is obtained using samples collected and analyzed off site or by using, a mobile pilot test bed and laboratory. An onboard gas chromatograph-mass spectroscopy (GC-MS) instrument is used for "sniffing" and sample collection and analysis. However, the most useful information for procurement purposes is obtained using olfactory methods or panel testing whereby a panel of at least four qualified persons determines the odor threshold at different dilutions. An example of an odor profile using gas chromatography analysis, the NOSE, a mobile pilot test bed, and typical pilot results are illustrated in Figure 2.

a. Odor profile - GC-MS results | b. The Centriair NOSE - Nuisance Odor Solution Evaluator | c. Odor (o.u./m^3) results from pilot testing

FIGURE 2. a) Example of odor profile as measured by GC-MS. b) The "NOSE"- a mobile test bed for unit process assessment. c) Typical pilot test data examining different unit processes

Facility Design Parameters

Strict performance criteria had been established by the facility owner and included a Not-to-Exceed odor concentration threshold at the exhaust stack and a required uptime during which the threshold odor concentrations could not be exceeded. However, the odor profile was not provided by the facility owner in the bid documents. Instead, the odor technology provider was left to rely on either his experience and/or on-site measurements conducted at his own expense. This is not optimal for the odor control equipment supplier but is not unusual as facility owners and/or their consultants are often well aware of available odor control technologies, performance characteristics, and the overall experience and reputation of service providers.

DISCUSSION

There is myriad of odor control technologies being offered to solve the problems emanating from the operation of wastewater collection systems and pumping stations, wastewater treatment plants, biosolids- and biogas production, etc. The problem for facility owners arises when evaluating different technologies in the absence of reliable performance data compounded by the fact that when data do exist the results are highly variable. However, both biological as well as physical-chemical odor control technologies are in widespread use in most EU countries and in the UK so facility owners and/or their consultants have ready access to such facilities close by.

Technology Evaluation

In one of the few published studies (Barr, 2013) comparing odor reduction technologies at POTWs in Sweden revealed that the average odor reduction, in units of o.u./m^3, was a meager 55 percent taken over all available different odor control technologies. This is not helpful for a facility owner when the odor generating facility is in close proximity to area residents, a situation where performance predictability and certainty are absolute requirements. A summary of the study results (Barr, 2013) is presented in Table 1.

TABLE 1. Comparison of Odor Control Technologies at POTWs in Sweden

Technology	Number of measurements	Number of facilities tested	Removal rate (average %)	Removal rate (median %)
Biofilter, in-ground	13	9	50	61
Bioscrubber	7	5	13	6
Activated Carbon	19	16	63	72
Thermal Oxidation	6	3	98	99
UV+Activated Carbon	10	4	86	98

The study identified a wide variation of odor removal rates among the installations tested. Some installations were very efficient, exceeding 90 percent removal, while other odor removal systems of the same general type did not remove any odor at all. From the facility owner's standpoint, most problematic is the wide range of observed outcomes when having to select an odor control technology for a specific application. The study authors identified several of reasons that contributed to the observed variability and included incorrect selection and/or sizing of the odor control technology. Here, Thermal Oxidation and UV+Activated Carbon were identified as the most reliable odor control technologies.

Performance Criteria

Given the location of the Nestle Purina facility and proximity to nearby sensitive receptors the client specified the threshold odor concentration at the fence line to be < 1 o.u./m^3, which corresponds to < 100 o.u./m^3 at the exhaust stack when allowing for an assumed dispersion factor of 100 that was determined using atmospheric modeling software.

In addition, the client required a redundancy or uptime with a target availability for the odor control system of 99.5 percent of the year, which corresponds to 363 days of operation without exceedances. All scheduled (and, when feasible, emergency) maintenance activities were to be carried out on either a non-operating unit with remaining units operating or within the specified time period. The approximate interval between scheduled maintenance and the time (duration) required to perform the associated maintenance tasks, and the start-up time needed to reach the design odor removal efficiency are summarized in Table 2 for different odor removal technologies. Again, these values can vary widely depending on system size, media type, organic and inorganic loading rates, and operating temperature.

TABLE 2. Approximate Time Interval and Duration for Scheduled Maintenance

Technology	Interval	Duration	Notes
Biofilters, in-ground	5 to 7 years	1 to 2 weeks	Acclimation of bioactive microrganisms – up to several weeks
Bioscrubbers, towers	12 to 30 months	2 to 4 days	Acclimation of bioactive microrganisms – up to several weeks
Activated Carbon	3 to 12 months	1 day	No start-up time
Thermal oxidation	varies	2 to 5 days	Downtime includes unit cool-down time
UV+Activated Carbon	12-24 months	1 day	No start-uptime

The design parameters for the Nestle Purina odor control system are summarized in Table 3.

TABLE 3. Design Parameters for Nestle Purina Odor Control Facility

Parameter	Quantity	Units
Flow	20,000 (12,500)	m^3/h (CFM)
Not-to-Exceed Odor Concentration (stack)	100	o.u./m^3
Uptime (redundancy)	99.5	percent
Validation Protocol	EN13725:2007	

The specified uptime of 99.5 percent leaves only 44 hours or less than two days per year for planned or emergency maintenance. The selection criteria for the odor control system also included minimizing life-cycle costs and having readily available access to spare parts and repair/maintenance services.

Selected Odor Control Technology

Given the strict performance criteria and based on life-cycle-cost analysis the facility owner selected an UV Oxidation+Activated Carbon odor control system with ancillary equipment including tank covers, stainless steel ducting and building air handling equipment to maintain overall building air quality. The equipment specifications are summarized in Table 4.

TABLE 4. Design Specifications

Component	Description
UV Reactor	Flow 20,000 m^3/h (12,500 CFM)
	Installed Power 9.5 kW
	Lamp life ~16,000 hours
	Automatic Clean-in-Place (CIP) system for lamp cleaning
	UV system can be expanded 40% for unanticipated future changes
Activated Carbon Bed	Dual carbon bed (minimize pressure drop)
	Non impregnated "standard" activated carbon
	Carbon life 1 to 2 years
	Carbon bed can be expanded 40% for unanticipated future changes
	Pressure drop <3.82 in. H_2O (950 Pa)
Fan	20,000 m3/h (12,500 CFM)
	Installed power 25 kW (running load 19.5 kW)

An added feature was the low overall weight of the odor control system, which allowed for placing the unit on the roof (see Figure 3) and thus freeing up valuable space outside the building

FIGURE 3. Nestle Purina with Roof Mounted Odor Control System

Commissioning and Validation Testing

This facility owner had strict performance requirements defined in units of odor units per cubic meter [o.u./m^3] and uptime requirements. The contract stipulated validation testing to be performed using olfactory panel testing in accordance with the standard EN13725:2007 by an independent odor laboratory, here the Olfactometric Research Laboratory, Wroclaw University of Science and Technology, Poland (Olfactometric Research Laboratory, 2016). The source odor concentrations, a blend of vented air from different locations within the waste treatment plant, and treated air odor concentrations were measured at two instances, once during facility commissioning and once during validation testing. The results are presented in Table 5.

TABLE 5. Measured Odor Concentrations of Untreated and Treated Air During Facility Commissioning and Validation

Combined Odor Source	Commissioning 5-12-2016		Validation 6-30-2016	
	Untreated Air	Treated Air	Untreated Air	Treated Air
Sludge Processing Flotation Unit Room Sludge Tanks	12,147 o.u./m^3	80 o.u./m^3	10,461 o.u./m^3	50 o.u./m^3

Odor removal exceeded 99 percent in both tests, but more important was that odor emissions were consistently less than 100 o.u./m^3 at the stack, which was the contractually specified odor threshold value. At this site, this translates to a fence-line odor concentration of less than 0.25 to 0.5 o.u./m^3 – well within the owner defined odor emission threshold for the most sensitive receptors at the facility fenceline.

The odor removal rates achieved here using UV Oxidation+Activated Carbon are comparable to those obtained using thermal oxidation but are attained at a fraction of the cost. The annual energy use for the UV Oxidation+Activated Carbon odor control system is 0.82 MWh/yr. compared to a Regenerative Thermal Oxidizer (RTO) operating at 95% efficiency that has an annual energy demand of 2.29 MWh/yr., or about nine times higher.

CONCLUSION

The procurement of the Nestle Purina odor control system was guided by strict performance criteria and life cycle cost. Odor emission requirements included air flow, threshold concentration, and uptime, as well as applicable validation procedures, e.g. EN13725, that were specified at the project bidding stage. Given the strict emissions requirements combined with a specified uptime called for a robust and easily maintained odor technology solution. Odor concentrations were measured by a third party at the source and at the exhaust stack to verify compliance with contractual odor emission requirements.

EN13725, as a unifying testing standard, produces confidence in using odor testing for performance measures and for comparing performance of different facilities, industry wide. It is argued that an objective odor standard would also provide industries with benchmarks against which they can gauge their success at preventing or mitigating odor impacts and for facility owners to gauge the effectiveness of odor control technologies. Strict and enforceable odor limits are useful criteria for the selection and procurement of odor control facilities. It is good for the facility owner and straightforward for the project bidder.

REFERENCES

Barr, Sten- Åke et.al., 2013, WR55 Evaluation and Recommendation for Odour Control Equipment at Waste Refinery Plants/Waste Water Treatment Plants, SP Sveriges Tekniska Forskningsinstitut, Box 857, 501 15 Borås Sweden, (http://wasterefinery.se/tidigareprojekt)

Olfactometric Research Laboratory, Research Report (treated air), 2016, W7/Z3/2016/13, Wroclaw University of Science and Technology, Environmental Engineering Department, Unit of Ecologistics and Environmental Risk Management, Plac Grunwaldzki 9, 50-377 Wrocław, Building D-2, room 17, Tel. +48 713.202.560.

Olfactometric Research Laboratory, Research Report (treated air), 2016, W7/Z3/1351/13, Wroclaw University of Science and Technology, Environmental Engineering Department, Unit of Ecologistics and Environmental Risk Management, Plac Grunwaldzki 9, 50-377 Wrocław, Building D-2, room 17, Tel. +48 713.202.560.

Olfactometric Research Laboratory, Research Report (untreated air), 2016, W7/Z3/101/13, Wroclaw University of Science and Technology, Environmental Engineering Department, Unit of Ecologistics and Environmental Risk Management, Plac Grunwaldzki 9, 50-377 Wrocław, Building D-2, room 17, Tel. +48 713.202.560.

Testing the DMS Capacity of Specialized Adsorptive Media

Ryan McKenna[1], Matthew Ward[2], Eyasu Yilma[3]
1. Hazen and Sawyer, 10002 Princess Palm Avenue, Registry One Suite 200, Tampa, FL 33619;
2. CH2M, 3221 Rigsbee Road, Chapel Hill, NC 27514;
3. DC Water, 301 Bryant Street NW, Washington, DC 20001

ABSTRACT

DC Water, in conjunction with other jurisdictions discharging to the Potomac Interceptor (PI), constructed six odor control facilities (PI OCFs) for the long-term abatement of odors along the main sections of the sewer. These facilities were designed to pull odorous air from the sewer air space with a fan, and pass the foul air stream through a radial vessel of activated carbon prior to discharging the treated air into the atmosphere. However, soon after the startup of the PI OCFs, it became apparent that nearby receptors at several of the facilities could detect exhaust odors as nuisance odor complaints began to be filed. Being responsive to community concerns, DC Water initiated an investigation to determine the nature and cause for the off-site nuisance odor complaints. This revealed that although H_2S and most of the other odorous wastewater compounds were being removed, the activated carbon was not very effective at removing some of the reduced sulfur compounds, especially dimethyl sulfide (DMS).

This paper is an update of the optimization efforts that have been conducted since the topic was reported at this conference in 2014 and 2016. Additionally, this paper discusses the development of a modified test based on the American Society for Testing and Materials (ASTM) D6646-03 – *Determination of the Accelerated Hydrogen Sulfide Breakthrough Capacity of Granular and Pelletized Activated Carbon*, that was used to evaluate and compare the capacity of various types of permanganate-impregnated media for the removal of DMS in the polishing layer of the odor control facility treatment vessels.

KEYWORDS: Odor control, dimethyl sulfide, DMS, DMS capacity, testing, permanganate media

INTRODUCTION

Odors are generated naturally by microorganisms within the wastewater collection system – where there's wastewater, there are inevitably odors too. And in communities around the world, development and collection system infrastructure are constantly coming in closer and closer proximity, bringing more and more receptors in the vicinity of these odors. Furthermore, the public's patience for odors in our communities seems to be dwindling as well. As utilities strive to be good neighbors, it is important to balance community tolerance to odors with best available treatment technologies and practices.

The Potomac Interceptor (PI) is a portion of the District of Columbia Water and Sewer Authority (DC Water) collection system that is over 45 miles (72 km) in length with diameters ranging from 36-inch to 96-inches (0.8-meters to 2.3-meters). The PI conveys around 50 million gallons of wastewater per day (mgd) (2.2 cubic meters per second) to the Blue Plains Advanced

Wastewater Treatment Plant for state-of-the-art treatment before discharging into the Potomac River.

Several jurisdictions discharge into the PI system, including Loudoun and Fairfax Counties in Virginia, Montgomery County in Maryland, and the District of Columbia (Figure 1). As the wastewater is collected in the local community sewers and transported through the DC Water interceptor, biological activity consumes the oxygen. Anaerobic conditions transform sulfates into dissolved hydrogen sulfide (H_2S), resulting in the release of nuisance odors. Many factors influencing the generation of H_2S within the PI are outside of DC Water's control such as temperature, organic (Biochemical Oxygen Demand - BOD) load, dissolved oxygen concentration, sulfate concentration, detention time, etc. Furthermore, the H_2S gas that is released is also responsible for creating corrosive conditions to exposed metal and concrete surfaces within the collection system, which can dramatically reduce the lifetime expectancy of the infrastructure.

DC Water, in conjunction with other jurisdictions discharging to the PI, constructed six odor control facilities (PI OCFs) for the long-term abatement of odors along the main sections of the sewer. These facilities were designed to pull odorous air from the sewer air space with a fan, and pass the foul air stream through a radial vessel of activated carbon prior to discharging the treated air into the atmosphere.

One of most widely-used methods to control odors is adsorption by activated carbon. Activated carbon has a microporous structure, a high surface area, and a high degree of surface reactivity that allows it to effectively adsorb a variety of compounds. Its advantages also include consistent performance and a mechanically-simple process.

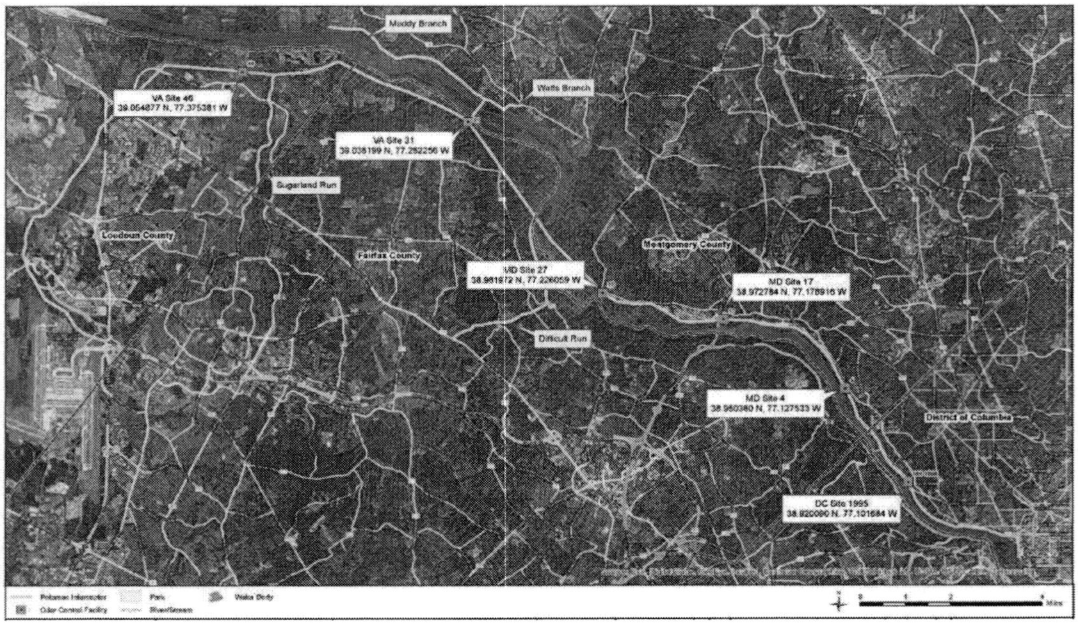

Figure 1 - Map of the Potomac Interceptor (PI) Sewer System.

However, soon after the startup of the PI OCFs (Figure 2), it became apparent that nearby receptors at several of the facilities could detect exhaust odors as nuisance odor complaints began to be filed. Being responsive to community concerns, DC Water initiated an investigation to determine the nature and cause for the off-site nuisance odor complaints. Influent and exhaust air samples were collected from the activated carbon odor control systems and sent to an independent laboratory for analysis. This revealed that although H_2S and most of the other odorous wastewater compounds were being removed, the activated carbon was not very effective at removing some of the reduced sulfur compounds, especially dimethyl sulfide (DMS).

Figure 2 – "Site 27" Odor Control Facility

DMS has an odor that is often described as being like that of decaying cabbage or rotting vegetables. According to published odor threshold concentrations, the average person can detect DMS at a concentration of less than 1 part per billion (ppb). The lab data showed that the DMS in the system's exhaust far exceeded this value, generally by orders of magnitude. This set in motion a search for the potential source of DMS in the wastewater of the PI, and more importantly, ways to mitigate the nuisance odor impact on the surrounding communities.

This paper is an update of the optimization efforts that have been conducted since the topic was reported at this conference in 2014 and 2016. Additionally, this paper discusses the development of a modified test based on the American Society for Testing and Materials (ASTM) D6646-03 – *Determination of the Accelerated Hydrogen Sulfide Breakthrough Capacity of Granular and Pelletized Activated Carbon*, that was used to evaluate and compare the capacity of various types of permanganate-impregnated media for the removal of DMS in the polishing layer of the odor control facility treatment vessels.

Initial Media Pilot Testing

It was previously reported that a pilot scale 208-liter (55-gallon) drum was used to initially test alternative media, including coconut- and coal-based carbons in catalytic and virgin forms, as well as a potassium permanganate-impregnated media (Pope, Lucas, 2014). The pilot unit exhaust was consistently analyzed for reduced sulfur compounds by an independent laboratory, and odor parameters (strength, intensity, character, hedonic tone, and persistence) were evaluated by a separate independent odor panel. Additionally, data was collected for other parameters such as volatile organic compound concentrations, relative humidity, and air flow.

The carbons were evaluated over a two- to three-week period, unless the data demonstrated that one was not effective prior to reaching the end of the three-week test period. Additionally, potassium permanganate-impregnated zeolite was placed as a polishing layer on top of the carbon media after the two to three-week period to assess how effective it was in further reducing odors from the pilot scale exhaust.

According to the initial pilot testing, the coal-based carbon with the potassium permanganate-impregnated zeolite polishing layer had the best chance of achieving the desired objective of no odor impact at the restaurant across the street. But the capacity of the potassium permanganate-impregnated media had not been fully tested and the media life was uncertain. In addition, the radial configuration of the carbon bed made installing a polishing layer of permanganate media unfeasible without extensive modifications to the interior of the vessel.

Facility Optimization

Based on a thorough study of wastewater characteristics, inlet and exhaust compounds, system configuration, media pilot testing, and facility operation and maintenance practices, an Optimization Plan was developed for the six PI OCFs. The Optimization Plan included the following components in an effort to mitigate potential odor impacts to the surrounding communities:

- Dual-media Vessel Retrofit
- Additional Media Pilot Testing
- Regular Media Monitoring
- Influent Airflow Control
- Maintenance Management Manual
- Differential Pressure Testing
- Liquid-phase Treatment Pilot Study

Dual-media Vessel Retrofit

While activated carbon has been a "work horse" of the odor industry for decades, reduced sulfur compounds (DMS, dimethyl disulfide, mercaptans, etc.) are less readily adsorbed into the carbon's pore spaces. Retrofitting the existing radial vessels so that an additional screen could be inserted into the annular media bed would allow for the installation of two separate layers of media. The first layer was recommended to be a pelletized, high-H_2S capacity coal-based carbon to remove the H_2S loading entering the treatment system. This carbon had proven to be successful at removing H_2S in the original PI OCF vessels. The second layer was recommended to be a specialized media to polish the remaining organo-sulfurs (DMS, DMDS, mercaptans, etc.).

The radial configuration of the carbon bed made installing a polishing layer of polishing media difficult without significant modifications to the vessels, including the lids, access hatches, exhaust stacks, and internal screens. However, providing an initial bed to remove the H_2S, coupled with a polishing layer of specialized media to remove the remaining organosulfur compounds, would greatly enhance overall system performance and media life (Figure 3).

DC Water worked with the equipment manufacturer to determine if retrofitting the vessel to allow for a dual bed configuration was feasible. The system manufacturer proposed a cost-effective plan that involved laminating base support rings, utilizing modular intermediate screen pieces, screen support tabs, and modifications to the access hatches of the vessel lid. Vessel retrofits for the sixth and final odor control facility were completed in December of 2016.

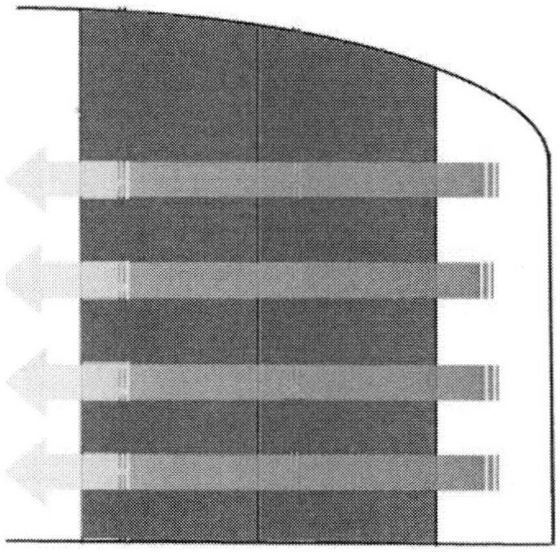

Figure 3 – Dual Media Radial Flow Vessel

Media Pilot Testing

The original pelletized, high-H_2S capacity anthracite-based carbon was effective at removing H_2S at the various PI odor control facilities. However, identifying the most effective, cost-efficient polishing media was desired for use in the retrofitted dual-media layer vessel configuration. This would optimize the media life and system performance of the OCFs – particularly with respect to DMS removal.

DC Water conducted a second pilot test to compare the performance of various types of specialized media side-by-side to determine if a more desirable alternative exhaust polishing product is available. The following specialized media was selected for the second round of pilot testing:

- 6% potassium permanganate-impregnated zeolite
- 6% sodium permanganate-impregnated activated alumina
- 12% sodium permanganate-impregnated activated alumina
- Potassium iodide-impregnated carbon pellet
- 12% permanganate-impregnated carbon pellet
- A new proprietary "ferritic-based" media designed to specifically target DMS/DMDS

The test was conducted in a two-stage configuration to replicate the dual media beds of the modified PI OCF vessels. Foul air from the interceptor first passed through a single bed of high-H_2S capacity carbon, followed by the various columns of specialized polishing media (Figure 4). Samples were collected monthly and sent to an independent laboratory for ASTM D5504-12 – *Standard Test Method for Determination of Sulfur Compounds by Gas Chromatography and Chemiluminescence* testing for sulfur-based compounds. While the pilot test was cut short due to the accumulation of moisture in the first few columns (in retrospect, a moisture trap would have been good!), some inferences could be made from the data that was able to be collected. The best performing polishing media with respect to DMS breakthrough (the primary compound of interest) appeared to be the permanganate-impregnated media.

Figure 4 – Media Pilot Test Apparatus

Regular Media Monitoring

In order to proactively assess whether breakthrough of the media was occurring in the full-scale PI OCFs, periodic media monitoring was also recommended as part of the Optimization Plan. This consisted of collecting influent and exhaust samples from each location every two months utilizing the ASTM D5504-12 test for 20 sulfur compounds (including H_2S, DMS, DMDS, methyl mercaptan, carbonyl sulfide, carbon disulfide, etc.). Sampling results are compiled and tracked for media replacement forecasting purposes, and DMS breakthrough initiates a media replacement event.

Influent Airflow Control

Airflows to the OCF treatment vessels were found to be above the design parameters, based on inlet testing at each facility with a hot wire anemometer. This increases the velocity and odor compound loading to the media, and reduces retention time and media life. It was found that the existing manual influent damper handwheels did not lock in place properly and "drifted" open over time. To address this, it was recommended that variable frequency drives be installed for each of the PI OCF fans for the purpose of periodic testing of airflow and adjustment of fan speed.

Maintenance Management Manual

The original O&M material was extensive and cumbersome, consisting of 35 separate manuals with a combined total of 7,700 pages of information. In order to facilitate maintenance activities and scheduling, required routine preventive maintenance activities were extracted into a stand-alone document and condensed into a more user-friendly manual.

This included compiling a list of odor control components that are required to be maintained, as well as identifying the specific tasks and task frequencies for each component. Additionally, a schedule of the required routine preventative maintenance activities was developed, presenting specific future dates when activities would need to be performed (and input into DC Water's asset management and scheduling tool, Maximo).

Differential Pressure Testing

Differential pressure testing in manholes along the PI is currently being performed to further aid in optimizing odor control system performance and media life. The differential pressure testing consists of operating one PI OCF fan at a time while recording the differential pressure between the sewer headspace and the ambient air at manholes along the PI. This is providing insight regarding the zones of influence of the various OCFs which will be used to potentially reduce airflows at certain OCFs, reducing energy costs and increasing media life. Testing efforts are currently underway (winter 2018) to take advantage of low temperature/odor conditions in the winter months.

Liquid-phase Treatment Pilot Study

Corrosion and odor mitigation via liquid phase treatment would decrease the generation of H_2S and slow the impacts of corrosion, reducing costly rehabilitation of the Potomac Interceptor and extending odor control facility media life. A pilot study is planned for spring/summer of 2018 to help define actual required chemical dosing rates, frequencies, costs, infrastructure, and other information needed to assess the feasibility of long-term implementation of liquid phase treatment for this specific application.

DMS Capacity Test

Based on the results of the two rounds of media pilot testing, permanganate-impregnated media was recommended for use in the polishing layers of the PI OCFs. Permanganate media utilizes oxidation-adsorption processes to indiscriminately oxidize the adsorbed odor compounds on the media surface. Like activated carbon, physical and chemical properties of permanganate media such as particle size, surface area, porosity, and permanganate concentration are important (but not well understood) factors affecting performance.

There are several commercially available permanganate-impregnated media products (Figures 5 and 6), including products utilizing differing substrates (e.g. zeolite, activated alumina), differing forms of permanganate (e.g. sodium permanganate, potassium permanganate), and differing concentrations of oxidant (e.g. 6%, 8%, 12%). There is also a balance between the amount of oxidant utilized in a media and the pore space available for adsorption. The pore volume available for physical adsorption is reduced by the addition of the permanganate oxidant.

Figure 5 – Hydrosil HS-600 Permanganate-Impregnated Zeolite (Photo courtesy of Hydrosil)

Figure 6 – Pure Air PA-8 Permanganate-Impregnated Activated Alumina (Photo courtesy of Pure Air Filtration)

Media replacement is often a significant operational cost. As DC Water began the procurement process for a media replacement contractor for fiscal year 2018, selecting the most efficient polishing media available was desired – taking into consideration both treatment performance as well as cost.

Accelerated Adsorption Capacity Test

The use of an accelerated adsorption capacity method is a way to estimate the adsorption capacity of activated carbon for a specific compound. In odor control applications, the standard test method is ASTM D6646-03 – *Determination of the Accelerated Hydrogen Sulfide Breakthrough Capacity of Granular and Pelletized Activated Carbon*. This standard method is commonly used by vendors to demonstrate the adsorption effectiveness of their activated carbon on an "apples to apples" basis.

H_2S has historically been the target compound in both liquid-phase and vapor-phase wastewater odor control. And rightly so, as it is frequently present in high concentrations, and has a very low, single digit ppb level, odor detection threshold. The ASTM D6646-03 method determines the adsorption capacity of the tested activated carbon at the breakthrough point under the specified laboratory test conditions using a controlled temperature (around 25 deg. C), a humidified air stream (relative humidity ~ 85%), containing 1% H_2S (10,000 ppmv), with a contact time of 4.8 sec. The air stream is passed through a carbon bed until 50 ppm breakthrough of H_2S is observed in the exhaust, which (represents a 99.5% removal efficiency). The H_2S adsorption capacity of the carbon is then calculated on a per unit volume basis.

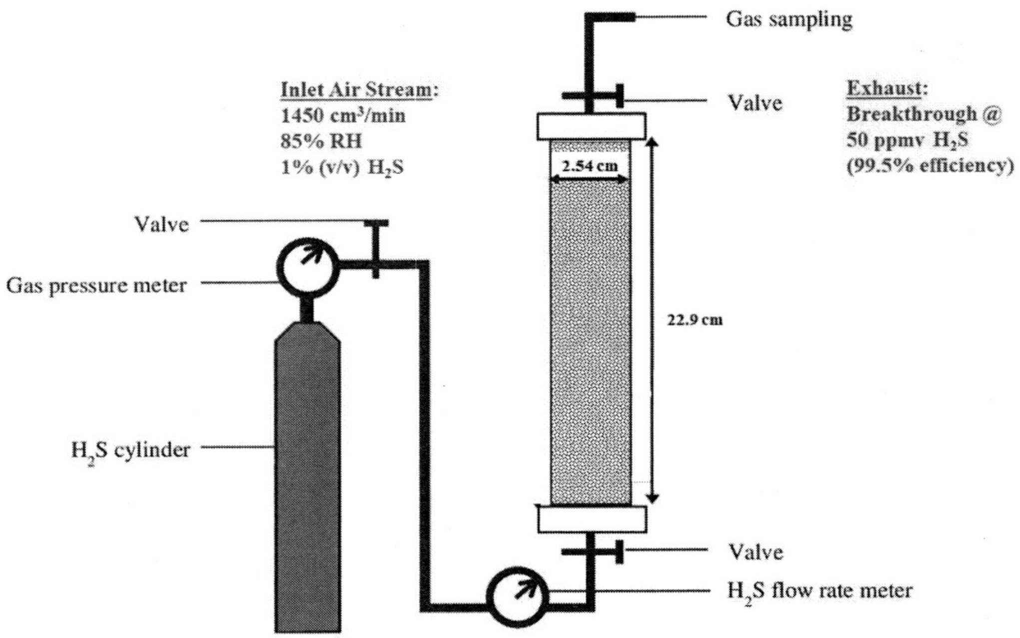

Figure 7 – ASTM D6646-03 Carbon H_2S Capacity Test Set-up

The advantages of this method are that it is a straightforward, standardized testing protocol and it is less time/labor intensive than developing adsorption isotherms. Unfortunately, the method also has some limitations. First, it only evaluates the adsorption capacity of a single compound (H_2S), and in reality, sewer gas is made up of many compounds. The potential for multi-compound adsorption effects isn't assessed. Also, the test does not simulate actual operating conditions of full-scale systems (which also differ from site to site). The H_2S concentration used for ASTM D6646-03 is 10,000 ppmv – orders of magnitude higher than the ppm to thousand ppm range routinely encountered for wastewater infrastructure. This difference favors activated carbon with rapid removal of H_2S over carbons with slower kinetics (Le-Minh, et al, 2014).

So, the actual lifespan of the activated carbon in a full-scale system is sometimes quite different than the predicted lifespan using the accelerated adsorption capacity method. Regardless, the benefits of this simple, short-duration, standard method have led to its common use in the industry.

However, DMS removal capacity data is rarely (if ever) provided by the manufacturers of activated carbon, and more importantly for DC Water, permanganate media. With the variability in media substrates, oxidant types, and permanganate concentrations, what product would be the most cost-effective long-term DMS removal solution for DC Water's PI OCFs? How could the various media types be evaluated on an "apples-to-apples" basis under the same conditions?

DC Water reached out to several independent laboratories across the country and was able to identify a lab willing to coordinate on the development of a modified ASTM D6646-03 test using DMS gas instead of H_2S gas:

Engineering Performance Solutions
3161 St. Johns Bluff Road S., Suite 3
Jacksonville, FL 32246

In order to determine the capacity of these specialized media for DMS, the following parameters were used:

Table 1 – Modified ASTM D6646-03 DMS Test Parameters

Parameter	Value
Dimethyl Sulfide Conc., ppm	50
Bed Diameter, cm	3.8
Bed Depth, cm	22.9
Gas Flow, L/min	3.253
Face Velocity, cm/sec	4.78
Relative Humidity, %	87
Breakthrough, ppm	1.0

As part of the media replacement contractor procurement process, vendors were required to send media samples to the independent laboratory to be tested under these conditions and a breakthrough loading was determined in units of grams of DMS per grams of sample:

- Hydrosil HS600 – 6% potassium permanganate-impregnated zeolite
- Nichem PPM8 – 8% potassium permanganate-impregnated activated alumina
- Purafil SP-12 – 12% sodium permanganate-impregnated activated alumina
- PureAir PA-6 – 6% potassium permanganate-impregnated activated alumina
- PureAir PA-8 – 8% potassium permanganate-impregnated activated alumina

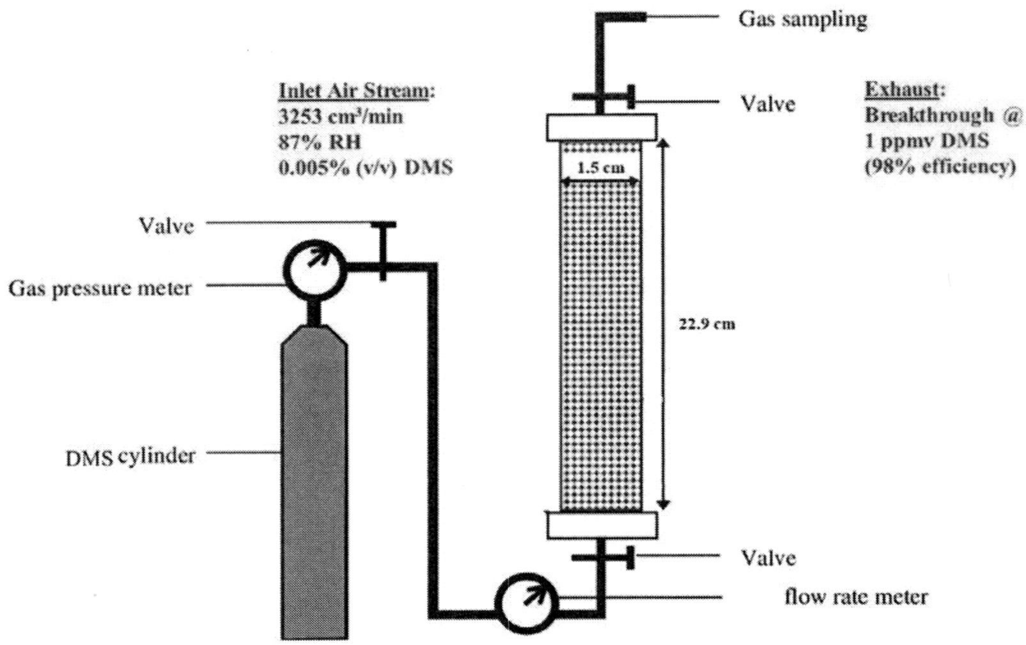

Figure 8 – Modified ASTM D6646-03 DMS Capacity Test Set-up

RESULTS

Based on the densities of the various media, each loading value was converted from the lab-provided grams of DMS per grams of sample to grams of DMS per cubic centimeter of media, which are the units that are typically presented on industry product data sheets. For the five media tested, the DMS capacity was found to range from 0.718 to 1.051 g DMS/cc media, as can be seen in Table 2 below:

Table 2 – DMS Capacity and Cost Efficiency Rankings

	Media				
Parameter	**W**	**X**	**Y1**	**Y2**	**Z**
Media Density (lb/ft3)	55	50	50	50	60
Media Density (g/cc)	0.88	0.80	0.80	0.80	0.96
Breakthrough Loading (g DMS/g Sample)[1]	1.187	1.19	1.312	0.897	0.967
DMS Capacity (g DMS/cc media)	1.046	0.953	1.051	0.718	0.929
DMS Capacity Ranking	2	3	1	5	4
Media cost ($/ft3)	$98.00	$140.00	$115.00	$110.00	$110.00
Media cost ($/cc)	$0.00346	$0.00494	$0.00406	$0.00388	$0.00388
Cost Efficiency ($/g DMS removed)	**$0.00331**	**$0.00519**	**$0.00386**	**$0.00541**	**$0.00418**
Cost Efficiency Ranking	**1**	**4**	**2**	**5**	**3**

Notes:
1. As measured by independent lab with modified ASTM D6646-03 test for DMS capacity.

However, just as the substrate type, oxidant type, oxidant concentration, and DMS capacity varies by product, cost varies as well. A product may remove 30% more of a compound when compared to another product, but if it's 40% more expensive, then the capacity benefit is moot. Because the most cost-effective media is desired for long-term use at DC Water's PI OCFs, a cost efficiency parameter was used to compare the products ($/g DMS removed).

The manufacturers and product names have not been disclosed in order to conceal pricing data. However, it can be seen that when cost efficiency was included in the analysis, the rankings shifted as compared to the DMS capacity rankings alone.

Modifying the widely-used accelerated adsorption capacity test method ASTM D6646-03 for DMS capacity instead of H_2S capacity provided valuable insight into an "apples to apples" comparison of various specialized media types for the polishing layer of DC Water's PI OCFs. This simple, short-duration, standardized method – while not perfect – may be of use in the industry for other odor control projects and applications, as focus moves from the well-understood treatment of H_2S to the less understood treatment of other odorous compounds.

REFERENCES

1. Pope, R.J., Lucas, B. (2014). "I'll Take a Breath of Clean Air with My Entrée – Sewer Odors and Neighborhood Restaurants Don't Mix". Proceedings of the 2014 Water Environment Federation Odors and Air Pollution Conference. Miami, FL.

2. McKenna, R., Pope, R., Reeves, J. (2016). "DMS CSI: A Multi-pronged Investigation for Comprehensive Odor Control." Proceedings of the 2016 Water Environment Federation Odors and Air Pollution Conference. Milwaukee, WI.

3. Le-Minh, N., Sivret, E., Stuetz, R. (2014). "Review of the Application of Activated Carbon for Sewer Odour Abatement." UNSW Water Research Centre – School of Civil and Environmental Engineering, The University of New South Wales.

Odor Practices in the Collection System and WRRF: Results from the 2016 Odor Benchmarking Survey

Carla D. Dillon[1], Dennis Froehlich[2], Eric Hansen[3], Mark S. Holstad[4], Ray Porter[5], Philip Wolstenholme[6]

1 City of El Monte; cdillon@elmonteca.gov*
2 Pima County Regional Wastewater Reclamation; Dennis.Froehlich@pima.gov
3 Evoqua Water Technologies; eric.c.hansen@evoqua.com
4 Albuquerque Bernalillo County Water Utility Authority; mholstad@abcwua.org
5 Porter Odor Science; ray.porter-ocm@comcast.net
6 Brown and Caldwell; pwolstenholme@brwncald.com

ABSTRACT
The Air Quality & Odor Control (AQ&OC) Committee of the Water Environment Federation (WEF) initiated an industry-wide survey to benchmark odor control practices in wastewater agencies. The survey covered many topics specific to odors including management strategies, agency characteristics, staffing, treatment, maintenance, design and construction, compliance, spending, good-neighbor strategies, and future needs/gaps in the field. This paper presents results and analyses from the survey.

KEYWORDS
Odor control, odor management, benchmarking, level of service, air treatment, vapor phase odor control, liquid phase odor control, hydrogen sulfide, H_2S.

INTRODUCTION
Benchmarking is valuable to organizations as a means to compare practices of other organizations and influence organizational growth. Members of the Water Environment Federation (WEF) Air Quality & Odor Control (AQ&OC) Committee determined that benchmarking with a focus on odors would provide a valuable reference point for all sectors including owners/operators, consultants, and suppliers. A subcommittee was formed to develop a benchmarking survey.

The benchmarking project objective was to gather plant and collection system odor control best practices from North American municipal wastewater utilities. The best practices will include policies, strategies, operational and capital costs, levels of service, and staffing. The end product will serve as a resource for other organizations to establish levels of service.

METHODOLOGY
The Benchmarking subcommittee of the WEF Air Quality & Odor Control Committee held regular conference calls beginning in Spring of 2015 to start the benchmarking effort. In addition WEF staff participated in all calls and meetings and served as the technical lead for the survey. The eight-member subcommittee remained intact throughout the process.

The Benchmarking Committee defined overarching issues such as the sub-categories within the larger goal, what metrics and units of measure to use, how to receive and organize data, and how to work within the group to post questions and to review. Categories of questions were agreed and the task of developing appropriate questions began. The categories are presented in Table 1.

Table 1: Survey Categories

Item	Category
1	Plant, Collection System Data and statistics
2	Full time equivalent (FTE) personnel allocated for odor control
3	Action levels (levels of service)
4	Outreach efforts related to odor control
5	O&M Costs per year a. Spent on vapor phase/yr b. Spent on liquid phase/yr
6	Design Standards
7	Business practices – include organizational structure, whether drivers are regulatory, internal, or customers
8	Operations (including energy) a. Technologies used: (Liquid phase, vapor Phase or both) also could be operational/infrastructure changes i. Type of liquid phase used ii. Type of vapor phase used
9	Regulatory Compliance
10	Maintenance and System integrity
11	Capital Investment

Developing well-crafted questions is difficult enough, but first agreement was sought on what was important to quantify and the units to use. The odor control issues, methodologies, and approaches are different at a water resource recovery facility (WRRF) versus in the collection system, and yet many attributes are very similar. Since each utility is organized differently, questioning structures that worked for one utility may not work for another utility. One such structure is the number of WRRFs owned-operated by each respondent. To address this, the survey allowed responses for up to ten WRRFs.

The subcommittee also tried to balance questions that would extract enough detail yet allow the survey to be completed in a reasonable amount of time. The committee recognized that those filling out the survey would be very busy and may stop answering and not participate if the survey took too much time. While the option was made available to save the survey and complete at a later time, most respondents may only be willing to answer the survey in one sitting.

The subcommittee utilized several questions constructed to gain the maximum responses and to gain the maximum number of specific values. First, the survey tool allowed the respondent to skip a group of questions negated by a prior answer. Since the survey addressed both the collection system and plant issues the respondent would likely be more knowledgeable about one or the other but not necessarily both. So, if a respondent indicated that they were not knowledgeable or willing to answer questions about the collection system, then 58 questions were skipped about the collection system and collection system odor control. This reduced the number of questions presented to considerably less than the maximum of 347.

Next, range answers were allowed so the respondent, who may not have specific data to memory, would feel competent to answer without any research. Finally, specific values were requested and frequently provided to allow rigorous statistical analysis.

The survey was ultimately structured in the following six sections:

A. Strategic Efforts, Organizational Structure, and Outreach
B. Regulatory Compliance
C. Collection System and Collection System Odor Management
D. Odor Management – WRRF Information
E. Maintenance and System Integrity
F. Capital Improvements in Odor Control (Costs and Design Criteria)

The WEF utilizes SurveyMonkey® under the Select monthly plan and the subcommittee opted to use this tool. The SurveyMonkey® software was found to be easy to work with in setting up the various categories, skip logic, etc. It is likely that this survey will be refined and repeated. The WEF committee member loaded the questions into SurveyMonkey® and in March 2016, the beta-version of the questionnaire was deployed. Members of the committee completed the survey and provided feedback for issues that were noted. The Beta version was open for about three and a half months when it was replaced by the official survey. The Odor Benchmarking Survey was launched in June 2016 and announced via WEFCOM, which is an online community hosted by the Water Environment Federation. The survey was further promoted in several ways such as the WEF Water Log e-mail newsletter and the AQ&OC Committee. Preliminary results were presented at WEFTEC 2016. The survey was closed in December 2016.

After closing the survey, results were screened to include owner-operators only, which resulted in 50 respondents. Initially there were 90 submissions; however, many were beta testers or those wanting to see what the survey was about, but not owner-operators. Methods used for analyses are described throughout the paper.

RESULTS

The survey participants were primarily from the United States with some respondents also from Canada, New Zealand, and Australia. Temperature of the wastewater ranged from 10C (50F) to greater than 30C (86F) indicating geographic and climate variability. Participants had both combined and separated storm sewers with approximately 80% of participants having separate systems. The service area populations ranged from 15,000 to 6 million. Participants identified having one or more of the following types of pipes: force mains, interceptors, trunks, mains, laterals/branches, and lengths ranged from 69.2 km (43 miles) to approximately 22,530 km (14,000 miles). Dry weather treatment/conveyance ranged from 1.893 megaliters per day (ML/D) (0.5 million gallons per day (MGD) to 2,612 ML/D (690 MGD). Average BOD ranged from 96 milligrams per liter (mg/L) to 500 mg/L. TSS ranged from 130 to 785 mg/L. The range of responses indicate a variety of participants based upon volume treated, characteristics of wastewater, miles of sewer, population, combined versus separate storm sewers, etc.

Odor Management: Organizational Resources, Regulatory Compliance, and Community Engagement

Many questions were presented to understand organizational management of odors, drivers for decisions, and community involvement. In understanding odor management, all of these factors, and more are important.

Drivers

Survey participants were asked to rank the top drivers for making decisions related to odor control and management. This was a forced choice ranking with three options: regulatory, external customers, and internal goals. 27 responders (59%) selected external customers as their top driver for making odor related decisions. This was followed by regulatory factors (27%) then internal goals (13%). Since this question did not allow for selections other than these three, a free form question allowed for input. Of those that responded with other drivers, there were multiple responses for cost (25%), corrosion control (20%), and being a good neighbor (20%). Additional drivers included citizen lawsuits, safety of workers, and system integrity.

Organizations may have different triggers for taking action to respond to odors. The free form responses can be categorized into numeric triggers, no nuisance, none, and other. Some responders indicated triggers that fall into one or more categories, as shown in Figure 1.

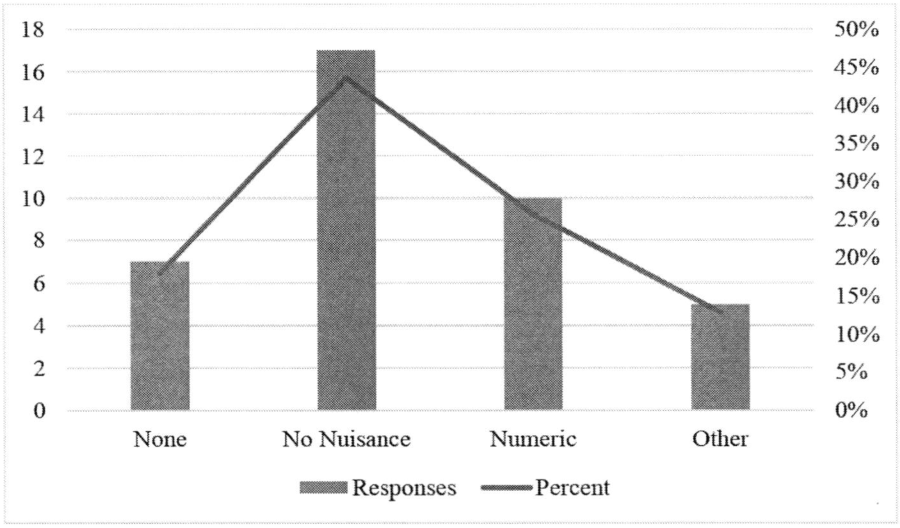

Figure 1: Response to Triggers

Causing *no nuisance* was the highest proportion of answers for triggers to respond. Answers included 'no nuisance,' 'respond to any complaint,' 'resolve all complaints,' and 'if you smell it we will try to fix it.' Of those with numeric triggers, 40% cited 10 ppb for hydrogen sulfide but also had additional criteria such as hydrogen sulfide for a specific duration or peaks at the receptor. Source or scrubber emissions was listed by 20% of responders as a trigger, fenceline hydrogen sulfide was indicated as a trigger by an even smaller group, 20% of those with numeric triggers. Other triggers included ambient levels within the facility, headspace hydrogen sulfide, dissolved sulfides in the collection system, and odor units in dilutions to threshold. None of the responders with combined storm and sanitary sewers had numeric action levels (i.e. measured H_2S) for responding to odors. Future surveys should include a breakout of triggers by response type such as adjustment to treatment, initiation of a capital project, contact customer to better understand unique drivers.

Types of Responses/Outreach
When asked what outreach activities are undertaken related to odor, respondents were able to select all those that apply. The top responses selected were site visits (72%), a standardized complaint form (54%), a complaint hotline (54%), facility tours (50%), trained customer representative (39%), and having an odor complaint response time goal (37%). Participants were also permitted to enter their own text. Other responses included extensive odor patrol and inspections by third parties.

Time
In a free-text format, participants were asked if they had a response time goal, and if so, what. The predominant answer fell within one hour, some stating immediately, within ½ hour, or within one hour, Figure 2. The next prevalent response was within a day or 24 hours. A few responses indicated different goals for investigating versus contacting the complainant. Future surveys should separate the time goals for investigating versus follow up with the customer/complainant.

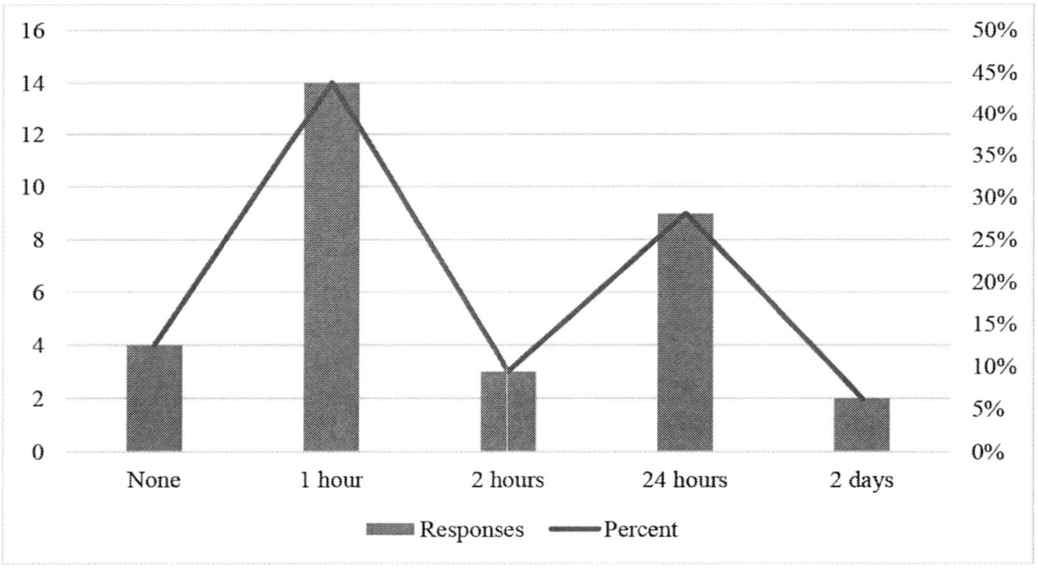

Figure 2: Complaint Response Time

Community Engagement during Design and Construction of Odor Control Projects
Survey participants were asked if they engage community members during the design and construction of odor control capital projects. Responses were similar for both phases, Figure 3. Only two participants responded that they always include the community. For construction, ten responders indicated they sometimes involve the community and for design, eleven. Fourteen

responders indicated they do not involve the community for design and sixteen for construction. In future surveys this involvement should be further explored.

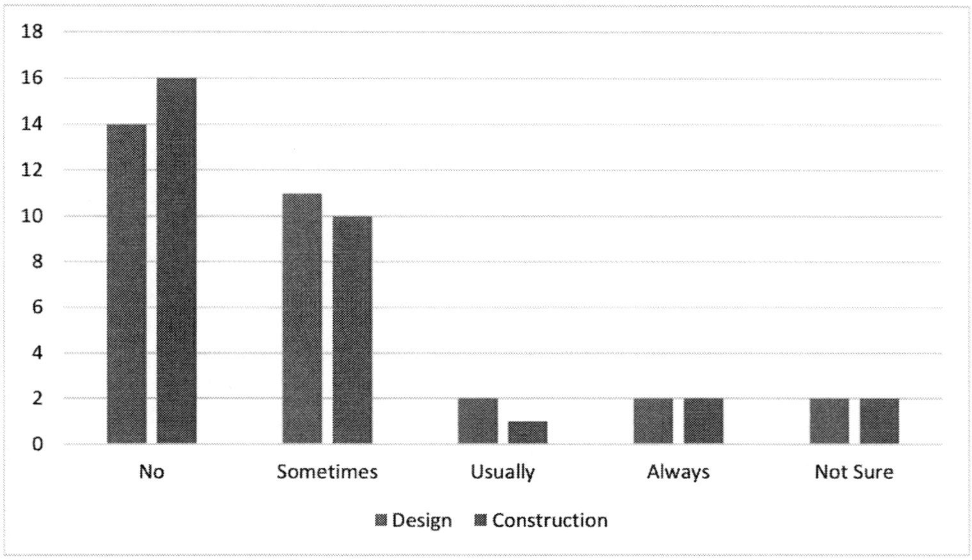

Figure 3: Community Involvement

Organizational Factors

Organizational factors were explored related to odor control activities including whether there was a dedicated group for odor control, how many full-time equivalents (FTEs) were devoted to odor control, and if the collections and WRRFs were considered holistically.

The survey asked about teams and the number of staff working on odor control. Of those responding, 39% have teams dedicated to odor control. 52% of respondents do not have an odor control team. 9% did not know if there was a team.

To understand how many people had some or any responsibilities related to odor control and full time equivalents, the question asked for the number of people with some odor control responsibilities in a selection of ranges. The highest response grouping had 61% of responders indicating their organization has 1-5 people with some responsibilities for odor management or control. 17% of responders indicated they have 6-10 employees with responsibilities for odor management or control; 9% have 10-20 employees and 13% have over 20 employees with responsibilities for odor management or control.

Four of the six responders with over 20 people having odor control responsibilities were from large service areas, with over 1,500,000 people in their service area. While size of the service area may be a factor, other factors that were not asked may also drive the need for devoting staff to odor control purposes. These may include design of conveyance system (slope, detention time in pipes, lift/pump stations), proportion of high strength discharges, sensitivity of surrounding community to name a few.

The survey asked how many full-time equivalents (FTEs) were devoted to odor control and management if they had a dedicated odor control team. The response was free-field text. The responses ranged from 0 to 20, which is consistent with the prior question where the majority indicated that they have 1-5 employees working on odor control. Organizations that serve larger populations were more likely to have dedicated teams for odor control. All those who serve more than 900,000 residents indicate they have teams dedicated to odor control, Figure 4.

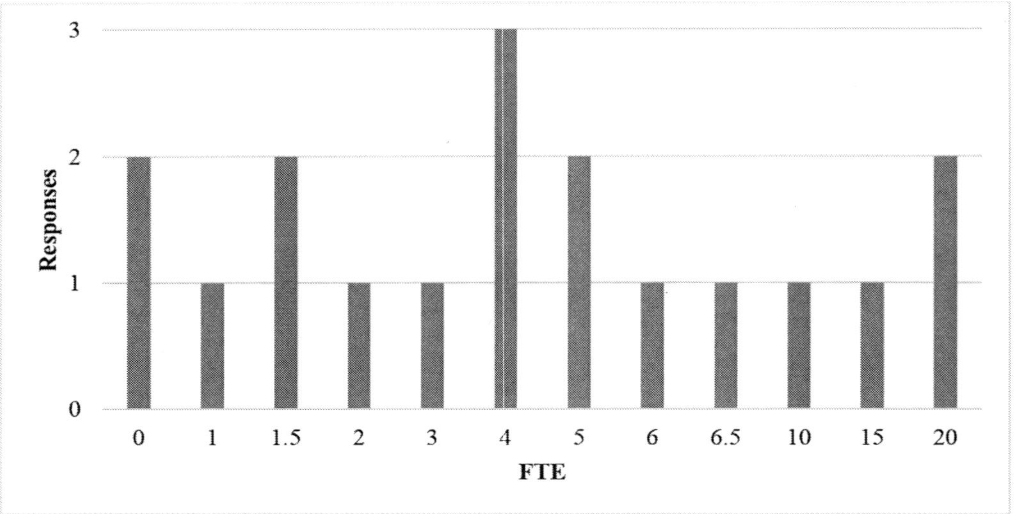

Figure 4: Full Time Equivalents Devoted to Odor Control

As treatment in the collection system can often affect the WRRF, participants were asked if they consider the collection system and WRRFs holistically in treatment. Just under half of the responders indicated they consider the two systems holistically and of those 67% have liquid phase odor control in the collection system. The potential for downstream affects may heighten the awareness of the inter-relatedness of the collection system and WRRFs.

Compliance and Odor Goals

Air permits may be required for odor control equipment or regulations may apply to the WRRF as a whole. 60% of respondents indicated that they are required to comply with an air permit or other air or odor regulations, 35% responded that they did not, and 5% did not know if their organization was required to comply with an air permit. Regulations will vary by jurisdiction and the type of air treatment and wastewater processes may trigger different regulations.

Air regulations may be a numerical standard, such as limiting the emissions of hydrogen sulfide or volatile organic compounds. Regulations may also restrict air emissions that cause a nuisance or impact the quality of life. Based upon the survey, 44% indicate they are required to comply with both a quality of life and numerical standards. Having both requirements may influence the organization's response to odor complaints and internal goals.

Oftentimes a nuisance may be caused even if an agency meets its numerical limits. For that reason, some organizations may establish more stringent standards. Based upon the survey, 37%

of responders have established more stingent internal emission goals. 86% of those who have more stringent goals, they either have a quality of life requirement or both quality of life and numeric goal. The remaining 14% of those with more stringent internal goals indicated they have no regulatory requirement for air emissions. In a future survey, it would be valuable to explore the reasons for establishing a more stringent organizational goal.

Collection System Odor Control

There are two primary methods to treat odors, particularly hydrogen sulfide (H_2S) and associated reduced sulfur compounds in municipal wastewater collection systems. The first is vapor phase odor control (VPOC), which is the method of extracting odorous air from a defined point in the collection system and passing it through one of many types of vapor phase technologies to remove the odorous compounds and exhaust the air into the atmosphere. The most widely used VPOC technologies are:

- Activated Carbon Adsorbers
- Bioscrubbers
- Biofilter Units
- Chemical Scrubbers
- Iron Sponge
- Ionization Treatment Units

The second method to treat for odors in the collection system is liquid phase odor control (LPOC). This is the method of adding a chemical treatment solution directly to the wastewater to either prevent the formation of H_2S and/or other reduced sulfur compounds or remove existing compounds from the wastewater. The most commonly used LPOC technologies are:

- Nitrates
- pH shift/Alkaline solutions
- Iron salts
- Hydrogen Peroxide
- Enzymes
- Air/Oxygen Injection
- Bleach/Chlorine

Based on the responses of the Benchmarking Study, the most widely used VPOC technologies in order from greatest to least are: activated carbon units, bioscrubbers, biofilters, chemical scrubbers, and iron sponges. Figure 5 displays the number of units and the reported total airflow flow in cubic feet per minute of the reported VPOC technologies of the respondents.
No respondents are currently using ionization.

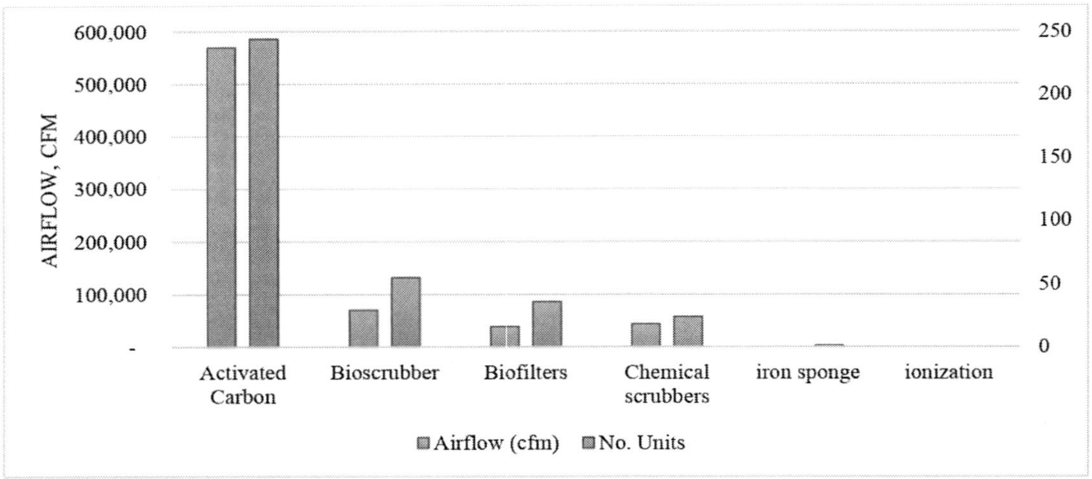

Figure 5: Collection System Vapor Phase Odor Control Distribution

Conversely the most widely used LPOC technologies in order from greatest to least are: nitrates, pH shift/alkaline, bleach/chlorine, hydrogen peroxide, and three tied (air/oxygen injection, iron salts and enzymes), Figure 6.

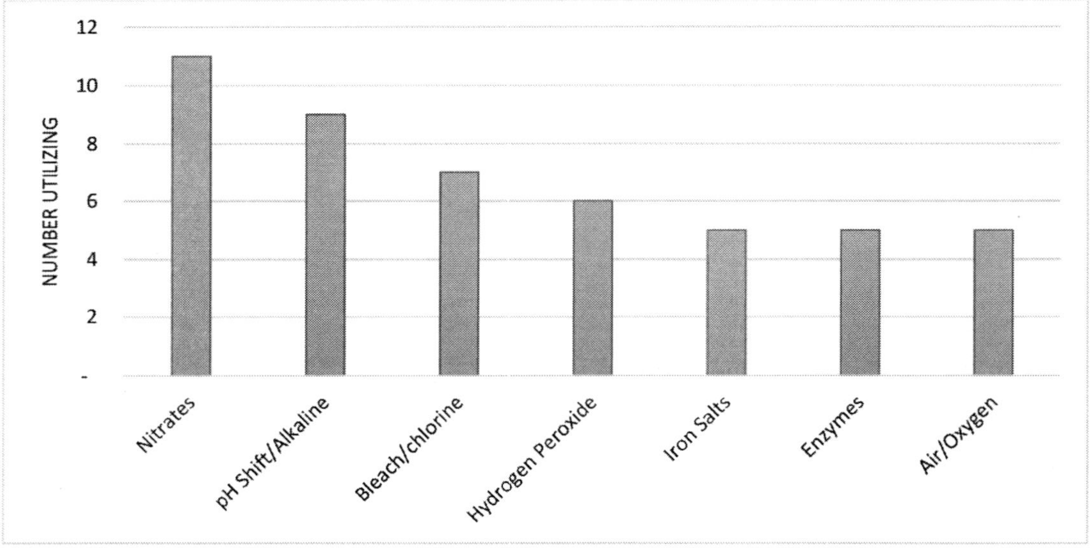

Figure 6: Collection System Liquid Phase Odor Control

For LPOC, the survey asked about the percent of sewage treated by length of pipes and then by flow. The responses indicate differences that are worth noting. Six of the responders (24%) indicated they treat more than 50% of the sewage with LPOC based on flow. No responders indicated that they treat more than 50% by length of pipe. At the low end of the range, 18 responders (69%) indicated they treat less than 10% of the pipes by length, yet 10 responders (40%) treat with LOC less than 10% by volume. Distinguishing the basis is important as seen with the differences in response. This may represent some responders selecting pipes with greater flow to create the biggest reduction of odor or it may indicate that high flow pipes cause the most significant odor generation. Responses are shown in Figure 7.

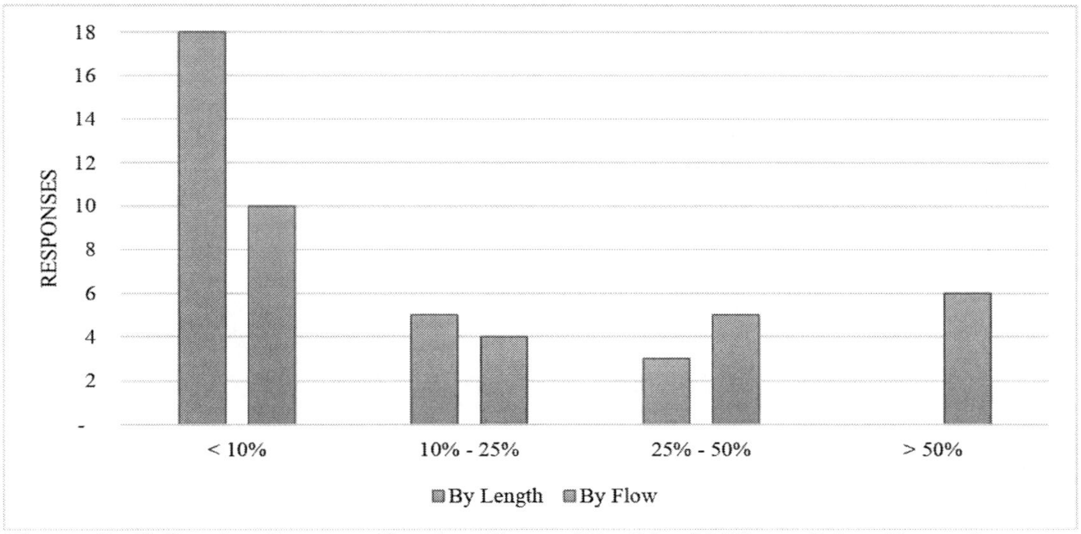

Figure 7: Collection System - Portion Treated by Liquid Phase Odor Control

Water Resource Recovery Facilities Odor Control

Of the utilities that responded to the survey, 33 provided information from their WRRFs. Several provided information for multiple treatment works. Thus, information was obtained for 49 WRRFs as part of this survey.

For many WRRFs, the management of odor emissions was siting the facility away from population centers or providing an adequate buffer around the facility to allow for fugitive odors to disperse. As population centers expanded, neighbors moved next door to WRRFs or buffer zones become public open space with playing fields and nature paths. Figure 8 shows, 76% WWRFs have, neighbors sensitive to odors within one-half mile of their facility, and over 40% of the participants identified receptors were located within 0.1 mile.

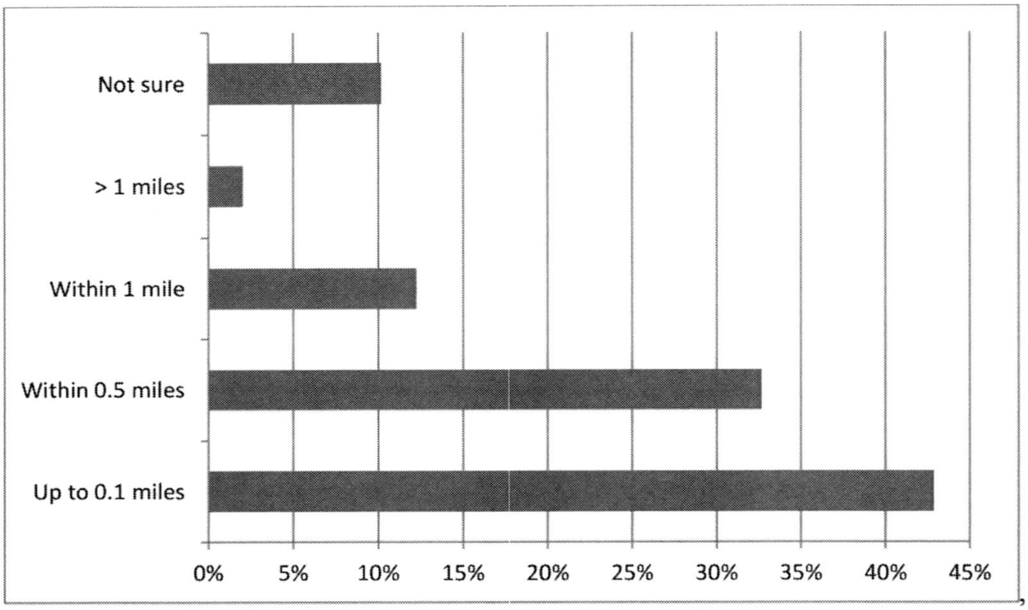

Figure 8: Distance from the WRRF to Nearest Neighbor

To track the odor generation potential at a WRRF, the operational staff can monitor dissolved hydrogen sulfide in the liquid phase or hydrogen sulfide gas in the headspace of tanks to exhausts from process units. Most (71 %) do not monitor for hydrogen sulfide in the liquid phase or do so only if necessary, Figure 9.

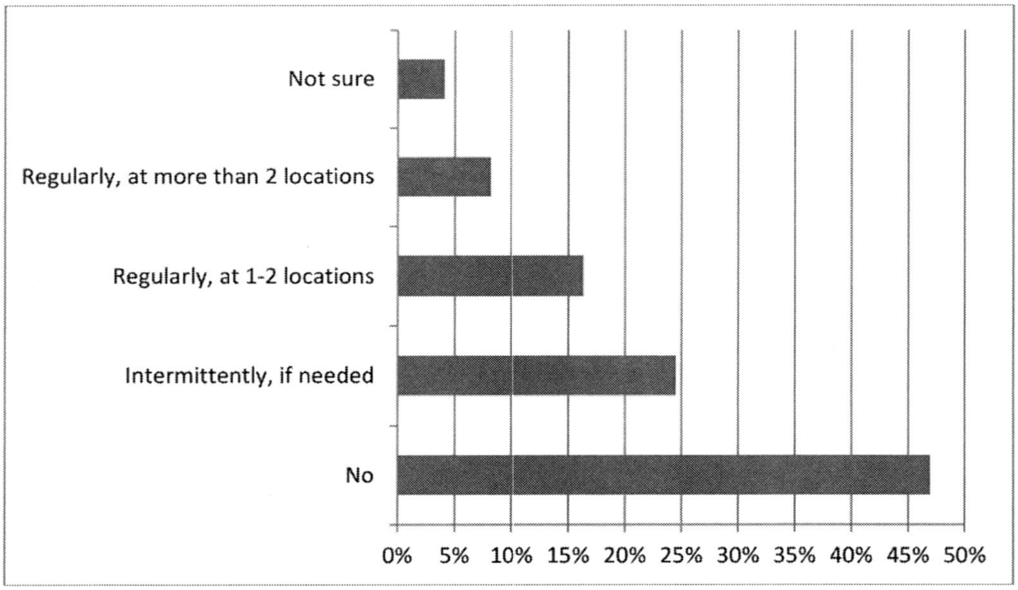

Figure 9: Frequency at Which Sulfide is Monitored in the Liquid Phase

It is more common for operational staff to monitor hydrogen sulfide (H_2S) in the gaseous phase. More than 60% of the facilities monitor H_2S for staff safety, to optimize process operations or monitor odor concentrations. About 15% of the facilities monitor H_2S concentrations as a requirement of their operating permit, Figure 10.

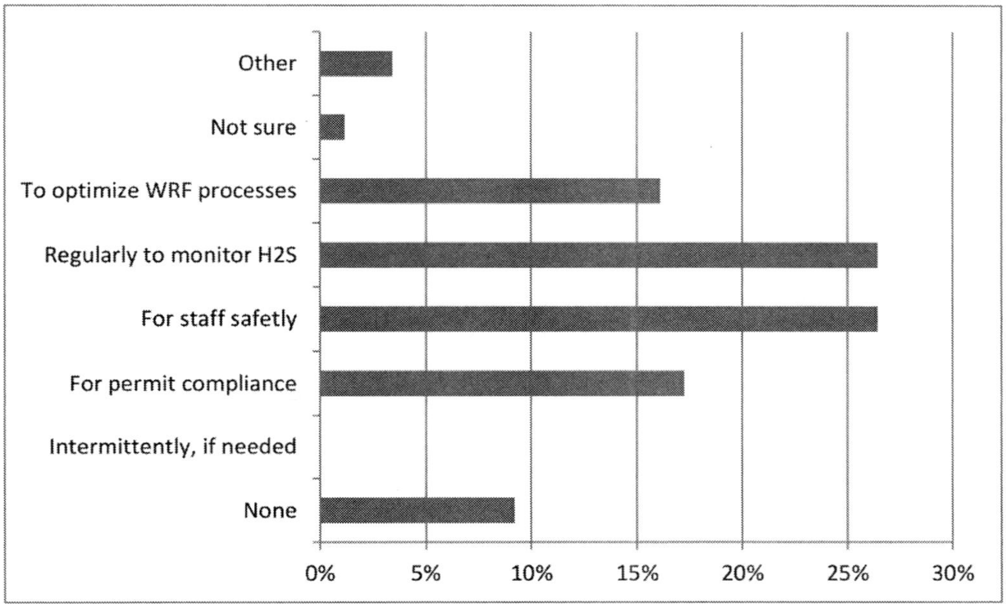

Figure 10: Purpose for Monitoring H_2S Concentrations in the Gas Phase

To reduce H_2S and related odor emissions, an operator may control H_2S in the liquid phase or capture and treat odors in the gas phase. The approaches are similar to those used to manage odors in the collections system.

Liquid Phase Approaches

For most process units, no liquid treatment is used. Upstream liquid phase treatment is most common approach for managing dissolved sulfide concentrations at the WRRF. Iron salts and hydrogen peroxide are the more common compounds used to reduce dissolved sulfides. The alkaline agent, sodium hydroxide, is used to raise the pH and increase the solubility of sulfide keeping it in the solution. Air and oxygen serve as inhibitors to the strict anaerobes, which form sulfides from sulfates. Permanganate is also a strong oxidant used to minimize emissions of reduced sulfur compounds.

Liquid phase treatment chemicals are not used in secondary treatment processes. In some cases, returning activated sludge to the head of a treatment plant can suppress sulfide concentrations and reduce potential odors in the headworks. Nitrates, alum and enzymes are rarely used in the treatment of odors at the WRRF. Liquid phase treatment for WRRFs are presented in Figure 11.

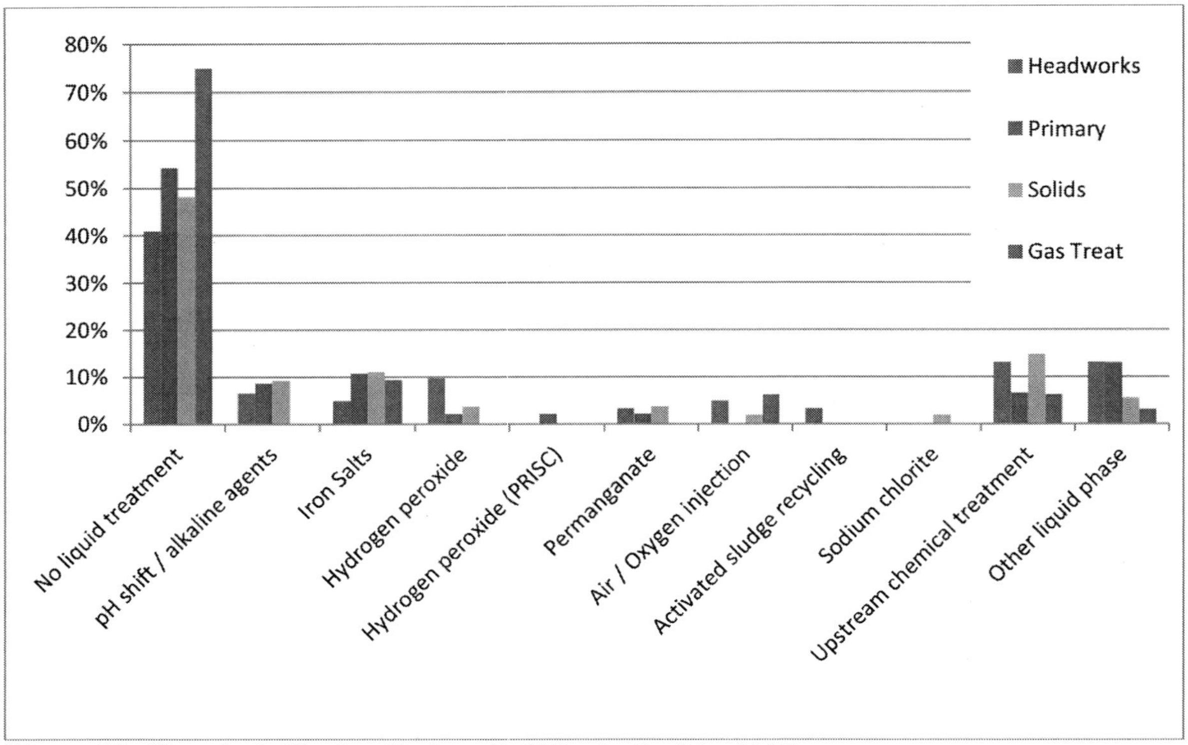

Figure 11: Chemicals Used to Reduce Sulfides in the Liquid Phase

Gas Phase Approaches
Before the odors in a gas stream can be treated it must be captured and ducted to the treatment unit. Table 2 summarizes whether a process unit is covered, ventilated, and/or treated.

Table 2: Process Units that are Covered, Ventilated and Treated

	Headworks	Primary	Secondary	Solids Processing	Gas Treatment
Mostly Uncovered	18%	36%	80%	22%	12%
Mostly Covered but no ventilation or treatment	4%	6%	4%	4%	9%
Mostly covered and ventilated	24%	4%	0%	24%	19%
Mostly covered, ventilated, and treated	20%	19%	2%	18%	7%
Mostly covered, ventilated, treated, and treatment exhaust is monitored	31%	21%	10%	22%	9%
Not sure	2%	4%	2%	2%	2%
Do not have this process	0%	9%	2%	6%	42%

The headworks are mostly (75%) covered and ventilated and 51% receive treatment for their odorous emissions. It is less likely that a primary unit will be covered as compared to headworks, but if a full or partial cover is installed, the capture exhaust gas is likely to receive treatment. Secondary process units are most likely not covered. Typically, they are large and have a high volume of air passing through them. Some trickling filter system may be covered and possibly treated. Almost 65% of the solids processing area are covered, and ventilated. Over 40% of the solids processes also received treatment.

There are a number of factors that go into the selection of an odor treatment technology. Life cycle cost is certainly a factor, but space available to install a system, chemical composition of the odorous gases, and intermittency of operations are also important in selecting the appropriate technology. Table 3 summarizes the technologies used by the facilities in this survey.

Table 3: Odor Control Systems Used by Facilities in This Survey

	Headworks	Primary	Secondary	Solids Processing	Gas Treatment
Biofilter	32%	35%	13%	28%	38%
Biotrickling filter	19%	13%		4%	
Activated carbon	27%	22%	50%	24%	50%
Chemical scrubber	19%	26%	25%	36%	13%
Oxidation				4%	
Other	3%	4%	13%	4%	

Air Flow Rate

Critical factors for sizing an odor control system include knowing the design air flow rate, peak loading rate, and composition of key odorants. While air flow rate is proportional to surface area, which is a function of wastewater design flow rate, it is also important to know the headspace and air exchange rate. The headspace can be quite large if it is intended for human occupancy. This may also require a high air exchange rate (12 air changes per hour) to meet fire safety codes (NFPA 820). More closely fitted covers, fully or partially covering the liquid surface, may have a lower air exchange rate (3 to 6 air changes per hour) depending on the H_2S concentration in the headspace and the corrosion resistance of the cover materials.

Unfortunately, information about the nature of the covers from treated units and air exchange rates was not requested in the survey. Further not all the facilities that reported that process areas were ventilated and treated provided air flow rates for their treatment units. In some cases, a treatment unit might serve more than one process area, so some apportionment might have been required. Nevertheless, a comparison of air flow rate and wastewater flow rate are summarized in the figures below (Figures 12, 13, and 14).

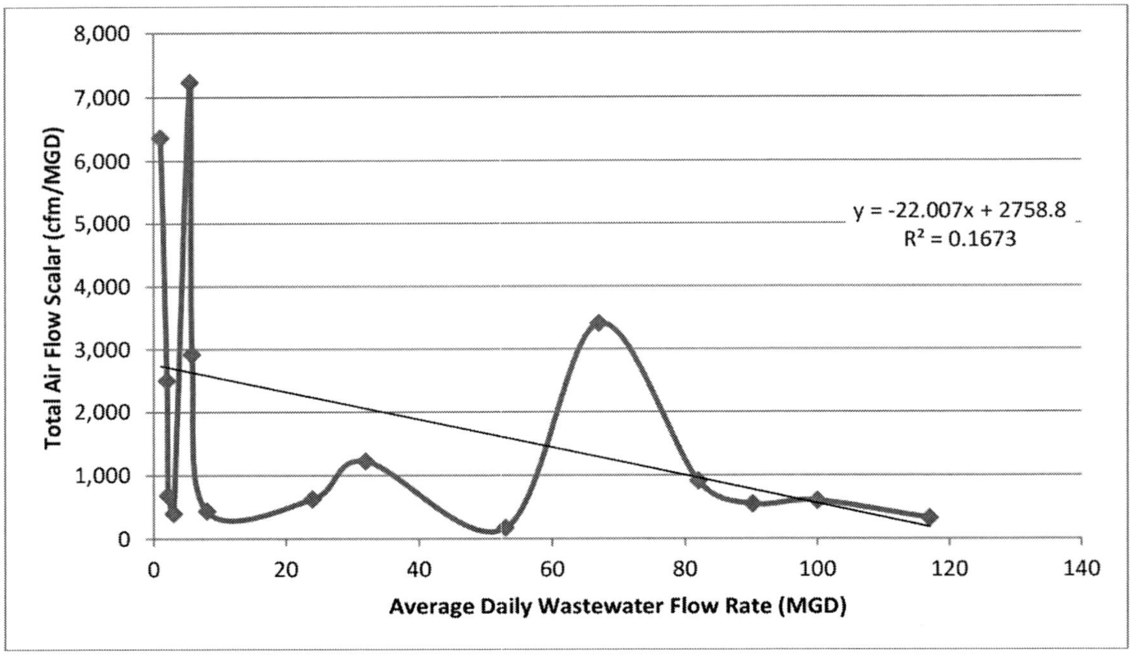

Figure 12: Air Flow Scalar Headworks (cfm/MGD)

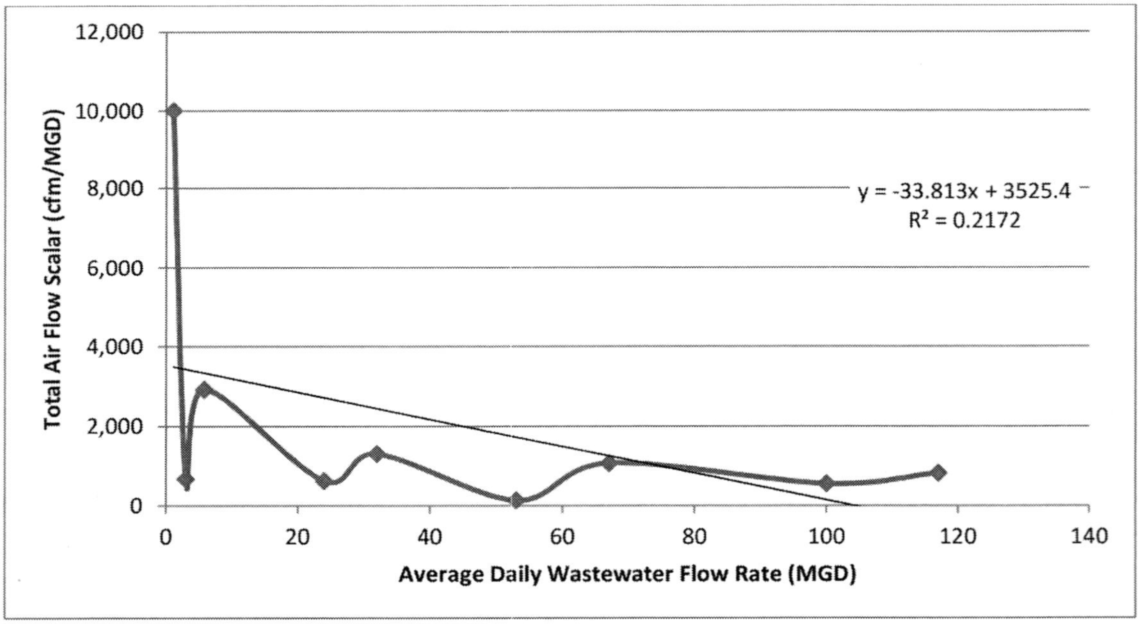

Figure 13: Air Flow Scalar Primary (cfm/MGD)

These figures apply to facilities with solely sanitary sewer collection system. Only one facility with a combined sewer system provided air flow information. The design capacity of a WRRF with a combined sewer system would likely have a larger odor control system air flow rate to capture the peak wastewater flows, not adequately represented by the average daily wastewater flow rate.

Secondary treatment processes are not usually covered, ventilated and treated. Only three facilities provided air flow information for secondary process units. There is limited usefulness in extrapolating from this small set of data.

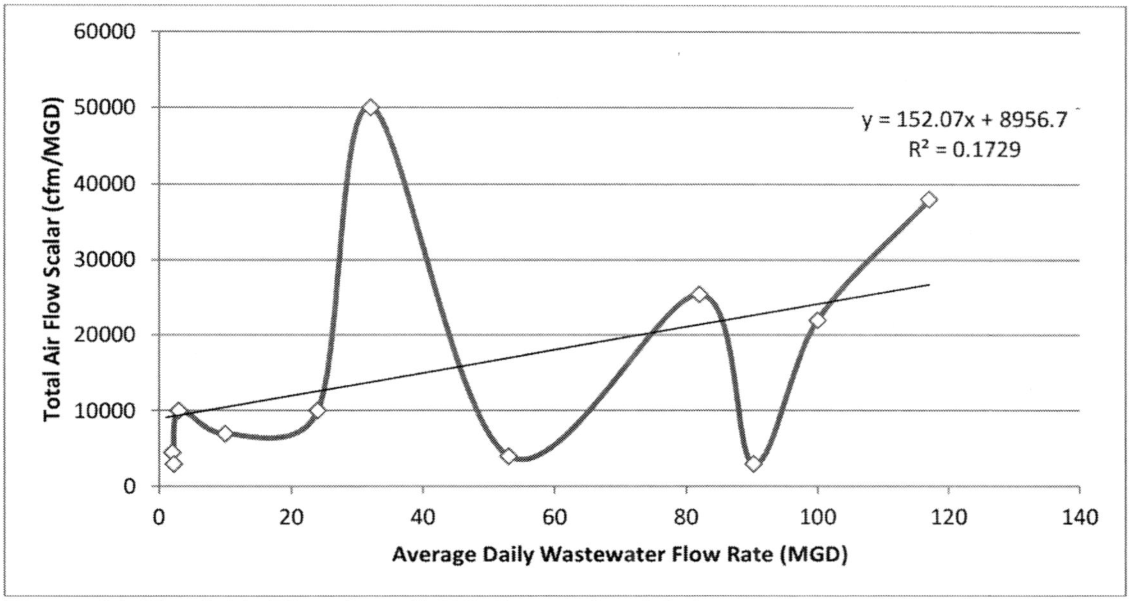

Figure 14: Total Air Flow from Solids Processing (cfm)

While there is a generally favorable trend in the graphs and positive correlation, it is clear that design air flow rate cannot be determined for average daily wastewater flow alone.

Maintenance, Operation, and Capital Investment

In this survey, the community size served by wastewater organizations spanned over four orders of magnitude, therefore, numeric benchmarks must be functions of system scale.

Wastewater infrastructure regional planning typically employs a combination of current connections (customers) with a combination of development permits and transportation data to estimate future wastewater flows based on per capita or per residence/parcel information. Sub-basin population and hence flow estimates are based on residences per acre estimates, average persons per residence, average gallons per person per day, and engineering formulae for diurnal peak flows and wet weather flows.

Wastewater system flow, in millions of gallons per day, results directly from population served. Sewer system extent in miles indirectly results from population served and some population density or aggregate residences per acre. With this in mind, system scale variables include population served, miles of sewer, and system wastewater flow (or average dry weather flow at the WRF). This benchmark analysis uses both population served and the product of flow and miles of sewer, with the unit: "MGD*Mile". The sewer miles versus population served is presented in Figure 15. The analysis in this section combined sewers due to limited sample size.

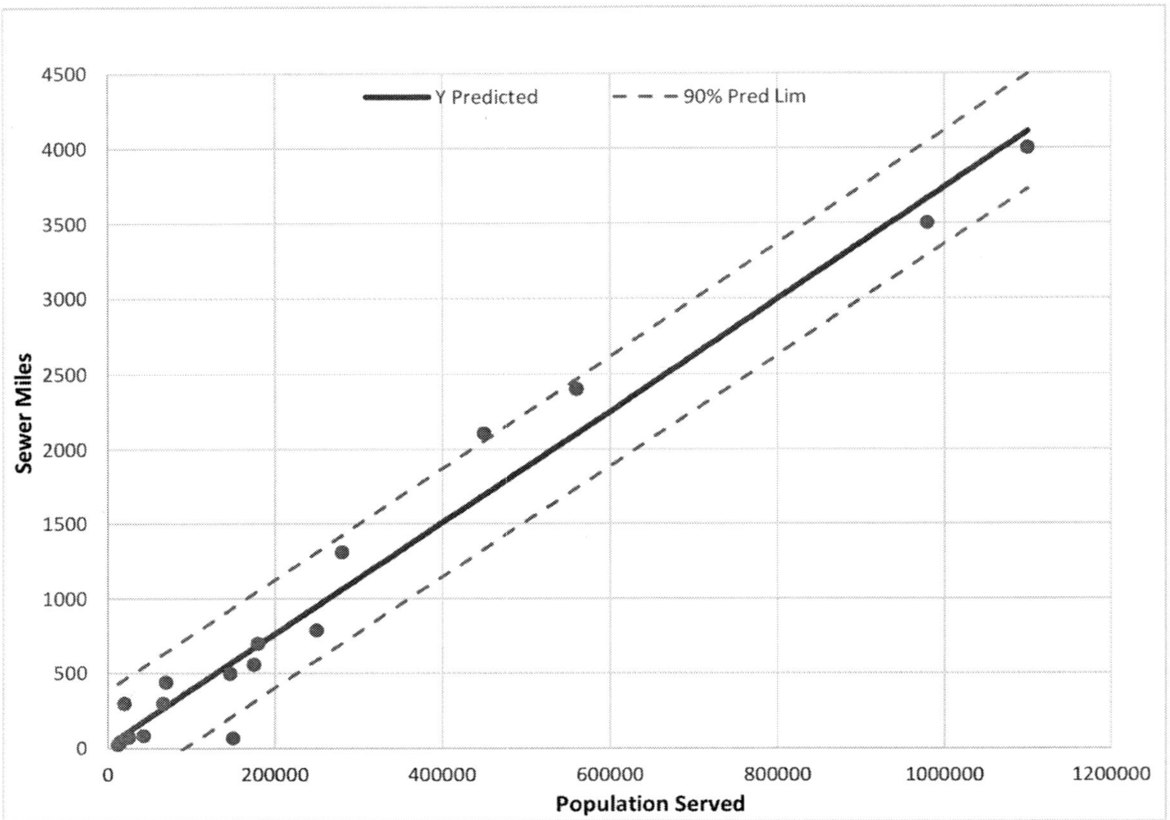

Figure 15: Sewer Miles and Population Served

O&M Costs versus Population Served and System Scale

For each participating organization, operating and maintenance (O&M) costs for vapor and chemical treatment systems in the conveyance system and in each WRRF was regressed against the population served and an aggregate variable for that represents system scale: the product of average dry weather flow and the conveyance system extent expressed in miles of sewer (mgd*mile).

Since the survey requested costs/expenses in bracketed ranges, this posed challenges for a detailed evaluation. A unique approach was employed to better understand the relationships. The following procedure was used to reduce the data set and find the central tendency:

1) For survey responses that were data ranges (costs), each record was provided for upper and lower values for the bounds, as well as a median value for the range.
2) This created three values for each record (e.g. population value). This was done by copying the records (e.g. population values) twice and appending these values as rows added to the bottom of the original list.

3) With the three columns of low, median, and high values (costs) in three columns to the right of the "population" record value, the median and high values were cut and paste, appending them to the bottom of the low value column.
4) There were now two columns: one population values repeated three times and one column with the low values, the median values, and the high values.
5) These two columns were sorted based on the population.
6) These two columns were graphed on a scatter plot then the linear trend line added with equation and r-squared value.
7) From high population values to low population values, one or two of the three values per "population" value were removed, based on an increase or decrease in the r-square value.
8) For zero values that may have also been no-data, all zero values were removed unless the population was less than 50,000.

In future surveys, actual dollars spent should be requested to allow for more rigorous analysis. For conveyance system and aggregated wastewater reclamation facilities vapor and chemical odor control O&M costs are separately graphed against population served and "mgd*mile". A linear regression of each data set produces the benchmark: a linear equation of the specific O&M for the average municipality of that system scale, Figures 16 to 23. Included in the regression are the 90% prediction interval and the survey data points.

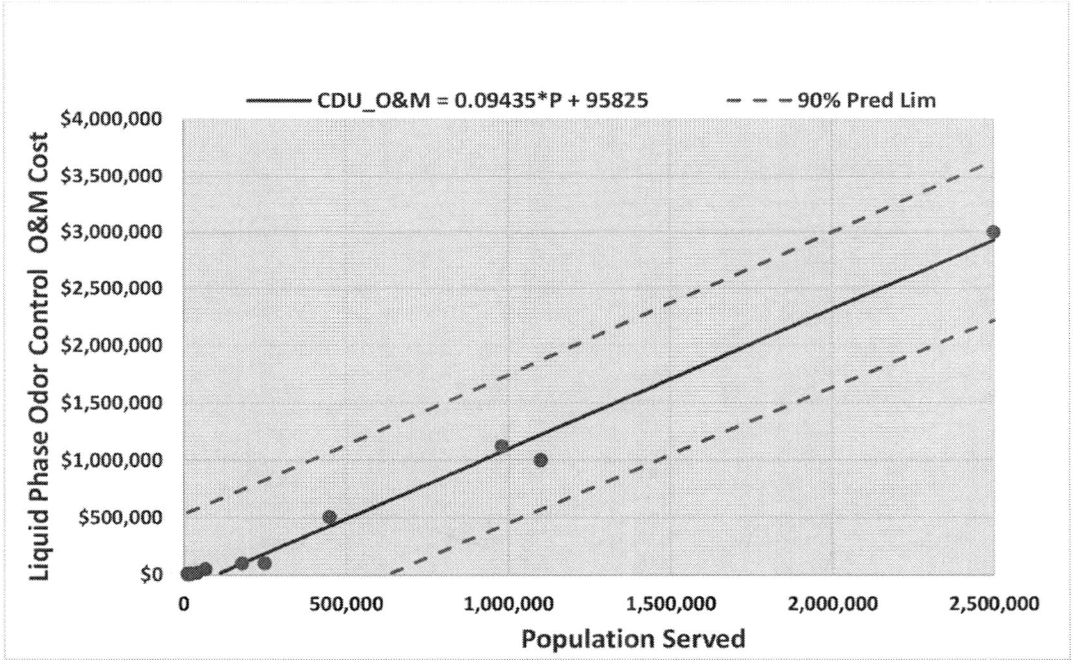

Figure 16: Collection System Liquid Phase Odor Control O&M Cost v Population Served

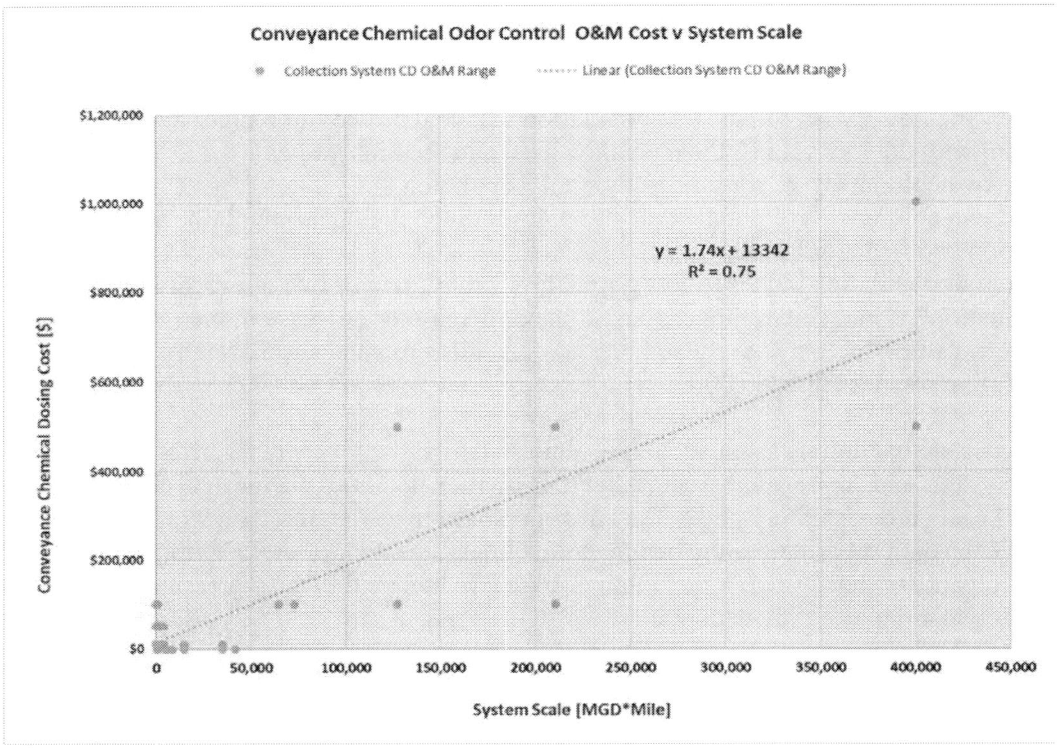

Figure 17: Collection System Liquid Phase Odor Control O&M Cost v System Scale

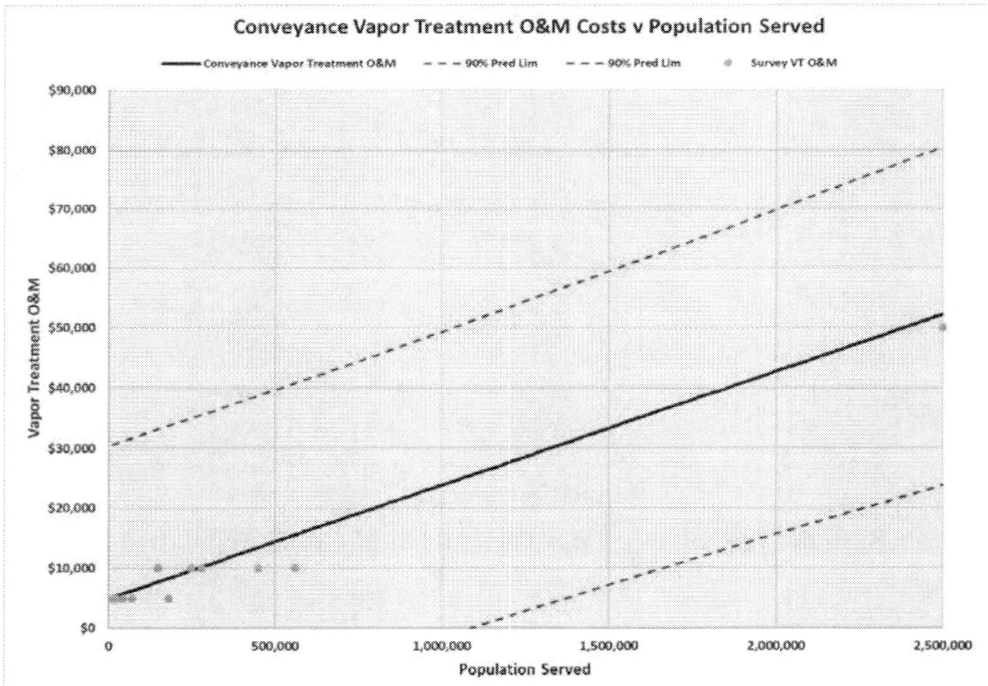

Figure 18: Collection System Vapor Phase Treatment O&M Costs v Population Served

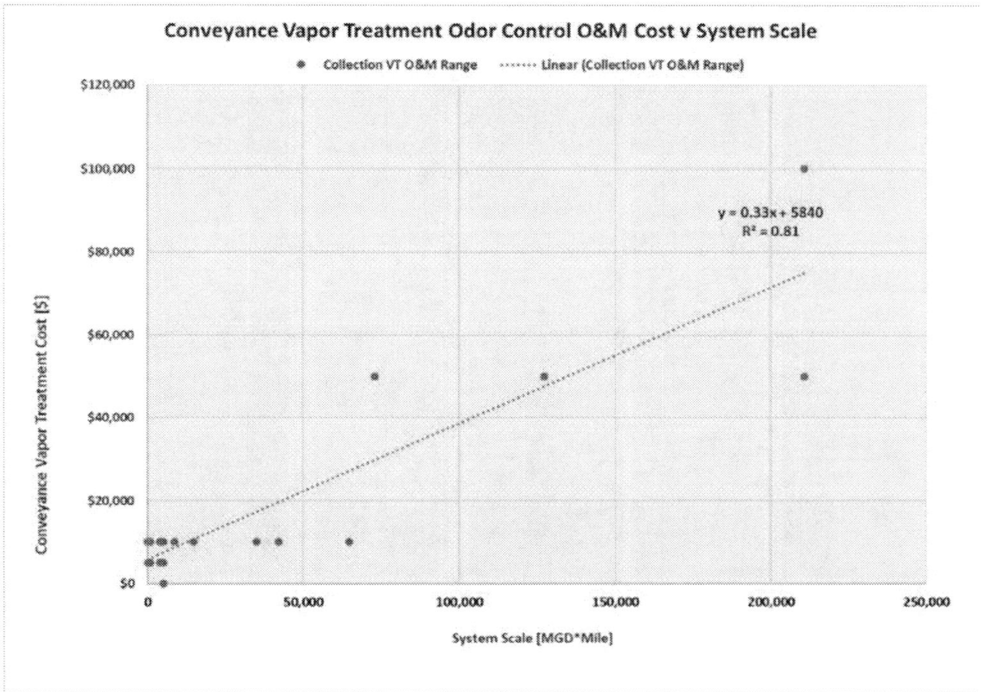

Figure 19: Collection System Vapor Phase Treatment O&M Costs v System Scale

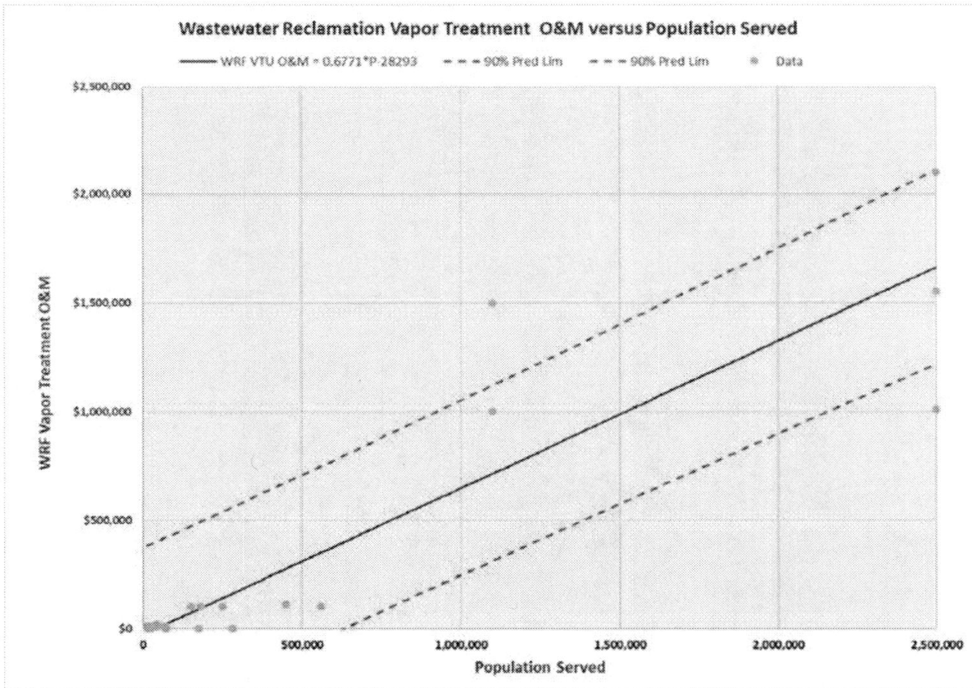

Figure 20: WRRF Vapor Phase O&M v Population Served

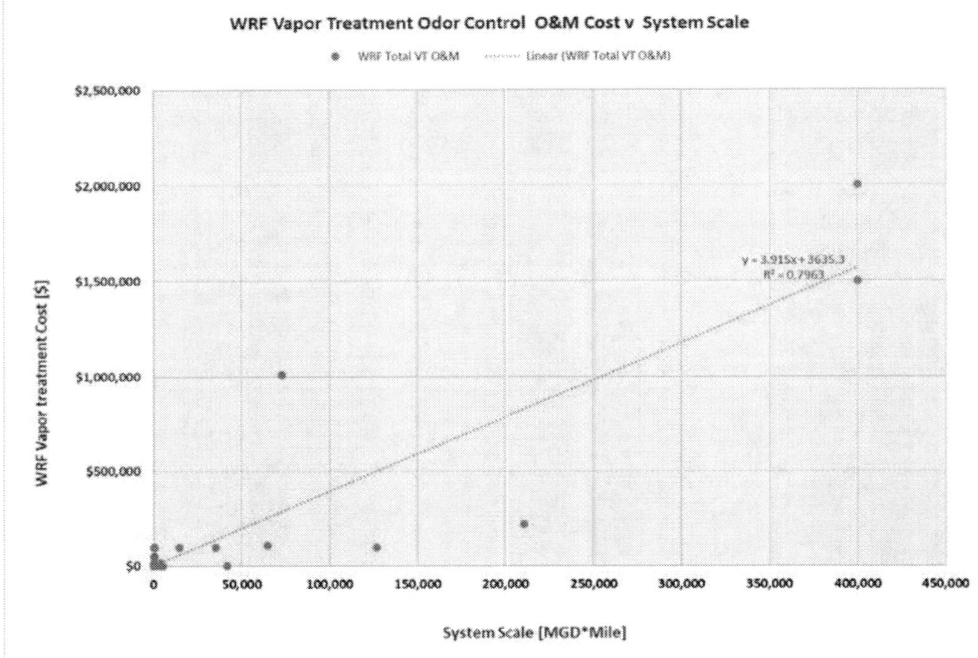

Figure 21: WRRF Vapor Phase O&M v System Scale

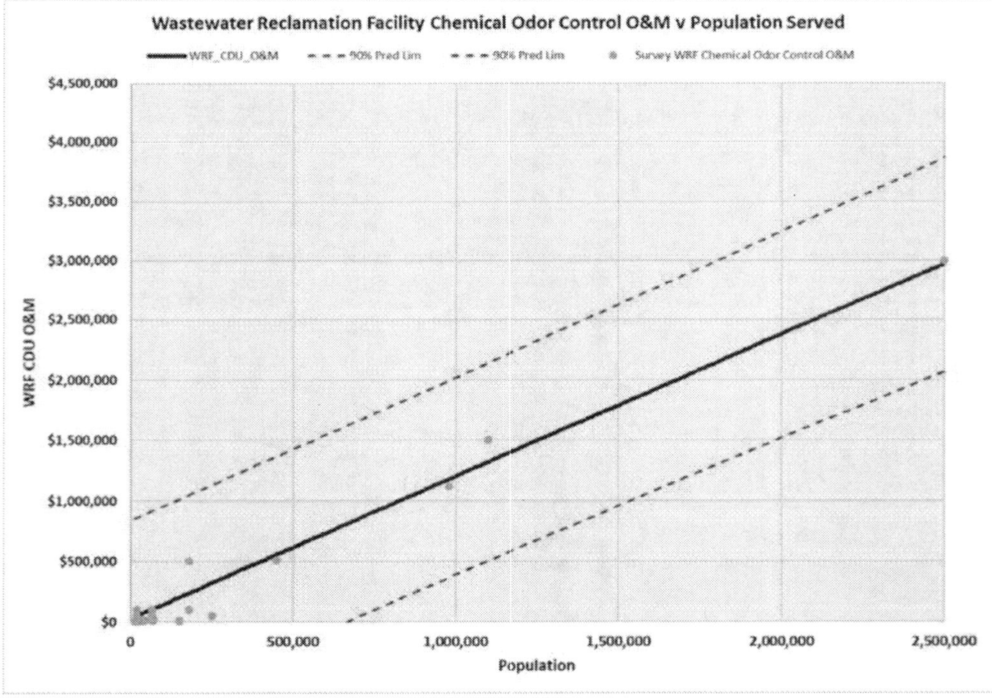

Figure 22: WRRF Liquid Phase O&M v Population Served

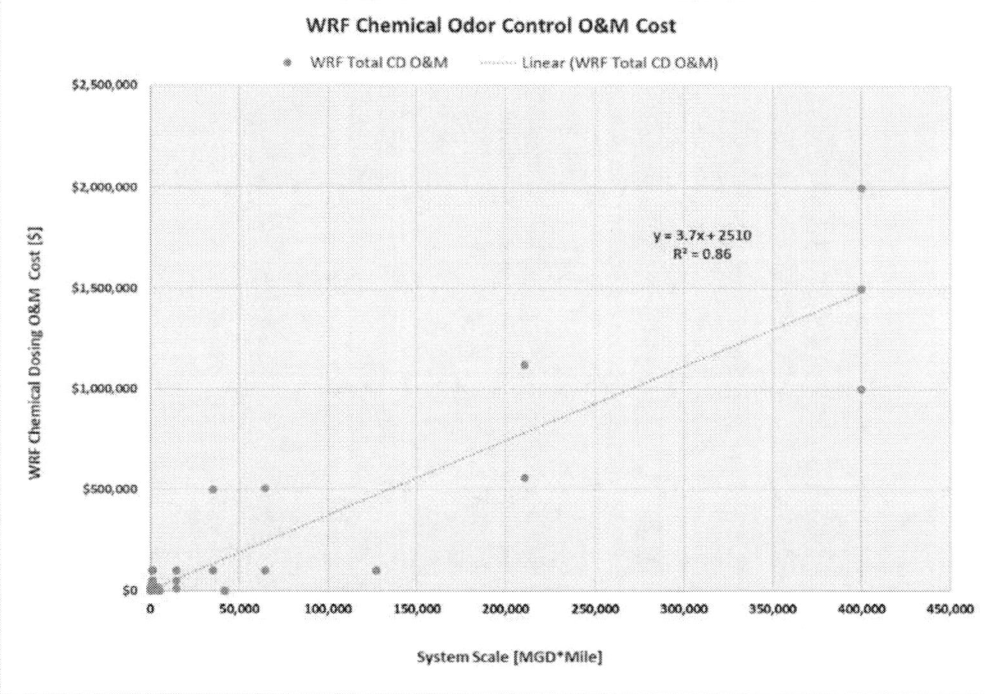

Figure 23: WRRF Liquid Phase O&M v System Scale

Capital versus Population Served and System Scale

For capital improvement expenditures, the sum of each responders' 5-year odor control capital program was graphed versus population served and system scale in "mgd*mile". This analysis assumes the approximation that both the population served and the system scale was constant during the 5-year capital program at the value reported for the fifth year of the capital program. The capital program cost for each municipality is graphed versus the "population served" and "mgd*mile". The linear regression on this data set produces the linear benchmark for 5-year capital expenditure for the average municipality odor control program in terms of service population and system scale, "mgd*mile" (Figures 24 and 25). Included with these graphs are the 90% prediction intervals for the benchmark.

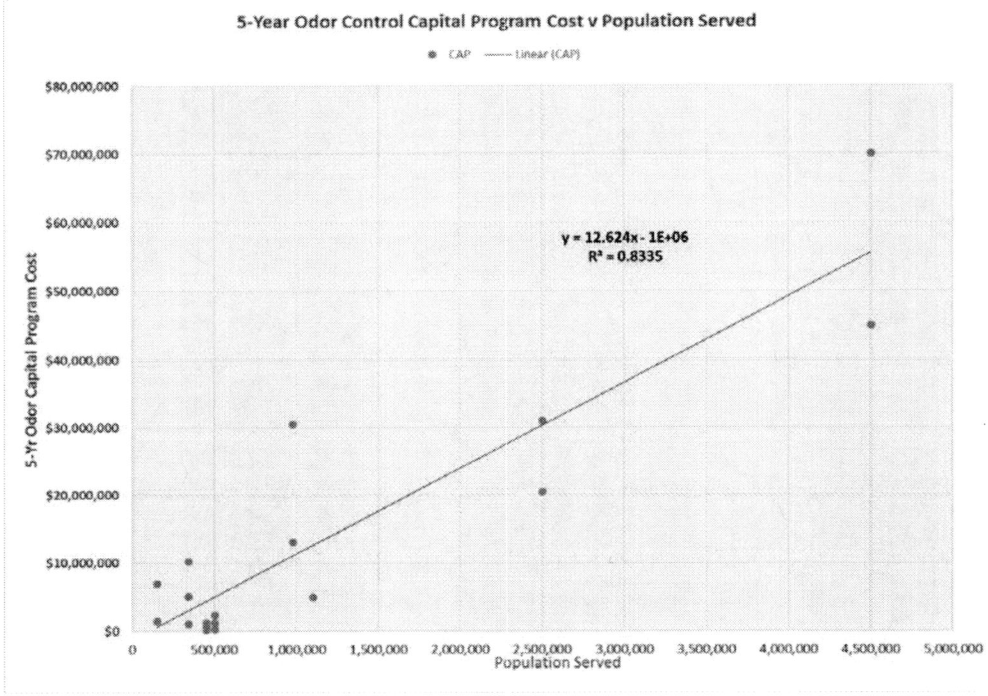

Figure 24: 5-Year Odor Control Capital Program Costs v Population Served

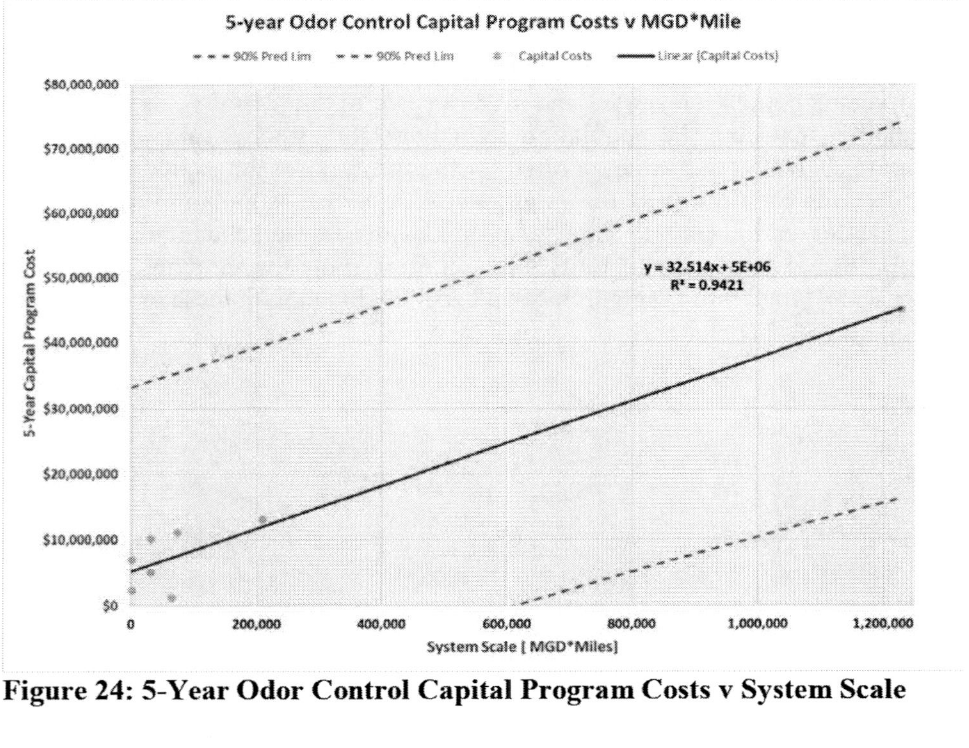

Figure 24: 5-Year Odor Control Capital Program Costs v System Scale

Design Standards
Nine responders (31%) indicated their organization has established odor control system design standards. Eight responders (89%) indicated having preferred odor control technologies.

Maintenance and Records
Conducting routine inspections and maintenance of odor control equipment provides assurance that it is operating as designed and ninety five percent of responders take these pro-active steps. Of those that record monitoring data, 56% directly record all or some of it electronically without an intermediate paper step.

Conclusions

This survey provides wide-sweeping data about odor control for collection systems and WRRFs. In many cases, there were insufficient data to draw correlations. It is recommended that an odor benchmarking study be conducted every three years continuing to improve the methods and questions. Recommendations for future studies include:

- Break out odor triggers by response type such as adjustment to treatment, initiation of a capital project, customer contact to better understand unique drivers,
- Separate the time goals for investigating versus follow up with the customer/complainant,
- Further explore community involvement in capital project design and construction,
- Explore factors that drive the need for devoting staff to odor control such as design of conveyance system (slope, detention time in pipes, lift/pump stations), proportion of high strength discharges, sensitivity of surrounding community,
- Question the reasons for establishing a more stringent organizational goal for odor control beyond permit requirements,
- Continued use of both flow and length of pipe for LPOC in the collection system,
- In future surveys, actual dollars spent, as opposed to ranges, should be requested to allow for more rigorous analysis,
- Request more information about the nature of the covers from treated units and air exchange rates,
- Ask why not monitor for hydrogen sulfide,
- Look at industry impact as a percent of load on odor,
- Ask about inflow and infiltration as part of flow,
- Ask what metric is used to judge effectiveness,
- Ask why certain treatment was selected,
- Look at life cycle costs,
- Correlate number of employees to effectiveness, and
- Reach out to engage new survey responders.

Acknowledgements
The authors would like to recognize and thank WEF staff, Rebecca Arvin-Cólon who made the survey possible. In addition, Richard J. Pope, AQ&OC committee member, was part of the benchmarking committee providing significant input in the survey development and evaluation.

How To Boil Down OCSD's Odor Control Master Plan Into One Page

Tadeo G. Vitko

Orange County Sanitation District
10844 Ellis Avenue
Fountain Valley, CA. 92708, USA

Abstract

This paper extracts the most relevant information from OCSD's 2017 Odor Control Master Plan. It demonstrates how one work table is sufficient to show which odorants are emitted beyond nuisance levels and require abatement. Also, how it can serve as a future design criteria for odor control at a water resource recovery plant.

Key Words

Odors, Foul Air, Odor Nuisance, Odor Control Master Plan, Design Criteria, Sampling Guidelines, Treatment Guidelines

1. Introduction

Odor characterization involves identification and quantification of odorants. It is only when odorants are characterized at a particular source that an effective odor abatement plan and an odor treatment technology can be properly defined for that particular source (Vitko et al., 2014). There are three basic ways to determine odorants; analytical, sensory, and combined methods. There are many different analytical methodologies to quantify odorants. Each analytical methodology is dependent on the family of compounds being analyzed. Table 1 summarizes the

Table 1 - Odorants Identified at OCSD Wastewater Treatment Process Areas Compared to their Sampling Techniques, Analytical Techniques, and Method Reporting Limit (Vitko et al., 2014)

Odorant	Odor Threshold Concentration[1] (ppbV)	Sampling Technique[2]	Analytical Technique[2]	Method[3] Reporting Limit (ppbV)
Acetic Acid	1,019	1.0 L/min rate for 100 – 200 min. in sorbent tube.	TO-11A Method by GC/MS.	9.2 - 4.6
Butyric Acid	0.33			0.76 - 0.38
Isobutyric Acid	8,108			0.34 – 0.68
Ethyl Mercaptan	0.013	Grab sample into 1 L Tedlar bag.	ASTM D5504 Method by Modified GC/SCD with sulfur chemiluminescence detection.	5.0
Methyl Mercaptan	0.020			5.0
Dimethyl Disulfide	0.026			2.5
Dimethyl Sulfide	0.99			5.0
Hydrogen Sulfide	0.5			5.0
Dimethyl Amine	20.55	1.0 L/min rate for 100 – 200 min. in sorbent tube. *0.2 L/min rate.	CAS 101/TO-17 Method by Modified GC/NPD *OSHA 60M.	5.6 - 2.8
Ethyl Amine	264			6.0 - 3.0
Trimethyl Amine	0.33			4.1 - 2.1
Methyl Amine*	19.89			0.0078 - 0.0039
Ammonia	38.28	0.5 L/min rate for 60 min.	OSHA ID-188/ID-164.	0.26
D/T, R/T, I, HT[4]	N/A	Grab sample into 1 L Tedlar bag.	ASTM E679 & EN13725	N/A
OPM[5]	N/A	Grab sample into 10 L Teflon bag	Std. Method 2170 applied to air	N/A
MIB[5] IPMP[5] Skatole[5] Indole[5]	0.020 0.004 0.02 0.5	Grab sample into 10 L Teflon bag	GC/MS	0.04 0.1 0.15 0.25

[1] Ruth, J (1986). [2] ALS sample and analysis techniques. [3] A range in MRL indicates the doubling of volume captured in sorbent tubes. [4] St. Croix Sensory. [5] Suffet et al. (2016).

odorants identified at Orange County Sanitation District (OCSD) waste resource recovery plants along with their particular odor threshold concentration (OTC), sampling technique, analytical technique, and associated method reporting limit (MRL). The importance of OTCs and MRLs are key to understanding odors and the importance of sensory analyses in the determination of odors.

There are two different sensory methodologies used in odor control. One is widely used and is known as the odor threshold method (ASTM E679, 2011; EN13725, 2003; ASTM E544, 2010). It uses an olfactometer to precisely dilute air samples to a known ratio. The dilution to threshold (D/T) is a measure of the total odor of a sample at its detection level. The recognition to threshold (R/T) is a measure of the total odor of a sample at its recognition level. The other sensory method is known as the odor profile method (OPM), which is a variation of the threshold method modified to identify odors at a particular process. The OPM identifies the different odor characters (how odors smell like) present in an undiluted sample and reports them in terms of the odor intensity for each particular and identified odor character. Odor intensity is based on a nine-scale sugar standards (APHA, 2012).

Combined methods use the same select assessors used for OPM (ASTM Method STP 440, 1968) to evaluate the odors while the sample is being analyzed by gas chromatography (GC). This is also known as GC sensory analysis. This method enables the simultaneous identification of the odorant by chromatography as it is being sensed by the assessors. This procedure helps identify the odorant present in a sample even when it is not apparent by simple view in the chromatograph. Once identified by GC sensory method, the odorant can be properly quantified by a standard method or by a modified standard method. The only limitation of the GC sensory analysis, as for all analytical methods, is when the odorants being tested have their concentrations below their particular MRLs. When this happens, the odorants are invisible to the method being used to detect and quantify them. Odorants that have very low OTCs (below their MRLs) tend to fall in this category. In these cases, the only available and viable methodology to approximately quantify odorants is the use of the OPM.

2. Objective

OCSD recently completed its Odor Control Master Plan (OCSD, 2017). The objective of the Odor Control Master Plan (OCMP) was to determine the foul air treatment level required to obtain zero odor incidents and events under normal operating conditions. To achieve this goal the OCMP set out to: 1) identify and determine what odorants cause odors, 2) at what concentrations these identified odorants cause a nuisance, 3) what is the extent of the modeled odor impact under current and proposed conditions, 4) what odor abatement technologies are more effective at reducing the identified odorants, and 5) what is the cost associated with the proposed odor abatement technologies for single and dual stage applications. The objective of this paper is to condensate the OCMP and tabulate key information into a one page table matrix. This matrix establishes standard testing techniques for any future plant project requiring odor control and defines the design criteria for existing plant process areas requiring odor control.

3. Method and Materials

OCSD is the third largest wastewater treatment facility west of the Mississippi River. It is located in the county of Orange in Southern California. OCSD has a service area of 463 square miles and provides sewerage for a population of 2.6 million. The total average influent flow is 180 million gallons per day (MGD) that is treated in two separate plants located 5 miles apart. Plant 1, with a headworks capacity of 280 MGD, is located in the city of Fountain Valley and receives an average of 130 MGD. Plant 2, with a headworks capacity of 340 MGD, is located in the city of Huntington Beach and receives an average of 50 MGD. Plant 1 is a wastewater reclamation facility, as its entire secondary effluent is further treated by the Orange County Water District to produce 100 MGD of potable water, which is used to replenish the groundwater in Orange County, which is known as the Groundwater Replenishment System (GWRS). There are plans to expand GWRS to include all reclaimable flows from Plant 2.

Population growth and urban development in Southern California soon began surrounding both plants with residences and commercial facilities. As this encroachment advanced, odors became an issue. OCSD addressed it by covering all major odor-producing wastewater treatment plant process areas, providing negative pressure, conveying foul air in ducts, and installing air scrubbing facilities. In spite of all these odor control efforts, OCSD keeps getting odor complaints at both plants from its neighbors (OCSD, 2017). There is one key factor that contributes to these odor complaints. Besides the occasional operational and climatic variables that periodically exacerbates the odor impact beyond OCSD fence lines, the reason odor complaints persist is that the existent odor scrubbing technologies mainly target the removal of hydrogen sulfide (Gabriel et al., 2004; Wang et al., 2014; Shammay et al., 2016).

The OCMP has demonstrated that besides hydrogen sulfide, there are eight other odorants being consistently emitted from the different plant treatment process areas (Table 2) and that the current treatment systems and technologies are insufficient to remove these odorants from the foul air. Surprisingly, hydrogen sulfide is not the odorant that tops the list, but methyl mercaptan.

Table 2 – Most Detectable Odorants at OCSD with Respect to Process Area

Odorant	Trunk-lines	Head-works	Primary	Trickling Filters	Activated Sludge	DAFTs	Dewa-tering	Truck-loading
Dimethyl Disulfide	X	X						X
Dimethyl Sulfide								X
Hydrogen Sulfide	X	X	X	X	X	X		X
Methyl Mercaptan	X	X	X	X	X	X		X
Ammonia							X	X
MIB				X	X		X	
IPMP		X	X	X	X	X	X	
Skatole							X	X
Indole								

NOTES: MIB = 2-Methyl Isoborneol and IPMP = 2-Isopropyl-3-Methoxypyrazine

In an effort to improve odor reduction, the existing packed bed chemical scrubbers at OCSD over time have been operated in different modes, each scrubber and mode with a different hydrogen sulfide removal efficiency, as shown in Figure 1. All obtained hydrogen sulfide removal efficiencies, calculated from continuous H_2S monitors, are below 90%, which is insufficient to reduce hydrogen sulfide below nuisance levels, much less to reduce the other eight odorants below nuisance levels. The reason for this low efficiency has been found to be the operation outside optimum pH range and excessive makeup water.

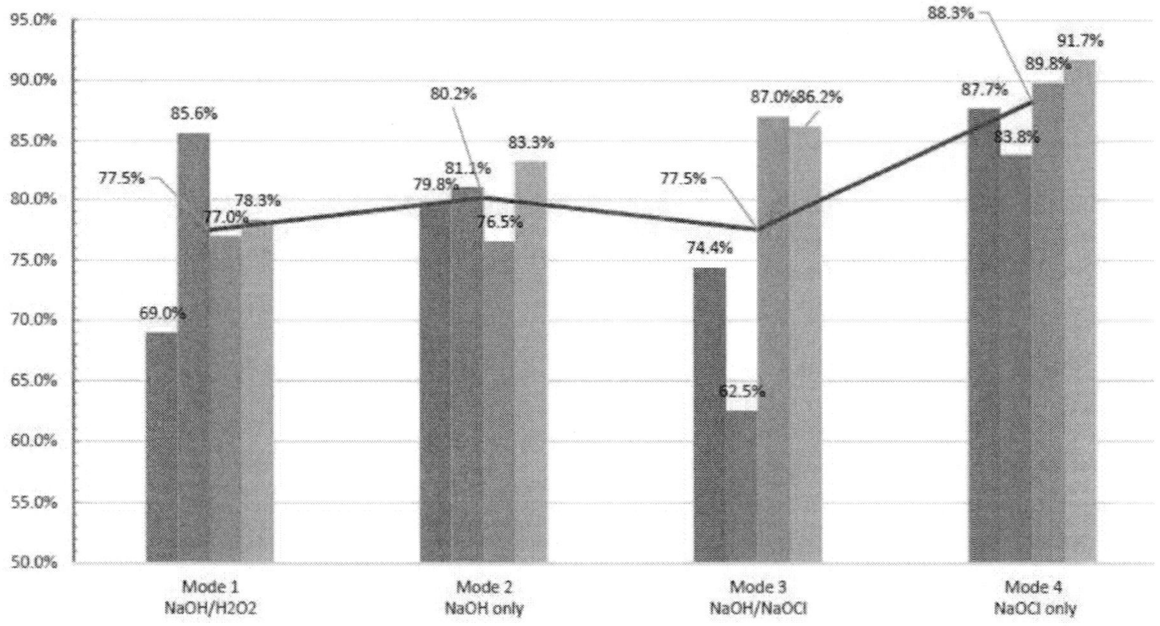

Figure 1 – Percent Efficiency of the Four OCSD Headworks Chemical Scrubbers Operated in Different Modes from 2009 to 2014 (OCSD, 2017).

4. Methodology Used

Sixteen monitoring stations from eight odorous process areas had their foul air repetitively sampled at each plant using Standard Methods. Some process areas had duplicate locations and all foul air treatment systems were sampled at their inlet and outlet locations to verify their efficiency in removing the nine identified odorants. All sample points were chosen at positive pressure locations (to avoid possible dilution with ambient air), except for those process areas that had no odor control. In those cases, the foul air was sampled using EPA flux chambers with nitrogen sweep gas. Sample repetitions included bias-avoiding schemes, such as sampling at different days of the week and changing the order of the sampling source. The sampling was done during August and September, 2013 for Plant 1 (a total of 5 repetitions per monitoring station) and October and November, 2013 for Plant 2 (a total of 4 repetitions per monitoring station). The sampling period was between 7:00 AM and 1:00 PM in order to ensure the laboratory analysis was done within the prescribed holding times. The average ambient temperature during the sampling period was about 70°F (21°C). All results were later corrected for daily and seasonal variations, obtained from continuous hydrogen sulfide monitors, in order to represent a worse-case discharge situation (Vitko et al., 2016-b).

What follows is a description of the methodology used broken down by the objectives of the project:

a. Description of Field Sampling Methods

The sampling methods depended on the odorant being sampled are shown in Table 1. Samples for reduced sulfides and odor threshold consisted of a grab sample in 1 Liter Tedlar bags using vacuum containers. Samples for OPM and GC Sensory consisted of a grab sample in 10 Liter Tedlar bags[1] using vacuum containers. Samples for carboxylic acids, ammonia, and amines, consisted of sorption of a regulated flow into sorbent tubes to concentrate the odorants in a media for subsequent extraction and quantification. The sample time was later doubled to

[1] During Phase II of the project, it was found that Teflon bags were more adequate than Tedlar bags for the determination of skatole as it virtually "disappeared" by either adsorption to or diffusion through the bag material (Zhou et al., 2016).

verify if this increased the probability of detection for these compounds. However, this effort did not change the frequency of carboxylic acids or amines in the samples indicating that these compounds were not consistent odorants. Ammonia, on the other hand, was found at high concentrations only in the solids dewatering and solids truck loading plant processes.

b. Description of Laboratory Techniques

The laboratory methods utilized depended on the odorant sampled, as shown in Table 1. Standard analytical methods utilize mass spectrometry (MS) and/or gas chromatography (GC) for the determination of reduced sulfur, carboxylic acids, and amines. Ammonia was determined by an ion-specific electrode. Combined methods utilize GC sensory for the determination of odorants such as skatole, indole, 2-methyl-isoborneol (MIB), and 2-isopropyl-3-methoxypyrazine (IPMP). These odorants were found to be prevalent in foul air at most plant process areas, as shown in Table 2. It was found that the analyses for skatole and indole do not fully represent the fecal odor category as low analytical results were obtained, while the OPM shows high detections for this family of odorants.

c. Description of Sensory Techniques

Threshold sensory analyses are performed by a group of 4 to 10 screened panelists (meeting ASTM E253, 2015) that are asked to choose which one of the three presentations, one diluted odor sample and two blanks, has an odor, even if no apparent difference is found between the three. This is known as the "triangular forced choice method" (ASTM E679, 2011). The samples are presented to the panelists using an olfactometer at a given flowrate, known as "presentation flowrate", which in this case was 20 L/sec. The panelists are given samples in batches of increased concentrations (each at lower dilutions) and asked to either guess, detect, or recognize the odorous sample in what is known as "ascending concentration series". The dilution is the ratio between the presentation flowrate and the sample flowrate. The geometric mean of the last dilution ratio, where the odor was not detected, and the dilution ratio where the odor was first detected, is the odor threshold for that sample. A D/T value is reported from the results of the test when 50% of the panelists first detect an odor at a particular dilution factor. D/T represents the total odor at an OPM equivalent intensity of one (I = 1). Similarly, when 50% of the panelists are able to recognize (identify or describe) the odor, that dilution factor is known as the recognition threshold (R/T). R/T represents the total odor at an OPM equivalent intensity of four (I = 4). OCSD established the OPM equivalent intensity of three (I = 3) as that particular odorant nuisance action level (Figure 2). Whenever an odorant is shown to disperse beyond OCSD fence lines at the concentration equivalent to I = 3, the corresponding concentration at the plant process stack (based on the AERMOD dispersion model) must be met to ensure a nuisance free operation (OCSD, 2016).

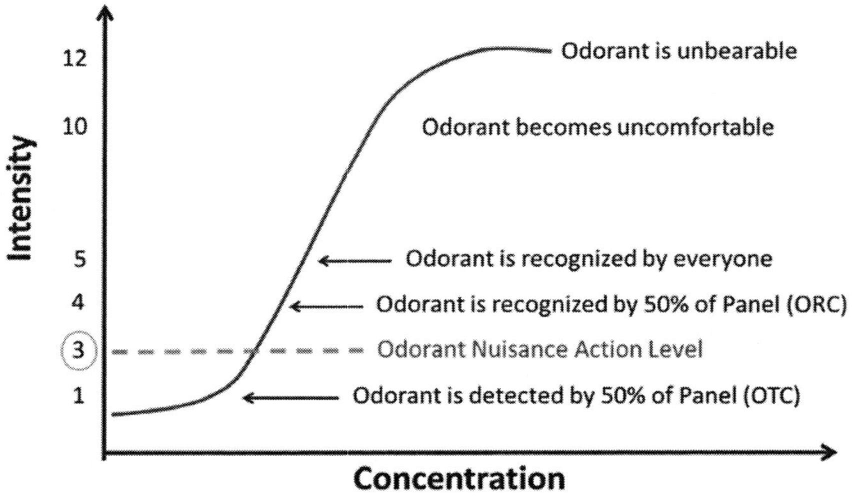

Figure 2 – Relationship Between an Odorant Intensity and Concentration and Its Threshold (Vitko et al., 2014)

OPM is a sensory method for determining odorants present in samples. The method is similar to the recognition threshold method, but without the use of an olfactometer, as the samples are not presented in a diluted form. Because a real-life odor sample is a mixture of odorants, the odor panelists are trained to identify multiple odor characters (how each odor smells like) and their respective intensities (how strong each odor feels like) in a single sample. They are trained to distinguish odor mixtures of 2, 3, and 4 different odorants from standards before assessing a real-life foul air sample (Gostelow et al., 2001; Khiari et al., 1992). The stock odorants are chosen based on the odors associated with the waste stream to be tested. In this case stock odorants used those compounds identified in the Wastewater Odor Wheel (Burlingame et al., 2004), which evolved to the Wastewater Odor Peel (Figure 3) in an attempt at giving odors a third dimension in trying to explain odors as "peeling an onion". The order of the layers is theorized to be according to each odorant's OTC, which explains the concept that odors can be masked by other odorants and become apparent when those odorants masking them are removed (foul air treatment) and/or become sufficiently diluted (dispersion plume). This mechanism explains why public odor complaints persist and often qualifies them as offensive (e.g. rotten vegetables, caused by methyl mercaptan or musty/moldy, caused by MIB and/or IPMP).

The odor intensity in OPM is evaluated using a modification of the Standard Method 2170 (2012), Flavor Profile Analysis (FPA) method (APHA, 2012) using a sugar scale as shown in Table 3. Curren et al. (2014) demonstrated that FPA and the Butanol method (ASTM E544, 2004) are similar and valid methods for evaluating odor intensities, FPA being the easiest. If a given character is reported by greater than 50% of the panel, the odor intensity of each character is reported as its average and the calculated standard deviation. If the odor character is reported by less than 50% of the panel, it is considered an odor note.

Table 3 – Flavor Profile Method Odor Scale for Intensity and Fenceline Effects

Intensity Scale	% Sugar in Water	Associated Odor Strength	Expected Fenceline Odor Effects From Wastewater Treatment Plants
0	0	Odor free	None
1		Threshold	None
2		Very Weak	Unlikely Off-Site Odors
4	5	Weak	Probable Off-Site Odors
6		Weak-Moderate	Definite Off-Site Odors
8	10	Moderate	Definite Off-Site Odors
10		Moderate-Strong	Definite Off-Site Odors
12	15	Very Strong	Definite Off-Site Odors

5. Results

The description of how the results were obtained, a discussion about the reasoning used in the calculation, and the presentation of the results are broken down by the objectives of the project:

1) Identify and determine what odorants cause the odors

All odorants have a specific odor character (how they smell), method reporting limit (MRL), and odor threshold concentration (OTC). Table 4 shows the list of odorants detected at OCSD with specific characteristics organized by odor families. Each odor family is color-coded to distinguish among them. This color code is used in the Wastewater Odor Peel and OPM. Sulfur-based odorants are shown in green, nitrogen-based odorants are in purple, carboxylic acid odorants are in dark red, volatile organic carbons (VOC's) are in gray, musty/moldy odorants in blue and fecal odorants in orange. Note that the yellow highlighted odorants in this list have their OTCs considerably lower than their MRLs. When these compounds are reported as "non detect" or ND, the OPM analysis is the only available technique to determine if any of these odorants are

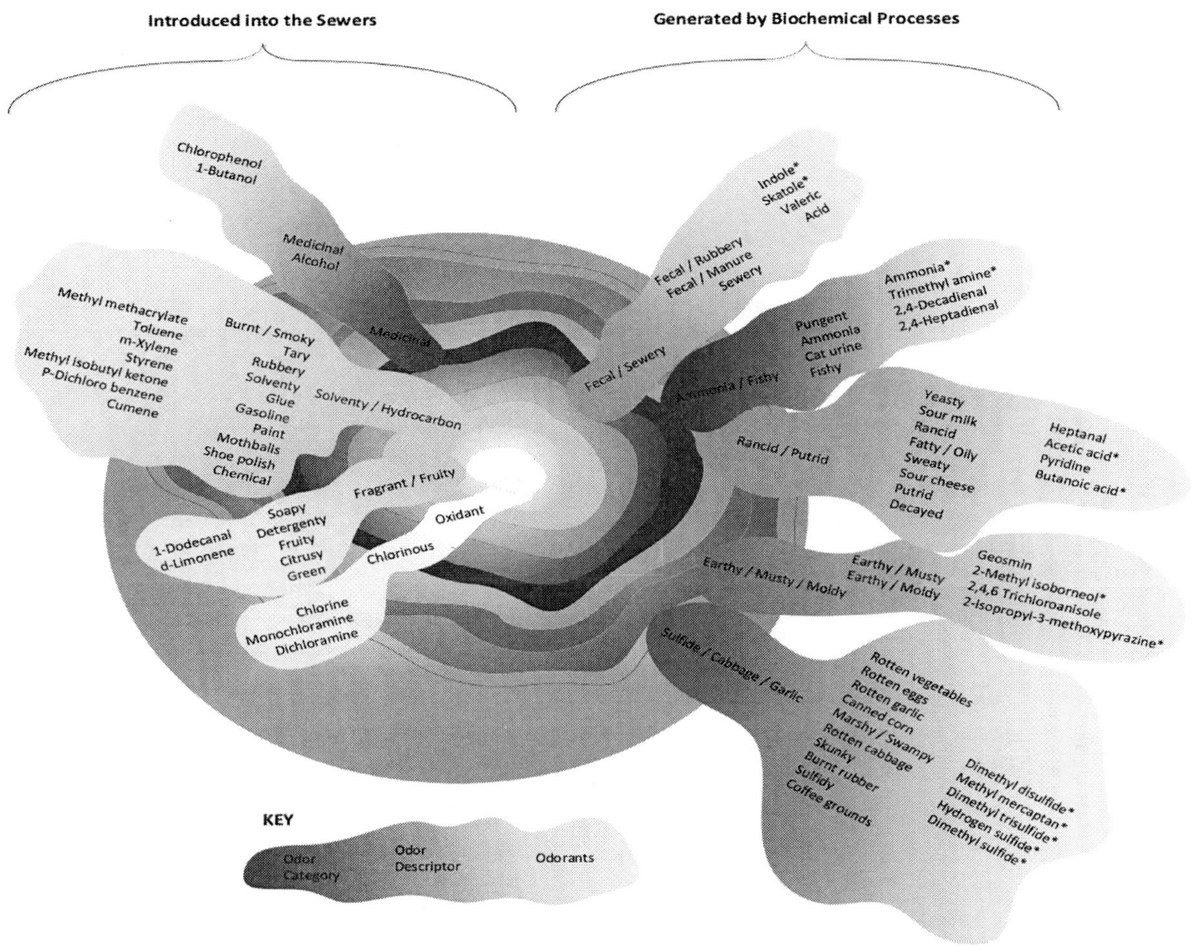

Figure 3 – The Wastewater Odor Peel Identifies Odor Category, Odor Descriptor, and Odorants in Foul Air at Wastewater Treatment Plants Process Areas (* Detected at OCSD)

present in a sample. This illustrates how much odor information may get lost if we only rely on standard analytical results to describe odors. For example, the human nose is 10 times more sensitive than the lab method for hydrogen sulfide, but is 62 times more sensitive for methyl mercaptan, and 384 times more sensitive for ethyl mercaptan. The nine most consistently detected odorants have been marked with an asterisk (*).

OTCs in the literature vary widely depending upon how they were measured. OTCs represent the dilutions with clean air for any particular odorant from a pure standard so that 50% of the panelists begin to detect an odor. The OTC result, as of any sensory analysis, is principally

Table 4 – Comparing MRLs and OTCs for Odorants Detected at OCSD Wastewater Treatment Plants

Odorant	Characteristic	MRL (ppbV)	OTC (ppbV)
Dimethyl Disulfide*	Rotten garlic	2.5	0.2
Dimethyl Sulfide*	Canned Corn	5.0	3.0
Ethyl Mercaptan	Rotten cabbage	5.0	0.013
Hydrogen Sulfide*	Rotten eggs	5.0	0.5
Methyl Mercaptan*	Rotten vegetable	5.0	0.08
Trimethylamine	Pungent, Fishy	2.1	0.33
Ammonia*	Pungent, Irritating	0.26	1,000
Acetic Acid	Vinegar	4.0	1,019
Butyric Acid	Rancid	0.3	0.33
Valeric Acid	Sweat	0.3	0.62
Acetone	Solvent	5.5	19,798
Carbon disulfide	Ether-like	8.0	7.81
Methyl Ethyl Ketone	Butterscotch	4.9	250
Tetrahydrofuran	Solvent	3.8	2,502
Toluene	Solvent	3.0	2,131
Methyl Isoborneol*	Musty/Moldy	0.04	0.020
Isopropyl Methoxypyrazine*	Musty/Moldy	0.1	0.004
Skatole*	Fecal	0.15	0.018
Indole*	Fecal	0.25	0.5

* Most detectable odorants. OTC determined by OCSD odor panel in 2015
Highlighted OTCs indicate that these odorants require OPM to determine their concentrations when below the MRL

affected by the conditions in the environment at the time of the test and the age and gender of the panelists (Vitko et al., 2016-a). To ensure reliable OTC results, it is crucial that the test be conducted by a minimum of 12 panelists that can pass the UPSIT test (Doty et al., 1984), that the panelist be evenly composed of both genders and all age groups, and that 50% of them have no particular experience with OPM.

Furthermore, it is very important that the same panelists used for the determination of OTCs participate in the OPM work. This will ensure consistency in the results. At OCSD this meant that some OTCs were somewhat different than the literature, but still within the reported range in the literature (Suffet et al., 2016).

2) At what concentrations these identified odorants cause a nuisance

As shown in Figure 2, odorants become a nuisance somewhere between their OTC (odor intensity 1) and their recognition threshold (odor intensity 4). The OCMP determined that an odor intensity 3 whould be a safe level in between these two intensities. Using odorants from laboratory quality stocks and OCSD's olfactometer, a panel (as described above) determined the number of dilutions required for each identified odorant to reach its nuisance level. With this information the concentration in ppbV could be determined. Table 5 summarizes the odor intensity 3 (I = 3) concentrations for each odorant. These became the nuisance criteria values.

3) What is the extent of the modeled odor impact under current and proposed conditions

A comprehensive AERMOD air dispersion model was developed to determine the theoretical odorant target concentration (TOTC) at each odorous plant process area. By knowing the nuisance criteria for each identified odorant, TOTCs are calculated by multiplying the location dilution factor (LDF) by the nuisance criteria. When Design is greater than TOTC, it means that this particular odorant requires abatement to avoid nuisance beyond the fenceline.

Table 5 – Nuisance Concentrations for the Identified Odorants

Odorant	Characteristic	OTC (ppbV)	Nuisance (ppbV)
Dimethyl Disulfide	Rotten garlic	0.2	0.77
Dimethyl Sulfide	Canned Corn	3.0	7.9
Hydrogen Sulfide	Rotten eggs	0.5	1.3
Methyl Mercaptan	Rotten vegetable	0.08	0.22
Ammonia	Pungent, Irritating	1,000	4,900
Methyl Isoborneol	Musty/Moldy	0.020	0.060
Isopropyl Methoxypyrazine	Musty/Moldy	0.004	0.035
Skatole	Fecal	0.018	0.037
Indole	Fecal	0.5	1.1

NOTE: Nuisance is achieved when the concentration of the odorant is equivalent to an odor intensity of three (I = 3).

4) What odor abatement technologies are more effective at reducing the identified odorants

An eight-technology pilot unit (See Figure 4) was used to test side by side the removal efficiencies of the nine odorants in the foul air from the incoming trunklines at Plant 2. As shown, the pilot unit is not of laboratory scale, but about $1/20^{th}$ scale of typical scrubbers at OCSD. The technologies were chosen after a thorough literature search of the most promising candidates. Technologies were tested at two different EBGRTs for the five biological systems and run to failure for the three activated carbon systems.

Figure 4 – The OCMP Pilot Unit Simultaneously Tested the Removal of Odorants by Five Different Biological Technologies and Three Activated Carbon Units

6. Discussion

The matrix (See Table 6) consists of a list of all the foul air generating plant process areas at both OCSD treatment plants with their particular specifications of nuisance criteria concentrations and airflows followed by the calculated location dilution factor (LDF) and the design information about the nine "most detectable" odorants. Each odorant has subdivisions for the maximum obtained odorant concentration (Max), the theoretical odorant target concentration (TOTC), and the design odorant concentration (Design).

Nuisance criteria is determined by olfactometry and the odor panel using stock odorants of known concentration presented at several dilutions to construct a sufficient relationship of odor intensity and odor concentration. LDFs is determined by the air dispersion model. They are specific to the plant location of the source of foul air and depict the number of dilutions that an odorant would experience between being emitted at that particular source and arriving at the plant fence line's prevailing wind under the worse-case weather conditions.

Max data is determined during the repetitive and multi-day sampling that was done to characterize each of the foul air sources at all plant process areas at both plants. This sampling was done at all plant process areas at different days of the week to avoid any bias in the results. All foul air sources were so sampled and analyzed at least four separate times. Max uses the highest concentration detected.

TOTCs is determined by calculation and is equivalent to the LDF times the nuisance concentration.

Design data is determined by multiplying the Max data by a corrective factor representing the daily and seasonal reduced sulfur and fecal results of foul air variations for Headworks and Primary. The data from the continuous monitoring of hydrogen sulfide was used as a reference point and compared with the hydrogen sulfide data obtained from the samples. This was done to calculate a daily correction factor (to correct the results to the observed daily peak at that sampling day) and to calculate a seasonal correction factor (to correct the results to the observed 2013 summer peak). For all other odorants and sources, a standard grab sample corrective factor of 2 was used.

The odorant removal rate table summarizes what was obtained with the pilot unit tested with Plant 2 trunkline foul air (collections entering the treatment plant) and compares removal efficiencies with data from the existing chemical scrubbers (both old packed bed chemical scrubbers installed in the 1990's and new dual bed chemical scrubbers installed in the 2010's) as well as data from the existing 2-second dual-bed foam cube biotrickling filters also installed in the 2010's. The odorant removal rate table demonstrates the limitations of most technologies in removing methyl mercaptan, which not only is the most important odorant at OCSD, but it is a key odorant to remove if the second stage is activated carbon, as it tends to dissociate into dimethyl disulfide in the presence of moisture. If methyl mercaptan, which smells like rotten vegetables is not removed by at least 90% in the first stage of treatment, the undesirable outcome will be a malodorous rotten garlic smell in the second stage activated carbon system.

Table 6 – Odor Control Design Criteria Matrix

(As reflected by the plant conditions prevalent during 2013 when the odor assessment was made)

Odorant (ppbV)			Hydrogen Sulfide			Methyl Mercaptan			Dimethyl Sulfide			Dimethyl Disulfide			Ammonia			Methyl Isoborneol			Isopropyl Methoxypyrazine			Skatole			Indole		
Nuisance Citeria (ppbV)				1.3			0.22			7.9			0.77			4,900			0.06			0.035			0.037			1.1	
Plant 1	Airflow (cfm)	LDF	Max	TOTC	Design	Max	TOTC	Design	Max	TOTC	Design	Max	TOTC	Design	Max	TOTC	Design	Max	TOTC	Design	Max	TOTC	Design	Max	TOTC	Design	Max	TOTC	Design
Wastehauler	300	636	19,000	827	38,000	2,400	140	4,800	7	5,024	13	2.5	490	5	0	0	0	0.1	38	0.1	0.1	22	0.1	0.5	24	0.9	3	700	7
Headworks	96,000	82	4,800	107	18,720	330	18	1,287	35	648	137	26	63	101	0	0	0	0.1	5	0.3	0.3	3	0.6	0.7	3	3	5	90	18
Primary	106,680	159	2,000	207	9,900	150	35	743	24	1,256	119	3.0	122	15	470	+	2,327	0.1	10	0.3	0.2	6	0.3	0.3	6	1	2	175	11
Trickling Filter*	395,810	59	15	77	62	16	13	66	34	466	140	2.5	45	10	320	+	1,322	0.8	4	2	0.1	2	0.2	0.2	2	0.7	0.4	65	2
Activated Sludge-2	78,000	53	8.2	69	34	12	12	50	7	419	29	2.5	41	10	0	0	0	0.4	3	0.7	0.1	2	0.2	0.2	2	0.7	2	58	2
Activated Sludge-1	10,000	17	8.2	22	34	12	4	50	7	134	29	2.5	13	10	0	0	0	0.4	1	1	0.1	1	0.2	0.2	1	1	2	19	7
DAFTs	50,000	101	5.00	131	10	5	22	10	5	798	10	2.5	78	5	0	0	0	0.1	6	0.2	0.1	4	0.1	0.2	4	0.4	1	111	3
Dewatering	61,376	117	13	152	26	5	26	10	5	924	10	2.5	90	5	3,300	+	6,600	0.1	7	0.2	0.1	4	0.1	0.2	4	0.4	6	129	12
Truckloading	13,375	357	52	464	104	390	79	780	520	2,820	1,040	27	275	54	7,900	+	15,800	0.2	21	0.4	0.3	12	0.3	0.3	13	0.7	3	393	5
Plant 2																													
Trunklines**	40,000	184	7,000	239	14,000	920	40	1,840	68	1,454	136	98	142	196	0	0	0	0.1	11	0.1	0.1	6	0.2	0.1	7	0.2	0.3	202	1
Headworks	188,400	98	600	127	6,600	180	22	1,980	21	774	231	19	75	196	0	+	0	0.1	6	0.2	0.1	3	0.1	0.5	4	5	3	108	37
Primary (NSC)	106,680	105	2,100	137	2,625	100	23	125	10	830	12	3	81	209	260	+	0	0.1	6	0.2	0.1	4	0.1	0.6	4	0.7	4	116	5
Primary (SSC)	53,340	197	2,100	256	2,625	100	43	125	10	1,556	12	3	152	3	260	+	325	0.1	12	0.2	0.1	7	0.1	0.6	7	0.7	4	217	5
Trickling Filter	33,000	362	2,300	471	2,990	230	80	299	160	2,860	208	12	279	3	0	0	0.00	0.3	22	0.5	0.4	13	0.7	0.1	13	0.1	3	398	4
TF Contact Basin	5,600	41	5	53	7	41	9	53	30	324	39	3	32	16	0	0	0.00	0.3	2	0.6	0.9	1	2	0.4	2	0.5	3	45	4
TF Mixed Liquor	3,600	156	5	203	7	28	34	36	13	1,232	17	3	120	3	1,500	+	1,950	0.4	9	0.8	2	5	3	0.2	6	0.2	1	172	2
Activated Sludge	500	589	85	766	111	79	130	103	45	4,653	59	3	454	3	2,100	+	2,730	0.3	35	0.7	0.2	21	0.4	0.2	22	0.3	2	648	2
DAFTs	35,000	62	370	81	740	110	14	220	61	490	122	3	48	3	2,000	+	4,000	0.1	4	0.1	0.1	2	0.1	0.5	2	0.9	3	68	7
Dewatering	81,000	133	32	173	64	5	29	10	5	1,051	10	3	102	5	2,800	+	5,600	0.1	8	0.1	0.0	5	0.0	0.7	5	1	5	146	9
Truckloading	2,280	423	82	550	164	84	93	168	230	3,342	460	15	326	30	29,000	+	58,000	0.1	25	0.2	0.1	15	0.2	0.1	16	0.2	4	465	8

NOTES:

Nuisance Criteria - the concentration of each odorant at which it becomes a nuisance. This has been determined to be equivalent to an OPM odor intensity of 3 (I = 3).

* The four stacks at the two P1 Trickling Filters discharge a total of 48,000 cfm. However, the 43,260 ft2 of open tops generate an estimated additional emission of 347,810 cfm.

** These have no possibility of discharge to the atmosphere, but have been included here to simulate the odorants being emitted at the collections system (i.e. to design foul air treatment systems at pump stations). However, a new location dilution factor would have to be determined.

+ Ammonia has a very high odor threshold concentration. As a result TOTCs are too large to indicate. However, SCAQMD requires Ammonia emissions to not exceed 5,000 ppbV daily average at Dewatering and Truckloading.

LDF = location dilution factor calculated by the air dispersion model which does not take into account possible synergistic effects of other sources.

Max = maximum observed odorant concentration from multiple testing of the foul air at the source (ppbV).

TOTC = theoretical odorant target concentration from modelling. This is the concentration that needs to be achieved at the source to avoid nuisance at the fenceline (ppbV).

Design = adjusted odorant emissions for daily and seasonal variations (ppbV).

The highlighted figures represent the odorants and its sources that require abatement (Design > TOTC). However, modelling may show that odor abatement is needed at certain non-highlighted locations as well.

Fecal odorants (Skatole and Indole) appear to underrepresent this category of odors. Analytical results do not reflect the sensory Odor Profile Method results. It is possible that another compound exists that smells like fecal.

Whenever a plant process change occurs, a new characterization of the foul air must be conducted and a new data analysis must be prepared to determine the extent of the nuisance and the treatment needed. An adequate foul air characterization includes:

Odorant	OTC (ppbV)	Sampling Technique	Analytical Technique	MRL (ppbV)
Hydrogen Sulfide	0.51	Grab sample into 1 L Tedlar bag	ASTM D5504 Method by Modified GC/MS with sulfur chemiluminescence	5.0
Methyl Mercaptan	0.077			5.0
Dimethyl Disulfide	0.22			2.5
Dimethyl Sulfide	3.0			5.0
Ammonia	1,300	0.5 L/min rate for 60 min.	OSHA ID-188/ID-164	0.26
D/T & R/T	N/A	Grab sample into 1L Tedlar bag	ASTM E679 & EN13725	N/A
OPM	N/A	Grab sample into 10L Teflon bag	Standard Method No. 2170 applied to air	N/A
Methyl Isoborneol	0.02	Grab sample into 10L Teflon bag	GC/MS	0.06
Isopropyl Metoxypyrazine	0.004			0.06
Skatole	0.02			0.11
Indole	0.5			0.12

NOTES:

Tedlar bags have been shown to sorb fecal odorants and to impart a residual "chemical" odor to the samples. If possible use Teflon bags for all sampling instead.

OTC = Odor Threshold Concentration - the concentration at which each lab quality odorant from a standard is detected by 50% of the panelist.

MRL = Method Reporting Limit - the minimum detectable concentration by the referenced Standard Method for the determination of that odorant.

D/T = Dilution to Threshold - number of dilutions needed to render the foul air barely detectable by 50% of the panelists. Equivalent to an OPM odor intensity of 1 (I = 1).

R/T = Recognition to Threshold - number of dilutions needed to render the foul air barely recognizable by 50% of the panelists. Equivalent to an OPM odor intensity of 4 (I = 4).

OPM = Odor Profile Method - sensory method that determines the concentration of odorants in foul air and reports them in units of odor intensity.

By the use of the Weber-Fechner relationship the OPM odor intensities can be translated into odor concentrations.

When the odorants to be removed and the TOTCs have been established, it is then necessary to determine which odor batement technology would be most effective:

Technology	Odorant Removal Rate (%)					
	H2S	MM	DMS	DMDS	Musty	Fecal
Chemical Scrubber (current old single stage)	92	62	44	66.5	5	45
Chemical Scrubber (current new 2-stage)	98	97	72	75	63	80
Bioscrubber w/foam cubes @ 2 sec. EBGRT (current Biotower)	76	14	15	42	38	7
Biotrickling Filter w/scructured foam @ 15 sec. EBGRT	99.9	21	17	31	15	76
Biotrickling Filter w/structured foam @ 30 sec. EBGRT	99.9	78.5	42	84	38	90
Biofilter w/engineered media @ 30 sec. EBGRT	99.9	91	66	97	5	88
Biofilter w/engineered. media @ 45 sec. EBGRT	99.9	98.7	92	95.5	31	92.5
Activated Carbon (virgin coconut, granular)	98.7	92.7	25	-141	-25	84

NOTES:

Chemical scrubber = caustic/bleach mode with scrubbant pH > 9 and makeup water rate of 5 to 8 gpm.

Bioscrubber = recirculating mode at 1 gpm operated with scrubbant pH < 2 with 3.5 gpm makeup water.

Biotrickling Filter = foam & reticulated foam arrangement, once-through at 0.8 gpm makeup water for 5 minutes at 20 minute interv

Biofilter = lava rock or engineered media, once-through mode at 0.8 gpm makeup water for 5 minutes at 20 minute intervals.

Activated Carbon operated with raw foul air (worse-case condition). Note transformation of MM into DMDS.

Literature Cited

APHA, 2012. Standard methods for the examination of water and wastewater: Method 2170, 22nd ed. American Public Health Association (APHA), American Water Works Association (AWWA), and Water Environment Federation (WEF), Washington, DC. USA.

ASTM E253, 2015. Standard Terminology Relating to Sensory Evaluation of Materials and Products. American Society for Testing and Materials, West Conshohoken, PA.

ASTM E544, 2004. Standard practices for referencing suprathreshold odor intensity ASTM Standard. American Society for Testing and Materials, West Conshohoken, PA.

ASTM E544, 2010. Standard practices for referencing suprathreshold odor intensity. American Society for Testing and Materials, West Conshohocken, PA.

ASTM E679, 2011. Standard practice for determination of odor and taste thresholds by a forced-choice ascending concentration series method of limits. American Society for Testing and Materials, West Conshohocken, PA. pp. 1-7.

ASTM STP 440 Method, 1968. Correlation of subjective-objective methods in the study of odors and taste. American Society for Testing and Materials, West Conshohocken, PA. pp.

Burlingame, G. A., Suffet, I. H., Khiari, D., and Bruchet, A., 2004. Development of an odor wheel classification scheme for wastewater. Water Science and Technology, 49: 201-209.

Curren, J., C. L. Snyder, S. Abraham, and I. H. Suffet. Comparison of two standard odor intensity evaluation methods for odor problems in air or water. Water Science & Technology, 69.1: 142-146, 2014.

Doty, R. L., Shaman, P., Kimmelman, C. P., and Dann, M. S., 1984. University of Pennsylvania smell identification test: a rapid quantitative olfactory test for the clinic. Laryngoscope, 94: 176-178.

EN13725, 2003. Air quality – determination of odor concentration by dynamic olfactometry. Original ion French. European Commission on Normalization, Brussels, Belgium.

Gabriel, D., Cox, H. H. J., and Deshusses, M. A. 2004. Conversion of full-scale wet scrubbers to biotrickling filters for H_2S control at publicly owned treatment works. Journal of Environmental Engineering. Vol. 130, No. 10. 1110-1117.

Gostelow, P., Parsons, S. A., Stuetz, R. M., 2001. Odor measurement for sewage treatment works. Water Research, 35: 579-597.

Khiari, D., Brenner, L., Burlingame, G. A., and Suffet, I. H., 1992. Sensory gas chromatography for evaluation of taste and odor events in drinking water. Water Science and Technology, 25: 97-104.

OCSD, 2017. Odor Control Master Plan. Orange County Sanitation District, Fountain Valley, CA. CH2M-Hill Corp.

Ruth, J. H., 1986. Odor thresholds and irritation levels of several chemical substances: a review. American Industrial Hygiene Association Journal, 47: A-142-151.

Shammay, A., Sivret, E. C., Le-Minh, N., Lebrero, R., Evanson, I., and Stuetz, R. M. 2016. Review of odour abatement in sewer networks. Journal of Environmental Chemical Engineering. Vol. 4, Issue 4, Part A. 3866-3881.

Standard Method 2170: the Flavor Profile Analysis Method, APHA, et al., "Standard Methods of Water and Wastewater" 2012.

Suffet, I. H., Zhou, Y., and Vitko, T., 2016. How does dynamic olfactometry really work? Water Environment Federation, Odors and Air Pollutants 2016, Milwaukee, WI.

Vitko, T., I. H. Suffet, Y. Zhou, S. Abrahams. 2014 OCSD Odor Control Master Plan. Water Environment Federation. Odors and Air Pollutants 2014, Miami, Florida.

Vitko, T., Cowden, S., Erdal, Z., Witherspoon, J., and Suffet, I. H., 2016-a. Innovative odor mapping and management method sets the stage for targeted foul air treatment. Water Environment Federation, Odors and Air Pollutants 2016, Milwaukee, WI.

Vitko, T., Suffet, I. H., and Zhou, Y., 2016-b. How OCSD identified the most detectable odorants. Water Environment Federation, Odors and Air Pollutants 2016, Milwaukee, WI.

Wang, B., E. C. Sivret, G. Parcsi, X. Wang, N. M. Le, S. Kenny, H. Bustamante, and R. M. Stuetz. Is H_2S a suitable indicator for odor abatement performance of sewer odors? Water Science & Technology, 69.1: 92-98, 2014.

Zhou Y., Braithwaite, S., Vitko, T., and I. H. Suffet, 2016. Investigation of odorant losses in Tedlar, multi-layer foil, and Teflon sampling bags by solid-phase micro extraction with GC-MS analysis. Water Environment Federation Technologies. Creating the Future of Water, New Orleans.

El Paso Implements Odor Control Improvements to Remain a "Good Neighbor"

Richard J. Pope, P.E., BCEE
Anthony L. Chacón, P.E.

ABSTRACT

Located in central El Paso, TX, with a cluster of homes 450 feet from its Headworks, a residential subdivision 1,200 feet from its Dewatering Building, the international border with Mexico on its southern boundary, with long interceptors, and warm weather, El Paso Water Utilities' (EPW's) Haskell WWTP has a long-running, agile posture on odor control. As Haskell's major odor control facilities approached 20 years of service in 2012, EPW initiated investigations into the status of the existing odor control systems to assess their current performance and determine whether the technology was able to meet the odor control demands of the short and long term. On-site odor studies identified the Headworks Building, Dewatering Building, Grit Building, and Primary Clarifiers as having the highest current offsite odor impact. Technology upgrades, moving from the existing mist wet scrubbers to biotrickling filters, were recommended. By 2016, capital upgrades to those four odor control facilities were planned, and 2 of the 4 sites were already under construction when a spat of neighborhood odor complaints from around the collection system and the facility prompted EPW to respond with focused interim O&M mitigation measures and community outreach.

KEYWORDS

Odor Control, Retrofit Odor Control, Bioscrubber, Chemical Addition, Ferrous Chloride, Magnesium Hydroxide, Sodium Hypochlorite, Wastewater Treatment Plant Odor Control, VOCs, Odor Dispersion Modeling, Air Permits, Good Neighbor, Odor Monitoring

INTRODUCTION

El Paso Water (EPW) retained Hazen to design upgrades to the odor control systems and associated appurtenances at the Haskell R. Street Wastewater Treatment plant (HSWWTP)'s Headworks, Dewatering Building and Return Activated Sludge (RAS) Aeration Channel. Previous odor studies at HSWWTP during 2012 and 2013 identified the Headworks and the Dewatering buildings along with the Grit Building and Primary Clarifiers as having some of the foulest odors with the highest offsite impact to the surrounding neighborhoods. EPW has since installed one biotrickling filter at the Aerated Grit Building and has purchased three more to be installed at the Primary Clarifiers. The goal of this project is to improve the odor control capabilities of the Headworks and Dewatering buildings, as well as to control odors within the RAS Aeration Channel in order to mitigate offsite odor impact and to protect plant personnel's health and safety.

Treatment Facility Approaches

During the study phase of this project, Hazen explored building and field modifications which may be required for the installation of new odor control systems and support facilities, and established field constrains impacting equipment layout / installation, including all civil, mechanical, structural, architectural, instrumentation, electrical, and HVAC equipment. Additionally, Hazen performed bench scale testing to determine the most viable chemical addition odor control strategy to address collections system odors in the HSWWTP service area as well as to mitigate odor release during ongoing construction work at the facility. These efforts highlighted to following issues:

- The Headworks and Dewatering buildings currently have a chemical wet scrubber system that uses water-based solutions of sodium hydroxide and sodium hypochlorite to treat the buildings' foul air. The manufacturer of neither system remains in business, so obtaining parts has been increasingly challenging for EPW. The two granular activated carbon odor control units that serve the RAS Channel are in good condition and will only require periodic replacement of the carbon media.
- The Headworks and Dewatering buildings' existing make-up air units (MAU's) and foul air exhaust fans are showing signs of corrosion and are not performing per the original design due to their poor conditions.
- Ongoing projects at HSWWTP pose an increased odor release risk. Odor control chemical addition at the facility has been recommended as an effort to mitigate offsite odor impact during ongoing and future construction projects.
- EPW has received odor complaints within the HSWWTP service area related to the collection system. The numerous complaints warranted an investigation in order to appreciate their severity and to quickly determine whether implementing a collection system chemical addition strategy is warranted.
- This memo describes field investigations findings, presents the chemical addition bench scale test results, presents air dispersion modeling results of baseline and future odor conditions, and provides odor control recommendations as well as a preliminary opinion of probable cost. Additionally, the air dispersion modeling results are described herein.

Smoke Testing

The assessment of the Dewatering Building ventilation system was accomplished through flow testing of ventilation equipment, which was discussed in the previous section, and through smoke testing. Smoke testing is used for visualization of air movement. The smoke inspection was completed by introducing non-toxic smoke into the odor controlled areas and observing how the smoke is captured, contained and transported to the existing odor control systems. Two smoke tests were conducted in the Belt Filter Press Room. Smoke testing in the Dewatering Building revealed that there were many areas where air remained stagnant, particularly above the belt filter presses.

Odor Characterization

Hazen conducted an assessment of current odor characteristics and conditions, and investigated the existing facilities to determine if conditions were adequate to support the new odor control system. During the field investigation, air and wastewater samples were collected and evaluated, and the ductwork, building structural conditions, HVAC, and odor control system conditions were inspected. Additionally, smoke testing and air balancing were performed in order to evaluate the existing odor control capabilities to capture, contain, and convey. Based on these assessments, Hazen recommend improvements. Based on previous odor control studies, the dewatering building appears to be the fourth major contributor to offsite odors. The building has a distinct solids type odor profile which is less hydrogen sulfide related. This is confirmed by the hydrogen sulfide gas levels measured in the belt press room using Odalog monitors.

Collection System Approaches

EPW's operational mitigation measures were to identify and optimize chemical addition upstream in the collection system and attempt to optimize the mist wet scrubber odor control systems at the facility Headworks and Dewatering Buildings that had not yet been replaced. In fact, the wastewater phase chemical addition intervention. phase intervention was already envisioned as a necessary long term odor control step when planned odor-control construction delays compromised containment and treatment of odorous air at the facility and forced it to be implemented sooner. EPW jar tested the effectiveness of several chemical agents, including: hydrogen peroxide; sodium hypochlorite; magnesium hydroxide; and ferrous chloride. The jar test procedure varied the chemical doses on samples collected from the collection system and facility influent to determine the optimal dose of each chemical required for H_2S reductions and the cost-effective chemical to achieve the maximum reduction. EPW then studied H_2S and wastewater sulfide data collected over time along with potential accessible sites that could accommodate chemical addition equipment without burdening the community. The selected chemical addition odor control strategy feeds ferrous chloride into the facility's two main interceptors.

Water Reclamation Facility Approaches

EPW's maintenance mitigation program was designed to inspect the existing odor control facilities against a checklist of mid-term maintenance procedures, procedures that under normal circumstances are reasonably deferred until capital upgrade. This new maintenance regimen prioritized repairs that would make the greatest odor reductions with less cost and/or less time. The attention illuminated other routine maintenance that was promoted in the plant's CMMS. During the field investigations, odor control ductwork, building structural conditions, HVAC and odor control system conditions were inspected. Additionally smoke testing and building air balancing were performed in order to evaluate the existing odor controls strengths, weaknesses, and capabilities, and propose recommended improvements.

Dispersion Modeling

The modeling performed included an evaluation of existing conditions at HWWTP (Base scenario), and an evaluation of the facility with all planned upgrades in place (Future scenario) in comparison to the Texas Administrative Code pertaining to the allowable impact due to emissions of H2S from a facility on residential, business, or commercial property. The dispersion modeling analysis of the HWWTP H2S emissions with the implementation of new and recommended improvements, such as the installation of biotrickling filters at the Grit Building,

Primary Clarifiers and Headworks Building, along with carbon vessels at the Dewatering Building and replacement of spent carbon at the RAS Mixing Channels determined that the offsite impacts from sources at HWWTP will be below the TCEQ 0.08 ppm threshold by an order of magnitude. Because offsite impacts do not exceed the odor standard according to TCEQ, no further measures are recommended at this time to comply with air treatment goals.

Methodology

Dispersion modeling of facility sources was performed using the USEPA and TCEQ-recommended AERMOD dispersion model to predict the potential for H2S impacts offsite from sources at HWWTP for the 30-minute averaging time. AERMOD is an advanced state-of-the-art Gaussian plume model that incorporates updated treatments of the boundary layer theory, turbulence, and terrain. AERMOD can be used to predict ambient air impacts from multiple source types, including areas sources (e.g., open tanks), point sources (e.g. odor control exhaust stacks), and volume sources. AERMOD can also consider the effects of building downwash, stack-tip downwash, and gradual plume rise. The most recent and updated regulatory version of AERMOD (version 16216r), was used for the present study. The modeling methodology was performed following regulatory modeling guidance recommended by the Texas Commission on Environmental Quality (TCEQ) and the United States Environmental Protection Agency (USEPA).

The dispersion modeling performed for the HWWTP utilized information provided in the 2012 AERMOD study (CDM Smith, 2013), including the initial facility setup and source parameters, which were then revised and updated with data gathered from the odor sampling at the HWWTP conducted from November to December 2016. In the present study, H2S emissions emitted from each onsite process were modeled to evaluate the potential H2S offsite impact individually and in combination (i.e., as the whole plant) with respect to the TCEQ requirement under TAC Title 30 Rule§112.31 that offsite H2S impacts to the surrounding residential, commercial, and business properties not exceed 0.080 ppm averaged over any 30-minute period. Following TCEQ guidance, the 1-hour averaged model output was used for comparison to the 30-minute standard.

Model Input

AERMOD requires information about the emission sources, representative surface and upper air meteorology, site data and characteristics (building downwash, urban/rural, etc.), terrain, and receptor data. The model input from the 2012 modeling study was reviewed and specific model input was updated for the present analysis. The present study evaluated the current status of the facility (Base scenario) and also the status of the facility for each proposed odor control and modification put in place (Grit Facilities, Addition of Chemical to Wastewater, Primary Clarifiers, Headworks, Aeration Channel Cover and Dewatering, and Future scenario) with respect to TCEQ regulations.

Community Outreach

EPW's community outreach consisted of publicizing the schedule for odor-reducing collection system and plant modifications and construction. To ensure consistent and sustained messaging, the project entailed regular all-hands status meetings including senior administration personnel from operations and engineering. EPW conducted dispersion modeling to help identify key facility odor sources and their off-site odor impacts, as well as to target its messaging, to project odor impact improvements, and to provide a visual communication tool for "before and after" messaging.

CONCLUSION

Developing an odor master plan for an existing wastewater treatment plant can be completed effectively and successfully when you pay attention to the details when planning the study. Starting with defining the goals and objectives of the study helps to formulate the road map on how to arrive at the finish feeling and looking like a "Good Neighbor". That road map begins with the basics of finding where the odors are and how strong they are. That data is key to creating the foundation for the odor master plan. Employing available tools such as odor modeling helps to independently determine the impact of the odor sources and the mitigation requirements that will be needed to comply with off-site impact criteria. Designing and constructing the retrofit technologies without impacting the normal plant operations and the health and safety of the staff requires special planning and staging. A MOPO plan forces the engineer and plant staff to sit down and carefully consider the steps that will be necessary to install the odor control technologies and to include contingencies. These steps were taken successfully for the HWWTP.

Quality Control of the Containment of Gases in Covered Processes

N.J.R. Kraakman[1], G. Hamer[2], R. Gonzalez-Abraham[1], J. Cesca[2], J. Witherspoon[3]
[1] CH2M/JACOBS, Portland, OR, USA
[2] CH2M/JACOBS, Sydney, NSW, Australia
[3] CH2M/JACOBS, San Francisco, USA

ABSTRACT

Fugitive emissions such as odorous gases are difficult to identify and quantify and can be a significant concern for an odor-producing facility. The emissions from odor-generating processes that are exposed to the ambient air are usually contained with covers. Containment covers inevitably have openings for access requirements and instrumentation and as a result of the wear and tear of covers and their seals over time. While some methodologies have been developed to quantify fugitive emissions, there is no standard method to quantify the capture efficiency of containments at covered processes.

This paper explores the use of a new method to quantify the containment of odor emissions at a water recovering facility (WRF) treating municipal and industrial wastewaters. The method predicts the "Capture Efficiency" (CE), which is defined as the percentage of time that an air extraction system (ventilation system and covers) eliminates the release of fugitive gas emissions. The odor Capture Efficiencies at the covered process units of the WRF are determined three times over a period of several years. This paper demonstrates that the method can successfully be used to verify, optimize and maintain ventilation systems controlling fugitive emissions such as odors from covered processes.

KEYWORDS: Ventilation, Cover systems, Fugitive Emissions, Air Extraction, Capture Efficiency, Odor.

INTRODUCTION

Containment Covers

The emissions from odor-producing processes that are exposed to the ambient air are usually contained with fiberglass reinforced plastic, aluminum, concrete or fabric covers. It is assumed that once a source (e.g. a tank or channel) is covered and connected to an air extraction system, the gases are contained and so no longer a concern. However, as every practitioner of odor control knows, the complete containment of volatile gases is not fully solved with covers and extraction systems and is typically subject to changing conditions like wind speed.

In addition, containment covers inevitably have openings for access requirements and instrumentation to provide the facility's operator with visibility and accessibility. Operators use their senses to assess the influent characteristics and the wastewater treatment performance based on factors such as turbidity, color and smell, while the water levels can change diurnally, seasonally and during storm events. Also, scum can accumulate in a routine fashion and diffused air systems can create recognizable patterns on the surface of aerated channels or tanks. Thus, maintaining operability is usually the biggest hurdle to overcome when designing and installing covers. Furthermore, covers are easily subject to being partly or fully left open during and after an inspection or a maintenance activity, directly losing effectiveness of the ventilation system and resulting in increased risk of health issues, odor nuisance and localized corrosion.

Finally, openings in the covers can result from the wear and tear of covers and seals over time and the number of methods to properly assess containment covers is limited.

Forces Creating Fugitive Emissions

The containment of volatile gases is a function of the negative pressure under a cover. Enough negative pressure must be created to negate the following forces to prevent the escape of gases from under the cover:

> ***Wind*** - The pressure difference created by air movement over the surface of a cover with an opening is usually the most important force to overcome when designing a cover and extraction system. Wind speed over a site has a large influence on the magnitude of leakage from covers and buildings. The wind conditions over a site are a function of the local environment (topography, ground roughness and nearby obstacles).
>
> ***Buoyancy*** - The buoyancy caused by the density difference between the inside and outside of a tank or process unit is the second most important cause of fugitive emissions from a covered process. This is also known as the stack effect. For some applications, this force can be neglected as it will be relatively small when the temperature difference is less than 10 degrees Celsius (6° F) for ambient temperature around 20°C (68 °F). For cold weather conditions, relatively warm process temperatures, or for large distances between the liquid surface and the air vent height, this force increases and will have to be taken into account (e.g. relatively warm digested sludge in a sludge holding tank and buildings containing large thermal loads).
>
> ***Displacement*** - The third force is caused by air or liquid forced into the covered process units (e.g. aeration in tanks or the pumping of liquid into a tank causing the liquid level to increase). The liquid level inside tanks changes when there are differences in the filling and emptying rates. McKetta (1998) notes that odor complaints can be correlated to tank filling events in some cases.

Air needs to be extracted from the covered sources to overcome the sum of the above three forces that will otherwise result in gases escaping from under covers. The total amount of extracted foul air that will have to be extracted (and treated) can be minimized by using properly sealing covers. The effectiveness of covers to contain fugitive emissions is directly related to the size and number of gaps in the covers. The smaller the sum of all opening of gaps in the cover, the more effective a ventilation system can be in maintaining the required negative pressure to overcome the sum of the above three forces.

In summary, the factors that influence fugitive air leaks from covered sources are: wind, buoyancy, displacement, net air extraction and the area of openings for the air to escape.

Quantifying Containment of Fugitive Emissions

When gaseous emissions for a facility need to be considered, it is common practice to focus only on stack and area sources (Bokowa and Liu, 2009), ignoring fugitive emissions due to the difficulty in their identification and quantification. However, this does not mean that fugitive emissions are of minor importance. It is acknowledged (e.g. Schlegelmilch, Streese and Stegmann, 2005) and has been demonstrated (Bokowa and Liu, 2009) that fugitive emissions can be a significant concern for an odor-producing facility. If these emissions are to be dismissed, they first need to be quantified.

As there is no standard method to calculating containment efficiency, quantifying fugitive emissions is the more common approach. To qualify fugitive emissions, various methods have been developed over time. One approach is to measure the concentration downwind (e.g. using field olfactometry) and then back-calculate the emission rate, while taking into account the weather condition at the time of the measurement. Bokowa and Liu (2009) proposed a method to estimate fugitive emission rates that involves air sampling in the wind cavity of a structure associated with a fugitive source and the use of a dispersion model. For an enclosed area, such as a building or a tank, tracer gases such as sulphur hexafluoride, nitrous oxide, or carbon dioxide are proposed. These are non-toxic, chemically inactive, and are not absorbed to building surfaces. The method involves creating a uniform concentration of the tracer gas in the enclosed area, and measuring the concentration decline over time. This allows the air renewal rate to be calculated (Ritcher and Frechen, 2009; Sivil and Hobson, 2009). The air renewal rate can be taken as the fugitive air flow rate from the enclosed area. If the area is connected to a ventilation system, the difference between the air renewal rate and the extraction rate would be the fugitive emission rate. The outcome of these proposed measurements is directly related to the weather conditions, especially wind at the time of the measurement. In addition, the methods explained above involve some degree of specialist technical skill and can be complicated to apply.

This paper explores the use of a new method that is simpler and takes into consideration the changing weather conditions. The containment of odorous gases at covered process units is quantified at a water recovering facility (WRF) treating municipal and industrial wastewaters. The odor Capture Efficiencies at the covered process units of the WRF are determined several times over a period of three years and used as quality control method to verify, optimize and maintain the odor control ventilation system.

METHODOLOGY

Capture Efficiency

As discussed previously in Kraakman et al., (2016) "Capture Efficiency" (CE) is defined as the percentage of time that an air extraction system (ventilation and covers) eliminates fugitive gas emissions from covered processes.

The CE can be predicted and requires relatively few input parameters: the measurement of the negative pressure under the cover under no wind conditions, regional wind speed data, and the topographical conditions around the covered process unit to determine the effect of wind velocity acting on the covers. The air extraction rate at the covered process unit is also measured. The CE of an individual covered process unit is determined as follows:

step 1: The differential pressure over the cover of the process unit (dP) under no wind or calm wind conditions is measured, as well as the foul air extraction rate from the process unit (Q_g).

step 2: The total area of opening (A) or the so-called total leakage area is calculated using Bernoulli's equation:

$$A = Q_g / (C_d * (2 * dP/p)^{1/2}) \quad (2)$$

using a nominal value for the discharge coefficient (C_d) of 0.60 - 0.65.

step 3: The average velocity across the openings (V) is calculated by using equation (3):

$$V = Q_g / A \quad (3)$$

Note: This is the average air velocity across the opening of the cover, caused by all forces except wind. The main force will be the extraction by the fan and could involve the buoyancy or displacement forces.

step 4: Wind speed (V_w) up to where the air velocity across the opening negates the effect of wind above the cover (V_o) is calculated. This is the wind speed at which air containment will be lost due to the pressure difference created by air movement above the cover, using the equation derived by Cadee and Wallis (2007):

$$V_w = V_0 / (C_d * C_{wo}^{1/2}) \quad (4)$$

C_{wo} is the global wind pressure coefficient, which is defined as the fraction of the dynamic pressure of the free wind acting on the outside of a structure (see addendum Negating the Effect of Wind on Covers) as is calculated as:

$$C_{wo} = C_p * (K * z^a)^2 \quad (5)$$

The external pressure coefficient (C_p) is typically 0.9 (AS/NZ 1170.2, 2002). The height of covers above ground level is *z*. The coefficients to include the effect of height and topography on the wind velocity acting on the covers (K and a) are selected from **Table A-1** (see addendum Negating the Effect of Wind on Covers).

step 5: The percent of the time that the air velocity across the opening of the cover negates the effect of wind (V_w) due to factors, such as foul extraction, is derived. The percent of the time that the effect of the local wind is negated is called the CE of an individual covered process unit.

Figure 1 shows the regional wind data, consisting of the statistical percentile of the wind speed (10-minute average is recommended), against the CE. The resulting CE is specific for that covered process unit under the local range of wind conditions.

For example, when the differential pressure under a cover is negative 15 Pascals (measured during calm or no wind conditions) and the extraction rate is 0.5 m^3/s, a total leakage area of 0.2 m^2 is calculated using equation 2. The influence of wind on the containment is determined when the cover height (e.g. 2 metres) and the topographical condition are defined (e.g. *Open Country - many windbreaks* from **Table A-1** in the addendum) and by using equations 3 to 5. In this case, the wind speed (V_W) at which containment of gases will be lost due to the pressure difference created by air movement over the cover is predicted to be about 8.8 m/s. The cover at a location with regional wind characteristics as shown in **Figure 1** results in an odor CE of 95% (95% of the time the foul air will be contained under the cover). However, when the differential pressure over the cover is smaller, (e.g. negative 5 Pascals), the wind speed (V_W) at which containment of gasses will be lost, due to the pressure difference created by air movement over the cover, is only 5.1 m/s. In this instance, the odor CE will only be about 60%, meaning 60% of the time the foul air under the cover will be contained under the range of local wind conditions.

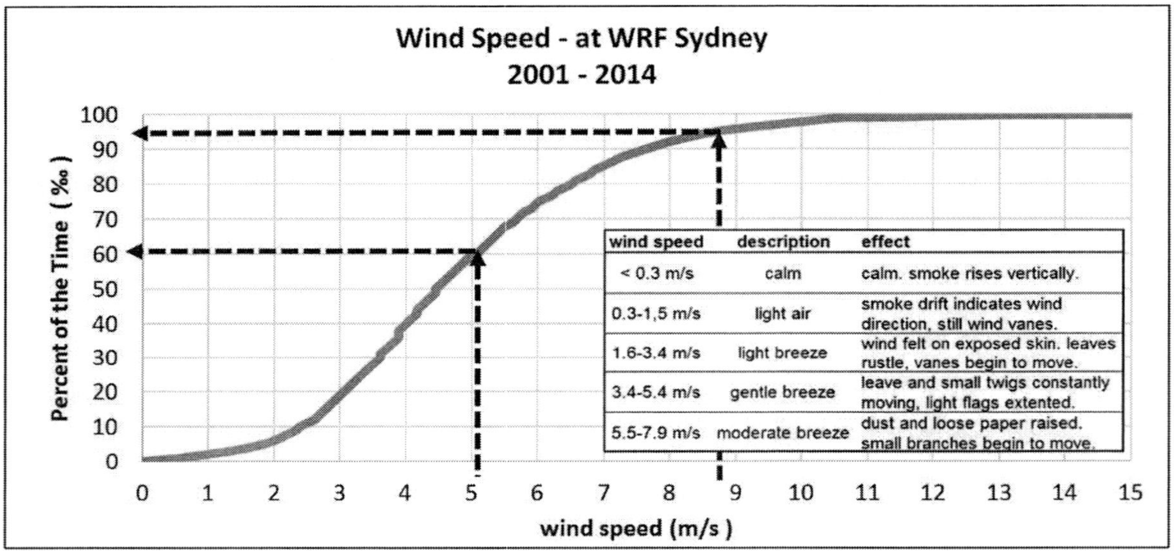

Figure 1: The wind speed for Sydney, Australia, near the Water Recovery Facility as percentage of the time measured over the period 2001 – 2014.

The impact of wind speed (V_W) on a cover is influenced by the topographical conditions. Open spaces with minimal obstructions will have a larger impact than when the wind is slowed down by surrounding obstacles like buildings and trees. **Figure 2** shows the required negative pressures under a cover to overcome the effect of wind for three different typical topographical conditions around WRFs under wind conditions similar to shown in Figure 1. The figure shows the negative pressure required for three topographical conditions when the effect of wind is the force that must be negated. At negative 15 Pascals, the odor CE is more than 99% for the topographical conditions defined as *Rough Country* (typical topographical conditions for inland WWTPs near a small town). This odor CE is only about 87.5% for the topographical conditions defined as *Open Country – few wind breaks* (typical topographical conditions for WRFs along the coastline).

Figure 2 shows the required negative pressure under a cover to overcome the effect of wind under wind conditions similar to Sydney. As previously discussed and quantified in Kraakman et al., (2016), the required negative pressure under a cover is expected to be higher when wind conditions are different and when other forces such as buoyancy and displacement are applicable.

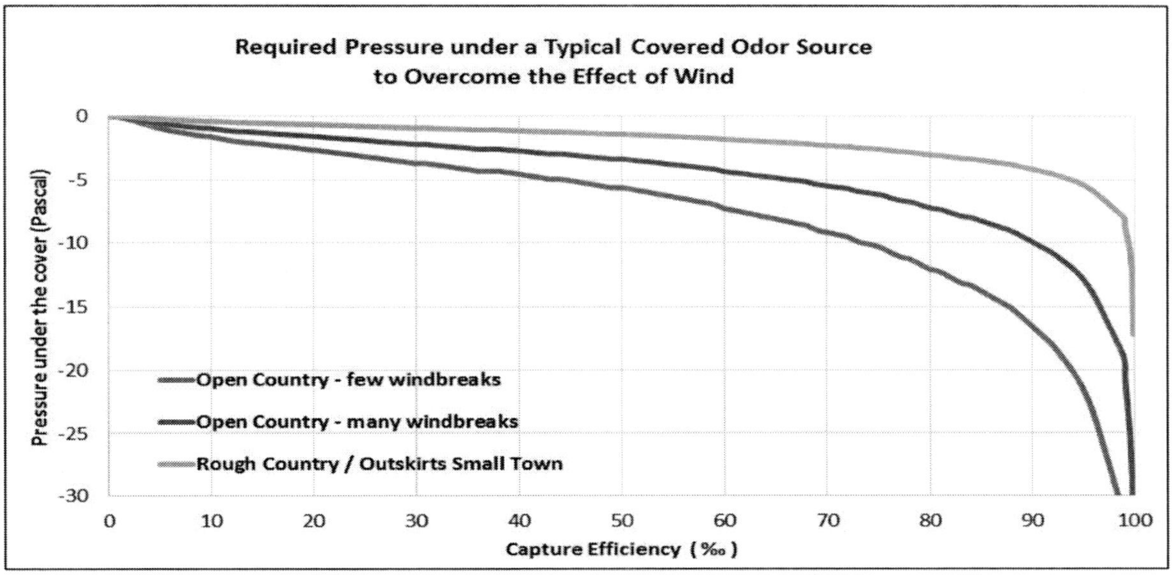

Figure 2: The required negative pressure under a cover for typical topographical conditions around water recovery facilities when only the effect of wind is to be negated.

In conclusion, the wind data together with the local topographical location forms an important basis to determine the required negative pressures under covers to overcome the effect of wind. The overall negative pressure required under covers to eliminate foul air escaping is determined by other multiple site-specific conditions (buoyancy, displacement, and cover leakage area. Finally, the wind speed at which air containment will be lost due to the pressure difference created by air movement above the cover and the CE are estimated.

The process to determine the CE of an individual covered process unit under the local range of wind conditions is visualized in **Figure 3**.

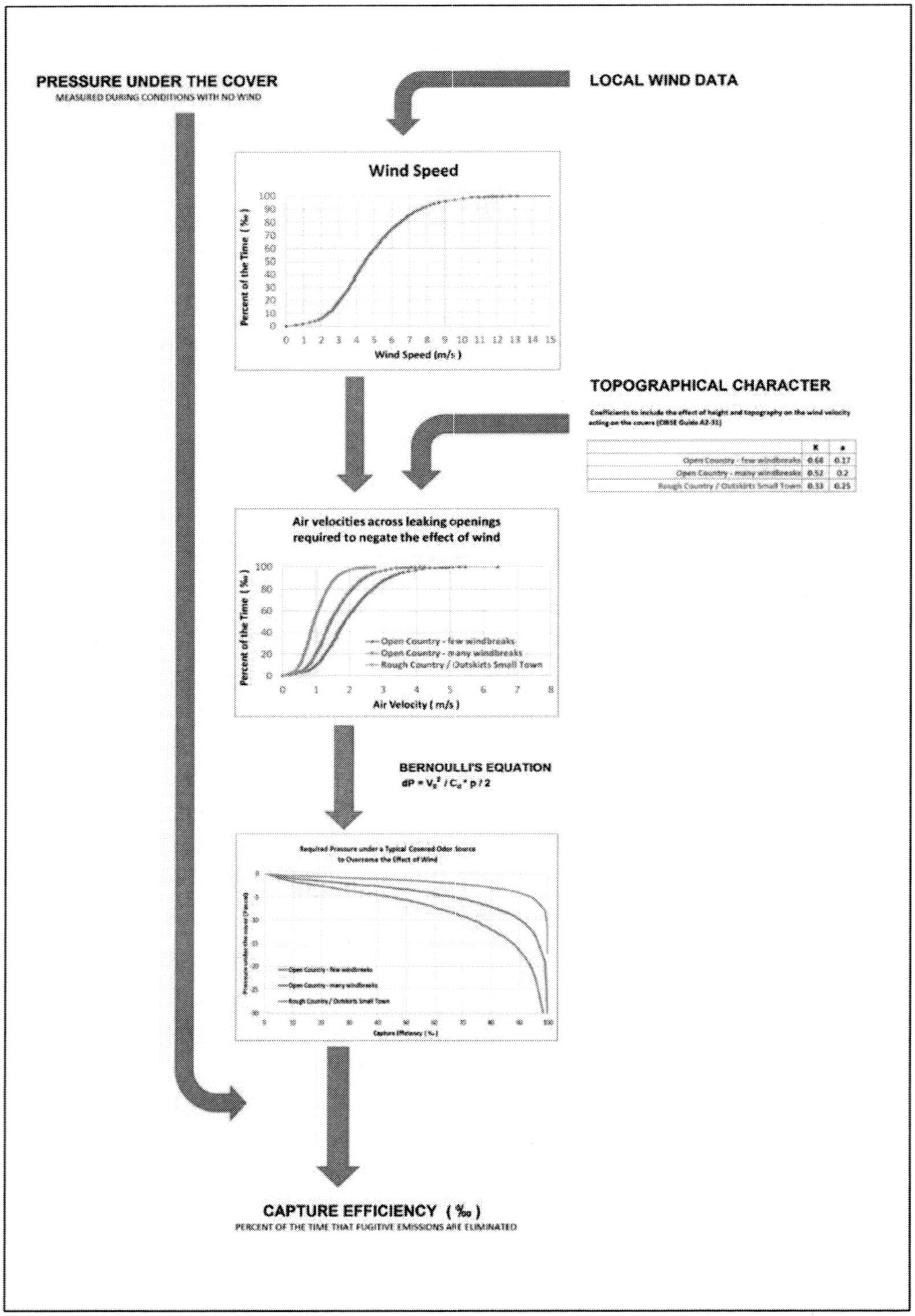

Figure 3: Summary of the method to determine the Capture Efficiency of a covered and ventilated process unit.

Description of the Ventilation System

The studied ventilation system was part of an odor control system treating 45,000 m^3/h (26,500 cfm) of foul air from different process areas of a WRF treating 54 million litres per day (14.3 MGD) of municipal and industrial wastewater in Sydney, Australia. Foul air was extracted from a covered area totaling 2,989 m^2 (105,556 scf), which includes the Inlet Works, the Primary Settling Tanks (PSTs), the Overflow Storage Area and the Interstage Pumping Station. Concerns about the potential impact of odors on a new residential development near the WRF led to a yearly quality control of all potential site odor sources, including the foul air capture performance of the foul air extraction system.

Multiple types of covers were used (see **Table 1**). Some areas required regular access and needed to be trafficable, while other areas did not require access nor need to be trafficable. The covers were divided up during the construction stage into two types to set the minimum design requirements and to set guidelines for operation and maintenance activities after installation. Type 1 covers are used for frequent inspection and are easy removable. As these covers are frequently removed for inspection, each cover must have its own identity and has a rubber seal. They are built to the needs of the operations crew and the standard practice is to inspect rubber seals every of year and replace when necessary. Type 2 covers are permanent, fixed and not frequently used for inspection. Type 2 fixed covers are not used as a general means of regular operator checks or entry for inspecting any assets under the covers. The maintenance crew removes the covers when required and the standard practice is that any rubber seals, metal strips and permanent sealing are to be renewed when replacing these covers.

The design ventilation rates of the ventilation system were selected to effectively capture odors, and to keep the undercover environment less corrosive for materials such as concrete and steel. The minimum ventilation rate at a process unit is based on multiple criteria, including: a) the concentration of hydrogen sulphide, b) the effectiveness of the covered area seals, c) the number of air changes per hour (ACH) and, d) the air sweep velocity (m/s) in channels. The ventilation rate per unit surface area ($m^3/m^2/h$) and extracted foul air to wastewater flow ratio (unitless) are also used to benchmark and verify the design.

The design basis for the air extraction from the different process units is illustrated in **Table 2**. Design criteria were established to achieve near-complete odor capture while creating enough ventilation under the cover to protect against microbial induced corrosion. The air changes per hour is relatively high at the screening area due to large openings that are typical for screens and high hydrogen sulphide concentrations. The ventilation air velocity over the liquid surface, the so-called sweep velocity, under the covers is typically around 0.10-0.15 m/s. However, in the PSTs a relatively high sweep velocity of 0.20 m/s is used to reduce the risk of concrete corrosion as no corrosion resistant coating of the concrete was applied in the PSTs.

Table 1: The characteristics of the covers at the different covered process units.

Process Unit	cover design	cover material	cover type[1)]	trafficable	inspection hatches
Inlet channel	flat	aluminium	2	Yes	1
around screen (stone trap)	flat	aluminium	1	Yes	5
around screen	flat	aluminium	2	Yes	10 (2 per screen)
grit tanks	barrel arch	FRP	2	No	16 (4 per grit tank)
wet weather by-pass channel	flat	FRP	2	No	4
PST inlet channel	flat	aluminium	2	Yes	24 (3 per channel)
PST drive platform	flat (hinged)	aluminium	1	Yes	16 (2 per channel)
PSTs/Launders	retractable fabric	fabric	1	No	56
PSTs/Launders	flat	aluminium	2	Yes	96
Launders outlet	flat	aluminium	2	Yes	16 (2 each channel)
effluent channel	flat	aluminium	2	Yes	32
wet weather storage area	retractable fabric[2)]	fabric	1	No	2
wet weather storage area	flat	aluminium	2	Yes	2
interstage pump station	flat	aluminium	2	Yes	4

1) Type 1 covers are easy removable, not fixed and used for frequent inspection and Type 2 covers are permanent, fixed and not frequently used for inspection.

Table 2: Design basis of the ventilation rate at the different covered processes.

Process Units	Sweep velocity (m/s)	Air Changes (h^{-1})	Ventilation flux ($m^3/m^2/h$)	Gas to Liquid ratio (-)
Inlet channel	0.15	46	64	1.9
Screenings Channel/Screens	NA	63	99	2.9
Outlet Screen Channel	0.15	46	71	2.5
Grit Tanks	0.1	16	13	1.6
PST Inlet Channel	0.15	34	34	1
PSTs	0.2	13	12	18.5
PST Launders	0.1	33	26	6.8
PST Outlet Channel	0.15	27 - 30	53 - 59	1.2 - 2.2
Interstage Overflow Area	NA	8	16	3.7 - 5.5
Interstage Pumping Station	NA	8	16	0.6

After installation, all covers were smoke tested to ensure large gaps were absent. During the commissioning stage, the negative pressure under the covers was measured during no wind conditions. The ventilation system (covers, airflow balancing dampers, ductwork, ventilation compartment curtains, fans) is inspected yearly since commissioning and corrective actions are taken to ensure the system performs optimally. Examples of corrective actions are, the removal of debris under cover hatches, repair or replacement of rubber seals, cleaning of duct drains, repair of cover inspection hatch closure mechanisms, and adjustment of dampers.

RESULTS & DISCUSSION

The quality of the ventilation system was measured three times over a period of three years. The Capture Efficiency (CE) of the individual covered process units under the local range of wind conditions are determined as descripted. For this, the negative pressure of all covered process units was measured under the "no wind" condition. The constant airflow extracted during the measurements was 45,000 Nm^3/s. The measured negative pressures and the predicted CEs are summarized in **Table 3**. The overall CE combines the Capture Efficiency of the individual process units and considers the footprint of each process unit. The footprint is used to weight the CEs of each process unit and calculate the overall CE.

Table 3: The negative pressure and predicted Capture Efficiencies at the covered process units.

Process Units	footprint (m2)	Nov-14 measured pressure (Pa)	Nov-14 Odor Capture Efficiency (‰)	Dec-16 measured pressure (Pa)	Dec-16 Odor Capture Efficiency (‰)	Nov-17 measured pressure (Pa)	Nov-17 Odor Capture Efficiency (‰)
Inlet Channel	33	-4.5	61.0	-4.7	63.0	-4.1	57.0
Screenings Channel	30	-4.9	65.0	-5.0	65.0	-4.2	59.0
Outlet Screen Channel	40	-5.6	70.0	-4.9	65.0	-4.9	65.0
Grit Tanks	140	-9.0	87.0	-7.3	80.0	-7.1	79.0
PST Inlet Channel	32	-9.4	88.0	-7.5	81.0	-8.7	86.0
PSTs	2299	-21.4	99.1	-21.3	99.1	-19.6	99.0
PST Outlet Channel	67	-20.4	99.0	-19.6	99.0	-18.7	98.0
Interstage Overflow Area	257	-22.9	99.3	-20.6	99.1	-18.1	98.0
Interstage Pumping Station	91	-20.1	99.0	-19.4	99.0	-18.6	98.0
Overall	**2989**		**97.3**		**96.8**		**96.5**

Many of the covers were newly installed as part of the odor control improvement works at the plant just prior 2014. In November 2014, the overall Capture Efficiency was around 97.3 percent. Two years later the overall Capture Efficiency slightly decreased to 96.8 percent and three years later to 96.5 percent.

It can be concluded that fugitive odor emissions from the Inlet Works (Inlet Channel, Screenings Channel, Outlet Screen Channel) is eliminated only about 60 to 70 percent of the time. The negative pressure was about negative 5 Pascals and becoming slightly lower over time (see also **Figure 4**).

The PSTs and Interstage Overflow Area and Interstage Pumping Area had significantly lower pressures with less than negative 20 Pascals eliminating fugitive emissions more of 99 percent of the time. These areas also became slightly lower in negative pressure over time, but still resulting in a Capture Efficiency of 98 % (**Figure 5**).

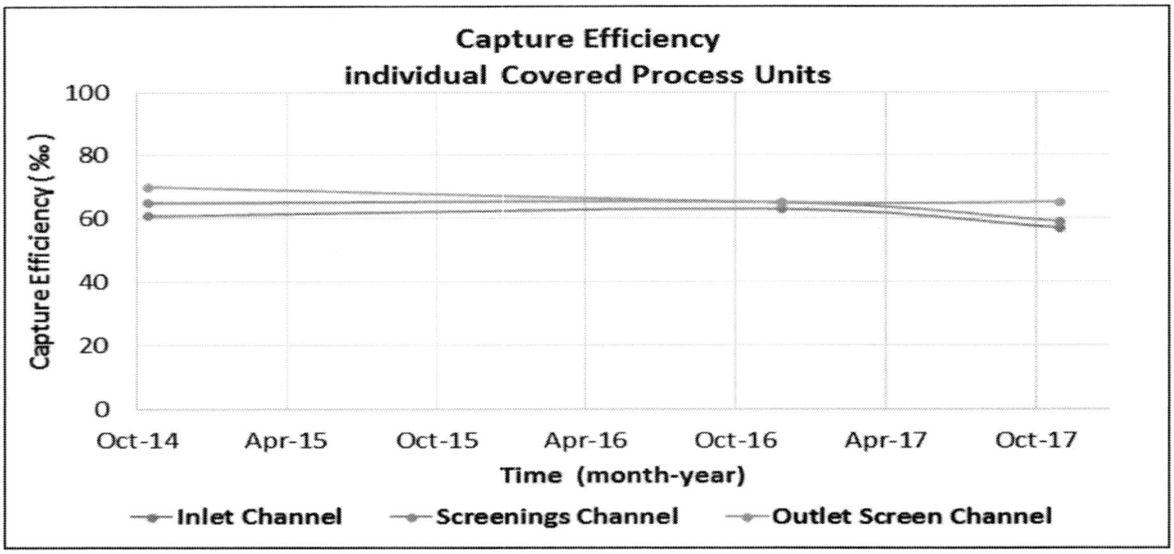

Figure 4: The Capture Efficiency at the Inlet Works.

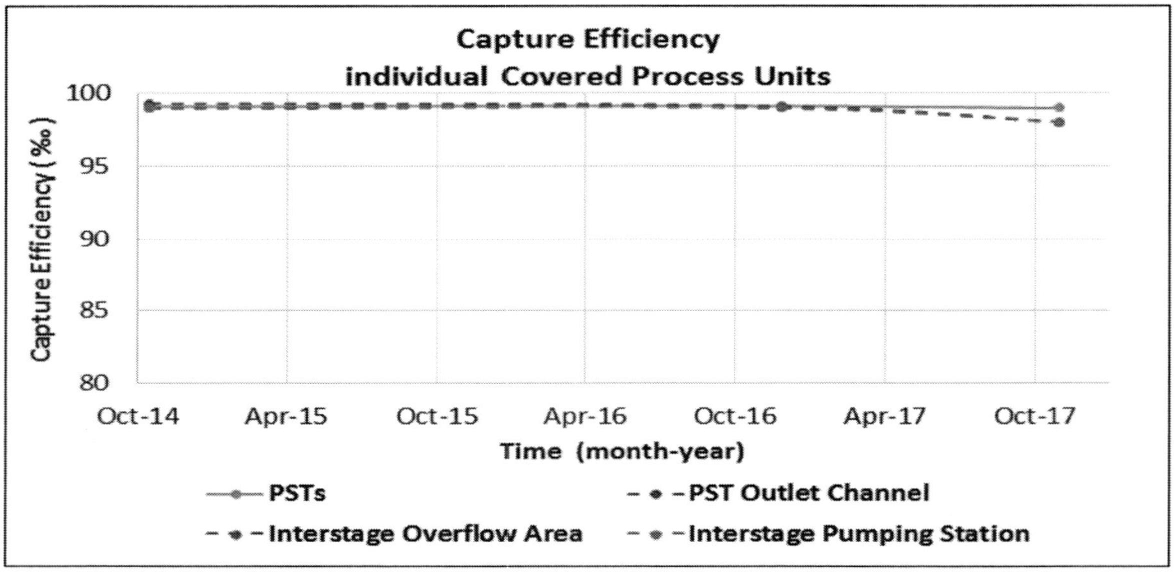

Figure 5: The Capture Efficiency at the PSTs and Interstage Areas.

The negative pressure under the covers of the Inlet Works is significantly lower than the other covered process units at the facility. A possible explanation is that most of the old covers at the Inlet Works were re-used with only the rubber seals being replaced during the odor control improvement works at the plant just prior 2014.

The covered area of the Inlet Works is relatively small compared to the other covered areas with only 103 m^2 (3,637 scf and less than 4%) of the total covered surface areas of 2,989 m^2 (105,555 scf). At a moderate wind breeze, a slight odor has been observed at the Inlet Works, with no odors being observed at that time around the PSTs and Interstage Areas confirming the measurements above. The areas around the PSTs and Interstage Areas include some of the most turbulent processes with relatively high hydrogen sulphide concentrations generated (e.g. Launders and PST effluent channel).

The least negative pressures (i.e. weaker containment) were obtained in the highest ventilated areas where existing covers were used (Inlet Works). On the other hand, with the relatively lower ventilation rates, the covers around the PSTs and Interstage Areas would maintain a more negative pressure (i.e. stronger containment) under the covers for these process units. Ventilation rate rules-of-thumb, such as the number of Air Changes per Hour, are therefore not a guarantee that fugitive emissions are being prevented. A certain number of Air Changes per Hour for a specific process unit preventing fugitive emissions could generate fugitive emission at another process unit when different covers are installed. In this case, the most negative pressures were measured at the areas with the lowest ventilation rates (PSTs and Interstage Overflow & Pumping Area).

The measured negative pressures and predicted Capture Efficiencies of this study are in line with the negative pressure reported of at least -7 Pascals required under low wind speeds (Degremont Handbook, 2007), and more than -20 Pascals required under more extreme wind conditions (Cadee and Wallis, 2007).

The concept of Capture Efficiency of a ventilation system can be used for the technical design of extraction systems as well as for cost optimization to obtain the most cost-efficient containment of volatile emissions (Kraakman et al., 2016). As shown in this paper, it can also be used as quality control method for existing cover systems and their ventilation to verify, optimize and maintain containment of fugitive emissions over time.

CONCLUSIONS

"Capture Efficiency" can be used as a quality control method for existing ventilation systems at covered process units to verify, optimize and maintain containment of fugitive emissions over time.

The Capture Efficiency can be predicted using data that is typically collected at a WRF, or that can be measured with a basic handheld manometer. The CE requires three input parameters: the measurement of the negative pressure under the cover under no wind conditions, regional wind speed data and the topographical conditions around the covered process unit to estimate the effect of wind velocity acting on the covers.

The overall Capture Efficiency of the ventilation system considered in this case study was initially around 97.3 percent and slightly decreased to 96.5 percent over a period of three years. The Inlet Works had the least negative pressure (i.e. weakest containment) despite the relatively high ventilation rates (ACH), resulting in a predicted Capture Efficiency of 60-70 percent, while the large areas of the PSTs and Interstage Overflow & Pumping Area had significantly lower negative pressures (i.e. strongest containment) resulting in a predicted Capture Efficiency of 99 percent. The results are in-line with onsite odor observations and reported negative pressures required under covers.

REFERENCES:

AS/NZ 1170.2 (2002). Structural Design Actions – Wind Action.

Bokowa, A., & Liu, H. (2009). Quantification of Odour Emissions from Fugitive Sources. *Odours and VOCs: Measurement, Regulation and Control Techniques*, *31*, 123.

Cadee, K. and I. Wallis (2007). Odour Containment and Ventilation at Perth's Major WWTPs. Water, March: 54-60.

CIBSE (1999). Guide A2-31: Environmental Design. Chartered Institute of Building Services Engineers.

Degremont Handbook (2007) 7th English edition.

Kraakman, N.J.R., Cesca, J., Hamer, G., Witherspoon, J. (2016) Design and Implementation of Air Extraction using the *Odour Capture Performance* for the Containment of Odours. 7th IWA Conference on Odours and Air Emissions. March 2017. Paris, France.

McKetta, J.J, 1998. "Encyclopedia of Chemical Processing and Design: Volume 65 -- Waste: Nuclear Reprocessing and Treatment Technologies to Wastewater Treatment: Multilateral Approach", CRC Press, Aug 21, 1998.

Schlegelmilch, M., Streese, J., & Stegmann, R. (2005). Odour management and treatment technologies: an overview. *Waste management*, *25*(9), 928-939.

Sivil, D. and Hobson, J. A. (2009) Odour, Covering and Ventilation. *Water Science & Technology*, 59 (7): 1377–1384.

ADDENDUM

Negating the Effect of Wind on Covers

Containment of fugitive emissions is achieved by maintaining a negative pressure over the cover of a process unit.

$$dP = P_{inside} - P_{outside} \quad (6)$$

An important force to overcome (for maintaining containment) is the pressure difference created by air movement over the surface of the cover.

$$dP_{wind} = P_{inside} - P_{outside} \quad (7)$$

The pressure on the outside surface of the cover ($P_{outside}$) may be calculated as:

$$P_{outside} = P_{atmosphere} - C_p * p * [K_w * V_w]^2 / 2 \quad (8)$$

with C_p the external pressure coefficient (-), V_w the free wind speed measured at the standard height of 10 meters (m/s) and K_w the velocity coefficient related to the topography and height of a structure (-).

Thus when combining equations (7) and (8)

$$dP_{wind} = P_{inside} - P_{atmosphere} - C_p * p * [K_w * V_w]^2 / 2 \quad (9)$$

$$dP_{wind} = dP_{measured\ over\ cover} - C_p * p * [K_w * V_w]^2 / 2 \quad (10)$$

The velocity across an opening (V) may be derived from Bernoulli's equation:

$$V = C_d*(2*dP/p)^{1/2} \quad (11)$$

The velocity (V_0) across an opening that negates the effect of wind on the cover ($dP_{wind} = 0$) is given by combining the last two equations above:

$$V_0 = C_d * C_p^{1/2} * K_w * V_w \quad (12)$$

or

$$\boldsymbol{V_o = C_d * C_{wo}^{1/2} * V_w} \quad (13)$$

with the global wind pressure coefficient (C_{wo}), defined as the fraction of the dynamic pressure of the free wind acting on the outside of a structure. Defining further:

$$C_{wo} = C_p * (K * z^a)^2 \quad \text{and} \quad K_w = K * z^a$$

with the external pressure coefficient (C_p) being typically 0.9 (AS/NZ 1170.2,2002), and the height of covers above ground level (z) is specific to each case and the coefficients to include the effect of height and topography on the wind velocity acting on the covers (K and a) are referenced from Table A-1.

Table A-1: The coefficients to include the effect of height and topography on the wind velocity acting on the covers (CIBSE, 1999).

Topographical condition	K	a
Open Country - few windbreaks	0.68	0.17
Open Country - many windbreaks	0.52	0.2
Rough Country / Outskirts Small Town	0.33	0.25
City Centre	0.21	0.33

Odor Control Scrubber System Planning and Design with Limited Source Specific Data

Dirk Apgar, PE, PMP
King County Wastewater Treatment Division
201 South Jackson Street
Seattle, Washington 98104

ABSTRACT

The design of gas phase odor control scrubber systems should be based on known parameters. These include the overall odor concentration, odorous compounds present and their concentrations, the airflow rate to be treated and the location, height and diameter of the exhaust stack as well as adjacent building dimensions, historic meteorology and local topography. However, these parameters are not always known in the planning phase of a project and many still many may not be specifically known when designs must be produced. This is particularly true on budget or time limited projects when resources are not available to conduct odor source data collection. In these cases it is possible to move forward with limited specific knowledge by using information available in literature, accepted modeling methods and conservative assumptions. This paper will describe sufficient methods that will allow for early project planning and the design of single point of emission odor scrubber systems. These information sources, methods and a discussion of how to conservatively apply them is presented.

KEY WORDS: Odor, Modeling, Hydrogen Sulfide, Scrubber

INTRODUCTION

In the ideal case, decisions about the need for, and design of, odor control systems would be based on complete knowledge of all important parameters. However, even in the best real-world case, decisions must always be made with incomplete knowledge and in some cases with very limited knowledge. These situations are typical during very early planning phases for new wastewater facilities or when limited budget and/or schedule prevent the gathering of data. However, it is still possible to move forward with planning and even design of relatively simple odor emission sources or control systems.

It would always be best to obtain information on odorous compounds present and their concentrations through source testing. However, without source specific data sufficient information can often be gathered from the literature on wastewater odors such as conference papers and works of the Water Environment Federation, American Society of Civil Engineers and others. Using these sources and making conservative estimates can provide sufficient information and justification to move forward with planning and design of scrubber systems that provide robust odor removal and control.

Airflow rates for treatment can be developed based on the process area room ventilation rates required to maintain a hazardous area classification, worker safety and comfort, and to prevent the escape of untreated foul air. Once the airflow is estimated, dispersion modeling can be conducted to estimate the amount of atmospheric dilution that will occur between the point of emission and downwind impacts. This allows the maximum allowable exhaust stack odor concentration to be calculated and the evaluation of alternative technologies that can be used to reduce the source odor intensity prior to emission to the atmosphere.

INFORMATION NEEDED FOR PLANNING AND/OR DESIGN

In order to investigate the need for and/or plan and design an odor scrubber system basic information is required. This information includes the following:

- Odorous compounds present
- Concentrations of odorous compounds present
- Odor concentration in terms of Dilutions-to-Threshold (D/T)
- Volumetric airflow requiring treatment
- Location, elevation and diameter of the stack at the point of emissions
- Geometry of buildings at the release location
- Meteorology at the release location
- Local topography
- The acceptable downwind impact

If no source specific information is available information in the literature can be used with suitably conservative assumptions to proceed with evaluation and design of your project.

Odorous compounds potentially present

It is preferable to have information available on the speciation and concentration of odorous compounds present in the foul air source of concern. Source sampling and analysis by methods such as ASTM D-5504 for reduced sulfur compounds and the Environmental Protection Agency's method TO-15 for volatile organic compounds are particularly useful to obtain a list of compounds present and their concentrations (at the time of sampling). Monitoring of hydrogen sulfide concentrations using commercially available data logging devices can be very useful to obtain an understanding of the hydrogen sulfide (H_2S) concentration and variability over time. If it is not possible to obtain data using these methods, there are many sources of information on the species of odorous compounds present in wastewater facilities. Several of these that have proven useful are:

- Water Environment Federation Manual of Practice 25 – Control of Odors and Emissions from Wastewater Treatment Plants (WEF, 2004)

- American Society of Civil Engineers Manual of Practice 69 – Sulfide in Wastewater Collection and Treatment Systems (ASCE, 1989)
- American Society of Civil Engineers Manual of Practice 82 – Odor Control in Wastewater Treatment Plants (ASCE, 1995)
- WERF Study – Minimization of Odor and Corrosion in Collection Systems (WERF, 2007)

These sources are useful when considering the compounds that may be present and provide a reasonable first estimate. While each of these sources lists compounds that may be present, and additional information such as detection and recognitions thresholds, none provide information on potential or even typical concentrations of those compounds found in samples of wastewater fouled air. Because odorous compound concentrations very substantially between even very similar processes it is not practical to describe typical compound-specific concentration values. However, overall odor and H_2S concentrations can be sufficient for planning and design. A method of estimating H_2S concentrations is presented later in this paper. One paper that provides estimates of odor concentrations for various wastewater processes based on years of sample analysis has proven very useful. It is entitled Odor Threshold Emission Factors for Common WTP Processes (McGinley, 2008).

The McGinley paper provides odor data for many wastewater processes based on laboratory analysis of multiple samples from similar sources at different facilities. The mean, minimum, maximum, 1st and 3rd quartile DT values are reported and provide a defendable basis for planning or design of an odor control system when source specific data is unavailable. Figure 1 below is a recreation of Table 2 from the McGinley paper.

Figure 1: Odor values based on laboratory analysis of samples from conveyance system facilities (McGinley 2008).

Code	Category	n	Geo. Mean	Min.	Max.	1st Quart.	Median	3rd Quart.
120	Pump Stations	15	639	35	5,488	175	1,100	1,988
130	Wet Wells	17	2,245	338	33,000	1,116	2,963	3,650

If no odor data can be obtained for the source of concern, odor concentrations for the source category from the McGinleys' paper can be used. When considering a source of odors, the median odor concentration can be assumed for a typical process under normal conditions and the 3rd quartile value could be used as a more conservative value or as a likely maximum (McGinley, 2008). In hot, dry climates where hydrogen sulfide is more readily generated a higher value may be considered than if the source of interest is in a cool wet climate. If more conservatism is desired a higher value between the 3rd quartile and maximum reported may be assumed. This could lead to an overly conservative estimate and increase cost of an installed odor control system unnecessarily. Ultimately, it is up to those planning or designing the odor control system to decide the level of conservatism warranted based on the risk tolerance of the owner/operator.

Estimation of Foul Airflow Rate Requiring Treatment

Several issues must be considered when deciding how much airflow to evacuate and treat with an odor control system. In no particular order of importance these include:

- Hazardous area (sometimes referred to as electrical) classification
- Worker safety/hazardous gas exposure
- Corrosion control
- Condensation control
- Airflow into the classified space that must also be treated

Hazardous Area Classification

The National Fire Protection Association (NFPA) publishes NFPA 820 – Standard for Fire Protection in Wastewater Treatment and Collection Facilities (NFPA, 2016). As the name states, its primary focus is protection of the facility (and presumably occupants) from fire. It is not meant to be an industrial hygiene guide to protect workers from hazardous atmospheres. NFPA 820 is a "Standard", and not part of an official code unless specifically adopted directly or by reference by the legal authority having jurisdiction. However, regardless of its formal adoption it does serve as a means of selecting minimum airflows to evacuate and treat depending on the hazardous area classification desired and the type of equipment that will be installed in the facility. These minimum airflows may need to be increased due to other considerations and will be discussed below.

NFPA 820 provides a systematic way to select a minimum airflow to evacuate from an area by stepping through the standard's tables. The users must familiarize themselves with the text also to verify proper interpretation. The basic method is to identify the source requiring foul air evacuation and the desired classification and, using the standard's tables, find the necessary airflow rate. The rate is typically twelve, six or no air changes per hour (AC/hr.) depending on the classification. In addition to evacuating foul air, fresh supply air must be provided for some areas. The reader must familiarize themselves with NFPA 820 and other jurisdictional codes to ensure they are designing to local legal requirements.

As an example, consider a conveyance system pump station wet well with internal dimensions of 10ft X 20ft X 15ft = 3,000 cubic feet. For this to be considered a Class 1/Division 2/Group D area, 12 air changes per hour would be required. The volumetric flow rate, Q, would be calculated as in equation 1 below.

$$Q = 3{,}000\ cubic\ feet * \frac{12}{hr} * \frac{1hr}{60min} = 600cfm \qquad \textbf{Eq. 1}$$

In this simple case, maintaining 12 (or more) air changes per hour would allow the use of equipment that was not explosion proof. If a lower airflow was desired explosion proof equipment would be required and it must be maintained in that condition. This usually requires a higher initial costs and more ongoing maintenance. The owner of the facility should be made aware of these facts so they can make an informed decision regarding design options and cost ramifications.

Worker Hazardous Gas Exposure

As stated above, NFPA 820 is not an industrial hygiene guide meant to help protect workers from hazardous atmospheres or workplace dangers other than fires. As such, the minimum airflows found using the document may need to be increased if dangerous levels of H_2S or other gases may be present. Calculating such an airflow can be particularly difficult if no empirical data for the area is available because it is very difficult to predict the flux of hazardous gas from wastewater into an airspace. The fact that there is little control of the variables that promote the production of sulfide and what materials may be discharged into a sewer results in the inability to accurately predict the presence of hazardous atmospheres. As a general rule, H_2S should be kept at concentrations less than 10 PPMv but the reader is cautioned to verify local rules and regulations when considering a target concentration for design purposes. It is recommended that continuous gas monitoring for H_2S, CO, O_2 and Lower Explosion Limit (LEL) be provided that will allow sufficient warning of hazardous conditions for the area. If continuous fixed monitoring and warning is not available, workers should carry personal monitors that can detect these gases and warn them of hazardous conditions.

Corrosion Control

Hydrogen sulfide gas is oxidized to sulfuric acid by thiobacillus bacteria living in moist surfaces above wastewater. The acid reacts with the matrix of concrete commonly composed of calcium silicate hydrate gel (CSHG), calcium carbonate and unreacted calcium hydroxide in concrete structures to form gypsum ($CaSO_4$) as shown in Equation 2A through 2C below (WEF, 2001).

$$H_2SO_4 + CaSi \rightarrow CaSO_4 + Si + 2H^+ \quad \textbf{Eq. 2A}$$

$$H_2SO_4 + CaCO_3 \rightarrow CaSO_4 + H_2CO_3 \quad \textbf{Eq. 2B}$$

$$H_2SO_4 + Ca(OH)_2 \rightarrow CaSO_4 + 2H_2O \quad \textbf{Eq. 2C}$$

The rate of corrosion is dependent on the amount of H_2S present so lowering its concentration by dilution with fresh air from outside the wastewater system (or preventing its formation by liquid phase chemical addition) can slow down or eliminate the problem. As mentioned above, it is difficult to accurately predict the amount of H_2S present so it is difficult to predict the amount of additional airflow required to keep corrosion rates acceptably low. Imperfect methods of predicting sulfide production in wastewater systems and gas phase concentrations are available. One method is described later in this paper. Actual gas monitoring should be conducted after a system is in operation to verify actual gas concentrations and vigilance must follow to observe actual rates of corrosion. If corrosion rates are found to be unacceptably high increases in airflow or other methods of sulfide and corrosion control may be required. These include chemical addition to wastewater for sulfide control and coatings or linings to protect concrete from contact with the sulfuric acid produced.

Condensation Control

In addition to increasing the airflow to lower the possibility of hazardous gas exposure or corrosion, airflows may need to be increased above NFPA 820 requirements to lower humidity and/or temperature levels within the ventilated classified area. If humidity levels are too high, condensation can result in wet surfaces that can cause slip hazards and a generally unpleasant working environment.

Airflow into the Classified Area

Flowing wastewater will cause air movement due to the viscous drag between the two medium. Various methods can be used to predict the air movement based on wastewater flows (WERF 2007). Work by Pescod and Price (Pescod, 1982) provides such a method which is sufficient for most simple sewers when no empirical information exists. This work is frequently used to describe the sewer airflow phenomenon, at least qualitatively.

NFPA 820 only requires certain air change rates be maintained for a specific hazardous area classification. However, if wastewater fouled air is also entering the area from a sewer or other source it is prudent to also add that airflow to the evacuation rate.

The movement of water in a pipe will cause the air within a sewer to accelerate to an equilibrium velocity. The velocity will reach this equilibrium when the viscous drag force exerted on the air by the water is equal to the drag force exerted on the moving air by the sewer walls and obstacles within the sewer. Where the main obstacle to the moving air is the sewer wall, the air velocity will be at a maximum at the water surface and a minimum at the pipe wall.

The work of Pescod and Price had limitations due to the measurements in a relatively small pipe (300 mm, 11.8 in) and relatively low water flow velocities along with considerable scatter in their data (WERF, 2007). The real value of their studies is that they established a relationship between average air velocity to the water surface velocity and both the air space perimeter and air space hydraulic radius (WERF, 2007). Their work is generally attributed with establishing a rule of thumb, suggesting that the average air velocity in a half full pipe (depth/Diameter = d/D = 0.5) will be between 35% and 50% of the water's surface velocity. Because surface velocity is difficult to predict, the mean velocity, as predicted by Manning's equation, has been used as a surrogate value.

The rule-of-thumb that has been established as a result of Pescod and Price's work indicates the sewer air velocity will be half that of the water velocity. This likely over estimates the true air velocity at other values of d/D. This is because the d/D of 0.5 that they used to measure air velocities provides the greatest area of interface between the air and water and thus the most area over which the viscous drag between the two medium can be exerted.

To predict the wastewater velocity in a pipe we can use Manning's equation for the calculation of open channel flow velocity, shown in Equation 3 below, for a circular pipe shown in Figure 2.

- $V = \frac{\{R_h\}^{2/3} * \{s\}^{1/2}}{n}$ (this is in SI unit version) **Eq. 3**

Where

- s = Slope of the energy line. For long, constant slope pipes the physical slope is close enough
- n = Pipe roughness
  Concrete about 0.012, PVC about 0.010
- R_h = Hydraulic radius = $\frac{D}{4}\left\{1 - \frac{\sin(\alpha)}{\alpha}\right\}$
  (Janna, 1983)
- $\alpha = 2\pi - 2\arccos\left(\frac{2d-D}{D}\right)$

Figure 2: Partially filled pipe

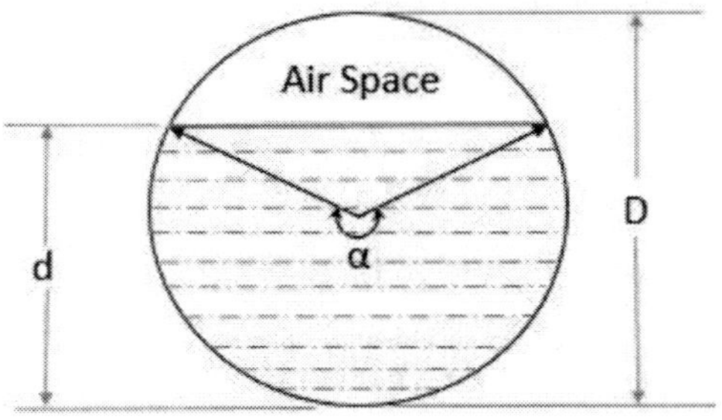

Assuming the depth/pipe diameter = 0.5 for a 0.3m pipe,

$$\alpha = 2\pi - 2\arccos\left(\frac{2(.15m) - 0.3m}{0.3m}\right)$$

$$\alpha = 2\pi - 2\arccos\left(\frac{0.3m - 0.3m}{0.3m}\right) = 2\pi - 2\frac{\pi}{2} = \pi$$

With $\alpha = \pi = 180^o$,

$$R_h = \text{Hydraulic radius} = \frac{0.3}{4}\left\{1 - \frac{\sin(\pi)}{\pi}\right\} = 0.075m\left\{1 - \frac{0}{\pi}\right\} = 0.075m$$

Assuming the slope of the energy line is equal to the pipe slope of 0.02 and the pipe roughness is equal to 0.012, then

$$V_{water} = \frac{\{0.075\}^{2/3} * \{0.02\}^{1/2}}{0.012}$$

$$V_{water} = \frac{\{0.075\}^{2/3} * \{0.02\}^{1/2}}{0.012} = 2.1m/s$$

$$V_{water} = \frac{2.1m}{sec} * \frac{3.28ft}{m}$$

$$V_{water} = 6.9ft/sec$$

Now that we have an estimate of the mean water velocity we can apply the rule of thumb that the mean air velocity will be half that of the water velocity.

$$V_{air} = \frac{1}{2}V_{water} = \frac{1}{2} * 6.9\frac{ft}{sec} = 3.45\frac{ft}{sec}$$

The sewer airflow is then found by multiplying the air velocity by the cross section of the pipe above the water line as shown below. Since d/D = 0.5 in our example 0.3 meter (11.8") pipe the air space is simply half the cross sectional area of the pipe.

$$Q_{air} = V_{water} * \frac{1}{2}\left\{\frac{\pi D^2}{4}\right\} = 3.45\frac{ft}{sec} * \frac{1}{2}\left\{\frac{\pi(11.8"/(12in/ft))^2}{4}\right\} * \frac{60\ sec}{1\ min} \approx 79\ cfm$$

Now that we have an estimate of the airflow rate from the sewer, we can add it to the airflow rate required by NFPA 820 to maintain our hazardous area classification thus

$$Q_{air\ total} = Q_{NFPA\ 820} + Q_{sewer\ air}$$

$$Q_{air\ total} = 600\text{ cfm} + 79\text{ cfm}$$

$$Q_{air\ total} = 679\text{ cfm} \approx 700\text{ cfm}$$

Given the level of accuracy of these calculations it is appropriate and prudent to round up to 700 cfm.

Sulfide Production and Hydrogen Sulfide Gas Phase Estimation

Hydrogen sulfide gas concentration has a significant impact on the amount of consumable media and/or chemicals that must be accounted for in the design of odor control systems. There is no perfect method for predicting the rate of sulfide production and the resulting gas phase concentration. However, the method described in the American Society of Civil Engineers Manual of Practice 69 (ASCE, 1989) based on the method described by Pomeroy and Parkhust (Pomeroy, 1977) provides a reasonable method. It should be noted that it is important to use these two references together as there are several errata or typographical errors in the ASCE publication that can cause confusion. Below, the present author has attempted to clarify the method after combing through both publications.

Pomeroy and Parkhurst developed an empirical relationship between relatively easy to obtain wastewater parameters and sulfide production. While more accurate models may have been available at the time of their work these were dependent on data that was hard or impossible to obtain in a reasonable way when working in a real-world active sewer system. The methodology Pomeroy and Parkhurst produced is fairly simple and allows for sulfide concentration levels to be predicted for both surcharged sewers, including siphons and force mains, and partially filled gravity sewers.

For filled pipes Pomeroy and Parkhurst described the rate of change in sulfide concentration in Equation 4 below (Pomeroy, 1977).

$$\frac{d[S]}{dt} = 3.28M[EBOD][1 + 0.48r]r^{-1} \qquad \textbf{Eq. 4}$$

Where

S = Total sulfide concentration in mg/L

M = Specific sulfide flux coefficient = $3x10^{-4}$ m/hr. (ASCE, 1989).

$EBOD$ = Effective BOD_5 in mg/L = $BOD_5*(1.07)^{T-20}$, where T = temperature degrees centigrade

r = Hydraulic Radius in feet (diameter/4 for a full pipe)

3.28 = conversion factor for converting meters to feet (ASCE, 1989).

t = time in hours

Equation 4 can be rewritten to show that the production of sulfide occurs partially in the slime layer on the pipe wall and partially within the bulk flow of the wastewater as shown in Equations 5 and 6 below (Pomeroy, 1977).

$$\frac{d[S]}{dt} = [sulfide\ from\ slime\ layer] + [sulfide\ from\ bulk\ flow] \qquad \textbf{Eq. 5}$$

$$\frac{d[S]}{dt} = \left[\frac{3.28M[EBOD]}{r}\right] + [1.57M[EBOD]] \qquad \textbf{Eq. 6}$$

For partially filled pipes, Pomeroy and Parkhust described sulfide concentration rate of change as being a function of sulfide production and sulfide loss (presumably to the sewer air as H_2S) as shown in Equation 7 below (Pomeroy, 1977).

$$\frac{d[S]}{dt} = M'[EBOD]r^{-1} - N(sv)^{\frac{3}{8}}[S]d_m^{-1} \qquad \textbf{Eq. 7}$$

Where:

M' = Sulfide flux coefficient = $0.4x10^{-3}$m/hr. (ASCE, 1989) (differs from M for filled pipes in Eq. 4 and Eq. 6 above)

N = Sulfide loss coefficient = 0.64 (ASCE, 1989).

s = slope of hydraulic conduit (Pomeroy, 1977). It should be noted that in the referenced paper, s is said to be the slope of the conduit and elsewhere as the slope of the hydraulic energy line. In a long reach of constant sewer slope, the two should be essentially equal.

v = stream velocity (m/sec)

S = Total sulfide concentration in mg/L

d_m = mean hydraulic depth (cross sectional area of stream divided by the surface width)

$d_m = \frac{D}{8}\left(\frac{\alpha - sin(\alpha)}{sin(\alpha/2)}\right)$ (Janna, 1983)

When the production of sulfide is equal to the loss, there is no longer a change in sulfide concentration with time and Eq. 7 (ASCE, 1989) can be written as

$$M'[EBOD]r^{-1} = N(sv)^{3/8}[S]_{lim}d_m^{-1} \quad \textbf{Eq. 8}$$

And, therefore,

$$[S]_{lim} = \frac{M'[EBOD]d_m}{N(sv)^{3/8}r} \quad \textbf{Eq. 9}$$

Using these simple (relatively) equations, one can readily estimate the liquid sulfide concentrations using the default values of $M' = 0.32x10^{-3}$ m/hr, and $N = 0.64$ (ASCE, 1989).

The downstream sulfide concentration S_2 can be calculated by Equation 10 (ASCE, 1989) below.

$$S_2 = [S]_{lim} - \left\{\frac{[S]_{lim} - S_1}{log^{-1}\left\{\frac{(sv)^{3/8}\Delta t}{1.15 d_m}\right\}}\right\} \quad \textbf{Eq. 10}$$

Where S_1 and S_2 are the upstream and downstream concentrations respectively.

Once a sulfide concentration is estimated the gas phase concentration can then be estimated as well but not with any great accuracy. Even with very good data from an active sewer it is often difficult to show good correlation between liquid and gas phase sulfide concentrations. Because sewers are very dynamic in terms infiltration of ambient air and expulsion of sewer air equilibrium conditions seldom, if ever, exist and Henry's law equilibrium between the liquid and gas phase sulfide concentrations cannot be simply applied. The concentration of H_2S can vary widely from high concentrations in areas of extreme turbulence to concentrations as low as 5 or 10% of the Henry's law predicted concentrations (Speece, 2008). For planning and simple system designs 25% of the Henry's Law equilibrium concentration would not be unreasonable to assume.

Odor Impact Estimation Using Airflow and Odor Concentration Values

With both the odor concentration and the amount of airflow that must be treated estimated it is possible to estimate the impact of the emissions on downwind receptors by using atmospheric dispersion modeling. Simple atmospheric dispersion models used for estimating pollutant impacts (including odors) are based on the concept of Gaussian dispersion as depicted in Figure 3 below. The dispersion and impact can be calculated using Equation 11 (WEF, 2004) below.

$$C = \frac{Q}{2\pi\sigma_y\sigma_z} exp\left(-\frac{1}{2}\frac{y^2}{\sigma_y^2}\right)\left\{exp\left(-\frac{1}{2}\frac{(z-H)}{\sigma_z^2}\right)\right\}\left\{exp\left(-\frac{1}{2}\frac{(z+H)}{\sigma_z^2}\right)\right\} \quad \textbf{Eq. 11}$$

Where

C is the downwind concentration

σ_y and σ_z are the horizontal and vertical dispersion coefficients respectively

H is the "Effective" stack height (sum of actual stack height and plume rise)

Figure 3: Gaussian plume horizontal and vertical dispersion

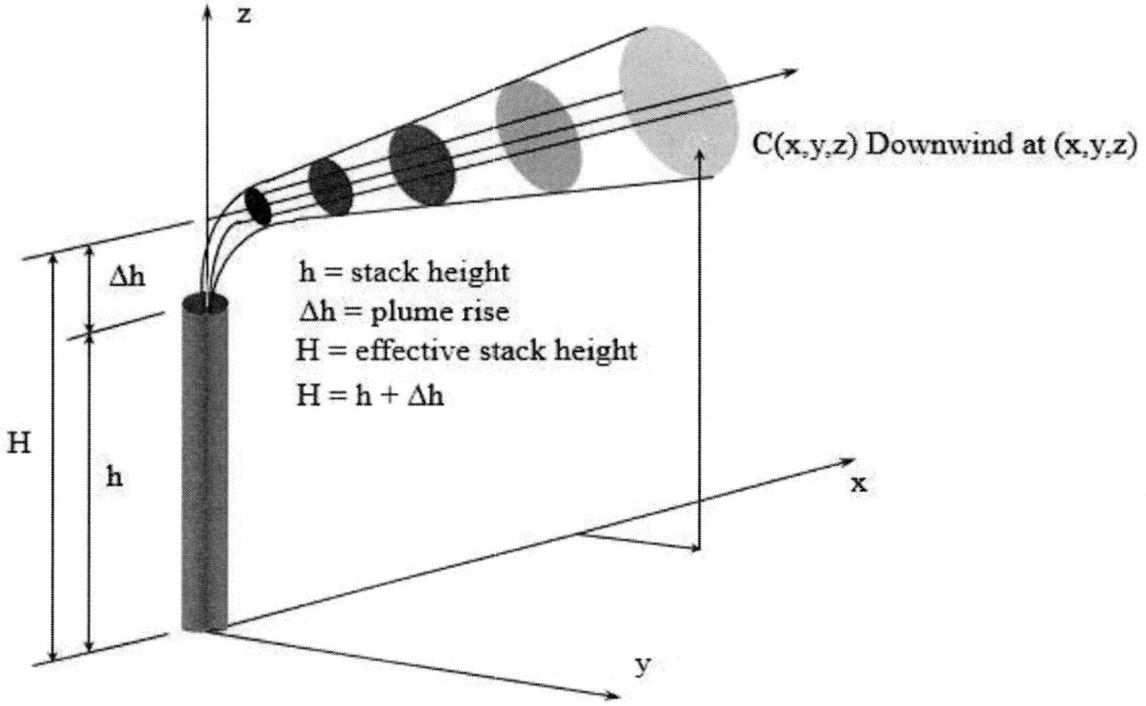

The effective stack height is the actual stack height with elevation added to account for plume rise due to buoyancy resulting from the plume's temperature difference with the ambient atmosphere and momentum due to its initial velocity.

The reader is advised to consult WEF's Manual of Practice 25 for additional information if this method of calculation is desired. It should be noted that this method is cumbersome for calculating impacts at more than a few locations. This is due to the need to find values for σ_y and σ_z for each location and each atmospheric stability class that may occur. Figures 4 is a table for selecting the atmospheric stability class based on the time of day, wind speed and solar radiation. Once the stability class is known, the vertical dispersion coefficient, σ_z, can be found from the chart presented as Figure 5. The horizontal dispersion coefficient, σ_y, can be found using a similar appropriate chart. In addition to determining the stability class and the dispersion coefficients, the calculation made using Equation 11 requires that the effective stack height (H) be calculated. This requires knowledge of the actual stack height and its diameter, vertical velocity of the emitted air stream and its temperature and the wind speed. Obviously all these calculations would have to be repeated for each alternative scenario considered and must be done with care to avoid introducing errors.

Figure 4: Table for selecting Atmospheric Stability Class (WEF, 2004)

	Day			Night	
	Incoming Solar Radiation			Thinly Overcast	
Wind Speed, 10 m (m/sec)	Strong	Moderate	Slight	>4/8 Cloud	<3/8 Cloud
<2	A	A–B	B	E	F
2–3	A–B	B	C	D	E
3–5	B	B–C	C	D	D
>6	C	D	D	D	D

Figure 5: Vertical Atmospheric Dispersion Coefficient (Wark, 1981)

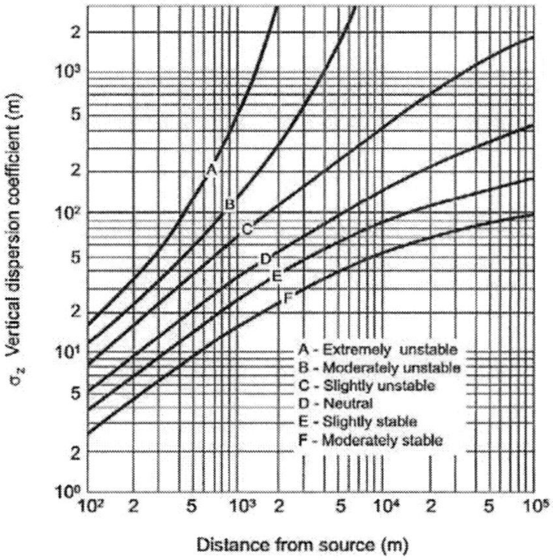

Fortunately, computerized atmospheric dispersion models allow a far easier means of estimating downwind impacts. For initial planning level analysis when little is defined with regard to actual emission stack heights or geometry and adjacent structures or when performing alternatives analysis of simple odor sources screening level dispersion models may be used. The AERSCREEN model is the screening version of popular and more sophisticated AERMOD model and EPA's SCREEN3 is the screening version of the Industrial Source Complex 3 model. SCREEN3 is an old DOS (disk operating system) based program that is relatively simple to use and, with some familiarity, the user can develop impact estimates in several hours for multiple emission scenarios. More sophisticated models, such as, AERMOD require substantially more time to develop. All of these models and documentation can be found on the Environmental Protection Agency's web site.

Of course, the point of modeling is to determine if a pollutant emission from a source will cause an unacceptable downwind impact. Since we are interested in odors, the acceptable downwind concentration must be determined whether it is in the interest of simply being a good neighbor or because of a regulatory requirement. Because odors can cause a nuisance with just a few minutes of impact it is appropriate that a short averaging time be considered. For the purpose of discussion in this paper an acceptable odor impact of 5 D/T and a 3-minute averaging time will be used.

Most models require the input of the mass emission rate (gram/sec) of pollutant and the volumetric emission rate (m^3/sec) of the carrier gas as inputs. Typical output includes mass concentrations ($\mu g/m^3$) over a 1-hour averaging time at points of downwind impact. With these inputs and outputs, the amount of dilution between the emission point and the point of impact can be calculated. A method of converting a 1-hour average to a shorter averaging time is discussed later in this paper.

Because we have methods of estimating the odor concentration at the point of emission as described earlier in this paper, once the amount of atmospheric dilution is known the downwind impact can be easily calculated to determine if odor control is required or if an odor control system will be adequate.

Table 1 below shows some of the typical output from the SCREEN3 model. The most important information for our purposes appears in the first two columns that show the concentration (CONC) at the point of impact and distance (DIST) to that point.

Table 1: SCREEN3 Atmospheric Dispersion Model Output (Partial)

DIST	**CONC**	STAB	U10M	USTK	MIX HT	PLUME HT	σ_y	σ_z	DOWN WASH
(M)	**(UG/M^3)**		(M/S)	(M/S)	(M)	(M)	(M)	(M)	
1	**0**	0	0	0	0	0	0	0	NA
100	**2150**	6	3	3	10000	6.7	4.07	4.83	SS
200	**1242**	6	3	3	10000	6.7	7.73	6.31	SS
300	**896**	6	2.5	2.5	10000	7.67	11.23	7.3	SS
400	**676.7**	6	2.5	2.5	10000	7.67	14.64	8.64	SS
500	**561.2**	6	2	2	10000	9.58	17.97	9.43	SS
MAXIMUM 1-HR CONCENTRATION AT OR BEYOND 1. M:									
17	**5995**	6	4	4	10000	5.86	0.82	2.84	SS

In this example the mass emission rate used in the modeling run was 1 gram per second. It is not important what emission rate is used for this method because, as will be demonstrated below, it is only the relation between the emitted mass concentration and the reported downwind impact that is needed to estimate atmospheric dilution of the emitted odors. It should be noted that the SCREEN3 model does not provide the direction from the point of emission to the point of impact.

As stated earlier, the SCREEN3 model predicts a one-hour concentration and in our example it is desired to avoid a shorter term 3-minute average concentration. Fortunately there is a simple method of converting between different averaging times known as a 1/5th power law (WEF, 2004) which is shown as equation 13 below.

$$C_1 = C_0 \left(\frac{t_0}{t_1}\right)^{1/5} \qquad \textbf{Eq. 13}$$

Where C_1 is the concentration at averaging time t_1 and C_0 is the concentration at averaging time t_0.

In our case we want to convert a 1-hour averaged concentration, C_0, to a 3-minute concentration, C_1, so, substituting like unit values for the two different averaging times into Eq. 13 we find

$$C_{3\,min} = C_{60\,min}\left(\frac{60\ min}{3\ min}\right)^{1/5} = C_0(20)^{1/5} = 1.82C_{60\,min}$$

In Eq. 12 above then, the impact concentration must be increased by the time average conversion factor calculated using Equation 13 using the method shown below in Equation 14.

$$Dilution\ Factor_{3\,min} = \frac{Emitted\ Concentration}{Impact\ Concentration_{1\,hr} * Conversion\ Factor_{3min}} \qquad \textbf{Eq. 14}$$

Earlier in the paper the volumetric emission rate was determined to be 700 cfm and the SCREEN3 output indicates that the maximum 1-hour impact would be 5,995μg/m³ (Table 1). Therefore, using equation 14, the atmospheric dilution factor is

$$Dilution\ Factor_{3\,minute} = \frac{\left(\frac{1\frac{gram}{sec}}{700\frac{ft^3}{min}}\right) * \left(\frac{35.3ft^3}{m^3}\right) * \left(\frac{60sec}{min}\right)}{\left(5{,}995\frac{\mu g}{m^3}\right) * \left(\frac{1gram}{1{,}000{,}000\mu g}\right) * 1.82} = \frac{3.0\ gram/m^3}{1.091x10^{-2}gram/m^3}$$

$$Dilution\ Factor_{3\,minute} \approx 275$$

Now that we have our dilution factor we can compare it to the likely source odor concentration from in the McGinleys' paper to determine if it is likely that the untreated air emission would cause an odor nuisance. In the case of a conveyance system wet well shown in Figure 1, the odor concentration of the 3rd quartile of the wet well source category is 3,650 D/T. Therefore, that concentration would have to be diluted 3,647 times to reach our limit of 3 D/T. Because the atmospheric dispersion alone will provide only about 275 dilutions, this would not be sufficient to dilute the odor to our 3 D/T target and a nuisance would likely be caused.

It is now possible to calculate how efficient an odor control scrubber would need to be in order to reduce the stack emission odor concentration to our 3 D/T target. To do this we perform a simple efficiency calculation as shown in Equation 15 below.

$$N = \frac{Emitted\ Concentration - Dilution_{atm} - Impact\ Target}{Emitted\ Concentration} * 100\% \qquad \textbf{Eq. 15}$$

$$N = \frac{3{,}650 - 275 - 3}{3{,}650} * 100\% = 92.4\%$$

Now that the required overall odor control efficiency of the scrubber system is known and you have an estimate of H_2S concentrations that could be present in your foul air stream, enough is known to consider what type of scrubbing technology would be appropriate and sufficient.

CONCLUSION

As stated at the beginning of this paper, in the ideal case decisions about the need for, and design of, odor control systems would be based on complete knowledge of all important parameters but, with little, if any, source specific information it is still possible to move forward with planning wastewater facilities and design of relatively simple single sources of odor source and/or treatment.

As shown, the methods described in this paper are sufficient to estimate

- Which odorous compounds would likely be present
- Concentrations of H_2S present
- Overall odor concentration as described by Dilutions-to-Threshold (D/T)
- Volumetric airflow requiring treatment
- Likelihood of causing an odor nuisance
- The overall odor removal efficiency of an odor control system

With this information reasonable decisions can be made regarding the necessity of providing odor control for existing or planned wastewater facilities. Selection of odor control technologies and sizing of odor control system components can be made and the design of robust odor control scrubbing systems can be developed.

REFERENCES:

American Society of Civil Engineers, Sulfide in Wastewater Collection and Treatment Systems, 1989.

Pomeroy, R. D., Parkhurst, J. D., "The Forecasting of Sulfide Buildup Rates in Sewers", Prog. Wat. Tech., Vol. 9, Pergamon Press, London, England, 1977, pp. 621 – 628.

Water Environment Federation, Control of Odors and Emissions from Wastewater Treatment Plants, Manual of Practice No. 25, 2004.

McGinley, M. A., McGinley, C. M., "Odor Threshold Emission Factors for Common WTP Processes", Water Environment Federation Odors and Air Emissions 2008 Conference Proceedings, Phoenix, AZ: 6-9 April 2008.

Water Environment Research Foundation, "Minimization of Odors and Corrosion in Collection Systems (Phase 1)", 2007.

Pescod, M. B., and Price, A. C., "Major Factors in Sewer Ventilation", Journal of the Water Pollution Control Federation, Vol. 54, No. 4, Apr. 1982, pp 385-397.

Speece, R. E. (2008). Anaerobic Biotechnology and Odor/Corrosion Control for Municipalities and Industries. Nashville, TN, Archae Press.

"Odor and Corrosion – Prediction and Control in Collection Systems and Wastewater Treatment Plants", Water Environment Federation (WEF), 2001.

Janna, W.S. (1983). Introduction to Fluid Mechanics. Belmont, CA, PWS Publishers.

NFPA 820 – Standard for Fire Protection in Wastewater Treatment and Collection Facilities, National Fire Protection Association, Quincy, MA. 2016

Odor Control in Wastewater Treatment Plants, American Society of Civil Engineers Manual of Practice 82, 1995

Wark, K., Warner, C.F. (1981). Air Pollution – Its Origin and Control, Second Edition, New York, NY, Harper and Row, Publishers, Inc.

Implementing an Odor Control Program for a Centralized Sludge Processing and Loading Facility

Corresponding Author:
James S. McMillen, P.E.
Perkins Engineering Consultants, Inc.
6001 Interstate 20 West, Suite 219
Arlington, Texas 76017
jmcmillen@perkinsconsultants.com

Coauthors:
William Gase, Director of Wastewater Treatment, City of Garland, Texas
Toshio Shimada, Ph.D, P.E., Carollo Engineers, Inc.
Charlotte G. Smith, P.E., Perkins Engineering Consultants, Inc.

ABSTRACT

The City of Garland, Texas needed to upgrade the solids processing facilities at their Rowlett Creek Wastewater Treatment Plant. The upgrades included replacing belt filter presses (BFPs) with centrifuges; replacing gravity belt thickeners with rotary drum thickeners (RDTs); converting anaerobic digesters to mixed sludge holding tanks; and other improvements. By upgrading the sludge processing equipment, the City was able to minimize the cost of odor control for these facilities.

The equipment upgrades and process modifications required varying methods of odor control, including biotrickling filters, carbon adsorbers, and organic media biofilters. With the change in processes, determining the appropriate design odorant concentrations required innovative field testing. Dispersion modeling was used to determine the effectiveness of carbon adsorbers and stack discharge prior to bidding so that the City could make an informed decision on certain additive alternate bid items. The design process included prequalifying odor control manufacturers and interviews to select vendors for the systems.

This paper discusses the interaction with the biosolids process design; minimization of odor control costs; determining appropriate odorant concentrations and design airflows; technology selection and design; construction and performance testing.

KEYWORDS

Odor Control; Biotrickling Filter; Carbon; Solids; Centrifuge; Rotary Drum Thickener; Silo; Lift Station

INTRODUCTION

The City of Garland, Texas owns and operates two wastewater treatment plants (Rowlett Creek WWTP and Duck Creek WWTP) serving customers in the Dallas-Fort Worth

Metroplex. The Rowlett Creek WWTP serves the east side of Garland including the cities of Rowlett and Sachse. RCWWTP is a Trickling Filter/Activated Sludge (TF/AS) plant and is permitted for an annual average daily flow of 1.05 m^3/s (24 MGD) and peak two-hour flow of 1.88 m^3/s (43 MGD). All biosolids from the Duck Creek WWTP and RCWWTP are processed at the RCWWTP with ultimate disposal in the City of Garlands' landfill.

In the mid-1980's, the plant was expanded and new dewatering facilities were constructed. These facilities included anaerobic digesters, dissolved air floatation thickeners (DAFTs), and three 2-meter belt BFPs. In 2007 the DAFTs were replaced with two 2-meter gravity belt thickeners.

By 2014, the existing BFPs were near capacity and were undergoing frequent maintenance. After a study and pilot testing were performed, the City elected to convert the anaerobic digester to sludge holding tanks; replace the gravity belt thickeners with RDTs; replace the BFPs with centrifuges; and install screw conveyors to transport dewatered solids to a new cake storage silo and truck loading facility. Odor control facilities were required for the new facilities as well as the existing lift stations. Figure 1 shows an aerial of the RCWWTP before the current project.

Figure 1 Rowlett Creek WWTP Preconstruction Aerial Photograph

PROJECT DESCRIPTION AND BACKGROUND

In 2003, the City developed the "Odor Abatement Evaluation and Master Plan" (Master Plan) to evaluate and plan odor control improvements. Concurrent with the development of the Master Plan, the City constructed odor control facilities for the primary clarifiers. In 2010, a comprehensive Biosolids Improvements Study was completed for the City. The Biosolids Study pointed out that the digestion and dewatering systems at the Rowlett Creek Plant were nearing the end of their useful lives and recommended improvements.

In 2011, the City updated its Odor Abatement Evaluation and Master Plan. As part of the this update the plant-wide air dispersion model was updated to reflect 2010 conditions. In 2012, the City commissioned an evaluation of the feasibility of decommissioning its anaerobic digesters. The evaluation recommended converting the anaerobic digesters to aerobic sludge holding tanks. In 2013, the Centrifuge Evaluation recommended replacing the BFPs with higher capacity centrifuges to meet plant ultimate buildout capacity.

In 2013, the City began design of the improvements to the solids processing facilities and associated odor control improvements which were to include the lift stations. Odor control technologies are not a case where "one-size-fits-all". Each technology has a situation for which it is best suited depending on the wastewater characteristics, odor characteristics, site requirements, and operator preferences. To customize treatment, each area (lift stations, cake storage and loading facilities, solids handling equipment, and the sludge holding tanks) was evaluated separately to determine the appropriate odor control technology for that area.

ODOR CONTROL TECHNOLOGIES

There are multiple odor control technologies available, including liquid phase (chemical addition), vapor phase (biological treatment, chemical scrubber, carbon adsorption, and others), and containment (covers). To avoid introducing new chemical hazards to the facility, the City elected to remove chemical scrubbers from evaluation. The City operates a wood chip media biofilter for odor control at the preliminary and primary treatment areas but did not have land area available at the RCWWTP for a second biofilter. For the odor control on this project, biological and carbon systems were evaluated.

Bioscrubbers/Biotrickling Filters

Biotrickling filters (BTF's) and bioscrubbers are typically constructed from FRP towers with media inside. Foul air enters the BTF at the base of the tower where a sump helps maintain humidity. Air passes up through the media while recirculated water flows down through the media. Bacteria grows on the media (for a BTF) or in the liquid (for a bioscrubbers) to consume odorous compounds in the air. Cleaned air exhausts from the top of the tower and liquid drains from the sump area. Recirculation and fresh water irrigation rates can be adjusted to encourage the growth of different biological compounds that target hydrogen sulfide and other odorous compounds. Sizing of BTFs or bioscrubbers depends on the air flow rate, compounds in the foul air stream, elimination capacity of the media and the empty bed residence time of the vessel.

Carbon Adsorbers

Carbon adsorbers are FRP vessels containing activated carbon. Foul enter the vessel and passes through the carbon media. Compounds in the air stream are adsorbed into the pores of the carbon and remain in the media as the cleaned air exhausts. Specific kinds of carbon can be used to target different odor compounds and there are several varieties of media with varying capacities. Three types of adsorbers are typically used: deep bed, dual bed and radial flow. Deep bed adsorbers have a single bed of carbon and foul air usually enters at the base of the vessel and flow upward through the media. Dual bed adsorbers have two beds of media. Foul air enters in the middle of the vessel and flows upward and downward through the media. At the top bed air exhausts directly to an exhaust stack; at to bottom bed, an exhaust stack penetrates both layers of media. Since carbon adsorber sizing depends partly on the face velocity of the air stream on the media, a dual bed adsorber can provide two surfaces and treat higher airflows. A radial flow adsorber has an annular space filled with media. Flow enters the side of the vessel and passes through the media bed toward the center and is exhausted. A radial bed has the greatest surface area and can treat much higher airflows in smaller footprint, albeit with a taller vessel.

TESTING AND EVALUATION

Early in the design, it became clear that separate odor control facilities would be needed at each area. This was due to site requirements and the expected varying makeup of the airstreams that would need customized treatment. Areas were separated into four systems, one each for the lift stations, solids processing equipment, cake loading facility, and sludge holding tanks. Since only the lift stations were currently constructed and operating, small scale pilot testing was needed to determine design values for the other areas.

At the two lift stations, the Rowlett Lift Station and the Transfer Lift Station, hydrogen sulfide monitors (H_2S) logged concentrations for approximately one week. Both lift stations had fairly low average concentrations (9 ppm and 2 ppm respectively) but exhibited peaks up to 35 ppm. Data was recorded in the spring; vapor phase hydrogen sulfide concentrations typically peak in the late summer for the project location so higher design values were selected.

To determine design odorant levels in the future sludge holding tanks, sludge samples were taken from the sludge transfer line that conveys sludge from the DCWWTP to the RCWWTP for processing. A small aeration chamber was constructed so that the sludge sample could be aerated while the headspace air was collected for testing. Two samples were tested for sulfur compounds in accordance with ASTM D5504. Results indicated the presence of methyl mercaptan, ethyl mercaptan, dimethyl sulfide, carbon disulfide, and hydrogen sulfide. H_2S concentrations ranged from 200 ppm at the beginning of aeration to 32 ppm after several minutes of aeration.

Since the City did not operate any centrifuges, RDTs, or other high solids processing equipment, field testing to determine design values was not practical. Based on literature published by the Water Environment Federation and other sources, as well experience on similar designs, it was anticipated that volatile sulfur compounds would make up most of the odorants in the air stream.

DISPERSION MODELING

The American Meteorological Society (AMS)/Environmental Protection Agency (EPA) Regulatory Model (AERMOD) was used produce isopleths using odor units for the existing and proposed treatment units. AERMOD is a steady-state plume dispersion model which simulates transport and dispersion of multiple point (scrubber stacks or vents), area (uncovered process units) or volume (open doors, etc.) emission sources. Typically, a combination of area and point sources are used to model odors from wastewater treatment facilities.

Odor dispersion modeling can be a helpful tool to illustrate the reduction of the area of odor impact from different odor control options. An odor dispersion model shows how odors generated at a wastewater plant are transported to the fence line and offsite. The model predicts the intensity, frequency, and spatial extent of nuisance odors generated by the treatment facilities. Air dispersion model output is most useful for illustrating the differences between odor control strategies. Dispersion modeling was conducted to illustrate the reduction in odor impact by controlling odors at the influent lift stations and solids handling facilities. Dispersion modeling was also used to illustrate the impact of the additive alternates of the additional two aerated sludge holding tanks and carbon polishers at the lift stations and aerated sludge holding tanks.

For the existing odorous treatment processes, OU values measured as part of an earlier study were used for dispersion model input. For the proposed odor treatment technologies, the OU values at the stack discharges were derived from current performance guarantees from odor control equipment manufacturers, as well as data collected at similar facilities and engineering judgment. The assumed OU levels for the discharges of the biotrickling filters, carbon adsorbers, and carbon polishers are 400, 150, and 150, respectively.

Dispersion modeling showed that for the current project, adding odor control to the solids handling areas in conjunction with the new solids handling equipment would greatly reduce the offsite odor impact. Figure 2 shows an example of the dispersion model isopleth.

Additonal modeling efforts showed that the height of the biotrickling filter exhaust stack played a key role in offsite impacts. Stack heights above 30-feet provided a greatly reduced area of impact. It was also shown that, under the assumptions used for the model, adding carbon polishing units on the exhaust of the biotrickling filters resulted in minimal changes to impact area. Carbon units were ultimately included in the project to guard against process upsets.

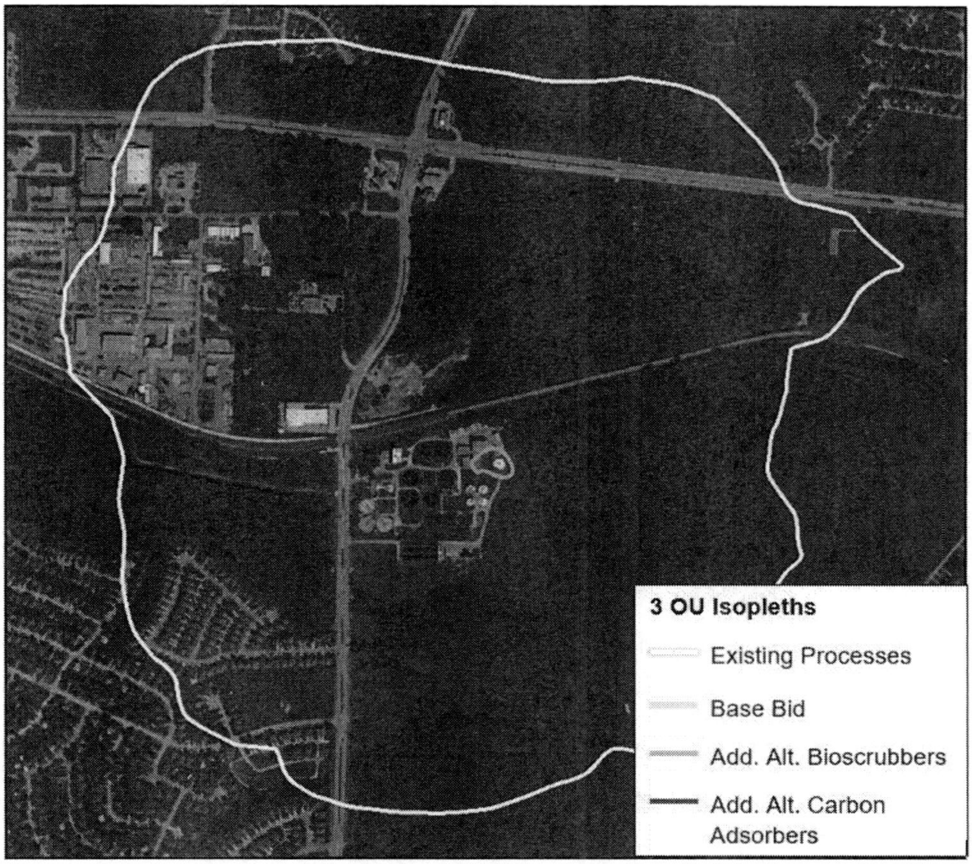

Figure 2 Dispersion Model – 3 OU Isopleths for Solids Processing Only

ODOR CONTROL DESIGN

Four odor control systems were designed at the RCWWTP, with one system treating each of the lift stations, solids processing equipment, cake loading bay, and sludge holding tanks.

Lift Station Odor Control

Located adjacent to the fence line, two lift stations serve the RCWWTP. The Rowlett Lift Station serves as an influent lift station; the Transfer Lift Station sends high flows to the DCWWTP. The Transfer Lift Station consists of a single large wetwell with submersible pumps, hatches for access, and ventilators. The Rowlett Lift Station has channels feeding the wetwell, a fiberglass reinforced plastic (FRP) cover over the wetwell, and a separate pump discharge chamber. The pump discharge chamber has an elevated concrete slab cover with hatches and a gooseneck vent.

Regulations in the State of Texas require lift station wet wells to have 12 air changes per hour which was provided by the existing supply ventilators. Additional air is exhausted to maintain a slight vacuum on the wet wells to prevent fugitive emissions. Airflows for odor control total 371 m^3/min (13,100) cfm for the lift stations, with air pulled from the wet wells, channels, and discharge chamber. A two biotrickling filters were designed to operate in parallel followed by an activated carbon adsorber polisher. The biotrickling filter were designed to treat H_2S concentrations that average 50 ppm and peak at 100 ppm and remove 99% of H_2S with the carbon adsorber removing remaining H_2S and non-H_2S odorants. Figure 3 shows the lift stations system under construction. In the figure, the carbon adsorber is behind the two biotrickling filter towers.

Figure 3 Lift Station Odor Control System

Cake Loading Bay Odor Control

Situated below the live bottom cake storage silo, the truck loading bay provides a fully enclosed space to load dewatered solids for transport to the City's landfill. As a truck approaches the loading bay, a high-speed overhead door opens. After the truck enters the building, the door closes and provides an enclosed loading area. Since the enclosed building will be occupied by personnel, a fresh air supply fan provides 20 air changes per hour to the bay. Slightly more air is exhausted to the 198 m^3/min (7,000 cfm) carbon adsorber to keep the building under negative pressure. Activated carbon treatment was selected because the system operates intermittently and the odor airstream would consist primarily of non-H_2S compounds. Figure 4 shows the dual bed activated carbon adsorber.

Figure 4 Cake Loading Bay Carbon Adsorber

Solids Processing Equipment Odor Control

Solids processing equipment includes three centrifuges, two RDTs, a cake storage silo, and interconnecting shaft less conveyors. Three centrifuges in the dewatering building each have an odor control airflow of 5.66 m^3/min (200 cfm) per centrifuge (and one future centrifuge) with the duct connection at the liquid drain line. The centrifuges are fully enclosed and the building has an independent HVAC supply and exhaust system. By treating only the air at the enclosed equipment, the entire room can be exhausted to atmosphere which greatly reduces the odor control equipment costs. Two RDTs are installed in the existing gravity belt thickener building. Like the centrifuges, each thickener exhausts 5.66 m^3/min (200 cfm) directly from the drum and the room air has an independent HVAC system.

Screw conveyors transfer dewatered sludge from the centrifuges to the storage silo. The screw conveyors are 24-in. diameter shaftless conveyors. An airflow of 11.32 m^3/min (400 cfm) was selected for the horizontal conveyors, which is exhausted at two locations (5.66 m^3/min or 200 cfm each). The vertical conveyor is located between the two horizontal conveyors and air is exhausted through the horizontal conveyors.

The fully enclosed sludge silo has a total volume of 214 m^3 (280 cubic yards) with an outloading capacity of 75 m^3/hour (2,650 cubic feet per hour). Although the sludge volume in the silo will vary as sludge is added and removed, for odor control purposes a total empty volume of the silo was assumed. Airflow was calculated at 21 m^3/min (750 cfm) for 6 air changes per hour. A vent with a gravity inlet damper is included on the silo to avoid pulling a strong vacuum on the silo.

A mixture of sulfur compounds from the solids handling equipment was anticipated. For the low concentrations of hydrogen sulfide ad non-H_2S compounds, a 67 m^3/min (2,350 cfm) deep bed carbon adsorber was selected. Figure 5 shows the carbon adsorber.

Figure 5 Solids Processing Equipment Deep Bed Carbon Adsorber

<u>Sludge Holding Tanks Odor Control</u>

Two existing digesters were converted to mixed sludge holding tanks and fully covered with flat aluminum covers. Airflow of approximately 71 m^3/min (2,500 cfm) is added through large bubble diffusers for mixing, although the airflow will vary based on sludge volume. Mixing air comes from a fresh air supply that must be exhausted to the odor control system in addition to airflows needed to maintain a slight vacuum on the system. It was determined that an airflow of 127 m^3/min (4,500 cfm) per sludge holding tank would be needed. Bench scale tests on aerated sludge form the plant indicated the presence of hydrogen sulfide, methyl mercaptan, ethyl mercaptan, dimethyl sulfide and carbon disulfide. To treat the hydrogen sulfide and non-H_2S compounds, a 255 m^3/min (9,000 cfm) biotrickling filter with a 12 second empty bed residence time was installed. A carbon adsorber was also installed to polish the airstream and capture any untreated compounds. Initially only two digesters were converted to sludge holding tanks, so the initial installation provided space for a future biotrickling filter to be added when the remaining two tanks are converted. Figure 6 shows the covered sludge holdings tanks, biotrickling filter and carbon polisher (left side).

Figure 6 Sludge Holding Tanks, Biotrickling Filter and Carbon Adsorber

Odor Control Equipment Performance

Performance testing was conducted on the biotrickling filters and carbon adsorbers; only data for the BTF performance was available at the time this paper was prepared. At the sludge holding tank BTF, H_2S average removal efficiency averaged 99.96% with inlet concentrations ranging from 50 ppm to 200 ppm. Outlet concentrations ranged from non-detectable to 2 ppm, with peaks corresponding to inlet concentration spikes. At the lift station BTF, removal efficiency average 99.51 ppm with inlet H_2S concentrations ranging from 5 ppm to 125 ppm. Data from the carbon adsorber units was not available at the time this paper was prepared.

The District of Barriere, British Columbia Implements an Innovative Hydraulic Odor Control System.

Richard St-Aubin, P. Eng,, National Sales Manager – Municipal Systems, IPEX USA LLC.,

Abstract

The District of Barriere is located in the interior of the Province of British Columbia in Western Canada. While forestry is an important economic driver for Barriere, Its rugged location in the mountains of BC draws skiers in the winter and hikers in the summer. Barriere is also home to a new and innovative aeration and odor control system - one of the first of its kind ever installed in North America.

In 2012, the District solved an odor problem at a septage receiving station by installing a new aeration device called a Vortex Force aerator. The results were immediate as the odor problem quickly dissipated. A later evaluation by a hydraulic consultant found that the device was extremely effective in increasing dissolved oxygen (DO) levels in the septage tank. Based on these results, the District installed five more Vortex Force devices to support a new solar aquatics based treatment process in its new treatment plant.

This paper will fully describe the theory behind the new Vortex Force aerator, and results of both laboratory and field testing will be presented. Finally, details of the installations in Barriere will be presented, including the energy savings achieved and some of the other, more intangible benefits of using this new approach to aeration and odor control.

Causes of Odor and Corrosion in Municipal Wastewater Systems

The control of hydrogen sulfide and other odorous gases are a constant concern for operators of sewage collections and treatment systems. These natural by-products of sewage decomposition not only generate complaints from residents and businesses located near problem areas in a system, they are also extremely destructive. Corrosion from hydrogen sulfide attack is one of the primary causes of failed sewer pipes, manholes and other system components.

Virtually all municipal wastewater contains some amount of elemental sulfur, and therefore the potential for hydrogen sulfide generation. When wastewater is in an aerobic state (containing relatively high concentrations of dissolved oxygen) sulfur exists in the form of ionized H_2SO_4. This is because the aerobic bacteria that populate the slime layer on the submerged surfaces of waste water pipes use the dissolved oxygen in the waste water for their respiration process. The respiration of these bacteria generates a large portion of the biochemical oxygen demand (BOD) of the wastewater.

As a result of this process, the DO content of the waste water falls as time progresses. The rate of DO decay is dependent on many factors, however given enough time, the DO in the water water flow will eventually be depleted. When this occurs, the bacterial population in the slime layer of the sewer pipe changes from aerobic bacterial that metabolize DO, to anaerobic bacteria that use the H_2SO_4 ions in the waste water for their respiration process. By stripping the oxygen from the H_2SO_4, we are left with higher concentrations of H_2S, hydrogen sulfide, in the wastewater. Unfortunately, hydrogen sulfide is not particularly soluble in water, so at any areas of turbulence H_2S gas is expelled from the water flow, much like CO_2 is released when shaking up a can of soda.

The unsubmerged areas of the sewer pipes and manholes typically harbour bacteria that can metabolize H_2S gas. These bacteria convert the gas bask into H2SO4, commonly known as sulfuric acid. It is this sulfuric acid that attacks the Portland cement matrix of concrete sewer system components, converting it into calcium sulfate (gypsum). This can be easily seen as a fine, white powder on the affected components in a system. The hydrogen sulfide cycle summarized above (often called biogenic corrosion) is a well known and documented process, and is shown in Figure 1.

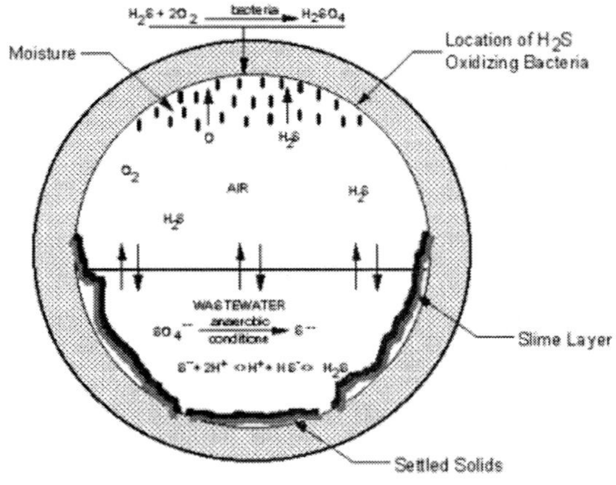

Figure 1-The Sulfide Cycle

Control of Hydrogen Sulfide (H_2S) in Wastewater

Given the serious consequences associated with uncontrolled H_2S gas generation, there have been many different methods and products developed to mitigate the problem. These include precipitating sulfur out of the wastewater flow, controlling the pH of the wastewater flow, controlled air stripping (and subsequent air-phase treatment) of the H_2S and others. However, one of the most common methods of dealing with H_2S is to oxidize it while it is still in the waste water (ie: before it is released into the sewer atmosphere). This process not only destroys the H2S that is in the water, but also changes the anaerobic conditions that are the root cause of H_2S generation.

The oxidation reaction can be accomplished through the addition of oxygen based chemical compounds, injection of pure oxygen, or the vigorous mixing of air into the flow. A typical design guide is to provide 5 mg O_2 per mg of dissolved sulfide[i]. Other sources state that dissolved sulfides are oxidized whenever dissolved oxygen concentrations exceed 1 mg/L[ii]. The reaction is as follows[iii]:

$$2O_2 + 2HS^- \rightarrow S_2O_3^{2-} + H_2O$$

While there are a number of different methods of increasing the dissolved oxygen content of the wastewater, in this paper a method of using the energy of fast flowing wastewater to vigorously mix air into the flow and thus diffuse oxygen into the water will be described.

The Vortex Force Aerator

The Vortex Force is a device that was designed by Dr. Eugene Natarius, a retired municipal engineer who worked for Minneapolis/ St-Paul's Metropolitan Council for many years. During his time there, Dr. Natarius invented the Vortex Drop Insert – an engineered drop structure that uses flow energy from sewer drops to control odor and corrosion in sewer collection systems. The Metropolitan Council completed a trial installation of Dr. Natarius's design in 1998, the successful results of which can be reviewed in a paper presented at WEFTEC in 2000[iv]. Since then, the design has been used in over 300 installations worldwide. The Vortex Drop also won a Technical Innovation Award from the American Public Works Association.

However, the Vortex Drop has a number of limitations; it requires a drop of at least six feet to be effective, and high velocity flows can be challenging. As a result, Dr. Natarius designed a new device designed to work with pipes flowing full, and that would require less of a drop.

Simply described, the Vortex Force is an aeration device that mixes air into a zone of high turbulence in a sewage flow.

This is accomplished in a rather unique fashion: the flow is split into two separate streams and directed into two vortex forms, oriented in opposite directions (See Figure 2). Both of these vortex forms outlet into the drop shaft, creating an area of extreme turbulence. Air is drawn into the device via the top air vent through a siphoning action, and quickly enters the turbulent zone. The turbulence breaks the air into very small bubbles, dramatically increasing the surface area of the air in contact with the water. As the oxygen diffuses from the air into the flow, sulfides and other odorous and corrosive compounds in the sewage are oxidized.

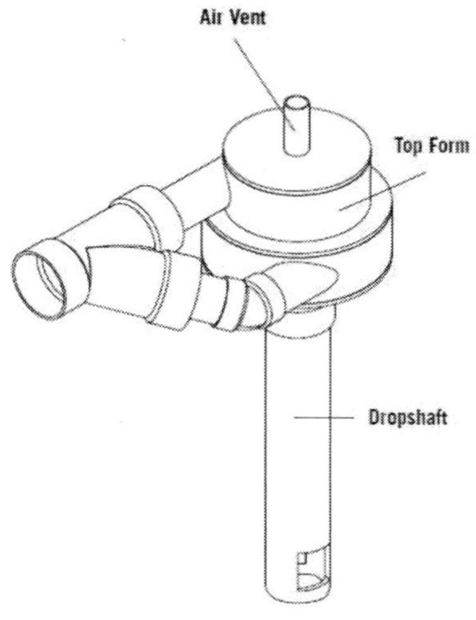

Figure 2-Vortex Force Diagram

First Installation in the District of Barriere

The initial design was validated using CFD software, and a number of lab tests were conducted using tap water, however in order to properly assess the performance of the new device it was necessary to test it with wastewater.

The District of Barriere, British Columbia is located about 66 kilometers north of the City of Kamloops in the interior of the province. It is a small town with a population of just over 3,000, with forestry, tourism and agriculture as its main economic drivers.

Figure 3-The District of Barriere, BC

In 2012, Mr. Doug Borrill, the District's Water Treatment Specialist, was looking for a solution to the problem of a 38,000 gallon sewage tank located on the south side of the Town. The tank received municipal sewage from the Town, as well as septage from a nearby trailer park, which was causing significant airborne odor issues. He was considering various options, including chemical injection to mitigate the odor problem, when he read about the vortex aerator in an industry publication. A design was proposed, accepted and the prototype was installed. A recirculating pipe loop was constructed, connecting the device to a small 2 horsepower pump. In this way the contents of the tank could be continuously recirculated through the device.

The results were evident soon after installation – the odors disappeared and the device ran smoothly and efficiently with a minimum of maintenance. The device was monitored over the following year and there were no operational problems.

Figure 4-Vortex Force Aerator in Recirculation Loop.

In 2015 IPEX, the manufacturer of the vortex device, commissioned Northwest Hydraulic Consultants Ltd. (Vancouver, BC) to complete a field evaluation of the device's aeration performance and efficiency. While it was clear that the unit was working as expected and attenuating any odors, the objective of the study was to quantify the effectiveness of the device as an aerator.

The test was set up as shown in Figure 5 and conducted on March 16th, 2015 and the results proved that the device was effective in elevating dissolved oxygen levels. The tank was left undisturbed for two weeks before the test was run to ensure that the DO levels were virtually zero. The pump was turned on at a flow of approximately 9.5 L/s (152 gal/min) and in less than one hour the measured DO level in the tank had increased to approximately 5 mg/L. The report concluded that the upper limit for the device would be about 6 mg/L.[v]

Photograph 1 – Test setup

Photograph 2 - DVA In-situ

Figure 5-Test Setup

Laboratory Testing

In addition to the field testing completed in Barriere, Northwest Hydraulics also undertook a laboratory based study of the device. A small prototype was manufactured using clear acrylic material for the top vortex forms and clear PVC pipe for the drop. The clear materials allowed the mixing of the air and water to be observed directly during the study. This unit was installed in their laboratory, and studied under varying flow conditions and drop heights. The focus of this work was the hydraulic performance of the device as it was not possible to accurately measure the increase in DO in this phase, due to the lack of an effective way to remove DO at the returned water sump.

Figure 6-Experimental Setup

The study successfully defined the operational limits of the device. The operational limit was defined as the point at which water begins to exit the device from the top air vent rather than travelling down the drop pipe. The most important finding was that the device needs an elevation change of approximately 1 metre (3 feet) in order to draw in sufficient air for effective mixing.

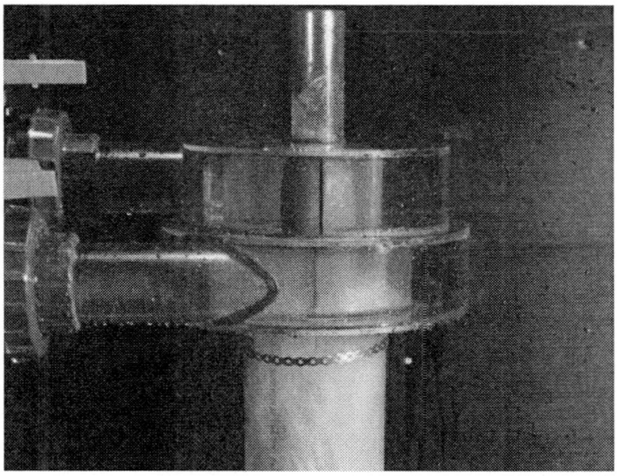

Figure 7-Close up of Turbulent Zone

Barriere Builds a New Sewage Treatment Plant

During testing of the Vortex Force Aerator in 2015, planning was already underway for the construction of a new sewage treatment plant for the Town. The project called for the installation of an innovative solar aquatics based treatment system that will allow the District to remove old and failing septic systems and reclaim the water for community uses[vi].

Figure 8-District of Barriere Treatment Plant

The new system uses a series of tanks that are populated with specific species of plants and algae designed to metabolize the various nutrients in the sewage. The flora in each treatment stage change with the sewage composition as the process advances. The final product is surface level discharge quality reclaimed water[vii].

High levels of dissolved oxygen are required at all parts of the process in order to ensure that the plants remain healthy. The original design called for the installation of a 15 hp (11 kW) compressor and fine bubble aeration system, however given the results from the initial Vortex Force installation at the septage receiving station, it was decided to substitute Vortex Force devices for the fine bubble aeration system. There were two main advantages to this approach:

- The Vortex Force units were driven by 2 hp (3 kW) pumps, rather than the15 hp (11 kW) compressor required for the fine bubble aeration system. This allowed significant energy savings throughout the process. Given that energy is generally the largest operational cost for a treatment facility, this is an important benefit.
- The facility is located near the centre of the District, in a commercial area. The compressors would have generated significant noise that could have been a potential problem for surrounding businesses. The Vortex Force units, and the pumps required to drive them are virtually silent.

Figure 9-Treatment Stream (Plant was not fully operational for this picture)

The first stage of the process is a large sewage tank (the round grey structure shown at the upper right of the plant in Figure 8) that receives pumped effluent from the septage station and other sources in the District. The flow runs through a Vortex Force device as it enters the tank, increasing its DO content to approximately 4 mg/L. From there, it is pumped into the plant itself, which is housed in a greenhouse to support the plants required for the treatment process.

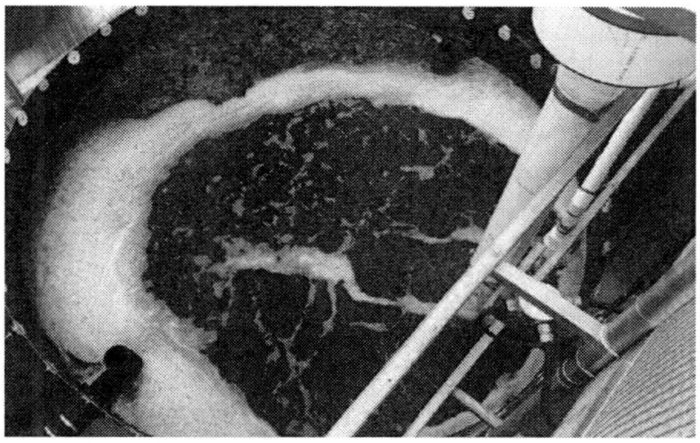
Figure 10-Close up of Aeration in Tank

The flow is then divided into two separate streams, as there are two parallel treatment streams in the plant. Figure 9 shows one of the treatment streams, however the plant was not fully operational when the picture was taken. The flow then goes through a second device, which boosts the DO to between 6 and 8 mg/L, which is sufficient to support the solar aquatic process. The sewage flows through a series of four tanks, each populated by a specific mix of plants and algae designed to metabolize the various nutrients in the sewage flow. The flow is pumped through a third device as it enters the fourth treatment tank, in order to boost the DO for the last stage of the process. The sewage exits the fourth and final tank as fully reclaimed water, suitable for surface irrigation.

Conclusions

The results of both laboratory and field testing confirm that the Vortex Force device is an effective aerator that can be used to boost DO levels in various types of effluent. As the first municipality to use the device in its septage receiving station and as a component of it s new, innovative treatment plant, the District has shown that other municipalities can benefit by using this new technology. Municipalities that are plagued with odor and corrosion issues in their sewage pumping stations or at forcemain outlets could benefit from this method of boosting dissolved oxygen.

---

[i] "Interim Odor Control Report", Prepared by Webster Environmental Associates for the Indianapolis Department of Public Works, Indianapolis, IN, April 1987
[ii] "Design Manual: Odor and Corrosion Control in Sanitary Sewage Systems and Treatment Plants", EPA, 1985
[iii] "An Overview of Methods and Approaches for Estimating and Solving Odor and Corrosion Problems in Collection Systems", Jim Joyce, PE, Odor and Corrosion Technology Consultants, Inc., Published in WEF Manual: "Odor and Corrosion – Prediction and Control"
[iv] "The Vortex Drop Structure Implementation for Odor and Corrosion Control", William P. Moeller Jr., P.E., Dr. Eugene Natarius. P.E., Proceedings WEFTEC 2000.
[v] Final Report – Double Vortex Aerator Field Test Results, Northwest Hydraulic Consultants Ltd, April 1, 2015.
[vi] www.ecotek.ca, EcoTek is the developer of the Solar Aquatics system used in Barriere's new plant.
[vii] This is a very simplified description of the process. More detailed information is available at www.ecotek.ca.

Wait, Why Do I Need Odor Control at a Farm-Based Digester Project?

Michael Lannan*, Tech Environmental, Inc., mlannan@techenv.com
Matthew Riegert, Tech Environmental, Inc.

ABSTRACT

Anaerobic digestion, by design, consumes compounds that are odorous. At first glance, digestate has less odor loading potential than raw manure. Therefore, there is no need for odor control at a digester project when the farm traditionally had open manure pits or lagoons. Unfortunately, odor loading potential alone does not drive odor nuisance potential. The odor nuisance potential changes when the odor loading changes, as well as the odor character, the hedonic tone (relative pleasantness), the reaction rate, and other factors. This paper discusses how exploring odor control is necessary, even at a farm based digester with traditional open manure storage. And, although there is very little odor potential from the biogas itself, there is off-gassing odor potential. These are new sources of odor potential and they cannot simply be dismissed as "less than before" without consideration.

KEYWORDS: Anaerobic digestion, odor control, digester, land application, digestate, dairy farms

INTRODUCTION

Most dairy farms in New England, and most other parts of the country, are family-owned and often require multiple income streams to make ends meet. So, besides milking cows, these farms typically have large areas of land dedicated to growing corn, hay, and other forms of agriculture. And, it is also now common to see a pumpkin patch, corn maze, hay ride, and/or other dairy products, such as cheese and ice cream, available for purchase. In this vein, many New England dairy farmers have recently welcomed installations of anaerobic digestion facilities onto their family's land. Those facilities typically accept processed manure, created by the on-site milking cows, and biosolids or food waste, from off-site sources, to generate sustainable renewable energy and valuable byproducts, including fertilizer and organic soil amendments.

But, this paper is about odor, and odor issues have always been common at dairy farms. Traditionally, manure is the major source of those odors, and dairy farmers have been controlling odors from manure as long as they have been milking cows, especially those farmers with nearby neighbors, or those who live in an area with an odor regulation. However, with the installation of an anaerobic digestion facility, many of these farmers have needed to consider even more odor control. It may seem silly that a farm that used to simply apply manure to its fields now has to consider odor control for an anaerobic digester, but after going through the process a number of times, some of it actually makes sense. The rest, well, it is politics.

This paper will present an overview of farm-based anaerobic digester systems, and the sources of odors associated with them. This paper will also discuss the exchange of odor sources with the installation said systems, and which odorants are reduced, and which odorants are increased.

Lastly, this paper will discuss what drivers justify a need for odor control at farm-based digester projects, and when odor control may not be needed.

FARM-BASED DIGESTER SYSTEMS

Automated, continuous mix, high solids agricultural anaerobic digestion technology had been proven to work in hundreds of applications in Europe and Japan before being attempted in the United States, but over the last decade that has changed. These facilities typically accept and process manure produced on-site and homogenized food waste derived feedstock, or biosolids, to generate sustainable renewable energy and produce valuable byproducts, which include fertilizer and organic soil amendments. The combination of food waste or biosolids and manure provides the necessary energy capacity to make a successful project that can be operated remotely in a cost effective manner. Traditionally, manure-to-energy projects in the U.S. operate with marginal energy output, so there is insufficient energy value to justify a sophisticated biogas-to-energy process. Many of the historical manure-to-energy facilities, therefore, were built on a shoe-string by the farmers themselves. While these systems function adequately, there are long-term operation and capital equipment concerns. The use of food waste feedstock or biosolids is a great example of resource optimization. Instead of being managed in a landfill, these materials are returned and processed by the farm, capturing energy, valuable organic material and nutrients, and creating a sustainable nutrient cycle.

A typical installation can be divided into the following areas:

1. Receiving and feedstock tanks (hydrolysis and acidification);
2. Digester tank (mesophilic fermentation);
3. Effluent management (dewatering & fertilizer storage); and
4. Gas processing (generator & flare).

Receiving and Feedstock Tanks

Typically, food waste or biosolids is delivered to the facility receiving tanks by truck and pumped through a "rock catcher" to filter out unwanted solid materials from the biomass. It is then mixed with recirculated biomass and incoming liquid manure, from nearby barns, and passed through a grinder/macerator becoming further refined and pumped into a feedstock tank. Typically, a pipe will connect the head space in the feedstock tank with the digester tank gas storage to maintain equalized gas pressure in the system. This has the added benefit of controlling odor from the feedstock tank. This also provides an area for displaced gas due to filling/discharging or temperature expansion of contained material, which would otherwise be released to the atmosphere.

Digester Tank

The digester tank provides the mean hydraulic retention time necessary to allow methanogenic bacteria to convert the organic biomass into biogas. The digestate is heated using a heat exchanger supplied by waste heat from the generator. The conditioned biomass from the feedstock tank is then fed into the digester tank at a turbulence zone created by a mixer to minimize the time required to obtain a complete mix. Gas storage in the digester tank is accomplished with a single membrane roof system. The membrane is supported by biogas pressure. The membrane inflates and deflates according to the amount of biogas in the system. Safe pressure levels are maintained by a water trap mounted to the tank. This prevents excessive pressure or vacuum in the system.

Effluent Management

Traditionally, manure from a farm is stored in a lagoon and then directly land applied in the spring and fall. But, at a farm with an anaerobic digester, the digested material is separated via a dewatering plate process. The dewatered solid material is used as farm bedding materials, plant pots, or other beneficial reuse options. The liquid is stored on-site and used as a liquid fertilizer. The liquid fertilizer drastically reduces the odor potential historically observed when the manure is directly applied in the spring and fall. And, it also allows for a more normalized application throughout the growing season, since the liquid fertilizer is applied as crops grow. Normalizing applications further reduces odor potential. The liquid fertilizer also offsets much of the farmer's costs of traditional commercial fertilizers.

Gas Processing

The biogas created by the digester is typically used as fuel in a generator set where electricity is produced for use on the farm or provided to the electric grid via net metering. Heat from the generator is typically used to maintain mesophilic temperature in the digester as well as a local heat source for barns or homes. A flare is typically also installed to combust gas when the generator set is down for maintenance or as an emergency.

Figure 1. The digester at Jordan Dairy Farms was the first of its kind in Massachusetts

ODOR SOURCES & CONTROL TECHNOLOGIES

Farm-based anaerobic digester system are typically designed to maximize the capture of biogas to capitalize on energy production. This operational goal inherently leads to little-to-no fugitive emissions and a great reduction in odor potential. But, odor control has been warranted in certain cases.

Receiving Tanks

The only component of digester systems not completely enclosed is the receiving tank(s), which have headspaces not connected directly to the biogas storage area. Any tank "breathing" from filling operations of the receiving tanks should consider odor treatment. Truck deliveries of feedstocks occur at least daily, but with trucks being on-site only temporarily, and with the actual feedstock discharge process lasting only a matter of minutes. However, this intermittent source has been problematic for some farms in less rural environments, and in those situations odor control systems have been successful in treating those discharges. But, in more rural areas where sources, such as open manure pits, exist with no odor concerns, odor control has not been necessary to prevent change to the existing odor potential of the farm.

The emissions from these tanks are generally a mix of volatile organic compounds, fatty acids, reduced nitrogen compounds, and some reduced sulfur compounds. And, the type of successful odor control, if necessary, can vary from passive single stage activated carbon, to lead-lag activated carbon with a blower, to biological systems. Typically, a passive activated carbon system has been adequate to control these odors. In cases where the digester system and receiving tanks was closer to a neighbor, or property line, an intermittent ventilated activated carbon system was warranted to minimize fugitives during feedstock discharges. At another location, a biofilter was installed because the receiving tanks had a larger capacity for feedstocks, were in close proximity to a neighbor, and it was necessary to have continuous odor control.

Regardless of the technology, facility operating permits generally require "sniff tests" of the system outlets to monitor performance, or, in some cases, monitoring with gas detection tubes. When testing with gas detection tubes, hydrogen sulfide has typically been chosen as the surrogate for total odor with a threshold of five (5) parts per million by volume (ppmv) determining when a change of the media is required. The hydrogen sulfide emitted by the receiving tanks is low, and is proven by the longevity of the life that these systems have been in service without media change-out determined by a "sniff test" or breakthrough greater than 5 ppmv of hydrogen sulfide. The carbon systems deployed at these facilities have operated for well over a year before requiring change-out.

Feedstock Tank

The best odor control for a feedstock tank has been connecting the head space of the feedstock tank with the inflatable digester tank gas storage to control displaced air during filling/discharging and/or temperature expansion of the feedstock materials. A pressure relief valve is typically installed on the feedstock tank to ensure safe pressure levels are maintained, but it should never see an active release unless both the flare and engine are out of service. This rare condition could only occur for a very limited time since a reduction in feedstock would occur simultaneous with any significant generator set malfunction. Consequently, odor potential from this relief valve would be extremely low and generally does not warrant control.

Digester Tank

Like the feedstock tank, the digester tank is also completely sealed to maximize the capture of biogas to capitalize on energy production. So, odors from the digester tank are generally not a concern. But, there is the potential for an emergency release, which could create the potential for an odor, and health concern. This situation is rare, but it has occurred at one (1) farm where an element of the feedstocks caused foaming in the digester resulting in biogas being blocked from both the engine and the emergency flare. Gas pressure continued to increase in the digester and some of the biogas was eventually vented to the atmosphere to prevent damage to the digester tank. Although not ideal, this situation is very rare, and was a lesson learned for the facility operator. And, no formal odor complaints resulted from the event, likely due to the short duration of the release and the significant dispersion associated with it. The facility operator self-reported the release to the regulator per a condition of their operating permit.

Effluent Management

Nuisance odors from dewatering of biosolids at a wastewater treatment plant are a common problem, but the difference at a farm-based digester is that the digestate is, or at least should be, mostly oxidized. The result is that odors from the dewatering process are generally not a significant concern. None of the farm-based anaerobic digesters reviewed have included odor control on the effluent management processes.

Gas Processing

Releases of hydrogen sulfide, and other oxygen-deficient odorants, from the engine and flare are generally not an odor concern due to the high destruction efficiencies of those combustion devices, but the resulting formation of sulfur dioxide is worth noting. Sulfur dioxide is produced by the combustion sulfur, and thus the combustion of biogas containing hydrogen sulfide will produce sulfur dioxide and its characteristic "burning match" odor. Sulfur dioxide can also be a severe irritant at high enough concentrations and could result in a nuisance condition, particularly if emitted from a stack with human exposure directly downwind. Minimizing the impacts of sulfur dioxide is achieved by minimizing the concentrations of hydrogen sulfide in the biogas produced in the digester.

Hydrogen sulfide will always be present in the biogas created by anaerobic digesters, although the concentrations will vary with the feedstocks. The concentrations of hydrogen sulfide have typically been less than 500 ppmv at the farm-based digesters systems reviewed, but at times have had peaks as high as 4,000 ppmv. Some farms have installed biogas pretreatment systems to address this odor concern, but generally their primary motivation of these pretreatment systems is to satisfy regulations to minimize sulfur dioxide for regional and national concerns. Regardless, these pretreatment systems have limited the release of sulfur dioxide and have reduced the odor potential from the combustion of biogas at farms.

Use of iron oxides (i.e. iron sponge) technology has been the most successful hydrogen sulfide pretreatment system at the farms reviewed. This technology is proven, and is capable of reducing hydrogen sulfide at some of the very high peak concentrations (i.e. greater than 1,000 ppmv) experienced at the farms. The technology is relatively maintenance free, and if sized properly, can allow for a significant duration of time between change-outs.

Figure 2. An iron sponge biogas pretreatment system at a farm-based digester

Land Application

Undigested manure at a traditional dairy farm is usually stored in open lagoons and is land applied in the spring and fall emitting odorous ammonia, volatile organic compounds, especially fatty acids, and reduced sulfur compounds, which can be an odor nuisance to neighbors. The liquid fertilizer produced from digested manure has a reduced odor potential in comparison with the practice of using raw cow manure from the open lagoon. The end result is a major reduction of odor complaints associated with land application at the farms with digester systems. Digestate contains all the nitrogen, phosphorus and potassium from the original feedstocks, and thus, has high value as an organic fertilizer. But, the digestate also contains any material that has not fully biodegraded in the digester and was converted into biogas. Consequently, there have been odor issues at some farms land applying digestate from their farm-based digester.

At one farm, odor complaints arose from their nearest neighbors after land application of digestate, but the complaints did not persist. Although, at another farm odor complaints were more persistent and occurred as much as three (3) to four (4) miles away from the farm and in two (2) different towns after land application of digestate from their digester. In both of these cases, it was determined that the material in the digester was not allowed enough time necessary to allow the methanogenic bacteria to fully convert the organic biomass into biogas. The digestate removed from the digesters were not fully decomposed and still possessed many of the same odorants that are emitted from the spreading of raw manure.

The reason for the problematic digestate in these two (2) examples is unknown. It is possible that a change in feedstocks required a longer detention time than the facility was typically utilizing. And, it is also possible that the retention time was reduced to make room in the receiving tanks for more material. Regardless, additional processes could be explored to minimize the potential for land applying immaturely digested material. Direct injection of digestate into the soil when land applying would also reduce odors.

Digestate, done right (i.e., with residence time long enough to extract the optimal energy), is typically less odorous than raw manure. Yet, odor complaints have been obtained from areas with land applied digestate. How can this be? Simply put, it gets back to what individuals deem a "normal" or, to use a double negative, a not unpleasant smell. It is not that manure odor is necessarily pleasant, but it is something that many people grew up knowing, and acclimating. It can remind individuals of summertime, harvest time, or childhood, in general, when time seemed endless. They did not, and therefore, do not associate digestate with these memories. As a result, their tolerance is much lower.

Manure odor contains three dominant odorants, "two volatile fatty acids—isovaleric acid and butyric acid—and the aromatic compound 4-methylphenol were responsible for over two-thirds of detectable beef manure odors. Most of these odors were released within 24 hours after manure was applied to the soil" (https://www.ars.usda.gov/news-events/news/research-news/2014/sniffing-out-the-source-of-beef-manure-odor/). None of these compounds have a positive hedonic tone (relative pleasantness), however, combined together they form the dominant smell of manure.

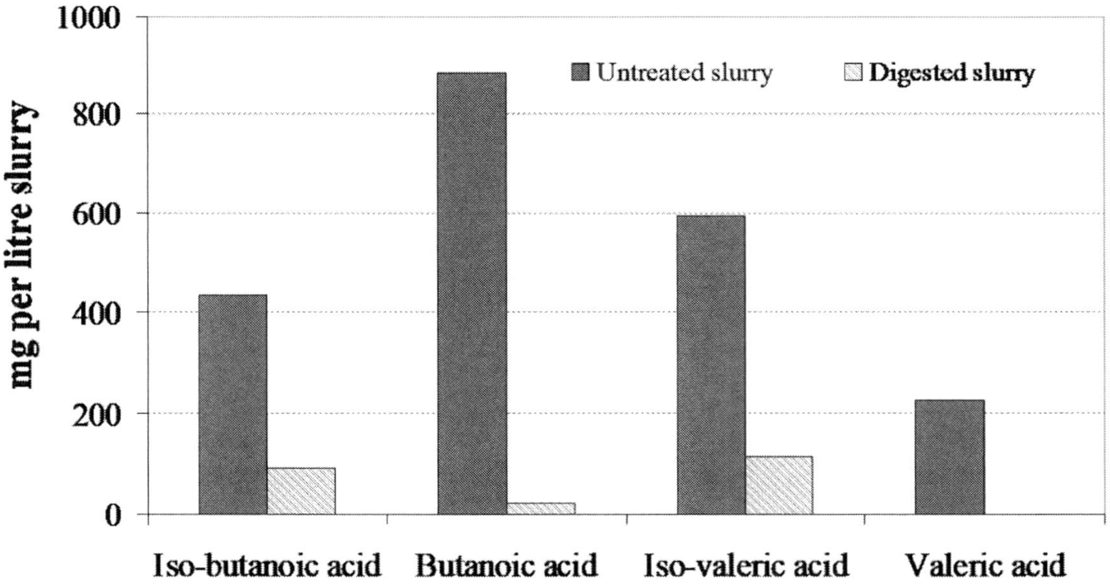

Figure 3: Concentration of volatile fatty acids in untreated slurry and digested slurry (Hansen et al., 2004)

These fatty acid compounds are reduced in digestion. As one can see from Figure 3, there is a sharp drop in volatile fatty acids of concerns with manure, as expected, since they are a good food source for the microorganisms in anaerobic digestion. Clearly, the dominant odors now change.

Although these three compounds dominate in odor, there were a number of other compounds, 135 identified in the USDA study. The other compounds consist of the typical odorants of concern from wastewater, composting, solid waste degradation, etc. These compounds are reduced sulfur and reduced nitrogen compounds. The reduced nitrogen compounds are commonly ammonia and the reduced sulfur compounds and are commonly methyl or dimethyl chain sulfur compound, such as methyl mercaptans and dimethyl sulfides. Once the manure odor is gone, the traditional odorants now will dominate, so in some ways, the distinctive manure odor masks the smell of traditional odorants of concern.

Furthermore, the oxidation rate and the degree to which manure will stabilize and a digestate will stabilize once land applied are different. The digestate has active biology that can react more quickly, causing higher initial peaks of odor release. So, although the overall odor potential is less, the odor character and the odor release rates can cause more odor concerns with digestate than from manure land application. Unfortunately, people often associate these compounds with wastewater, garbage, sewers, etc.

The shift in character and release rate are important considerations, since digestate has the added advantage of being applied during the full growing season, and not just before and after crop harvesting, as with manure, when less people are generally outdoors. This potential odor shift is manageable, but it was a surprise for some farmers at first that simply thought, "Hey, the digestate odor is less, so my odor nuisance potential is less."

CONCLUSIONS

It may seem silly that a farm that used to simply apply manure to its fields now has to consider odor control for biosolids and manure, or a food waste and manure digester, but after going through the process a number of times, it actually makes sense.

Odor control may be required, even though odor potential is going down. Manure odor is different than digestate odor. The change may be acceptable as a "new farm odor" over time, but initially, it may smell more like a waste processing facility to neighbors, and not the traditional "folksy" farm smell.

The change in odor coupled with the new industrial look at the farm can change public perception as well. Tolerance for odor is a funny thing. While rural neighbors have a high tolerance for traditional farm smells (and many local laws exclude farm odors from nuisance regulations), once the facility is no longer only a farm, their tolerance may go down.

And lastly, farming is traditionally a daytime activity. Yes, they often start early in the morning, but it is quiet at night. Anaerobic digestion is a continuous, 24-hour a day operation. Other potential nuisances, such as noise, can affect a farm's tolerance for odor too. So, essentially the answer is "Yes, even though the farm had open pits for lagoons previously, changes made by adding anaerobic digestion could require a facility to reduce odors to less than those present before the farm was just a farm." The good news is that many farms have significant buffer between them and their neighbors, so these changes will need to be explored, but may not require significant odor control requirements. The key here is odor control at farm based digesters is not, one size fits all. It is dependent on many very local factors.

REFERENCES

Hansen, M.N., Birkmose, T., Mortensen, B., Skaaning, K. (2004). "Miljøeffekter af bioforgasning og sepa-rering af gylle". Grøn Viden, Markbrug, nr. 296.

Identification of Musty Odor Nuisance at Wastewater Treatment Plants

Yubin Zhou[1]; Tadeo Vitko[2]; S. Cowden[3]; I.H. (Mel) Suffet[1]

[1]Department of Environmental Health Sciences
School of Public Health
UCLA
Los Angeles CA 90095

[2]Orange County Sanitation District
10844 Ellis Avenue
Fountain Valley, CA 92708

JACOBS

1100 NE Circle Blv'd, Suite 300
Corvallis, OR 97330

Corresponding Author: Mel Suffet

Email: msuffet@ucla.edu

Cell: +1 310-601-0971

Address: Department of Environmental Health Sciences, School of Public Health, UCLA, Los Angeles CA 90095

Abstract

Complaints from neighbors around wastewater treatment plants (WWTPs) are primarily about the odors emitted from treatment processes. A study of the nuisance odors at two WWTPs located at the Orange County Sanitation District (OCSD) in Fountain Valley, CA indicated that musty type odors were actually present in the odors released. These odors have been known from studies to occur from aerobic sites at natural and drinking water locations. These types of odors have not been studied as an odor nuisance from WWTPs. OCSD operated a variety of air treatment technologies (bioscrubbers, chemical scrubbers, biofilters, and activated carbon filters). These odor control technologies at the OCSD WWTPs as presently operated failed to remove most of the musty odor and thus the musty odor causing chemicals.

A Gas Chromatography-Mass Spectrometry (GC-MS) system was applied to investigate the chemical sources of these odors. 2-Methylisoborneol (MIB) and 2-isopropyl-3-methoxypyrazine (IPMP), that have been reported to cause musty odor in drinking water supplies, were identified by GC-MS as principal contributors to the musty nuisance odor at the WWTPs odor sources. The odor threshold concentrations (OTCs) were measured and determined to be 0.1 ng/L (0.014 ppbV) for MIB and 0.02 ng/L (0.003 ppbV) for IPMP. Rotten vegetable and fecal odors apparently mask odor panel identification of the true intensity of the musty odor.

Keywords

Odor Nuisance; Musty Odors; Chemicals Causing Musty Odors, Treatment of Musty Odors

Abbreviations

- Wastewater treatment plants ---------------------------------- WWTPs
- Orange County Sanitation District --------------------------- OCSD
- 2-Methylisoborneol --- MIB
- 2-isopropyl-3-methoxypyrazine ------------------------------- IPMP
- Gas Chromatography with Mass Spectrometry ------------- GC-MS
- Odor threshold concentrations ---------------------------------- OTCs
- Volatile organic compounds ------------------------------------ VOCs
- Odor profile method --- OPM
- Odor activity value -- OAV
- Flavor profile analysis -- FPA
- Solid phase micro-extraction ------------------------------------ SPME
- Granular activated carbon -- GAC

1. Introduction

Complaints from neighbors around wastewater treatment plants (WWTPs) are primarily about the odors emitted from treatment processes. Usually complaints are from anaerobic treatment processes (Hwang et al., 1994). Volatile sulfur compounds (VSCs), especially hydrogen sulfide are considered as the primary cause of odor emissions from WWTPs (Fisher et al., 2017). Typical odors from VSCs are rotten eggs (hydrogen sulfides) (Van Langenhove et al., 1985; Zarra et al., 2008) and rotten vegetable (methyl sulfides) (Zarra et al., 2008). Therefore, in many odor studies at WWTPs, the control of hydrogen sulfide is a surrogate for odor control (Fisher et al., 2017). However, other volatile organic compounds (VOCs) can also contribute to the odor emissions, but have been paid little attention. For example, Zhou et al., (2016) reported the fecal odor as a major odor from WWTPs. Two fecal odorants indole and skatole were identified as chemicals that were present that have very low odor threshold concentrations (OTCs) at these WWTPs. When these chemicals were removed, the fecal odor dissipated.

Musty odors has been detected by OCSD staff at the OCSD WWTPs and residents near the WWTPs that were studied (Vitko, 2014). However, musty odors were not determined as an odorant of concern at the WWTPs. OCSD operated a variety of air treatment technologies (bioscrubbers, chemical scrubbers, biofilters, and activated carbon filters) to minimize odor nuisances from different unit operations. An initial study of the nuisance odors at 32 WWTP process locations at the two Orange County Sanitation District (OCSD) WWTPs in Southern California indicate that the musty/moldy/earthy group herein called musty odors identified by the Wastewater Wheel were actually present in odors released (Vitko et al., 2014). These types of odors have not been investigated as an odor nuisance from WWTPs.

Musty/earthy odors are the most common descriptors for odor complaints in the drinking water industry from aerobic natural water locations (Suffet et al., 2018). These odor types from studies at natural and drinking water locations occur from aerobic sites (Mackay et al., 2008, Yu et al., 2014; Joe et al., 2007). The Drinking Water Taste and Odor Wheel (Suffet et al., 2017) identifies - Trans-1,10-dimethyl-trans-9-decalol (geosmin), an earthy odor, 2-methylisoborneol (MIB), a musty odor, 2-isopropyl-3-methoxypyrazine (IPMP), an earthy/musty odor and 2,3,6-trichloroanisole (TCA), a musty/moldy/cork odor as primary causes of earthy/musty/moldy type odors. For example, the major source of MIB in drinking water is primarily from the metabolism and biodegradation of algae, such as, cyanobacteria (Izaguirre et al., 1982) and IPMP has been detected as a result of the metabolism of algae (Peter et al., 2009). Musty nuisance odors have also been found in wine (Ortega et al., 2001), indoor air (Ömür-Özbeket et al., 2011) and pulp mill waste (Ferguson et al., 1995), but little attention has been given to the earthy/musty/moldy odors at WWTPs. The Wastewater Odor Wheel notes musty/earthy odors as a possibility, but without identification of a specific chemical (Burlingame, 1999; Burlingame et al., 2004; Suffet and Rosenfeld, 2007; Burlingame, 2009). Recently, MIB has been detected in the effluent water of municipal wastewater, with a concentration of up to 30 ng/L (Agus et al., 2012). It was not identified in the air from different WWTP processes.

The OTC of the known earthy/musty/moldy odors in drinking water have been determined. At about 20 °C, the OTC of MIB in water is reported ranging from 5 to 15 ng/L as generally compiled by Mallevialle and Suffet (1987) and Suffet et al. (1995a). A recent study (Yu et al., 2014) found a wide variation in MIB odor sensitivity between eight different drinking waters using Flavor Profile Analysis (FPA) (APHA, 2012) with trained odor panels in China. The OTCs in water ranged from 7 to 20 ng/L for samples at 45 °C. The OTC in water was determined by FPA to be 2.1 ng/L for MIB at 45 °C in USA studies in Philadelphia PA and at the Metropolitan Water District of Southern California (Mackey *et al.*, 2008). Drinking water consumers usually complain when the levels of MIB reach 10 ng/L (Suffet et al., 2008). The OTC of MIB in air varies from 0.013-0.07 ppbV and the OTC of IPMP in air varies from 0.0008 to 0.008 ppbV as reported in literature (Ömür-Özbeket et al., 2011; Sala et al., 2004). Thus, it was strongly suspected these odorants would be odor nuisance at WWTPs in air at the ppbV level.

The objectives of this study at the two WWTPs of the OCSD were: 1) to observe if the musty group odor nuisance was present by odor panel analysis, 2) to determine the musty group chemical odorants found at different WWTP sources, 3) to evaluate the effect of current treatments of off-gases for removal of the musty group odors by odor panel analysis 4) to evaluate the effect of an activated carbon pilot plant treatment of off-gases for removal of the musty group odors by an odor panel analysis 5) to determine the OTCs of the musty odorants present and quantify the chemical concentration of the musty odorants at the WWTPs.

2. Methodology

In order to control odors, it is necessary to have a consistent language for their definition and an understanding of the possible chemical compounds responsible. The Wastewater Odor Wheel and the odor profile method (OPM) (Burlingame, 1999; Burlingame et al., 2004; Suffet and Rosenfeld, 2007; Burlingame, 2009) were developed to categorize potential wastewater odors and their intensities by odor panels as well as to help identify the chemical odorants that define those odors characteristics by chemical analysis methods.

Specific chemical causing odor nuisances can be evaluated by the ratio of the concentration of the odorants divided by its odor threshold concentration (OTC) of the offending compound, the so-called Odor Activity Value (OAV) (Feilberg et al., 2010). The Weber-Fechner curves which relates the OTC in air to the odor intensity of a chemical concentration in air must be defined to evaluate the odor nuisance. Weber observed that increased stimuli is relative to the previous amount experience and his student, Fechner (1859) described this relationship that has been used by Greenman et al., (2005); Munoz et al., (2010); and Suffet et al., (1995a). The intensity of an odor characteristic is a measure of its odor strength, which is related to the log of its concentration via the Weber-Fechner law Equation (1):

$$I = k \, Log(C/C_o) \quad (1)$$

The intensity (I) is proportional to the logarithm (base 10) of the concentration (C) relative to a previous concentration (C_o). This has previously been demonstrated for the fecal odorant skatole (Zhou et al., 2016).

Figure 1 shows the Weber–Fechner Curve. For each odor character, the OPM method develops an odor intensity that is related to log concentration of the odorous chemical. The odor intensity scale has 7 points from 1 to 12; 1 (threshold), 2, 4 (recognition), 6, 8, 10, and 12. At the odor threshold intensity of 1, an odor panelist can state that the odor is different than clean air. Only at an odor intensity of 4 can the panelist recognize odor characteristic. An average odor intensity of 3 is considered as an action level for drinking water plants to avoid odor nuisance complaints (Burlingame, 2004). Thus, a calculated average panel odor threshold of 3 was used as an action level warning to the WWTP that an odor nuisance can occur from a particular type of odor. The OTC is formally described as the point where 50% of a population (minimum of 4 panelist) can define a difference between pure air and air containing an odorant. The odor recognition threshold is where 50% of an odor panel can define the type of odor from the odor wheel.

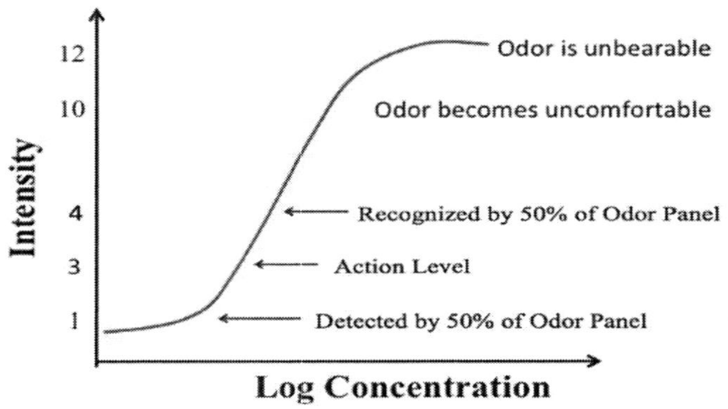

Figure 1. Intensity Scale for Each Odor by Weber-Fechner Curve

2.1 Odor Profile Method

The odor panel used the Odor Profile Method (OPM) with the Wastewater Odor Wheel to describe odor characters and the FPA Intensity Scale for each odorant to report odor intensity (Burlingame, 1999; Curren et al., 2014). The OPM is based upon the FPA Standard Method #2170, (APHA, 2012) that is used as sensory panel methods in drinking water odor control studies (Suffet et al., 1987). A minimum of 4 trained panelists were used to analyze air samples at UCLA as well as 14 consumer panelists at OCSD.

2.2 Solid Phase Microextraction and Gas Chromatography-Mass Spectrometry (GC-MS) of Earthy/Musty/Moldy Odorants at the OCSD WWTP

A solid phase micro-extraction (SPME) method (Godayol et al., 2013) that was used to identify and quantify skatole and indole (Zhou et al., 2016) as odor nuisances at WWTPs was adapted for analysis of the musty group of odorants. A SPME fiber of DVB/CAR/PDMS was injected into a 10-L sample bag and exposed for 20 minutes at room temperature. The SPME fiber was then injected onto a Varian 450 gas chromatograph (Varian Inc., Palo Alto, California) through a splitless #1177 liquid-injector port at 250 °C followed by a Varian 220 mass spectrometer (SGE Analytical Science, Austin, Texas) with 99.9999% helium as the carrier gas. The gas chromatography (GC) system was chosen based upon the broad spectrum GC system used for the indole and skatole method (Zhou et al., 2016). The GC was equipped with an Agilent DB-5MS column (length 60 m, inner diameter 0.25 mm, film thickness 0.25 μm) and held at an initial temperature of 65 °C with a ramp of 9.5 °C/min to 160 °C and then 12 °C/min to 240 °C. The carrier-gas flow rate was 1.0 mL/min through the column. The retention time and mass spectrum of standards of MIB, geosmin and IPMP were determined and standard curves were prepared.

The ion-trap mass spectrometer used selective ion monitoring for the primary range of m/z units of the two musty group odorants found: MIB (56-58, 70-72 and 84-86 m/z from 14.50 to 17.00 minutes) and IPMP (108-110, 123-125, 136-138, and 150-153 m/z from 14.40 to 15.00 minutes). The detection limit for MIB was 0.4 ng/L in air. The detection limit for IPMP was 0.4 ng/L in air. The SPME recovery for MIB and IPMP were 92% for MIB and 90% for IPMP. It should be noted that these levels are significantly greater that the OTC levels reported later in this paper. The concentrations measured in field samples were not corrected for recovery.

2.3 Odor Threshold Concentration (OTC)

The OTCs of odorants in this study were determined by trained odor panelists by the OPM (Burlingame, 1999) using the ASTM Method 679-04 (2011), a forced choice triangle test in an ascending concentration series with dilutions made by an AC'SCENT olfactometer (St. Croix Sensory, Lakewood, MN). An olfactometer is an air-dilution instrument that precisely dilutes a given odorous compound with pure air so that the exact concentration is known. The bag with odorants was then attached to an olfactometer, which delivered dilutions at known ratios to panelists. The panelists were forced to choose between two blanks and an odorous sample in the triangle test. The panelists were then asked to provide an odor-intensity rating for each dilution using the OPM scale. OTCs were then calculated by plotting the average odor intensity of at least 4 panelist for a given concentration versus the logarithm (base 10) of the concentration (in ng/m^3), performing a least-squared linear regression, and calculating the concentration that would give an intensity of 1 (threshold) using the linear equation.

2.4 Sampling Location

Samples were collected at 16 locations at each of the OCSD WWTPs in 2013. Plant 1 is located in Fountain Valley, CA and Plant 2 in Huntington Beach, CA. The sampling locations studied were from:

<u>Plant 1</u>

Headwork Scrubbers, Primary Scrubbers, Waste Hauler Biofilter, Trickling Filters, Activated Sludge, Dewatering, And Truck Loading Facilities;

Plant 2
Trunkline Scrubber, Headwork Scrubber, Primaries Scrubber, Trickling Filter Scrubber, Trickling Filter Contact Basin, Trickling Filter Mixed Liquor Channel, Activated Sludge Discharge Channel, Dewatering And Truck Loading Scrubber. Subsequent samples were collected from 2014-2015 from a pilot air scrubbing system study using Trunkline foul air (at the start of the Headworks system at Plant 2).

Pilot Plant
Carbon Column (CARB 3) Danasorb DAS- 100C Virgin Coconut Shell Granular Activated Carbon (GAC) (COC –A60). Conditions: Air Flow Rate = 120 cfm, Empty Bed Contact Time =3.5 seconds

2.5 Sample Collection Method

Air samples were collected at existing positive pressure air collection ducts in 10 L Tedlar or Teflon Bags, except for those processes with open surfaces exposed to the atmosphere. At the latter locations, such as: activated sludge discharge channel, trickling filter media surface, trickling filter reactor basin, and trickling filter mixed liquor channel, a flux chamber with sweep gas was used to fill the bags (EPA, 1986). Where the WWTP process area foul air was treated with a scrubber, sampling was done before (IN) and after (OUT) of the scrubber unit.

3. Results and Discussion

3.1 Odor Nuisances Determined by the OPM

The OPM determined many nuisance odors at the 32 WWTP locations at the two OCSD WWTPs. In order to better characterize the odors identified at OCSD, odors were grouped into 4 broader categories in this study (Vitko et al., 2014):

Rotten vegetable/canned corn/sulfide (rotten vegetable group),
Rotten garlic (rotten garlic group),
Musty/moldy/earthy (musty group),
Fecal/sewery/manure/rubbery (fecal group)

This general nomenclature is based on the Wastewater Odor Wheel (Burlingame, 1994). Since an average odor intensity of 3 was considered as an action level for drinking water plants to avoid odor nuisance complaints (Burlingame, 2004), it was used as guideline for odor nuisances in this study as defined by the Weber-Fechner Curve (Figure 1).

3.2 Musty Group Odor Found at the WWTPs of the OCSD

Table 1 showed that the musty odor group was reported by the OPM at over half of the locations in both WWTPs. The intensities by the OPM of musty odors at Plant 2 were higher than those at Plant 1. The OPM values that are reported are the average of 4 grab samples of 10L taken in the morning at each location on different days from August to October in 2013. Thus, the average and standard deviation shows the variability of the OPM data at each location.

Plant 1	Pretreated Trunklines & Bin Loading Bldg.	Bar-screens & Grit Chamber	Primaries	Waste-Hauler	Trickling Filter	Activated Sludge	DAFTs	Dewatering	Truck Loading	
Scrubber Treatment	Chemical	Chemical	Chemical	Biofilter	None	None	Chemical	None	None	
Earthy/ Musty/ Moldy Odor Group	1.3±1.9 (in)	1.3±1.2 (in)	0.9±1.5 (in)	0 (in)	3.3±0.6	2.8± 1.8	2.7±1.1	2.9±1.7	2.2±2.3	
	2.0±1.2 (out)	1.5±1.5 (out)	0.5±0.8 (out)	0.5±1.0 (out)						
Plant 2	**Trunklines**	**Pretreated Trunklines & Headworks**	**Primaries**	**Trickling Filter**	**Trickling Filter Contact Basin**	**Trickling Filter Mixed Liquor**	**DAFTs**	**Activated Sludge Discharge Channel**	**Dewatering**	**Truck Loading**
Scrubbing Technology	Bioscrubber	Bioscrubber & Chemical	Chemical	GAC	None	None	Chemical	None	None	GAC
Earthy/ Musty/ Moldy Odors	2.9±1.1 (in)	1.5±1.8 (in)	0.5±1.0 (in)	2.9±2.5 (in)	4.8±0.9	4.5±1.5	1.4±1.6	4.5±1.2	1.1±1.3	2.4±1.8 (in)
	1.8±1.4 (out)	1.4±1.1 (out)	0 (out)	1.0±1.2 (out)						1.7±2.0 (out)

Table 1: **Average ± the Standard Deviation of Four Odor Profile Method's Individual Odor Intensities of Earthy/Musty/Moldy Odor Group at Unit Operations at the OCSD WWTP by an Expert Panel.**

NOTE 1 - Present odor treatment operations are included.
NOTE 2 - Other Odors were present (not shown) and thus these odor intensity values are minimum values as masking of these odors could occur from the rotten garlic, rotten vegetable and fecal groups.

The OPM results indicate the highest intensities of musty odors were primarily from aerobic secondary wastewater biological treatments unit operations, such as activated sludge and trickling filters. Musty odors were observed from the unit operations of Dewatering (Intensity = 2.9) and Trickling Filter (Intensity = 3.3) at Plant 1 and Trickling Filter Contact Basin (Intensity = 4.8) and the Activated Sludge Discharge Channel (Intensity = 4.5) as well as the Trickling Filter Mixed Liquor Channel (Intensity = 4.5) at Plant 2. It is noted that with the addition of appropriate grab samples peaking factors risk is higher at these locations.

It is important to note noted that other odors were present but not shown on Table 1. The rotten vegetable and fecal odor groups were dominant at all locations at both plants and could mask the total musty odor. Even though masking probably occurred especially at the locations before the Tricking Filters, musty odors were detected as part of the OPM odor characteristics at these locations. Thus, these musty group odor intensity values are minimum values as masking of the musty odors could occur from the rotten garlic, rotten vegetable and fecal odorants.

3.3 Effectiveness of the Present Musty Odor Treatment Determined by OPM

Four different odor treatment technologies were in use at OCSD: biofilters (natural media), bioscrubbers (biotrickling filters with synthetic media), packed bed chemical scrubbers with caustic and/or bleach and activated carbon filters (Vitko et al., 2014). The present efficiencies of these treatment technologies after a long time of service were evaluated by comparing the inlet and outlet intensities of musty odors based upon the data of Table 1, that reported OPM panel analysis values as the average of 4 samples taken at each location. Thus, the average and standard deviation shows the variability of the OPM data at each location. The rotten vegetable and fecal odor groups were dominant at all locations at both plants, but musty odors were detected as part of the OPM odor characteristics.

The input and output of 4 treatment processes were investigated in Plant 1. Figure 2 shows that the 3 primary chemical scrubbers that were operated in the caustic/bleach mode (at pH = 8-9) at the time, performed best for musty odor control, removing up to 44% of the OPM odor intensity. This is surprising since it is generally accepted that chlorine does not oxidize musty odorants as MIB and geosmin under water treatment conditions (Srinivasan and Sorial, 2011). Thus, it is hypothesized that 1) dissolution in water is the primary mechanism of removal of the musty chemicals or 2) this data may indicate masking of the inlet odor by high concentrations of fecal and rotten vegetable odors that would make the inlet musty odor lower than it is chemically whereas the outlet odor would <u>not</u> be as strongly masked as the fecal and rotten vegetable odors are decreased. The mechanism for removal of musty odorants on a true chemical basis in the chemical scrubbers should be investigated.

Figure 2 shows that the biological treatment technologies of a biofilter, operating at the Plant 1 Wastehauler Station actually created musty odors. The general findings in water treatment technology was that biological treatment do not remove musty odors (Srinivasan and Sorial, 2011). Thus, the second hypothesis above for chemical treatment probably explains the result, i.e. masking of the inlet odor by high concentrations of fecal and rotten vegetable odors would make the inlet odor lower than it is chemically whereas the outlet odor would not be as strongly masked as the fecal and rotten vegetable odors are decreased. The mechanism for removal of musty odorants by biological treatment should be investigated.

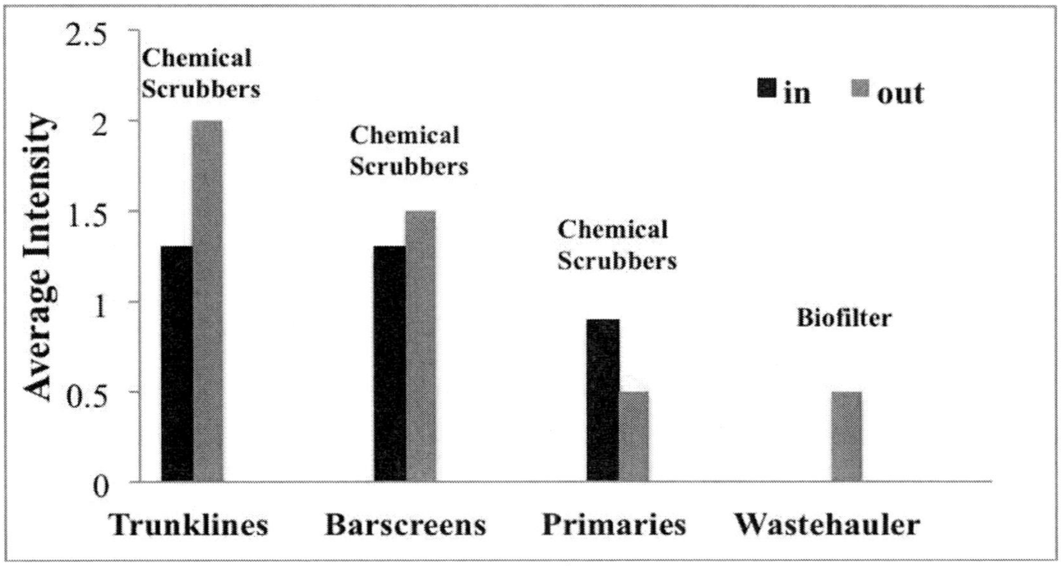

Figure 2. Air Treatment of Musty Odor Intensities "In" vs. "Out" of Treatment Plant Processes at Plant 1. Note: Headwork contains Barscreens.

In Plant 2, there are 5 plant processes that undergo treatment, using Chemical Scrubbers, Bioscrubbers, or GAC filters. The present efficiencies of these treatment technologies after a long time of service were evaluated by comparing the inlet and outlet intensities of musty odors based upon the data of Table 1, that reports OPM panel analysis values as the average of 4 samples taken at each location. Thus, the average and standard deviation shows the variability of the OPM data at each location. The rotten vegetable and fecal odor groups were dominant at all locations at both plants, but musty odors were detected as part of the OPM odor characteristics.

Figure 3 shows that based upon the OPM intensities, all treatments can remove part of the musty odor. The musty odors were removed 100% at Primary Chemical Scrubbers. This is similar to Plant 1 and remains surprising since it is generally accepted that chlorine does not oxidize MIB under water treatment conditions (Srinivasan and Sorial, 2011). Thus, it is again hypothesized as for Plant 1 that dissolution in water is the primary mechanism of removal of MIB or this data may indicate masking of the inlet odor by high concentrations of fecal and rotten vegetable odors would make the inlet odor appear lower than it is chemically whereas the outlet odor would not be as strongly masked as the fecal and rotten vegetable odors are decreased. Again, the mechanism for removal of MIB in chemical scrubbers is not known and should be investigated.

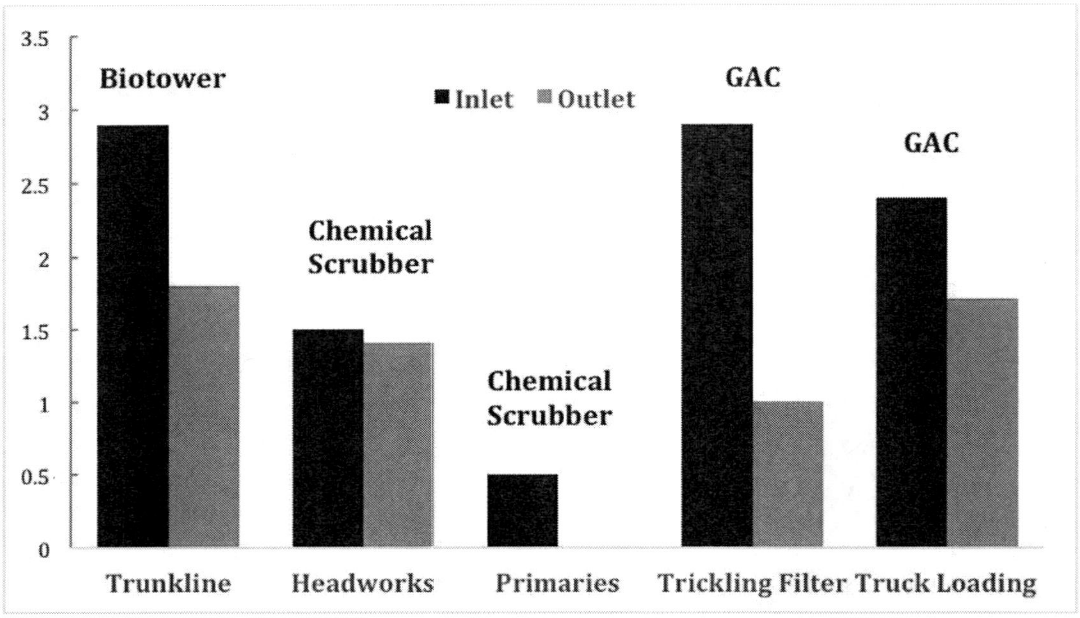

Figure 3. Air Treatment of Musty Odor Intensities "In" vs. "Out" of Treatment Plant Processes at Plant 2

In Plant 2, the Biotower at the Trunkline remove 38% of the musty odor intensities. This result was surprising. No explanation can be developed for this to occur at this time. More studies are needed.

GAC columns are widely used on taste and odor control for musty odorants as MIB and geosmin in drinking water (Suffet et al., 1995, Bruce et al., 2002). GAC filters at the Trickling Filter removed 65% of the musty odors while only 29% were removed by the GAC columns at the Truck Loading Facilities. The GAC systems at these locations in Plant 2 did not completely remove the musty odors. The probable reasons were that 1) the GAC may have been partial blinded by excess water 2) the GAC may have been saturated and desorbing odorants and/or 3) the masking effect discussed above is less from high concentrations of other odorants present. Here, the greater musty odor removal efficiency for the trickling filters would be considered more accurate since overall masking effects are lower for fecal and rotten vegetable odors at this location. Knowledge of the history of the musty input chemicals and competing chemicals in these column is not sufficient to completely understand the reason for the low removal by these GAC filters.

Thus, the current air treatment technologies of off-gases from WWTP processes as presently used at OCSD do not remove musty odors efficiently. The ability of the current foul air treatment technologies to remove the musty odors seemed to relate more to odorant loading of the GAC columns rather than specifically to the removal of musty odors. Musty odorants can be removed efficiently by activated carbon filtration at drinking water treatment plants (Suffet et al., 1995b). WWTPs do not currently treat musty odors in any targeted manner. GAC systems from start-up should be evaluated for musty odorants with the chemical matrix of the process to determine the useful efficiency of these systems. Advanced oxidation technologies using hydroxyl radical chemistry (e.g. ozone and UV-peroxide) have been found to be effective in removing musty odorants from water (Suffet et al., 1995b; Jung et al., 2004) and should be tested also as an alternative musty odorant removal method in air treatment systems of off-gases from WWTP processes.

3.4 Evaluate the effect of a GAC pilot plant treatment of off-gases for removal of the musty group odors by odor panel analysis.

Figure 4 shows that based upon the OPM intensities, the coconut carbon can remove the fecal odors and the D/T total odors. After a 20 % removal of the musty odor, the musty odor increases possibly as. the carbon became spent it also became wetter which could result in possible musty odor generation. However, it is hypothesized that this disparity occurred because inlet musty odors were very low and therefore removal rate accuracy was poor. This example is used to show the complications that can occur when odor characteristics are observed and not just D/T. Chemical analysis of the musty odorants would best evaluate the changes of the musty odor during treatment.

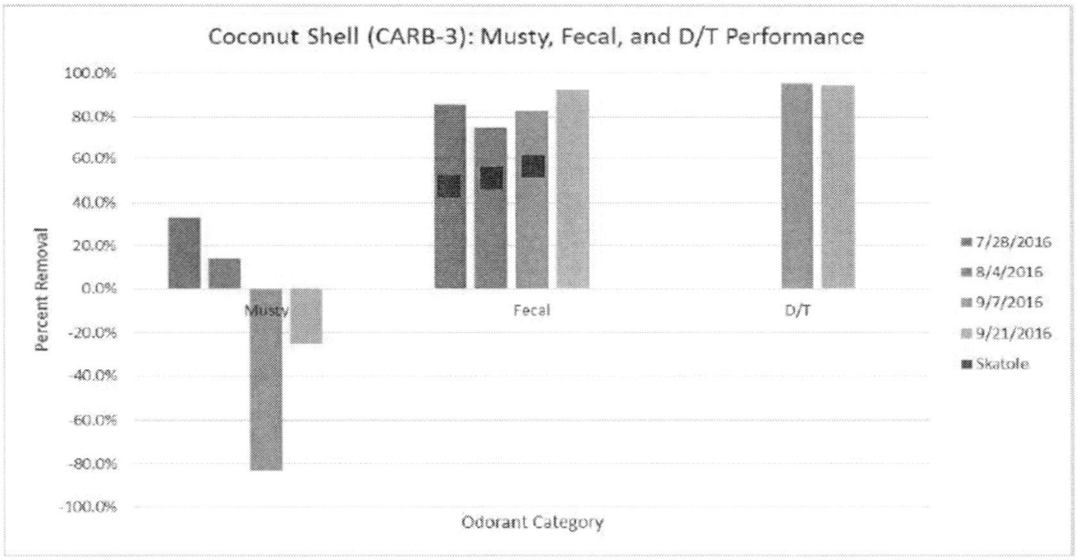

Figure 4. Air Treatment of Musty Odor Intensities "In" vs. "Out" of Pilot Treatment Plant Processes CARB 3 at Headwork. Carbon Column (CARB 3) Danasorb DAS- 100C Virgin Coconut Shell GAC (COC –A60). Conditions: Air Flow Rate = 120 cfm, Empty Bed Contact Time =3.5 sec.

3.5 Odor Threshold Concentration (OTC) of Musty Compounds in Air

The OTCs of MIB and IPMP in air samples were measured using the Olfactometer (Figure 5). The odor threshold of MIB was found to be 0.1 ng/L (0.02 ppbV). The odor threshold of IPMP was found to be 0.02 ng/L (0.004 ppbV). The odor threshold of MIB was found to be 0.1 ng/L (0.014 ppbV). The odor threshold of IPMP was found to be 0.02 ng/L (0.003 ppbV).

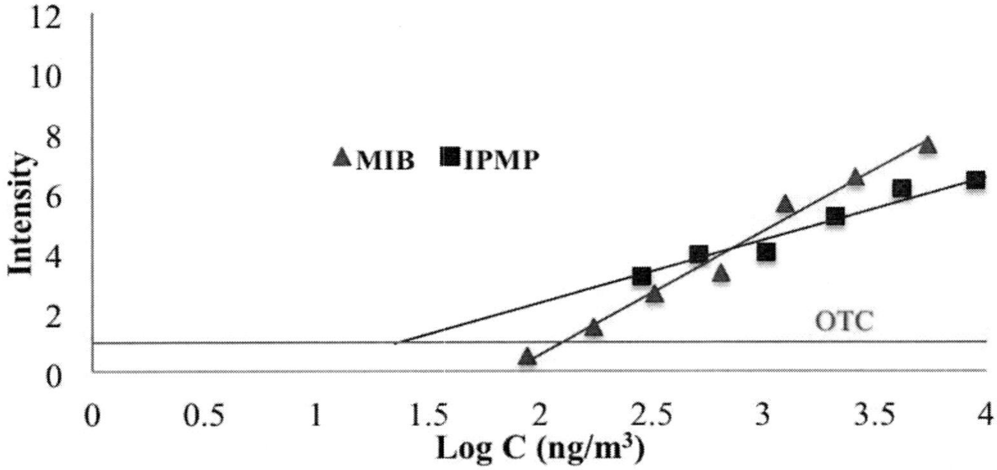

Figure 5. The Weber-Fechner Curves of MIB and IPMP

The OTC of MIB in air was reported as 0.09-0.47 ng/L (0.013 – 0.067 ppbV) in air (Ömür-Özbeket et al., 2011). The OTC of IPMP was reported as 0.005-0.05 ng/L (0.001 – 0.008 ppbV)in air (Sala et al., 2004). Both the OTCs of MIB and IPMP measured in this study were within the range of literature values in air. The difference between the literature and the results reported here may be due to the sensitivity of panel members, the methodologies used, the number, age, and sex of the panelists (Curren et al., 2014). The wide range of odor thresholds in the literature indicates variability in odor detection of different individual panelists. OTCs provide important information on the identification of the source of the odors as well as the importance of it odor nuisance potential.

3.6. Qualitative and Quantitative Analysis of Musty Odorants by GC-MS at the OCSD WWTPs

GC-MS scans of the samples from each OCSD location were evaluated for the four primary known earthy, musty and moldy odorants known: MIB, IPMP, geosmin and TCA. MIB (musty) was detected at 6 locations at Plant 1 and Plant 2 and IPMP (musty) were detected at 6 locations at Plant 1 and Plant 2. There was no detection of other earthy/musty/moldy odorants, such as geosmin or TCA in the samples. The concentrations of MIB and IPMP in air samples were quantified by GC-MS. The concentrations of MIB and IPMP and the intensities of the OPM of musty odors at 9 locations at both OCSD Plants are shown in Table 2. The concentration of MIB at Activated Sludge 2.5 ng/L [**0.36** ppbV) was the highest at Plant 1 and the concentrations at Mixed Liquor 2.8 ng/L (**0.40** ppbV) were the highest at Plant 2. The concentration of IPMP at the inlet of Headworks 1.8 ng/L (**0.28** ppbV) was the highest at Plant 1 and the concentrations at Trickling Filters 2.4 ng/L (**0.38** ppbV) were the highest at Plant 2.

Table 2 shows that the musty odorants IPMP and MIB were detected by the GC-MS where the musty odors were reported. This indicates a conclusive confirmation that MIB and IPMP would be the major musty odorants released from WWTPs. Table 2 shows also the importance of the odor nuisance potential by the ratio of the odor concentration to odor threshold concentration (C/OTC), i.e. the OAV in air. The OAV ratios are included for MIB and IPMP based on the OTCs measured in this study. The concentrations of MIB detected were higher than the concentrations of IPMP. However, the OAV ratio of IPMP (17-98) were higher the OAV ratio of MIB (4-20) due to the very lower OTC of IPMP. In this case, IPMP may contribute more to musty odors. However, both MIB and IPMP can likely lead to a musty odor nuisance off-site.

April 2016 Location	IPMP			MIB			Site OPM
	C (ng/L in air)	C (ppbV)	OAV = C/C	C (ng/L in air)	C (ppbV)	OAV = C/OTC	
P2 Trickling Filter	2.4 ±0.1	0.38 ±0.02	98	1.3 ±0.2	0.19 ±0.03	9	3.0
P2 Trickling Filter (Reactor A)				2.2 ±0.4	0.31 ±0.06	16	5.0
P2 Mixed Liquor West				2.8 ±0.3	0.40 ±0.04	20	2.0
P1 Headworks Inlet #1	1.8 ±0.1	0.28 ±0.02	77				Note
P1 Activated Sludge #5	0.4 ±0.0	0.06 ±0.0	17	2.5 ±0.3	0.36± 0.04	18	3.5
P1 Dewatering	0.5 ±0.1	0.08 ±0.02	20	1.7 ±0.3	0.24 ±0.04	12	3.5
P1 Daft Inlet	0.4 ±0.0	0.06 ±0.0	17				1.5
P2 Daft Inlet	0.6 ±0.4	0.09 ±0.06	25				1.0
P2 Trickling Filter #1				0.6 ±0.0	0.09 ±0.0	4	2.8

Table 2. Concentration of MIB and IPMP and OPM Results for one sampling set.
Note: Site OPM indicates minimal musty odors but may be masked by higher concentrations of Fecal and Rotten Vegetable Odorants

The concentrations of MIB and IPMP were used to calculate the expected musty odor intensities, using the Weber-Fechner curve, Figure 5. Table 3 shows the comparison of intensities of musty odors detected by OPM and calculated by Weber-Fechner Curves. The detected intensities were all lower than the expected values calculated from the concentration of either MIB or IPMP. The difference was most obvious at P1 Headworks Inlet, where strong fecal and rotten vegetable odors were shown. This strongly indicates that a masking effect for the OPM would occur where high fecal and rotten vegetable odors occur. The mechanism of masking effect of musty odors by the dominated odors needs to be investigated in future research. Therefore, if musty type odors are found by the OPM by odor panels, a chemical analysis for the musty odorants MIB and IPMP must be completed to evaluate the true effect of a musty odor nuisance off-site by the ratio of (C/OTC), i.e. the OAV in air at different WWTP sources.

Apr-16	Musty Odor Intensity Detected by OPM Panel	Calculated Musty Intensity	
Location		IPMP	MIB
P2 Trickling Filter	3.0	5.2	5.0
P2 Trickling Filter (Reactor A)	5.0		5.9
P2 Mixed Liquor West	2.0		6.4
P1 Headworks Inlet #1	Note	4.9	
P1 Activated Sludge #5	3.5	3.6	6.2
P1 Dewatering	3.5	3.8	5.5
P1 Daft Inlet	1.5	3.6	
P2 Daft Inlet	1.0	3.9	
P2 Trickling Filter #1	2.8		3.6

Table 3. Comparison of Intensities of Musty Odors Detected by OPM and Calculated by Weber-Fechner Curves.

4. Conclusions

- Musty type odors (Musty/moldy/earthy) have been detected by residents near the WWTPs that were studied. Musty odors were observed as odor nuisance from the two WWTPs of the OCSD by odor panel analysis at different WWTP locations.
- The intensity of the musty odor by the OPM at the WWTP were apparently lower as compared to the offsite locations by treatment processes at the WWTPs and masking by strong fecal and rotten vegetable odorants that were apparently diluted out quicker than the musty odorants.
- The musty odor was not removed by present long term used foul air treatment devices from the WWTP unit operations, as determined by odor panel analysis. Specifically, the biological treatment methods that were in use at OCSD did not remove musty odors well. GAC current treatment technologies of off-gasses from WWT processes as used at OCSD do not remove musty odors efficiently. GAC saturation and desorption is hypothesized to be the primary cause of the problem with present GAC columns. The wet scrubbers performed best to remove musty odors among current control technologies. Probably, dissolution in water is the primary mechanism of removal of the musty chemicals.
- The odorous chemicals MIB and IPMP were determined as two primary chemicals that cause the musty odor nuisance from WWTP unit operations. The OC/OTC (OAV value) data for MIB and IPMP confirm that musty odor group could lead to off-site nuisance from WWTPs.
- The OTCs of the musty odorants of MIB and IPMP were determined by the Weber-Fechner curve. The odor threshold of MIB is 0.1 ng/L (0.014 ppbV) and the odor

threshold of IPMP is 0.02 ng/L (0.003 ppbV) for our untrained panelists. These values are within the range of other OTC determined for these compounds in air.

- Rotten vegetable and fecal odors apparently mask the odor panel's identification by the OPM of the true intensity of musty odors. When rotten vegetable and fecal odors are removed efficiently or diluted out leaving the WWTP, musty odors may become major odor issues for WWTP locations as musty odorant have very low OTCs and have OAV ratios sufficient to be an off-site odor nuisance. Thus, if an odor panel at a WWTP identifies musty odors, a chemical analysis is needed to determine removal efficiency.
- The ratio of the odor concentration to OTC in air were determined at specific locations at the OCSD Plants and large OAV values indicate potential off-site odor nuisance.

References

Agus, E., Zhang, L., Sedlak, D., 2012. A framework for identifying characteristics odoring compunds in municipal wastewater effluent. Water Research, 46, 5970-5980.

APHA (American Public Health Association), 2012. *Standard Methods for the Examination of Water and Wastewater*: Method 2170, 22nd ed. APHA, the American Water Works Association (AWWA), and the Water Environment Federation (WEF), Washington, DC.

ASTM (2011) E679-04, 2011. Standard practice for determination of odor and taste thresholds by a forced-choice ascending concentration series method of limits, pp. 1-7, ASTM, West Conshohocken, PA, USA.

Bagheri, H., Salemi, A., 2006. Headspace solvent microextraction as a simple and highly sensitive sample pretreatment technique for ultra trace determination of geosmin in aquatic media. Journal of Separation Science 29, 57–65.

Bruce, D., Westerhoff, P., Brawley-Chesworth, A., 2002. Removal of 2-methylisoborneol and geosmin in surface water treatment plants in Arizona. J. Water Supply Res. Technol. 51(4), 183–197.

Burlingame, G.A., 1999. Odor profiling of environmental odors. Water Sci. Technol. 40(6), 31-38.

Burlingame, G.A., Suffet, I.H., Khiari, D., Bruchet, A.L., 2004. Development of an odor wheel classification scheme for wastewater. Water Sci. Technol. 49, 201–209.

Burlingame, G.A., 2009. A practical framework using odor survey data to prioritize nuisance odors. Water Sci.Technol. 59(3), 595-602.

Curren, J., Snyder, C.L.C., Abraham, S., Suffet, I.H., 2014. Comparison of two standard odor intensity evaluation methods for odor problems in air or water. Water Sci. Technol. 69(1), 142-146.

EPA (1986)/600/8-86/008. M.R. Kienbusch, Radian Corp. Measurement of gaseous emission rates from land surfaces using an emission isolation flux chamber.

Fechner, G., *Elemente de Psychophysik*, Thesis, Leipzig University, Germany.

Feilberg A., Liu, D., Adamsen, A.P., Hansen, M.J., Jonassen, K.E., 2010. Odorant emissions from intensive pig production measured by online proton-transfer-reaction mass spectrometry. Environ. Sci. Technol., 44 (2010), 5894-5900.

Ferguson, K., 1995. Mill odor control—A pulp and paper mills location and process determines how much end-of-pipe equipment it must use to control odorous emissions. Pulp Paper 69, 147–148.

Fisher, R.M., Le-Minh, N., Sivret, E.C., Alvarez-Gaitan, J.P., Moore, S.J., Stuetz, R.M., 2017. Distribution and sensorial relevance of volatile organic compounds emitted throughout wastewater biosolids processing. Science of The Total Environment 599–600, 663-670.

Godayol, A., Alonso, M., Sanchez, J.M., Antic_o, E., 2013. Odour-causing compounds in air samples: gas-liquid partition coefficients and determination using solid phase microextraction and GC with mass spectrometric detection. J. Sep. Sci. 36, 1045-1053.

Greenman, J., El-Maaytah, M., Duffield, J., Spencer, P., Rosenberg, M., Corry, D., Saad, S., Lenton, P., Majerus, G., Nachnani, S., 2005. Assessing the relationship between concentrations of malodor compounds and scores from judges. JADA. 136, 749-757.

Hwang, Y., Matsuo, T., Hanaki, K., Suzuki, N., 1994. Removal of odorous compounds in wastewater by using activated carbon, ozonation and aerated biofilter. Wat. Res. 28, 2309-2319.

Iwasaki,Y., Ishiguro,T., 1978. Measurement of odor by triangle odor bag method: Japan Society Atmospheric Environment 13(6), 34-39.

Izaguirre, G., Hwang C.J., Krasner S.W. and McGuire M.J. 1982. Geosmin and 2-methylisobomeol from cyanobacteria in three water supply systems. Appl. Envir. Microbial. 43, 708.

Joe, W.H., Choi I.C., Baek Y.A., Choi Y.J., Park G.S. and Yu M.J. 2007. Advanced treatment for taste and odour control in drinking water: case study of a pilot scale plant in Seoul, Korea. Water Sci. Technol. 55(5), 111-116.

Jun S.W., Baek, K.H., Yu, M.J., 2004 Treatment of taste and odor material by oxidation and adsorption. Water Sci. Technol. 49(9), 289-295.

Mackey E.D., Suffet I.H. and Booth S. 2008. A decision tool for earthy/musty taste and odor control. Water Works Association Research Foundation, Research Report #3032, Publ. AWWA Research Foundation, Denver, Colorado, USA.

Mallevialle J. and Suffet I.H. (1987). *Identification and Treatment of Tastes and Odors in Drinking Water,* ISBN 0-89867-392-5. AWWA Research Foundation and Lyonnaise Des Eaux Cooperative Research Report, American Water Works Assoc., Denver, Colo., USA.

Muñoz, R., Sivret, E.C., Parcsi, G., Lebrero, R., Wang, X., Suffet, I.H., Stuetz, R.M., 2010. Monitoring techniques for odour abatement assessment. Water Res. 44, 5129-5149.

Ömür-Özbeket P., Gallagher, D., Dietrich, A., 2011. Determining human exposure and sensory detection of odorous compounds released during showering. Environ. Sci. Technol. 45(2), 468–473.

Ortega. L., Lopez. R., Chaco, J., Ferreira, V., 2001. Use of solid-liquid distribution coefficients to determine retention properties of Porapak-Q resins: Determination of optimal conditions to isolate alkyl-methoxypyrazines and beta-damascenone from wine. Journal of Chromatography A 931, 31-39.

Peter A., Koster O., Schildknecht A., von Gunten U., 2009. Occurrence of dissolved and particle-bound taste and odor compounds in Swiss lake waters. Water Res. 43: 2191–2200.

Sala, C., Bustoa, O., Guascha, J., Zamorab, F., 2004. Factors affecting the presence of 3-alkyl-2-methoxypyrazines in grapes and wines: A review. Ph.D. Dissertation, University Rovira I Virgili, Tarragona, Spain.

Srinivasan, R., Sorial, G., 2011. Treatment of taste and odor causing compounds 2-methyliosborneol and geosmin in drinking water: A critical review. Journal of Environmental Science 23(1), 1-13.

Suffet, I.H., Mallevialle, J., Kawczynski, E., 1995a. *Advances in Taste-and-Odor Treatment and Control* (ed. by I. H. Suffet, J. Mallevialle and E. Kawczyski) American Water Works Research Foundation, American Water Works Association, Denver CO, ISBN # 0-89867-744-0.

Suffet, I.H., Ho J., Chou D., Khiari D. and Mallevialle J. (1995b). Taste-and-odor problems observed during drinking water treatment. In *Advances in Taste-and-Odor Treatment and Control* (ed. by I. H. Suffet, J. Mallevialle and E. Kawczyski), pp. 120. ISBN 0-89867-744-0. AWWA Research Foundation and Lyonnaise Des Eaux Cooperative Research Report, American Water Works Assoc., Denver, Colo., USA

Suffet, I.H., Mallevialle, J., 1987. *Identification and Treatment of Tastes and Odors in Drinking Water*. American Water Works Research Foundation, American Water Works Association, Denver CO, ISBN # 0-89867-392-5.

Suffet, I.H.; Burlingame, G.; Mackey, E. 2008. Controlling taste and odor events for the city of Philadelphia USA: a case study. Water Science and Technology: Water Supply, 8(2), 135-141.

Suffet, I.H., Rosenfeld, P., 2007. The anatomy of odour wheels for odours of drinking water, wastewater, compost and the urban environment, Water Science & Technology, 55(5), 335-344

Suffet, I.H., Braithwaite, S., Zhou, Y., Bruchet, A., 2018. The drinking water taste-and-odor wheel after 30 years. 2018, *In "Taste and Odour in Source and Drinking Water: Causes, Controls, and Consequences",* Editors: T-F. Lin, S. Watson, and I.H. (Mel) Suffet, Publisher, International Water Association, In Press, 2018.

St. Croix Sensory, 2005. AC'SCENT® International Olfactometer User Manual. Lakewood, MN.

Van Langenhove, H., Roelstraete, K., Schamp, N., Houtmeyers, J., 1985. GC-MS identification of odorous volatiles in wastewater. Water Res. 19 (5), 597-603.

Vitko, T., 2014. Private communication. .

Vitko, T., Suffet, I.H., Zhou, Y., Abraham, S.,2014 OCSD Odor control master plan, Proceedings: Odors and Air Pollutants. Water Environment Federation, Miami, FL.

Yu J., An W., Cao N., Yang M., Gu J., Zhang D. and Ly N., 2014. Quantitative method to determine the regional drinking water odorant regulation goals based on odor sensitivity distribution: Illustrated using 2-MIB. Journal of Environmental Sciences, 26, 1389-1394.

Zarra, T., Naddeo, V., Belgiorno, V., Reiser, M., Kranert, M., 2008. Odour monitoring of small wastewater treatment plant located in sensitive environment. Water Sci. Technol. 58 (1), 89-94.

Zhou, Y., Hallis, S., Vitko, T., Suffet, I.H., 2016. Identification,quantification and treatment of fecal odors released into the air at two wastewater treatment plants. Journal of Environmental Management, 180, 257-263.

Effects of Diurnal pH Variation in Sewer Process Modeling

Muriel Steele PhD, CH2M/Jacobs; Matthew Ward PE, CH2M/Jacobs; Jes Vollertsen PhD, Aalborg University

ABSTRACT

Hydrogen sulfide (H_2S) formed during microbial transformation of wastewater in sewer systems can cause odor and safety issues in and around collection systems. The release of H_2S is dependent, in part, on the pH of the wastewater due to the acid-base equilibrium reactions between H_2S and its conjugate bases and Henry's law which drives the equilibrium concentrations of H_2S in the aqueous and gaseous phases of a collection system. Despite the theoretical basis for a pH-H_2S relationship, these effects can be difficult to capture when modeling transformations in collection systems.

The work presented here demonstrates the pH-H_2S relationship through use of high-quality data obtained from a municipal client. An existing force main was modeled and calibrated using the diurnal pH curve derived from this data; the results show strong agreement between modeled and observed H_2S gaseous concentrations, particularly with respect to minimums and maximums observed over the day when diurnal variation in pH is considered. The calibration was applied to a planned force main and used to inform its construction and operation with respect to odor control along the length of the force main and at the headworks of the receiving treatment plant.

This modeling was performed using the WATS (Wastewater Aerobic/anaerobic Transformations in Sewers) model. WATS development is an ongoing process by researchers at Aalborg University in Denmark; CH2M is providing full-scale applications of the model and practical engineering knowledge to identify improvements that can be made to the model and its use.

KEYWORDS

Collection systems, hydrogen sulfide, modeling

INTRODUCTION

The WATS model was developed at the University of Aalborg in Denmark through pioneering research, sustained over the past several decades, targeting sewer biological, chemical, and physical processes within wastewater conveyance facilities, the understanding of which is key to solving problems related to odors and corrosion. The WATS model is continuously extended and improved as knowledge on in-sewer processes increases. It constitutes the backbone of other sewer process models and has, as such, set the paradigm for sewer process modelling. It exists as a mathematical model able to simulate extended sewer networks.

In brief, modeling with WATS typically aims at solving the following in-sewer problems and related management strategies:

- Concrete corrosion caused by H_2S
- Hydrogen sulfide impacts on human health
- Odor and volatile organic compound emissions

- Hydrogen sulfide and VOC control strategies
- Analysis of wastewater quality at inflows to wastewater treatment plants

With appropriate physical inputs and field sampling data, the model can be calibrated to the network in question and then used to simulate future conditions, identify prospective problem solving strategies, optimize those strategies, estimate costs for the best prospective options, and develop design guidance to inform capital improvement projects. The current state of the WATS model can be found in *Microbial and Chemical Process Engineering of Sewer Networks* (Hvitved-Jacobsen et al., 2013) and related journal publications. This book provides detailed descriptions and mathematical formulations of key processes included in the model

The application of study is a large municipality's planned pump station and pipeline. This rising main will transport approximately 121 MLD (32 mgd) of wastewater to an advanced water reclamation plant (AWRP), where it will be treated before being sent to another plant for further purification to use in an indirect potable reuse context. Construction will include a pump station and pipeline 18 kilometers (11 miles) long to the AWRP.

The paper summarizes the sewer process modeling efforts related to the planned pump station and pipeline (PPSP). A WATS model was created for an existing pump station (EPS) and force main and data was collected in order to calibrate the model to local conditions, flow variation, and wastewater characteristics. The 30% design plans for the PPSP were then used to construct a network in WATS and the calibrated parameters from the EPS system were applied. The resulting PPSP model was used to assess proposed odor control strategies and the effects on the downstream treatment plant, AWRP.

METHODOLOGY

WATS Model Setup and Calibration

The EPS force main is approximately 2,900 meters (9,500 feet) of 107 cm (42") steel pipe that extends from the EPS pump station to a tunnel that discharges into the AWRP. The network was constructed in the sewer process modeling software, WATS, using pipe geometry obtained from as-built drawings.

Flow data from the EPS pump station (upstream of the force main) were provided by the City, in one hour increments, for fiscal year 2015 (July 2014 – June 2015). A diurnal curve was developed using the average and standard deviation for each hour of the day. The flow pattern in the EPS force main shows a diurnal curve (Figure 1) typical of domestic wastewater with the lowest flows occurring between midnight and 8 AM and highest flows at midday and towards the late evening.

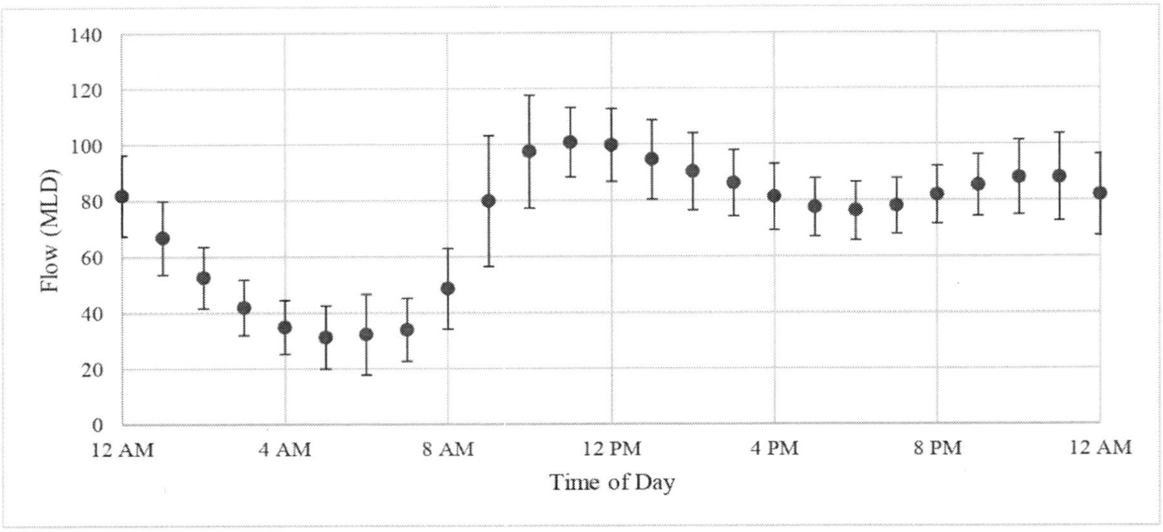

Figure 1. Diurnal Curve of Flow in EPS.
Average and standard deviation for one year's worth of data (FY15) are shown.

Wastewater samples from the EPS pump station wet well were analyzed to determine wastewater characteristics entering the force main. Total, dissolved, and flocked/filtered COD were measured, as well as volatile fatty acids (VFAs; products of fermentation reactions). These constituents were converted to COD-based WATS parameters as listed in Table 1.

Table 1. Wastewater COD Fractionation Based on Sampling Results

Analysis	WATS Parameter	Calculation	Value (mg COD/L)
Total COD (COD_{TOT})	Total COD		647.0
Dissolved COD (COD_{DIS})			214.0
Flocked/filtered COD (COD_{FF})			184.1
Volatile Fatty Acids (VFAs)	Fermentation Products (S_A)	*See Table 2*	137.8
	Fermentable Substrate (S_F)	COD_{FF} - S_A	46.3
	Heterotrophic biomass in bulk water (X_{Hw})[1]	COD_{TOT} * 0.0365	23.6
	Sulfide-forming biomass in bulk water (X_{Hsulf})[2]		0
	Fast Hydrolysable Substrate (X_{Sf})	COD_{DIS} - COD_{FF}	29.9
	Medium Hydrolysable Substrate (X_{Sf})[3]	(COD_{TOT} – COD_{DIS} – X_{Hw}) * 0.3	122.8

Table 1. Wastewater COD Fractionation Based on Sampling Results

Analysis	WATS Parameter	Calculation	Value (mg COD/L)
	Slowly Hydrolysable Substrate (X_{Sf})[3]	($COD_{TOT} - COD_{DIS} - X_{Hw}$) * 0.7	286.6

[1] Default WATS fraction for X_{Hw}/Total COD (0.0365 g biomass COD/g total COD) used.

[2] Assumed no sulfide-forming biomass enters system in bulk water; most of sulfide formation occurs in the biofilm/on the surface of the pipes.

[3] Default WATS fractions for medium- and slowly-hydrolysable substrate portions of non-biomass particulate COD used.

VFA concentrations were converted to concentrations as COD based on the theoretical COD of each compound, shown in Table 2.

Table 2. Theoretical COD of VFAs Sampled

	g COD/g VFA
Acetic acid	1.07
Butyric acid	1.82
Heptanoic acid	2.34
Hexanoic acid	2.21
Isobutyric acid	1.82
Isovaleric acid	2.04
4-Methylvaleric acid	2.21
Propionic acid	1.51
Valeric acid	2.04

In addition to COD components, other pertinent wastewater characteristics were measured. The constituents used in the EPS force main model are shown in Table 3.

Table 3. Non-COD Wastewater Characteristics Used in WATS Model

	As Reported by Laboratory		Converted to WATS Units	
Analysis	Concentration (mg/L)	StDev (mg/L)	Concentration (mmol/L)	StDev (mmol/L)
Alkalinity[1]	272.0	18.1	5.44	0.36
Ammonia[2]	45.3	4.3	3.24	0.31
Nitrate[3]	0.15	0.07		
Orthophosphate[4], Dissolved	3.4	0.6	0.11	0.02

Table 3. Non-COD Wastewater Characteristics Used in WATS Model

	As Reported by Laboratory		Converted to WATS Units	
Analysis	**Concentration (mg/L)**	**StDev (mg/L)**	**Concentration (mmol/L)**	**StDev (mmol/L)**

[1] Alkalinity reported as mg/L as $CaCO_3$ and converted to a molar concentration of $CO_{2,total}$

[2] Ammonia reported as mg-N/L and converted to molar concentration

[3] Nitrate concentration mg-N/L is equivalent to g-N/m^3 used in WATS model

[4] Dissolved orthophosphate reported as mg-P/L an converted to molar concentration

The model was calibrated using data sampled at a condo vault at the downstream end of the EPS force main, near the discharge to the tunnel, where Odalog (hydrogen sulfide; H_2S) and pH measurements were recorded continuously between June 5 and 11, 2017 (shown in Figure 2). The sampling results show a strong relationship between pH and H_2S gas concentration in the condo vault, where peak H_2S occurred concurrently with low wastewater pH measurements. The low-pH, high-H_2S periods were observed during low-flow periods within a diurnal flow pattern at the pump station. These low-pH, high-H_2S values were most likely due to a greater degree of wastewater transformation associated with longer detention times in the system. Therefore, it was determined that the daily variation in pH must be considered in the model to account for transformations that occurred upstream of the pump station.

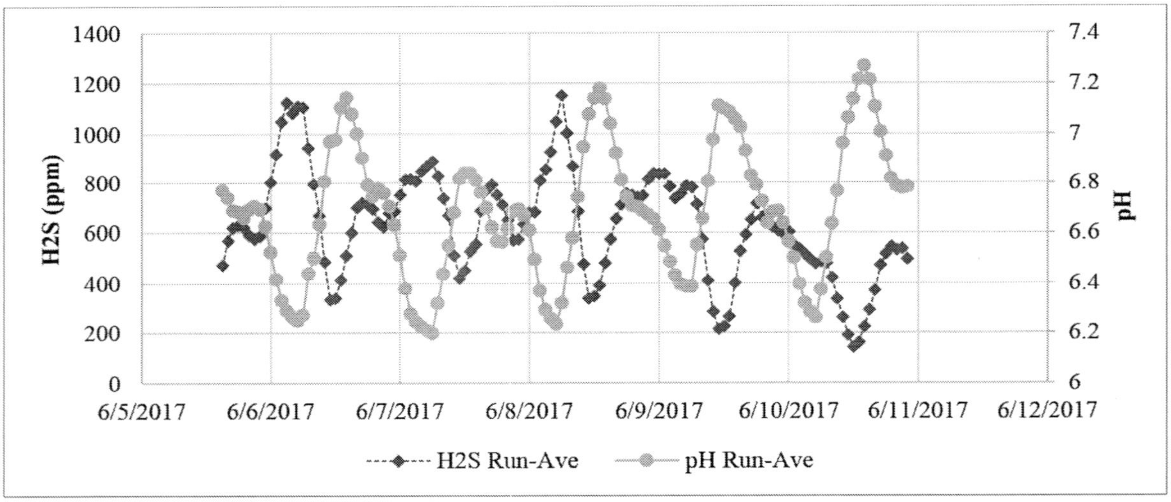

Figure 2. H_2S and pH measurements downstream of the EPS forcemain.
Data are shown using a 5-hour running average to reduce noise.

Diurnal nature of pH is less than that of flow (Figure 3) but does show a relationship to the diurnal flow curve. Variability in pH is highest when flow is low and thus retention time is longest. The pH also increases slightly after the period of low flow indicating a biological activity has occurred during detention. The dip in H_2S gas concentrations on the downstream end of the force main, which occurs after the morning low flow period is consistent with the subsequent increase in pH. The resulting diurnal H_2S curve (Figure 4) was used to calibrate the model.

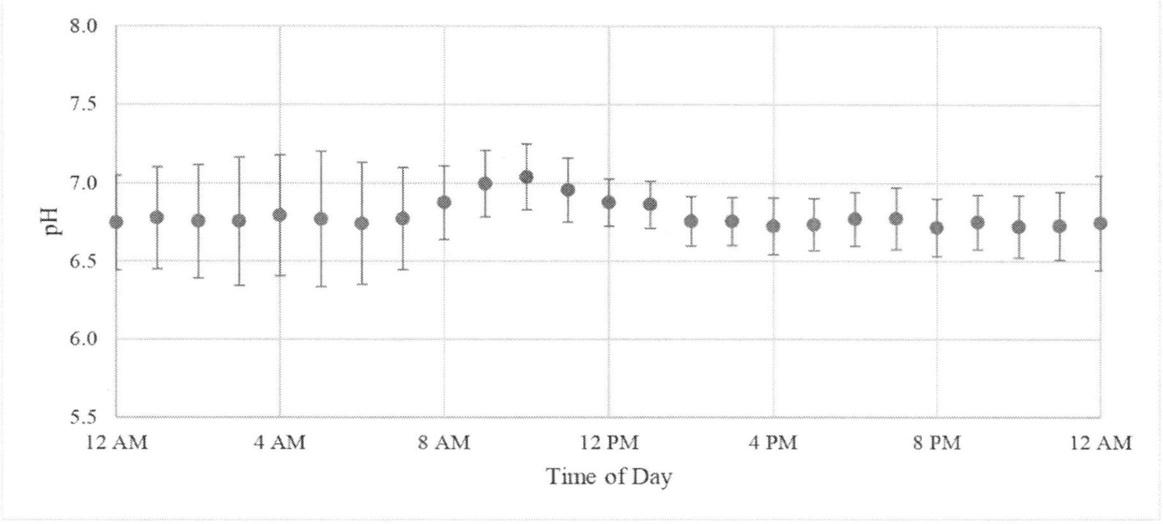

Figure 3. Daily pH Curve Upstream of the EPS Forecemain.
Average and standard deviation for one week's worth of data (June 4 – 11, 2017) are shown.

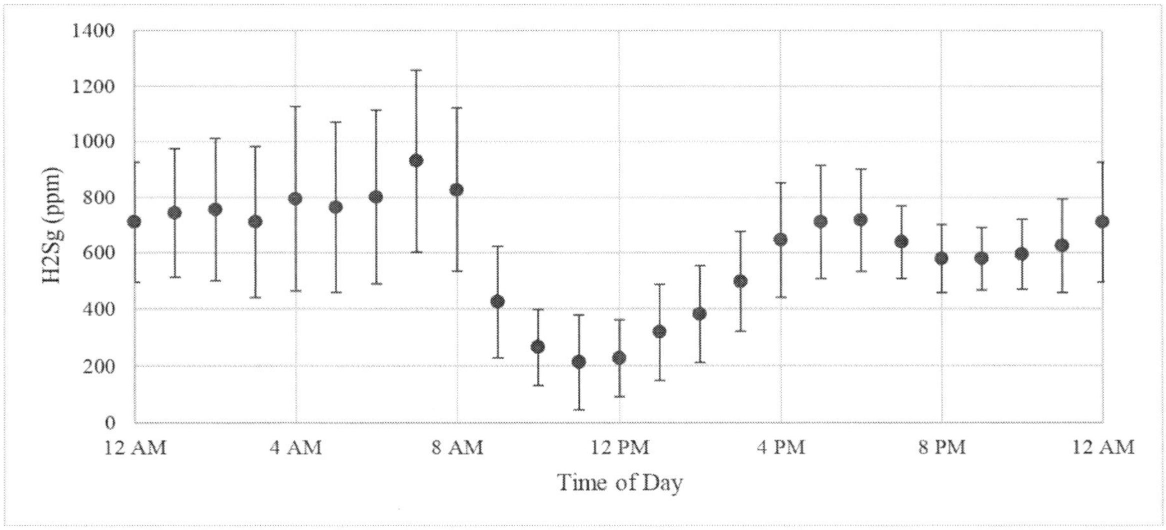

Figure 4. Diurnal Gaseous Hydrogen Sulfide (H_2S_g) Fluctuations at the Downstream End of EPS Forcemain.
Average and standard deviation for one week's worth of data (June 5 – 12, 2017) are shown.

In addition to Odalog data, grab samples were collected between June 5 and 8, 2017 and analyzed for total and dissolved sulfide (shown in Table 4). The results of this sampling effort showed high variability at each location, and dissolved sulfide concentrations at the condo manhole were less than could be explained by Henry's law and the measured H_2S Odalog data for the same time period. H_2S gas measurements are generally more reliable than sulfide grab samples due to potential volatilization of H_2S from aqueous samples; therefore, it was assumed the H_2S Odalog readings were more representative of actual conditions than the sulfide grab samples. Moreover, the configuration of the vault where Odalog data was recorded was likely to be close to equilibrium, so assumed equilibrium conditions were assumed for calculating H_2S gas concentrations.

Table 4. Dissolved and total sulfide sampling results
Showing average (range) of 4 samples collected June 5 – 8, 2017

Location	Total Sulfide, mg/L	Dissolved Sulfide, mg/L
EPS Wetwell (Upstream of FM)	0.53 (0.3 - 0.9)	0.03 (0.0 - 0.1)
Condo Manhole (Downstream of FM)	1.00 (0.3 - 1.3)	0.23 (0.1 - 0.3)
AWRP Influent	1.28 (0.7 - 2.5)	0.63 (0.1 - 1.5)

RESULTS

The EPS force main model from the wetwell to the condo manhole was run using the network constructed from as-built drawings, wastewater characteristics shown in Tables 1 and 3, and diurnal flow and pH patterns shown in Figures 1 and 3. Influent dissolved H_2S and the hydrogen sulfide formation rate constant in the biofilm (k_{H2S}) were varied to find agreement for this application between modeled and observed gas-phase H_2S concentrations at the manhole (Figure 5 and Figure 6, respectively). The peaks and valleys of the data were the targeted points for calibration; relative percent difference (RPD) of minimum and maximum value, excluding the first 12 hours of model results to allow for steady-state to be achieved, was used to determine best fit.

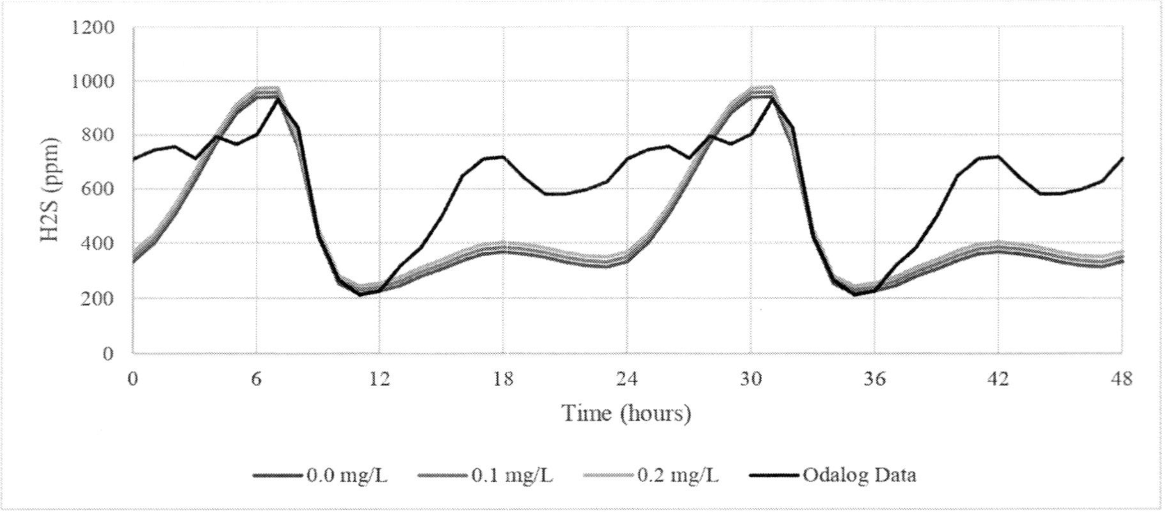

Figure 5. Modeled and observed gas-phase H_2S concentrations at various influent dissolved H_2S concentrations.
Hydrogen sulfide formation rate constant in the biofilm was 60 mg-H_2S/m^2-hr. Not all scenarios run are shown.

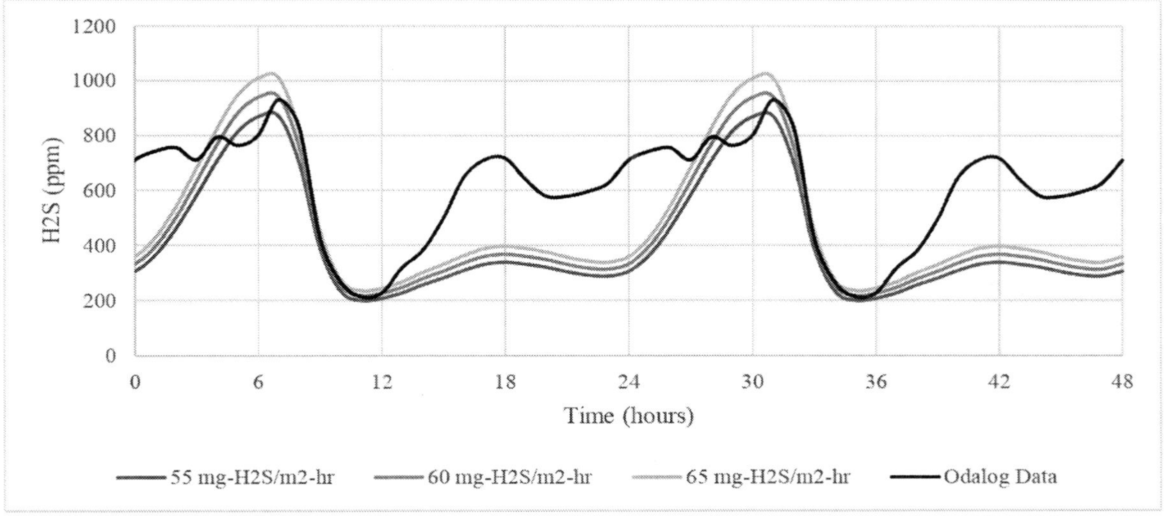

Figure 6. Modeled and observed gas-phase H_2S concentrations at various sulfide formation rates.
Influent dissolved H_2S concentration of 0 mg/L. Not all scenarios run are shown.

Table 5. Summary of calibration model runs showing minimum, maximum, and RPD
Selected calibration scenario shown in bold/italics

Calibration Scenario		Downstream H_2S Concentration (ppm)			
Influent Dissolved Sulfide (mg/L)	**k,H_2S (mg-H_2S/m^2-hr)**	**Minimum**	**RPD, min**	**Maximum**	**RPD, max**
0.0	60	216.9	1.9%	941.0	1.1%
0.0	55	199.3	6.6%	869.9	6.7%
0.0	65	234.4	9.6%	1011.0	8.3%
0.1	60	230.7	8.1%	957.8	2.9%
0.2	60	244.8	13.9%	974.9	4.6%
0.0	58	209.9	1.4%	912.7	2.0%
0.0	***59***	***213.4***	***0.3%***	***926.9***	***0.4%***
0.0	61	220.4	3.5%	955.1	2.6%
0.0	62	223.9	5.1%	969.1	4.0%
0.01	59	214.7	0.9%	928.5	0.2%
0.02	59	216.1	1.5%	930.2	0.1%
0.03	59	217.5	2.1%	931.8	0.1%
0.04	59	218.8	2.8%	933.5	0.3%
Data from Odalog		212.9		930.7	

The final values selected for the calibrated model are an influent dissolved H_2S concentration of 0 mg/L and a hydrogen sulfide formation rate constant in the biofilm of 59 mg-H_2S/m^2-hr (Figure 7) with RPDs in observed and modeled minimum and maximum values of 0.3% and 0.4%, respectively. These represent the lowest RPDs seen when dissolved H_2S was varied between 0.0 and 0.2 mg/L and formation rate was varied between 55 and 65 mg-H_2S/m^2-hr.

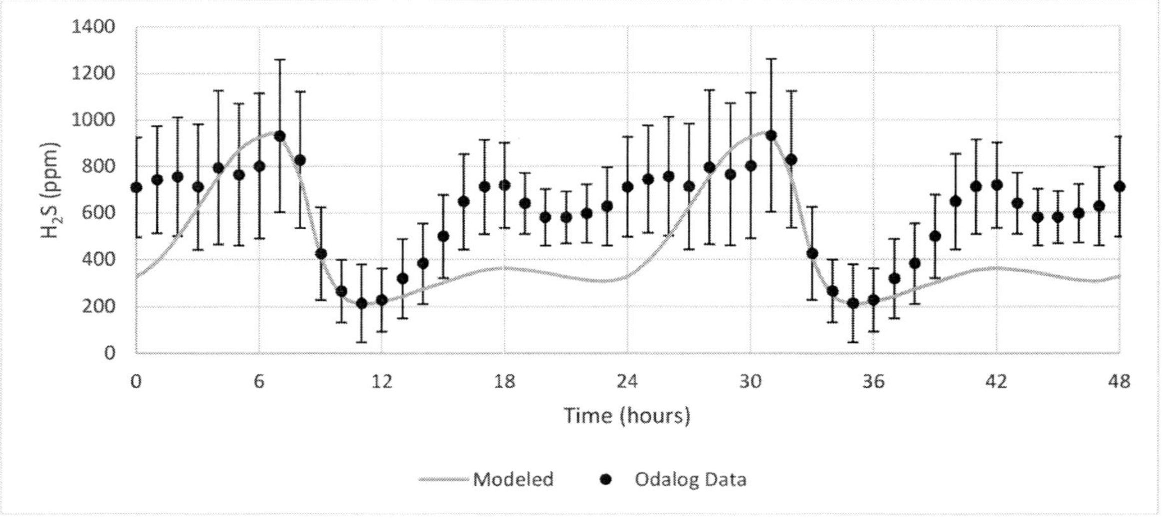

Figure 7. Modeled and observed gas-phase H_2S concentrations for calibrated model.
Influent dissolved H_2S concentration of 0 mg/L and hydrogen sulfide formation rate 59 mg-H_2S/m^2-hr were modeled. Odalog data is shown as average and standard deviation for each time period.

DISCUSSION AND CONCLUSIONS

As shown in Figure 7, the calibrated WATS process model could achieve relatively good agreement with field measured hydrogen sulfide concentrations when 0.0 mg/L and 59 mg-H_2S/m^2-hr were chosen in this situation. The model reflected the diurnal peaks and lows accurately. While the field-measured hydrogen sulfide showed greater diurnal variability than the model-predicted variability, the results were reasonable. Odalog data represent the gas-phase concentrations at an air-release valve and discharge of that gas from the sewer system is dependent on pressure at the valve and since this pressure dependence is not part of the WATS model, the resulting difference in discharge of gas from the system may explain the differences in variability between the data and modeling results. Therefore, the calibrated model provides a sufficient basis for evaluating sulfide control strategies for the new planned alignment, especially since the minimum and maximum H_2S concentrations will form the basis of design for any recommendations that follow.

Often when modeling collection systems, diurnal COD curves are assumed to explain diurnal variation in H_2S production, but this calibrated model demonstrates that the pH-H_2S relationship can be used in conjunction with high-quality data can be used to accurately model H_2S

production in force mains. This is advantageous because pH can be measured online and recorded in real-time, while COD can't.

References

Hvitved-Jacobsen, T., J. Vollertsen, A.H. Nielsen. 2013. *Microbial and Chemical Process Engineering of Sewer Networks.* Second Edition, CRC Press, ISBN 978-1-43-988177-4.

MULTIPLE TECHNOLOGY PILOT TESTING SUPPORTS SELECTION OF TARGETED FOUL AIR TREATMENT

Scott Cowden, CH2M 1100 NE Circle Blvd. Suite 300, Corvallis, OR 97330
Jay Witherspoon, CH2M 6 Hutton Center Drive, Suite 700, Santa Ana, CA 92707
Zeynep Erdal*, CH2M 6 Hutton Center Drive, Suite 700, Santa Ana, CA 92707
Carla Dillon, Orange County Sanitation District, 10844 Ellis Ave., Fountain Valley, CA 92708
Tadeo Vitko, Orange County Sanitation District, 10844 Ellis Ave., Fountain Valley, CA 92708
Sam Boswell, Daniel Company, 1939 W 11th Street Suite E, Upland, CA 91786
Yubin Zhou, UCLA Department of Environmental Health Sciences, School of Public Health, Los Angeles, CA 90095
I.H. (Mel) Suffet, UCLA Department of Environmental Health Sciences, School of Public Health, Los Angeles, CA 90095

* Present Address: AECOM, 999 Town and Country Road, Orange, CA 92868

Email: scowden@ch2m.com

ABSTRACT

The Orange County Sanitation District (OCSD) approached its Odor Control Master Plan (OCMP) from a unique and innovative perspective. In lieu of the traditional focus on hydrogen sulfide (H_2S) and general odor (dilutions-to-threshold [D/T]), OCSD conducted a wide sweep of sampling and analysis of all plant process areas to determine the top offenders at its Plant No.1 (P1) and Plant No.2 (P2) located in coastal Southern California. Key odorants were identified using chemical and sensory analytical methods, and using the Weber-Fechner relationship between each odorant concentration and their perceived intensity, maximum fenceline concentrations below nuisance were determined.

A comprehensive evaluation of foul air management technologies was conducted including a bibliographic search of all viable vapor phase odor control technologies followed by in-situ technology testing. Test data would be used for supporting the selection of targeted foul air treatment at specific P1 and P2 sources and for facilitating Capital Improvement Plan (CIP) decisions. Two approaches were considered for accomplishing this including remote installation technology testing (i.e.; testing of existing vapor phase systems at multiple remote sites) and OCSD pilot unit technology testing. Pilot unit testing conducted on OCSD premises was selected.

Pilot testing was conducted at trunk line scrubbers at P2 under two distinct test conditions for the following technologies:

- Biotower with reticulated polyurethane foam cubes
- Biotower with Matala® filter media mat and foam filter media mat
- Biofilter with engineered media
- Biofilter with bioshell media

- Biofilter with lava rock media
- High capacity carbon media
- Coconut shell carbon media
- Combination coconut shell carbon media + potassium permanganate impregnated activated carbon

Test conditions were varied to include bioscrubber configuration (recirculation mode), bio-trickling filter configuration (once-through irrigation mode), and different empty bed gas residence times (EBGRTs). In addition, tests were conducted for measuring biological acclimation duration, ability to react to spikes, and ability to react to short-term shutdowns and varying inlet loadings. All testing included both sensory and analytical methods for measuring sulfur compounds, ammonia, D/T, fecal, and musty constituents.

In-situ pilot unit testing proved to be invaluable for obtaining pertinent performance data under specific operating parameters and unique operating conditions. Pilot unit test results provided the necessary backbone for allowing OCSD to make incisive decisions regarding selection of targeted foul air treatment technologies as part of their OCMP effort and future CIP projects.

KEYWORDS: Vapor phase technologies, pilot testing, air dispersion model

INTRODUCTION

Orange County Sanitation District (OCSD) operates two wastewater treatment plants (WWTP's); Plant No. 1 (P1) located in Fountain Valley, California, and Plant No. 2 (P2) located in Huntington Beach, California. P1 currently treats approximately 133 million gallons per day (mgd) while P2 currently treats approximately 50 mgd. OCSD has always strived to be a good neighbor to the surrounding communities and, as such, has developed a 5-year Strategic Plan that calls for zero odor incidents/events under normal operating conditions at both plants P1 and P2. Although hydrogen sulfide (H_2S) and other specific odorants are generally controlled throughout the two plants, OCSD is not meeting these levels of service goals and therefore embarked on an update to their previous Odor Control Master Plan (OCMP) completed in 2002.

OCSD approached its updated OCMP from a unique and innovative perspective. In lieu of the traditional focus on H_2S and general odor (dilutions-to-threshold [D/T]), OCSD conducted a wide sweep of sampling and analysis of all plant process areas to determine the top offenders at P1 and P2. Key odorants were identified using chemical and sensory analytical methods. The following nine odorants, along with their odor characteristic, were identified (Suffet et al., 2016, Vitko et al., 2014, Vitko et al., 2016):

- Hydrogen Sulfide (H_2S): Rotten Egg
- Methyl Mercaptan (MM): Rotten Vegetable
- Dimethyl Disulfide (DMDS): Rotten Vegetable
- Dimethyl Sulfide (DMS): Canned Corn
- Ammonia (NH_3): Pungent
- 2-Methyl Isoborneol (MIB): Musty
- 2-Isopropyl-3-Methoxypyrazine (IPMP): Musty

- Skatole: Fecal
- Indole: Fecal

Air dispersion modeling was conducted to determine offsite impacts of each key odorant and mitigation measures were developed which focused on technology selection tailored to each required key odorant removal.

In an effort to ensure due diligence in selecting the preferred technology approach, a comprehensive evaluation of foul air management technologies was conducted including a bibliographic search of all viable vapor phase odor control technologies followed by in-situ technology testing. This paper focuses on the development, methodology, operation, and results of the comprehensive technology testing effort.

BIBLIOGRAPHIC RESEARCH OF FOUL AIR MANAGEMENT TECHNOLOGIES

There are many technologies used for the treatment of odors from wastewater treatment plants. A combination of multiple technologies is necessary to provide high levels of performance and removal of different odorants. For the OCSD OCMP effort, the universe of available technologies was considered. An initial screening to identify the most viable technologies to be carried forward was provided along with the rationale for removing those technologies that were not considered. A worldwide bibliographic search was conducted to identify the vapor phase odor control technologies and their specific odorant removal efficiencies for the nine "most detectable" odorants of interest for OCSD's P1 and P2.

The technologies continue to evolve with a shift towards green approaches, reduction or zero use of chemicals and emphasis on energy optimization, smaller footprints, and reduced loading rates. Figure 1 categorizes the most common odor control technologies.

INITIAL SCREENING OF FOUL AIR MANAGEMENT TECHNOLOGIES

All viable odor control vapor phase technologies included in Figure 1 were considered for evaluation. An initial screening step was used to vet out those technologies not considered viable for this project. Those technologies, along with justification for dismissal, are provided as follows:

- Activated sludge diffusion is not applicable to this application.
- Counteractants were screened out since the plants have very close proximity to neighbors and there could be mist issues onto adjoining property. Also, performance results are shown to be mixed.
- Dispersion fans were screened out since they would likely not meet the offsite odor goals.
- Hydroxyl ion radical spray was screened out due to poor past performance. A unit was previously installed at an OCSD pump station site but was taken out of service because it was believed to be unsafe, was not designed to handle the wetwell gas pressures, and was causing urethane and polyvinyl chloride coating to become brittle and crack.

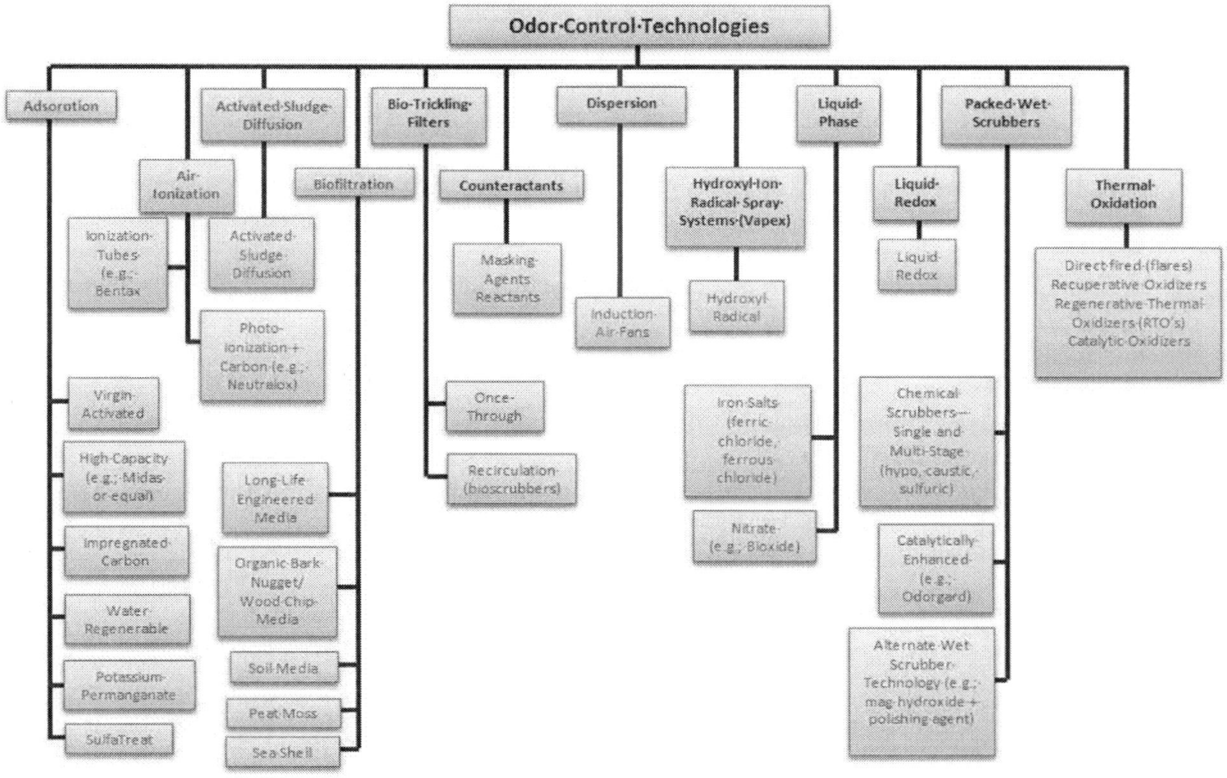

Figure 1. Universe of Odor Control Technologies

- Cross-flow packed bed chemical scrubbers and horizontal flow scrubbers were screened out given the advantages of the counter-current packed tower design
- Chemical mist scrubbers were screened out due to high operating costs and poor removal efficiency under varying loads
- Thermal oxidation was screened out for the following reasons:
  - — Regenerative thermal oxidizers (RTO) are effective for a treatment of a range of compounds, but expensive and complex and considered overkill
  - — Plasma arc destruction is effective for a range of compounds but not proven full scale for this application
  - — The use of existing plant combustion sources is only suitable for processing small volumes of air, and is restricted by operating times and processed gas volumes, resulting in limited potential.

Therefore, only the following vapor phase odor control technologies were carried forward for evaluation:

- Carbon Adsorption
- Air Ionization using UV-Photoionization (UVPI)
- Biofiltration using Engineered Media Biofilter

- Biotrickling filter (BTF) with once-through irrigation and recirculating type (Bioscrubber)
- Packed bed chemical scrubber

OCSD PILOT TESTING

A technology pilot testing effort was initiated to further evaluate the shortlisted vapor phase odor control technologies. Pilot testing was considered valuable for the following reasons:

- Provide representative performance data for key technologies for the nine "most detectable" odorants of interest
- Shed light into key operational parameters of each technology (e.g.; empty bed gas residence time [EBGRT])
- Provide insight into technology limitation and benefits (e.g.; required acclimation period for biological systems)
- Used to develop design criteria for the shortlisted vapor phase treatment technologies

Pilot Testing Approaches Considered

Two approaches were identified as possible scenarios for pilot testing. These included:

- Remote Installation Technology Testing: This approach would entail sampling selected technology vapor phase odor control systems at remote locations for obtaining performance data. This approach would have likely required Owner approval for modifying system operating parameters to meet those identified as key for the OCMP project.
- OCSD Pilot Unit Technology Testing: This approach would entail the manufacture of a tailored pilot unit complete with multiple technologies for testing of selected technologies operating at specific key operating parameters. The pilot unit would be placed at specific P1 or P2 source location(s) considered representative of favorable odorant loadings for testing.

The OCSD Pilot Unit Technology Testing approach was selected as the preferred approach for the following reasons:

- Odorant profile unique to OCSD allows for more representative performance data
- All technologies placed on a common skid allowing true side-by-side testing under identical loading conditions.
- Full flexibility provided for modifying operating parameters for each specific technology to meet project goals and objectives
- Sampling for all technologies can be conducted at a single location in relatively close proximity to air quality laboratories allowing for ease of sample collection, standardized and consistent hold times, and identical sampling methodologies.

Daniel Company was retained to manufacture the pilot unit. Daniel Company is a supplier of odor control systems with manufacturing facilities located in Southern California. Close proximity, familiarity with OCSD plants and processes, verbal commitment to this project, reputation excellence, and technical competence were all reasons why Daniel Company was selected.

Selection of Technologies

Technology selection for pilot unit implementation considered both shortlisted vapor phase technologies and OCSD knowledge base. For example, significant testing of existing OCSD chemical scrubbers at both P1 and P2 had already been completed under various operating parameters. Therefore, this technology was not considered critical for inclusion. In addition, attention was paid to technologies with multiple competitors, to reduce the use and selection of proprietary media and processes. Significant efforts were made to include the Air Ionization using UV-Photoionization (UVPI) technology into the pilot test unit. However, confidentiality concerns along with OCSD risk ultimately prevented its inclusion.

Technology selection ultimately came down to a combination of synthetic and natural media for biological treatment and various carbon medias. These are further defined as follows:

- Reticulated Polyurethane Foam Cubes: Foam cube media with a nominal 10 pores per inch of polyether foam and a nominal density of 2.0 pounds per cubic foot. The foam cubes were selected to operate as both a bioscrubber (water recirculation mode) and as a BTF (once through irrigation mode). Nutrients were provided to the foam cube biological treatment vessel, as the foam cubes have no inherent nutrient properties.

- Matala® Filter Media Mat and Foam Filter Media Mat: The Matala® media is a three-dimensional filter mat that promotes high organic loadings through a specific surface area of approximately 150 square meters per cubic meter. The Matala® media was coupled with a foam media mat, to increase the surface area of the media, promoting increased biofilm yields. The Matala® and foam media were selected to operate as both a bioscrubber (water recirculation mode) and as a BTF (once through irrigation mode). Nutrients were provided to the Matala® and foam biological treatment vessel, as the media have no inherent nutrient properties.

- Cell-Max by Global Environmental Solutions, Inc. (GES): Cell-Max media is an engineered, inorganic substrate. A proprietary formula of nutrients, adsorbents, minerals and food source is applied to the engineered Cell-Max media structure to support natural occurring selection of fungal and bacterial microorganisms for the capture and consumption of complex odors and VOCs. Cell-Max is light weight, rigid, and will not degrade in a low pH sulfuric environment. It has buffering capabilities which limits the low pH quality of the leachate. Supplemental nutrients were not provided, as the engineered media contains inherent nutrient properties. Multiple manufacturers of this media type exist including GES, Biorem, and ECS.

- Bioshells by Melcourt Industries: Bioshells are a shell-based biofilter medium with very high surface area properties. Bioshells are produced as a processed byproduct from the seafood industry. Supplemental nutrients were not provided, as the seashell media contains inherent nutrient properties. Multiple manufacturers of this media exist including Monashell by Bord na Mona and Bioshells from Melcourt.

- Lava Rock: Lava rock media, specifically Crater-Max™ media manufactured by GES, was selected. The porous Crater-Max™ media serves as part of the ecosystem for the growth of acidithiobacillus thiooxidans microorganisms. The media also contains iron content that in a low pH environment can create a chemistry of iron ions to support a colony of acidithiobacillus ferroxidans microorganisms. Supplemental nutrients were provided, as the lava rock media does not contain all nutrient properties required for optimal biological growth.

- AddSorb OX30, manufactured by the Jacobi Carbons Group: AddSorb OX30 is a performance activated carbon for the combined removal of malodors such as hydrogen sulfide, mercaptans, methyl sulfides and organics. It is considered a "high capacity" carbon. The manufacturing process of AddSorb OX30 develops a highly reactive and retentive internal pore structure without chemical impregnation. This feature permits the use of in-situ regeneration techniques with water. The theoretical H_2S removal capacity of AddSorb OX30 is greater than 0.30 g/cm^3 based on initial use. Regeneration does not provide full capacity restoration.

- Vapor Phase Coconut Shell Carbon, manufactured by Carbon Activated Corporation: Coconut shell carbon (Model Number COC-A 60) is a highly active granular activated carbon manufactured via high temperature steam activation from selected grades of coconut shell. Coconut shell carbons feature high surface areas, fine pore structures, and high packing density. The theoretical H_2S removal capacity of Vapor Phase Coconut Shell Carbon is greater than 0.04 g/cm^3.

- Potassium Permanganate (KMnO4) Impregnated Activated Carbon, manufactured by Carbon Activated Corporation, combined with Vapor Phase Coconut Shell Carbon (50:50 mix): Potassium permanganate, impregnated into activated carbon, is used to oxidize pollutants such as H_2S, mercaptans, sulfur dioxide, and various aldehydes. The pelletized activated carbon was impregnated with a minimum of 8 percent potassium permanganate. The theoretical H_2S removal capacity of Potassium Permanganate (KMnO4) Impregnated Activated Carbon is greater than 0.13 g/cm^3.

Pilot Unit Description

The pilot test unit (pilot) featured five biological treatment vessels and four carbon treatment vessels. Eight different technologies were selected and implemented, and the pilot was configured to allow maximum flexibility to include options for water recirculation and combining technologies for first and second stage treatment analysis.

The pilot was mounted on a flat-bed trailer. The overall dimensions of the pilot were approximately 6 feet 11-inches by 23 feet 9-inches. A detailed plan of the pilot test unit is provided in Figure 2. A photo of the unit is provided in Figure 3.

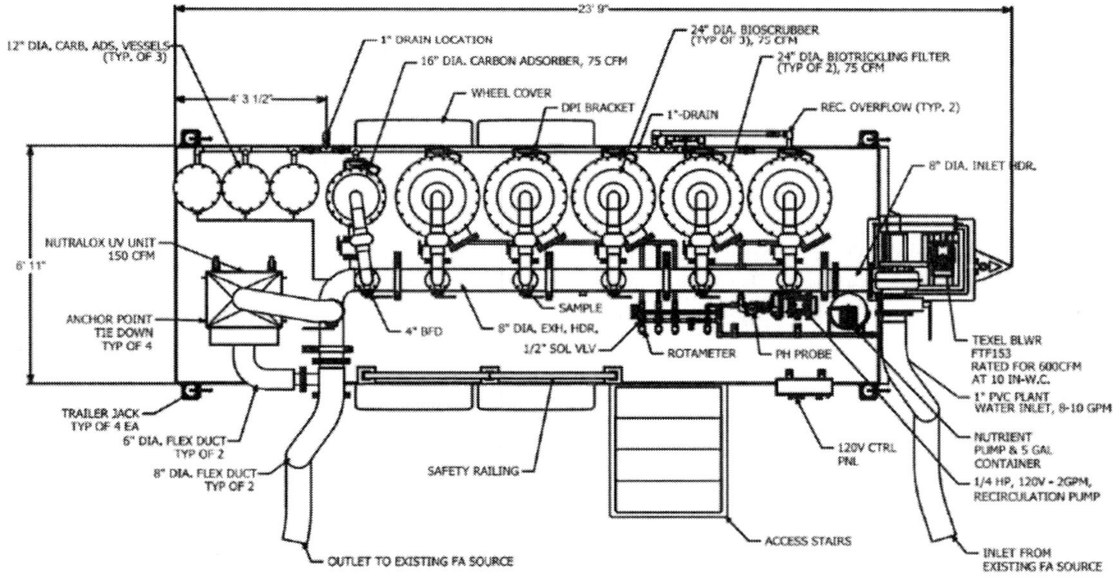

Figure 2. Plan View of Pilot Test Skid

Figure 3. Pilot Test Skid at Plant No. 2 Trunkline Scrubbers

Biological treatment vessels were approximately 9 feet tall and 2 feet in diameter. The single large carbon vessel was approximately 8 feet tall and 16-inches in diameter. The three small carbon vessels were approximately 6 feet tall and 1 foot in diameter. A 2-horsepower, 240-volt, 600-scfm, single-phase power extraction fan was utilized to supply raw foul air to the pilot test unit. A water recirculation pump and water control panel, complete with timer controlled solenoid valves, rotameters, and pH controlled water valves were included on the pilot test unit. Additionally, a nutrient feed system, complete with a 5-gallon feed tote and metering pump were also included on the pilot test unit.

The pilot test unit was configured to be operated under various operating parameters. The foul air ducting was configured to allow for simultaneous foul air treatment through each vessel on the pilot test unit. The water supply system was configured to allow the first two large biological treatment vessels to operate as both bioscrubbers, where the drainage water is recirculated through the system, and as BTF units, where the drainage water is removed and fresh water is used. The remaining three large biological treatment vessels were configured to operate only as biofilters.

The inlet ducting included a sampling port and an OdaLog port to measure inlet odor concentrations. All vessels on the pilot test unit included a sampling port and OdaLog port to measure the outlet odor concentrations. The biological treatment vessels included a sampling port at the midpoint of the media bed for testing removal efficiencies at a reduced EBGRT. Carbon vessels were provided with sample ports at three equidistant locations within the carbon bed to allow for testing removal efficiencies at reduced contact time conditions. The biological treatment vessels were provided with drain sampling ports for manually pulling a sample and checking pH. Drainage pH monitoring was conducted to ensure proper operating conditions, including verifying adequate makeup water was provided and for verifying acclimation status of the pilot test unit.

There was a concern with accurate pilot test results with regards to scale-up effects. Small scale vessels can experience wall effects and subsequent channeling which can skew results. For the scaled-down reactor vessel sizes of 12, 16, and 24-inch diameter, the following actions to minimize or eliminate scale-up issues were taken:

- Air Distribution: Because pressure drop through the newer media tend to be low, and velocities into the bottom of the reactor can be relatively high, air can impact the far end of the reactor vessel and want to channel up the wall between the media and the wall. This "wall effect," which does not have to be limited to startup and clean media conditions, can result in inaccurate results, especially when dealing with lower odorant concentrations. It is imperative that proper air distribution be provided. To this end, the following features were incorporated to the pilot system:
  - A perforated plate was provided just below the media to provide adequate backpressure and ensure uniform air distribution across the bed.
  - Space was provided between intermediate media layers to allow redistribution of air and pressure equalization between intermediate layers. In addition, sample ports were provided at these intermediate zones to allow for sampling and evaluation of distribution and treatment performance of the system for the specific odorants along the height of the reactor.
  - One- to 2-inch wide rings were provided along the perimeter inside wall of each vessel within each media bed to provide a torturous path for air passage along the edge of the wall, to prevent short-circuiting and inflated wall effect along the height of the vessels.

Velocities entering the bottom of the vessel were limited to a maximum of 800 fpm (4 mps) through air flow rate control and sizing of the unit components.

- Water Distribution System: With a single central nozzle located above each media bed, water distribution was provided uniformly to each vessel as needed. Plugging and spray pattern of the nozzles was checked throughout the pilot test period.

Identification labels for vessels pertaining to specific technologies are summarized in Table 1.

Table 1. Pilot Unit Vessel Identification Labels

Vessel Identification	Technology Description
BTF-1	Biodan-100Z Jurgen Loy Reticulated Polyurethane Media
BTF-2	Biodan-100R Structured Media (Matala® and Foam)
BSBR-1	Biodan-100C Global Environmental System Coated Engineered Media
BSBR-2	Biodan 100S Melcourt Seashell Media
BSBR-3	Biodan-100L Global Environmental System Lava Rock Media
CARB-1	Danadsorb DAS-100 High Capacity Carbon (Addsorb OX30)
CARB-2	Danasorb DAS-100J High Capacity Carbon (Addsorb OX30)
CARB-3	Danasorb DAS-100C Virgin Coconut Shell GAC (COC-A60)
CARB-4	Danasorb DAS-100K 50% Virgin Coconut and 50% Potassium Permanganate

Selection of Operating Parameters

Criteria for selection of operating parameters included:

- Parameters selected should be within normal industry standard ranges. For example, selecting BTF EBGRT of 60 seconds would be considered atypical and therefore not selected for pilot testing.
- Parameters selected should be physically achievable. For example, selecting an EBGRT of 90 seconds for biofilters would generally require a footprint size considered nonviable for P1 and P2 due to limited available real estate.
- Two different operating parameters were selected for each biological technology. Due to limited scope, only a limited number of different operating parameters could be selected. Therefore, care was taken to select those considered to provide maximum benefit.

The pilot was configured to two unique operating conditions. During the first operating condition, the first two biological treatment vessels (BTF-1 and BTF-2) operated as bioscrubbers, which featured drainage water recirculation. The addition of makeup flush water was pH controlled, with a set point of a pH of 1.8. When the pH of BTF-1 and BTF-2 dropped below 1.8, the makeup flush valve turned on to add supplemental water, raising the pH above 1.8. CARB-1 (the large carbon) was offline during both the first and second operating conditions to save the carbon life for the potential use as a second stage carbon treatment system.

Under the second operating condition, BTF-1 and BTF-2 were adjusted from bioscrubbers (with water recirculation) to BTFs.

A summary of the operating conditions and airflow rates is provided in Table 2.

Table 2. Pilot Operating Conditions

Name / ID	First Condition (6/20/2016 – 8/12/2016)		Second Condition (8/12/2016 – 10/13/2016)	
	Air Flow Rate	EBGRT	Air Flow Rate	ERBT
BTF-1[a,b]	75 cfm	15 seconds	37.5 cfm	30 seconds
BTF-2[a,b]	75 cfm	15 seconds	37.5 cfm	30 seconds
BSBR-1	37.5 cfm	30 seconds	25 cfm	45 seconds
BSBR-2	37.5 cfm	30 seconds	25 cfm	45 seconds
BSBR-3	37.5 cfm	30 seconds	25 cfm	45 seconds
CARB-2	120 cfm	3.5 seconds	120 cfm	3.5 seconds
CARB-3		3.5 seconds		3.5 seconds
CARB-4		3.5 seconds		3.5 seconds

Notes:
[a] Recirculation pump operating during first condition, recirculation pump shut off during second condition.
[b] Nutrient feed system operating during both the first condition and second condition.

Pilot Testing Location

The location of the pilot was selected based on three parameters, including strength and characteristic of odors from a foul air stream, availability of space for the pilot, and the availability of utilities, such as potable water and electricity, for the operation of the pilot.

Odor sources that exhibited higher odorant concentrations and greater range of odorants were ranked higher when determining the ideal pilot test location. Higher odorant concentrations were preferred, as odorant concentrations near the laboratory MRL values can skew results. Pilot testing results may be skewed when inlet odorant concentrations are low and the treatment removal performance levels are high (i.e., outlet levels are low and fall below the MRL value for that odorant). A foul air source exhibiting a broader range of odorants is ideal for testing the "peeling the onion" effect (Suffet et al., 2016), as when stronger odorants are removed initially, remaining untreated odorants can become more prevalent.

Ultimately, the pilot was placed at the P2 Trunkline Scrubbers. This location was selected with input from OCSD staff due to its high odor loadings, the mixture of odorants of concern, available pilot test unit space, and proximity to utilities required for the operation of the pilot.

Pilot Operation

The pilot was mobilized and placed adjacent to the trunkline bioscrubbers at OCSD's P2 for treating un-treated trunkline foul air upstream of bioscrubbers. All installation procedures, including utility hookups, foul air inlet and outlet hookups, and functional testing were completed by OCSD staff or under the supervision of OCSD staff. Subsequent to the arrival of the pilot, utility hookups, including water, drainage, electricity, and electrical grounding were completed. Flexible ducting was used to connect the inlet foul air source as well as the outlet treated air discharge to the outlet of the trunkline scrubbers for further treatment by the headworks scrubbers.

Upon the mobilization and installation of the pilot, functional testing was then conducted to ensure the proper operation of all systems on the pilot test unit. Upon the completion of all connections and functional testing, the pilot was then air balanced, using a hot wire anemometer and an air foil. Air balancing was an iterative process, and was fine-tuned by adjusting the air damper on the blower and the ball valves on the discharge side of each treatment vessel included on the pilot. Upon the completion of all connections, functional testing, and air balancing, the acclimation phase was entered. The pilot was monitored daily to ensure proper acclimation of the system.

Upon the completion of all testing and analysis under the first operating condition, the pilot was transitioned to the second operating condition examined under this study. The EBGRTs were adjusted through an iterative process, and was fine-tuned by adjusting the air damper on the blower and the ball valves on the discharge side of each treatment vessel. Upon the completion of all air balancing, the pilot unit was then allowed to acclimate to the new operating condition.

Shutdowns and Stress Events

Two major pilot shutdowns occurred during the operation of the pilot in August 2016. The first major shutdown lasted three days due to water leakage from the water supply pump. The unit was turned back on after 72 hours, providing air flow and water flow to the pilot. This unplanned stress event revealed the resilience of the unit. The second major shutdown occurred in October and lasted for approximately 24 hours. Results and impacts to the pilot for both shutdown events are discussed herein.

High and Low Loading Events

The pilot experienced both extreme high and low H_2S loadings throughout its operation. During the first operating condition, in August 2016, the H_2S loading increased from approximately 10 ppm to 40 ppm. During this period, performance of all biological units, except for BSBR-2 were monitored. BSBR-2 was not monitored during this spike, as the outlet OdaLog was not operational during this time. Pilot test unit performance and robustness to high loading spikes are discussed herein.

Chemical dosing upstream of the trunkline scrubbers occurred at the end of September 2016. H_2S inlet concentrations steadily dropped until it reached its lowest measured value of 0.48 ppm after approximately one week. Inlet H_2S concentrations did not return to average loading values until approximately 19 days. Pilot test unit performance and robustness to low loadings are discussed herein.

Daily Monitoring

The pilot test unit was monitored daily during operation, Monday through Friday. The following monitoring activities were conducted:

- The makeup water flow to each vessel was observed and recorded, including the timing of the water delivery.
- When operational, the recirculation water flow to BTF-1 and BTF-2 was observed and recorded.

- The inlet and outlet of each treatment vessel was monitored through the use of a hand held and portable Jerome Hydrogen Sulfide Analyzer (Jerome). The Jerome was then regenerated daily after monitoring activities.
- The pH was observed and recorded through the use of colorimetric test strips.
- The pressure drop through the bed was monitored and recorded using an installed pressure sensor on each treatment vessel.
- OdaLog performance was observed daily, and batteries to each OdaLog unit were changed on a weekly basis. Any accumulated water from the water traps were drained on a daily basis.
- MM and DMS was monitored weekly through the use of colorimetric tubes.
- The air flow rate, measured through velocity readings were verified and adjusted weekly, as needed. Velocity measurements were taken through a hot wire anemometer and an air foil.

Testing Procedures and Methodologies

Foul air samples were collected at the inlet and outlet of each treatment technology to determine the effectiveness in reducing particular odorants using standardized methods. Sample collection methodology and analyses conformed to industry best practice and approved Standard Methods for the Examination of Water and Wastewater, and U.S. Environmental Protection Agency (EPA) protocols.

Odorous air samples for musty and fecal odorant analysis, as well as Odor Profile Method (OPM), were collected and transported the same day to University of California Los Angeles (UCLA) for immediate analysis. Similarly, odorous air samples for reduced sulfur odorant analysis were collected and transported same day to ALS Environmental Laboratories for immediate analysis. It was imperative that the time between sample collection and analysis be minimized to prevent sample decay. To that end, all samples were driven to the laboratories immediately after the samples were collected to minimize hold times and associated decay effects. Odor threshold (odor panel) samples were shipped overnight to St. Croix Sensory for analysis within all EPA approved holding times.

Air samples were collected using a vacuum chamber conforming to EPA-approved design parameters and sampling methodologies. Grab samples were collected into Teflon bags from the treatment system ducting using sample taps at steady-state conditions. Air sample bags were filled by connecting the sampler to duct ports and filling the bag, collecting a 1- to- 10-liter sample at the ambient humidity levels. All sample bags were preconditioned.

PILOT TEST RESULTS

Four official sampling events were conducted, with two official sampling events conducted per operating condition. During an official sampling event, samples were taken for each technology and the inlet and analyzed for reduced sulfur compounds, fecal, and musty compounds. Samples were also analyzed using the OPM and D/T analysis.

Through the OPM analysis conducted, it was discovered that while overall removal rates of fecal compounds remained relatively high in each sample, removal of skatole remained much lower than the removal of fecal compounds in each treatment technology. This discrepancy suggests that the OPM method indicates that there is an unknown fecal compound that is being recognized and removed by the technologies.

Performance: BTF-1 Foam Cube

BTF-1, which utilized the reticulated foam cubes, received nutrients throughout the duration of the test. BTF-1 achieved relatively steady removal of H_2S under both EBGRTs tested. BTF-1 achieved greater removal of musty, fecal, and reduced sulfur compounds under the second operating condition. This may be expected as the EBGRT was increased from 15 seconds to 30 seconds from 15 seconds. See Figure 4 for H_2S performance results along with pH trending. See Figure 5 for reduced sulfur removal performance results. See Figure 6 for musty, fecal, and D/T performance results.

For BTF-1, pH remained steady between 2 and 3 but shifted upwards when inlet levels dropped in October as mentioned previously. Tabulated performance data for key sulfur-based odorants are summarized in Table 3.

Note that there are blue boxes in Figure 6 and other similar figures herein that are shown overlaid on top of the fecal performance bars. These represent specific removal efficiencies for skatole. This reveals the fact that skatole is believed to be a small component of all fecal odorants, some of which have not been identified. Therefore, skatole is believed to be poorly removed in comparison to other unknown fecal compounds.

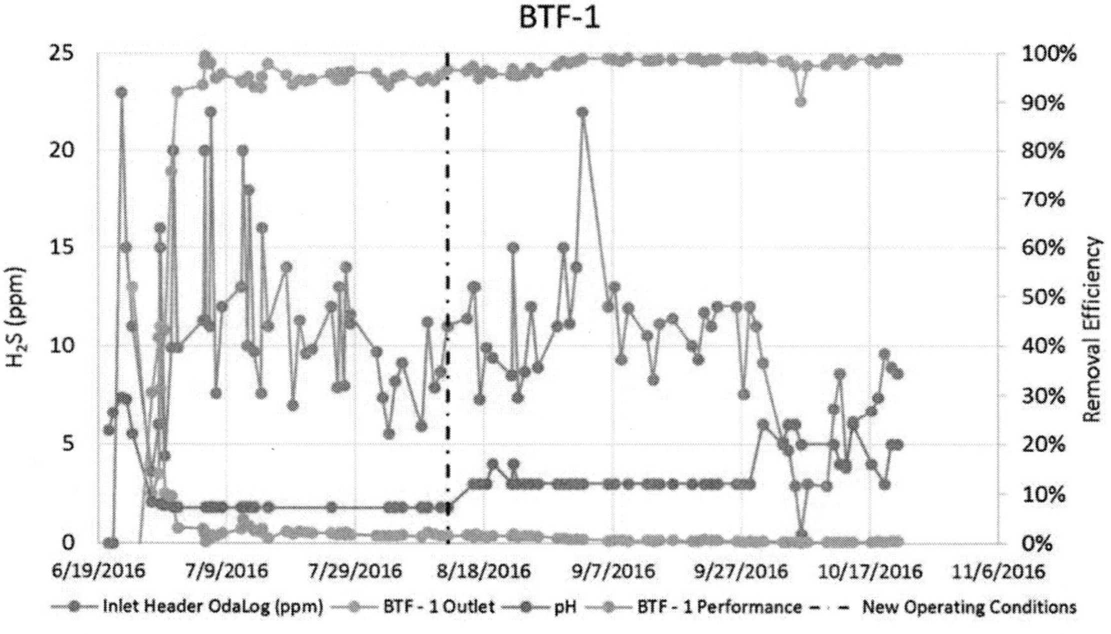

Figure 4. BTF-1 (Foam Cube) H_2S Performance Results and Associated pH Trending

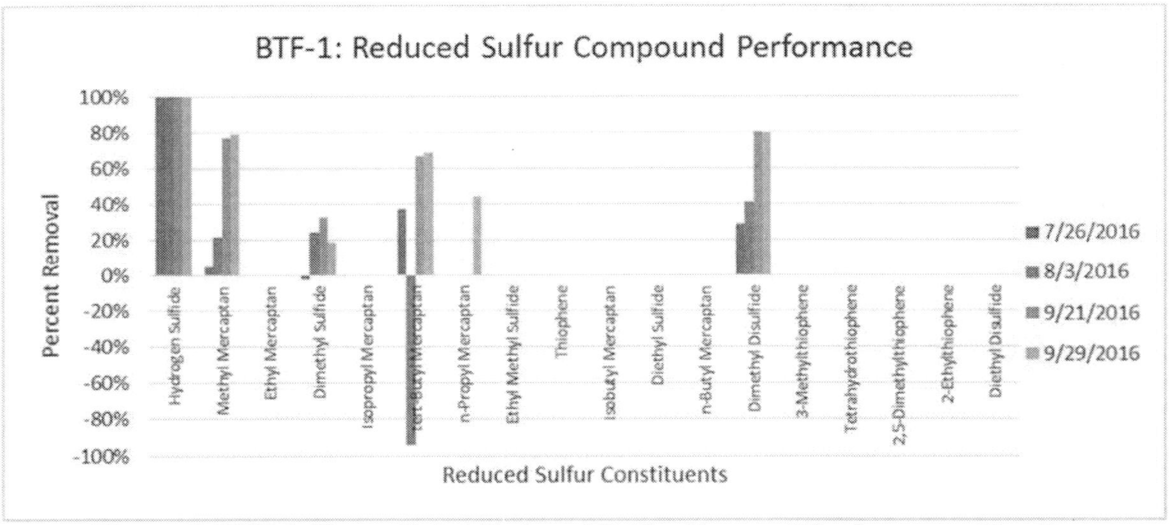

Figure 5. BTF-1 (Foam Cube) Reduced Sulfur Removal Performance Results

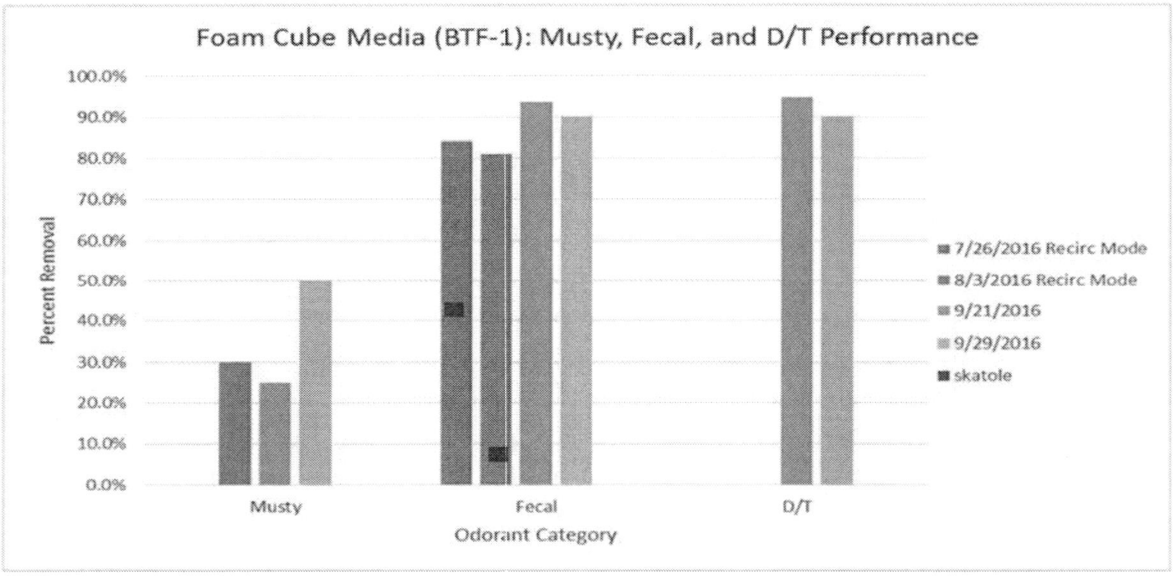

Figure 6. BTF-1 (Foam Cube) Musty, Fecal, and D/T Performance Results

Table 3. BTF-1 (Foam Cube) Sulfur-Based Odorant Performance Data

Constituent	First Condition (6/20/2016 – 8/12/2016) – Removal (%)		Second Condition (8/12/2016 – 10/13/2016) – Removal (%)	
	First Event	Second Event	First Event	Second Event
H_2S	99.93	99.91	99.96	99.87
MM	4.69	21.43	77.00	78.95
DMS	-1.96	24.29	32.39	18.33
DMDS	28.74	41.00	80.47	79.27

MM = methyl mercaptan, DMS = dimethyl sulfide, DMDS = dimethyl disulfide

Performance: BTF-2 Matala® Media

BTF-2, which utilized Matala® and foam mats, received nutrients throughout the duration of the test. BTF-2 achieved relatively steady removal of H_2S under both EBGRTs tested, and achieved greater removal rates than BTF-1 in general. Similar to BTF-1, BTF-2 achieved greater removal of musty, fecal, and reduced sulfur compounds under the second operating condition. See Figure 7 for H_2S performance results along with pH trending. See Figure 8 for reduced sulfur removal performance results. See Figure 9 for musty, fecal, and D/T performance results.

For BTF-2, pH remained steady between 2 and 3 but shifted upwards when inlet levels dropped in October as mentioned previously. Tabulated performance data for key sulfur-based odorants are summarized in Table 4.

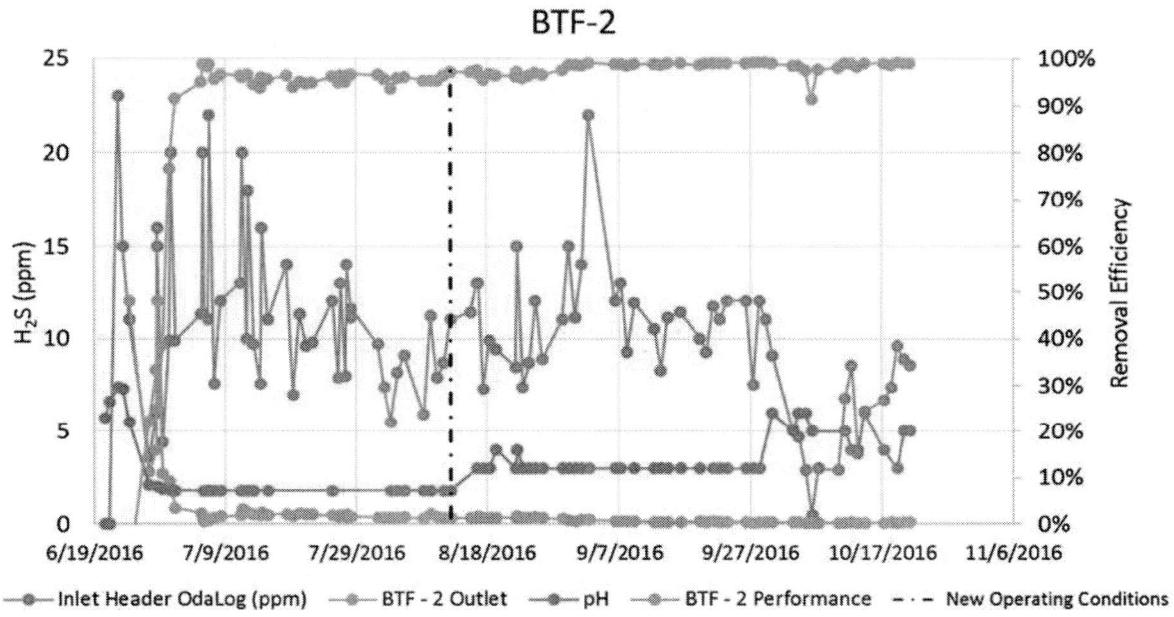

Figure 7. BTF-2 (Matala®) H_2S Performance Results and Associated pH Trending

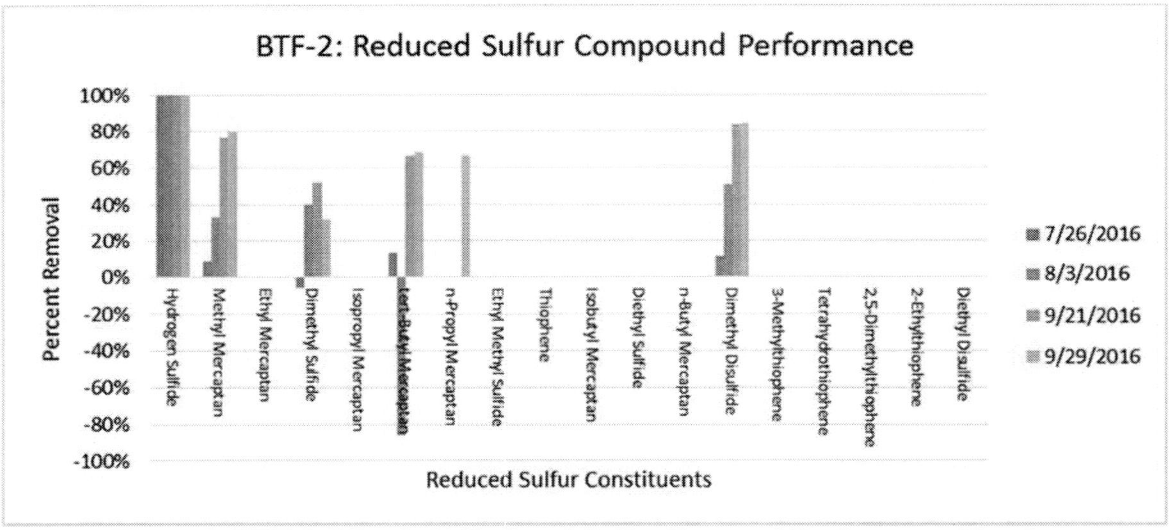

Figure 8. BTF-2 (Matala®) Reduced Sulfur Removal Performance Results

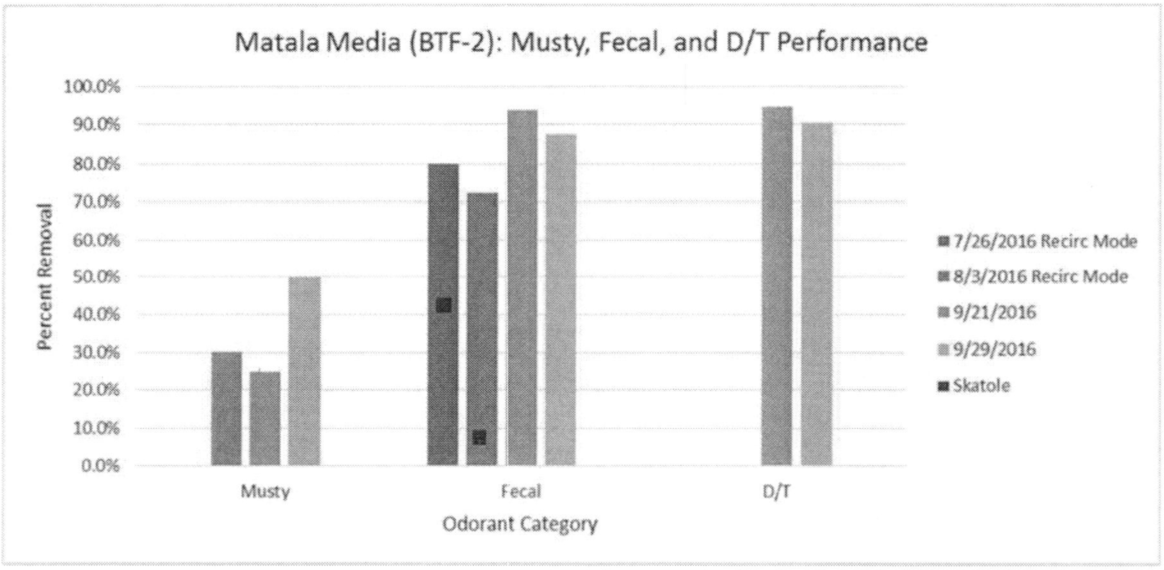

Figure 9. BTF-2 (Matala®) Musty, Fecal, and D/T Performance Results

Table 4. BTF-2 (Matala®) Sulfur-Based Odorant Performance Data

Constituent	First Condition (6/20/2016 – 8/12/2016) – Removal (%)		Second Condition (8/12/2016 – 10/13/2016) – Removal (%)	
	First Event	Second Event	First Event	Second Event
H_2S	99.93	99.93	99.91	99.96
MM	9.4	33.3	77.0	80.0
DMS	-6	40.0	52.1	31.7
DMDS	11.5	51.0	83.7	84.1

MM = methyl mercaptan, DMS = dimethyl sulfide, DMDS = dimethyl disulfide

Performance: BSBR-1 Cell-Max Engineered Media Biofilter
BSBR-1, which utilized the engineered media Cell-Max, did not receive nutrients throughout the duration of the test. BSBR-1 achieved steady removal of H_2S under both EBGRTs tested, and achieved greater removal rates than BTF-1 and BTF-2 in general. BSBR-1 achieved high removal of fecal and reduced sulfur compounds and reduced removal rates of musty compounds. Removal rates of most compounds were greater under the second operating condition. See Figure 10 for H_2S performance results along with pH trending. See Figure 11 for reduced sulfur removal performance results. See Figure 12 for musty, fecal, and D/T performance results.

For BSBR-1, pH remained steady around 5 for the duration of the test. Tabulated performance data for key sulfur-based odorants are summarized in Table 5.

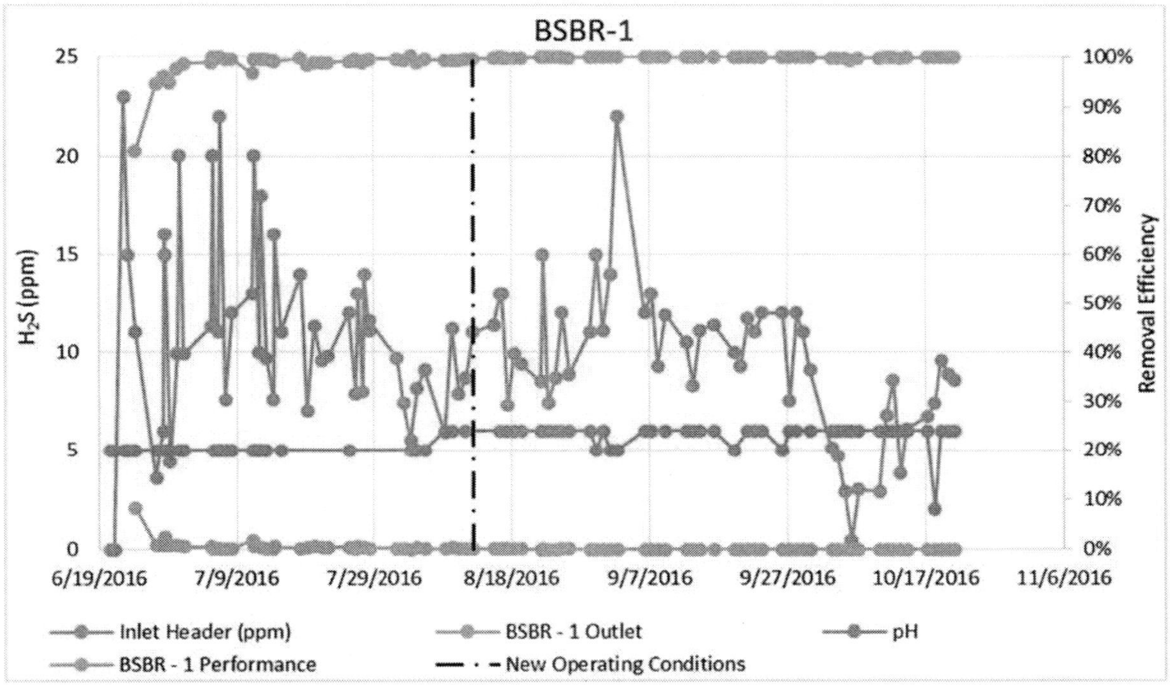

Figure 10. BSBR-1 (Cell-Max) H_2S Performance Results and Associated pH Trending

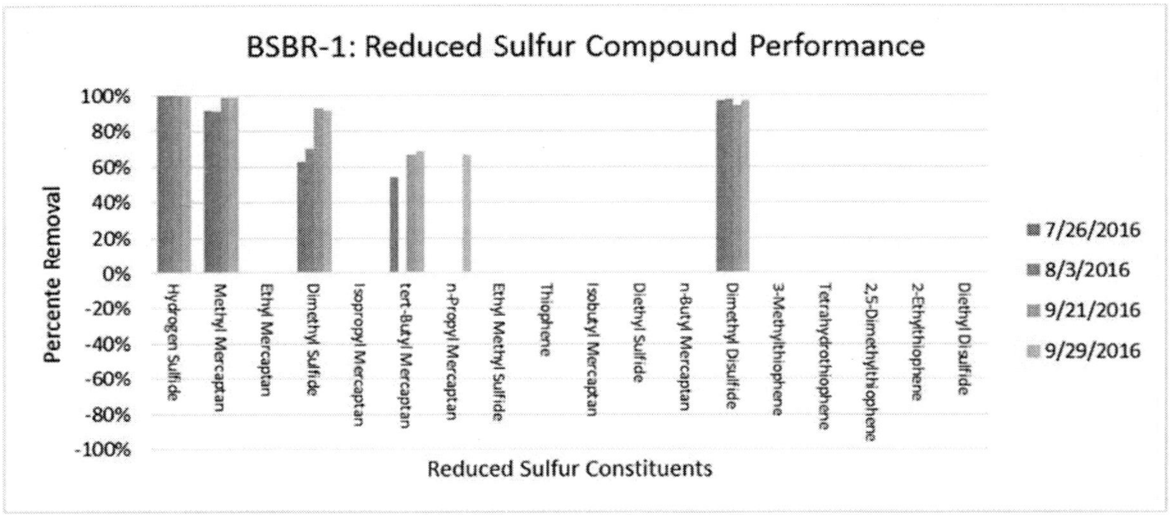

Figure 11. BSBR-1 (Cell-Max) Reduced Sulfur Removal Performance Results

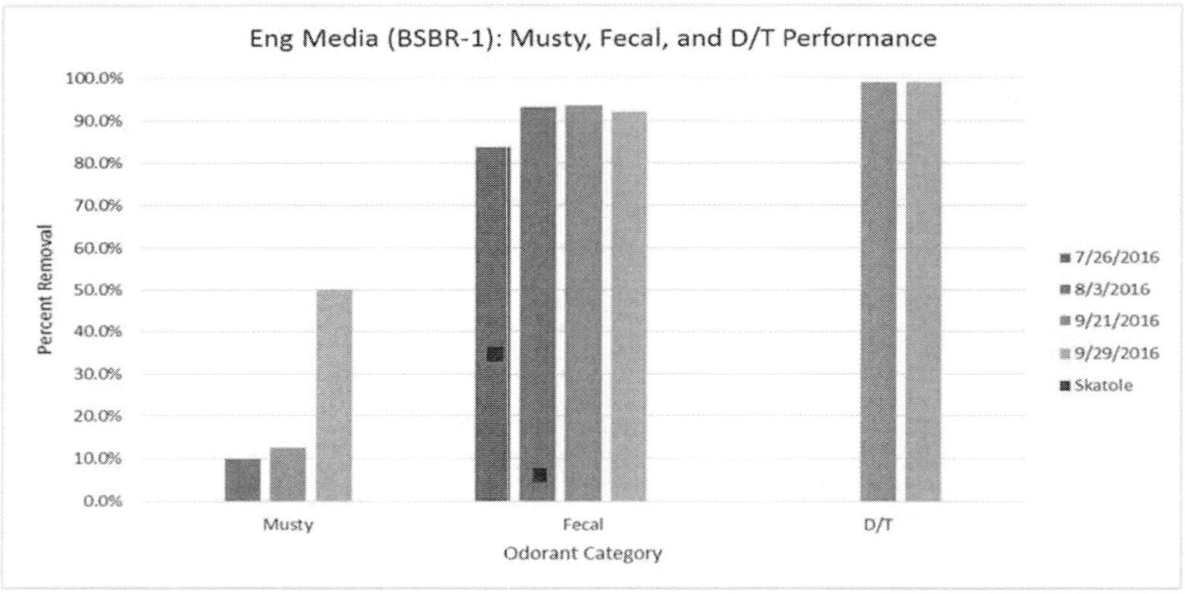

Figure 12. BSBR-1 (Cell-Max) Musty, Fecal, and D/T Performance Results

Table 5. BSBR-1 (Cell-Max) Sulfur-Based Odorant Performance Data

Constituent	First Condition (6/20/2016 – 8/12/2016) – Removal (%)		Second Condition (8/12/2016 – 10/13/2016) – Removal (%)	
	First Event	Second Event	First Event	Second Event
H_2S	99.93	99.93	99.97	99.96
MM	91.4	91.0	98.8	98.6
DMS	62.7	70.0	93.0	91.7
DMDS	97.0	97.5	94.2	97.0

MM = methyl mercaptan, DMS = dimethyl sulfide, DMDS = dimethyl disulfide

Performance: BSBR-2 Seashell Media Biofilter
BSBR-2, which utilized seashell media, did not receive nutrients throughout the duration of the test. BSBR-2 achieved steady removal of H_2S under both EBGRTs tested, and achieved greater removal rates than BTF-1, BTF-2, and approximately equal removal rates to BSBR-1. BSBR-2 achieved high removal of fecal and reduced sulfur compounds and mixed removal rates of musty compounds. Removal rates of most compounds were greater under the second operating condition. See Figure 13 for H_2S performance results along with pH trending. See Figure 14 for reduced sulfur removal performance results. See Figure 15 for musty, fecal, and D/T performance results.

For BSBR-2, pH remained steady around 6 for the duration of the test. Tabulated performance data for key sulfur-based odorants are summarized in Table 6.

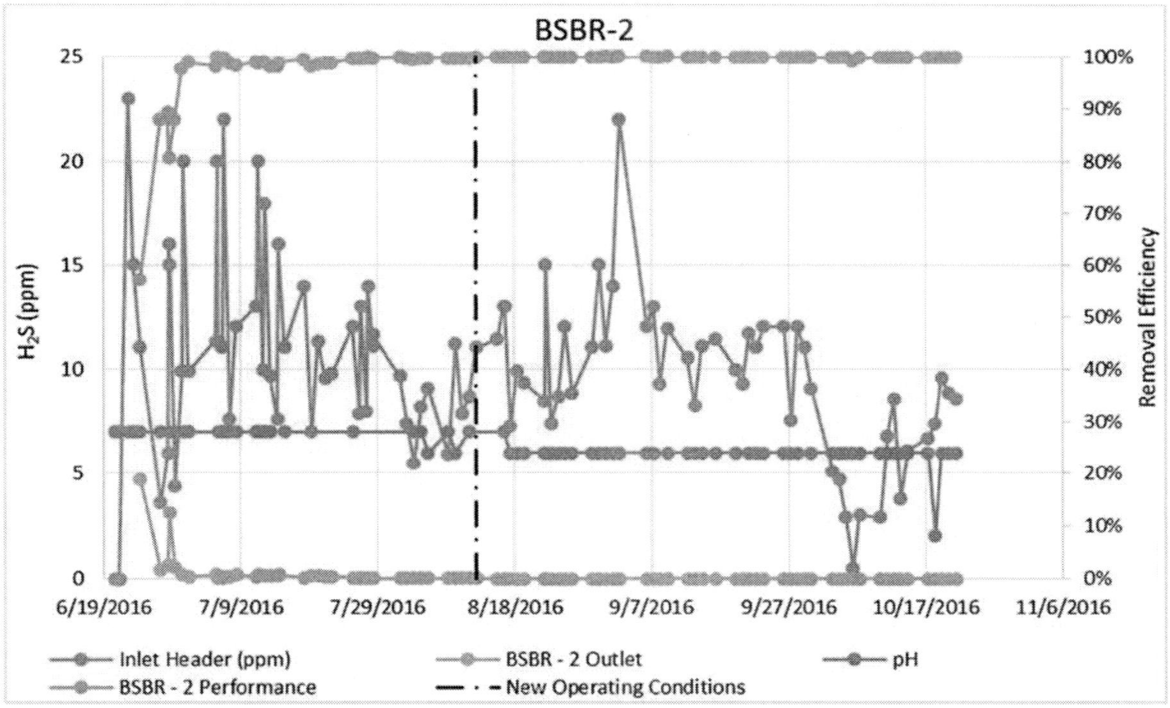

Figure 13. BSBR-2 (Seashell) H_2S Performance Results and Associated pH Trending

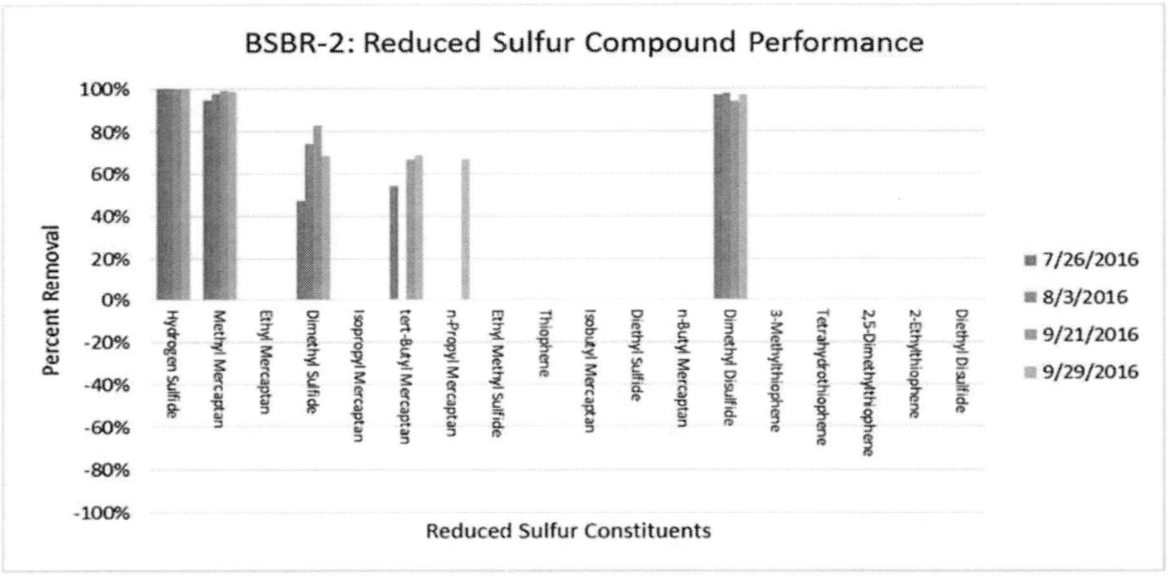

Figure 14. BSBR-2 (Seashell) Reduced Sulfur Removal Performance Results

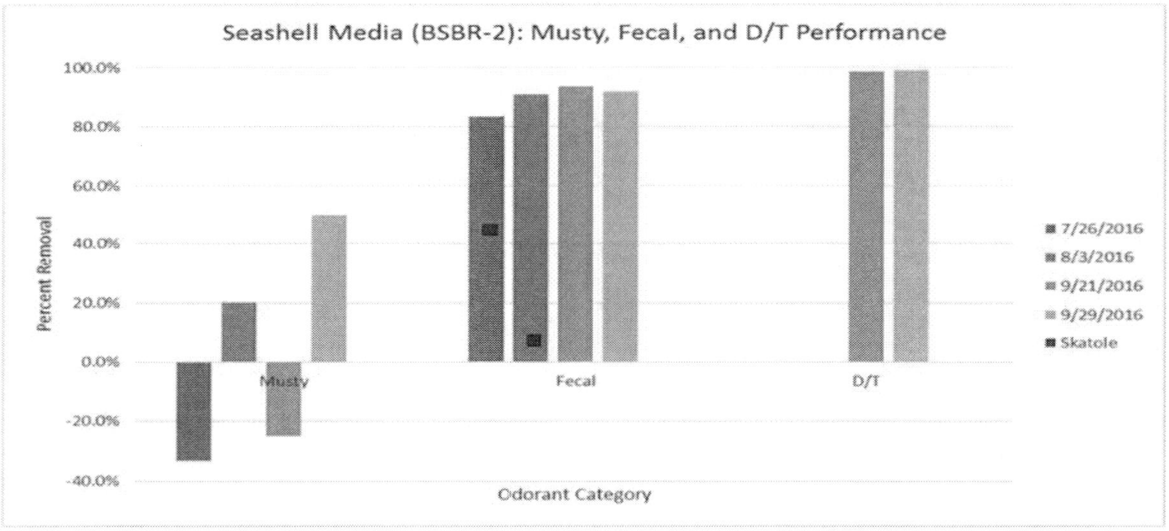

Figure 15. BSBR-2 (Seashell) Musty, Fecal, and D/T Performance Results

Table 6. BSBR-2 (Seashell) Sulfur-Based Odorant Performance Data

Constituent	First Condition (6/20/2016 – 8/12/2016) – Removal (%)		Second Condition (8/12/2016 – 10/13/2016) – Removal (%)	
	First Event	Second Event	First Event	Second Event
H_2S	99.93	99.93	99.95	99.96
MM	94.5	97.5	99.0	98.6
DMS	47.1	74.3	83.1	68.3
DMDS	97.1	97.5	94.2	97.0

MM = methyl mercaptan, DMS = dimethyl sulfide, DMDS = dimethyl disulfide

Performance: BSBR-3 Lava Rock Media Biofilter
BSBR-3, which utilized lava rock media, received nutrients throughout the duration of the test. BSBR-3 achieved high removal of H_2S under both EBGRTs tested, and achieved greater removal rates than BTF-1, BTF-2, and approximately equal removal rates to BSBR-1 and BSBR-2. BSBR-3 achieved high removal of fecal and reduced sulfur compounds and reduced removal rates of musty compounds. Removal rates of most compounds were greater under the second operating condition. See Figure 16 for H_2S performance results along with pH trending. See Figure 17 for reduced sulfur removal performance results. See Figure 18 for musty, fecal, and D/T performance results.

For BSBR-3, pH remained steady between 5 and 6 for the duration of the test. Tabulated performance data for key sulfur-based odorants are summarized in Table 7.

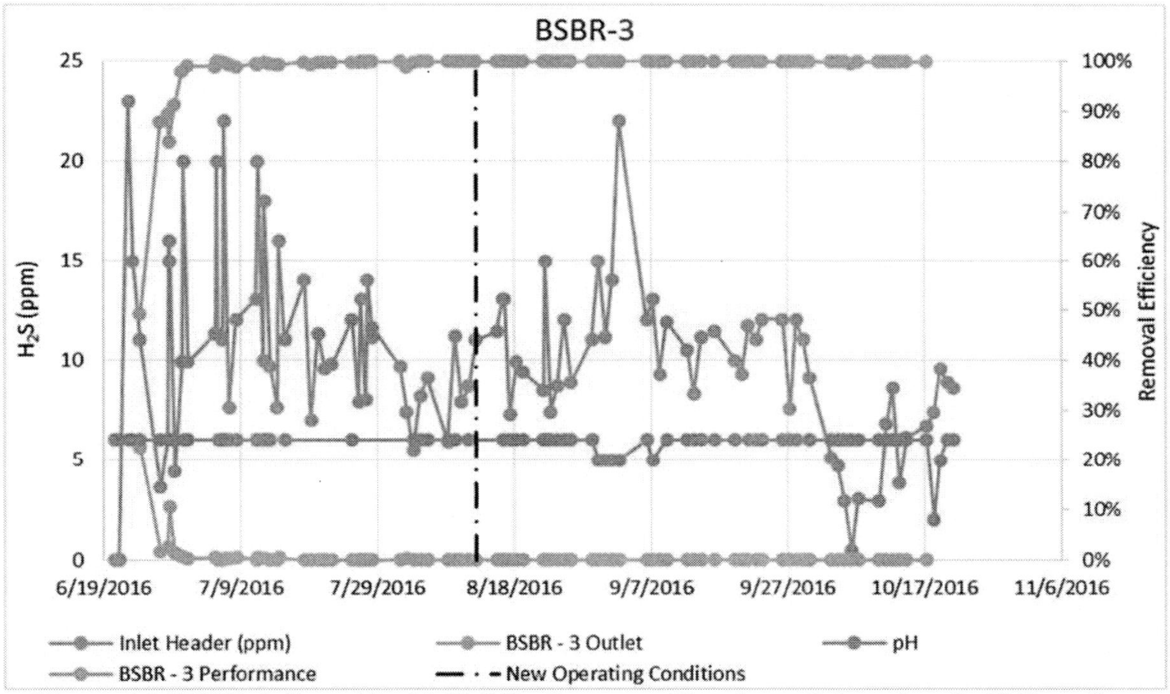

Figure 16. BSBR-3 (Lava Rock) H_2S Performance Results and Associated pH Trending

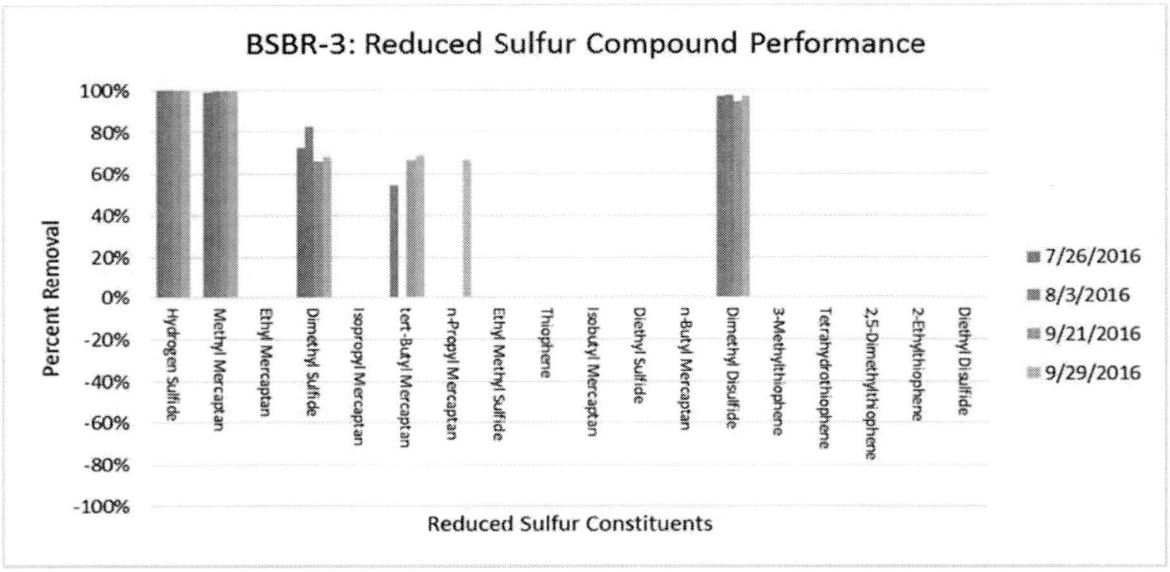

Figure 17. BSBR-3 (Lava Rock) Reduced Sulfur Removal Performance Results

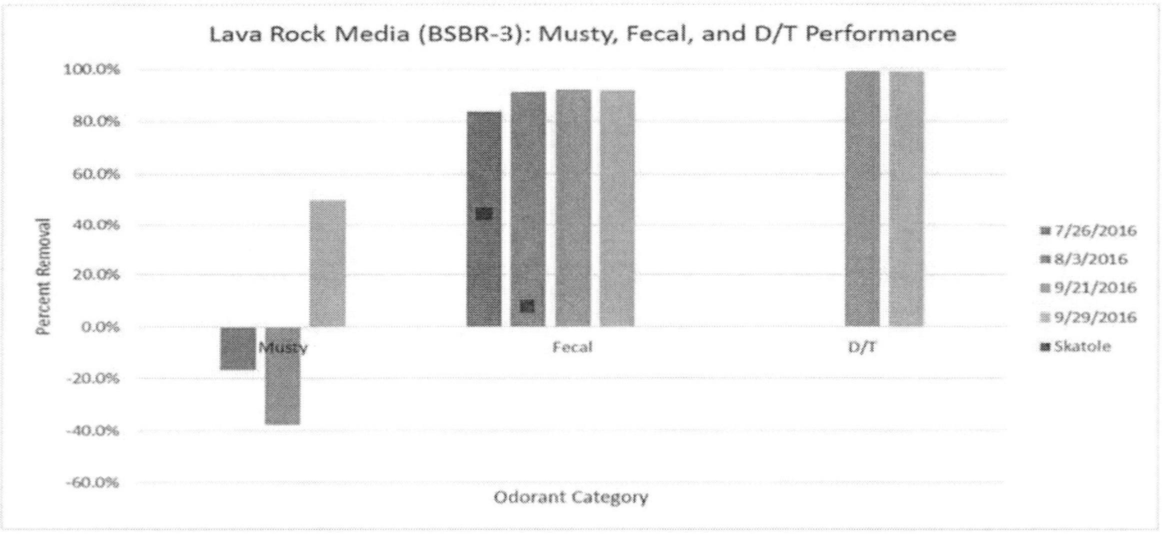

Figure 18. BSBR-3 (Lava Rock) Musty, Fecal, and D/T Performance Results

Table 7. BSBR-3 (Lava Rock) Sulfur-Based Odorant Performance Data

Constituent	First Condition (6/20/2016 – 8/12/2016) – Removal (%)		Second Condition (8/12/2016 – 10/13/2016) – Removal (%)	
	First Event	Second Event	First Event	Second Event
H_2S	99.93	99.93	99.97	99.96
MM	98.9	99.4	99.5	99.5
DMS	72.5	82.9	66.2	68.3
DMDS	97.1	97.5	94.2	97.0

MM = methyl mercaptan, DMS = dimethyl sulfide, DMDS = dimethyl disulfide

Performance: Dry Media Adsorption Medias

The carbon treatment vessels operated at a contact time of 3.5 seconds throughout the entire testing period. CARB-3, which contained the coconut shell carbon, achieved the greatest removal of H_2S throughout the duration of the test. CARB-4, which contained 50 percent coconut shell carbon and 50 percent potassium permanganate, achieved the second greatest performance initially, but then quickly declined. This decline is expected, as CARB-4 contained approximately half of the activated carbon as the other two vessels due to the inclusion of 50 percent potassium permanganate. CARB-2 performed the worst of the three carbons initially, but due to its high capacity, it performed at a relatively steady level throughout the duration of the test, and experienced minimal sharp declines in performance throughout the test. Figure 19 illustrates the H_2S removal performance trending results for all three carbon units.

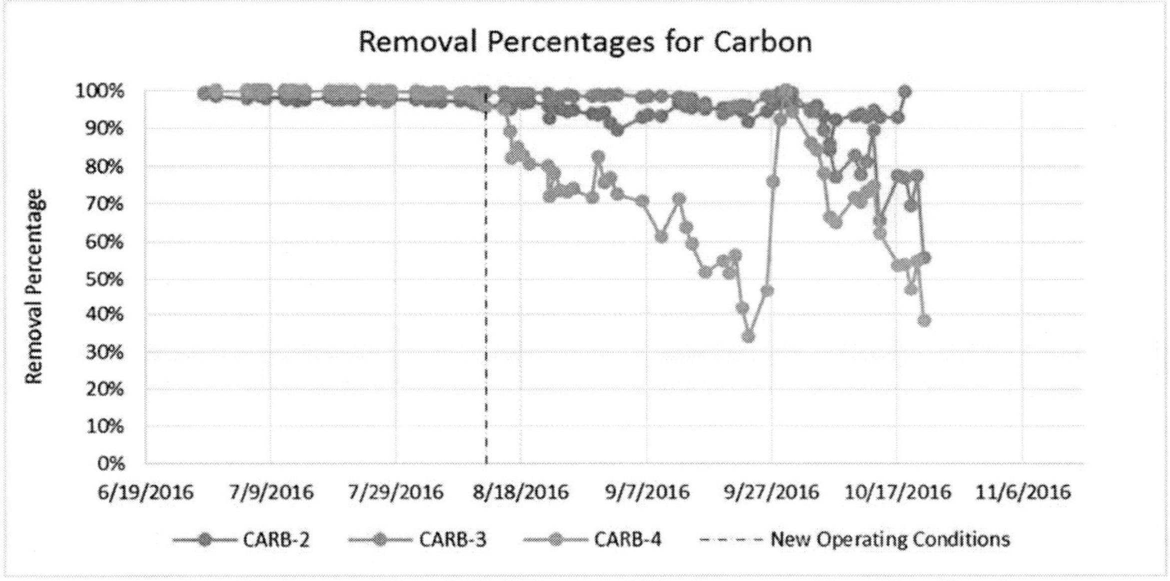

Figure 19. Carbon Vessel H_2S Removal Performance Trending Results

CARB-2 produced greater amounts of DMDS as the duration of the pilot unit testing increased and also produced DMS. MM removal also dropped off over time. CARB-2, CARB-3, and CARB-4 all achieved similar removal rates of fecal compounds although CARB-4 efficiency dropped off quickly. After approximately three months of testing, CARB-3 H_2S removal performance began to sharply decline for the remainder of the testing period. After approximately two months of testing, CARB-4 MM removal performance sharply declined. Figure 20 provides reduced sulfur removal performance results pertaining to CARB-3. Figure 21 provides musty, fecal, and D/T performance results pertaining to CARB-3.

Tabulated performance data for key sulfur-based odorants for CARB-3 are summarized in Table 8.

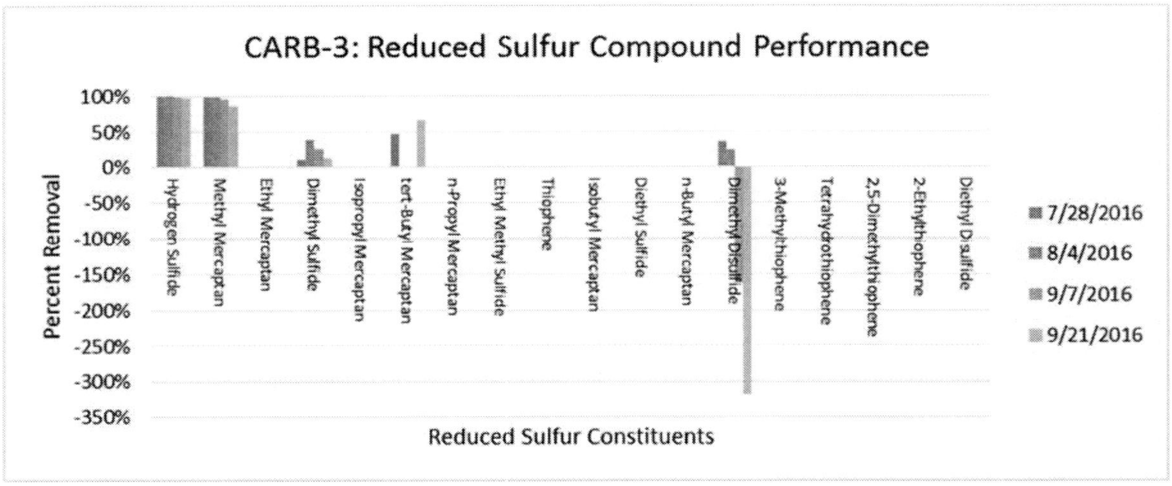

Figure 20. CARB-3 Reduced Sulfur Removal Performance Results

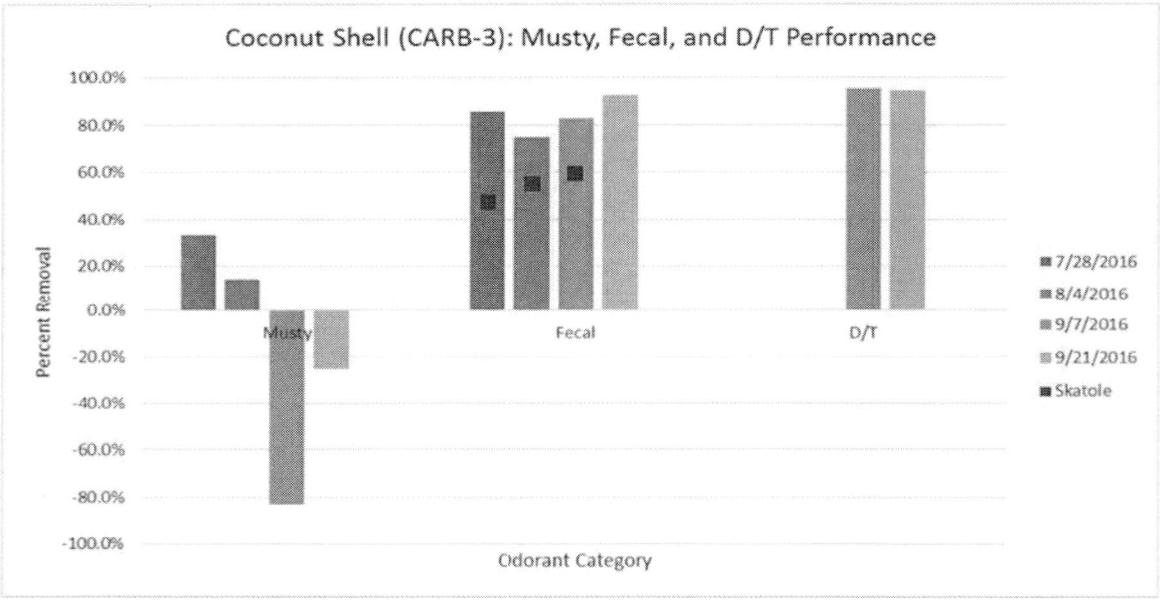

Figure 21. CARB-3 Musty, Fecal, and D/T Performance Results

Table 8. CARB-3 Sulfur-Based Odorant Performance Data

Constituent	First Condition (6/20/2016 – 8/12/2016) – Removal (%)		Second Condition (8/12/2016 – 10/13/2016) – Removal (%)	
	First Event	Second Event	First Event	Second Event
H_2S	99.94	99.95	99.45	97.53
MM	99.3	99.4	96.3	86.0
DMS	11.8	38.2	26.1	12.7
DMDS	35.2	26.6	-163.2	-318.6

MM = methyl mercaptan, DMS = dimethyl sulfide, DMDS = dimethyl disulfide

Acclimation

The inlet foul air connection was connected on June 20, 2016. Measurement of the biological units began on June 24, 2016. Acclimation and steady state performance for the BSBR units was achieved on July 1, 2016, yielding an acclimation time of approximately two weeks. BTF units achieved acclimation and steady state performance on July 14, 2016, yielding an acclimation time of approximately four weeks. Acclimation trending data related to H_2S removal performance is graphically illustrated for all biological technologies in Figure 22.

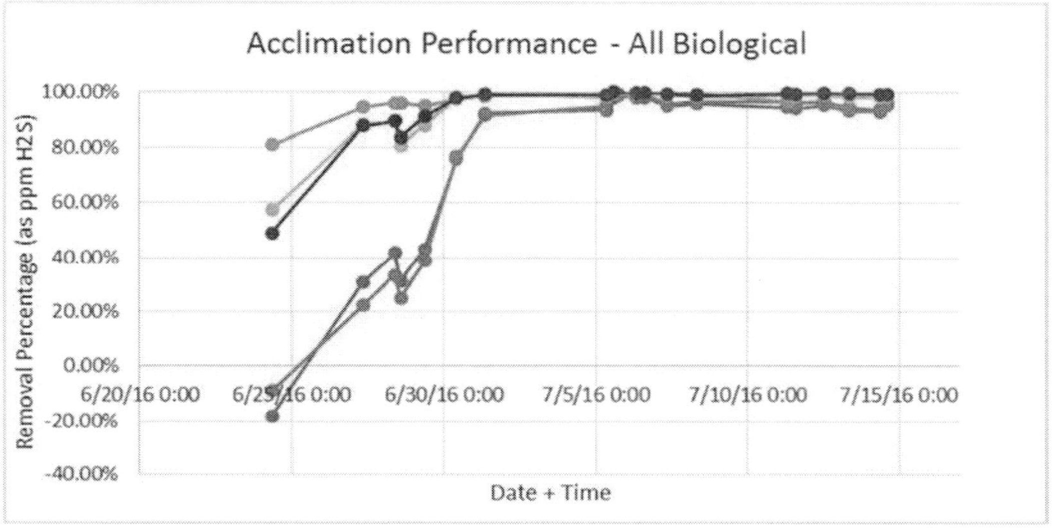

Figure 22. Acclimation Trending Results Based on H_2S Removal for Biological Technologies

Ability to React to Spikes

On August, 1, 2016, a spike in H_2S occurred over the course of approximately six hours. During the spike, H_2S concentrations increased from approximately 10 ppm to about 40 ppm over the course of thee hours. During this elevated H_2S loading event, removal performance of BTF-1, BTF-2, BSBR-1 and BSBR-3 were analyzed using OdaLog data. The OdaLog on BSBR-2 was not operational at this time. See Figure 23 below for performance results during this event.

All measured biological units reacted well to the spike, and did not exhibit large drop offs in removal performance. BSBR-3 achieved the greatest and steadiest performance throughout the duration of the H_2S spike.

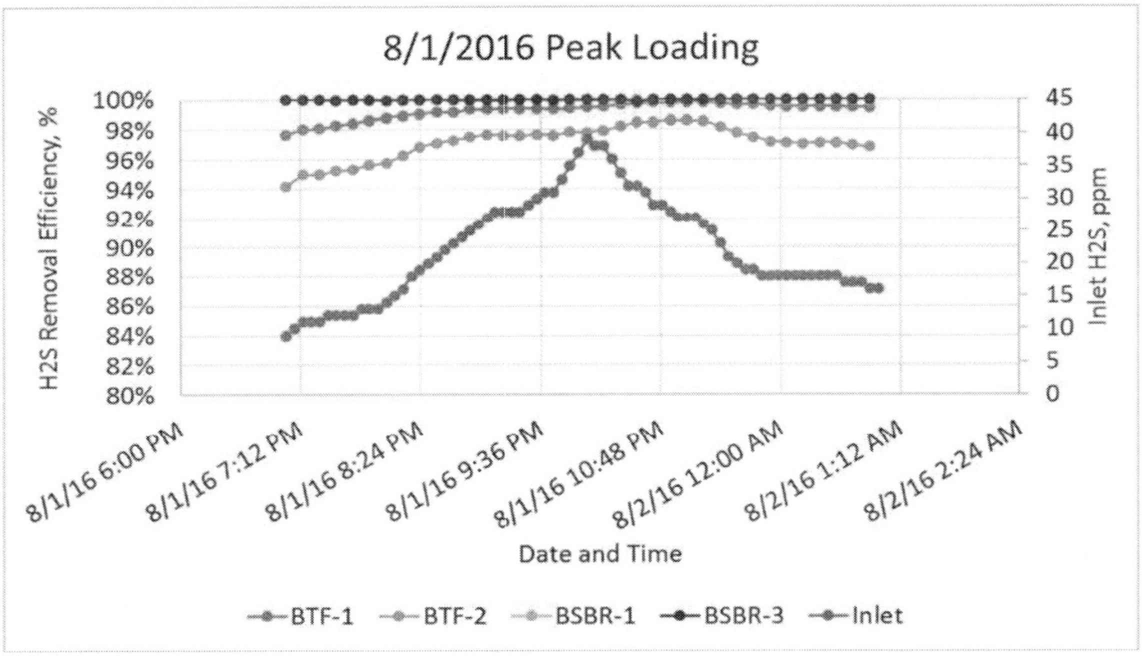

Figure 23. Biological Technology Performance Results during Odor Spike Event

Ability to React to Short-Term Shutdowns and Operating Condition Changes

An extended pilot shutdown occurred on August 12, 2016 in which the pilot was without water and inlet foul air for approximately 72 hours. Upon the restart up of the pilot, the system began operating under the second operating condition, detailed previously. The BSBR units showed increased robustness to the shutdown, and continued to achieve high level of H_2S removal. The BTF units required greater than 3 weeks to completely stabilize and acclimate due to the shut down and change in operating conditions. See Figure 24 below for performance results during this event.

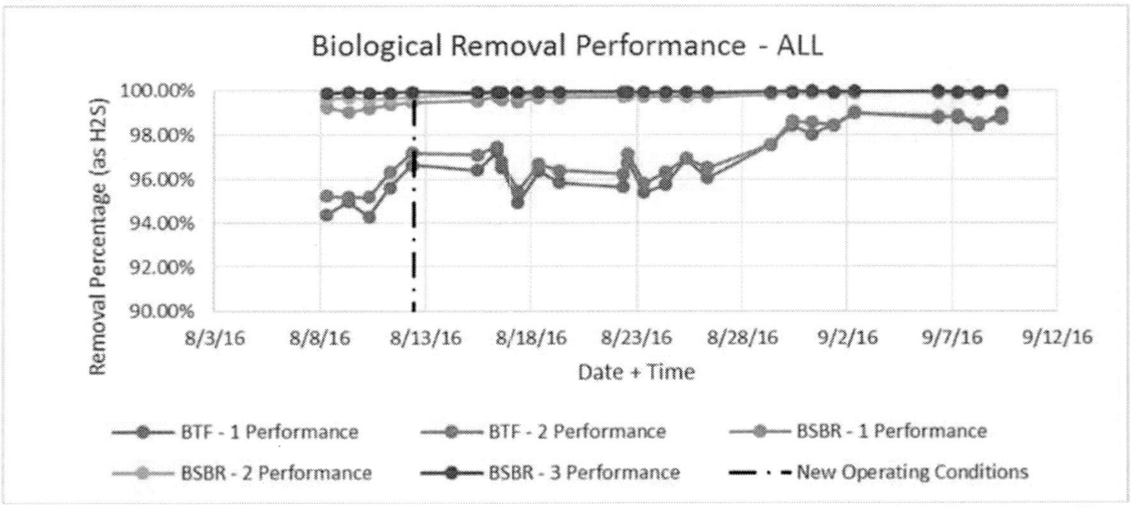

Figure 24. Biological Technology Performance Results during Extended Shutdown Event

Ability to React to Short-Term Shutdowns and Reduced Inlet Conditions
A short-term shutdown occurred on October 17, 2016, which resulted in the loss of both inlet air and water to the pilot for approximately 24 hours. In addition to the short-term shutdown of the pilot, chemical dosing occurred upstream of the trunkline scrubbers, and H_2S inlet concentrations began to decline starting on September 30, 2016, steadily dropping until it reached its lowest measured value of 0.48 ppm on October 6, 2016. Inlet H_2S concentrations did not return to average loading values until approximately October 19, 2016. See Figure 25 below for performance results during this event.

The reduced inlet loadings caused the performance of the BTF units to drop significantly. The BSBR units continued to show strong robustness to changes in inlet loading concentrations. The shutdown on October 17, 2016 caused a slight dip in BTF unit performance, while the BSBR units proved to be more resilient to short term disruptions in air and water feed.

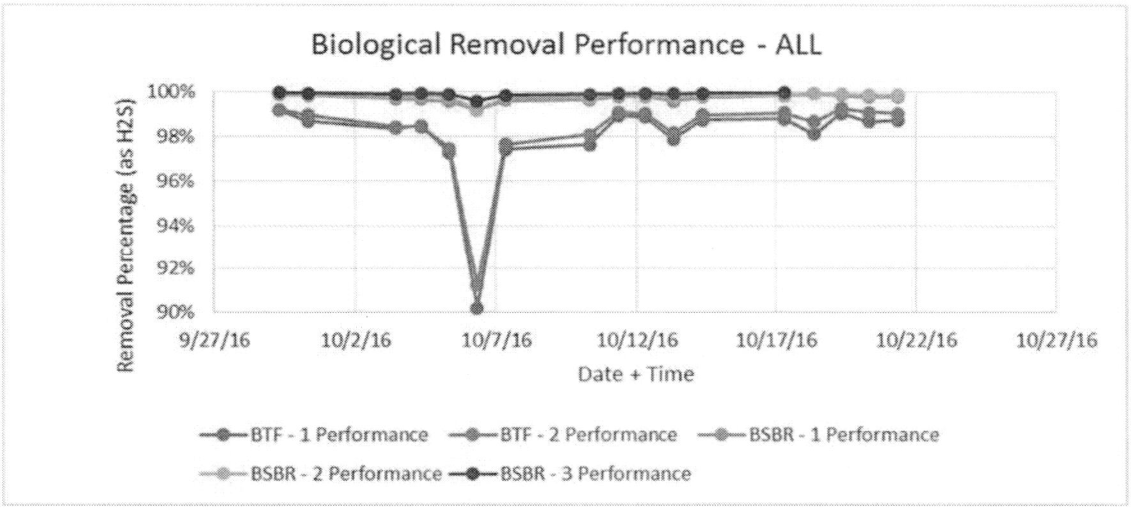

Figure 25. Biological Technology Performance Results during Short-Term Shutdown and Reduced Loading Event

FINDINGS AND CONCLUSIONS

The following is a summary of findings and conclusions:

- Biofilter technologies operated at 45 seconds EBGRT outperformed biotrickling filter and bioscrubber technologies with greater than 99.5 percent removal of H_2S and approximately 99 percent removal of MM.
- Coconut shell carbon appeared to outperform high capacity carbon and the coconut shell/potassium permanganate mixture over the duration of the test. However, the coconut shell/potassium permanganate mixture performed best initially but experienced a sharp drop-off in performance over time. No impact due to moisture was observed.
- All dry medias tested performed well with regards to removal of fecal odorants.
- Biotrickling filters outperformed bioscrubbers with regards to removal of organic reduced sulfur compounds.

- Biotrickling filter and biofilter technologies provided good (~ 90 percent) removal of fecal odorants while bioscrubber technologies provided moderate (~80 percent) removal of fecal odorants.
- Biofilter technologies appeared to exhibit superior robustness when compared to biotrickling filter and bioscrubber technologies under short-term shutdown conditions and reduced inlet loading conditions.
- Biofilter technologies required approximately half the time to achieve full acclimation when compared to biotrickling filter and bioscrubber technologies.
- The pilot results should be considered incomplete with regards to musty/moldy odorant performance removal. The technologies were tested with trunkline odors which exhibited high levels of sulfur-based compounds and fecal odorants but lower levels of musty/moldy compounds. Therefore, this source was not conducive to observe the removal of musty/moldy odors. Additional testing should be performed at a location where must/moldy odors are prevalent (e.g., P2 Trickling Filters). Furthermore, photocatalytic oxidation technologies should be considered for testing at this future location as this technology is considered a promising technology for musty/moldy removal.
- High-capacity carbon media and combined coconut shell/potassium permanganate media are not considered viable as primary treatment for OCSD but would be considered viable in a polishing application (i.e., second stage). Both these media types exhibited poor characteristics with high capacity carbon exhibiting poor DMS and DMDS removal and coconut shell/potassium permanganate exhibiting reduced media life.
- Seashell biofilter frequency of media change-out makes this technology less favorable when compared to lava rock and engineered media. An assumed media life of 3 to 5 years based on typical inlet loadings observed at P1 and P2 will result in significant plant disruption fairly regularly as well as relatively high cost related to change-out labor and media replacement.
- Where high H_2S and/or MM removal is required, biological technology mitigation measures generally provide the best fit. However, other factors including odor loading variability and robustness must be considered before selecting a biological approach.
- Proprietary issues, both favorable and unfavorable, related to media types should be considered before moving forward with a specific media type. For example, lava rock media can be procured as a nonproprietary product (i.e., is readily available). However, there is risk that a guaranteed removal performance may be unattainable for nonproprietary media and should be considered on a case-by-case basis.

Results of pilot testing were distilled into general technology performance categories considered viable for OCSD and fed into mitigation alternative analyses through iterative engineering/air dispersion modeling for meeting stringent offsite odor goals.

Based on the findings and conclusions provided herein, the following provides a summary of viable single stage technologies that were carried forward in the OCMP:

- Chemical Scrubbers (existing baseline)
- Chemical Scrubbers (new and optimized)
- Biological Systems:
    - BTF – 15 seconds EBGRT
    - BTF – 30 seconds EBGRT

    - Biofilter – 30 seconds EBGRT
    - Biofilter – 45 seconds EBGRT
- Dry Media Adsorption Systems – Coconut Shell

REFERENCES

Suffet, I.H.; Zhou, Y.; Vitko, T. "Determination of Odor Threshold Concentrations – How Does a Dynamic Olfactometer Really Work?" In: Proceedings of the Odors and Air Pollutants Meeting. Water Environment Federation, Milwaukee WI, USA. 2016.

Vitko, T.; Suffet, I.H.; Zhou, Y.; Abraham, S. "2014 OCSD Odor Control Master Plan". In: Proceedings of the Odors and Air Pollutants Meeting. Water Environment Federation, Miami, FL, USA. 2014.

Vitko, T.; Suffet, I. H.; Zhou, Y. "How OCSD Identified the Most Detectable Odorants". In: Proceedings of the Odors and Air Pollutants Meeting. Water Environment Federation, Milwaukee WI, USA. 2016.

Two-Stage Biological Odor Control System Treats WWTP Headworks Exhaust Air

Chester M. Morton, P.E., Clean Air & Water Engineering, PLLC
Tom Card, P.E., Environmental Management Consulting
Matthew Smith, P.E., Strand Associates, Inc.
Wesley J. Merkle, P.E., Clay Township Regional Waste District

ABSTRACT

A wastewater treatment plant located north of Indianapolis receives all of its wastewater via pump stations and force mains. The detention time in the force mains ranges from 1.5 to 5 hours. Exhaust air from the plant's headworks building and influent splitter box was being treated in an ozone system and carbon adsorber, respectively. Despite this treatment, odors from the plant were detectable in the neighborhood around the plant. A field study was conducted which characterized the inlet odors to the plant. Gas phase measurements of the exhaust air from the headworks building and splitter box were conducted which consisted of continuous hydrogen sulfide measurements using Odalog units, and grab samples were analyzed for reduced sulfur compounds and odor concentration. Liquid phase measurements were conducted of the influent wastewater which included dissolved sulfide, pH, oxidation-reduction-potential (ORP) and temperature. The analytical results indicated gas phase hydrogen sulfide concentrations exceeding 100 parts per million by volume (ppm_v), dissolved sulfide concentrations of up to 10 milligram per liter (mg/l) and odor concentrations greater than 60,000 odor units. A performance assessment was conducted of the ozone system and carbon adsorber. As a result of the field study and performance assessments it was determined that the existing odor control systems should be replaced. An alternative control technology evaluation was conducted. It was determined that a two-stage biological odor control system would be the most cost-effective alternative to treat the headworks building and splitter box exhaust air.

A two-stage biological odor control system was designed which consists of a bio-trickling filter followed by a biofilter. The system was designed so that a minimum biofilter inlet air temperature can be maintained by heating the bio-trickling air with heated recirculation water. The biofilter is an in-ground configuration that utilizes Biosorbens™ media and is covered. Construction of the odor control system is complete and was started-up in July 2017.

This paper will present the following:

- Results of the field study odor measurements
- Results and findings of the ozone system and carbon adsorber performance assessment
- Bio-trickling filter and biofilter process design summary
- Project costs
- Performance test results

INTRODUCTION

The Michigan Regional Wastewater Treatment Plant (MWWTP) is located in Zionsville, Indiana. The plant is operated by the Clay Township Regional Waste District (CTRWD). The plant was constructed in 1991 with a capacity 0.6-million gallons per day (mgd). It had multiple expansions and now has a capacity of 3.05-mgd. It is expected to be further expanded to 4.05-mgd to meet the population growth of its service area. All wastewater is pumped to the plant through multiple force mains. The collection system consists of gravity sewers and twelve pump stations (lift stations). A number of pump stations are in series resulting in wastewater travel times of up to 5-hours. Five of the secondary pump stations discharge to the MRWWTP. One of the six secondary pump stations discharges to the Carmel Wastewater Treatment Plant (Figure 1).

The force mains discharge to an influent wet well located in the Pretreatment Building (PTB) where the influent wastewater is screened by mechanical step screens and flow is measured in a Parshall flume. The plant's treatment processes include activated sludge consisting of an oxidation ditch and four vertical loop reactors, secondary clarification and UV disinfection. Biosolids are aerobically digested and dewatered in a belt filter press.

The MWWTP is located in a commercial/light industrial area with essentially no buffer between the plant and the community. The influent well and bar screen channel is covered and air from within these spaces was discharged to an ozone treatment unit. A splitter box receives wastewater from the PTB and distributes it to the secondary treatment processes. The splitter box is covered and air from within it was discharged to a carbon adsorption unit. The plant has received odor complaints since the ozone unit was installed in 2011.

Clean Air & Water Engineering, PLLC (CAWE) and Environmental Management Consulting (EMC) were retained by CTRWD in July 2015 and conducted the following:

- A field assessment and to characterize odors at the plant
- Assessed the performance of the existing odor control systems and determined their performance was inadequate
- Evaluated alternative odor control technologies and recommended a two-stage biological system
- Prepared the process design for the two-stage biological system and assisted Strand Associates prepare the construction contract documents for this system.
- Prepared the performance test protocol for the two-stage biological system.

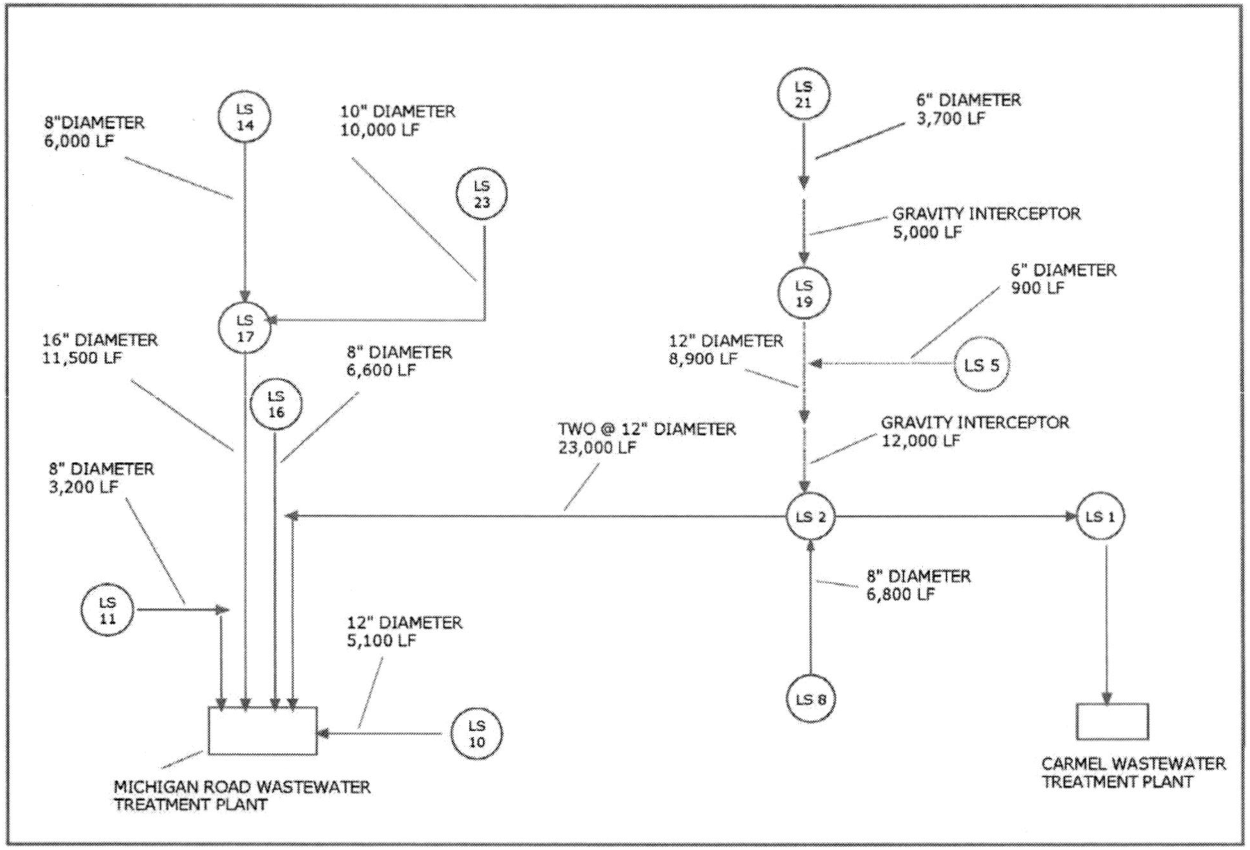

LS = Lift Station

Figure 1 – Zionsville Wastewater Collection System

FIELD MEASUREMENTS

Field measurements were made from August 3 to 5, 2015 at the following locations:

- Ozone Unit Inlet and Exhaust
- Influent Splitter Box Carbon Adsorber Unit Inlet and Exhaust

Air Sample Collection and Analytical Methods

Air samples were analyzed using two methods: by collecting air samples and sending them to off-site laboratories; and onsite using field measurement equipment.

Air Sample Collection Methods

Air samples were collected in Tedlar™ bags using an evacuated lung. An evacuated lung allows an air sample to be collected directly into a sample bag without contacting the internals of a pump and potentially altering the sample's chemical or odor characteristics. A sample bag is placed in the evacuation lung and is connected to a sample port. The lung cover is closed; a sample tube is connected to the external side of the sample port and is run to the sample source. A vacuum pump is connected to the evacuated lung which creates a vacuum in the lung and sample bag, and causes air to be drawn into the sample bag.

Air Analytical Methods

- Hydrogen sulfide. Hydrogen sulfide is the most dominant odor emitted from wastewater treatment plants. It has an odor detection concentration of 0.01 to 0.02 parts per million by volume (ppm_v) depending on the individual perceiving the odor and other odors that may be present. It is a gas that originates from bacteria that thrive in sewers, wastewater treatment plants, and biological solids. The bacteria, Desulfovibrio, convert sulfate, a naturally occurring ion in water, to sulfide under anaerobic conditions.

  Under typical wastewater conditions, sulfide exists in solution as hydrogen sulfide (H_2S) gas; and the dissolved ions, hydrosulfide (HS^-) and sulfide (S^{-2}), with their distribution pH dependent. At pH 7.0 there is a 50/50 distribution of hydrogen sulfide and hydrosulfide present, with an infinitesimal amount of sulfide. At lower pH values the percent of hydrogen sulfide increases. These bacteria are most active at wastewater temperatures greater than 68 °F.

  Continuous hydrogen sulfide measurements were taken using OdaLog units which are manufactured by App-Tek International. These units utilize an electro-chemical sensor, are intrinsically safe and have data-logging capabilities. The units were programmed to record a hydrogen sulfide reading every 5 minutes. It was intended to also use a Jerome meter to measure H_2S which is manufactured by Arizona Instruments. This device employs a gold film sensor to which H_2S absorbs causing its resistance to increase in proportion to the amount absorbed. It has a measurement range of 0.003 to 50 ppm_V. However, the first sample tested exceeded the Jerome Meter's measurement range and was therefore not used.

- Odor Parameters. Odors were measured using industry standard test methods[i]. All the odor parameters were quantified in a laboratory using an odor panel consisting of six to eight members.
- Reduced Sulfur Compounds. Reduced sulfur compounds are found in areas where hydrogen sulfide is present. These compounds have odor detection concentrations similar to hydrogen sulfide. For instance, the detection concentration for methyl mercaptan is 0.5 parts per billion by volume (ppb_V). The detection threshold for carbonyl sulfide is 100-ppb_V. These compounds were analyzed by gas chromatography that measures for twenty reduced sulfur compounds (ASTM Method D5504).

Liquid Phase Parameters

Wastewater samples were collected from a sample tap on three of the five force main pipes just upstream of the influent wet well. No flow was present in the force mains from Lift Stations 10 and 11 at the time of sampling. Each sample was analyzed for the following parameters.

- Total Dissolved Sulfide. The emission of hydrogen sulfide gas is the result of the volatilization of hydrogen sulfide in solution. The amount volatilized is dependent on the concentration of hydrogen sulfide in solution, which as noted above is a function of the pH of the solution. Greater turbulence and higher temperature also increases volatilization.

  Total dissolved sulfide was measured to verify gas phase measurements and to provide a basis for liquid phase H_2S control in the event the gas phase treatment requires augmentation with liquid phase treatment. Total sulfide measures hydrogen sulfide gas (H_2S); and the dissolved ions, hydrosulfide (HS^-) and sulfide (S^{-2}). Total sulfide was measured with a field kit that uses the methylene blue method as described in Standard Methods[ii]. This method has a detection limit of 0.1 mg/l.
- pH. As noted above, the pH of the wastewater determines the fraction of the three sulfide species that are present, H_2S, HS^- or S^{-2}. A portable pH meter was used to determine the pH of the water from which emissions were being measured.
- ORP. Oxidation Reduction Potential (ORP) is a general measure of the relative concentrations of oxidants (chlorine, oxygen, nitrates, sulfate) versus reductants (sulfides,

ammonia, organics) in an aqueous solution. The greater the ORP, the greater the oxidizing potential of the solution, the greater its ability to oxidize reduced compounds such as hydrogen sulfide. ORP measurements were conducted using a portable ORP meter.

Air Sample Results

The inlet and exhaust ozone system H_2S concentrations were measured using Odalog units during two time periods during 2015: August 4 to 5, and November 6 to 9. For the time period August 4 to 5 the inlet average and maximum H_2S concentrations were 61 and 133-ppm$_v$, respectively. The exhaust average and maximum H_2S concentrations were 51 and 113-ppm$_v$, respectively. Hydrogen sulfide concentrations greater than 1-ppm$_v$ in the ozone system exhaust are considered to have a high potential to be perceived off-site given its proximity to facility's fenceline which is the case for these measured values. The average removal efficiency over this time period was 18%.

For the time period November 6 to 9 the inlet average and maximum H_2S concentrations were 278 and 428-ppm$_v$, respectively. The exhaust average and maximum H_2S concentrations were 117 and 186-ppm$_v$, respectively. The average removal efficiency over this time period was 58%.

The reason for the increase in inlet H_2S concentrations measured in November as compared to the August sampling is believed to have been caused by a reduction in flow through the collection system's force mains, a lower cycling rate of the pump station's pumps and an increase of wastewater travel time from the pump stations to the plant.

The inlet and exhaust influent splitter box carbon adsorber unit H_2S concentrations were measured from August 4 to 5, 2015. The inlet average and maximum H_2S concentrations were 50 and 167-ppm$_v$, respectively. The exhaust average and maximum H_2S concentrations were 27 and 92-ppm$_v$, respectively. The average removal efficiency over this time period was 46%.

H_2S measurements were made in the Pretreatment Bldg. on the bar screen and dumpster levels from August 4 to 5, 2015. On the bar screen level the average and maximum H_2S concentrations were 3 and 24-ppm$_v$, respectively. The maximum hydrogen sulfide concentration measured on the bar screen level was the result of a power loss to the ozone unit which included its exhaust fan for approximately 1.5-hours. On the dumpster level the average and maximum H_2S concentrations were 3 and 5- ppm$_v$, respectively.

The air phase sample results measured with the OdaLog units are summarized in Table 1.

Table 1 – OdaLog Hydrogen Sulfide Measurement Summary

Location	Average (ppm_v)	Maximum (ppm_v)	Monitoring Date/Time
Ozone Unit Inlet	61	133	8/4/15 to 8/5/15
Ozone Unit Exhaust	51	113	8/4/15 to 8/5/15
Ozone Unit Inlet	278	428	11/6/15 to 11/9/15
Ozone Unit Exhaust	117	186	11/6/15 to 11/9/15
Pretreatment Bldg. Bar Screen Level	3	24	8/4/15 to 8/5/15
Pretreatment Bldg. Dumpster Level	3	5	8/5/15 to 8/5/15
Influent Splitter Box Carbon Adsorber Unit Inlet	50	167	8/4/15 to 8/5/15
Influent Splitter Box Carbon Adsorber Unit Exhaust	27	92	8/4/15 to 8/5/15

The reduced sulfur scan results indicated only hydrogen sulfide being present in the inlet sample and hydrogen sulfide and methyl mercaptan present in the exhaust. Methyl mercaptan and other reduced sulfur compounds may have been present in the ozone system inlet, and the inlet and exhaust of the influent splitter box carbon adsorber unit, but were not detected because the air samples had to be diluted to measure the hydrogen sulfide at a concentration that was compatible with the detector of the laboratory's gas chromatograph's detector. Diluting the samples may have reduced the concentration of other compounds present below their detection limit. It is also possible that the methyl mercaptan in the ozone system exhaust may have been caused by this compound having been formed in the ozone unit. The reduced sulfur scan results from the ozone unit inlet and exhaust and the carbon adsorber inlet and exhaust are summarized in Table 2.

Table 2 - Reduced Sulfur Scan Results

Location	Hydrogen Sulfide (ppm_v)	Methyl Mercaptan (ppm_v)
Ozone Unit Inlet	45.0	nd
Ozone Unit Exhaust	30.0	0.550
Influent Splitter Box Carbon Adsorber Unit Inlet	10.0	nd
Influent Splitter Box Carbon Adsorber Unit Exhaust	7.6	nd

nd = Non-Detect

The reported odor concentrations are high and exceed the dilution capacity of the olfactometer used by the odor laboratory. The high odor concentration of the samples is consistent with the measured hydrogen sulfide concentrations and indicates a high potential to cause odorous impacts offsite. The odor sample analysis results are summarized in Table 3.

Table 3 - Odor Analysis Results

Location	Detection Threshold	Recognition Threshold
Ozone Unit Inlet	>60,000	>60,000
Ozone Unit Exhaust	>60,000	>60,000
Influent Splitter Box Carbon Adsorber Inlet	>60,000	>60,000
Influent Splitter Box Carbon Adsorber Unit Exhaust	26,000	14,000

Liquid Phase Results

The total dissolved sulfide present in the force mains, 5.2 to 10.0 mg/l, are significant and consistent with the elevated gas phase hydrogen sulfide concentrations that were measured with the Odalog units. The pH values are typical for a wastewater. The ORP values ranged from -229 to -163 millivolts and are consistent with a wastewater with low dissolved oxygen and the presence of hydrogen sulfide. Table 4 summarizes the liquid phase measurements conducted on the force main sample lines in the Pretreatment Bldg.

Table 4 - Force Main Liquid Phase

Date	Time	Lift Station	Total Dissolved Sulfide	pH	ORP	Temperature
			(mg/l)	S.U.	mV	Deg F
8/3/2015	17:00	No. 2	9.5	7.57	-220	73.1
8/3/2015	17:15	No. 2	6.5	7.45	-209	70.8
8/3/2015	17:35	No. 16	8.0	7.3	-200	71.3
8/3/2015	17:50	No. 17	9.4	7.61	-205	68.3
8/3/2015	17:55	No. 10 & 11	No Flow			
8/4/2015	16:30	No. 2	5.2	7.43	-163	75.5
8/4/2015	16:50	No. 2	5.4	7.22	-199	75.5
8/4/2015	17:12	No. 16	Limited Flow			
8/4/2015	17:25	No. 10 & 11	No Flow			
8/4/2015	17:30	No. 17	10.0	7.57	-229	70.2

EXISTING ODOR TREATMENT EQUIPMENT ASSESSMENT

Existing Ozone Unit Assessment

The ozone unit (Figure 2) treating the Pretreatment Bldg. exhaust air had a 2,550-cubic meters per hour (m^3/hr) (1,500-cubic feet per minute (cfm)) fan which pulled 1,700-m3/hr (1,000-cfm) from the Pretreatment Bldg. and 850-m^3/hr (500-cfm) that was used to generate ozone. These two air streams combined and passed through the unit's reaction chamber.

Figure 2 - Ozone Unit

The contact chamber had a volume of 2.80-m^3 (99-ft^3) and an ozone generating capacity of 3 to 6 gram ozone per hour.

Given the chamber's volume and air flow rate, the unit provided a contact time of about 4 seconds. At a flow rate of 2,550-m^3 (1,500-cfm) and an ozone generation rate of 6 gram per hour, the unit had the capacity to generate 1-ppm_v of ozone.

An EPA document[iii] indicates that 1-ppm_v of ozone will oxidize 10-ppm_v of hydrogen sulfide and recommends a contact time of 30 to 40 seconds.

As noted in the Air Sample Results section above, the ozone unit's inlet hydrogen sulfide reached 428-ppm_v. Using the EPA document guidance the ozone unit should have the capacity to generate 43-ppm_v of ozone. The ozone unit was undersized based on its contact time and ozone generation capacity. The inadequate treatment capacity of the unit was also demonstrated by the unit's average H_2S removal efficiency of 18% and 58% in the August and November 2015 monitoring efforts, respectively. The inlet and exhaust odor concentrations were too high relative to the odor laboratory's olfactometer dilution capability to be measured in order to estimate odor removal efficiency.

Existing Influent Splitter Box Carbon Adsorber Assessment

The existing influent splitter box carbon adsorber unit (Figure 3) has the following characteristics:

- Air flow rate = 510-m^3/hr (300-cfm)
- Diameter = 12.2-cm (31-inch)
- Bed volume = 0.28-m^3 (10-ft^3)

The monitoring effort recorded an average inlet hydrogen sulfide concentration of 50-ppm_V. Using this concentration, the unit's air flow rate and assuming a high capacity carbon is used (e.g. Midas Carbon with a capacity of 0.3 g H_2S/cc carbon), the carbon life would be approximately 4 months. This would result in a carbon change-out frequency that is considered to be excessive.

This carbon adsorber was breaking through during the monitoring effort conducted in the August when the average exhaust H_2S concentrations was 27-ppm_v, and the average removal efficiency was 46%, indicating the carbon should have been replaced. The measured odor concentrations indicate minimal removal through the unit.

The reduced sulfur scan results indicate only hydrogen sulfide being present in the inlet and exhaust samples. The measured hydrogen sulfide concentrations are consistent with those measured with the OdaLog units.

Figure 3 - Splitter Box and Carbon Adsorber Unit

REPLACEMENT ODOR CONTROL SYSTEM DESIGN BASIS

It was determined that the ozone unit was undersized and was not providing effective odor control for the Pretreatment Building exhaust air stream. The influent splitter box carbon adsorber was not providing effective odor control because its carbon was exhausted and due for replacement. Given the relatively high hydrogen sulfide concentration of the influent splitter box and the carbon's finite hydrogen sulfide adsorption capacity, the carbon will require frequent replacement and is not an optimum control technology for this location.

As a result it was proposed that an alternative and common odor control technology be used to treat the exhaust air from the Pretreatment Building and Influent Splitter Box. Following is the basis of design for the replacement odor control system.

Air Flow Rate

The Pretreatment Building consists of an upper and lower level. The upper level contains the bar screens and influent channels and the lower level contains the screenings dumpster. Table 5 summarizes the exhaust air flow rates for spaces within the Pretreatment Building, including the influent splitter box.

Table 5 – Design Air Flow Rate

Location	m^3/hr (scfm)
Influent Wet Well and Bar Screen Channels	510 (300)
Bar Screen and Dumpster Levels	10,878 (6,400)
Influent Splitter Box	510 (300)
Total	**11,898 (7,000)**

These exhaust air flow rates result in 10 and 6 air changes per hour for the Pretreatment Building general space and for enclosed non-occupied spaces (e.g., covered channels, tanks), respectively. The Pretreatment Building has Class 1 Division 1 equipment which results in no required minimum air change rate with respect to National Fire Protection Association (NFPA) 820 guidance. The Pretreatment Building has a 10,198-m^3/hr (6,000-scfm) supply air system. Given the exhaust air flow rate from the building spaces of 11,388-m^3/hr (6,700-scfm), this results in the building being under a slight negative pressure.

Inlet Hydrogen Sulfide Concentration

The design odor control system inlet H_2S concentrations are summarized in Table 6 based on the field measurements conducted in August and November 2015 and a mass balance of odor emissions from the Pretreatment Bldg. and influent splitter box.

Table 6 - Odor Control System Inlet Hydrogen Sulfide Concentrations

Parameter	(ppm_v)
Average Inlet	65
Maximum Inlet	100

TECHNOLOGY EVALUATION

A technology screening was conducted to identify a single odor control system to replace the Pretreatment Building ozone system and the influent splitter box carbon adsorber. The technologies that were screened included: packed tower wet scrubbing, carbon adsorption, biological treatment, activated sludge scrubbing, bio-scrubbing and ionization. Packed tower wet scrubbing and biological treatment technologies consisting of biofilters and bio-trickling filters remained after the screening process. Preliminary designs were prepared that would provide the required performance and included construction cost estimates based in part on vendor quotes.

Present worth cost for bio-trickling filter designs ranged from $1.6 to 2.8 million. Present worth cost for biofilters ranged from $2.3 to 3.0 million. Present worth cost for a packed tower system was $3.6 million.

Ultimately a two-stage biological treatment system was chosen consisting of a bio-trickling filter followed by a biofilter with a capital cost of $1.9 million and an estimated present worth cost of $2.4 million.

ODOR CONTROL SYSTEM DESIGN

Odor Control System Design Summary

The odor control system treats exhaust air from the Pretreatment Building and the influent splitter box. The odor control technologies consist of a humidifier/bio-trickling filter followed by a biofilter. These treatment devices are designed to remove hydrogen sulfide and non-hydrogen sulfide odors from the exhaust air streams.

Fans and Humidifier/Biotrickling Filter

Air from the odor sources is collected and pushed through the odor control devices by a fan located upstream of the humidifier/biotrickling filter. Two fans are provided, with one serving as the duty fan and the second as a standby fan.

The humidifier/biotrickling filter is a vertical fiber-reinforced plastic (FRP) column filled with a proprietary media. It functions as a humidifier to saturate the air that is discharged to the biofilter and as a reactor to remove H_2S by autotrophic bacteria that grows on its media. Moisture and nutrients are provided to support the growth of autotrophic and other bacteria of this unit by circulating secondary effluent from a sump in the bottom of the column to a spray nozzle at the top of the column where the water is distributed over the media. Heat is added to the circulating water to offset heat lost through the evaporation process by passing it through a heat exchanger and to maintain a minimum temperature of 55 degree F.

Secondary effluent is continually added to the recirculation water, referred to as makeup water, to replace water that is evaporated as the inlet air is saturated. The makeup water is added at a rate slightly in excess of the requirement to replace evaporated water, and to maintain a pH between 1 and 2 which is optimum for the autotrophic microorganisms metabolizing hydrogen sulfide. The excess makeup water addition results in a blowdown stream. If the makeup water addition rate is too high the pH will be too high. Conversely, if too little water is added the pH will be too low. The recirculation water system components which include piping, circulation pumps, and heat exchanger are corrosion resistant to resist the effects of this low pH water. Exposed outdoor piping is insulated and heat traced to protect from the cold winter temperatures of the area.

Biofilter

The biofilter receives its inlet air from the humidifier/biotrickling filter. It is a concrete structure made up of three cells that contain the biofilter media. It serves to remove any H_2S and organic based odors not removed in the bio-trickling filter by bacteria that grow on its media. The three cells operate in parallel. Air from the humidifier/biotrickling filter flows to a three-branch manifold. Each branch has a damper and connects to an air trench in the bottom of each cell.

Perforated plastic floor plates (manufactured by Bac-Tee) are installed on the floor of the biofilter cells. Biofilter media (Biosorbens™ media manufactured by Biorem) is in place on the plastic floor plates and has a 15-year design life. Air flows upward from the air trench and through the plastic floor plates which provide even distribution of air through the media.

Each cell has a flat pultruded FRP cover with an exhaust stack, light-manway and access-manway. The covers will be removed when the biofilter media is to be replaced.

City water is applied as irrigation water to each biofilter cell to maintain a minimum media moisture content. Nutrients to support bacteria are leached from the media. Water application is controlled by a programmable logic controller and is distributed onto the media with irrigation spray nozzles. Extensive instrumentation was installed to continuously monitor system operations.

All interior surfaces of the biofilter cells are coated with a corrosion resistant coating.

Figure 4 is a process flow diagram of the biological odor control system. Table 7 is a design summary for the bio-trickling filter and biofilter. Figure 5 is photograph of the biological odor control system.

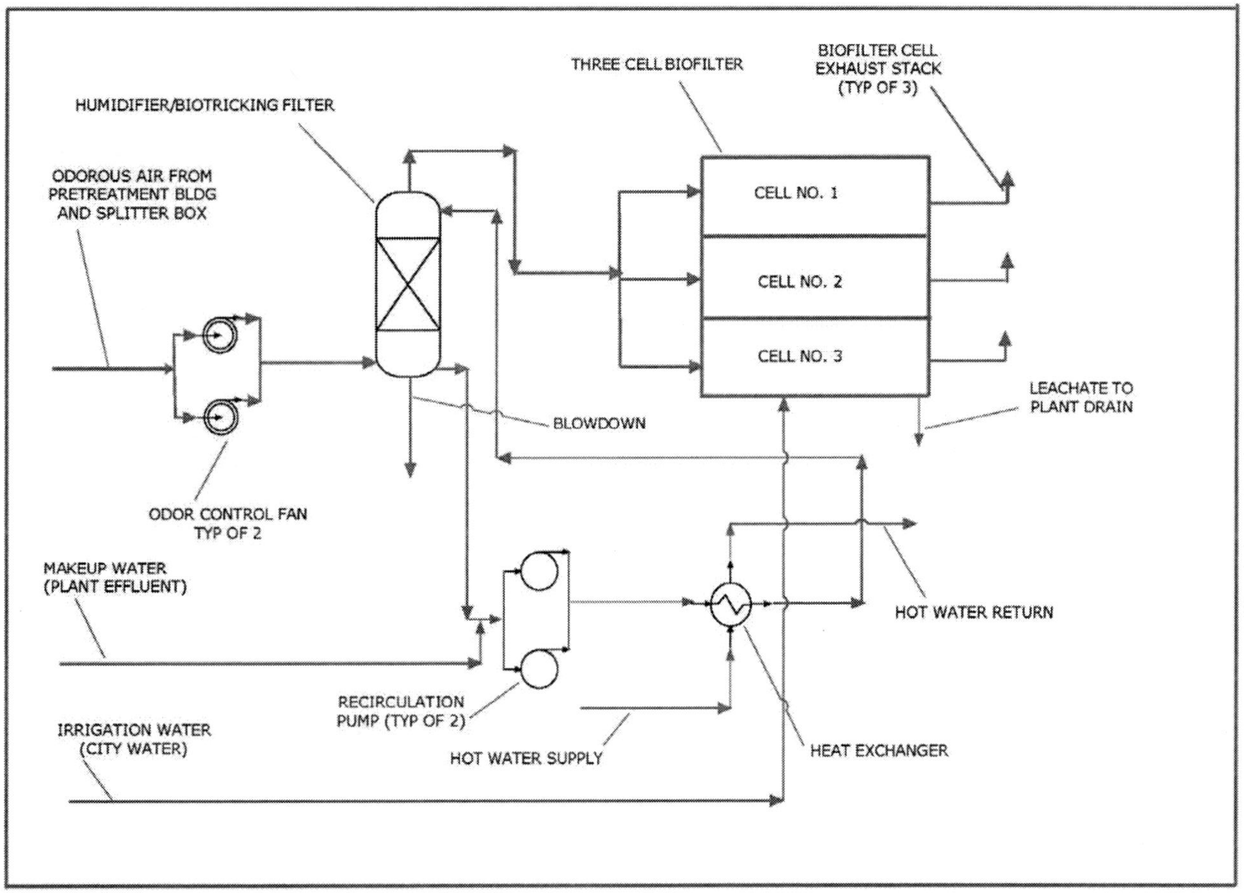

Figure 4 – Two-Stage Biological Odor Control System Process Flow Diagram

Table 7 - Biotrickling Filter and Biofilter Design Summary

Humidifier/Bio-Trickling Filter	
Parameter	**Value**
Quantity	1
Air Flow Rate, m3/hr (cfm)	11,898 (7,000)
Diameter, cm (ft)	305 (10)
Media Depth, cm (ft)	229 (7.5)
Contact time, sec	5
Recirculation rate, m^3/hr (gpm)	0.21 (55)
Maximum Media Pressure Drop, kPa (inch Water Gauge)	249 (1)
Heat Provided to Recirculation Water, KW (BTU/hr)	121 (415,000)
Biofilter	
Number of Cells	3
Cell Dimensions, m (ft)	427 X 427 (14 X 14)
Media Depth, cm (ft)	183 (6)
Contact Time (sec)	30
Maximum Media Pressure Drop, kPa (inch Water Gauge)	996 (4)

Figure 5 – Two-Stage Biological Odor Control System

PROJECT COSTS

The engineering costs for the project including field sampling/preliminary design, development of contract documents, bidding services, and services during construction were \$144,000. Construction costs were \$1.9 million. The unit cost for the system was \$170 per m^3/hr (\$289/cfm). Project costs are summarized in Table 8.

Table 8 - Two-Stage Biological Odor Control System Costs

Total Engineering	\$144,000
Construction	1,879,000
Total Project Cost	**\$2,023,000**
Air Flow Rate, m^3/hr (cfm)	11,898 (7,000)
Unit Cost, \$/m3/hr, (\$/cfm)	170 (289)

PERFORMANCE TEST

The 2-stage biological odor control system was started up in July 2017 and performance tested in September 2017. The duration of the performance test was 6-hours. Sample locations were the: bio-trickling filter inlet, bio-trickling filter exhaust and biofilter exhaust. Table 9 summarizes the performance test parameters. The performance test requirements were:

- 99% H_2S removal through the bio-trickling filter and biofilter
- 90% odor removal through the bio-trickling filter and biofilter

The H_2S removal performance test results are summarized as follows:

- The bio-trickling filter inlet H_2S concentrations ranged from 12 to 29 ppm_v
- Most of the bio-trickling filter and biofilter exhaust concentrations were below the detection limits of the Odalog unit and Jerome Meter.
- The odor control system met the performance requirements of greater than or equal to 99% removal of H_2S through the bio-trickling filter, not accounting for treatment through the biofilter.

The odor removal performance test results are summarized as follows and in Table 10:

- Inlet odor concentrations to the bio-trickling filter ranged from 43,000 to 46,000 odor units (OU)
- The odor concentration in the bio-trickling filter exhaust ranged from 3,200 to 8,900 OU.
- The odor concentration in the biofilter exhaust ranged from 25 to 45 OU.
- The percent odor removal for the BTF ranged from 79 to 89% removal. The percent odor removal for the biofilter and the overall system exceeded 99% removal.

Table 9 - Performance Test Parameters

Location	Parameter	Frequency	Measuring Device/Method
Bio-trickling Filter Inlet	H_2S	1 minute	0 – 200 ppm OdaLog Unit
Bio-trickling Filter Exhaust	H_2S	1 minute	0 – 50 ppm OdaLog Unit
Bio-trickling Filter Exhaust	H_2S	30 minutes	Jerome Meter
Biofilter Cell Nos. 1 – 3 Exhaust	H_2S	10 minutes	0 – 2 ppm OdaLog Unit
Biofilter Cell Nos. 1 – 3 Exhaust	H_2S	30 minutes	Jerome Meter
Bio-trickling Filter Inlet	Odor Concentration [a]	Once each hour in the last 3 hours of the test	Laboratory Odor Panel
Bio-trickling Filter Exhaust	Odor Concentration [a]	Once each hour in the last 3 hours of the test	Laboratory Odor Panel
Biofilter Cell Nos. 1 – 3 Exhaust	Odor Concentration [a]	Once each hour in the last 3 hours of the test	Laboratory Odor Panel
Bio-trickling Filter Exhaust	Temperature	15 minutes	Temperature sensor in BTF exhaust
Biofilter Cell Nos. 1 – 3 Exhaust	Temperature	15 minutes	Temperature sensor in biofilter
Biotrickling Filter	Pressure drop	15 minutes	Pressure differential gauge
Biofilter Cell Nos. 1 - 3	Pressure drop	15 minutes	Pressure differential gauge
Biotrickling Filter Blowdown	pH	15 minutes	pH meter

[a] ASTM 679-91, EN 13725:2003

Table 10 - Odor Removal Performance Summary

Perf. Test Hour No.	BTF Inlet (OU)	BTF Exhaust (OU)	BF Exhaust 1 (OU)	BF Exhaust 2 (OU)	BF Exhaust 3 (OU)	Avgerage BF Exhaust (OU)	BTF % Odor Removal	BF % Odor Removal	Overall % Odor Removal
4	43,000	8,900	30	30	30	30	79.3%	99.7%	99.9%
5	46,000	4,900	25	45	35	35	89.3%	99.3%	99.9%
6	43,000	3,200	25	40	40	35	92.6%	98.9%	99.9%

Other recorded parameters from the performance test are as follows:

- The pressure drop through the BTF was measured at 149-kPA (0.6-inch WG), approximately half the specified maximum. This reflects an intact media with no apparent degradation.
- The pressure drop through the biofilter media was 249-kPA (1-inch WG) which is well below the specified maximum of 996-kPA (4 inches WG) which is good and reflects a young media that has not broken down.
- The pH of the BTF blowdown ranged from 1.68 to 1.77 which is indicative of autotrophic bacteria (Thiobacillus) activity converting H_2S to sulfuric acid.
- The exhaust air temperature of the BTF and biofilter ranged from 58 to 59 degree F.

REFERENCES

[i] M.A. McGinley, C.M. McGinley, Developing a Credible Odor Monitoring Program, American Society of Agricultural Engineers (ASAE) 2004 Annual Conference, Ottawa, ON, Canada

[ii] Standard Methods for the Examination of Water and Wastewater, 22nd Edition. American Public Health Association, American Water Works Association, Water Environment Federation, 2012

[iii] USEPA, Odor and Corrosion Control in Sanitary Sewerage Systems and Treatment Plants, EPA/625/1-85/018, October 1985

Performance of Combined Two-Stage Biotower and Carbon System: When Great is not quite good enough.

Chris Easter[1] , Mark Feltner[2], Chris Wilson[2] , Andrew Keisel[2],

[1]HDR, 4470 Cox Road, Suite 200, Richmond Va. 23060
[2]Hampton Roads Sanitation District, 2401 G Ave, Newport News, VA 23602

ABSTRACT

This paper provides a case study where a combined two-stage odor control system, with a biotower followed by carbon, experienced breakthrough of an unusual and offensive odor. Overall the first stage biotower performance was excellent on hydrogen sulfide (H_2S) removal at nearly 100%, but key reduced sulfur organic based odor compounds began to breakthrough the carbon polishing stage after 8 months of operation. This paper focuses on the cause of this breakthrough and the recommended carbon replacement to address the issue.

The odor control system was installed in a pump station application on a combined gravity and force main sewer system in southeastern Virginia. The collection system was long and flat resulting in relatively high odor loads. The pump station was located in an urban community where both odor impact and visual aesthetics were important. The resulting design of the odor control system was a compact box shaped system adjacent to a new pump station architecturally similar to the adjacent homes with an architectural screening wall treatment so that the system is not visible from the street. The biotower/carbon system was close to the pump station building such that building downwash dispersion effects resulted.

This paper provides an extensive evaluation of odor causing compounds and how they changed from the inlet of the system to after the biotower and then after the carbon polishing step. Odor scans included:

- Odor Panel Analysis following ASTM E679
- Reduced Sulfur compound analysis following ASTM D 5504 using Gas Chromatograph (GC) Detection
- General Volatile Organic Compound (VOC) scan following EP T0-15 Method Using Mass Spec (MS) Analysis
- Carboxylic acid Analysis using sorption tubes and ALS Laboratory GC/MS
- Aldehyde Analysis Using sorption tubes and Following EPA Method T0-11A
- Amine Analysis using Sorption tubes and ALS Laboratory GC
- Field measurement of H_2S using Odalogs and Jerome H_2S meters

Five different carbon system suppliers were surveyed during the evaluation to solicit recommendations. Without identifying the suppliers, the five options are summarized in the paper identifying common themes amongst the recommendations. Additionally, a literature and experience survey was compiled. This effort identified two potential odor causes.

Multiple sources reported experience with conversion of methyl mercaptan (MM) to dimethyl disulfide (DMDS) in higher capacity catalytic carbons. The field data from this study reinforced the theme that MM that escaped the biotower was being removed in the carbon but the carbon exhaust was high in DMDS that was not in the biotower exhaust. This suggest conversion of the MM to DMDS which was concentrating in the carbon and breaking through with an offensive garlic and body odor like smell as observed to those in the field. Odor panel analysis reported it as an "animal" odor. Essentially the carbon exhaust had picked up a unique offensive smell that appeared to be associated with the DMDS build up and breakthrough.

The initial carbon combination had limited DMDS capacity. This was in part due to the initial carbon type selection and in part due to moisture effects carrying over from the biotower to the carbon. Field measurements showed that the biotower exhaust was very near 100% moisture saturation entering the bottom of the carbon bed. Field inspection on carbon replacement showed the lower levels of the carbon bed were very moist.

A second potential cause was also investigated. It was postulated that the lower levels of the carbon were becoming biologically active due to the moisture and that this biological activity was generating organic based odor compounds. Field test were completed to investigate this theory. These test included:

- Microscopic evaluation of the spent carbon
- Total Solids burn test to see if volatile biomass was present
- Protein evaluation of the carbon to look for the presence of biomass based protein
- DNA test to look for the presence of DNA suggesting biomass on the carbon

This paper reports on the findings of these evaluations. The findings do not support the likelihood for extensive biological growth in the carbon.

The following observations and conclusions are drawn from the available data:

- The biotower section was reliably removing nearly 100% of the inlet H_2S but was only removing 39% of the inlet MM. The limited MM removal may have been due to the relatively short biotower empty bed contact time of approximately 9 seconds.
- The initial carbon installation was overloaded by MM converting to DMDS. It is surmised that this occurred in the lower catalytic carbon layers of the carbon polishing stage. This was the dominant carbon layer with a high H_2S capacity catalytic carbon.
- The remaining carbon (top layer), which was virgin coconut shell carbon was eventually overloaded in DMDS resulting in breakthrough of a unique and offensive odor.
- Building downwash effects on the odor control system exhaust stack contributed to the issue with odors being detected approximately 50 to 100 feet downwind of the building. This was occurring even though the two-stage system was removing essentially 100% of the H_2S and providing 95% reduction in overall odors as measured by odor panel analysis.
- The replacement carbon shifted the carbon bed strategy towards carbon with a higher DMDS and organic compound capacity. On initial restart of the system the exhaust was essentially zero odor with no odors noticeable at the exhaust stack and zero odor

as measured by a low level Jerome H_2S meter. After 6 months of operation (at the time of this paper) the carbon exhaust was still odor free.

KEYWORDS
Biotower, carbon, hydrogen sulfide, methyl mercaptan, dimethyl disulfide

INTRODUCTION

This paper provides a case study where a combined two-stage odor control system, with a biotower followed by carbon, experienced early breakthrough of an unusual and offensive odor. Overall the first stage biotower performance was excellent on hydrogen sulfide (H_2S) removal at nearly 100%, but key reduced sulfur organic based odor compounds began to breakthrough the carbon polishing stage after only 8 months of operation which caused odor complaints by neighbors. This paper focuses on the cause of this breakthrough and the recommended carbon replacement to address the issue.

Background

The odor control system was a self contained two-stage system with a biotower in the first stage followed by an integral carbon section that was provided for carbon polishing as a second stage of treatment. Figure 1 shows views of the system:

Figure 1: Norchester Pump Station Odor Control System

The lower biotower section included internal plastic foam media with an intermittent spray system just above the media. Biological growth on the media consumes odorous compounds such as hydrogen sulfide (H_2S) in the odorous air. The spray keeps the environment moist for the biological system, provides nutrients to the biology, and washes out contaminants as the odorous compounds are removed.

For H_2S, the primary bacteria are dominated by autotrophic organisms (Thiobacillus) that oxidize the H_2S,which generates acidic (low pH) conditions. Heterotrophic bacteria are challenged by low pH environments caused by the autotrophic H_2S consuming population. If H_2S levels are high enough, then the biotower will be dominated by H_2S oxidizing bacteria. The biotower supplier claims that the "once through" spray allows the media to establish a gradient with H_2S consuming bacteria on the bottom and transitioning to heterotrophic bacteria in the upper levels of the media that will help remove other organic based compounds.

The Scrubbers Programmable Logic Controller (PLC) controls the spray cycle. Based on settings during the site visit, the PLC was set to initiate a 20 second spray cycle every five minutes. This is an adjustable spray cycle. This 20 second spray cycle was confirmed in the field. To some extent, the spray cycle is used to control the pH in the media such that the blow down pH is not too low. pH levels that are substantially below 2 are not recommended by the supplier because they feel this can interfere with the biological performance.

On initial start-up, the biotower was operated in a recycle mode with a temporary recycle pump. On start-up, the Biotower sump was seeded with mixed liquor from one of the HRSD wastewater treatment plants. Review of the HRSD start-up data shows that the Norchester Biotower was fully acclimated after approximately one month. After successful start-up, the recycle pump was removed and the system was operated as a "once through" spray with the biotower blowdown draining back to the pump station wet well.

During the spray cycle, nutrients (essentially dilute fertilizer) are injected into the spray water to help support the biological growth on the media. The vendor indicates that the empty bed retention time (EBRT) in the biotower media section was approximately nine seconds. This is considered reasonable for biotowers focused on H_2S removal. Longer detention times might be required for much higher H_2S levels or if broad spectrum odor removal is targeted for other non-H_2S compounds. Performance data will be reviewed in the following sections.

The carbon polishing section is located just above the spray nozzle assembly for the lower biotower section and provides 24 inches of carbon depth. The bulk of the carbon was a higher H_2S capacity carbon followed by an upper thinner layer of lower capacity virgin carbon. The high capacity carbon was a lignite based carbon. Review with the supplier indicated that it includes a metal oxide in the carbon matrix which increases H_2S capacity to approximately 0.2 grams/cc. The remaining upper layer of carbon was a coconut shell based virgin granular activated carbon with less H_2S capacity (0.03 gr/cc) but able to remove a wider range of the organic and VOC compounds.

Based on discussions with the system's supplier, the higher capacity H_2S carbon was intended to provide residual H_2S removal during start-up and thereafter while the virgin coconut shell carbon was provided to polish organic-based compounds. Based on the start-up data provided by HRSD, it is estimated that 10 percent of the H_2S capacity would have been used during the start-up process when the biotower was acclimating for H_2S removal. Since that acclimation, the biotower has removed approximately 99 percent or more of the H_2S allowing the carbon to act primarily as a polishing step for the other odorous compounds.

Field Odor Data and Site Visit Observations

An initial site visit was completed in April 2017. Key observations during the site visit were:

- The exhaust stack is relatively low compared to the building height (Figure 1). Building downwash effects were perceived with the exhaust from the odor control system being noticeable approximately 50 to100 feet downwind of the facility. Figure 2, from an ASHRAE HVAC manual, shows building eddy current downwash effect concepts. While the figures are for low stacks on top of buildings, the same effect is in play for Norchester Pump Station with the odor control stack very close to the building and exhausting below the top of the roof level. It is understood however that this is a compromise between aesthetic desires by HRSD and a tall stack that better promotes dispersion. As such the initial design did not provide an elevated stack based on Good Engineering Practice (GEP) guidance.

Figure 2: Building Downwash and Eddie Current Wake Effects

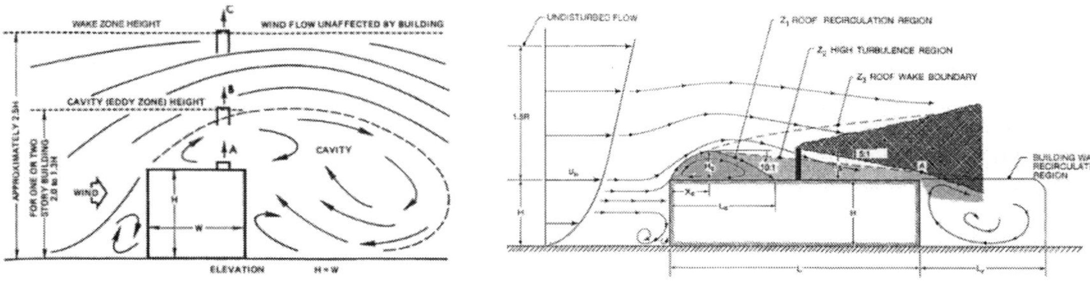

- Field observations suggested the downwind odor was an organic smell similar to body odor or a rotten garlic like smell. Some of the HRSD staff described it as dirty socks or urinal smell. Field odors from the biotower exhaust were inspected separately from the exhaust of the carbon. Observations from the field inspections were as follows:
  - The inlet H_2S to the biotower was high in the range of 20 to 30 parts per million (ppm) and smelled noticeably of rotten eggs.
  - The exhaust from the biotower section was much less odorous and smelled organic in nature similar to odors from typical wastewater plant aeration basins with a musty smell. While noticeable, the biotower exhaust was not offensive.
  - The exhaust from the carbon took on a peculiar organic smell that was more offensive as described above (body odor, rotten garlic, dirty socks, urinal). This was the same odor smelled downwind of the system at approximately 50 to 100 feet.
  - Field Jerome meter readings suggested that the exhaust from the carbon was higher in measureable odors than the biotower exhaust. The Jerome H_2S meter can cross read on other reduced sulfur organic compounds such as methyl mercaptan, dimethyl disulfide and others. Since H_2S removal was nearly complete, this potentially suggested the impact of organic based odor compounds being worse in the carbon exhaust.
- The area around the fan was noticeably odorous in H_2S odors. These odors were localized right at the fan but very noticeable. Figure 3 shows a photo of the Jerome H_2S analyzer measuring 3.2 ppm H_2S at the fan shaft seal location just inside the yellow safety cover over the drive shaft. Discussion with the fan supplier indicated the drive shaft seal was a Teflon plate sandwiched between FRP plates. Normally, while these are not complete seals, they are fairly effective. While noticeable next to the fan, this odor

was not the downwind odor that was noticeable outside the pump station fence line. Fan corrective measures by HRSD are presented latter in this paper.

Figure 3: Fan Shaft Seal H_2S measurements

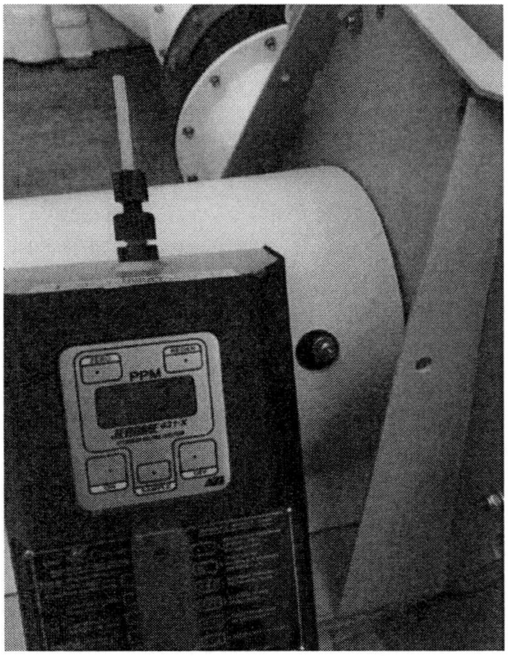

EVALUATION APPROACH

The goal of the evaluation was to understand the cause of the offensive odor breakthrough in order to help HRSD select the best path forward that avoided a repeat of the 8 month breakthrough. The evaluation included 1) odor scans to define odor causing compounds and how they changed through the 2-stage system and 2) Field Inspection of the spent carbon followed by laboratory evaluations to investigate the potential for biological growth in the carbon.

Odor scans included:

- Odor Panel Analysis following ASTM E679
- Reduced Sulfur compound analysis following ASTM D 5504 using Gas Chromatograph (GC) Detection
- General Volatile Organic Compound (VOC) scan following EP T0-15 Method Using Mass Spec (MS) Analysis
- Carboxylic acid Analysis using sorption tubes and ALS Laboratory GC/MS
- Aldehyde Analysis Using sorption tubes and Following EPA Method T0-11A
- Amine Analysis using Sorption tubes and ALS Laboratory GC
- Endole and Sactole scans following EPA Method T-017 by thermal desorption/gas chromatography

- Field measurement of H_2S using Odalogs and Jerome H_2S meters

Field Inspection testing of spent carbon included:

- Microscopic evaluation of the spent carbon
- Total Solids burn test to see if volatile biomass was present
- Protein evaluation of the carbon to look for the presence of biomass based protein using Navipure Soil Extraction Kit followed by Nanodrop (UV-Vis) protein assays
- DNA test to look for the presence of DNA suggesting biomass on the carbon by Soil Extraction methods using Genomic DNA Dneasy Soil Extraction Kits

RESULTS

Results are presented in two sections. First are the odor evaluation results of the system at the time when the carbon was exhibiting breakthrough of the unusual offensive odor. Second are field observations during the carbon change out and inspection of the spent carbon.

Odor Panel Evaluations

Odor panel evaluations were completed by St. Croix Sensory. These were done using Tedlar bag samples shipped overnight to the lab in Minnesota. We should note that during the overnight shipment that compounds captured in the bags can sometimes continue to react or change before the odor panel is exposed to the sample. That said, the results of the data can be summarized as follows:

- **Odor Control System Inlet**: The inlet was reported at 30,000 Dilution to Threshold (DT) with a negative hedonic tone of -5.6 out of potential -10. The sample was described as decay, sulfur, earthy and sulfidic vegetable. The dominant odors were decay and sulfur. This represents a strong and offensive odor source.
- **Biotower exhaust**: The biotower exhaust was reported at an average of 4,850 DT with -5.65 hedonic tone described as decay, sulfur, earth sulfidic vegetable, vegetation, and animal. While this is an 84 percent DT reduction, which is typical of biotower systems, the exhaust was still fairly high in DT concentration and negative in hedonic tone and therefore would be considered offensive. Staff notice that this was not entirely consistent with field observations which suggested the biotower section exhaust was a low level musty odor typical of aeration basins and not overly offensive.
- **Carbon exhaust**: The carbon exhaust averaged 1,467 DT with a - 4.93 hedonic tone described as decay, sulfur, animal, sulfidic vegetable. The animal descriptor became more consistent in the odor panelist for the carbon exhaust. This was consistent with the field odor observations. This will be discussed more in the following section on the reduced sulfur compounds.

Table 1 provides a summary of the Odor panel data just described.

Table 1: Odor Panel Data Summary

Sample Description	DT	RT	Intensity	Hedonic Tone	Comments
Inlet #1	30,000	18,000	120	-5.6	Decay, sulfidic vegetable, sulfur, earth
Bio Exhaust #1	5,200	3,700	110	-5.4	Decay, sulfidic vegetable, sulfur, earth, vegetation
Bio Exhaust #2	4,500	3,000	190	-5.9	Decay, sulfur, sulfidic vegetation, vegetation, earth animal
Carbon Final Exhaust #1	1,000	580	90	-5.0	Decay, sulfur, sulfidic vegetation, **animal**
Carbon Final Exhaust #2	1,400	930	90	-5.3	Decay, sulfur, sulfidic vegetation, **animal**
Carbon Final Exhaust #3	2,000	1,100	70	-4.5	Sulfur, decay, sulfidic vegetation, **animal**

Overall, this represents a 95 percent DT reduction through both stages, which is very good overall odor removal. However, the leftover hedonic tone is still negative and therefore problematic and appears focused on the carbon exhaust which had picked up a new odor description of "animal."

Also of interest is that the intensity was only reduced from 120 down to the range of 70 to 90. This is only a 30 percent reduction in intensity compared to 95 percent reduction in DT. Intensity is the odor panelist subjective comparison of the odor's intensity compared to an n-butanol sampling that the panelist are trained to estimate. The reduction of a perceived intensity of 120 ppm n-butanol equivalence down to only 70 to 90 suggest the panelist noticed a persistency in how the exhaust odor was perceived.

The combination of a persistent exhaust odorant with a negative hedonic tone and a relatively low stack exhaust, impacted by building downwash, creates a potential odor issue even when the scrubber is getting good 95 percent overall odor reduction. The two stage scrubbers was always meeting its specified design criteria for H_2S removal but still negatively impacting neighbors who passed through the downwash plume.

Odor Scans Evaluations

The Total reduced sulfur (TRS) GC/SC scan following ASTM 5504 evaluated for the following compounds:

- Hydrogen sulfide (H_2S)
- Carbonyl sulfide (CS)
- Methyl mercaptan (MM)
- Ethyl mercaptan
- Dimethyl sulfide (DMS)
- Carbon disulfide (CS2)
- Isopropyl mercaptan
- tert-Butyl mercaptan
- n-Propyl mercaptan
- Ethyl methyl sulfide
- Thiophene
- Isobutyl mercaptan
- Diethyl sulfide
- n-Butyl mercaptan
- Dimethyl disulfide (DMDS)
- 3-Methylthiophene
- Tetrahydrothiophene
- 2,5-Dimethylthiophene
- 2-Ethylthiophene
- Diethyl disulfide

The following six were detected:

- H_2S
- CS
- MM
- DMS
- CS2
- DMDS

Table 2 provides a summary of the TRS data by source.

Table 2:	DD 5504 TRS Scan Data Summary					
Source	**H2S**	**CS**	**MM**	**DMS**	**CS2**	**DMDS**
	ppb	ppb	ppb	ppb	ppb	ppb
Inlet	15000	18	350	14	3.1	0
BioExhaust	16	23	210	12	2.7	8.7
BioExhaust	10	41	220	14	3.5	13
Carbon	25	29	14	9.1	3.4	18
Carbon	140	33	25	13	3	19
Carbon	270	38	30	13	3.9	22

In general, the data suggest the following:

- The inlet air was high in H_2S at 15,000 parts per billion (ppb) or 15 ppm as well as fairly high in MM at 350 ppb (0.35 ppm). Since both of these compounds are detectable at 1 ppb these are significant contributors to the source odors.
- Much higher H_2S concentrations (40 PPM average, 100 PPM peak) have also been observed in the summer using Odalog H_2S Meters. We expect the methyl mercaptan to be proportionally higher during these periods as well.

- Over 99.9 percent of the H_2S was being removed in the biotower representing very good H_2S removal performance.
- Only 39 percent of the MM was being removed in the biotower section. This is somewhat low and likely caused by the short EBRT which is effective on H_2S but less effective for MM.
- No DMDS was detected in the inlet but DMDS was reported in the system exhaust.
- The system appears to be:
  - Allowing MM breakthrough past the carbon
  - Breaking through low level H_2S in the carbon after the biotower stage
  - Generating DMDS in the carbon after the biotower stage
  - Breaking through with low levels of CS
- Potential mechanisms for these observations will be discussed in the following sections.

Table 3 summarizes the key data from the EPA TO-15 scan. This scan looks for approximately 60 standard Volatile Organic compounds (VOCs) as well as reporting the top "Tentatively Identified Compounds (TICs)."

Table 3:	Volatile Organic Compunds By EPA T0-15											
Source	Chloromethane	Acetone	Methylene Chloride	MEK	Dichloroethene	Chloroform	Tetrahydrofuran	Bromodichloromethane	n-heptane	Toluene	Dichlorobenzene	Limonene
Inlet	ND	94	2.8	270	2.8	50	ND	2.2	ND	22	12	3.7
BioExhaust	ND	23	2.4	36	2.3	41	ND	2.5	ND	6.5	2	ND
BioExhaust	ND	63	2.6	130	3.6	50	ND	2.6	ND	6.5	3	ND
Carbon	ND	ND	3.2	ND	2.9	62	3.4	2.5	1.8	160	5	ND
Carbon	ND	ND	3.2	ND	2.6	46	3.9	2.5	1.4	130	5.1	ND
Carbon	5.7	ND	3.3	ND	2.8	61	3.8	2.6	1.2	99	6.8	ND

This table is limited to the standard T0-15 compounds that were reported. Additional TICs were reported in the exhaust for a variety of unidentified siloxane compounds and several TRS compounds. Review of the TRS data with the lab suggest the DD5504 scan is more accurate for the TRS compounds than the TO-15 method. As such, the DD5504 data was used for this evaluation for TRS compounds.

Data shown in Table 3 suggest the following:

- Although acetone was present, it was effectively removed by the combination biotower/carbon system.
- Methyl ethyl ketone (MEK) and Limonene was similarly removed.
- Methylene chloride, dichloroethene, chloroform, bromodichlormethane, and dichlorobenzene essentially passed through the scrubber unchanged. The levels are all low enough however that they should not be contributing significantly to exhaust odors based on reported detection levels for these compounds. They may however be competing for carbon capacity with the other odorous compounds such as MM and DMDS.
- The carbon seems to be generating tetrahydrofuran and toluene. This is particularly the case for toluene with the exhaust concentration being nearly six times the inlet concentration. The Toluene concentration in the exhaust may be at levels high enough to be detected. Human detection thresholds for Toluene are reported in the range of 0.02 to 4 ppm.

No aldehydes, amine or carboxylic acid compounds were detected in the exhaust. No Indole or Skatole were detected in the exhaust.

Spent Carbon Field Observations

Following the initial odor data evaluations the odor control system was inspected and the spent carbon replaced. Carbon replacement recommendation are provided in following sections. This section focuses on the spent carbon evaluations.

Spent Carbon Changeout

In August 2017 the spent carbon was removed. Figure 4 shows a photo of the carbon changeout in progress. Figure 5 shows key photos of the carbon that was removed.

Figure 4: Carbon Change Out (left to right: lid removal, spent carbon, vacc truck removal, supersack unloading)

Figure 5: Spent carbon Photos (left to right: clumped carbon from bottom layers just above the spray nozzles, moist catalyic carbon layer near mid point of the carbon bed, upper virgin carbon layer near the top of the carbon bed)

Field observations showed that the bottom of the spent carbon was wet and moisture saturated with portions of the carbon exhibiting clumps as shown in Figure 5. The middle layers of the carbon as much as 6 to 12 inches up into the bed were moist as shown above. The upper layers were less impacted by moisture effects. Field measurements during operation showed the inlet relatively humidity into the biotower and carbon at 100% (saturated) with exhaust at 85% measured coming out of the carbon exhaust stack.

Samples of the carbon were evaluated to determine Total Solids (TS) and Volatile Solids (VS) percentages. Figure 6 shows a comparison of the results. Based solely on this data the lower TS in the used carbon is likely due to the moisture content with the lowest TS value in the bottom layer of carbon near the biotower spray nozzles where 100% humidity was observed.

The VS is higher in the used carbon compared to new but this does not determine whether the VS is composed of biological growth on the carbon or adsorped materials in the carbon.

Figure 6: Comparison of Total Solids and Volatile Solids for New carbon compared to spent carbon at different depth in the spent carbon bed.

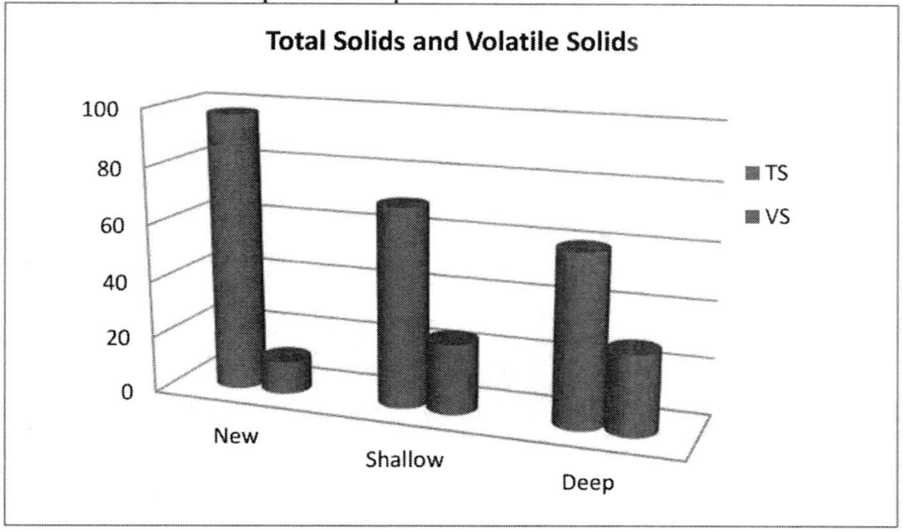

Microscopic inspection of the carbon did not show strong evidence of biological growth. In fact extensive views from many carbon samples did not exhibit any clear evidence of bacterial or other biological growth. Figure 7 shows photos comparing a suspension at 100X and 400X magnification of the lower levels of the spent carbon. This was the only microscopic observation showing potential filamentous bacteria.

Figure 7: Microscope photos comparing carbon suspensions (left: 100X Spent, right: 400 X spent carbon)

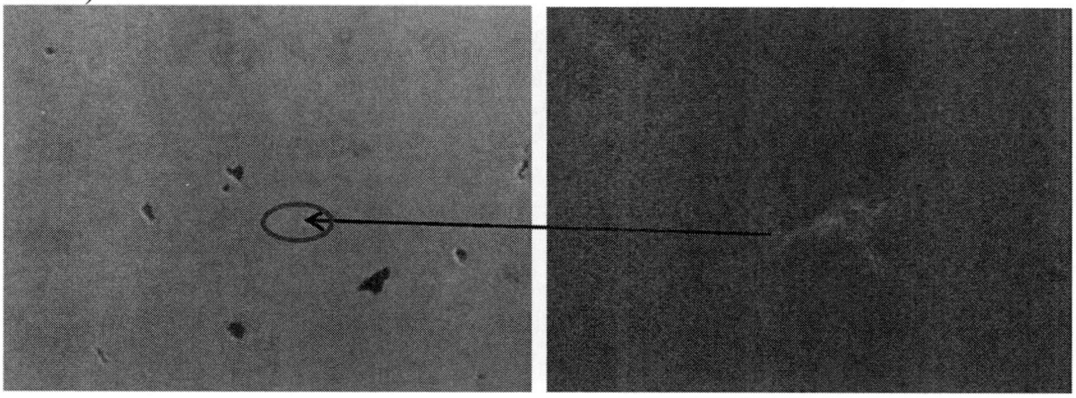

DNA evaluation did not detect DNA. Protein Evaluation suggested slightly higher protein levels in the bottom layers averaging 6.1 mg/ml compared to 3 mg/ml higher in the carbon media and 23.5 in the top of the bed. All of these were higher compared to 0.8 for unused virgin carbon. The data seems inconclusive in that:

- TS data can be explained by water and moisture content alone.

- VS data may suggest higher volatile content low in the media but this could be biological or just adsorbed VOCs in the carbon.
- Microcopic evaluation showed little to no indication of clear bacterial growth.
- DNA evaluation showed no detected DNA which means it was either zero or below the detection limit of the method.
- Protein evaluation suggest higher protein low in carbon and on the top surface.
    - The elevated low proteins could be low levels of biological growth on the carbon
    - The high surface proteins could be growth on the top media caused by residual light passing through the somewhat translucent FRP vessel creating conditions for algael growth.

Meaning of the Odor Data

This section provides a summary of the findings with regard to cause of the noticeable odors and measured odor data. The causes are broken into those where the author is more confident of the cause and the data compared to others that are postulated.

The first observation is that the odor character in the exhaust although reduced had shifted in character to include descriptors such as "animal." This seems consistent with the presence of DMDS in the exhaust which is not in the inlet air. A survey was undertaken with five different carbon suppliers and several other odor control specialist from other consulting firms along with a survey for literature. Multiple sources report the conversion of MM to DMDS in high capacity carbons as well as the concentration of DMDS until final breakthrough (1,2,3).

MM reportedly adsorbs and then in the presence of oxygen and moisture on the catalytic carbon converts to DMDS which has a greater affinity for carbon (1,2). Overtime the DMDS concentrates until carbon breakthrough occurs. Based on this observation supported by literature review any future carbon replacement should shift away from the catalytic higher H_2S capacity carbons and focus more on either broad spectrum virgin carbon with higher organic compound capacity or permanganate impregnated sorption systems manufactured to handle organic based difficult compounds. Figure 8 below shows a comparison of MM and DMDS. Displacing the hydrogen allows splicing of MM into DMDS.

Figure 8: Comparison of MM (left) and DMDS (right) Structures

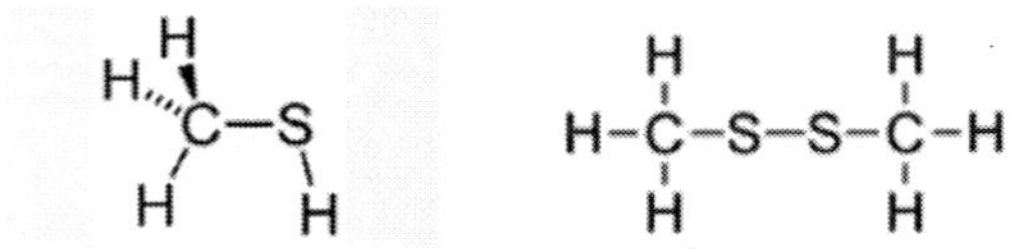

Reactions with the catalytic types of carbons is substantially a surface mediated reaction due to the catalytic nature of the carbon as compared to virgin carbon where organic compounds adhere to the deep pores in the carbon. The water has less impact on the surface reactions in catalytic carbons but can reduce capacity in deep pore carbons such as coconut shell. Moisture content is reported as a requirement for some of the higher performance carbons.

Since the biotower was fully acclimated the higher H_2S capacity is less important than the broad spectrum capacity for organics. In this case MM and DMDS are key compounds of concern.

The second observation is based on the data that suggest the carbon appears to be forming toluene. It was considered possible that the inlet edge of the carbon is moist or wet due to the fine mist spray carryover and humidity just below it and that portions of the carbon may have gone "biological".

It was also postulated that biological growth on the carbon could occupy pore space and could cause biological degradation within the carbon by decay of cell mass and built up contaminants. This might produce the toluene compounds. A similar phenomenon was observed for siloxanes reported at TICs in the T0-15 analysis. However HRSD completed siloxane evaluations that did not detect siloxanes suggesting the TIC observations for siloxane might be a false positive.

Alternatively, the toluene could just build over time and finally begin to breakthrough in mass but the siloxane and other compounds that are not reported in the inlet air seem to give credence to the potential for biologically active carbon. This was however not confirmed by the spent carbon analysis.

While the system did provide eight months of excellent performance, the existing carbon was compromised for odor removal after 8 months operation.

Corrective Measures

Since it had become clear that the existing carbon was spent, a key question was what type of second stage sorption material is best for the replacement. Various vendors were contacted and asked for their recommendations. These are summarized below followed by HDR's recommendation. Table 5 provides a summary of what was proposed by each vendor. The vendors are NOT named on purpose. The table is offered for comparison.

Table 4: Carbon System Supplier Proposals			
Supplier	Proposal	Cost Estimate	Comments
Supplier 1	3 layers with High H_2S Capacity carbon topped with Permanganate coated media and Coconut Shell Virgin Carbon	$10,163	Estimated life 2 years
Supplier 2	Layer of high H_2S carbon topped with a blended carbon/permanganate media	$13,100	Estimated life 2 years
Supplier 3	Coconut shell virgin carbon	$5,000	Estimated life 1 year
Supplier 4	Coconut shell carbon with potential thin bottom layer of catalytic water activated carbon	$4,000	Estimated life 2 years
Supplier 5	Coconut shell virgin carbon over a thin layer of lignite coal carbon that acts as a moisture barrier. Consider adding a mist mesh layer	No cost provided to date	No Estimated life provided to date

Although varied in cost and details a common theme recommended by the suppliers was a shift towards either more virgin carbon or impregnated media adsorption systems. One supplier also recommended retrofitting the existing system with a mesh layer as shown in Figure 9.

The original supplier expressed doubt as to the value of this retrofit and it was not implemented by HRSD because adding a mess pad retrofit under the carbon would represent research and development altering the original 2-stage system.

Figure 9: Mist Mesh (left) and Matala plastic media (right)

CONCLUSIONS AND RESULTS

Based on the site visit observations, data provided to date, and the investigation of options with other vendors, the following conclusions and recommendations were offered to HRSD:

- The existing carbon was compromised for odor removal having been overloaded with MM converting to DMDS. This is very likely due to the nature of the catalytic carbon layer.
- When acclimating initially the H_2S loading would have been high and might overwhelm lessor capacity carbon. It is estimated that 10 percent of the high capacity carbon was used during biological acclimation of the biotower. Now that the biotower is fully acclimated it is removing virtually all of the H_2S. As such, the replacement media should focus on the MM/DMDS challenge so high capacity H_2S carbon should not be the focus.
- The replacement carbon should be a layered approach with the bulk (60 percent or more) being lower costs virgin coconut shell.
    - HRSD should also include a thin layer of the catalytic carbon to catch the moisture before it reaches the virgin coconut shell carbon. Any MM or DMDS breakthrough in this thin layer would be treated in the thicker layer of the virgin carbon. The catalytic carbon would also help in conversion of MM to DMDS which has greater capacity in virgin carbons.
    - A thin permanganate impregnated layer can be provided on top if desired to enhance removal of difficult compounds such as CS and CS2.
    - In future HRSD should consider adding a thin mist eliminator layer on the bottom of the carbon to help remove droplet mist moisture before the carbon. The cost is minimal and would occupy only a small portion of the carbon space. Before this is pursued the media should be tested to show it can take the weight of 2 feet of carbon.
- The selected carbon vendor should be based on costs and estimates of projected carbon life.

- The selected vendor should be required to provide a guarantee period based on odor source loading as defined by the ongoing HRSD testing data. Vendors who will not provide at least a one-year guarantee should not be considered.
- The exhaust from the current system should be extended up to match or slightly exceed the peak of the adjacent building to help reduce downwash effects. While this is still not GEP for stack height, it would be a better compromise between risk of downwash impacts and aesthetics of the visible stack.
- The fan seal should be inspected and either replaced or tightened if required. Additionally, HRSD should consider retrofitting the fan with a fan Enclosure which would help contain odors and might be important to control fan noise to the adjacent property.

HRSD chose to implement carbon replacement shifting towards 20 % catalytic on the bottom with 80% virgin coconut on the top. This was the lowest cost approach. As of the time of this paper the new carbon system has been in service for 7 months and is running without issue and without noticeable odor breakthrough. HRSD continues to monitor the system for eventual breakthrough.

RECOGNITION

Special thanks to HRSD South Shore Operations personnel especially Mr. Kenneth "Buddy" Beaman and Claude "Travis" Stevens for their assistance and contributions on this project.

REFERENCES

Svetlana Bashkova, Andrewy Bagreev, and Teresa J. Bandosz : Catalytic Properties of Activated Carbon Surface in the Processs of Adsorption/oxidation of methyl mercaptan. Elsevier Catalysis Today. 2004

Svetlana Bashkova, Andrewy Bagreev, and Teresa J. Bandosz 2002: Activated Carbons as Adsorbents of Methyl Mercaptan. Fuel Chemistry Deivienv Preprints 2002.

Hisashi Tamai, HIsato Nagoya, Takeshi Shiono: Adsorption of Methyl Mercaptan on Surface Modified Activated Carbon. Graduate School of Engineering Hiroshima University Department of Chemistry.

Biotrickling Filtration of Sewage Plant Exhausts – Recirculation versus Once-through Irrigation

Derek S. Webb, Hadi Husain, Caleb Olesen

BIOREM Technologies, Inc.
Email: dwebb@biorem.biz

1 ABSTRACT

This work shows that the operation of a biotrickling filter (BTF) using once-through irrigation to achieve multiple benefits of high hydrogen sulfide removal and low exhaust odour (Total Odor) would require significantly higher empty bed residence time (EBRT) than for hydrogen sulfide removal only. Hydrogen sulfide elimination capacities for recirculation BTF are higher than those for once-through irrigation. Organic sulfur compounds (OSC) and odor removal in once-through irrigation BTF are at best comparable to recirculation BTF. If high hydrogen sulfide and OSC removal, and low Total Odor are desired, then other options, such as a recirculation BTF followed by a second stage of treatment should be considered.

These findings are based on a one-year long program in which sewage plant exhausts with high hydrogen sulfide peaks, and a consistent diurnal pattern were treated using a BTF pilot with three reactors. Capability existed to operate the pilots in series or parallel. Both once-through and recirculation irrigation could be tested. Hydrogen sulfide concentrations were continuously monitored in each process configuration, and the system was operated in each configuration for at least a month. In addition to continuous hydrogen sulfide monitoring, three sampling events were implemented to study removal of organic sulfur compounds and odor by both processes.

2 BACKGROUND

Biotrickling filtration is widely used for treatment of sewage plant exhaust streams with a broad range of hydrogen sulfide concentrations. Improvements in technology have resulted in reduced footprint and low cost of treatment. The process offers a significant cost saving over chemical scrubbing and physical adsorption, leading to its rapid acceptance by municipal utilities.

While there appears to be a significant amount of disagreement in the industry regarding nomenclature, for the purposes of this investigation and study we will define biotrickling filtration as an example of a fixed-film biological reactor. Bacteria are immobilized on the surface of an inert medium and then subsequently oxidize gaseous phase contaminants that have solubilized into the moisture layer or biofilm. Bioscrubbers differ from this process as the microbial cultures are freely suspended within the scrubbing liquid (suspended growth bioreactors).

Biotrickling filtration involves oxidation of hydrogen sulfide to sulfuric acid using autotrophic bacteria, *acidithiobacillus thiooxidans*. The bacteria uses carbon dioxide as a source of carbon, and exothermic oxidation of hydrogen sulfide as source of energy for growth, and the steps in hydrogen sulfide oxidation are well understood (Frederich et al, 2001). The process works well at a pH of 1.5 to 2, and a typical system consists of providing an inert packing surface on which the biomass is immobilized. Contaminated air is fed at the bottom and a low pH solution is recirculated to the top using a pump and sump counter flow arrangement. The only by-product, sulfuric acid, is discharged from the system using continuous blowdown from the recirculated fluid or as drain water in a once-through operating mode. The amount of waste liquid is a function of total hydrogen sulfide mass loading to the system.. It is typically a small fraction of recirculation rate and the process consumes little water for the recirculation mode of operation.

An operational change to the classical recirculation process is the use of once-through irrigation, which results in a variable pH, from neutral at the top to a low value, typically in the range of 2-3 at the bottom. Claimed advantages of once-through irrigation include:

1. Cost of a sump and recirculation system is avoided (but a pump is required for start-up).
2. Comparable or superior reaction kinetics to recirculation irrigation is claimed for hydrogen sulfide removal
3. Presence of a neutral zone at the top is claimed to provide superior odor and organic sulphur (OSC) removal.

A disadvantage of the once-through mode is significantly higher water consumption. This tends to be problematic for installations in arid areas and at waste water reclamation facilities where effluent is not readily available.

The current sales trend appears to promote the use of the once-through operating mode and for it to have a number of advantages such as higher hydrogen sulfide elimination capacities, high rates of OSC removal and low discharge odor concentrations. Claims are made that these results can be achieved efficiently under similar operating parameters required simply for hydrogen sulphide removal. (Woo & Singleton, 2017). However, limited published data to support this is available. For example, a comparative study of three different biotrickling filters, operating for 16 second residence time and fed with low hydrogens sulfide concentration, showed poor total odour removal (Easter, Quigley, Burrowes, Witherspoon, & Apgar, 2005). Although it is not stated, it is likely that one of these was operating in once-through irrigation mode. We found one reference (Johnson, le Roux, & Dunn, 2011) in which approximately 92% total odor removal from sludge storage exhausts was achieved using once-through irrigation. However, residence time and contaminant speciation was not available to make a full assessment of the process.

Biorem trialed once-through, intermittent irrigation on one of its large plants in the Middle East with a capacity of 150,000 m^3/h. Average daily removal efficiencies of 99% or greater could not be met consistently over a one month period. The system was then switched to recirculation irrigation and 99%+ removal efficiency was achievable. Figure 1 presents a typical performance for once-through irrigation, and Figure 2 presents typical performance of recirculation irrigation.

The paucity of published data, and our own experience indicated that a comparative evaluation of once-through irrigation process with recirculation for hydrogen sulfide kinetics is justified to help utilities and consultants in their decision making process. The primary purpose of the study was to investigate these claims and determine whether the microbial processes involved within a biotrickling filter system are optimized for odorant oxidation in one operating mode versus the other. Understanding the operating mechanisms for optimization allow for the most appropriate pairing of solution to application.

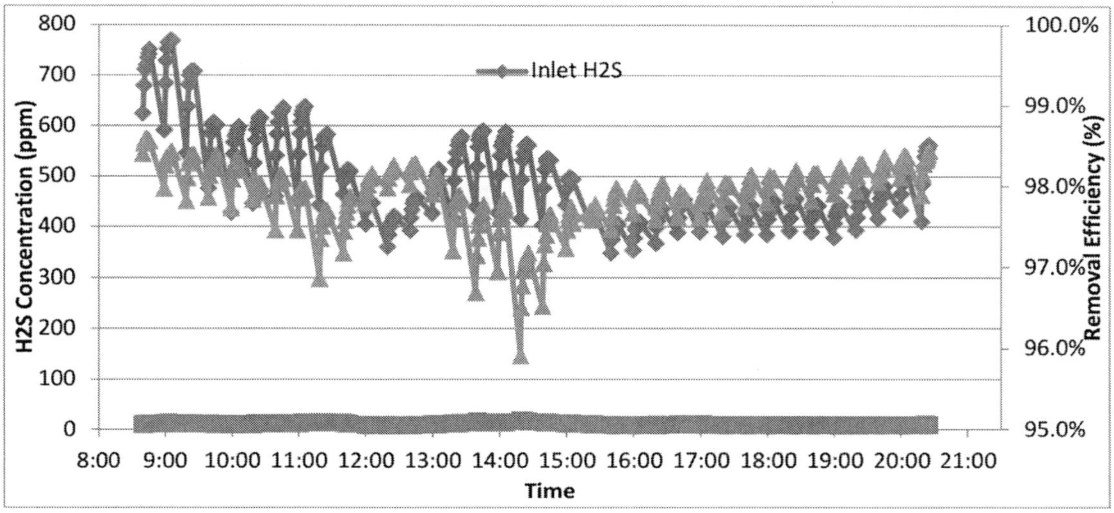

Figure 1: Once-through Intermittent Irrigation Performance for a 150,000 m³/h BTF

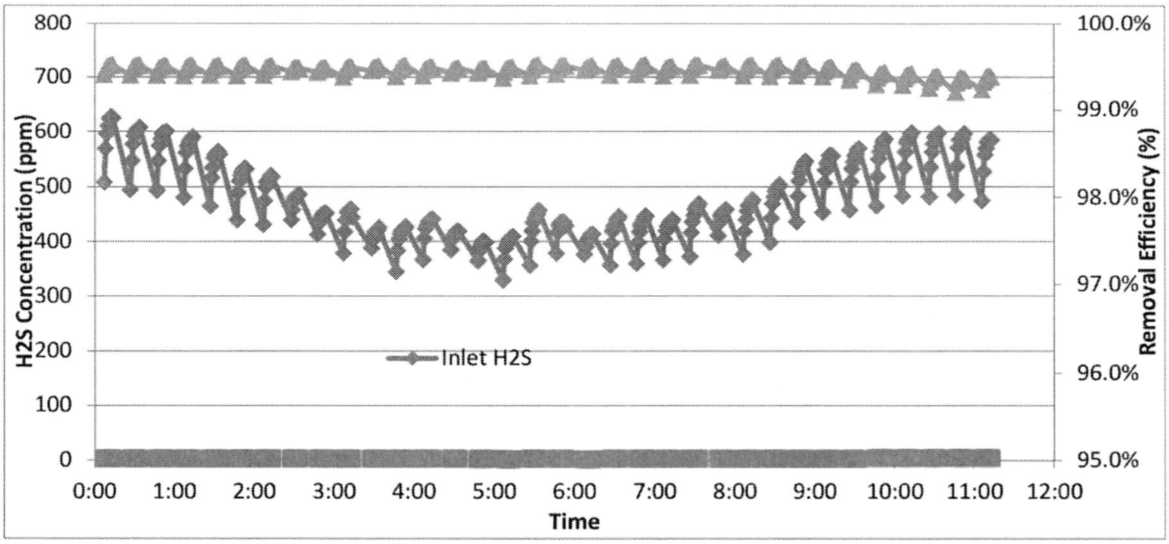

Figure 2: Recirculation Irrigation Performance for a 150,000 m³/h BTF

3 OBJECTIVES

Biorem conducted studies of the once-through and recirculation irrigation with the following objectives:

1. Compare kinetics of hydrogen sulfide removal for once-through irrigation and recirculation BTF.
2. Determine removal of OSCs in once-through irrigation at empty bed residence times (EBRT) determined to be appropriate based on the kinetics of hydrogen sulfide oxidation. That is, ensure that adequate residence time is allowed that high hydrogen sulfide removal is achieved yet a near neutral pH exists in a section of the reactor for OSC oxidation and odor treatment.
3. Compare this with OSC removal under recirculation at an EBRT optimum for hydrogen sulfide removal in this mode.
4. Use similar approach as Steps 2 and 3 for assessment of odor removal in a once-through biotrickling filter.

4 EXPERIMENTAL

A multi-vessel pilot, shown in Figure 3 was used. The pilot had three reactors (BTF1, BTF2 and BTF3), each with a capacity of 1.2 m^3 of packing capacity. Structured packing was used during initial phases of the program in each reactor. Several styles of structured (rolled) packing were used throughout the study to investigate the contribution of biological exchange area (BEA); total available surface area; porosity; and geometrical configuration. Random packing was used in the last phase of the program in each reactor. The pilot could be operated in a parallel mode with three side by side reactors, or a single series reactor. In the series configuration, all three reactors could be operated in recirculation mode, or BTF1 and BTF2 in recirculation mode and BTF3 in once-through, or BTF1 in recirculation and BTF2 and BTF3 in once-through mode. Other configurations were also possible if needed with some modifications.

Figure 3: Biorem biotrickling filtration pilot

A Florida municipal wastewater treatment plant site was selected for the following reasons:

1. Diurnal hydrogen sulfide concentration variations occurred throughout the year, in which high concentrations were typically present during a relatively consistent 12-hour period. This presented stable, yet real-life conditions for testing (see Figure 4).
2. Peak concentrations reached as high as 700 ppm_v providing the opportunity to test hydrogen sulfide removal kinetics over a broad concentration range.
3. A Biorem full scale recirculation BTF using random media and treating the same exhaust was available for use as a benchmark for comparative purposes.

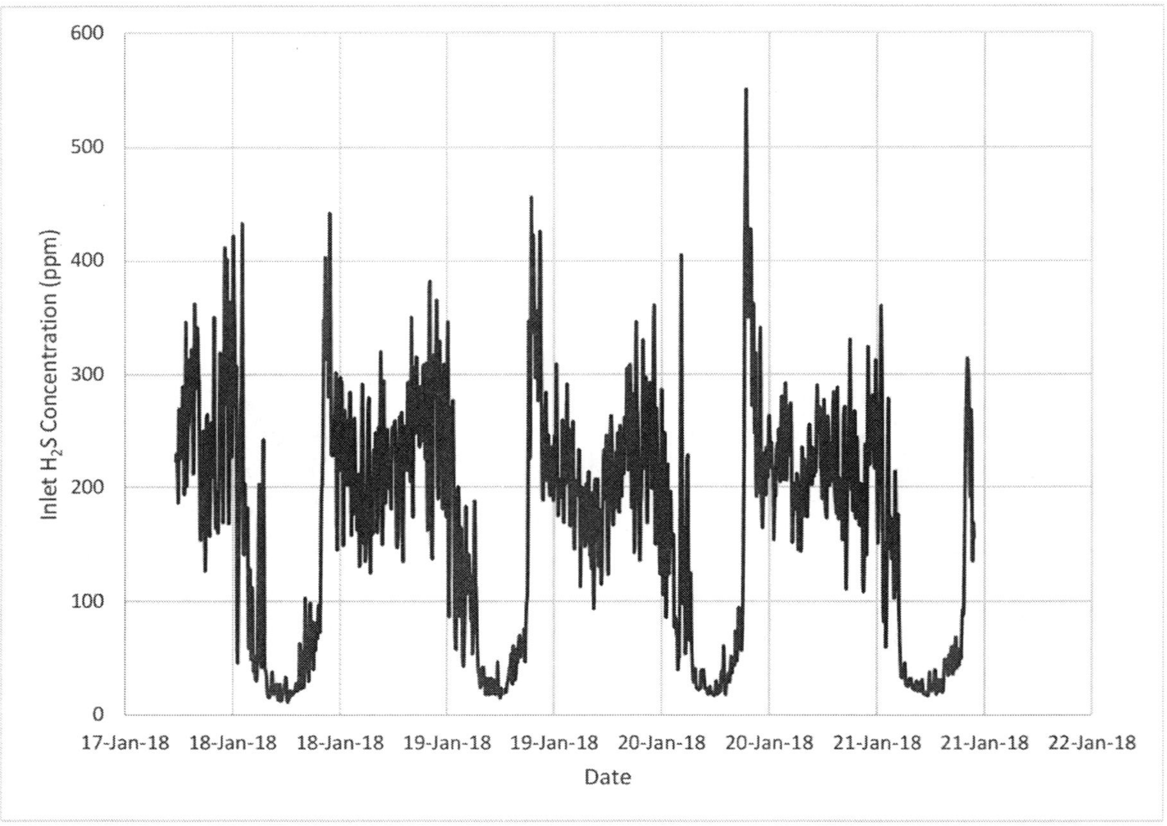

Figure 4: Typical diurnal conditions at the test location

A part of air was diverted from the full scale system to the pilot and was then returned for final treatment and discharge to meet permit conditions. The flow rate of air that was diverted depended on the pilot test conditions. Plant effluent water was used in the pilot for make-up during recirculation and for the once-through process. Nutrients were added on top of those present in the plant water to maintain the desired nitrogen and phosphorus concentrations. Typically, each operating configuration was investigated for at least a month. Hydrogen sulfide was monitored by Odalogs of appropriate ranges at inlet and outlet of each reactor. Temperature and recirculation rate was measured continuously; air flow rate and pH was monitored periodically as required.

5 RESULTS AND DISCUSSIONS

Hydrogen sulfide removal was studied throughout the program. Three campaigns were carried out for OSC and odor removal. Results are presented in this section.

5.1 Kinetics of Hydrogen Sulfide Removal

Figure 5 presents a plot of the hydrogen sulfide elimination capacity against inlet mass loading rate for once-through process and recirculation mode of operation. In the interest of keeping this paper focused on operational kinetics, not all of the data from the various trials of media type, geometry, flowrates and nutrient blends are being reported. Only the data from the optimized alternatives are being discussed in this paper. From the graph of this large data set, it is apparent that the once through mode of operation has a lower threshold for deviation from the complete removal demarcation line. Stated in another manner, the volume of media (or effective Empty Bed Retention Time) required for greater than 99% destruction of hydrogen sulphide is larger for once-through operation when compared with recirculation mode of operation. This trend was observed for all types of structured and random packing used during the study.

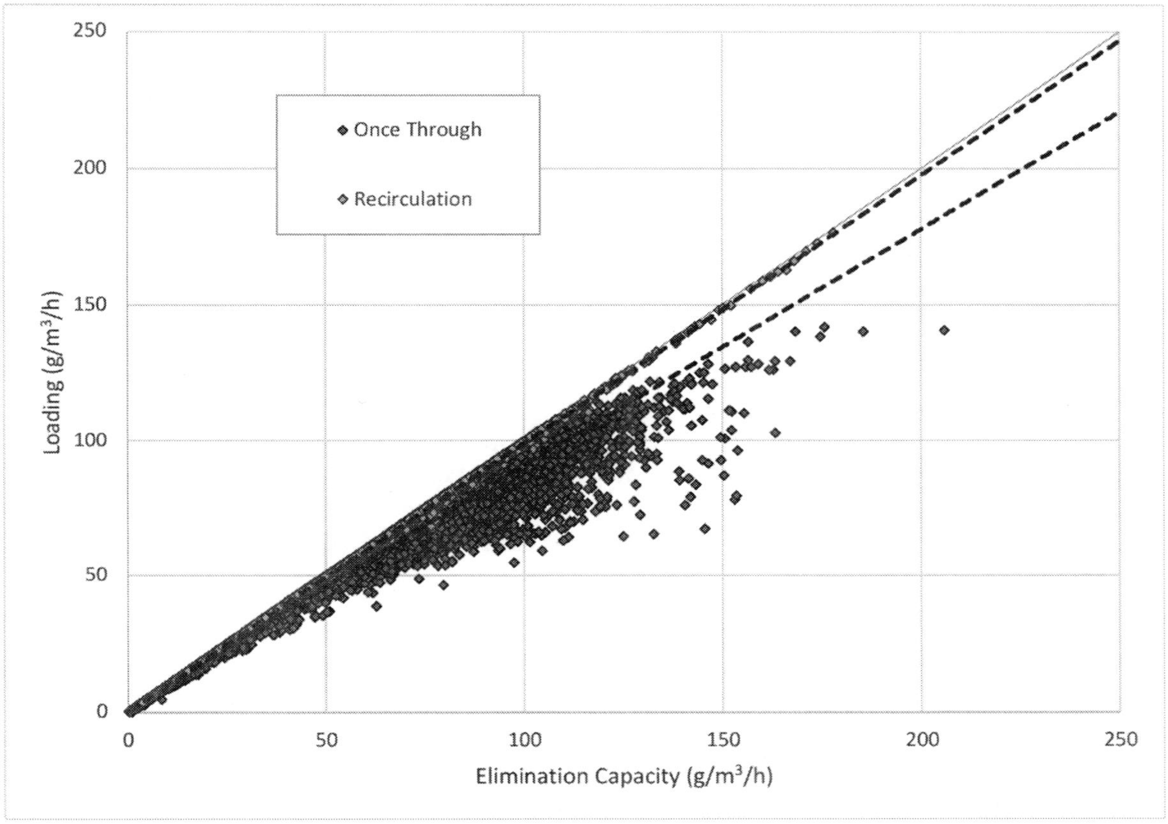

Figure 5: Hydrogen sulfide elimination capacity Once-through versus Recirculation

5.2 Organic Sulfur Compound and Odor Removal

Three sampling events were carried out to evaluate OSC and odor removal effectiveness during the two modes of operation. This is summarized in Table 1:

Table 1: OSC and odor sampling events

	Date	Media	Irrigation Operation		
			Stage 1	Stage 2	Stage 3
Sampling Event 1	Jul. 27, 2017	Structured	Recirculation	Recirculation	Once Through
Sampling Event 2*	Nov. 14, 2017	Structured	Once Through	Once Through	Once Through
Sampling Event 3	Jan. 10, 2018	Random	Recirculation	Recirculation	Recirculation

* During Sampling Event 2, a nearby full scale biotrickling filter employing recirculation and random media was also sampled for odor and OSC analysis

Results for the first test (Event 1), in which BTF1 and BTF2 used the recirculation process operating at a pH of 1.5, and BTF3 using once-through irrigation, are presented in Table 2. The bulk of hydrogen sulfide was removed in BTF1, and it was measured below detection limit at the BTF2 outlet.

BTF3 used an inert structured packing with a specific surface area >1,000 m^2/m^3 to ensure that very large surface was available for microbial growth. Despite the large effective surface area and the neutral operating environment, only 19% odor and 18% OSC removal was achieved. Interestingly, methyl mercaptan removal and odor removal in the neutral stage was comparable. It underlines the fact that bulk of the odor removal in a once-through BTF is due to hydrogens sulfide, and if significant odor is caused by OSC, high odor concentration in exhaust (Total Odor) will be present.

This is consistent with the finding of a study that 15-20s residence time is required for 95%+ OSC removal *even* in a high surface area engineered media biofilter specifically designed for OSC removal (Lee et al, 2012). In the above mentioned study, it was also indicated that an organic media biofilter, with a significantly longer residence time of up to 120s was unable to achieve 95% removal consistently. It is therefore concluded that an expectation of high odor and OSC removal, and low Total Odor within a short residence time once-through biotrickling filter using inert packing is most likely unrealistic.

Table 2: Sampling Event 1 Results

	Inlet	BTF1	BTF2	BTF3	Overall System
EBRT, s		4.0	4.0	3.0	11.0
H_2S, ppm	79.8	2.2	0.0	0.0	
H_2S Removal		97.2%	100%	NA	100%
MM, ppb	726	442	360	292	
MM Removal		39.1%	18.6%	18.9%	59.8%
Odor, D/T	>60,000	>60,000	31,000	25,000	
Odor Removal				19.4%	

Results of the second test are shown in The unit used for testing was the full-scale system at the facility which was connected to the same inlet source. Diurnal variations during the testing period resulted in testing be conducted on different mass loadings. Of note, different retention times were used in the side by side comparison due to equipment limitations. This will impact the elimination capacity reported in this part of the demonstration.

The second sampling event results demonstrate that in once-through operation, significant overall performance was achieved for the removal of hydrogen sulphide. This appears to be a function of the overall retention time used during the evaluation as inter-stage sampling clearly shows the structured packing operated in intermittent, once-through mode is struggling to achieve the 99% objective with over 16s of retention time. Methyl mercaptan removal rates are comparatively low during this phase of the study. This was an unexpected result as methyl mercaptan removal in biotrickling filters has been widely reported in previous studies.

The full-scale system demonstrated full removal of H2S and higher mercaptan destruction while operated in recirculation mode. It must be stated again that this system has a random packing for the microbial immobilization matrix and a higher retention time so that this data cannot be utilized for direct comparative purposes. The data does support previous work and understanding of standard biotrickling filters being operated in full recirculation mode.

Both modes of operation resulted in excellent overall Total Odor removal rates and is a function of the elevated inlet concentrations of hydrogen sulphide. Of note however, is that both biotrickling filter modes of operation resulted in still having a significant exhaust odor concentration despite high destruction rates. Outlet concentrations in these ranges would still be problematic for many applications in dense urban environments.

The exhaust odor concentration of the once through system has a significantly higher discharge level, due to the residual hydrogen sulphide and methyl mercaptan concentrations.

Table 3. In this test, the pilot was operated in a once-through mode, with a total residence time of 22s, which achieved a pH profile that would be expected in such a process (neutral at the top, dropping to a pH of 2-3 in water leaving BTF1). Side by side comparison with a random packed biotrickling filter operated in full recirculation mode was conducted in parallel. The unit used for testing was the full-scale system at the facility which was connected to the same inlet source. Diurnal variations during the testing period resulted in testing be conducted on different mass loadings. Of note, different retention times were used in the side by side comparison due to equipment limitations. This will impact the elimination capacity reported in this part of the demonstration.

The second sampling event results demonstrate that in once-through operation, significant overall performance was achieved for the removal of hydrogen sulphide. This appears to be a function of the overall retention time used during the evaluation as inter-stage sampling clearly shows the structured packing operated in intermittent, once-through mode is struggling to achieve the 99% objective with over 16s of retention time. Methyl mercaptan removal rates are comparatively low during this phase of the study. This was an unexpected result as methyl mercaptan removal in biotrickling filters has been widely reported in previous studies.

The full-scale system demonstrated full removal of H2S and higher mercaptan destruction while operated in recirculation mode. It must be stated again that this system has a random packing for the microbial immobilization matrix and a higher retention time so that this data cannot be utilized for direct comparative purposes. The data does support previous work and understanding of standard biotrickling filters being operated in full recirculation mode.

Both modes of operation resulted in excellent overall Total Odor removal rates and is a function of the elevated inlet concentrations of hydrogen sulphide. Of note however, is that both biotrickling filter modes of operation resulted in still having a significant exhaust odor concentration despite high destruction rates. Outlet concentrations in these ranges would still be problematic for many applications in dense urban environments.

The exhaust odor concentration of the once through system has a significantly higher discharge level, due to the residual hydrogen sulphide and methyl mercaptan concentrations.

Table 3: Sample Event 2 Results (Pilot System)

	Inlet	BTF1	BTF2	BTF3	Overall System
EBRT, s		8.1	8.1	6.1	22.3
H_2S, ppm	228	64	8.8	1.8	
H_2S Removal		71.9%	86.2%	79.5%	99.2%
MM, ppb	2,024	2,125	1,275	1,020	
MM Removal		-5.0%	40.0%	20.0%	49.6%
Odor, D/T	1,130,435	306,595	80,070	24,225	
Odor Removal		72.9%	73.9%	69.7%	97.9%

Table 4: Sample Event 2 Results (Full Scale System)

	Inlet	Outlet
EBRT, s		30
H_2S, ppm	159	0.0
H_2S Removal		100.0%
MM, ppb	1,567	486
MM Removal		69.0%
Odor, D/T	1,484,519	9,588
Odor Removal		99.4%

Results for Sample Event 3, conducted with random packing and recirculation are presented in Table 5. This test was carried out to provide further points of comparison between once-through and recirculation modes. Total residence time during this test was 18s, and the inlet hydrogen sulfide concentration was higher than the two previous runs. Despite this, removal of hydrogen sulfide and OSC was higher, and Total Odor significantly lower than the test results presented in Table 3. Table 6 presents this comparison. These results support other findings that even at a longer EBRT, once-through irrigation may not provide both high hydrogen sulfide removal and low Total Odor effectively to prevent off site odor complaints.

Table 5: Sampling Event 3 Results - Recirculation

	Inlet	BTF1	BTF2	BTF3	Overall System
EBRT, s		6.0	6.0	6.0	18.0
H_2S, ppm	296	4.1	2.9	0.125	
H_2S Removal		98.6%	29.3%	95.7%	100.0%
MM, ppb	1,164	393	339	306	
DMS, ppb	39	27	21	24	
DMDS, ppb	99	102	111	60	
Total OSCs, ppb	1,302	522	471	390	
Total OSC Removal		59.9%	9.8%	17.2%	70.0%
Odor Removal		98.3%	11.6%	58.7%	99.4%

Table 6: Once-through versus Recirculation for Comparable EBRT

	Sample Event 2 – Once-through irrigation	Sample Event 3 - Recirculation
Total EBRT, seconds	**22.3**	**18**
Hydrogen Sulfide Concentration, ppmv	**226**	**296**
Hydrogen Sulfide Removal, %	**99.2%**	**99.9%**
OSC Removal, %	**49.6%**	**70%**
Exhaust Total Odor, D/T	**24,225**	**1,207**

6 DISCUSSIONS AND CONCLUSIONS

The results of this study help to define criteria for the appropriate selection of operating modes for biotrickling filters based on application (ie source of odors or anticipated composition of odiferant compounds); inlet loading of hydrogen sulfide and required performance (location, proximity to neighbors, etc.).

A key finding of the study shows that once-through intermittent irrigation mode of operation is less efficient for the removal of hydrogen sulphide. This is independent of media type (structured or random). The technical explanation for this observed phenomenon is related to how the various microbial species adapt to changing biofilm conditions as a result of diurnal variations in inlet concentrations. Bacterial colonies that have been immobilised on smooth plastic surfaces tend to be very exposed and susceptible to changing conditions. Small changes in inlet hydrogen sulfide concentrations result in changes in acid production and the corresponding localized pH of the biofilm. Within a once-through biotrickling filter, the pH threshold continually moves up and down the vertical reactor, in synch with variable diurnal concentrations, and directly impacting microbial colonies. This creates inefficiencies as the biology tries to adapt to the changing conditions. In comparison, a full recirculation operation mode ensures that the full vertical profile of the reactor remains at a single constant pH.

The limited number of OSC and odor samples taken during the study impact the ability of the authors to make definite conclusions comparing the two modes of operation. However, the following statements can be made as observations:

1. Total odor removal is challenging in either mode of operation due to the limitations of the technology itself. If significant hydrogen sulfide is present in the air stream, high rates of removal are achievable, however, the resulting discharge concentrations may still be problematic in terms of offsite migration of residual odors.
2. Organic sulfur compound removal efficiencies of biotickling filters, operated in either recirculation or once-through mode are not comparable with a properly sized and designed biofilter (either engineered media or organic based blends). This is true even in the context of extended retention times utilized for the study.

While further work is required to refine the kinetic models on individual species of odiferant compounds, the results of the study tend to strongly indicate that the following guidance can be used for selecting the optimized technology and mode of operation for an application:

1. If hydrogen sulphide is the compound of concern for odor removal, a biotrickling filter operated in continuous recirculation mode is the preferred option. A once through mode of operation provides little additional benefit to the application.
2. If total odor removal is required, a biotrickling filter must be combined with another stage of treatment to ensure a low enough discharge concentration to prevent off site negative impacts to the surrounding community.
3. Sources of odors that are anticipated to have significant organic sulphur compounds (ie any source of air related to biosolids handling and treatment) should consider the use of alternative solutions rather than the selection of a biotrickling filter)
4. Elimination capacity is a useful indicator of the viability of a media or operating mode for a particular application.

7 REFERENCES

Easter, C., Quigley, C., Burrowes, P., Witherspoon, J., & Apgar, D. (2005). Odor and air emissions control using biotechnology for both collection and wastewater treatment systems. *Chemical Engineering Journal, 113*(2-3), 93-104.

Friedrich, c, Rathier, D, Bardischewsky, A, and Fischer, J, Applied and Environmental Microbiology, Vol67, No.7., 2001, pp: 2873-2882

Johnson, M., le Roux, L., & Dunn, W. (2011). *Use of Biotrickling Filters for Total Odor Control at Two Municipal WWTP's*. BioAir Solutions, LLC.

Lee, A., Morton, R., Tang, C., (2012) Proceedings of the Water Environment Federation, Odors and Air Pollutants, pp. 694-702(9)

Woo, M., & Singleton, B. (2017). *Breaking Down Biological Treatment of Odours - Design Considerations for Bioscrubbers and Biotrickling Filters*. CDM Smith.

Miniaturized biotrickling filters for process intensification of indoor air VOC treatment

Luis Rafael Lopez de León, Kelsey E. Deaton, Marc A. Deshusses

Department of Civil and Environmental Engineering 127C Hudson Hall; Box 90287; Duke University. Durham, NC 27708-0287. USA

ABSTRACT

Typical biotrickling filters and biofilters used for VOC control have treatment rates limited to 30 – 200 g/m^3-h. The low elimination capacities are due to common limitations of conventional biotrickling filters such as dilute substrate and moderate biomass density and activity, rendering biofilters and biotrickling filters non-ideal for treatment of indoor air pollutants. Here we report on the development and evaluation of micro-biotrickling filters (μBTFs) for the intensification of treatment of indoor VOC pollutants. The miniaturized reactor design, with maximized surface-to-volume ratios, allows for increased mixing and mass transfer of pollutants to the biofilm, which is expected to result in a greater degradation of VOCs. Furthermore, VOCs are fed to the μBTFs from a separate air pollutant concentrator, which will alleviate moisture management issues in the bioreactor and enable achievement of high microbial growth rates and pollutant elimination capacities. For this research, three different microbioreactor/micro-biotrickling filters were designed, constructed and their performance for removing vapors of toluene and methanol was assesses.The μBTFs were inoculated from pure culture of *Pseudomonas putida* previously grown in serum bottles. A test platform that allowed generating multiple air streams (for parallel testing) loaded with the VOC of interest was designed and multiple units were constructed to test the prototypes in parallel. The maximum elimination capacity (EC) achieved in each μBTF was quantified. Results showed that our miniaturized biotrickling filters are able to achieve ECs for methanol around 1000 g/m^3 h, 780 g/m^3 h and 12,600 g/m^3 h for the glass beads packed bed, polyurethane (PU) foam-packed bed and capillaries microbioreactor respectively. Toluene ECs were around 120 g/m^3 h, 250 g/m^3 h and 3050 g/m^3 h for the glass beads packed bed, PU packed bed and capillary microbioreactor respectively. Determination of kinetic parameters has been also conducted in parallel to inform the reactor design and a mathematical model of the process. Moreover, the interphase mass transfer coefficient (k_La) was determined in the each reactor prototype. Coefficients values were not statistically different for the glass beads and PU foam systems and very much in-line with earlier experiments in larger scales bioreactors. However, the capillaries bioreactor had mass transfer coefficients 13 to 17 times greater, suggesting that improved mass transfer could have contributed to the very high performance observed for both toluene and methanol vapor treatment in that bioreactor.

The results illustrate that the miniaturized biotrickling filters could be a promising component of novel technologies for indoor air pollution treatment.

KEYWORDS

Odor/Air Emission, Innovative Technologies, Microbioreactors, Biological Odor Control

1. INTRODUCTION

Taking into account that people spend 80-90% of their time indoors (Boor et al., 2017; Rudel and Perovich, 2009), indoor air quality matters as much as outdoor air quality. Poor indoor environmental air quality is an important public health concern worldwide (Wu et al., 2007). Indoor environments encompass homes and other residences, workplaces, transportation vehicles (including cars, trains, and planes) and other diverse enclosed settings where individuals spend a large portion of their day. There are different mixtures of pollutants and chemical agents in indoor environments that can have significant impact on human health, such as particulate matter, biological agents, and chemicals present in indoor air (Mitchell et al., 2007).

In particular, volatile organic compounds (VOCs) are a common class of indoor pollutants that are a major health concern. Short-term exposure to VOCs are known to cause respiratory irritation, headaches, dizziness, fatigue and nausea (Fiedler et al., 2005). Chronic health effects related to prolonged exposures to VOCs can cause asthma, damage to the liver, kidneys, and central nervous system (Rudel and Perovich, 2009; Rumchev, 2004; Wu et al., 2007). The generation of VOCs in indoor air are the result from consumer products, building materials, human metabolism and activities, pollution of outdoor air and more (EPA, 2011; Mitchell et al., 2007). According to one study, levels of most VOCs can be 5-10 times higher indoors than outdoors (Samet, 1988). Further, the EPA compiled the results of fifteen studies of North American residential air pollution (EPA, 2011). They found that VOCs range in concentration from 0.01 to 3.8 $\mu g/m^3$. Often, where outdoor air quality is good, providing a proper air exchange rate (AER) in indoor environments can be an adequate way to improve indoor air quality and reduce the concentration of VOCs generated indoors.

However, in closed environments without adequate ventilation of outdoor air and thus where the AER is null or very low, VOC levels can be much higher. Some of the worst air quality is found in enclosed spaces like factories and airplanes that do not have the capability for sufficient ventilation. For instance, toluene found at 0.03 to 1.9 $\mu g/m^3$ in residential settings, is present at one-hundred-fold higher levels in cabin of commercial aircraft 130 $\mu g/m^3$ (Waters et al., 2002). In non-civilian closed environments, such as the International Space Station (James et al., 2010), the levels of total VOCs are much higher, with total average VOC concentration of 14 mg/m^3. Even higher levels have been found on submarines. In a study of the air quality in Victoria Class submarines, the Canadian Royal Defense (Severs, 2006) found that levels as high as 67 mg/m^3. In such closed environments, since the AER is low, treatment of air is critical to mitigate adverse performance and health effects. Different air treatment methods exist. One of the most extensively studied gas-phase reactions are oxidation reactions involving oxidants like ozone (O_3) and free radicals like hydroxyls (OH^-) and nitrate (NO_3^-)(Wells et al., 2017). When O_3 and nitrite (NO_2^-) are present simultaneously, indoor NO_3^- may be the dominant indoor oxidant that effectively reacts with nearly all indoor VOCs (Mitchell et al., 2007). Surfaces also play an important role for indoor chemistry, due to large surface to volume rations in many indoor environments (3 $m^2\ m^{-3}$ vs 0.01 $m^2\ m^{-3}$), surfaces may be ideal for substance sorption and for water condensation (Mitchell et al., 2007). O_3 in air can react with unsaturated VOCs contained in surface coatings at a faster rate than when it reacts with the same compound in gas phase (Reiss et al., 1995). However, chemical oxidation processes can lead to the formation of oxygenated species (e.g. Formaldehyde) and secondary organic aerosols (SOA), compounds that can cause several health effects in the airways and the cardiovascular system (Nazaroff and Goldstein, 2015). While oxidants may be able to degrade VOCs, radicals like O_3 can be toxic on their own. An ozone generator with the intention of degrading indoor air pollutants could produce sufficient concentrations of ozone to become a health hazard.

There are several processes that can be employed to perform a more efficient VOC removal from indoor air such as Vacuum Ultraviolet (VUV) Photolysis(Huang et al., 2016), activated carbon fiber filters (ACF)(Sidheswaran et al., 2012) and photo-catalytic oxidation, with the latter being the most widespread technique for VOC removal in indoor air (Debono et al., 2017). The main drawback of some of these processes are that some of the oxidation reactions may be incomplete and reaction intermediates can be emitted even in the presence of typical indoor low concentrations of VOCs (Debono et al., 2013; Einaga et al., 2013).

Thus, there is a critical need for new technology that can effectively control VOCs in enclosed indoor environments. Biofiltration technologies, such as biofilters and biotrickling filters, have proved to be a very successful technology for treating contaminated gaseous streams at different scales, from laboratory scale studies (Dorado et al., 2012) up to full-scale implementation(San-Valero et al., 2017). The most successful application of biofiltration technologies have been for the treatment of dilute, high flow air steams containing odors, VOCs or other hazardous air pollutants (Gabriel and Deshusses, 2003; Guieysse et al., 2008; Kennes et al., 2009). The major advantage of air biofiltration is that air biotreatment is accomplished at low temperature, it produces no by-products such as NOx or SOx, does not require fossil fuel, and has lower operating and maintenance costs. A detailed review of the mechanisms, possibilities and limits of biofilters and biotrickling filters can be founded elsewhere (Cox and Deshusses, 1998, 2000).

One of the most relevant processes regarding biofiltration systems is the relationship between growth rate and pollutant degradation, since often biological stress or some form of nutrient limitation is imposed to limit growth and reduce operating costs or avoid excessive biomass growth and potential plugging of the bioreactor bed. The consequence of this choice is a lower Elimination Capacity (EC) of the system. The second critical point to understand is the importance of having a high interfacial area and effective gas-bacterial culture contact for rapid oxygen and pollutant mass transfer to the biofilm. Lastly, biological reaction rates are reduced at low concentrations; as a result, bioreactors that treat either low concentration or vey hydrophobic pollutants usually exhibit low volumetric performance. The research described here attempts to overcome the limitations of conventional biofiltration with the development of a novel high performance microbioreactors system. The use of microreactors has shown to have superior mass transfer and therefore higher reaction rates. Moreover microreactors offer a precise control of reaction parameters, contact time, shape and size of the interface between fluids (Yao et al., 2015). Low mass transfer and biodegradation rates of traditional biofiltration systems are overcome by the use of well-defined flow patterns and uniform interfacial areas in micro-bioreactors, that allows a precise tuning of the mass transfer processes and helps to have a more reliable and feasible prediction of mass transfer coefficients(Kashid and Agar, 2007). Moreover, microbioreactors working with specific VOCs or classes of VOCs are expected to ensure fast degradation rates due to highly active, specially acclimatized microbial cultures.

In this work, two concepts of micro-bioreactors will be pursued. The first one based on designs using conventional macro-scale gas-liquid bioreactor packing materials. The second configuration is a capillary-based bioreactor. The research focuses on demonstrating that mass transfer rates of oxygen and VOC degradation rates is greatly enhanced due to the very large surface-to-volume ratio achieved which each micro-bioreactor configuration and either the short diffusion lengths or intense mixing. The new microbioreactor system will rely on a pollutant microconcentrator developed by collaborators at the Warner Babcock Institute that will concentrate the VOCs from indoor air and release them in small volume air stream to the microbioreactors (not shown here).

2. MATERIALS AND METHODS

2.1. Micro-biotrickling Filters and Micro-Bioreactors Characteristics

Three different micro-bioreactor/micro-biotrickling filters were designed and constructed in order to evaluate their performance for removing toluene and methanol vapors. These two compounds were deemed suitable model pollutants for hydrophobic and hydrophilic VOCs, respectively. Also, there is a large body of published research on the biofiltration of these two compounds to use as a benchmark for assessing our results. The main characteristics of these bioreactors are reported on table 1, while pictures are shown in Fig. 1.

Table 1. Summary of the characteristics and experiments conducted with the micro-bioreactors prototypes

	PU Foam	Glass beads	Glass capillary
Description	Miniature packed bed of open pore polyurethane foam	Small bed packed with glass beads	Parallel glass capillaries for gas and liquid slug flow
Size (cm)	7 cm tall bed, 2.9 cm diameter	7 cm tall bed, 2.9 cm dia. Packed with 4 mm glass beads	19 capillaries, 1.1 mm ID, 10 cm in length
Reactor volume (ml)	30	30	1.90
Estimated interfacial area (m^2 m^{-3})	600	1500	3636
VOC tested	Methanol, Toluene	Methanol, Toluene	Methanol, Toluene

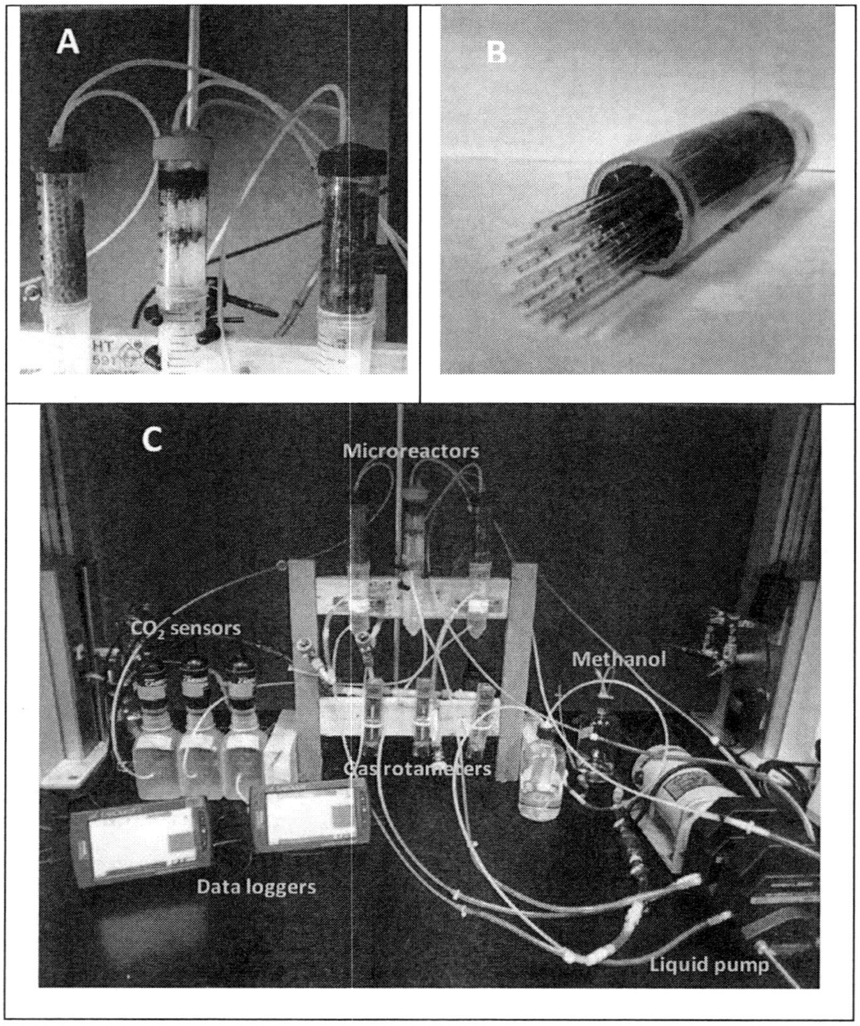

Figure 1. Pictures of the 3 different microbioreactor prototypes tested. Top, left to right: glass beads packed bed microbioreactor (note the brown color due to biofilm growth), glass capillary microbioreactor, PU foam microbiotrickling filter. To right: micro-pillar bioreactor (prior to connection, without bacterial cells). Bottom: picture of the test bench setup.

The microbioreactor/microbiotrickling filter designs aimed to maximize pollutant and oxygen mass transfer and the activity of the pollutant-degrading bacteria. Prototypes were built in-house using simple materials or 3D printed in the case of the micro-channels (capillaries) bioreactor. At the same time, a test platform that allowed generating multiple air streams loaded with the VOC of interest was designed and multiple units were constructed to test these prototypes in parallel. A picture of such test platform is shown in Fig. 1C and a flowsheet diagram for the test platform is showed in Figure 2.

2.2. Experimental Set Up For Liquid Film Mass Transfer Coefficients Under Abiotic Conditions

A set of experiments were carried out in order to determine the liquid interphase mass transfer coefficient (K_La) in each microbioreactor. Mass transfer coefficients are important to accurately model the bioreactor, to understand the fundamental principal of the interphase mass transfer in the system (i.e., determine the rate limiting step), and to define the possibilities of the bioreactor. Mass transfer coefficients were experimentally determined following the methodology detailed elsewhere (Kim and Deshusses, 2008). For the determination of K_La, absorption of oxygen, which is sparingly soluble in water, was used so the main resistance to mass transfer was in the liquid phase. To summarize the method, dissolved oxygen in the liquid phase was stripped prior to be introduced in the microbioreactor by continuously sparging nitrogen during the experiment. One deviation to the methodology applied by Kim and Deshusses, was that the liquid phase was passed by the packed bed only once, not being recirculated, using steady state outlet dissolved oxygen concentration to calculate K_La using the Equation 1:

$$K_La = \frac{F_L \cdot DO_{SS}}{V_b \cdot (DO_{SS} - DO_{sat})} \qquad \text{Equation (1)}$$

Where K_La stands for the liquid interphase mass transfer coefficient (s), F_L is the liquid flow rate (m^3/s), DO_{SS} the dissolved oxygen concentration at the steady state (g/m^3), DO_{sat} the dissolved oxygen concentration at saturation (g/m^3), V_b the volume of the reactor bed (m^3). The experiments were repeated over an array of relevant air and gas flow conditions as described in table 2.

Table 2. Experimental conditions for the determination of liquid film mass transfer coefficients

Packing materials tested	**Gas flow** *(ml/min)*	**Trickling velocity** *(m/h)*
Glass beads packed bed	190	0.16
		1.2
PU foam		2.2
Capillaries	190	6
		44
		79

2.3. Growth Media and Culture Conditions

A culture of *Pseudomonas putida* was inoculated into a 250 mL serum bottle with 50 mL autoclaved mineral medium(Fortin and Deshusses, 1999) and incubated at 20 °C on a shaker table at 200 RPM. Either 0.05 g/L toluene or 0.1 g/L methanol was added to the bottles. Methanol and toluene cultures were each grown in triplicate. Cultures were sequentially transferred every couple days to a new serum bottle with fresh media and substrate to maintain a high growth rate. Substrate concentrations were gradually increased to 0.4 g/L toluene or methanol. CSTRs were then set up to automate culture maintenance. One CSTR was set up for the toluene culture and one for the methanol culture. The CSTR set up consisted of a 500 mL glass stirred tank vessel with an inlet for mineral media, substrate, and air and outlets for reactor contents and air. Aliquots were periodically taken from the CSTRs for sampling of culture density which was measured by optical density (OD), and analysis of substrate concentration by gas chromatography (GC). Substrate concentrations were maintained at 13.8 g Methanol /L and 0.372 g Toluene /L. All cultures used in following experiments were obtained from the CSTRs.

2.4. Experimental Set Up For The Determination of VOC Degradation Biokinetics Parameters

Biokinetic parameters (essential for modeling our bioreactors and sizing biodegradation systems) were determined either by oxygen uptake rate (OUR) experiments or in direct measurements in CSTRs. In order to determine biokinetic parameters of our cultures, OUR was measured using respirometry tests, a technique defined as the measurement and interpretation of the biological oxygen consumption rate under well-defined experimental conditions (Spanjers and Vanrolleghem, 1995). The exogenous oxygen uptake rate (OUR) curves obtained after the addition of substrate pulses are related with the aerobic degradation processes, which allow the estimation of kinetic and stoichiometric parameters (Mora et al., 2014).

Respirometric tests were conducted in a microrespirometer coupled to a biological oxygen monitor (YSI 5300, YSI Incorporated). The micro respirometer set up consisted in a biological oxygen monitor, a micro oxygen chamber fitted with an data-logging oxygen probe and a magnetically stirrer. Aliquots of biomass from the corresponding CSTR (toluene or methanol) were withdrawn. The suspended biomass was centrifuged at 4000 rpm for 15 minutes and the supernatant was discarded and the biomass was resuspended in substrate-free mineral medium. The concentration of biomass was obtained from a correlation between the VSS and OD. After biomass was washed and the concentration was known. The biomass was poured into a vessel, were the liquid phase was gassed with air until reach saturation. Then 3 mL of biomass was poured into the micro oxygen chamber. Immediately the respirometer was spiked with a known substrate concentration. Concentration ranges test for both VOCs degrading cultures were from 0, which was to determine the endogenous respiration, until 36 mg COD L^{-1}, with increments of 2 mg COD L^{-1} until the concentration of 12 mg COD L^{-1} and then on with increments of 4 mg COD L^{-1}. The maximum specific growth rates, half-saturation constants and inhibition constants were determined by curve fitting to experimental respirometric profiles using Matlab. Differential equations were integrated with a routine with a variable step-Runge-Kutta method (ODE45), and the model was fitted using an optimization tool, fminsearch (Nelder-Mead-Simplex).

2.5. Operating Conditions for Micro-biotrickling Filters and Microbioreactors Testing

Experiments consisted primarily of operating the microbioreactors continuously for weeks at a time and monitoring the performance in terms of removal and elimination capacity of toluene and methanol. For that purpose, a controlled humidified clean air stream (1) and a second controlled air stream (2) that passed through saturated headspace of bottle containing pure liquid VOC (toluene or methanol) (3). The two air streams were mixed in order to create an air stream loaded with VOC (3). Mass flow controllers (MFC) were used to regulate each air stream in order to obtain the desired concentrations. The polluted air stream (4) was homogeneously distributed into the three microbioreactors by the use of gas flow controllers (5). For the glass beads (6) and PU foam (7) reactors, gas was fed at the bottom of the microbioreactor, flowing in counter-current with the liquid phase. For the glass capillary microbioreactor (8), gas phase was fed at the top, flowing co-current with the liquid phase. The liquid phase of mineral media was continuously recycled (9) from bottom-to-top of the microbioreactors at a trickling liquid velocity (TLV) of 1 m h^{-1} and 44 m h^{-1} for the glass beads and PU foam microbioreactors and capillary microbioreactor, respectively. After biological degradation in the microbioreactor, the treated air stream leaves the reactor. The outlet gas stream was monitored using on-line CO_2 NDIR probes to measure CO_2 production (10), as a measure of VOC mineralization, a common practice in biofiltration technologies to monitor biological activity (Dorado et al., 2012).

The pollutant-degrading bacterial culture was inoculated in to the glass beads and PU open pore foam microbioreactors, and allowed to grow as an immobilized biofilm on the packing material. For the glass capillary microbioreactor, a known concentration of the culture was added to the liquid phase before each experiment. Biomass accumulation for the glass beads and PU open pore foam microbioreactors was observed visually and by weighing selected bioreactors over time.

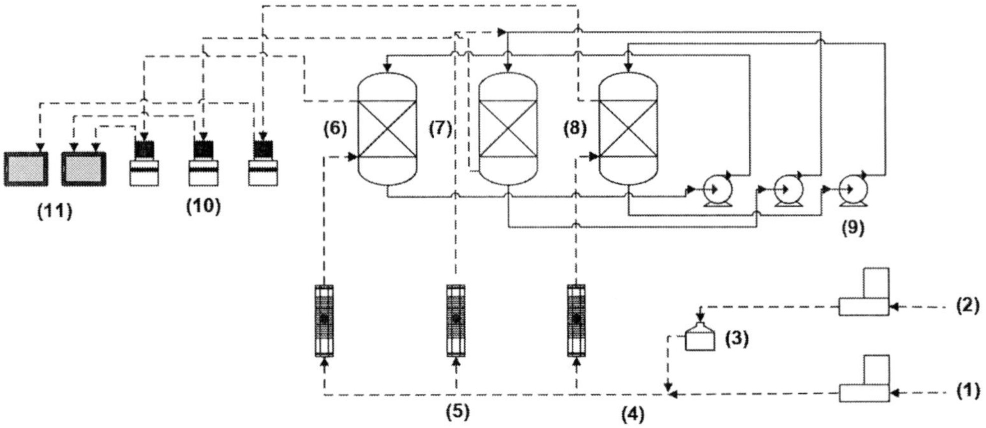

Figure 2: Flowsheet for the test platform

2.6. Experimental Conditions For The Evaluation of the Contribution of the Bed and Ancillary Liquid Recycle to the Overall Elimination Capacity

Additional experiments were conducted to determine the extent of pollutant removal in the liquid reservoir/liquid recycle tank of the microbioreactors. This was to a) ascertain that removal of the pollutant was primarily in the bed or capillaries, b) to justify the normalization of the rate by the bed or capillary volume, and c) ensure the performance of the prototypes can be accurately predicted using the bed or capillary volume. The methodology consisted of separating the liquid immediately after the bed and separately measuring CO_2 (as a measure of VOC mineralization) evolving from the bioreactor bed and CO_2 evolving from the recycled liquid during operation.

3. RESULTS AND DISCUSSION

3.1. Determination of Liquid Film Mass Transfer Coefficient

The effect of liquid velocity on the liquid film mass transfer coefficient was evaluated for the three different microbioreactors. Mass transfer coefficients obtained for the glass beads and PU foam packed bed microbioreactors, are showed in figure 3A wile results obtained for the glass capillary microbioreactor are showed in figure 3B. The mass transfer coefficient values for the glass beads and PU foam systems were not statistically different. Values were very much in-line with those of earlier experiments in a larger bioreactors determined in previous works (Kim and Deshusses, 2008). Interestingly, the glass capillary microbioreactor had mass transfer coefficients 13 to 17 times greater than the other two microbioreactors. The efficient mass transfer in the capillaries bioreactor certainly contributed to the very high performance observed for both toluene and methanol vapor treatment, as it will be later discussed. Mass transfer coefficients here determined for the glass capillary microbioreactor are in agreement with other gas-liquid glass capillary microbioreactors (Berčič and Pintar, 1997; van Baten and Krishna, 2004; Yue et al., 2007).

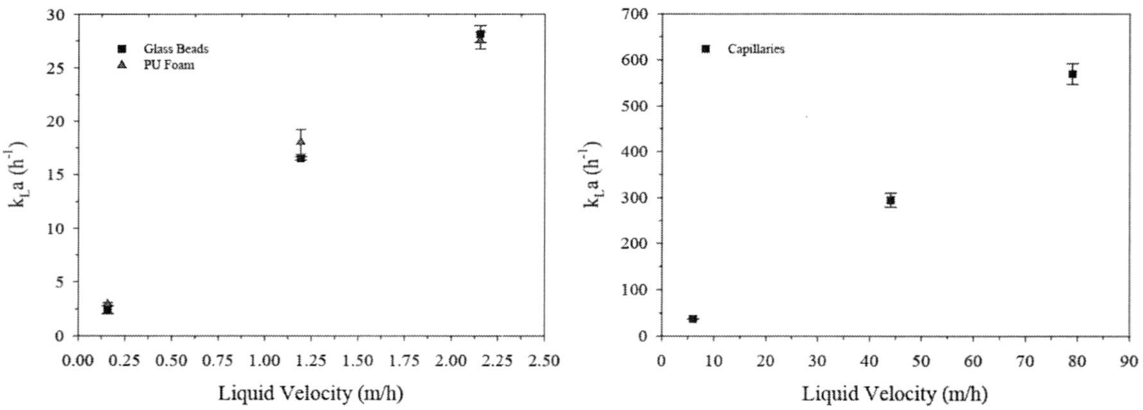

Figure 3. Overall mass transfer coefficient (k_La) in the glass beads and PU foam (left), and in the capillaries microbioreactors (right), as a function of the liquid velocity. Note the different Y-scale between the reactors.

3.2. Determination of VOC Degradation Biokinetics

Results for the kinetics of methanol and toluene biodegradation are shown in Figure 4, while biokinetic parameters are listed in table 3 and 4.

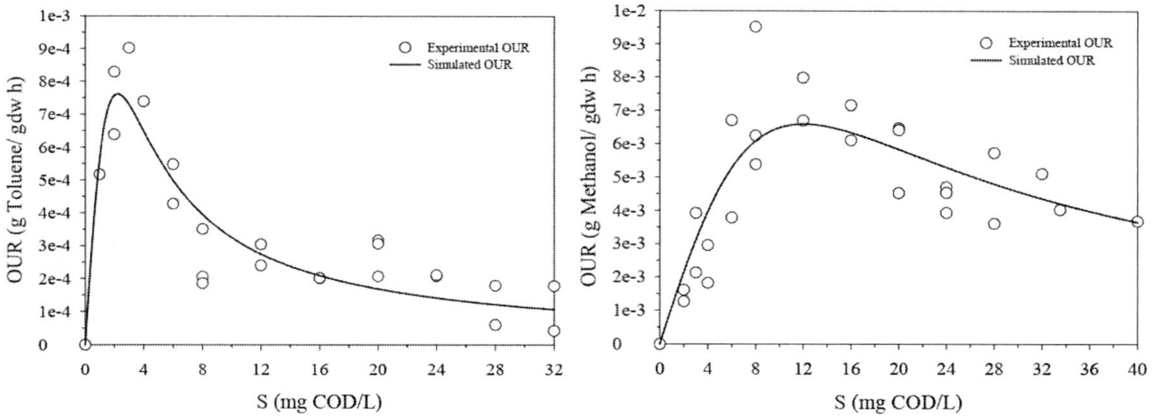

Figure 4. (A and B). Experimental and model-fitted biodegradation of toluene (left) and methanol (right)

The results show that biodegradation followed a Haldane kinetic model, i.e, inhibition by the substrate at high concentrations. The effect was more pronounced for toluene than for methanol, a finding consistent with the fact that toluene is more hydrophobic and more toxic to microorganisms than methanol. The substrate inhibition can have important implications for the microbioreactors as the plan is to operate them with high pollutant concentrations obtained from the microconcentrator. Thus, those systems will need careful feeding to avoid inhibitory concentrations. In addition, the fast biodegradation rates obtained (as reported in table 3 and 4) combined with the high cell density will result in a rapid sink for the pollutant, effectively protecting the pollutant-degrading bacteria from excessively high concentrations. Examination of table 3 and 4 reveals that the values obtained are in agreement with those reported in the literature (Littlejohns and Daugulis, 2008; Reardon et al., 2002; Shareefdeen et al., 1993)

Table 3: Biokinetic parameters determined for methanol-degrading bacteria and comparison with range of previously reported values

Parameter		Value	Units	Method of determination	Earlier reports
Endogenous decay	bh	0.03	d^{-1}	OUR endogenous respiration	0.04-0.10
Half-saturation constant	Ks	0.67	g COD L^{-1}	OUR tests, model fitting	0.63
Haldane inhibitory constant	Ki	2.15E-04	g COD L^{-1}	OUR tests, model fitting	2.15E-02
Specific growth rate	μ_{max}	0.082	h^{-1}	OUR tests, model fitting	0.1
Biomass-to-substrate yield	Y x/s	0.082	g VSS g COD^{-1}	CSTR experiment,	0.1-0.4
Biomass-to-substrate yield	Y x/s	0.055	g VSS g $MeOH^{-1}$	CSTR experiment	
Oxygen-to-substrate yield	Y o/s	1.44	g O2 g $MeOH^{-1}$	OUR test	1.5
Biomass concentration	X	1.70	g VSS L^{-1}	Measured	

Table 4: Biokinetic parameters determined for toluene-degrading bacteria and comparison with range of previously reported values

Parameter		Value	Units	Method of determination	Earlier reports
Endogenous decay	bh	0.02	d^{-1}	OUR endogenous respiration	0.04-0.10
Half-saturation constant	Ks	0.09	g COD L^{-1}	OUR tests, model fitting	0.03
Haldane inhibitory constant	Ki	5.28E-05	g COD L^{-1}	OUR tests, model fitting	7.89E-02
Specific growth rate	μ_{max}	0.28	h^{-1}	OUR tests, model fitting	0.53
Biomass-to-substrate yield	Y x/s	0.58	g VSS g COD^{-1}	CSTR experiment,	1.01
Biomass-to-substrate yield	Y x/s	1.80	g VSS g $C7H8^{-1}$	CSTR experiment	
Oxygen-to-substrate yield	Y o/s	3.56	g O2 g $C7H8^{-1}$	OUR test	3.13
Biomass concentration	X	0.67	g VSS L^{-1}	Measured	

3.3. Microbioreactors Performance: Elimination Capacity Under Different Loading Rate Conditions

Toluene and methanol vapors were treated in three different microbioreactor prototypes. Results reported in figure 5 and 6 shows the typical saturation (i.e., the rate reaching the steady maximum value at high loading) generally seen in conventional biofiltration and biotrickling studies (López et al., 2016; Montebello et al., 2014). Performance was the lowest for the glass beads, followed by the PU foam micro-biotrickling filter. Still, the EC for toluene in these two systems ranged from 60 to 260 g m^{-3} h^{-1}. This is much higher than most conventional biofilters and biotrickling filters which generally exhibit ECs in the range of systems capacity of 20 to 80 g m^{-3} h^{-1} for toluene biodegradation (Deshusses, 1997; Rene et al., 2017). The removal rates of toluene in the capillary channel microbioreactor were extremely high, with EC values between 1000 and 4200 g m^{-3} h^{-1}. Note that these high values are due in part to the fact that we are treating a large amount of gas in a small bioreactor, hence normalizing by the volume (1.9 mL) of the capillaries (and not the entire volume of liquid). However, this normalization is consistent with the fact that the reaction primarily occurs in the capillaries as was determined in separate experiments (see section 3.4).

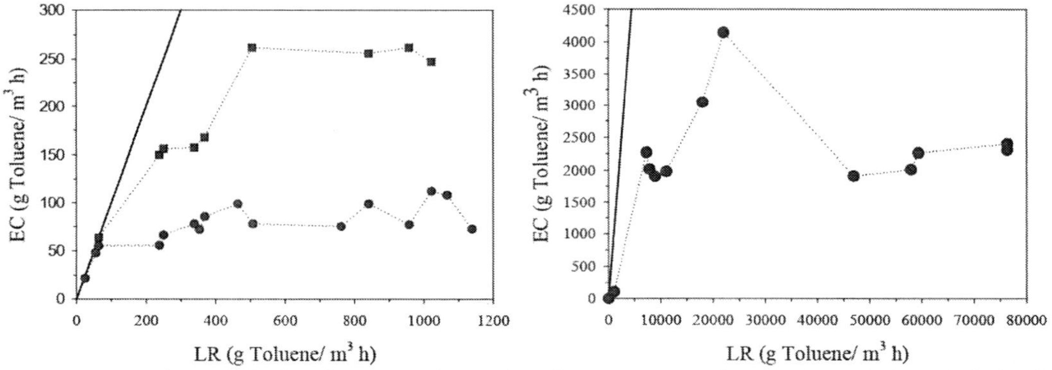

Figure 5. (A and B). Elimination capacity vs loading rate for toluene degradation: left, in the glass beads (blue circles) and the open pore PU foam (green squares) microbioreactors, and right, the micro-channels capillaries bioreactor. The solid line represents 100% removal. Note the much higher rate in the capillary bioreactor.

Different trends were observed for the removal of methanol vapors in the prototype microbioreactors, as shown in figure 6. The performance of the glass beads micro-biotrickling filter was higher than that of the PU foam micro-biotrickling filter at the lower loading, and about equivalent at the higher loading. This is probably due to the larger wetter area and liquid holdup in the glass bead microbioreactor combined with the fact that methanol is hydrophilic, and thus will adsorb readily in the free liquid. The rates observed (400 to 985 g m^{-3} h^{-1}) are 2 to 5 times higher than normally reported in the literature, again illustrating the performance gains obtained as a result of miniaturization and higher concentration feedings. Even higher rates were obtained in the glass capillary microbioreactor reaching an unprecedented rate in excess of 10,000 g m^{-3} h^{-1}. Further, 100% removal was achieved up to pollutant loads of 8000 g m^{-3} h^{-1} methanol.

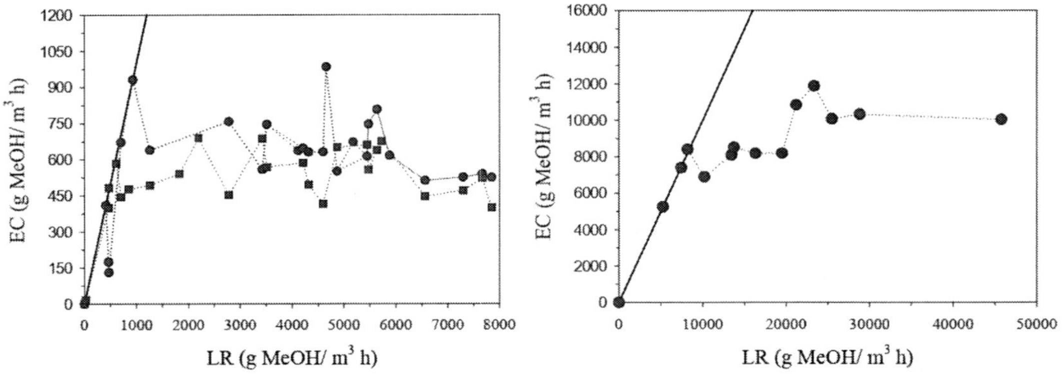

Figure 6. (A and B). Elimination capacity vs loading rate for methanol degradation: left, in the glass beads (blue circles) and the open pore PU foam (green squares) microbioreactors, and right, the micro-channels capillaries bioreactor.

After observing substantial substrate inhibition in oxygen uptake experiments, there was concern that methanol and toluene elimination capacities might be reduced as the loading increased. This is clearly not the case as shown in figure 5 and 6. A possible explanation is the rapid sink and slight mass transfer limitation, which locally reduced the concentration of the pollutant undergoing treatment in the microbioreactors.

3.4. Evaluation of the Contribution of the Bed and Ancillary Liquid Recycle to the Overall Elimination Capacity

Additional experiments were conducted to determine the extent of pollutant removal in the liquid reservoir/liquid recycle tank of the microbioreactors. The results of these experiments are presented in table 5 and show that for the toluene microbioreactors, the vast majority of the VOC that was removed, was indeed degraded in the bed or in the capillaries. This confirms that the outstanding performance of the prototype microbioreactors is due to the configuration of the miniaturized beds and not to the biodegradation capacity occurring in the ancillary vessels in the system. On the other hand, for methanol microbioreactors, the contribution of the ancillary vessels ranges between 15 and 25%, which is slightly higher than for the toluene microbioreactors. As it was above explained, methanol is a hydrophilic compound while toluene is more hydrophobic, and therefore it methanol is readily adsorbed in the free liquid. Consequently, the fraction of VOC removed in the ancillary vessel is higher than in those observed in toluene microbioreactors.

Table 5: Contribution of the bed and ancillary liquid recycle vessel to the overall elimination capacity for toluene and methanol degrading microbioreactors

VOC	Micro reactor	EC total (g/m^3 h)	EC liquid phase (g/m^3 h)	EC gas phase (g/m^3 h)	% Liquid phase	% Gas phase
Toluene	**Glass beads**	141	5.50	135.5	3.9	96.1
Toluene	**Capillaries**	1966	29	1937	1.6	98.4
Toluene	**PU Foam**	91	10	81	11	89
Methanol	**Glass beads**	466	71	395	15	85
Methanol	**Capillaries**	11877	2801	9076	24	76
Methanol	**PU Foam**	613	151	462	25	75

4. CONCLUSIONS

Novel microbioreactors for the removal of high loads of volatile organic compounds (VOCs) were successfully designed and operated. Results obtained here confirm that process intensification of VOC occurs when reducing the size of bioreactors. Miniature biotrickling filters and microbioreactors were 2-5 times more effective than conventional biotrickling filters, and our glass capillary microbioreactor had volumetric rates 40-80 times greater than conventional bioreactors. Mass transfer coefficients in the microbioreactors were quantified and were found to be particularly high in the glass capillary microbioreactors, providing a plausible explanation to the high VOC degradation rates that were observed. Overall, these results show that miniature bioreactors hold great promise for the treatment of volatile organics.

5. REFERENCES

Berčič, G., Pintar, A., 1997. The role of gas bubbles and liquid slug lengths on mass transport in the Taylor flow through capillaries. Chem. Eng. Sci. 52, 3709–3719.

Boor, B.E., Spilak, M.P., Laverge, J., Novoselac, A., Xu, Y., 2017. Human exposure to indoor air pollutants in sleep microenvironments: A literature review. Build. Environ. 125, 528–555.

Cox, H.H., Deshusses, M.A., 1998. Biological waste air treatment in biotrickling filters. Curr. Opin. Biotechnol. 9, 256–262.

Cox, H.H.J., Deshusses, M.A., 2000. Waste Gas Treatment in Biotrickling Filters 256–262.

Debono, O., Hequet, V., Le Coq, L., Locoge, N., Thevenet, F., 2017. VOC ternary mixture effect on ppb level photocatalytic oxidation: Removal kinetic, reaction intermediates and mineralization. Appl. Catal. B Environ. 218, 359–369.

Debono, O., Thévenet, F., Gravejat, P., Héquet, V., Raillard, C., Le Coq, L., Locoge, N., 2013. Gas phase photocatalytic oxidation of decane at ppb levels: Removal kinetics, reaction intermediates and carbon mass balance. J. Photochem. Photobiol. A Chem. 258, 17–29.

Deshusses, M.A., 1997. Biological waste air treatment in biofilters. Curr. Opin. Biotechnol. Opin. Biotechnol. 8, 335–339.

Dorado, A.D., Baeza, J.A., Lafuente, J., Gabriel, D., Gamisans, X., 2012. Biomass accumulation in a biofilter treating toluene at high loads - Part 1: Experimental performance from inoculation to clogging. Chem. Eng. J. 209, 661–669.

Einaga, H., Mochiduki, K., Teraoka, Y., 2013. Photocatalytic Oxidation Processes for Toluene Oxidation over TiO2 Catalysts. Catalysts 3, 219–231.

EPA, 2011. Background indoor air concentrations of volatile organic compounds in North American residences (1990-2005): a compilation of statistics for assessing vapor intrusion.

Fiedler, N., Laumbach, R., Kelly-McNeil, K., Lioy, P., Fan, Z.H., Zhang, J., Ottenweller, J., Ohman-Strickland, P., Kipen, H., 2005. Health effects of a mixture of indoor air volatile organics, their ozone oxidation products, and stress. Environ. Health Perspect. 113, 1542–1548.

Fortin, N.Y., Deshusses, M.A., 1999. Treatment of methyl tert-butyl ether vapors in biotrickling filters. 1. Reactor startup, steady-state performance, and culture characteristics. Environ. Sci. Technol. 33, 2980–2986.

Gabriel, D., Deshusses, M. a, 2003. Retrofitting existing chemical scrubbers to biotrickling filters for H2S emission control. Proc. Natl. Acad. Sci. U. S. A. 100, 6308–6312.

Guieysse, B., Hort, C., Platel, V., Munoz, R., Ondarts, M., Revah, S., 2008. Biological treatment of indoor air for VOC removal: Potential and challenges. Biotechnol. Adv. 26, 398–410.

Huang, H., Lu, H., Huang, H., Wang, L., Zhang, J., Leung, D.Y.C., 2016. Recent Development of VUV-Based Processes for Air Pollutant Degradation. Front. Environ. Sci. 4, 1–13.

James, J.T., Limero, T.F., Cheng, P.F., Vera, V.J. De, Hand, J., Macatangay, A., 2010. International Space Station Air Quality Assessed According to Toxicologically-Grouped Compounds, in: 40th International Conference on Environmental Systems.

Kashid, M.N., Agar, D.W., 2007. Hydrodynamics of liquid-liquid slug flow capillary microreactor: Flow regimes, slug size and pressure drop. Chem. Eng. J. 131, 1–13.

Kennes, C., Rene, E.R., Veiga, M.C., 2009. Bioprocesses for air pollution control. J. Chem. Technol. Biotechnol. 84, 1419–1436.

Kim, S., Deshusses, M.A., 2008. Determination of mass transfer coefficients for packing materials used in biofilters and biotrickling filters for air pollution control. 1. Experimental results. Chem. Eng. Sci. 63, 841–855.

Littlejohns, J. V., Daugulis, A.J., 2008. Kinetics and interactions of BTEX compounds during degradation by a bacterial consortium. Process Biochem. 43, 1068–1076.

López, L.R., Bezerra, T., Mora, M., Lafuente, J., Gabriel, D., 2016. Influence of trickling liquid velocity and flow pattern in the improvement of oxygen transport in aerobic biotrickling filters for biogas desulfurization. J. Chem. Technol. Biotechnol. 91, 1031–1039.

Mitchell, C.S., Zhang, J., Sigsgaard, T., Jantunen, M., Lioy, P.J., Samson, R., Karol, M.H., 2007. Current state of the science: Health effects and indoor environmental quality. Environ. Health Perspect. 115, 958–964.

Montebello, A.M., Mora, M., López, L.R., Bezerra, T., Gamisans, X., Lafuente, J., Baeza, M., Gabriel, D., 2014. Aerobic desulfurization of biogas by acidic biotrickling filtration in a randomly packed reactor. J. Hazard. Mater. 280, 200–208.

Mora, M., Fernández, M., Gómez, J.M., Cantero, D., Lafuente, J., Gamisans, X., Gabriel, D., 2014. Kinetic and stoichiometric characterization of anoxic sulfide oxidation by SO-NR mixed cultures from anoxic biotrickling filters. Appl. Microbiol. Biotechnol. 99, 77–87.

Nazaroff, W.W., Goldstein, A.H., 2015. Indoor chemistry: Research opportunities and challenges. Indoor Air 25, 357–361.

Reardon, K.F., Mosteller, D.C., Rogers, J.B., DuTeau, N.M., Kim, K.H., 2002. Biodegradation kinetics of aromatic hydrocarbon mixtures by pure and mixed bacterial cultures. Environ. Health Perspect. 110, 1005–1011.

Reiss, R., Barry Ryan, P.B., Tibbetts, S.J., Koutrakis, P., 1995. Measurement of organic acids, aldehydes, and ketones in residential environments and their relation to ozone. J. Air Waste Manag. Assoc. 45, 811–822.

Rene, E.R., Sergienko, N., Goswami, T., López, M.E., Kumar, G., Saratale, G.D., Venkatachalam, P., Pakshirajan, K., Swaminathan, T., 2017. Effects of concentration and gas flow rate on the removal of gas-phase toluene and xylene mixture in a compost biofilter. Bioresour. Technol.

Rudel, R.A., Perovich, L.J., 2009. Endocrine disrupting chemicals in indoor and outdoor air. Atmos. Environ. 43, 170–181.

Rumchev, K., 2004. Association of domestic exposure to volatile organic compounds with asthma in young children. Thorax 59, 746–751.

Samet, J., 1988. Environmental Controls and Lung Disease. Am. Thorac. Soc. 915–939.

San-Valero, P., Dorado, A.D., Martínez-Soria, V., Gabaldón, C., 2017. Biotrickling filter modeling for styrene abatement. Part 1: Model development, calibration and validation on an industrial scale. Chemosphere.

Severs, Y.D., 2006. A baseline air quality assessment onboard a Victoria class submarine: HMCS Windsor.

Shareefdeen, Z., Baltzis, B.C., Oh, Y. -S, Bartha, R., 1993. Biofiltration of methanol vapor. Biotechnol. Bioeng. 41, 512–524.

Sidheswaran, M.A., Destaillats, H., Sullivan, D.P., Cohn, S., Fisk, W.J., 2012. Energy efficient indoor VOC air cleaning with activated carbon fiber (ACF) filters. Build. Environ. 47, 357–367.

Spanjers, H., Vanrolleghem, P., 1995. Respirometry as a tool for rapid characterization of wastewater and activated sludge. Water Sci. Technol.

van Baten, J.M., Krishna, R., 2004. CFD simulations of mass transfer from Taylor bubbles rising in circular capillaries. Chem. Eng. Sci. 59, 2535–2545.

Waters, M., Bloom, T., Grajewski, B., Deddens, J., 2002. Measurements of Indoor Air Quality on Commercial Transport Aircraft. Indoor Air 782–787.

Wells, J.R., Schoemaecker, C., Carslaw, N., Waring, M.S., Ham, J.E., Nelissen, I., Wolkoff, P., 2017. Reactive indoor air chemistry and health—A workshop summary. Int. J. Hyg. Environ. Health.

Wu, F., Jacobs, D., Mitchell, C., Miller, D., Karol, M.H., 2007. Improving indoor environmental quality for public health: Impediments and policy recommendations. Environ. Health Perspect. 115, 953–957.

Yao, X., Zhang, Y., Du, L., Liu, J., Yao, J., 2015. Review of the applications of microreactors. Renew. Sustain. Energy Rev. 47, 519–539.

Yue, J., Chen, G., Yuan, Q., Luo, L., Gonthier, Y., 2007. Hydrodynamics and mass transfer characteristics in gas-liquid flow through a rectangular microchannel. Chem. Eng. Sci. 62, 2096–2108.

ACTIVATED CARBON AS A SUPPORT IN BIOTRICKLING FILTERS FOR THE REMOVAL OF SILOXANES IN SEWAGE SLUDGE BIOGAS

Eric Santos-Clotas, Alba Cabrera-Codony, Maria J. Marín*

LEQUIA, Institute of the Environment. University of Girona, Campus Montilivi, Girona, Catalonia, Spain

ABSTRACT

Biogas generated during anaerobic digestion in wastewater treatment plants and landfills is considered as a resource of renewable energy, and its use assists initiatives to reduce global warming. In order to be used as high qualify fuel upgrading processes are needed, such as siloxanes removal. They might be found at low concentrations yet they possess abrasive properties during combustion in engines, thereby decreasing the efficiency of energy recovery systems. The most widespread technology commercially available for siloxanes removal is adsorption onto porous materials, though their regeneration or replacement is energy-consuming and expensive. On the contrary, biotechnologies may reduce investment and operating costs, increment treatment capacities and require low energy and chemicals. Siloxanes, though, challenge biotechnologies due to their low solubility into water, which hinders their gas-liquid mass transfer.

Within this work, several batch studies were carried out suspending different activated carbons saturated with siloxanes in mineral medium. Siloxane mass transfer was investigated into ACs in presence of water, to select the most suitable AC for a biotrickling filter, being chemically activated carbons more suitable due to higher catalytic activity. A biotrickling filter was set up with and without the selected activated carbon in the packing bed, and were inoculated with enriched bacterial cultures isolated from sewage sludge from a wastewater treatment plant. A synthetic gas with siloxanes was treated counter currently to the mineral medium recirculation. The carbon proved to help the performance of the reactor increasing the removal efficiency, though its effect did not last long enough.

KEYWORDS

Biogas upgrading, Siloxanes, Activated carbon, Biotechnology

INTRODUCTION

The biogas generated during anaerobic digestion in wastewater treatment plants (WWTPs) and landfills is renewable energy source, since it can be used as high quality fuel. However, upgrading processes are needed, such as siloxane removal. Siloxanes are usually found at low concentrations yet during combustion of biogas they are converted into silica, which causes severe damage to the energy recovery systems (Soreanu et al., 2011).

Adsorption into activated carbon (AC) is the most widespread technology to remove siloxanes from biogas, although AC replacement is expensive. Biological technologies may reduce investment and operating costs, increment treatment capacities and require low energy. The biological degradation of pollutants is a well-known process in aqueous media. However, there is a lack of literature on biodegradation of low water-soluble pollutants from gas effluents (e.g. siloxanes). A few articles are found studying siloxanes biodegradation in biotrickling filters (BTF), most of them achieving rather low removal efficiencies even at large empty bed contact times (EBCT) (Popat and Deshusses, 2008). Transferring these gaseous apolar compounds to the liquid phase is of paramount importance since it challenges their biological treatment.

AC addition to a BTF might have an important role as not only a biomass support, but also enhancing the availability of the pollutants to the biofilm. Some ACs have shown catalytic activity towards the ring opening of cyclic siloxanes (A Cabrera-Codony et al., 2015), leading to the formation of silanediols, which are water-soluble compounds. Therefore, the end goal was to enhance the performance of BTF to remove siloxanes biologically by adding an appropriate AC that catalyzes the ring-opening of siloxanes to transform them into more bioaccessible compounds.

The aim of the present work was to investigate the elimination of siloxanes from biogas by means of a biotrickling filter. Since low mass transfer of siloxanes has already been reported, this research also aimed at studying the combination of the biotechnology and adsorption into activated carbon in order to improve the transfer of siloxanes from the gas stream to the liquid phase and so it could be biodegraded. For this purpose, it was also important to select an appropriate AC, so investigating their capacity of capturing siloxanes and their catalytic activity was another objective of this paper.

METHODOLOGY

Materials Physical and Chemical Characterization

A set of four different commercial activated carbons was used for studying their capacity to adsorb and transform siloxanes. For this purpose, the materials were characterized in term of physical and chemical properties.

Textural characterization

The surface areas and pore volumes of all ACs were calculated from the N_2 and CO_2 adsorption−desorption isotherms, performed at −196 °C in a Micromeritics ASAP 2420 volumetric adsorption system and at 0 °C in a Quantachrome Nova Station A, respectively. Prior to measurement, samples were outgassed overnight by heating at 523 K under vacuum. Specific surface areas (S_{BET}) were calculated using the Brunauer−Emmett−Teller (BET) method, taking 16.2 nm^2 for the cross-sectional area of the nitrogen adsorbed molecule. Total micropore volumes (VDR_{N2}) were assessed by applying the Dubinin–Radushkevich (DR) equation to the suitable adsorption data. Total pore volumes (V_t) were determined by the amount of N_2 adsorbed at $p/p0 = 0.99$. Mesopore volumes (V_{meso}) were calculated by subtracting the VDR_{N2} micropore volumes from the total pore volumes. Narrow (<0.7 nm) micropore volumes (VDR_{CO2}) were obtained from the CO_2 isotherms using the DR method.

Thermal-Programmed Desorption (TPD)

AC samples were heated at 10 °C min^{-1} under a He flow of 50 mL min^{-1} in the quartz reactor of an Autochem II apparatus (Micromeritics). Desorbed gases were analyzed with an Omnistar (Pfeiffer Vacuum) mass spectrometer. TPD curves were obtained by plotting desorbed CO and CO_2 (μmol g^{-1} s^{-1}) versus the temperature (°C). TPD curves were deconvoluted using the OriginPro software, following the criteria reported in previous studies (Figueiredo et al., 1999).

pH Slurry

A total of 0.50 g of AC were suspended in 25 mL of Milli-Q water and left stirring for 24 h. The pH was then measured with a Crison micro pH 2000 pHmeter.

Batch Experiments

Adsorption onto ACs is the most widespread technology to abate siloxanes, although usually requires a pretreatment step in order to remove all the humidity since it can notably decrease the adsorption capacity (Cabrera-Codony et al., 2014). As in a biotrickling filter mineral medium is being continuously recirculated, dry conditions are not expected. For this purpose, batch experiments aim to investigate the capacity of adsorbing siloxanes even though the carbon is covered in water, as it would be in a reactor.

The batch test consisted in placing 1 g of each carbon in headspace bottles and proper sealed with septum. Humidity was provided by the addition of 3 mL of mineral medium. The target siloxane in this work was octamethylcyclotetrasiloxane (D4, 98%, Sigma Aldrich), pure liquid siloxane was injected through the septum of the bottle to be adsorbed into the AC. The bottles were left for 96h in an orbital mixer at 150 rpm for adsorption to happen. Blank tests were prepared without activated carbon.

At the end of the experiment, the activated carbons were analyzed to determine the quantity of siloxane adsorbed and the formation of by-products. The analytical procedure followed for detecting these compounds was the one indicated by Varaprath and Lehmann (Varaprath and Lehmann, 1997). Extractions with anhydrous tetrahydrofuran (THF, 99%, Sigma Aldrich) were carried out by injecting 20 mL of solvent through the septum, vortexed for 2 min and shacken in an orbital mixer for 30 min. Solvent suspensions were centrifuged at 6000 rpm for 5 min (EBA 21, Hettich Zentrifugen) and later analyzed in a gas chromatograph coupled to a mass spectrometry detector (5977E series GC-MSD system, Agilent Technologies) equipped with a capillary column HP-5ms Ultra Inert (Agilent Technologies).

Biotrickling Filter Setup

A lab-scale biotrickling filter was designed and constructed for this study. The BTF consisted in a column made up in transparent Plexiglas with an inner diameter of 6 cm and a total height of 46 cm. The packing bed initially consisted in lava rock (8-12 mm), and later a small layer of activated carbon was added at the top of the lava rock; the height of the packing material occupied 33 cm of the reactor bed.

A synthetic gas stream of air containing 50 mg m^{-3} of D4 was produced by infusing liquid D4 by a syringe pump (Harvard Apparatus) to an air current regulated by a mass flow controller (Alicat Scientific). Gas samples at the inlet and outlet of the reactor were taken with Tedlar® sampling bags and measured by a GC/MS equipped with a gas valve.

The BTF was first inoculated with 300 mL consisting of enriched siloxanes-degrading cultures diluted in mineral medium (MM) isolated from sewage sludge from a wastewater treatment plant. MM was continuously recirculated using a peristaltic pump (Watson Marlow) and sprayed to the packing bed from the top of the reactor.

Scanning Electron Microscope

In order to prove that biomass was growing attached to the lava rock in the bed of the column; some samples were taken after a period of time for observing it with Scanning Electron Microscope (SEM).

Examinations of the samples were carried out with a scanning electron microscope (S-4100 FE-SEM Hitachi, Japan). Digital images were collected and processed by Quartz PCI program.

RESULTS AND DISCUSSION

Adsorbents Characterization

For exploring the potential benefits of using activated carbons in a BTF, four different AC were evaluated. Their chemical and textural properties are summarized in Table 1.

AC-1 and AC-2 ACs are chemically activated carbons with phosphoric acid, while AC-3 and AC-4 are steam activated. As expected, the pH of the slurry was in good agreement with these facts, showing acid pH for chemically activated carbons, and basic pH for steam-ACs. High acidic pH are often encountered in chemically activated carbons due to the activating agent remaining inside the nanopores (Puziy and Tascón, 2012).

The amounts of CO and CO_2 released from the carbons obtained by the TPD profiles were investigated. Steam activated carbons released low amount of CO and CO_2 indicating a low surface functionalization. On the other hand, phosphoric acid-AC presented larger pore volume (Vt) and higher textural development than steam-AC. At the same time, they displayed higher amounts of oxygen functional groups.

Batch Tests with Activated Carbons

The results of the tetrahydrofuran suspensions from each headspace bottle are showed in Figure 1, where dark grey bars correspond to the percentage of D4 recovered (i.e. not transformed) after 96 h of experiment in contrast to the unrecovered fraction in light grey bars (i.e. D4 by-products), which was not extracted as D4 siloxane.

Steam-AC displayed higher D4 recoveries, AC-3 having the highest percentage, indicating that a high quantity of the D4 initially injected remained adsorbed in the carbon. In the contrary, phosphoric activated carbons, AC-1 and AC-2, displayed really low D4 recoveries.

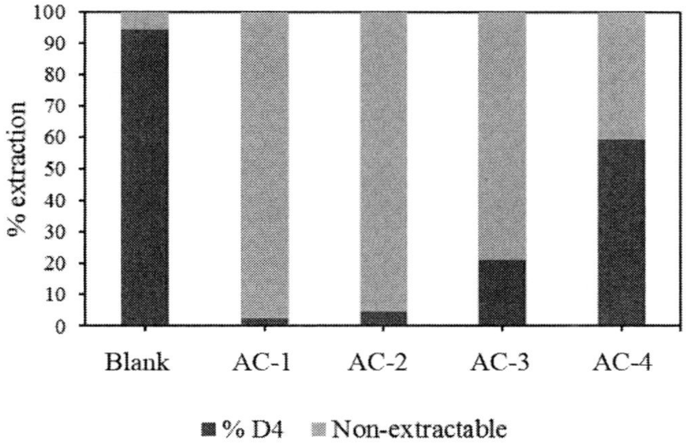

Figure 1. D4 extracted (%) from batch tests with different AC.

Within the resulting extractions with THF several transformation products were found. High molecular weight cyclic siloxanes such as decamethylcyclopentasiloxane (D5) and dodecamethylcyclohexasiloxane (D6) and also α-ω-silanediols. These by-products were identified by means of mass spectrometry and are shown in an example chromatogram obtained (Figure 2). Other authors have reported the catalytic activity of AC towards the ring-opening of cyclic siloxanes and the formation of silanediols, which are more soluble compounds (Alba Cabrera-Codony et al., 2015). Representing these silanediols, in Figure 3, three of them were detected; dimethylsilanediol, tetramethyl-1,3-disiloxanediol, and hexamethyl-1,5-trisiloxanediol.

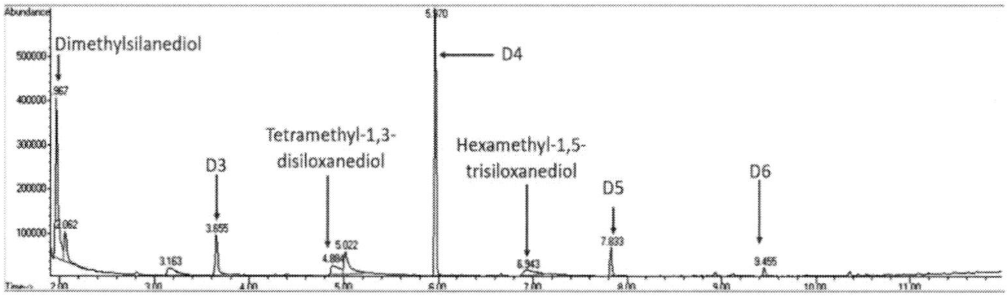

Figure 2. Chromatogram identifying the by-products of D4 detected in THF extractions.

Since some of these compounds are not commercially available they could not be quantified. However, in order to compare their presence, GC/MS area units were used in Figure 3 for obtaining a qualitative idea on the amount of by-products for each carbon. Phosphoric AC, which showed lower D4 recoveries, demonstrated a higher quantity of silanediols, especially dimethylsilanediol. This confirmed that chemically activated carbons presented higher catalytic activity than steam-AC, demonstrating that most of the D4 initially injected was being transformed into other by-products.

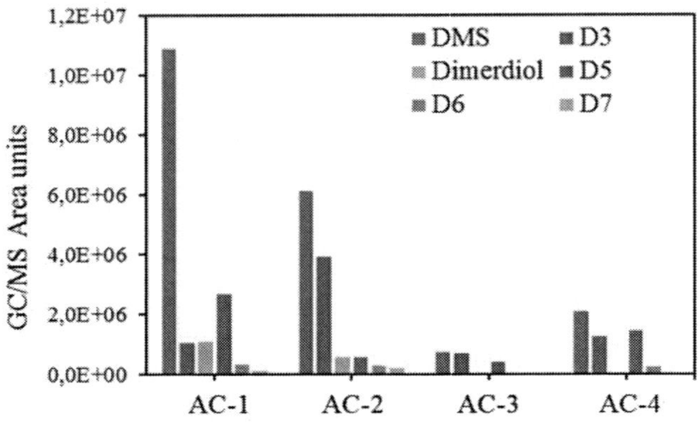

Figure 3. By-products identified from D4 transformation.

At the light of the results concerning carbon characterization and batch experiments, chemically AC demonstrated to have a higher surface functionalization as well as a higher catalytic activity towards the ring-opening of the cyclic siloxane D4. Siloxanes being transformed into more soluble compounds (e.g. silanediols) would increase the mass transfer of siloxanes, initially with a hydrophobic character, from the gas current to the liquid phase. Thus, they might be a potential solution for biotechnologies to overcome mass transfer issues.

Reactor Performance

After the inoculation with the biologic culture, the BTF was operated during a month for the acclimation of the microbiological community within the packing bed of the BTF. From day 40 to 50 the BTF was operated in an EBCT of 10 min and an gas flow of 100 mL/min, achieving an average D4 removal of 11%. Liquid-to-gas flow ratio was thought to play an important role in the performance of other BTFs in the literature. Thus, liquid flow was decreased at day 43, and the removal efficiency (RE) slightly increased. These results follow the trend pointed out by other reports on how a lower irrigation system maintains adequate moisture content in the packing material providing better environmental conditions for the microorganisms (Sakuma et al., 2009).

Apart from lowering the MM irrigation, within the stage III the EBCT was increased, which might have helped improving the performance of the reactor. These results are in good agreement with others previously reported of 40% RE at 20 min EBCT (Popat and Deshusses, 2008).

Finally, activated carbon (chemically AC) was added at the top of the lava rock and its performance was monitored for 15 days. RE slowly increased at day 36, but it decreased back some days. The carbon height in the column was a few cm, so the gas residence through the carbon material was so low.

CONCLUSIONS

D4 from the gas stream was partially removed biologically in a BTF inoculated with an enriched culture. Siloxanes were partially removed after one month acclimation period. The irrigation flow showed to benefit the system when it was decreased, offering more favorable conditions to the microorganisms in the reactor.

In order to enhance D4 biological removal in the BTF, increasing the transference from the gas to the water phase is mandatory. Phosphoric acid-AC, with high amounts of acidic surface sites, have been proved to catalyze D4 hydrolysis reactions, leading to water-soluble silanediols.

REFERENCES

Cabrera-Codony, A., Gonzalez-Olmos, R., Martín, M.J., 2015. Regeneration of siloxane-exhausted activated carbon by advanced oxidation processes. J. Hazard. Mater. https://doi.org/10.1016/j.jhazmat.2014.11.053

Cabrera-Codony, A., Gonzalez-Olmos, R., Martín, M.J., 2015. Regeneration of siloxane-exhausted activated carbon by advanced oxidation processes. J. Hazard. Mater. 285, 501–8. https://doi.org/10.1016/j.jhazmat.2014.11.053

Cabrera-Codony, A., Montes-Morán, M.A., Sánchez-Polo, M., Martín, M.J., Gonzalez-Olmos, R., 2014. Biogas upgrading: Optimal activated carbon properties for siloxane removal. Environ. Sci. Technol. 48. https://doi.org/10.1021/es501274a

Figueiredo, J.., Pereira, M.F.., Freitas, M.M.., Órfão, J.J.., 1999. Modification of the surface chemistry of activated carbons. Carbon N. Y. 37, 1379–1389. https://doi.org/10.1016/S0008-6223(98)00333-9

Popat, S.C., Deshusses, M. a., 2008. Biological removal of siloxanes from landfill and digester gases: Opportunities and challenges. Environ. Sci. Technol. 42, 8510–8515. https://doi.org/10.1021/es801320w

Puziy, A.M., Tascón, J.M.D., 2012. Adsorption by Phosphorus-Containing Carbons, in: Novel Carbon Adsorbents. pp. 245–267. https://doi.org/10.1016/B978-0-08-097744-7.00008-9

Sakuma, T., Hattori, T., Deshusses, M.A., 2009. The effects of a lower irrigation system on pollutant removal and on the microflora of a biofilter. Environ. Technol. 30, 621–627. https://doi.org/10.1080/09593330902838106

Soreanu, G., Béland, M., Falletta, P., Edmonson, K., Svoboda, L., Al-Jamal, M., Seto, P., 2011. Approaches concerning siloxane removal from biogas - A review. Can. Biosyst. Eng. / Le Genie des Biosyst. au Canada 53.

Varaprath, S., Lehmann, R.G., 1997. Speciation and Quantitation of Degradation Products of Silicones (Silane/Siloxane Diols) by Gas Chromatography-Mass Spectrometry and Stability of Dimethylsilanediol. J. Environ. Polym. Degrad. 5.

Trickling Filters Foul Air Treatment with Chemical Scrubbers and Removal Efficiencies for Methyl Mercapatan and Dimethyl Sulfide

May Kyi, P.E. and Richard Birdsell, P.E.
Orange County Sanitation District, Fountain Valley, California, United States of America

ABSTRACT

In 2014, I published an article discussing the trickling filters (TF) foul air characteristics and different activated carbon media and their efficiencies to treat the TF process foul airs at the Orange County Sanitation District (OCSD) wastewater treatment plant in Huntington Beach, California. It was also presented at the WEF Odors and Air Pollutant Conference in 2014. The article stated that activated carbon media was not able to remove dimethyl sulfide, dimethyl disulfide, and carbon disulfide (or some Volatile sulfur compound (VSCs) or reduced sulfur compounds - RSCs) from the highly saturated foul airs that are generated from the secondary treatment process. Furthermore, frequent carbon media replacement has become an economic burden for OCSD. Therefore, OCSD has installed an intermediate treatment system (chemical scrubbers) prior to the carbon scrubbers to treat the foul airs. The new odor control system is arranged such that the foul air can be discharged with or without the carbon media treatment. The chemical scrubber technology is selected based on the space availability, costs to operate and maintain the odor control system, OCSD's past experience with chemical scrubbers, and treatment efficiency for the foul air inlet concentrations at the secondary TF process. In general, chemical scrubbers are primarily designed to treat hydrogen sulfide, however, the project team considered possibilities of chemical scrubbers treating other odors that might have impacted the neighbors.

This paper will provide odor analysis results from the chemical scrubbers and their treatment efficiencies. The data will consistently show that chemicals scrubbers, with pH and ORP at the OCSD normal operating control levels, can effectively remove hydrogen sulfide (H2S), methyl mercaptan (MM), and dimethyl sulfides (DMS), but are inefficient in removing carbonyl sulfide (COS) and carbon disulfide (CDS).

KEYWORDS

Activated Carbon, Scrubber, OCSD, Trickling Filters, Wastewater Treatment, odor control, Hydrogen Sulfides and Methyl Mercaptan removal efficiencies, Chemical scrubbers

INTRODUCTION

The secondary treatment process (TF/SC process) in this article is located at OCSD's Plant No. 2, Huntington Beach, California; and is treating primary effluent with a combined daily average flow of 60 million gallons per day (mgd). The TF/SC process is also situated near the residential area (Figure 1) and the carbon scrubbers (A, B and C) were initially designed to treat the foul air exhausted from the TF process. However, the carbon medias were generating garlic-like odors at the discharge (outlet) stack. Therefore, three chemical scrubbers have been recently installed at three TFs facilities treating 11,000 cubic feet per minute (cfm) of air from each TF, totaling 33,000 cfm of foul air. This odor control system has been designed such that the treated air from the newly installed chemical scrubbers can be polished with the carbon vessels that were previously intended to treat the TF exhaust air or can discharge directly to the atmosphere, meaning operating the system with no carbon media in the carbon vessel (Figure 2). During the performance testing period, the air samples were taken at the inlet and outlet of chemical scrubbers and analyzed by the independent laboratory for specific RSCs. This paper has included all RSCs, whose concentrations are above Method Reporting Limit (MRL) or detectable by the referenced method. Other RSCs that are not mentioned in this paper are not detectable by the laboratory analysis.

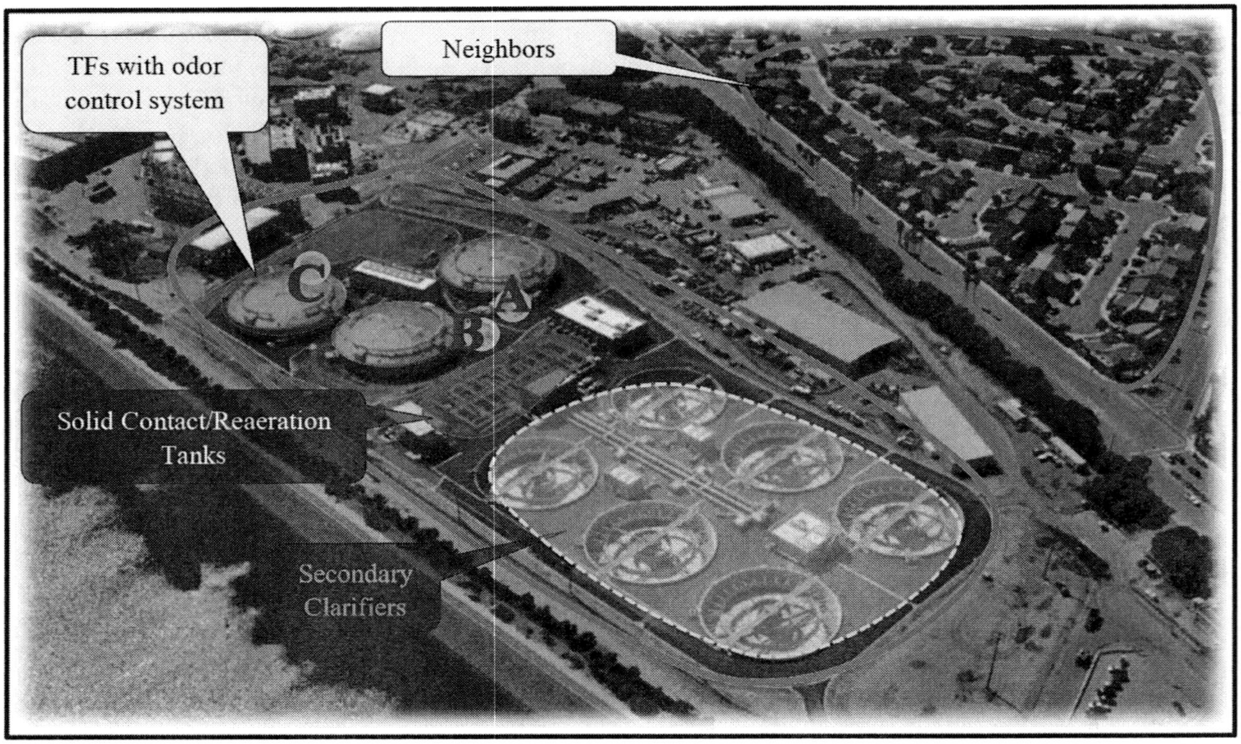

Figure 1: Trickling Filter Process Area and its Proximity to the Neighbors

DESIGN MATRIX AND CRITERIA

Finding the best odor control technology that fits the available space, while ensuring the new system to perform/work well with the existing system, operating the existing system during construction and minimizing the impact to plant operations, and achieving the desired efficiency, is the most challenging process for this project. First, the project team developed the primary objectives of the project and justification for selecting the chemical scrubber technology.

The objective is to mitigate the garlic-like odor, suspected to be DMS or dimethyl disulfide – (DMDS), created when the highly saturated foul air comes in contact with the carbon media. The laboratory results confirmed the higher DMDS concentration at the outlet of the carbon vessel, while the inlet concentration is non-detect. Some scholarly researchers stated that H2S can be methylated to form DMS, and sequentially to form DMDS (Higgins et al., 2002); hence, carbon manufacturers believed that the methylation process might have occurred at the OCSD carbon scrubber odor control system. Due to the DMDS breakthrough, the carbon media replacement frequency has become as often as every few months, which became an economic burden to OCSD. The annual cost for carbon media and replacement is approximately $700,000. In order to mitigate nuisance odor and reduce the operating cost, the project team began exploring different odor control alternatives to treat the foul air generated by the secondary process (TF).

First, the project team established the design (odor control technology) alternatives based on the South Coast Air Quality Management District (SCAQMD) requirements, removal of nuisance odors in addition to H2S, project duration, space requirement, maintenance requirement, cost effectiveness, and operational flexibility. The odor control technologies evaluated were biotrickling filters, biofilter, and chemical scrubbers (both low profile and vertical towers). However, biofilters and biotrickling filters were eliminated based on the required long empty bed residence time (EBRT) to achieve the efficiency, which will need a larger footprint. One scholarly article stated that typical EBRTs may range from 50 sec to 70 sec for biofilters and 12 sec to 16 sec for biotrickling filters to effectively remove the target compounds (Estrada et al., 2011). The space available for the future odor control at the existing facility is limited as it needs to fit on the available concrete pad (Figure 2). Another concern is – providing sufficient H2S to the system for the biological bacteria to thrive due to the low inlet H2S (0.5 to 5 ppm) from the secondary (TF) process, which does not have high sulfide contents.

Figure 2: Chemical Scrubber (new in 2017) and Carbon scrubber (Existing) Train

Therefore, the project team decided to install chemical scrubbers for the foul air treatment. The vertical chemical scrubbers not only fit the available space dedicated for the future odor control system, they provide some operational flexibilities to achieve the required treatment efficiency by controlling pH and oxidation reduction potentials (ORP) values. Table 1 shows the chemical scrubber design parameters used for the new system.

The foul air from each TF process unit is treated through the odor control system that consists of a demister (new), a chemical scrubber (new), another demister (existing), and a carbon scrubber (existing and optional) prior to the discharge stack (See Figure 2a). The SCAQMD permit allows for the carbon scrubber to be operated without the carbon media if the chemical scrubber can meet the required permit limit, which is less than 1.0 ppmv at the discharge stack to the atmosphere.

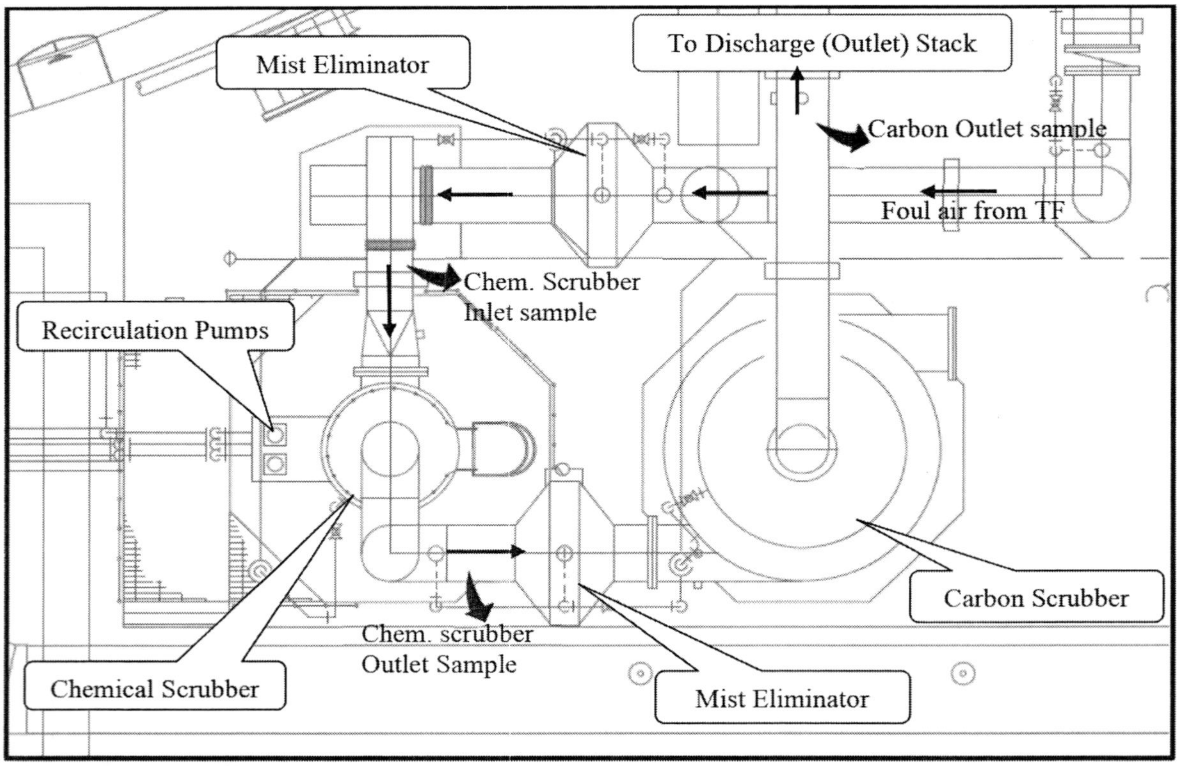

Figure 2a: Plan View for Chemical and Carbon Scrubbers in series

Table 1: Chemical Scrubber Design Parameters

Parameter	Value
Daily average inlet H2S concentration	1.0 ppm_v
Peak inlet H2S concentration	7.0 ppm_v
Target removal of H2S	99 % removal of inlet concentration, or < 0.1
Number of scrubbers	3
Scrubber diameter, feet	6, minimum
Scrubber packing height, feet	8, minimum
Maximum air flow, cfm (per scrubber)	11,000
Number of recirculation pumps (per scrubber)	2 (1 duty+1 standby)
Exhaust duct diameter, inch (use existing)	30
Maximum scrubber height, feet	23
Liquid Loading Rate	7.1 gallons per minute / square foot
Gas Loading Rate	1,754 pounds per square foot / hour
Gas Velocity	389 feet per minute

RESULTS

The new chemical scrubbers were commissioned in May 2017. This section will provide and discuss the results from scrubber performance testing plus other additional sample results. The air samples were collected at the inlets and outlets of the recently installed chemical scrubbers and outlets of the carbon scrubbers (Figure 2a). They were sent to an independent laboratory for reduced sulfur compound (ASTM D 5504-08), and St. Croix Sensory for detention threshold (D/T) and recognition threshold (ASTM E679 & EN13725) from the chemical scrubber outlets. However, OCSD has not determined the fence line D/T values based on the D/T at the discharge stack with some air modeling analysis. Therefore, this paper does not focus or analyze the D/T values.

The odor control systems for three separate TFs units are labelled as chemical (or) carbon scrubbers "A", "B", and "C". All three chemical scrubbers pH and ORPs were monitored and kept between 9 and 9.5 for pH and 600 to 750 mVs for ORP per manufacturer's design recommendations. Table 2 represents the data that were collected prior to the official performance testing period in a tabular form. Some of the non-detect (ND) constituents have been left out for clarity, meaning only five (5) RSCs in the table have been consistently detected by the lab under the standard test method.

Table 2: Sample Results (RSCs) prior to Performance Acceptance Test

Date: 6/19/2017	**Chemical Scrubber A Time: Between 11:00 am and 1:00 pm (Unit: ppbv)**					
	Sample 1		**Sample 2**		**Sample 3**	
Constituents	**Inlet 1**	**Outlet 1**	**Inlet 2**	**Outlet 2**	**Inlet 3**	**Outlet 3**
Hydrogen Sulfide	81	ND	130	ND	86	ND
Carbonyl Sulfide	22	15	20	19	21	18
Methyl Mercaptan	34	5.7	24	ND	32	ND
Dimethyl Sulfide	28	ND	22	ND	23	ND
Carbon Disulfide	14	8.7	13	12	13	12
Date: 6/21/2017	**Chemical Scrubber C Time: Between 10:00 am and 12:00 pm (Unit: ppbv)**					
	Sample 1		**Sample 2**		**Sample 3**	
Constituents	**Inlet 1**	**Outlet 1**	**Inlet 2**	**Outlet 2**	**Inlet 3**	**Outlet 3**
Hydrogen Sulfide	170	ND	230	ND	240	5.6
Carbonyl Sulfide	38	33	40	41	34	32
Methyl Mercaptan	14	ND	14	ND	16	ND
Dimethyl Sulfide	24	ND	25	ND	31	ND
Carbon Disulfide	16	14	15	16	18	18

On 6/21/2017 at the chemical scrubber "C", the carbon scrubber outlets were also sampled and the sample results consistently showed that carbon media was not able to remove both Carbonyl Sulfides and Carbon Disulfides as their inlet values are almost the same as the outlet values. The results are discussed later in this paper.

The data shown in the table below (Table 3) are representative of the performance acceptance tests that were conducted in 6/29/2017. Again, pH and ORPs ranges are kept similar for all the sampling events and all the samples were collected in the morning for uniformity. The air samples from Scrubber "B" were collected a couple of weeks later than Scrubbers "A" and "C" due to the samples being contaminated on 6/29/2017. Figures 3, 4 and 5 below represent the chemical scrubbers removal efficiencies for the constituents listed in the tables.

Table 3: Performance Acceptance Test Results

Date: 6/29/2017	**Chemical Scrubber A Time: Between 10:00 am and 1:00 pm (Unit: ppbv)**			
	Sample 1		**Sample 2**	
Constituents	**Inlet**	**Outlet**	**Inlet**	**Outlet**
Hydrogen Sulfide	280	ND	340	ND
Carbonyl Sulfide	68	58	67	58
Methyl Mercaptan	16	ND	16	ND
Dimethyl Sulfide	26	ND	31	ND
Carbon Disulfide	24	22	28	25
Date: 7/x/2017	**Chemical Scrubber B Time: Between 10:00 am and 1:00 pm (Unit: ppbv)**			
Constituents	**Inlet 1**	**Outlet 1**	**Outlet 2**	
Hydrogen Sulfide	210	ND	ND	
Carbonyl Sulfide	30	25	21	
Methyl Mercaptan	24	ND	ND	
Dimethyl Sulfide	33	ND	ND	
Carbon Disulfide	12	10	8.3	
Date: 6/29/2017	**Chemical Scrubber C Time: Between 10:00 am and 1:00 pm (Unit: ppbv)**			
	Sample 1		**Sample 2**	
Constituents	**Inlet**	**Outlet**	**Inlet**	**Outlet**
Hydrogen Sulfide	140	ND	130	ND
Carbonyl Sulfide	53	47	45	44
Methyl Mercaptan	9.5	ND	8.7	ND
Dimethyl Sulfide	24	ND	25	ND
Carbon Disulfide	19	17	19	19

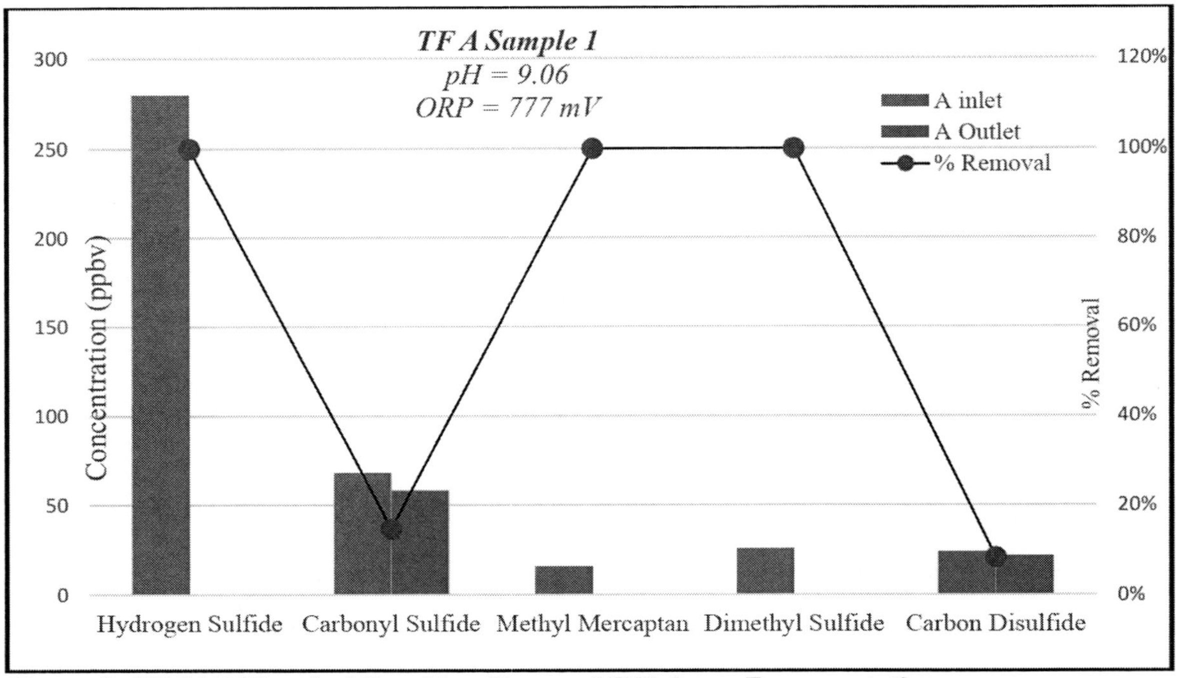

Figure 3: TF "A" Chemical Scrubber Removal Efficiency Representation

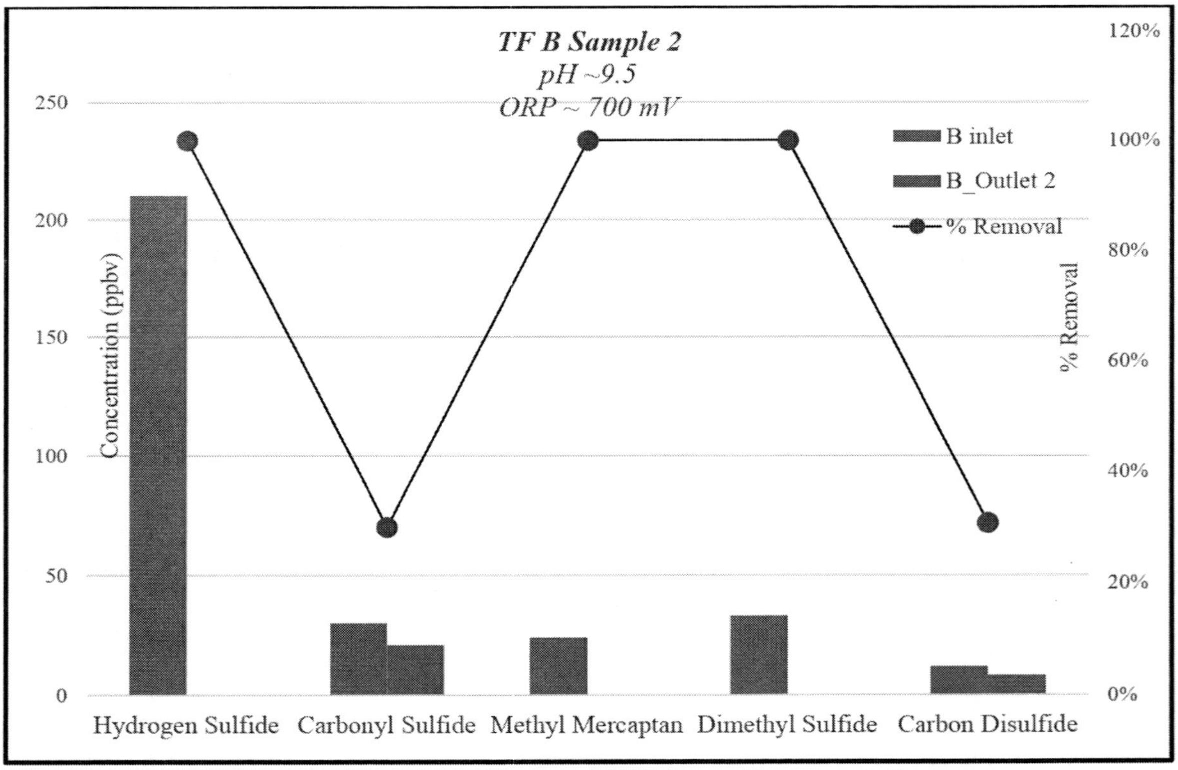

Figure 4: TF "B" Chemical Scrubber Removal Efficiency Representation

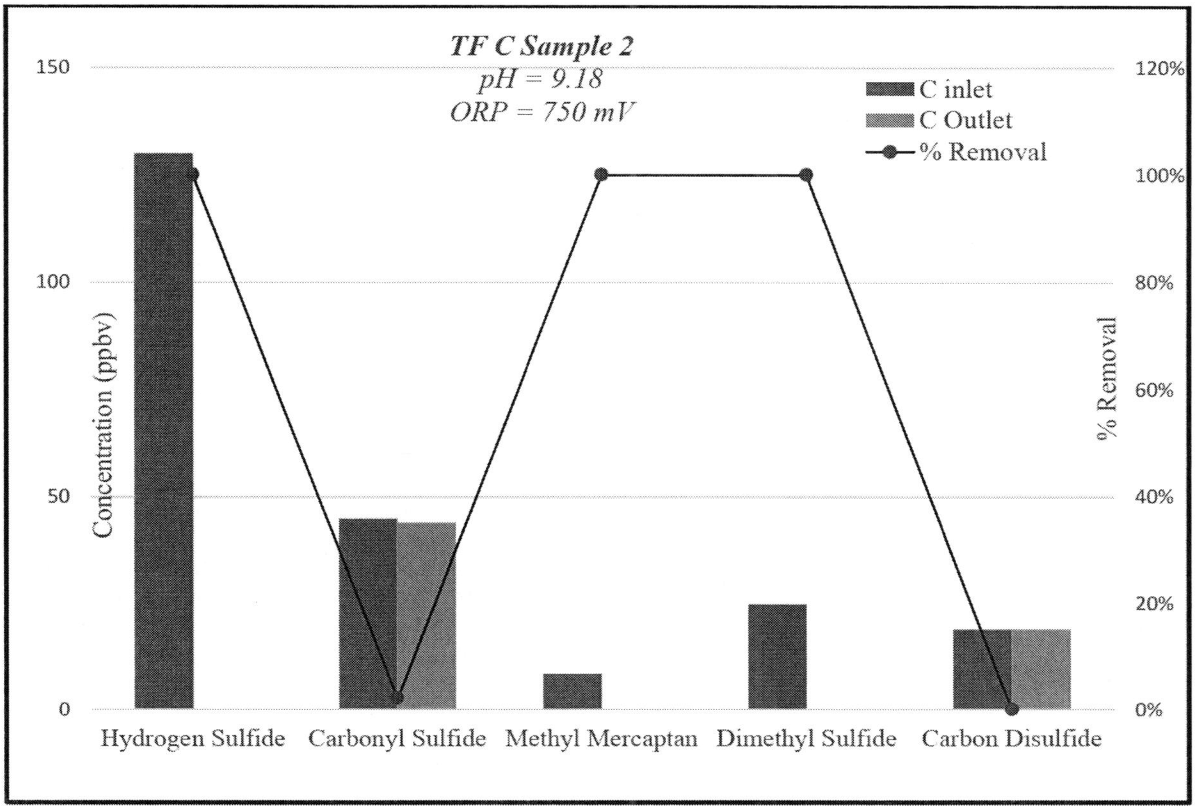

Figure 5: TF "C" Chemical Scrubber Removal Efficiency Representation

ANALYSIS/DISCUSSION

As stated earlier in this article, the new chemical scrubbers are installed prior to the existing carbon scrubbers to remove nuisance odors as well as to be cost effective. The chemical scrubbers are designed to remove only H2S per SCAQMD permit requirement. However, the foul air samples are sent to the lab to also determine the chemical scrubber removal efficiencies of other prominent RSCs that are often seen in the secondary wastewater treatment process.

One of the scholarly articles, "Biotrickling Filter and Chemical Scrubber Combination ….Perth's Stringent Odor Control Requirement", which was published in WEFTEC 2009 issue, stated that the two stage chemical scrubber, caustic followed by caustic & bleach, was able to remove Methyl Mercaptan and Dimethyl Sulfide (Van Durme & et. al). The findings in this paper are suggesting that one-stage chemical scrubbers, using both caustic & bleach in the same sump, are able to remove H2S, MM, DS very well, with an exception of 06/19/2017 sample 1 MM result, which can be considered as an anomaly, given that all other results are consistent. In the removal efficiency calculation, the non-detect results are considered as "zero" outlet concentrations, therefore, 100% efficiency may be slightly misleading in the graphical representations. However, the data consistently show that COS and CDS are not removed by chemical scrubbers. Furthermore, on

6/21/2017, the outlets of the carbon scrubbers (polishing scrubber at this point) were sampled and all three sets of samples from the carbon scrubber outlets showed inefficient removal of COS and CDS constituents. (see Table 4)

The United States Environmental Protection Agency (EPA) reference guide, published in March 1992, provided some odor thresholds and odor characteristics for the air pollutants. The thresholds for COS is 0.1 ppmv (0.0001 ppbv) and CDS is 0.16 ppmv (0.00016 ppbv), and their odor characteristics are described to be unpleasant, sulfide, vegetable, and medicinal. Some Odor Wheels identify sulfur-related compounds as having "rotten cabbage" and/or "garlic-like" characteristics. Due to the very low detection threshold level, the sample results indicate that the sulfur-related odors, COS and CDS, may be detected at the carbon scrubber outlet regardless of the carbon polishing stage.

Table 4: Chemical Scrubber Inlets and Carbon Scrubber Outlets Sample Results

Date: 6/21/2017	**Inlets Chemical Scrubber "C" and Outlets Carbon Scrubber "C" Time: Between 10:00 am and 12:00 pm (Unit: ppbv)**					
	Sample 1		**Sample 2**		**Sample 3**	
Constituents	**Chem. In**	**Car. Out**	**Chem. In**	**Car. Out**	**Chem. In**	**Car. Out**
Hydrogen Sulfide	170	ND	230	ND	240	ND
Carbonyl Sulfide	38	31	40	36	34	29
Methyl Mercaptan	14	ND	14	ND	16	ND
Dimethyl Sulfide	24	5	25	ND	31	ND
Carbon Disulfide	16	15	15	15	18	16

FINAL MODE OF OPERATION

Since the new chemical scrubbers are meeting the treatment requirements, the project team recommended to operate without using the carbon media as a polishing stage, which will be the most cost-effective mode. One major challenge with chemical system is - using excessive amount of caustic to maintain pH to be above 9.0. As a result, the valves, flow meters, and recirculation pumps become foul-packed after a couple of months of operating the chemical scrubbers. The OCSD staff noticed the irregular pumping patterns at all recirculation pumps. After acid-washing, and eliminating caustic dosing the system was back to operating well. Therefore, OCSD staff has discontinued using caustic, and due to very low H2S inlet from the TF process, bleach seems to be treating H2S well without the caustic addition. In a near future, OCSD staff will be analyzing the removal efficiencies on these RSCs with the bleach-only mode operation.

ACKNOWLEDGEMENT

This paper was supported by the Operations staff and Process Engineering staff from OCSD. The operation adjustments were implemented based on the multiple communications and decisions made between the project team and the Operations division; and the foul air samples were collected with collaborative efforts from the OCSD Process Engineering staff. The authors also would like to recognize the exceptional efforts from Dave Mazzarella (SCADA & Control System), GHD USA, and Jeff Jones & ECS Solutions teams for their cooperation, insights, and diligence in getting the project completed.

REFERENCES

Estrada, Jose M., Kraakman, N.J.R. Bart, Muñoz, Raul, and Lebrero, Raquel. (2010) A Comparative Analysis of Odour Treatment Technologies in Wastewater Treatment Plants. Environ. Sci. Technol. 2011, 45, 1100-1106.

Van Durme, Gayle P., Shaw, Andrew, Nichols, Cora, Cadee, Keith, McNeil. (2009) Biotrickling Filter and Chemical Scrubber Combination Meets Perth's Stringent Odor Control Requirements. WEFTEC 2009

Relationship between hydrogen sulphide and odour units for London sewage

Sofia Georgaki[1]*, Dejan Vernon[2], Rodrigo Baur[1], Donal Black[1], Bill Desing[3], David Crawford[1]

[1]CH2M now JACOBS, UK
[2]Thames Water Utilities Limited, UK
[3]CH2M now JACOBS, USA
*Corresponding author, e-mail: sofia.georgaki@ch2m.com

ABSTRACT
The development of Thames Tideway Tunnel (TTT) aims to capture and convey to treatment combined sewage which presently overflows into the tidal River Thames. The 25km long tunnel will pass beneath central London. An important consideration, is the incorporation within the design, of an air release treatment system, to mitigate the potential risk of odour nuisance in the vicinity of tunnel shafts. Hydrogen sulphide (H_2S) is a dominant odorant for many waste processes and is particularly associated with sewage treatment and is often used as a surrogate 'marker' compound to approximate odour levels in wastewater based odours. The TTT team have established H_2S generation and liberation models to estimate the levels of H_2S within the air released from the system after treatment. To support this approach, a pragmatic understanding of the sensory perception of odours likely to be generated by the system has been sought and the relationship between odour and H_2S concentrations, for the effluent samples collected from the TTT catchment, has been investigated. This relationship, which is presented in this paper, is based on testing samples of sewage at eight sites across London from November 2010 to October 2014. The relationship was derived by comparing the concentration of H_2S in the gaseous phase (mg/m^3) and laboratory olfactometry evaluation of the odour concentration (OU_E/m^3). The relationship between H_2S and odour units was found to vary depending on the concentration of H_2S that is present. At relatively low concentrations of H_2S ($<0.5 mg/m^3$), other odorous compounds found to be dominant, which became insignificant at increased concentrations of H_2S (ranging from0.5-5.0 mg/m^3).The results of the sampling programme, gave an acceptable correlation describing the relationship between H_2S and odour units for both high and low levels of H_2S. The impact of volatile organic compounds (VOCs) on the relationship was also investigated. The derived relationship has been applied to the odour measurement and dispersion modelling studies that were undertaken as part of the TTT project.

KEYWORDS
Combined sewer overflow, hydrogen sulphide, odour, odour measurement, odour units, olfactometry, wastewater.

INTRODUCTION
Odours are sensations resulting from the reception of a stimulus by the olfactory sensory system (WEF and ASCE, 1995). Human responses to an odour are highly subjective; different individuals have different perception to odours (Stuetz *et al.*, 2001). Odours in wastewater treatment arise from the anaerobic degradation of sewage during transportation and treatment. Other odours result from sources such as industrial wastewaters including solvents and volatile organic compounds or sulphurous effluents (Vincent and Hobson, 1998). Odours associated with wastewater emissions are made of an array of compounds including sulphurous, nitrogenous, acids, aldehydes and ketones. Sulphurous compounds cause the majority of odours in wastewater treatment with H_2S being the most significant. H_2S is formed either from the reduction of sulphate or the desulphurisation of organic compounds containing sulphur. Although odours from sewage treatment have been a concern for many years, odour complaints near treatment works and throughout sewer networks, in recent years, have increased. This is due to a number of reasons including increased housing, environmental legislation requiring new or upgraded facilities, and increased awareness of consumer rights (Toogood and Hobson, 1987; Balling and Reynolds, 1980). Controlling the odours from sewage treatment facilities and networks, has become the biggest challenge for the water industry and an important consideration in the design and operation of sewerage systems.

It is generally accepted, that in order for effective odour control measures to be implemented, quantification of the problem is required (Hobson, 1995). The perception of odours varies from person to person and therefore are difficult to fully measure. This is because many odorous emissions consist of a number of individual odorous components which makes the odour of the total mixture very difficult to be predicted. Consequently there is no universally accepted method available for the quantification of odours (Koe, 1989; Jiang, 1996).

Odorous compounds can be measured using analytical and sensory methods. Analytical measurements are associated with odorants which are the compounds that cause odour while sensory methods are associated with odours which represent the perceived effect of the odorant on the sense of smell (Gostelow and Parsons, 2000).

Analytical measurements refer to the physical or chemical properties of odorants, the most common being the odorant concentration. Analytical measurements offer several advantages such as objectivity, repeatability and accuracy. They can also be related directly to theoretical models describing chemical transformations or mass transfer. Some of the disadvantages of the analytical methods include:

- Most environmental odours are complex mixtures of dozens of components. This complicates analysis considerably, which usually requires a separation technique prior to analysis.
- Odorants may be present in very small concentrations. The limit of analytical detection may be below the threshold of smell. Non-odorous compounds will also be present in the sample in much larger concentrations than the odorants.
- It is difficult to relate analytical measurements to the intensity of odour as perceived by a human observer. This is especially the case for mixtures, as interactions between different odorants may lead to synergistic and antagonistic effects.

A key issue in analytical measurements, is the relationship between the concentration of the odorants and their effect on odours as perceived by humans. Where one or a few odorants are identified as dominant for certain odour sources, relationships can be formed between odorant and odour concentration. Several authors (Gostelow and Parsons 2000; Hobbs *et al.*, 1999; Noble *et al.*, 2001) have performed odorant /odour correlations for a range of wastes. Some of the issues discussed above, can be avoided by the application of sensory measurements.

Sensory measurements employ the human nose (usually multiple noses) as the odour detector and hence they relate directly to the odour properties as perceived by humans. Parameters which may be measured subjectively include odour character, hedonic tone and intensity. Interpretation of subjective sensory odour measurements is difficult due to inherent variation of odour perception even for well trained personnel (Koe, 1989). Olfactometry is the most common sensory method that is applied for the identification of odour concentration. It measures the odour concentration by diluting a sample with odour-free air until is reduced to a threshold concentration at which nothing can be detected (by the nose). The concentration of the odour is then expressed as the dilution required for this to be achieved. The main advantage of sensory measurements is that they require no special equipment and can be obtained quickly at low cost. Tables 1a and 1b set out the advantages and the disadvantages and the potential applications of the different sensory and analytical techniques available.

There are several factors that affect the perception of an odour, one of which is the variability in the sense of smell between different observers. Factors such as the order in which the samples are presented and the flow rate of the gas stream are also important. Previous studies using analytical and sensory measurements have shown that the main barrier in linking the two types of measurement is the effect of mixtures. Research by Laska and Hudson, (1991), who undertook work on mixtures of two to twelve odorants, suggested that odorants are additive. According to the authors, a mixture of odorants will have a stronger odour than any of the component odorants alone. However, the degree of variability is different for each mixture.

Further investigations by Laing *et al.*, (1994), suggested that odorants are partially additive, whereby the intensities of the mixtures were less than suggested by a simple addition of the individual component intensities. The intensity of the mixtures found to be close to that of the dominant contaminant could give a good indication of the overall odour concentration. Recent studies on odour mixtures have suggested a range of effects from simple additivity (Patterson *et al.*, 1993), synergistic additivity (Laska and Hudson, 1991), and partial additivity or suppression (Laing *et al.*, 1994; Hobbs *et al.*, 2001). Where one or few odorants are identified as dominant for certain odour sources, relationships can be formed between odorant and odour concentration. Several authors have performed odorant/odour correlations for a range of wastes (Gostelow and Parsons, 2000; Hobbs *et al.*, 1999, 2001; Noble *et al.*, 2001), however, there is no information available in literature on correlations between odorant and odour for a large CSO control storage tunnel system such as TTT.

MATERIALS & METHODS

This section describes the materials and methods used for the gas phase analysis of odour from the combined sewage samples, obtained from the catchment area that will serve the Thames Tideway Tunnel (TTT). The sites where dry weather flow (DWF) samples were collected are:

1. North of river Thames from West to East including:
    - Acton storm tanks manhole on Stanford Brook foul sewer
    - Western pumping station inlet chamber on Low level sewer No1
    - Wick Lane inflow chamber High level sewer
    - Abbey Mills pumping station F inlet chamber Low level sewers No1 and 2

2. South of river Thames from West to East including:
    - Heathwall pumping station foul sewer inlet
    - Earl pumping station foul wet well
    - Greenwich pumping station wet well
    - Charlton chalk pits overflow weir chamber on High level No 2

Odour concentrations, for the gas phase samples collected were determined using threshold olfactometry. Olfactometry is a sensory method that is based on the dilution of a sample with odour-free air until it is reduced to its threshold concentration. A simple schematic of a dynamic olfactometer is shown in Figure 1.

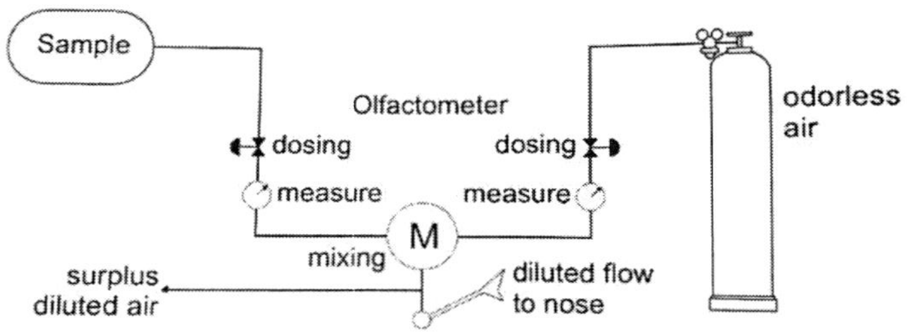

Figure 1. Flow schematic of dynamic olfactometer (Frechen, 1994).

The key elements of a dynamic olfactometer are:

- Air pumps for drawing in the odorous samples and providing the flow rate range.
- Flow meters for accurate measurement of air flow rates.
- Mixing chambers to mix the flows of odorous and odour-free air.
- Deodorising chambers, usually employing activated carbon, or an alternative supply of odour-free air.
- Control valves for accurate control of air flow rate.
- Connecting tubes for conveyance of flows between various parts of the olfactometer and to the sniffing port.
- Sniffing port, usually a face mask, hood or funnel where the sample is delivered to the observer's nose.

Because olfactometry cannot practically be performed on site, odour samples need to be collected, preserved and stored to allow transportation from site to laboratory. On this basis, the samples were collected under dry weather conditions, using a number of covered buckets or an ISCO peristaltic pump. The sampling equipment was supplied by Environmontel. Twelve sets of grab samples were taken at each site, made up of three sets of four samples noted as A, B, C, and D as listed in Table 1. These samples (approximate volume of 26 litres each) were collected during each visit and processed by Odournet, UK. Some of the samples were fixed at site using sodium carbonate (Na_2CO_3) and zinc acetate ($Zn(O_2CCH_3)_2$) preservatives. Sample analysis for dissolved sulphide (DS) was undertaken using an electrode provided by Odournet UK and laboratory testing at both Thames Water and Seven Trent test facilities.

Table 1. List of samples analyzed at each site.

Sample name	Fixed sample	DS analysis	Redox, pH, Temp	Gas phase analysis
A1, A2, A3 each 24.5L	No	Yes (Odournet UK by electrodes in lab) before and after dilution with de-ionized water	Yes	Yes (Odournet UK) H_2S Jerome analyzer, olfactometry, GC-MS, Hedonic tone
B1, B2, B3 each 0.5L	Yes	Yes (Odournet UK) (electrodes in lab)	Yes	No
C1, C2, C3 each 0.5L	Yes	Yes (TWL)	No	No
D1, D2, D3 each 0.5L	Yes	Yes (TWL and STL)	No	No

The locations at which the samples were collected are shown in Figure 2.

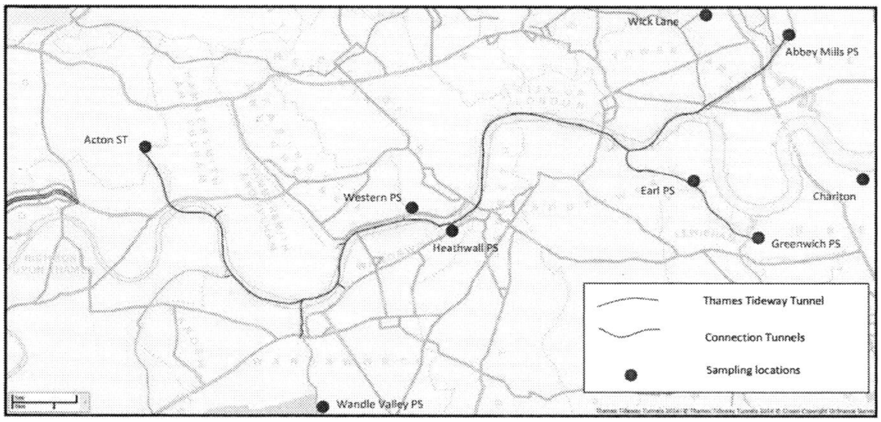

Figure 2. Map of the locations at which the sewage samples were collected.

Gas phase analysis was undertaken by transferring the liquid samples into a nalophan bag, where H_2S in the headspace was monitored until the concentration was stabilised. Bag materials are an important factor in sample preservation. The material should be odour free and impermeable, and should not encourage the adsorption of odorants. Several other materials are available, including tedlar, Teflon, nalophan, polyterephthalic ester, co-polymer, Mylar, saran and polyethylene. Gas phase odour concentration was determined using olfactometry analysis to BSEN13752. To establish the constituents of the odour in the samples, Gas Chromatography-Mass Spectroscopy (GC-MS) was applied. Gas chromatography relies on the differential adsorption of the sample components on the wall of a long chromatographic column, such that the different molecules emerge from the end of the column at different times. The H_2S concentration of the samples was measured using the Jerome H_2S analyser. The degree of odour offensiveness in the samples was obtained using the Hedonic Tone test. Typically this will be represented using a numeric scale depicting most pleasant at one end and most unpleasant at the other. Often, negative values are used to represent unpleasant odours and positive values to represent pleasant odours.

RESULTS & DISCUSSION

Experimental observations

Given the potential advantages of H_2S as an indicator of odour, correlations have been performed between H_2S and odour concentration, as measured by olfactometry, for a number of samples, at a total of eight sites around London. It is generally accepted that there is a direct relationship between gas phase H_2S and liquid phase DS. This results from the fact that H_2S is a weak dibasic acid that dissociates in water:

$$H_2S = H^+ + HS^- = HS^- + S^{2-} \quad (1)$$

At neutral pH, approximately 50% of the total H_2S is present in its molecular form and it is in this state that odour problems occur. Above pH 7, the majority of sulphide is present in ionic form, either as HS^- or S^{2-}. If pH is increased further, to pH 13, a second cross over occurs and approximately 50% of the sulphide present is in each ionic form.

The DS concentration in the liquid phase was determined from a number of test results using various methods (primarily electrode probe and laboratory testing). Some of the results that were obtained using the electrode probe found to be significantly different to the results obtained from the laboratory testing and were discarded (concentration difference >0.5mg/m^3). This could be due to potential errors in the testing procedure which include: 1. agitation during sampling which could release volatile sulphides from the liquid to the gas phase, 2. partially filled sampling bottles resulting in transfer of mass from the liquid into the gas phase, 3. the travel time and potential temperature effects during transportation of the samples, 4. possible errors during laboratory testing/olfactometry analysis of the samples.

The remaining samples were further tested using olfactometry and the relationship betweenH_2S concentration and odour concentration was explored. The results of this analysis showed that most samples were within the 95th percentile range and therefore were considered as acceptable data points for the derivation of the correlations describing the relationship between H_2S and odour concentrations. A range of H_2S concentrations between 2.9µg/m^3 (equivalent to 35OU_E/m^3) and 4,020µg/m^3 (equivalent to 16,040 OU_E/m^3) were examined.

H_2S/ Odour correlations

The equation that describes the relationship between H_2S and odour concentration is:

$$C_{(OU)} = mC^{n}_{(H2S)} \quad (2)$$

Where, $C_{(OU)}$ is the odour concentration in OU_E/m^3 and $C_{(H2S)}$ is the concentration of H_2S in mg/m^3. The values of the parameters m and n differ according to the odour composition. This study investigates the relationship between H_2S and odour concentration for H_2S concentrations in less than 0.5 mg/m^3 and in the range between 0.5 mg/m^3 and 5 mg/m^3 respectively. With these conditions the general equation (2) becomes:

For H_2S concentration less than 0.5mg/m^3,

$$C_{(OU)} = 12{,}000C_{(H2S)} \quad (3)$$

For H_2S concentrations greater than 0.5 mg/m^3 but less than 5 mg/m^3,

$$C_{(OU)} = 8{,}000C^{0.5}_{(H2S)} \quad (4)$$

The results of the samples for H_2S concentrations less than 0.5mg/m^3 are shown in Figure 3.

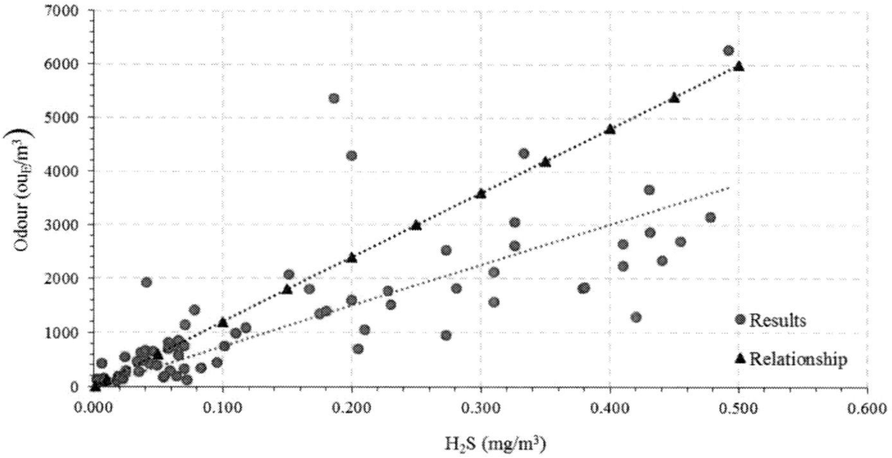

Figure 3. Relationship between H_2S and odour concentration for $H_2S < 0.5$ mg/m^3.

The relationship between H_2S and odour concentrations is expressed by a linear correlation as it would be expected for a pure H_2S odour sample. In this case, the parameter m is equal to the inverse of the H_2S threshold concentration while n is equal to 1. Previous studies (Bonin *et al.*, 1990), stated that threshold concentrations for pure H_2S are in the range between 0.0001-0.02 ppm, which suggests that the values of the parameter m could range from 50-10,000 (Bonin *et al.,* 1990). In this study, m values between 4,500 and 7,500 were obtained. It should be noted however, that the relationship between H_2S and odour concentration is a conservative approach and applies a boundary rather than a best fit solution. In this study, the value of n was around 0.4, which indicates that the contribution of H_2S to odour was suppressed by the presence of other odorants. The contribution of other odorants to odour becomes more significant at concentrations greater than 0.5 mg/m. Figure 4 illustrates this effect.

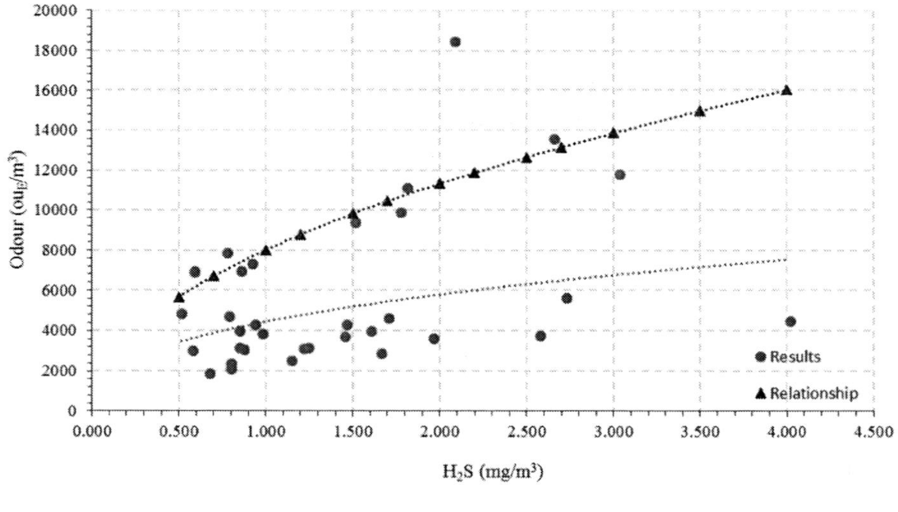

Figure 4. Relationship between H_2S and odour units for $H_2S > 0.5$ mg/m^3.

The gradients of the plots for the two sets of data ranged between 0.15-0.6 which suggest that 15-60% of odour can be explained by H_2S. The remaining percentage of odour could be due to other factors such as the contribution of other odorants, or errors in measurements including both sensory and analytical methods. Best fit results were obtained at low H_2S concentrations (< 5 mg/m^3) where 45 % of the odour concentration was due to other odorants contribution or measurement errors.

To determine the contribution of other odorants on the odour concentration, analytical measurements of volatile organic compounds (VOCs) were undertaken. The results of these studies are plotted in Figure 5. The results showed that the relationship between H_2S and odour concentration varies and is dependent on the concentration of H_2S. As expected, at relatively low H_2S concentrations, there is some contribution of VOCs to total odour. It should be noted, that the derived correlations describing the relationship between H_2S and odour concentration are valid for H_2S levels up to 5 mg/m^3.

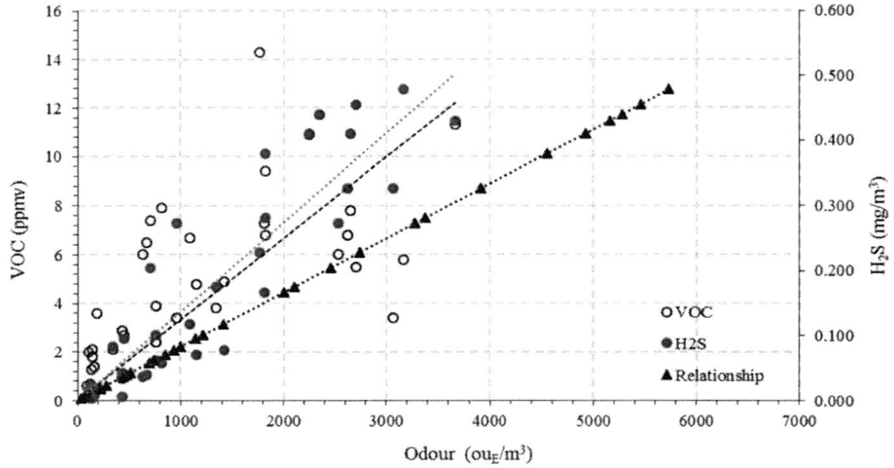

Figure 5. Comparison between VOC and H_2S surrogates. (H_2S concentrations <0.5 mg/m^3).

CONCLUSIONS

The derived relationships between H_2S and odour concentration, indicate that generally H_2S is an acceptable indicator for odour generated in sewerage system. The relationships derived are applicable for H_2S levels less than 5 mg/m^3, which indicates that analytical and sensory measurements can be linked for certain odour types. The experimental observations showed that there is a strong relationship between H_2S and odour units for H_2S concentrations less than 0.5 mg/m^3. At H_2S concentrations greater than 0.5 mg/m^3 a weak relationship was determined. This is possibly due to errors in the sampling methods or the contribution of other odorants to the total odour concentration. The derived correlation described by equation (1), could accurately predict the odour concentration, provided that H_2S is the dominant contributor to odour. The relationships derived could be applied to the conversion of H_2S monitored values to odour units and therefore in determining the value of odour units at release points from the tunnel system to demonstrate compliance with the regulatory guidance.

AKNOWLEDGEMENTS

The authors would like to thank the Thames Water Utilities Limited (TWUL) for funding this component of the TTT project and Odournet UK for the olfactory tests of the samples collected.

REFERENCES

Balling, R.C. and Reynolds, C.E., (1980). The diurnal flux in the dispersion of sewage plant odors. *Water, Air and Soil Pollution,* **14**, 323-329.

Frechen, F. B., (1994). Odour emissions of wastewater treatment plants: Recent German experiences. Water Science and Technology, **30** (4), 35-46.

Gostelow, P. and Parsons S.A., (2000). Sewage treatment works odour measurement. *Water Science and Technology*, **41** (6), 33-40.

Hobson, J., (1995). The odour potential: a new tool for odour management. *Journal of the Chartered Institution of Water and Environmental Management,* **9**, 458-463.

Jiang, J.K., (1996).Concentration measurement by dynamic olfactometer. *Water Environment and Technology*, 55-58.

Koe, L.C.C., (1989). Sewage odors quantification. Cheremisinoff, P.N. (eds.), *Encyclopaedia of Environmental Control Technology Volume 3: Wastewater Treatment Technology*, Houston, Gulf Publishing Company, 423-446.

Laing, D.G., Eddy, A., Best, D.J., (1994). Perceptual characteristics of binary, trinary and quaternary odor mixtures consisting of unpleasant constituents. *Physiology and Behaviour,* **56**, 81-93.

Laska, M. and Hudson, R., (1991). A comparison of the detection thresholds of odour mixtures and their components. *Chemical Sciences*, **16**, 651-662.

Stuetz, R. M., Gostelow, P. and Burgess, J. E., (2001). Odou perception. In: Odours in wastewater treatment: measurement, modelling and control, edited by Stuetz R.M. ad Frechen, F. B. IWA publishing (2001), New York (ISBN: 1 900222 46 9).

Toogood, S. and Hobson, J., (1987). On the scent? (Five little maxims for odour control at sewage works). *Water Sciences*, February, 58-60.

Vincent, A. and Hobson, J., (1998). Odour control. *CIWEM Monographs on Best Practice No. 2, London: Chartered Institute of Water and Environmental Management.*

Water Environment Federation (WEF) and American Society of Civil Engineers (ASCE), (1995). Odour control in wastewater treatment plants, WEF manual of practice – 22.

Odor Investigations - What Equipment Do I Need and What Will the Data Mean?

Richard J. Pope, PE, BCEE

Abstract
This paper and presentation is intended to provide the WRF staff and others with the basic fundamental understanding of the equipment needed to conduct a holistic odor investigation and how to make sense of the data collected to initially rank the WRF odor sources.

Keywords
Odor monitoring, odor equipment, monitoring plan, H_2S monitors, odor criteria

Background

In an attempt to reduce operating budgets, Water Reclamation Facility (WRF) management is asking staff to do more for less, and that includes reducing the need to retain odor assessment specialists that assist in monitoring nuisance odors and surrogate wastewater and physical parameters around their facility unit operations to determine key odor sources, performance effectiveness of existing odor control systems, off-site impacts, etc. As a result, more and more WRF's are looking to conduct nuisance odor investigations using their own staff. Although many have observed these odor assessment specialists conduct investigations at their facilities, most do not have the equipment to conduct even the most basic nuisance odor monitoring investigation. In addition, once the odor data is collected what does the information mean and how can it be used to develop a strategy to mitigate WRF nuisance odors?

This is a fundamental odor monitoring equipment presentation that includes a discussion of what the data collected means by identifying ranges of values that are of low, medium and high concern. Monitoring odors should consist of a holistic approach to understanding not only the direct vapor phase odor parameters, like hydrogen sulfide (H_2S) for example, but also the indirect parameters, like wastewater sulfide, that give us insight into the potential for odor production and release to atmosphere. Accordingly, odor monitoring should be conducted in three categories:

- Air
- Wastewater
- Physical

The equipment needed for the air and wastewater categories can be further subdivided into:

- Sample collection equipment
- Sample container devices (Sample containers may not be necessary for all situations)
- Sample analysis equipment

Typical standard equipment that is associated with each subcategory will be presented, with a few to several options being provided. Proven, simple, hand crafted devices will be illustrated when appropriate. The equipment choice is left up to the WRF staff to decide based on the criteria they have established as part of their odor investigation goals and objectives and the amount of money that is available for purchasing equipment.

The physical category consists of evaluating conditions and operations at the WRF that based on experience will indicate the potential and/or presence of odorous conditions, such as:

- Concrete wall pH where low levels indicate the presence of H_2S
- Observations of surface corrosion, wastewater color, foaming, etc.
- Smoke testing

The equipment needed to monitor for the physical category will also be presented.

Keep in mind that the best air phase field instrument is still the human nose! In many cases it has a lower detection level than equivalent analytical equipment, and it can evaluate combined odors. In addition, the human nose is the same instrument that the community residents are basing their judgements about the quality of the odors emanating from the WRF. So, taking advantage of the cheapest analytical piece of equipment that is available 24/7 is highly recommended.

Tables will be presented that look at the data collected and provide typical ranges of values. The ranges will generally be divided into three divisions:

- Low impact or low odor concern
- Moderate impact or moderate odor concern
- High impact or high odor concern

Converting the data into rates that is intended to help normalize the data from diverse sources and can be used to initially rank the sources. This approach represents a rough cut at odor source ranking. A more definitive and formal approach (Which will not be discussed in detail in this paper/presentation.) to ranking the WRF odor sources is the application of an air dispersion model.

This paper and presentation is intended to provide the WRF staff and others with the fundamental understanding of the equipment needed to conduct a holistic odor investigation and how to make sense of the data collected to initially rank the WRF odor sources.

Introduction

The on-site odor investigation begins with asking a series of important and basic questions that will help lead us to the resources that are necessary to conduct an effective and focused holistic monitoring study. The questions are:

Question #1 - Why Do We Need to Conduct this Odor Investigation?

There is a plethora of reasons why an odor investigation is necessary. Some of them are pro-active, understanding ahead of time that current or future conditions at the wastewater facilities will become a problem and trying to get ahead of that anticipated potential odor and/or corrosion problem. The rest are generally reactive, responding to outside forces that are highlighting the fact, real and/or perceived, that the wastewater facilities are an off-site nuisance odor source or the cause for the collapsed sewer. Whatever the reason may be, it is important to understand the "driving force" for wanting to conduct an odor investigation:

- Is the driving force the need to be pro-active?
- Is the response to regulatory limits that are being exceeded, notices of violation based on quality of life judgements, or calls to come in and sit down with the regulators to discuss nuisance odor issues at the facility?
- Have the number of odor complaints receive from the nearby community skyrocketed?

- Has legal action been brought against the wastewater facility to cease and desist releasing odors?
- Is the road collapse due to corrosion in the ruptured interceptor sewer?

The response relative to the type of odor investigation required must be appropriate to the driving force! For example, the level of attention regarding calibrations, chain of custody forms, industry defined procedures, etc. and the number of corroborating samples collected at each location would be less for a pro-active investigation than it would be for the situation where the information will be used in a court of law. The odor investigation should be designed to address the ultimate need for the information/data collected.

Question #2 - What Are My Goals?

After figuring out that an odor investigation is necessary, the next step is to recognize what my goals for the investigation are. Understanding what I hope to get out of the investigation will provide a critical path to help get there. For example, the goal may be to reduce odor complaints by finding out where the odors are being released and how to mitigate them. In mitigating the odors is the goal to achieve no odors at the odor source, fence line, nearest sensitive receptor, or closest residence. The approach to addressing any of these may require a different strategy and monitoring plan. It is always good to understand my end-point, so I can create a path and monitor it to make sure that I get there.

Question #3 - What Resources Do I Have?

Available resources are a key piece of the puzzle when planning an odor investigation. The main resources that are necessary include:

- Equipment – The meters, monitors, samplers, laboratories, health and safety, etc. necessary to perform the odor investigation.
- Time – With their current work load requirements do I have the staff time to dedicate to conducting the investigation?
- Staff – Are there adequate staff motivated to conduct the investigation? Are outside resources necessary?
- Experience – Have the staff performed the tasks and worked with the equipment that are required? Do they need training? Is a trainer necessary?
- Money – What is available?

Where am I resource limited? A gap analysis is imperative to assess what resources I have and what resources I need. Do I need to enlist outside resources (For example: rent monitoring equipment to be used for a limited timeframe; sub-contract with an independent laboratory to analyze/evaluate samples; retain the services of an outside expert(s) to manage or assist in field monitoring; etc.) to shore up gaps in the resources I have? It is important to make certain that the proposed odor investigation plan can be accomplished with the current in-house and external resources. And the bottom line question is – What can I afford? The ultimate monitoring plan will be designed to accomplish what is feasible with the resources that are available. If the resources prove to be limited this may mean adjusting the goals.

Question #4 - What are the Odor Investigation Basic Questions?

When developing the odor investigation plan the six basic questions that form the plan outline include:

- Who – Who is going to physically make-up the monitoring team? Will it be all staff or a combination of staff and outside resources? The monitoring team members must be dedicated, reliable, capable and committed to the successful completion of the project. Provide for some contingency team members that can step in when vacation, personal matters, medical issues, etc. may interfere with their ability to complete the assignment.

- What – Decide what the important parameters are that need to be included in the monitoring plan. A more complete odor investigation plan takes a holistic approach by realizing that monitoring air phase parameters alone will yield limited information about the potential of the unit process to emit nuisance odors. The recommended holistic approach includes monitoring for air phase; wastewater/liquid phase; solids phase; physical, and others. The **WHAT** basic question is the focus of this manuscript.
- When – There are a few timing questions that should be defined: the time of the year to conduct the investigation; and the time of day. The warmest time of the year is usually the worst-case condition for the production and release of odors as the microbiology attributed with the formation of odors is usually at its peak during warm weather. Those conditions will vary from region to region across the globe. Local meteorological records can assist with this determination. From a daily perspective, odors occur around the clock and are not known to follow a normal work schedule of 9 am to 5 pm! The odor monitoring plan should conduct some preliminary testing to determine the diurnal pattern of worst case odors using a tracer parameter like hydrogen sulfide in the vapor phase. The monitoring plan can then be scheduled to attempt to collect samples during the peak odor conditions during the day.
- Where – Select all the facility locations where there is a potential for odors to be emitted. This includes all potential sources: large and small; exhaust stacks with high flow volumes and vents with low or passive flows; enclosed areas and equipment; areas that might not ordinarily be considered like overflow boxes, annular spaces around floating covers on anaerobic digesters; odor control equipment (Just because an odor control system is in place and operating does not mean that it is working efficiently and effectively!); etc. Based on the available resources the full list may need to be adjusted, either before starting or during the investigation as screening and/or initial monitoring may demonstrate that a particular location is not an off-site odor source and can be eliminated from further monitoring. The sample locations also must include monitoring around the facility fence line and in the surrounding community. Whereas the heart of the odor investigation focuses on the potential odor locations at the facility, it is also very important to understand where along the fence line and in the community the facility-based odors can be perceived and to determine the paths of odor migration from the facility to the community. What plant-based nuisance odors are detected at the fence line/community? This information is invaluable when it comes to evaluating off-site odor impacts and the controls that can mitigate them.
- Why - The odor investigation is being conducted to identify odor sources, collect data on the parameters and concentrations being emitted and to use that information to estimate the off-site impact by desk-top evaluations or using odor dispersion models. An odor dispersion model creates a physical and operational simulation of the facility and its odor emissions given the local meteorology and

calculates the off-site impact of each individual odor source, as well as the combined impact of all the sources together, at specific receptors defined by the modeler. Subsequently, the degree of odor control required can be determined.

- How – The how is the odor investigation monitoring plan that is developed to incorporate the above five W's along with the equipment, procedures, and schedule to follow when performing the investigation. The monitoring plan is the road map that will lead the team to a successful investigation and to meeting the project goals.

This manuscript will focus on determining where the odors are and how strong they are by first presenting the industry standard list of parameters, and then the devices that can be used to collect/monitor/measure the odors. Finally, guidance will be presented that will help the user make sense of the data that is collected by highlighting common ranges for the various parameters and just how important, or not, those concentrations are relative to off-site odor impact or odor potential.

Odor Monitoring Parameters

When planning for an odor monitoring investigation it would seem intuitive that air phase parameters would be high on the monitoring plan list. Air phase measurements represent a direct impact to the nearby community. Not so intuitive are wastewater parameters and physical attributes. Nevertheless, these other two categories provide a wealth of information relative to the odor potential of the facility to generate, retain, control, and/or release odors and are almost as important as the vapor phase parameters when establishing a short or long-term plan to mitigate facility-based nuisance odors. Accordingly, the wastewater nuisance odor industry recommends taking a holistic approach to monitoring odors at a facility that includes monitoring: air phase; wastewater phase; physical/other. The subsequent sections will look at each phase, individually.

Air Phase

The list of air parameters that are commonly included in an odor investigation consist of the following:

- Hydrogen Sulfide (H_2S)
- Reduced Sulfur Compounds
- Amines
- Volatile Organic Compounds (VOCs)
- Ammonia
- Odor Parameters
  - Odor Strength
  - Odor Intensity
  - Odor Character
  - Hedonic Tone
  - Odor Persistence
- Electronic Nose
- Differential Pressure
- Air Flow

Not all these parameters are monitored in the field. Many are evaluated off-site at a laboratory. However, air samples are collected at the locations according to proper protocol and forwarded to the laboratory within the specific timeframes that are allowed according to the procedure by which they are to be evaluated.

Hydrogen Sulfide (H_2S)
Hydrogen Sulfide is the most important parameter to be monitored. It represents the most common odorous compound associated with wastewater and is almost always present in some concentration when monitoring wastewater facilities. It has a characteristic rotten egg scent and can be detected at concentrations in the air that are as low as one (1) part per billion (ppb) by volume in air and is persistent once it is in the atmosphere (Persistence means that it can travel extended distances and be diluted with odor free air in the atmosphere and yet it will lose very little in intensity, i.e., how strong it is perceived by the average individual.). Hydrogen sulfide is also closely linked with corrosion of exposed and unprotected metal and concrete surfaces. Hydrogen sulfide is directly corrosive to metal surfaces, and when converted by microorganisms in the headspace of wastewater facilities to sulfuric acid, it is corrosive to both metals and concrete. Hydrogen sulfide can also be detrimental to worker health and safety in the parts per million (ppm) concentration. For example, the Occupational Safety and Health Administration (OSHA) has a ceiling limit for H_2S of 20 ppm and the immediately dangerous to life and health (IDLH) is 100 ppm. It can be monitored with hand-held devices, as well as continuous data logging devices with varying sample intervals and deployment times. All these attributes support H_2S as a good candidate as a tracer compound regarding odors. However, do not be misled. Hydrogen sulfide is not the only nuisance odor associated with wastewater, and other compounds should be accounted for during an odor investigation.
Examples of H_2S monitoring devices are presented below. No endorsement of any particular brand/make of monitoring devices is intended or inferred by illustrating common devices or categories of devices displayed herein.

Figure 1 displays a portable, hand-held, H_2S monitor.

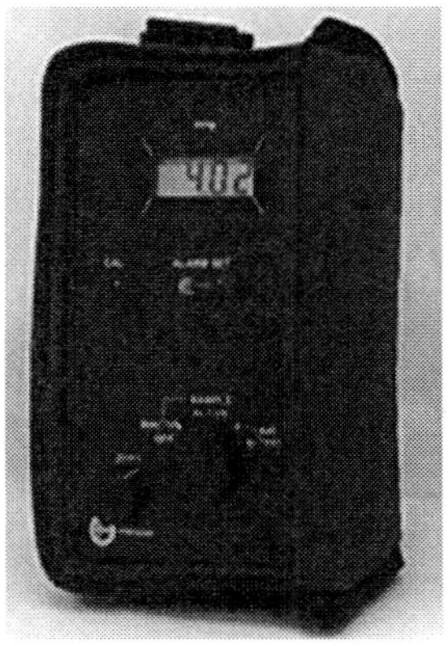

Figure 1 – H_2S Monitor

This unit uses an electrochemical sensor to measure H_2S and is capable of recording levels that range from 0.1 ppm to 1,000 ppm or greater. An internal air pump collects the sample and introduces it to the electrochemical sensor. This rechargeable battery-operated unit generally has enough of a charge to last for an eight-hour day. Because of the tendency of the electrochemical sensor to drift, the unit should be calibrated daily with a known concentration of H_2S gas that is selected to be within the range of anticipated concentrations that will be monitored during the day. Bottled gas with known concentrations of H_2S are available on the internet from gas suppliers. This type of unit is the workhorse of H_2S odor monitoring equipment as it has a good analytical range that is typical of wastewater facilities with odor problems and the internal pump allows for the direct measurement of odor sources.

The two H_2S odor monitoring units shown in **Figure 2** are very popular continuous, data logging, H_2S devices that are intrinsically safe (Meaning that they can safely be placed in locations like sewers and/wet wells.). They have internal batteries that allow them to operate in position for an extended period. Some can be left in place for upwards of a week. These units introduce the odorous air sample across a diffusion membrane. Just by being placed in a location where odors are located, the odors will diffuse across the membrane to the H_2S sensor. The sensor is also an electrochemical sensor, similar, but smaller, to that used in the device displayed in Figure 1. Many different H_2S ranges are available that bridge a low of 0.01 ppm to a high of 2,000 ppm. The sample interval can also be selected, from seconds to minutes. The integral eye hook allows the units to easily be suspended in the ambient air to collect the data from hard to reach locations. Recording minute-by-minute H_2S data helps the owners to better understand the diurnal variability of the H H_2S odors associated with their wastewater facilities, as well as how the concentrations will vary over the course of a week. The data is downloaded from the units either by an infrared reader or by Bluetooth technology and can be viewed in tabular form or by graph. These units have fast become the standard method for monitoring H_2S in collection systems and covered or enclosed units at waste H_2S water facilities. Because the entry into the unit is through the diffusion membrane on the bottom underside of the unit, the devices must avoid being submerged. In addition, the diffusion membrane must not be placed in the direct path of an air flow. The air flow will carry more H_2S to the electrochemical sensor than the unit would expect through diffusion, and so the recorded value will be higher.

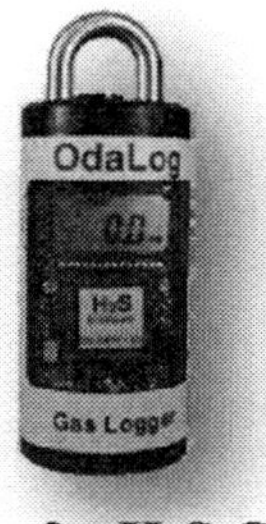

Figure 2 - H_2S Continuous Data-Logging Monitors

The H_2S monitor presented in **Figure 3** is a portable, hand-held, low range unit. It has an internal battery and air pump to collect samples.

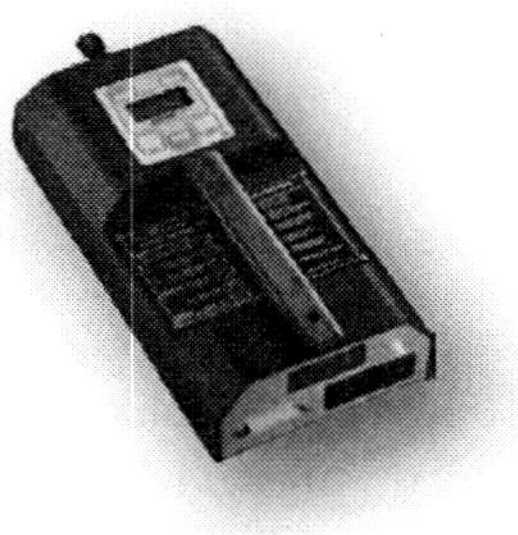

Figure 3 – Low Level Hand-Held H_2S Monitor

Hydrogen sulfide is analyzed by a gold filament sensor. Hydrogen sulfide plates out on the gold filament and the increase in resistance attributed to the additional H_2S relates to the mass that accumulated. Given the mass and the flow of the air that passed through the chamber, the concentration of H_2S can be determined. The gold filament has a limited capacity and needs to be regenerated when it gets full. This can happen during the day and would depend on the number of samples and the concentration. Regeneration takes less than 30 minutes and is accomplished by simply plugging the unit into a wall outlet. However, it can interrupt the monitoring investigation. This device is the only hand-held unit that is capable of monitoring for levels of H_2S in the single digit ppb. The device can measure from 0.003 ppm (3 ppb) to 50 ppm. The very low level makes this a perfect fit for monitoring the facility fence line and the community, as well as the exhaust from odor control exhausts.

Another H_2S monitoring device that is not as popular in the wastewater marketplace is the lead acetate paper cassette unit (see **Figure 4**). Lead acetate paper will react in the presence of H_2S and the paper will turn from brown to gray to black depending on the concentration of H_2S.

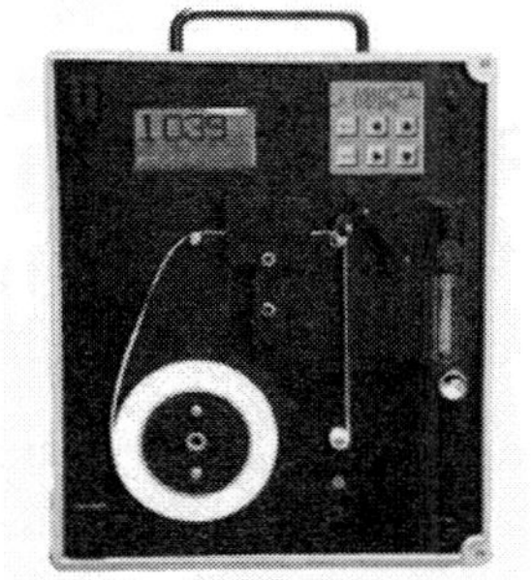

Figure 4 – Lead Acetate/Silver Nitrate H_2S Monitor

This device impinges the odorous air flow against the leas acetate paper. The stain created is then read by a photo eye and the concentration is calculated. The typical concentration range for this device is 0.01 ppm to 1,000 ppm or greater. This device suffers less from outside interferences associated with other reduced sulfur compounds. It has an internal air pump that pulls the sample and delivers it to the lead acetate paper. This device generally operates in a fixed location and is not designed to be dragged around a facility nor to withstand the corrosive and humid environments of most collection system and enclosed wastewater facilities. The lead acetate tapes have a limited shelf life (one to two years) but are sufficiently long for typical service. The monitors operate while being plugged in and/or by rechargeable battery.

Following the same principle of H_2S reacting with the lead acetate paper presented above and displayed in Figure 4, **Figure 5** displays lead acetate paper that will discolor in the presence of H_2S.

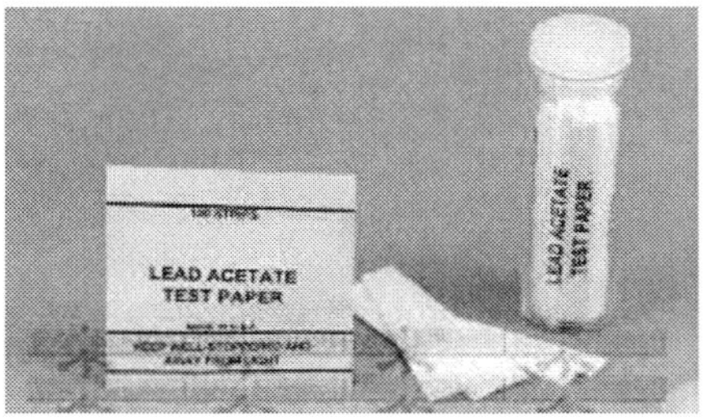

Figure 5 – Lead Acetate Paper for H_2S Detection

These lead acetate paper strips are used as general indicators of the presence of H_2S. They are placed in locations where the presence of H_2S needs to be verified. They can be used in the short term, for example, placing them in the air stream in which you may want to know whether H_2S is present. They are more commonly applied for long term use by leaving them in an area to test for the presence of H_2S over any length of time. In the past and currently these or similar strips have been placed in varying types of enclosures and attached to the sampling taps on carbon vessels (see **Figure 6**). The tap is opened to allow air to escape from the vessel and pass through the enclosure fitted with the lead acetate strip or round shape.

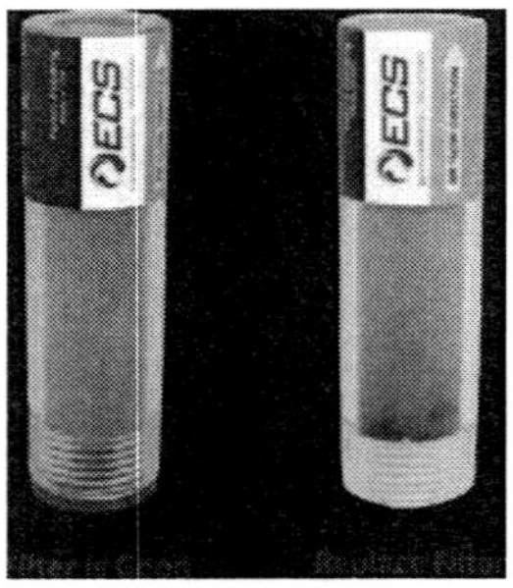

Figure 6 – Lead Acetate Paper Monitoring Device for Carbon Vessels

When H_2S gets to the sample tap the lead acetate paper will discolor and staff will know that the carbon bed up to that tap has reached saturation. You may find your own preferred use for these lead acetate strips.

Hydrogen sulfide can also be monitored using detection tubes, as shown in **Figure 7**. Detection tubes are glass tubes that are filled with media that will react with H_2S. The presence of H_2S will create a stain in the media. The concentration of H_2S will be determined by the media in the tube and the amount of air that flows through the tube.

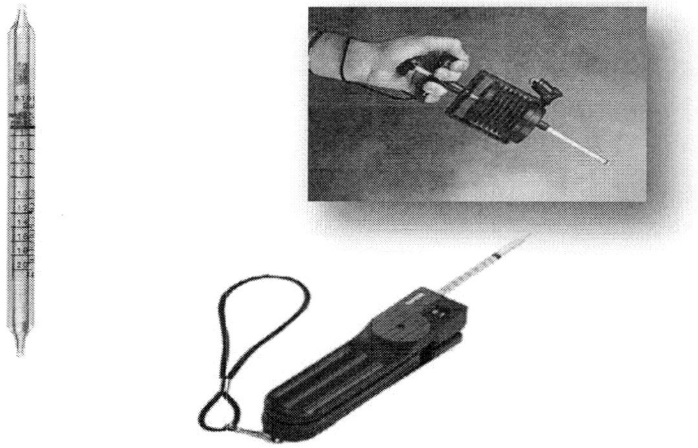

Figure 7 – H2S Detection Tube & Devices

To use a detection tube, both ends are broken off to allow air to pass through the tube. One end, indicated by a directional flow arrow imprinted on the side of the tube, is inserted into a hand pump. The other end is placed where the sample of air to be tested is located. The hand pump can be a bellows or piston type pump. The number of pump strokes are listed by the manufacturer for the specific tube and concentration range desired. The concentration graduations on the tube are also stipulated by the manufacturer. Detection tubes are used for grab samples and the results provide an order of magnitude indication of the H_2S concentration. For more refined concentrations, devices presented previously are more commonly used to assess H_2S levels at wastewater facilities.

The tubes are used once and disposed of properly. Detection tubes are popular when no other more refined monitors are available. They are also used for the initial screening of wastewater facilities. The detection tubes do have a limited shelf life, so attention should be focused on the expiration date.

Detection tubes are also available to monitor other odorous wastewater compounds such as ammonia, mercaptans, amines, VOCs, etc. **Figure 8** shows a variety of detection tubes that can also be used.

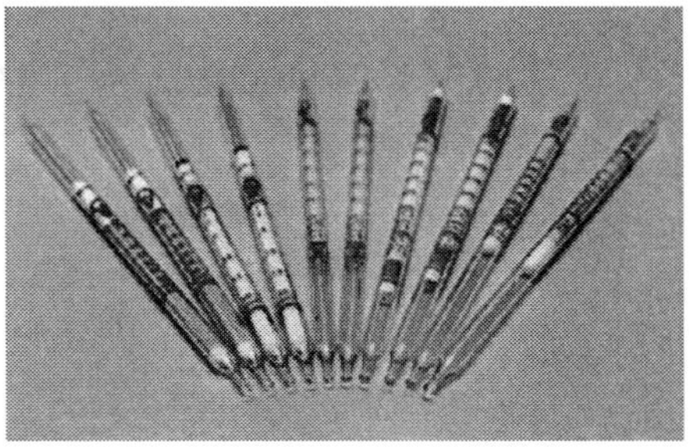

Figure 8 – Air Phase Detection Tubes

Reduced Sulfur Compounds
Next to H_2S the reduced sulfur compounds (RSC) have the lowest odor threshold concentrations and are important to identify and quantify during an odor investigation. These compounds are not measured in the field and require a sophisticated analytical process to determine their concentrations. Samples are collected in the field, as discussed later, and sent to an independent laboratory for analysis.

Ammonia
Figure 9 presents a hand-held ammonia monitor that can be used in the field to monitor grab samples. Equipped with an internal pump and battery, this monitor can measure ppm concentrations of ammonia in the air. Ammonia is not found as often in wastewater facilities and when it is it is more often associated with solids handling operations. The odor threshold level for ammonia in the published literature is in the ppm range with those industry critiqued threshold studies indicating that the detection threshold level is 17 ppm. At a detection threshold level that is four orders of magnitude greater than a compound such as H_2S, the concentrations of ammonia must be significantly greater to represent an off-site impact. Such levels can be found at wastewater facilities and it is a device such as this ppm unit that can be used to monitor for it.

When a monitor such as this is not available, a more practical way to screen for ammonia may be to use detection tubes which are available for ammonia.

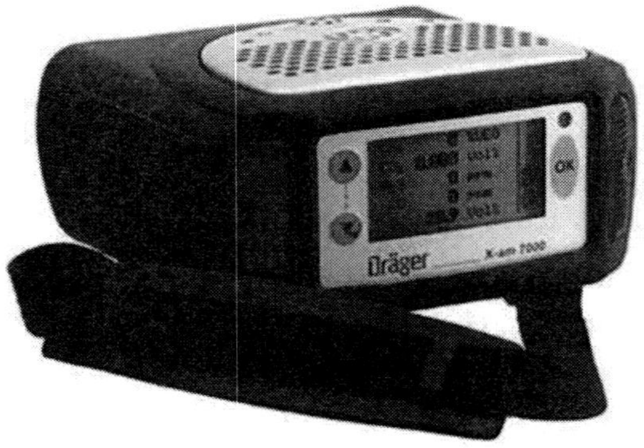

Figure 9 – Ammonia Monitor

Field Odor Strength

When it comes to monitoring for odor parameters it is generally best to collect samples and send them to an independent odor laboratory within 24-hours of collection where they can be evaluated by a trained odor panel. However, an individual conducting an odor investigation can be trained to assess some of the odor parameters in the field. For example, **Figure 10** displays a field olfactometer device that can be used to determine field odor strength (Odor strength is a measure of the number of times an odor sample must be diluted with equal volumes of odor free air before the average person does not detect that an odor exists. The number of dilutions is referred to as odor units.) levels.

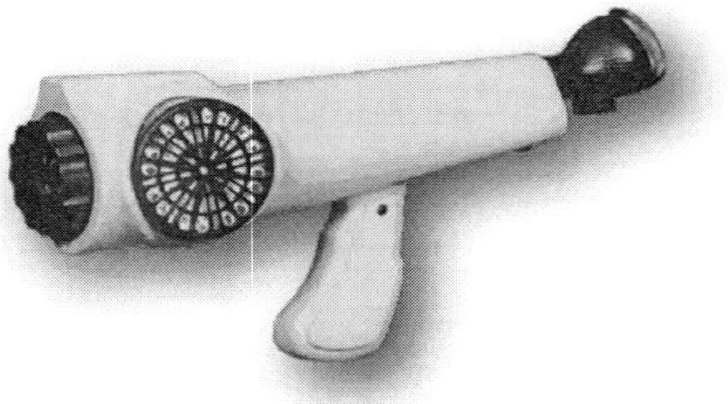

Figure 10 – Field Olfactometer

This device has a range of 2 to 60 odor units and is designed to evaluate odor strength in ambient air. The field observer securely places their nose in the nose port to the right and breathes in at a rate that is checked by the device (A green light is displayed when the correct volume of air is being applied.). A dial at the left front of the unit is turned to the various field odor strength levels provided until the correct value is observed. Space is provided between each field odor strength setting to pause and allow the observer to "cleanse their palate (nose)" by enabling them to breath in odor free air (The ambient air is put through a carbon canister to remove any odorous compounds.) before returning to evaluate subsequent dilutions. As a field-based monitoring unit, this device is normally used at fence line and community ambient air locations.

Field Odor Parameters
Some of the other odor parameters (odor intensity, odor character and hedonic tone) can also be evaluated anywhere in the field and at the facilities. The field investigator with training can be able to make observations on these three parameters. No instrument is required, just the observers nose. Accordingly, these observations are subjective but are a good way to screen the odor situation, understand specific odor characters, and potentially track the perceived odors back to a unit process source.

Electronic Nose
A more recent monitoring device that is still being evaluated to see how it can best be used to assist in an odor investigation is the electronic nose illustrated in **Figure 11**. This hand-held device does not monitor for a specific parameter, but rather consists of a neural network that can be calibrated to identify certain odors, where an odor is a combination of several compounds. As an example, we drink and smell coffee. Coffee is the odor, but it is made up of several compounds that when blended together gives us that aroma that many people crave.

Figure 11 – Electronic Nose

Typically, the unit is calibrated to odors at a wastewater facility and levels of concern are established. When that odor is subsequently detected it is compared to the calibration to assess potential nuisance odor issues. This monitoring device is currently being used at wastewater facilities in combination with an on-site meteorological tower and an odor dispersion model to provide real time off-site odor impacts of facility-based odors.

What Does the Data Mean?

After the data is collected, reduced, tabulated, graphed and reviewed it would be good to be able to compare it against general parameter criteria to understand where the facility that the data came from stands regarding off-site odor potential. **Table 1** and **Table 2** present typical concentrations and the level of importance for each of the parameters discussed above. Remember that these tables contain generalized information and that there are always exceptions to the rule.

Table 1 – What Do the Numbers Mean?
Air (2017)

Parameter	*Units*	*Concentrations*	*Importance*
Hydrogen Sulfide	Parts per million (PPM)	< 1 1 – 10 > 10	Low Moderate High
Reduced Sulfur Compounds	Parts per billion (PPB)	< 10 10 – 500 > 500	Low Moderate High
Ammonia	Parts per million (PPM)	< 10 > 10	Low Moderate - High
Amines	Parts per million (PPM)	< 10 > 10	Low Moderate - High
Volatile Organic Compounds	Parts per million (PPM)	< 10 > 10	Low Moderate - High

Table 2 – What Do the Numbers Mean?
Air (2017)

Parameter	Units	Concentration	Importance
Odor Strength	Odor Units (OU)	< 100 100 – 1,000 > 1,000	Low Moderate High
Odor Intensity	N-Butanol Scale (1 – 8)	1 – 2 3 – 5 6 – 8	Low Moderate High
Odor Persistence	Negative Slope	-0.5 to 0.0 (Flatter) < -0.5 (Steeper)	High Low

Other Air-Related Parameters
Other monitoring devices that measure separate air phase related parameters such as differential pressure and air flow are displayed in Figures 12, 13, 14, 15 and 16. **Figures 12** displays one type of differential pressure monitor that was not designed to take measurements in the corrosive environment of wastewater facilities.

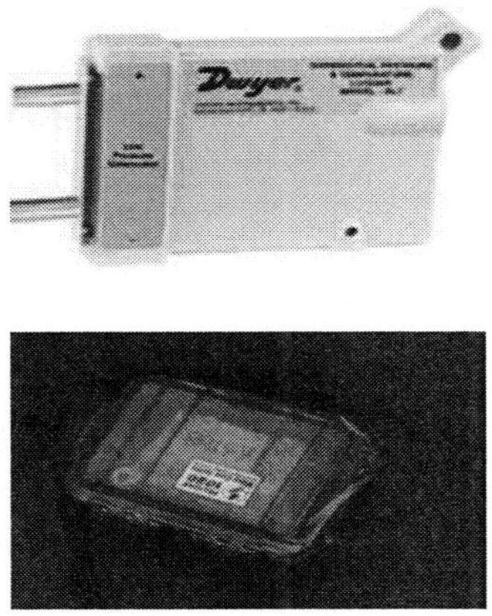

Figure 12 – Differential Pressure Data Logging Monitor & Case

To protect the device, it is placed in a plastic case shown below it. **Figure 13** illustrates another differential pressure monitoring device, but this unit is encased in a plexiglass enclosure that is integral and makes the unit intrinsically safe. Both devices are continuous data logging units and have two ports that tubing will be connected to. The end of one tube will be positioned in the headspace of the area to be monitored while the end of the other tube will be positioned to be open to the atmosphere. The differential pressure recorded will provide data that indicates whether the headspace of the area in question is negative (suggesting that the odor is contained) or positive (suggesting that the odor can be emitted to atmosphere through any crack, leak or vent). The information recorded by the devices can be downloaded and displayed in tabular or graphic form.

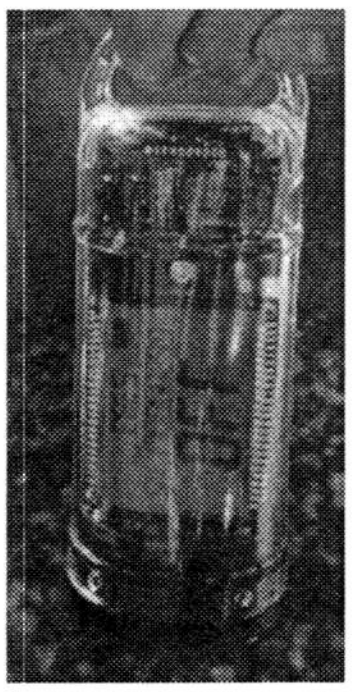

Figure 13 – Differential Pressure Data Logging Monitor & Water Trap

Figures 14, 15 and 16 present three different methods that can be used to monitor the velocity and subsequently the flow of air given the cross-sectional area of duct or structure that the air is flowing through. **Figure 14** represents one of the most basic methods for air velocity measurement using a pitot tube and gauge.

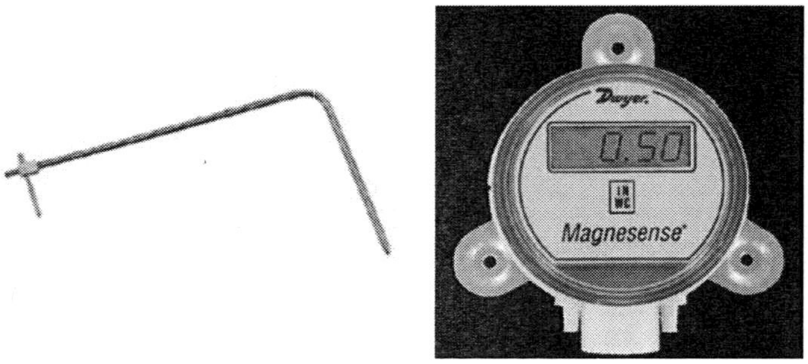

Figure 14 – Pitot Tube and Gauge

Figure 15 shows a vane anemometer with meter. The vane is located at the end of the tube to the right of the meter and can be extended in length to measure across the opening where the velocity and flow is to be determined. The velocity is visible on a digital readout on the face of the meter.

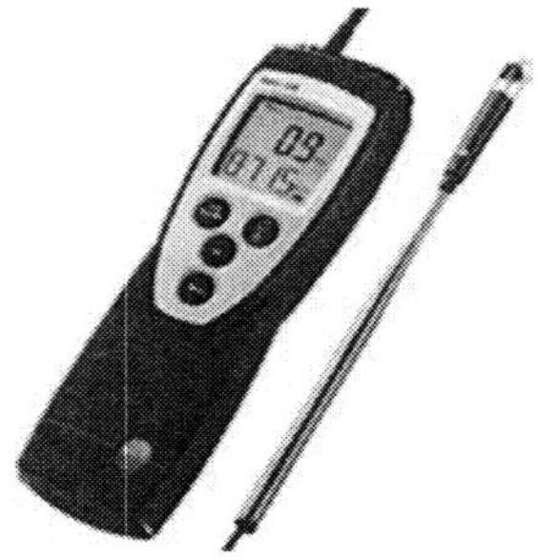

Figure 15 – Vane Anemometer

Figure 16 represents a hot wire anemometer which is applied in a similar fashion to the vane anemometer, but the velocity is determined by temperature differences across the hot wire. The velocity is also displayed digitally on the face of the meter. Any one of these devices can be used to estimate the air velocity and then flow in areas in question.

Figure 16 – Hot Wire Anemometer

Sampling Equipment

As noted above, many of the parameters of concern are not measured in the field. These parameters are collected in various containers and then sent to laboratories for evaluation. **Figures 17 and 18** present two similar devices that are used to collect air samples above a water or solids surface.

Figure 17 – Isolation Flux Chamber

Figure 18 – Isolation Flux Chambers

These devices are referred to as isolation flux chambers because they isolate a portion of the water/solids surface in question and prevent the local environment, like wind, from interfering with the collection of the sample. Odor free sweep air is also used to sweep and mix the sample within the chamber. The samples collected from the application of this flux chamber are placed in closed containers. The samples collected with the isolation flux chambers can also be converted to emission rates which are necessary for input to an odor dispersion model as well an initial screening to rank the odor sources at a facility.

The isolation flux chambers illustrated in Figures 17 and 18 can be rented or made. Not every facility has the resources to make these metal, and metal and plexiglass devices. Depending on the driving force behind the odor investigation, simpler devices may be considered. **Figure 19** shows a scaled down version of a surface isolating device referred to as a floating emission sampler. It can be constructed in a garage/basement out of PVC pipe, plexiglass panel, eye-hooks, and some rope.

Figure 19 – Floating Emission Sampler

A schematic of a floating emission sampler is presented in **Figure 20**. It accomplishes the concept of isolating a segment of the surface to be sampled and can be made to fit into areas where the full isolation flux chamber is too big.

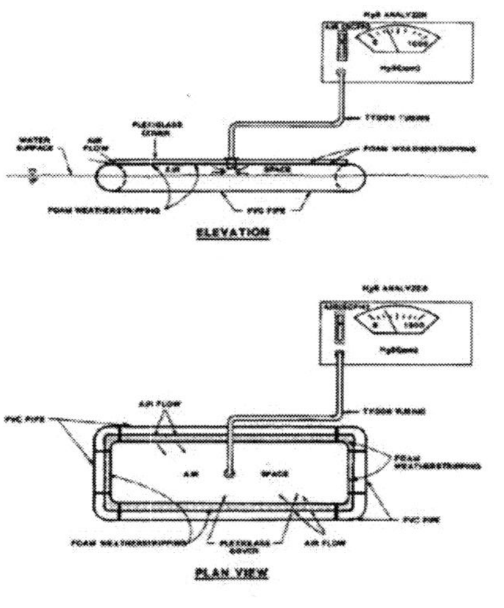

Figure 20 – Floating Emission Sampler Schematic

Whether the odorous air sample is collected from a surface isolation sampler, or directly from a stack, vent, gooseneck, or enclosed area some device is needed to move the air from the flux chamber or odor source to the container. If the positive pressure from the sampling location is not sufficient to be able to fill a container with the odorous air sample, another way to move the odorous air is necessary. An air pump is frowned upon to move the air into the container because of the potential of the pump to introduce additional odors to the sample. To avoid this potential contamination of the air sample an evacuation chamber is used, see Figure 21 and 22. The evacuation chamber in **Figure 21** can be purchased, rented, or constructed.

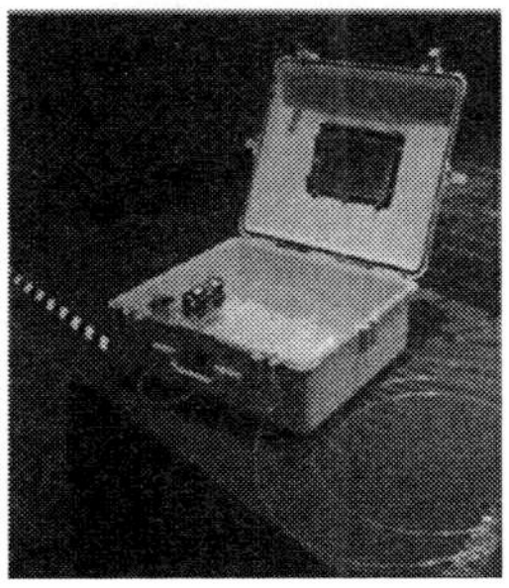

Figure 21 – Evacuation Chamber

Figure 22 is a simpler version using less expensive materials, is somewhat awkward to use, but can deliver the samples needed. The evacuation chamber consists of an air tight container where a sample tube from the odor source penetrates the wall of the container through a tight seal and attaches to the sample bag (Ex., Tedlar ®sample bag). An additional tube penetrates the wall of the container through a tight seal and attaches to an air pump. The evacuation chamber is closed, and the air pump is turned on to remove air from the inside of the evacuation chamber. As the air is removed from the chamber, odorous air is pulled through the odorous air sample line and fills the sample bag. The odorous air sample only passes through the sample tube and avoids any potential contamination from air pumps.

Figure 22 – Simple Evacuation Chambers

The bags that the odorous air samples are collected should be made of inert plastic material so that there is no loss of the odorous air through adhering to the walls of the sample bag. Figures 23 and 24 present examples of two of the most common inert plastic materials used today to make sample bags. **Figure 23** is an example of a Tedlar®, inert plastic bag, and **Figure 24** is a Nalophan®, inert plastic bag.

Figure 23 - Tedlar® Sample Bag

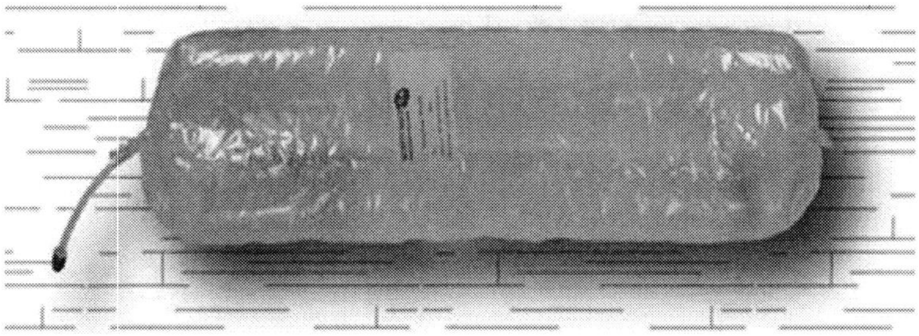

Figure 24 - Nalophan® Sample Bag

The sample bag should also be equipped with a sample tap that will allow for an air tight seal of the bag and a port that a tube can be connected to, so the bag can both be filled and exhausted. To be consistent with the inert plastic bag, the sample tap is these inert plastic bags the bags are filled with sample air, purged, and then filled again with sample air. The first bag fill is used to prepare the bag walls, as well as the walls of the sample tubing, with the sample odorous air. This way if there is any potential adhesion to the bag walls it will be accomplished with the first bag fill. In addition to inert plastic some samples are better stored in evacuated, sometimes glass lined (Necessary when analyzing for reduced sulfur compounds.), stainless steel canisters (see **Figure 25**). VOC samples are normally stored in stainless steel canisters. These canisters come pre-cleaned and are under negative pressure. When connected with the appropriate tubing and fittings the valve on the canister is opened to let the sample air into the canister. There is no ability to prepare the canister walls and so the tubing should be prepped before connecting to the canister. To prevent the sample from being collected too quickly, a flow controller can be connected to the canister valve to collect the sample more uniformly over an extended period.

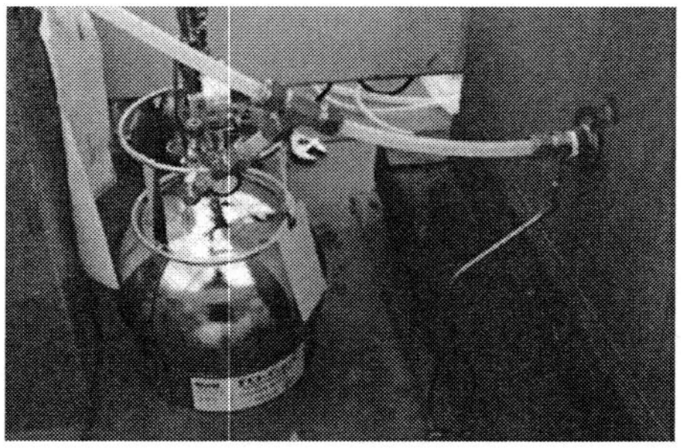

Figure 25 – Stainless Steel Canister

There are additional specialized devices, materials and procedures that are used to capture sample odorous air and send it to a laboratory for evaluation such as media tubes and impingers, but these devices are used for compounds that are not typically key nuisance odor compounds and so are not discussed further in this manuscript.

Tubing is used to convey the odorous air from the source to the container. Different tubing is used to accomplish this, and they are displayed in **Figure 26**.

Figure 26 – Air Sample Collection Tubing

For odor monitoring food grade Tygon® and Teflon® are typically used. When collecting VOCs stainless steel is more common. When setting up the odor monitoring plan it is always recommended that you have a conversation with the laboratory conducting the analytical work to see what they recommend.

Liquid Phase
The list of wastewater parameters that are commonly included in an odor investigation consist of the following:

- Sulfide (Total/Dissolved)
- Sulfate
- Temperature
- pH
- Oxidation Reduction Potential
- Dissolved Oxygen
- Biochemical Oxygen Demand (BOD)

Except for BOD, all the liquid phase parameters can be monitored in the field, at the sampling site. In fact, conducting the analysis in the field using a mobile laboratory precludes potential problems associated with labeling, preserving, holding times, transporting the samples, chain of custody forms, etc. In addition, the results of the analysis are known quickly, and generally in time to re-evaluate the sampling schedule and to re-test when the results do not seem to fit the location. The sampling and analytical techniques presented herein have been checked in the lab and tested in the field and have been found to be reliable.

Sulfide (Total/Dissolved)
Wastewater sulfide is a key parameter because it represents the potential of the wastewater to emit H_2S. What the wastewater is capable of emitting can be estimated by applying mass transfer equations, however, this approach is based on clean water and not wastewater. Nevertheless, Henry's coefficients can be applied to estimate H_2S emitted given the concentration of sulfide, temperature and pressure. In the case of sulfide, pH is also an important parameter. Sulfide is more soluble at higher pH ranges above neutrality and will strip out of solution quite readily at pH ranges below neutrality. Turbulence also plays a critical factor in the degree to which H_2S will strip out of solution because turbulence creates a significantly greater surface area over which the mass transfer of H_2S can occur.

Wastewater sulfide is monitored as total or dissolved. The more important parameter is dissolved sulfide because it represents the sulfide available to be stripped from solution. Total sulfide also measures the amount of sulfide that is tied up with other compounds in an organic or inorganic complex. This material will not be emitted from the wastewater as H_2S and so is not important from a nuisance odor nor a corrosive perspective. Accordingly, when conducting an odor monitoring investigation, be advised to analyze for dissolved sulfide whenever possible. Both total and dissolved are listed because the field kits that are used to measure wastewater sulfide generally take less time to analyze total sulfide than dissolved sulfide. **Figure 27** displays several field kits that are available for analyzing wastewater sulfides. They range from very simple with results that provide a range of concentrations to more detailed colorimetric tests that are based on the Standard Methods' Methylene Blue laboratory test and will define the concentration to tenths of a milligram per liter (mg/l).

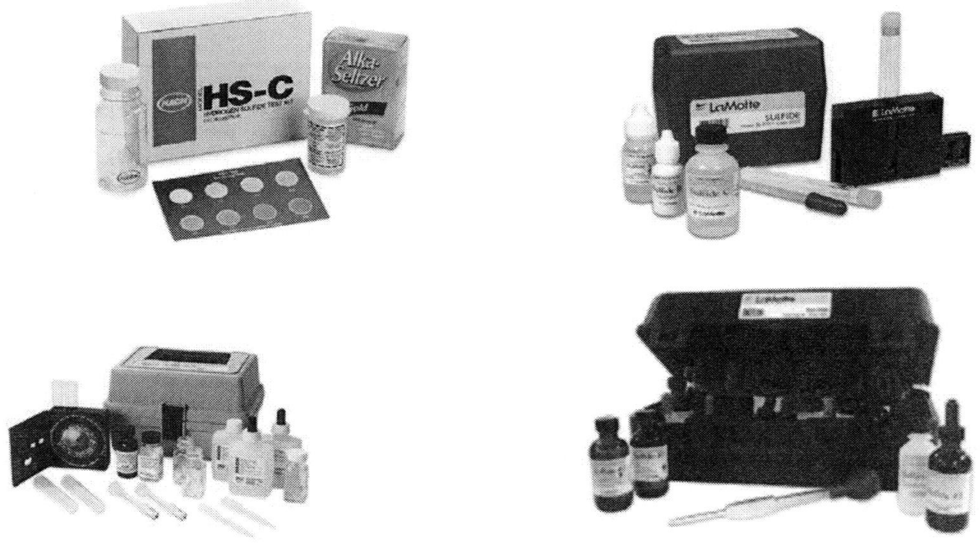

Figure 27 – Sulfide Test Kits

These test kits and others are used by many facilities in the industry. The selection is based on a personal choice, a recommendation, and/or cost. If field kits are not viable, wastewater samples can always be sent to the facility laboratory or an independent laboratory for more refined analysis with enhanced QA/QC.

Wastewater sulfides can also be analyzed by water detection tubes which are like the air detection tubes described earlier. The glass tube is filled with media that will stain in the presence of wastewater sulfide. The tube ends are removed, and the tube is placed in a container with wastewater sample in it. The tube media will discolor, and the length of the stain can be compared against the scale on the tube to determine the concentration. One drawback with these detection tubes is that they will not record sulfide levels below 1.0 mg/l. As with air detection tubes, they are not as precise as some of the field kits or a laboratory procedure, but they can be a good choice for screening elevated levels of wastewater sulfide. Understanding how these tubes can be used will help in determining whether they should be used.

Sulfate
Wastewater sulfate is the precursor to sulfide. Sulfate in the wastewater diffuses into the slime layer, along with other nutrients and dissolved oxygen, that forms on the pipe, channel or tank wall. When adequate oxygen is not available for the microorganisms that exist in the slime layer they begin to look for other ways to obtain oxygen. When sulfate is present these microorganisms will take the oxygen from the sulfate and release the sulfide. The sulfide then diffuses back out into the bulk wastewater. As a result, it is very useful to understand the concentration of sulfate in the wastewater. Like wastewater sulfide, field kits are available to analyze sulfate in the field.

Temperature, pH and ORP
Temperature, pH and ORP are all indicators of the potential of the wastewater to either produce odors or release them from the wastewater. With temperature, the warmer the wastewater the lower the solubility of dissolved oxygen and the more active microorganisms are.

The importance of pH to the solubility of wastewater sulfide was discussed above. ORP is another indicator of the aerobic or septic state of the wastewater, and the more septic the wastewater the greater the potential for generating sulfides. These parameters are monitored with hand-held probes and meters as shown in **Figure 28**. These monitoring units, equipped with batteries for use in the field, require routine calibration with buffer solutions as displayed in Figure 28. The probes should be maintained in a liquid solution between samples to protect the membranes. Readings are displayed on a digital screen on the face of the meter.

Figure 28 – pH and ORP Probes and Meters with Buffers

Dissolved Oxygen
Dissolved oxygen (DO) is an indicator of how aerobic or septic the wastewater is, which is directly proportional to the potential of the wastewater to generate H2S. Dissolved oxygen levels below 0.5 mg/l raise a flag that conditions may be right for the generation of H_2S. **Figure 29** presents devices that are used in the field to measure DO. In appearance they look similar to pH and ORP meters and probes, with the probes being the focus of the device and the most sensitive part. Also shown in Figure 29 is a field kit approach to measuring DO. Although less accurate, it would be less costly than purchasing a DO probe and meter. However, remember that almost all the devices presented in this manuscript can also be rented.

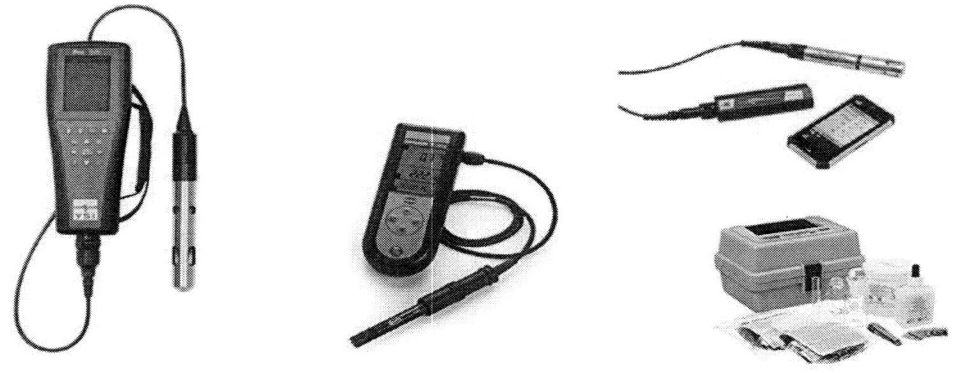

Figure 29 – Dissolved Oxygen Methods

Biochemical Oxygen Demand
Biochemical oxygen demand (BOD) represents the amount of food that is available to the microorganisms that are in the wastewater, the wall slime layer, and solids blankets across the wastewater facility. The more BOD available, the more microorganisms will exist to consume it.

The more microorganisms, the greater the demand for DO and the greater the chance that the DO level will drop below 0.5 mg/l creating septic conditions where sulfides will be formed. BOD can be an important driving force for the production of H_2S. Wastewater samples are collected and sent to a laboratory for analysis. At the laboratory the samples are analyzed for dissolved BOD.

What Does the Data Mean?
Table 3 contains a list of the concentration ranges that are important to the wastewater data collected.

Table 3 – What Do the Numbers Mean?
Wastewater (2017)

Parameter	Units	Concentration	Importance
Sulfate	mg/l mg/l mg/l	< 100 100 – 1,000 > 1,000	Low Moderate High
Sulfide	mg/l mg/l mg/l	< 0.2 0.2 to 1.0 > 1.0	Low Moderate High
Temperature	Celsius (°C) Celsius (°C) Celsius (°C)	< 18 18 to 23 24 to 30	Low Moderate High

pH	-- -- --	< 6 6 to 8 > 8	High Moderate Low
ORP	MilliVolts (mv) MilliVolts (mv) MilliVolts (mv)	< -200 100 to -200 > 100	High Moderate Low
Dissolved Oxygen	mg/l mg/l mg/l	< 0.5 0.5 to 2.0 > 2.0	Low Moderate High
BOD_5 (Dissolved)	mg/l mg/l mg/l	0 to 100 100 to 500 > 500	Low Moderate High

Physical Parameters
Two physical parameters that always seem to help understand odor conditions are surface pH and smoke tests. Surface pH is a process where litmus paper is used to evaluate the pH of the walls of structures, particularly manholes, wet wells and headworks structures where the humidity is high, and the walls are generally moist. The pH will indicate whether H_2S has been present in the headspace. In the presence of H_2S the pH of the wall will be low, below 3 to 4. If no H_2S has been present the pH will be near neutral.

Smoke testing is an excellent way to assess the tightness of a cover, enclosure, wet well, tank, etc. A smoke candle is lit, and the smoke will fill the space. The smoke will be contained by the ventilation within the space or it will be forced out through any crack, hole, opening, etc. The smoke helps in assessing the ability of the space to capture and contain odors. Areas where smoke is emitted is a release point that should be fixed. The smoke will also assist in evaluating the presence of dead zones in a building space by observing where the smoke lingers after the rest of the space has cleared out. Smoke testing must be a planned activity so that proper warnings can be made to nearby residences and/or staff. When performed in building spaces it is best to evacuate the building for although the smoke is non-toxic it is not good to breath in the particulates associated with the smoke.

Conclusion
Conducting an odor investigation takes a lot of planning and detailed understanding of the odor sources, the equipment available, how to use it, the staffing required, and what to do with the data afterwards. It may not be easy, but it is worth trying. Put together a plan and start monitoring. No plan is set in stone - and will need to be modified along the way. Let this manuscript be an initial guide.

Odor Strength – The Forgotten Odor Parameter

Monika Paduch, Former Secretary to odor groups at VDI, Dusseldorf, Germany
Carlos Diaz, Olores, Spain
Charles McGinley, St.Croix Sensory
Mike McGinley, St.Croix Sesnory, Inc.

ABSTRACT

After decades of applied olfactometry it is time that a fundamentally forgotten odor parameter of **Intensity** be highlighted and placed into the forefront of practice. The distinction needs to be made between odor concentration and odor intensity. The terminology difference is not just a minor linguistic matter: Odor Concentration (e.g. odor units) versus Odor Intensity. Further, there needs to be clarification between the stimulus of odorant(s) concentration and the effect, which is odor sensation.

Scientists and engineers in environmental protection started to quantify the properties of odor in order to avoid odor nuisance near emission sources. Methods that were developed became standards, i.e. ASTM E679 and EN 13725, and became the acknowledged basis of "odor" quantification with trained sensory assessors (a.k.a. panelists or judges).

The "problem" becomes obvious with odor abatement limits set as odor concentration without regard for the odor intensity of the limits that are set. In terms of olfactometry as applied in Europe (*Weber-Fechner*), I is the odor intensity (strength of odor sensation) and c and c_0 are the odorants concentration or the threshold concentration of odorants, respectively.

$$I = k_W \cdot \lg (c/c_0) \qquad (1)$$

The coefficient k_W is a measure of the slope of the curve (function) depicting the typical and individual relationship between odor intensity and odorant concentration of a defined source or production cycle (e.g. olfactometric fingerprint of an emission). It emphasizes the relationship between stimulus (odorants) and effect (odor intensity) with the odorant index as a mathematical link to avoid misinterpretation of olfactometric results and limit values.

The knowledge of this background requires not only revision of the interpretation of odorants abatement (commonly called odor abatement), but also of the necessary degree of gas cleaning efficiency to meet any emission limits and their results. The sections of this manuscript serve to become aware of the difference between stimuli (concentration) and effects (intensity) and the importance to distinguish them in terminology.

KEYWORDS: intensity, odor, odour, stimulus, threshold, olfactometry

INTRODUCTION

During the last decades, the assessment of odorous emissions from facilities was mainly based on standardized odor *threshold* determinations (ASTM 679-4 and EN 13725). The results are quoted as the Detection Threshold (DT) or odorant index (Odor Level) or odor units per cubic meter (ou/m³), depending on the measurement method and standard.

Definitions

Detection threshold (DT)

ASTM E679-04: "The lowest concentration of a substance in a medium relating to the lowest physical intensity at which a stimulus is detected as determined by the *best-estimate criterion,* … an interpolated concentration;…. It is the geometric mean of the last missed concentration and the next (adjacent) higher concentration."

EN 13725:2003: "(for an environmental sample) Dilution factor at which the sample has a probability of 0.5 of being detected under the conditions of the test"

Odor concentration

Commonly used to describe the dimensionless odor detection threshold value (or dilution-to-threshold value for field measurements).

EN 13725: "Number of European odor units in a cubic meter of gas (OU/m^3) at standard conditions"

Odor Index

Tenfold common logarithm of the ratio of a quantity **x** of substance A and the pertinent reference quantity **y** of substance A; a dimensionless number quoted in decibels for olfactometry (dBod).

NOTE: This definition also applies to the ***odor(ants) level*** following German standard DIN5493-2, clause 6.

Quoting the strength of the stimuli in decibels instead of in Dilution-to-Threshold values or Odor Units per m^3 is a practice in Japan and occasionally a practice in the USA. The advantage was detailed a decade ago (C. McGinley 2006).
NOTE: Remember that decibels are not a dimension or unit. They just identify the preceding ciphers as common logarithms of transformed antilog quantities.

One should bear in mind, however, that this is the quantification of the stimulus, not of the effect of odorants (mixtures). Other odor properties such as hedonic odor tone and odor quality were considered to characterize an odor situation as well as the impact duration and frequency. One odor property was mostly neglected because the quantification of the odor stimulus in terms of odorant concentration or odorant index was regarded a sufficient sensory parameter for many applications. The forgotten odor property is the odor intensity, the strength of the *suprathreshold* perception of odorants.

Importance of Unambiguous Terminology

Odor is a sensation, i.e. the effect of odorants in air in such an amount (concentration = quantity per unit of volume) that triggers the human sense of smell. Therefore, a distinction has to be made between parameters belonging to a stimulus and to the strength of perception of that stimulus (see Table 1). The authors emphasized the importance of precise terminology already in the past (*Charles McGinley* 2006, Monika Paduch 2017).

The odor threshold represents at a time the result of the determination of the stimulus strength and the starting point of the determination of the suprathreshold strength of perception of the odorants sample.

TABLE 1. Terminological distinction between quantities/properties of stimuli and effects (sensations)

Stimuli = Odorants	**Effects (Sensations = Odors)**
Source:	***Sensitivity threshold*:**
Odorants emission	Odor perception
Odorants concentration c_{od}	Odor threshold
Odorant index	Frequency of odor perception
Odor(ant) level L_{od}	
Volume flow of odorants $\dot{V}_{od}$	***Identification threshold*:**
Odorants abatement (technical degree of reduction in odorants) η_{od}	Odor identification (Identification of the odor quality and/or of the pertinent source of odorants)
Atmosphere:	
Dispersion of odorants	***Strength of odor perception*:**
(Plume of odorants)	Odor intensity response (verbal scale)
	Odor intensity I_{od}
***Ambient air*:**	Reduction in odor intensity response
Odorants in ambient air	
Plume of odorants	***Other odor properties*:**
	Hedonic odor tone (pleasant – unpleasant scale), odor quality (type of odor)

It is a common agreement that the aim of odorants abatement is to protect the neighborhood from intolerable SUPRATHRESHOLD odor impact. Therefore, the relationship between the stimulus concentration and the resulting odor intensity should be known in order to take well-founded decisions on the measures to be taken in order to improve an unsatisfactory odor situation.

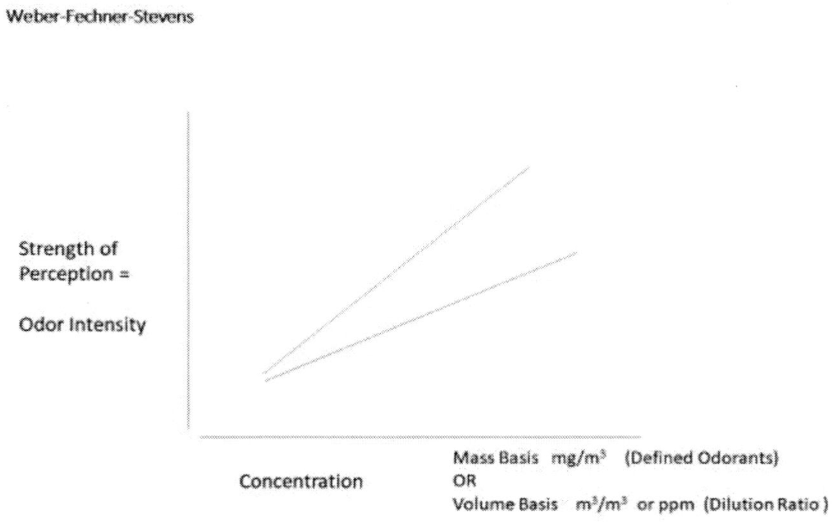

Figure 1. Fictitious 2 examples of the relationship between concentrations of defined odorants and the strength of perception.

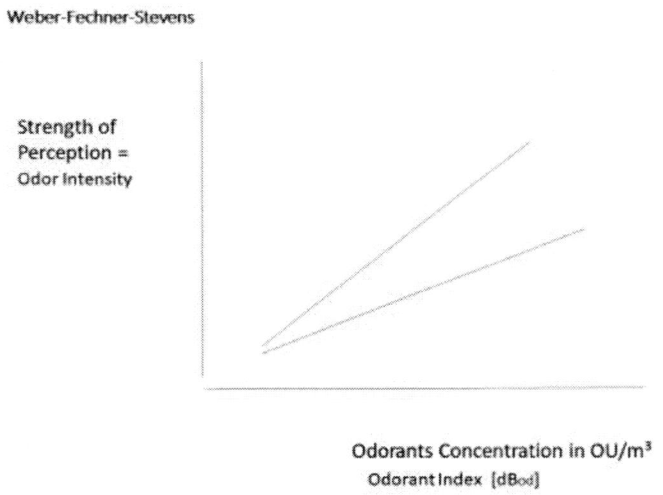

Figure 2. Fictitious 2 examples of the relationship between the odorants concentration in OU/m^3 or the odorant index as dB_{od} and the resulting strength of perception.

Methods of Odor Intensity Determination

There are two different methods known how to approach the parameter odor intensity: the standard ASTM E544-10 and the German guideline VDI 3882 Part 1 (plus amendment). The former is based on the theory of *S.S. Stevens*, the latter favors that of *G.T. Fechner* and *E.H. Weber*. A brief comparison of theories is given in Appendix A of VDI 3882-1. Both standardized methods aim at a maximum of objectivity.

ASTM E544-10 refers to the odor of any material. It invites panelists to compare the stimulus under investigation with two defined mass concentrations of n-butanol. The panelists have no time limit for decision, and adaptation to the stimuli is tolerated. The result is a correlation of stimuli with a sequence of reproducible n-butanol mass concentrations that lack a direct relationship to the human sensation expressed in unambiguous words/terms easily comprehensible by everyone.

VDI 3882-1 (under revision) filled this gap by creating a verbal scale for the intensity response (see Table 2). The guideline deals explicitly with odorants as air pollutants. The term intensity in this context refers to psychophysics, not to physical (instrumental) measurements such as those for sound intensity or radiation intensity. The distances between the six steps of the category scale are assumed to be equal. The terms were chosen with the help of psychologists. They are easily comprehensible and mark clear differences in the strength of perception. The comparable sensitivity of the selected panelists supports the assumption that their concept of e.g. a faint, moderate or very strong odor is equal.

TABLE 2: Odor intensity response as per VDI 3882 Part 1

6	extremely strong
5	very strong
4	strong
3	moderate (distinct)
2	faint
1	very faint
0	no odour

The maximum objectivity of the measurement results is achieved by the presentation method. Only the suprathreshold original stimulus is presented with the request for a spontaneous, quick answer without reflection or comparison. This is in accordance with the definition of objectivity in ISO 5492 on sensory analyses. VDI 3882-1 dates back to 1992 and does not yet consider the progress made in olfactometry up to 2017 regarding e.g. sampling technique, panel selection and overall repeatability of the olfactometric system. Nevertheless, the basic prescriptions of method are still valid. The following sections are based on the verbal intensity scale.

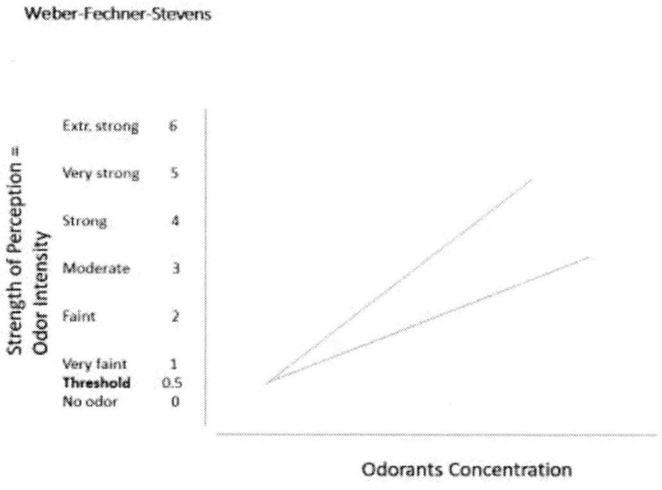

Figure 3. Fictitious 2 examples of the relationship between any odorants concentration (defined substances or undefined mixtures) and the 7-steps intensity scale as per guideline VDI 3882-1

The evaluation of the intensity responses from the panelists is similar to the evaluation of threshold measurements, i.e. the 50-percentile of responses in each category is determined and plotted over the logarithm of the concentrations presented to the panelists. By convention, the 50-percentile is defined to be the intensity category Iod + 0.5 to mark the center between two intensity categories. 50% each of the responses are then allotted to the adjacent intensity categories.

In order to adapt the odor intensity measurement to the results already gained by applying ASTM E 544-10 and n-butanol mass concentrations, a response scale was developed that facilitates this adaptation.

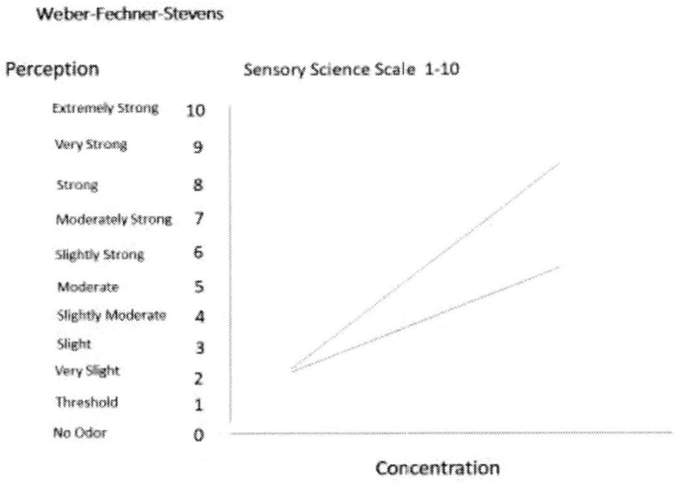

Figure 4. Fictitious 2 examples of the relationship between any odorants concentration (defined substances or undefined mixtures) and the 11-steps intensity scale developed by Sensory Science

Relationship Between Supra-threshold Concentrations and Perceived Intensity

Experience has shown that the course of the intensity responses versus the increase in odor concentration or odorant index or OU/m^3) differs as a function of the type of odorants mixture (or defined odorant). *Schulz et al.* (2002) published some experiments.

With these differences in mind, it is obvious that in practice, every threshold measurement ought to be followed by the suprathreshold intensity determination so that the characteristic **emission fingerprint** of the facility under investigation can be plotted in a diagram. Thus, the decision-makers can easily identify the **range** of odor concentrations belonging to the same intensity category. A few data from literature were taken to demonstrate the utility of an intensity measurement.

Case Example 1

Several wastewater treatment plants (WWTP) of different sizes were investigated in Italy (Capelli et al., 2009). The aim was to apply a tool for the assessment and prediction of odor impacts which is the Odor Emission Factor (OEF) (Sironi et al., 2005). The OEF in OU_E/m^3 was defined as the ratio of the odorant emission rate (OER) in OU_E/year and the annual treatment capacity C in m^3/year of the WWTP.

OEF = OER/C (2)

The following considerations are based on the results found for waste water arrival and secondary sedimentation:

Table 3: Waste water arrival – Odor(ant) concentration

1	Arithmetic mean value of the *logarithm* of the odor(ant) concentration; log c_{od}	3.354
2	Standard deviation of log c_{od} (item 1)	± 1.325
3	Odorant index of item 1	33.54 dB_{od}
4	Standard deviation of odorant index (item 3)	± 13.25 dB_{od}
5	Geometric mean value of the odor(ant) concentrations	$2.26 \cdot 10^3$ OU_E/m^3

Table 4: Secondary sedimentation – Odor(ant) concentration

1	Arithmetic mean value of the *logarithm* of the odor(ant) concentrations (c_{od})	2.086
2	Standard deviation of log c_{od} (item 1)	± 0.408
3	Odorant index of item 1	20.86 dB_{od}
4	Standard deviation of odorant index (item 3)	± 4.08 dB_{od}
5	Geometric mean value of the odor(ant) concentrations	$1.22 \cdot 10^2$ OU_E /m³

Although it is in common use, the term odor concentration does not refer to the sensation *odor*, but to the stimulus, i.e. *odorants*. As the "odor emission factor OEF" depends on the parameter "odor emission concentration", both terms need adjustment to clarify that they describe stimuli, not effects.

When looking at the odorant concentration values or the odorant indices of the waste water arrival and the secondary sedimentation, we cannot make any conclusions to the resulting strength of perception (odor intensity) that the mean values will provoke in trained panelists. Even if we find an equivalent n-butanol mass concentration as per ASTM E544-10, we can only guess what that figure would mean in terms of genuine sensational response.

This is where the necessity of an intensity scale becomes obvious.

Case Example 2

Assumption: The seller of odor abatement equipment offers his customer a system that would yield an "odor" reduction of η = 99 %. The customer will expect equipment that will reduce the "odor" emanating from his plant down to a value next to nothing. This expectation may be true for usual pollution abatement when a reduction e.g. of the mass concentration of a defined substance is concerned. But whenever sensory measurements are involved, the relationships between stimulus and effect are not proportional but logarithmic. When comparing the crude and clean gas concentrations of a source in OU_E/m^3 (e.g. 400,000

OU_E/m^3 versus 4,000 OU_E/m^3), which mathematically is the correct **odorants abatement** of $\eta = 99$ %, the remaining clean gas concentration of 4,000 OU_E/m^3 may not at all meet the requirements of **odor abatement** necessary for the environment.

Here again, the underlying error is obvious. One should be very skeptical about statements with percentages in olfactometry. Although speaking about $\eta = 99$ % **odor** abatement, the seller probably means "reduction in **odorant** concentration" (strength of stimulus) because he cannot give any other warranty. The customer understands "reduction in the perceivable **odor intensity"** (strength of effect) from his source. Misunderstandings like this one underline the importance of being precise in terminology. Moreover, they reveal the gap in knowledge concerning the real strength of perception (odor intensity).

Case Example 3

In 1993, *T.H. Misselbrook* et al. published the results of concentration and intensity measurements gained from untreated pig slurry applied on grassland and from the exhaust air of broiler houses. The concentrations (in OU/m^3) were measured with a forced-choice dynamic olfactometer, and the intensity measurements were based on the 7-point scale as per VDI 3882-1 (see Table 2). Table 5 shows the results based on Equation (1)

$$I = k_W \cdot \lg (c/c_0) \qquad (1)$$

with the addition of the pertinent odor indices.

TABLE 5: ***Olfactometric results by Misselbrook (1993) accomplished with odorant indices***

Intensity category	**Pig slurry: Odorant concen-tration in OU/m³**	**Pig slurry: Odorant index (decibels)**	**Broiler houses: Odorant concen-tration in OU/m³**	**Broiler houses: Odorant index (decibels)**
0	0 – 1.1	0 – 0.4	0 – 1.2	0 – 0.79
1	1.1 – 4.5	0.4 – 6.5	1.2 – 3.3	0.79 – 5.19
2	4.5 – 18.8	6.5 – 12.7	3.3 – 8.8	5.19 – 9.44
3	18.8 – 78.9	12.7 – 19.0	8.8 – 23.4	9.44 – 13.7
4	78.9 – 331	19.0 – 25.2	23.4 – 62.6	13.7 – 19.0
5	331 – 1390	25.2 – 31.4	62.6 – 167	19.0 – 22.2
6	> 1390	> 31.4	> 167	> 22.2

These values prove that a whole range of odorant stimuli result in the same odor intensity. Therefore, if the strength of perception (intensity category) does not change while the odorant concentration may vary to a certain extent, then the emission limitation philosophy has to be reconsidered. One of the consequences of these new aspects is the fact that any waste gas cleaning equipment will most probably be a waste of money if the clean gas concentration achieves only the same category of odor intensity response as the crude gas concentration. This is also true for intensity scales other than that of VDI 3882-1.

CONCLUSIONS AND PERSPECTIVES

The improved knowledge of the intensity and concentration relationship requires a revision and rethinking of:

- the suitability of rigid emission limits in terms of concentrations or odor(ants) indices or odorant levels,
- the interpretation of odorants abatement (commonly called odor abatement) with regard to the resulting odor intensity,
- the necessary order of magnitude of waste gas cleaning efficiency to meet any emission limits and their admissible spread of results,
- the suitability to exchange the concentration limit at the source for an intensity limit in ambient air,
- the suitability to exchange the efficiency of a reduction in odorants concentration for the efficiency of a reduction in odor intensity,
- the extension of field measurements from D/T to include intensity measurements,
- the adaptation of odor dispersion calculation to suprathreshold intensities as results instead of ambient odor concentrations,

Further research work will demonstrate how trained citizens can conduct odor monitoring in the neighborhood using a verbal intensity scale.

REFERENCES

ASTM E544-10 Standard Practices for Referencing Suprathreshold Odor Intensity. ASTM International, West Conshohocken, PA 19428-2959, USA (2010)

ASTM 679-04 Standard Practice for Determination of Odor and Taste Thresholds by a Forced-Choice Ascending Concentration Series Method of Limits. ASTM International, West Conshohocken, PA 19428-2959, USA (2004)

Capelli, Laura, Selena Sironi, Renato Del Rosso and Paolo Céntola: Predicting odor emissions from wastewater treatment plants by means of odor emission factors. Water Research 43 (2009), pp. 1977-1985

DIN 5493-2:1994 (in German): Logarithmic quantities and units; Logarithmic ratios, levels, in nepers and decibels. Berlin, Beuth Verlag

EN 13725 (2003): Air quality - Determination of odor concentration by dynamic olfactometry (plus corrigenda (2006)). European Committee for Standardization, Brussels

EN ISO 5492: Sensory analysis - Vocabulary (ISO 5492:2008), plus Amendment 1 (2017)

Fechner, G.T.: Elemente der Psychophysik. Leipzig 1860

McGinley, Charles, and Michael McGinley: An odor index Scale for Policy and Decision Making Using Ambient and Source Odor Concentrations. WEF/AWMA Specialty Conference *Odor and Air Emissions* 2006

Misselbrook, T.H., C.R. Clarkson and B.F. Pain: Relationship between concentration and intensity of odours for pig slurry and broiler houses. J. agric. Engng. Res. (1993) 55, pp. 163-169

Paduch, M.: The odor level L_{od}: the key parameter to distinguish *odorants* abatement from *odor* abatement (A discussion paper). German version: Der Geruchspegel: der Schlüssel zur Unterscheidung von Geruchsstoffminderung und Geruchsminderung (Ein Diskussionsbeitrag). Olores.org 2017

Schulz, Terry, A. Balch and S. Bowly: Odor Intensity Measurement: An Overview of its Potential for Use in Odor Impact Assessment and Control. Clean Air and Environmental Quality, Vol 36 No. 3 (Aug. 2002), pp 38-41

Sironi, S., L. Capelli, P. Céntola, R. Del Rosso and M. Il Grande: Odor emission factors for the prediction of Italian MSW landfills odor impact. Atmospheric Environment 39 (29), pp.5387-5394

Stevens, S.S.: Psychophysics. Introduction to its perceptual, neural, and social prospects. New York, J. Wiley & Sons, 1975

VDI 3882-1 (bilingual E/G) Olfactometry – Determination of Odor Intensity (plus corrigenda). Berlin, Beuth Verlag, 1992 (under revision)

Weber, E. H.: De pulsu, resorptione, auditu et tactu. Annotationes anatomicae et physiologicae. Leipzig, Verlag C.F. Koehler, 1834

CONDITION ASSESSMENT PRACTICES FOR FRP ODOR CONTROL DUCTWORK AND EQUIPMENT IN WASTEWATER TREATMENT PLANTS

By:

Randy Nixon, *President/Principal*
Douglas Sherman, P.E., *Principal Consultant*
CORROSION PROBE, INC.

Abstract

Fiberglass reinforced plastic (FRP) odor control ductwork and equipment, if properly designed, constructed and operated, should perform well for up to 50 years of service life in municipal wastewater treatment plants. However, the use of the wrong resin or curing practices can result in chemical breakdown of the corrosion barrier. Improper joining work at hand laid-up butt-and-strap joints can make the structural laminate vulnerable to chemical attack and loss of structural integrity. Additionally, excessive grit carryover from grit-removal units can cause significant corrosion barrier damage to FRP ductwork, resulting ductwork leakage and failure. These are only a few of the potential problems that can reduce the expected useful service life of FRP odor control equipment.

This paper will discuss condition assessment methods recommended for FRP ductwork and equipment based on matching the typical damage mechanisms or defects that occur to the best-suited condition assessment methods.

Assessment methods, including internal and external visual inspection, laminate sequence sampling and testing, resin type confirmation, microscopic examination, glass content testing, hardness measurement, physical property testing, and laminate thickness measurement, will be discussed and demonstrated with photographs. The importance of each assessment task will be explained, and the comparison of condition findings to the design and performance specification requirements for quality will be emphasized.

Introduction

Fiberglass reinforced plastic (FRP) is the most widely and successfully used material of construction for odor control ductwork and equipment in municipal wastewater applications. If the appropriate materials are selected and the fabrication and installation of this equipment and ductwork are properly specified and executed, FRP provides exemplary performance in these service conditions. Achieving the requisite quality for such outstanding material performance requires attention to detail in design, fabrication and installation. Once in service, FRP ductwork and equipment, including scrubbers, carbon absorbers, and other FRP equipment, needs to be periodically assessed to help assure its expected in-service performance. This paper presents the typical damage mechanisms and causes of premature degradation associated with FRP odor control ductwork and equipment. The discussion then turns to matching the appropriate condition assessment methods to those expected and commonplace damage mechanisms and defects.

Most FRP ductwork and equipment deterioration in odor control systems involves one or more of the following causal factors or damage mechanisms:

- Improper material selection, especially the FRP resin type.
- Inadequate design considerations for structural support.
- Poor fabrication practices.
- Improper field joining practices.
- Operational conditions that exceed the design criteria, and which promote premature degradation in the FRP laminate.

This paper provides technical guidance on where and how to inspect for deterioration or damage in FRP odor control equipment considering each of the causal mechanisms or contributing factors listed above. Examples of each type of degradation are explained and depicted for reference. In particular, assessment methods including both inspection watch-outs and laboratory tests are described, which provide the necessary insight into the expected service life of these important FRP assets.

Typical FRP Laminate Construction Methods

FRP material is made by mixing and combining liquid resin and fibrous reinforcing components to produce a cured, solid material with the desired properties. Unlike with metal alloys and thermoplastics, the material is created as the equipment is fabricated. A major consequence is that the properties of the FRP laminate materials can vary, depending on how much care and control are exercised when the laminate is constructed. Chemical resistance is provided by the resin matrix; strength is provided by the oriented, encapsulated fibers of glass. Components of the laminate include a glass or synthetic veil, chopped (fiberglass) strand mat (CSM), and filament-wound strands of glass called woven roving (WR).

The two most commonplace FRP construction methods for odor control equipment are as follows:

Hand Lay-Up (HLU) or Spray-Up (SU) Construction. This method is also called Contact-Molded Construction. HLU and SU laminates are made in round, square, and rectangular shapes, and are also used to make transitional shapes for round-to-square connections. **Figure 1** shows a typical cross-sectional view of HLU laminate construction. HLU construction consists of alternating layers of resin-impregnated CSM and WR in a predetermined sequence to create a solid laminate (after the resin cures) with the specified thickness and mechanical properties. The corrosion barrier is made using a glass or synthetic veil for resin-rich retention.

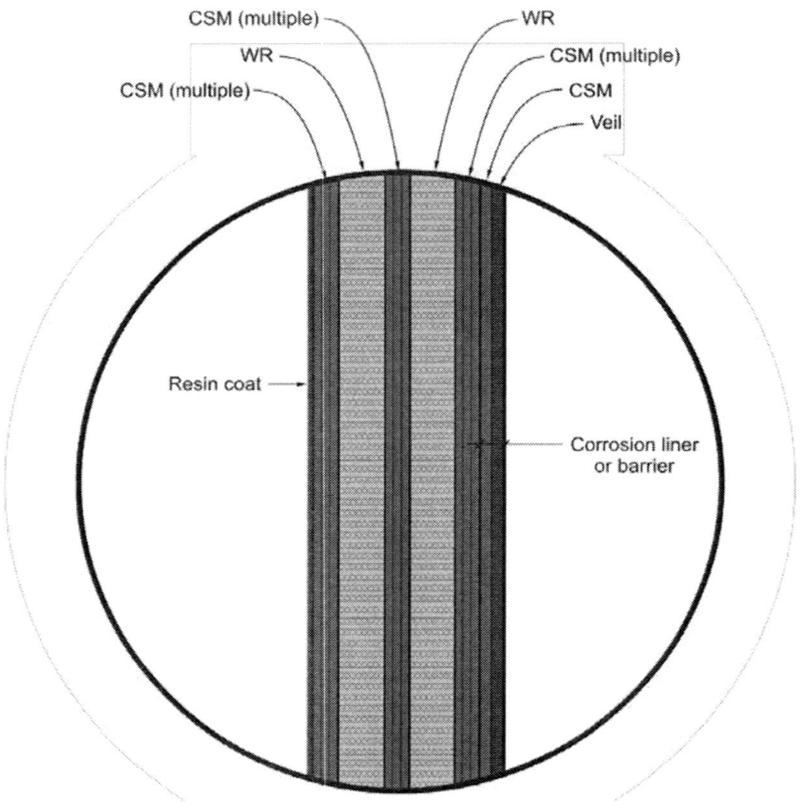

Approximate Resin Content Profile

CSM - 60% to 70%
WR - 40% to 60% (HLU)
Inner Surface CSM - 90%
Veil - 90%

Figure 1. Contact-Molded FRP (Hand Lay-up).

Filament-Wound (FW) Construction. FW construction is widely used for both ductwork and cylindrical tank construction. **Figure 2** depicts a typical FW laminate construction cross-section. This method involves automated helical winding of continuous fiber-glass threads wet with resin onto a cylinder, to produce cylindrical "cans" of specified thickness. FW construction consists of continuous WR for the structural wall. Winding angle determines axial and hoop strengths. The corrosion barrier or liner for FW equipment is typically made by hand lay-up or other contact-molded methods.

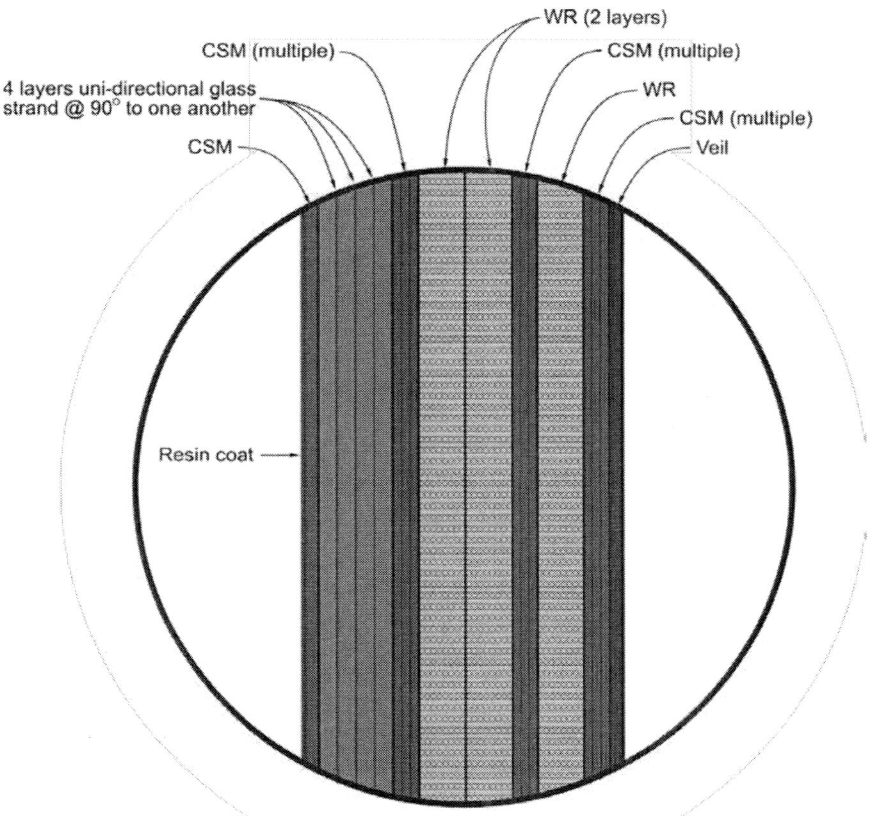

Approximate Resin Content Profile
CSM - 60% to 70%
WR - 40% to 60% (HLU)
Helical Wound Roving - 40%
Inner Surface CSM - 90%
Veil - 90%

Figure 2. Filament-Wound FRP.

Most scrubbers and carbon absorbers are constructed using contact-molded fabrication methods due to their geometry. This includes HLU and SU fabrication techniques.

Their higher glass content gives FW laminates higher strength-to-thickness ratios, but lower chemical resistance and higher permeability than HLU laminates.

For both construction methods, the corrosion barrier or liner's inner layer typically has >80% resin content (the veil-reinforced inner surface often has >90% resin). The structural laminate or wall generally contains 30 to 70% resin depending on the laminate design and how the structure was constructed.

All FRP laminate construction (whether FW or HLU) includes two major construction elements:

- The structural wall laminate that provides the shape and strength of the structure. Depending on the structural calculations for a given ductwork or equipment component, the number and sequence of the various reinforced layers of WR and CSM will vary.
- The corrosion liner, or barrier, which protects the process side of the structural wall from chemical attack or penetration by liquid or gaseous process contaminants or corrodents.

The greatest differences between the structural wall laminate and the corrosion liner include their resin-to-glass reinforcement ratios and the types of glass (or alternate synthetic material) reinforcement used within the two elements.

The corrosion barrier or liner typically consists of the inner surface, which has a high resin content and is reinforced with a glass or synthetic veil (typically 0.01 to 0.02 inch thick at the interior layer), and layers of CSM and resin (typically 0.1 inch thick).

The structural wall is reinforced with CSM (or strands when sprayed-up), WR, or combinations of these reinforcement types. An exterior corrosion layer is often included on the structural wall to protect against corrosive spillage or leakage from the equipment and ultraviolet radiation (UV) attack. This layer generally is built with CSM and resin, similar to the interior layer of the corrosion barrier or liner, but without a veil. This is often standard on well-built FRP equipment and ductwork.

Joining Methods for FRP Ductwork and Equipment

Three methods of joining FRP parts or components are most commonly used in odor control equipment. They are as follows:

Butt Joints (also called butt-and-strap joints). Components are butted together and are over-wrapped by HLU methods to form a splint type of joining. This requires careful cleaning and abrasion of the parent FRP component surfaces to be joined to ensure good bonding of the resin and reinforcement. Adhesion of wet FRP resin to cured resins relies on mechanical adhesion, rather than chemical adhesion. This is because the resins used are thermosetting resins and not thermoplastic resins. Resin and glass reinforcement are laid-up across the butted surfaces and subsequently sealed with resin overlapping onto the existing corrosion liner.

Bell-and-Spigot Joints. These joints are made for fitting one end of the component into the expanded end of the other component to which it is joined. The mating surfaces are first abraded to assure good mechanical adhesion. The joint is then filled with resin-based putty or paste and the parts are pushed together. After the excess paste is removed, it is typical to over-wrap the joint with resin and reinforcement using the HLU process, much like in the butt joint method.

Flanged Joints. Flanged joints are prefabricated for easy installation and to facilitate future access for in-service inspection or removal for maintenance. Flanged joints are generally used for ductwork manways, pipe nozzle connections, or nozzles for connections to scrubbers or absorbers. Flanged joints are constructed by HLU methods in the shop. When large flanged connections are needed, reinforcement plates of FRP are required on the vessel, tank, or ductwork wall at such connections. This helps avoid over-stressing the connection at the structure wall. Additionally, gussets are frequently used to add reinforcement to nozzles made in structures or vessel walls, again to prevent overstressing problems.

Common Damage Mechanisms and FRP Defects

Improper Resin Material Selection or Curing Chemical Degradation. The typical operating condition for the interior of FRP odor control equipment involves exposure of the FRP surfaces to varying concentrations of hydrogen sulfide (H_2S) gas, methane gas, carbon dioxide gas, various mercaptans, and wastewater liquid and biosolids particulate from airborne carry-over with air flow. Because H_2S gas concentrations can be quite high, the FRP process-side surfaces are also exposed to dilute sulfuric acid formed by biogenic sulfide acid formation. It is widely agreed by materials engineers that the sulfuric acid concentration in these applications peaks at approximately 7.5%.

Empirically, it is well-known that brominated bisphenol A epoxy vinyl ester resins perform extremely well in odor control ductwork and associated foul air treatment equipment. This is the commonly specified resin type for this type of odor control equipment. Resins with less robust chemical resistance should not be used in these service conditions. Hence, the resins used in the FRP odor control equipment should be checked periodically to ensure there are no signs of resin chemical attack or breakdown.

If the FRP resin system used was improperly selected or cured, various forms of defects may be present. One should look for:

- Softening or weakening of the resin.
- Craze cracking of the resin in the corrosion liner.
- Delamination of the corrosion barrier from the structural laminate.
- Dissolution of the resin. This will be manifested by exposure of fiberglass reinforcement, which has been disengaged due to resin degradation.
- Blisters, delamination, and swelling of the laminate may also be tell-tales of chemical attack or breakdown. This is often the result of chemical permeation of the resin, which can occur if the resin is insufficiently corrosion resistant or is improperly cured. **Photo 1** shows blistering of an improperly cured resin in the corrosion barrier in odor control ductwork.

Photo 1. Blistering of an improperly cured resin in the corrosion barrier

Probing of the cured resin to discern if it is hard and well-cured is also essential to this assessment, as is looking for color changes in the FRP resin—another tell-tale of chemical degradation.

If the resin does appear to be softened or permeated and blistered, samples of the laminate should be removed for laboratory analysis. Chemical testing of the resin used in the degraded laminate is most expeditiously done using Fournier Transform Infrared Spectrographic (FT-IR) methods. FT-IR spectra of the degraded resin in the FRP equipment and a known exemplar for the appropriate resin (brominated bisphenol A epoxy vinyl ester) are compared using a computerized database for the exemplar. Refer to **Figure 3**, below, which shows a close spectral fit between an in-service FRP ductwork resin and the appropriately specified resin system. This comparison indicated that the appropriate resin was used in the laminate construction being evaluated.

Mechanical Damage. This is a common issue in FRP odor control equipment and ductwork. FRP, unlike metals, has little capability for deformation under load, therefore it does not yield like metals. Rather, overstress or mechanical impact damage generally result in fracturing of the laminate. To assess FRP equipment for mechanical damage, one should inspect areas that are subjected to high stresses. This includes looking for cracks in the following areas:

- Gussets
- Flanges for nozzles
- Interfaces between metal mechanical supports and the FRP (e.g., hangers, saddle supports, etc.)

- Knuckles at the bottoms of tanks or vessels
- Lifting lugs
- Metal embedments or FRP attachments for ladders or baffles, pipe supports, etc. (can be internal or external)
- Hold-down lug connections

In addition, star cracks, scratches or gouges, and depressions in the laminate are likely indicators of impact damage.

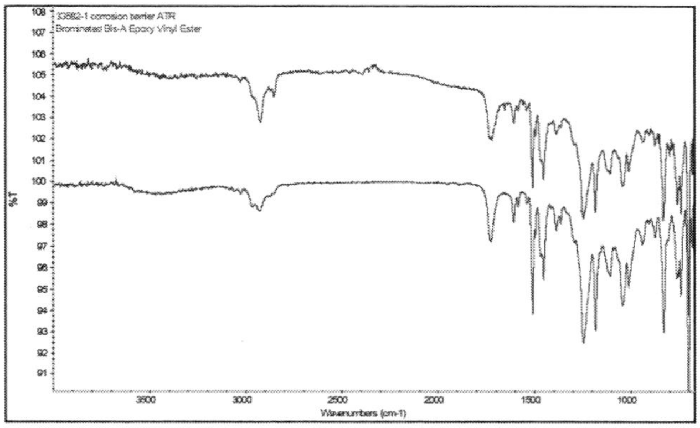

Figure 3A. FTIR-ATR spectra of FRP core inside surface (blue) and for reference sample of brominated bisphenol A epoxy vinyl ester resin (red).

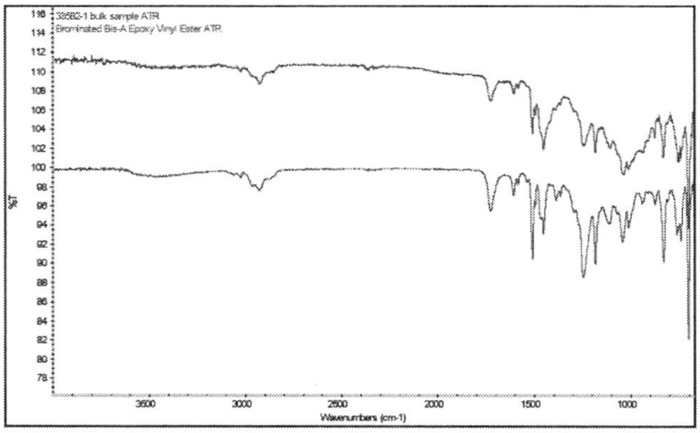

Figure 3B. FTIR-ATR spectra of FRP core outside surface (blue) and for reference sample of brominated bisphenol A epoxy vinyl ester resin (red).

Environmental Exposure Damage. Exposure to direct sunlight, which contains UV radiation, causes most organic materials to degrade to some extent. With FRP, sunlight exposure does cause the resin to embrittle and deteriorate with time. UV-resistant additives are used in FRP, and these materials help its UV resistance, but not for a long period of time. Typically, coating the FRP with UV-resistant polyurethane or other specialty coatings is required to prevent long-term UV degradation.

When UV breakdown occurs, the FRP experiences resin bloom. The white show-through color of the glass fiber will be visible initially, followed ultimately by fiber exposure.

Shop Fabrication Defects. In FRP equipment, fabrication defects can include gel coat delamination on either side of secondary joints (laid-up in the field) or the effects of using an insufficient number of the right laminate layers of reinforcement in sequence, as designed. This results in insufficient thickness of the FRP laminate. Fabrication defects also include dry pockets (insufficient resin impregnation), inadequate sealing of cut edges, unintended inclusion of materials in the laminate (such as duct tape), misalignment of components during lay-up, or failure to remove the Mylar film used as the parting or bond breaker on the mold for contact-molded FRP equipment. These types of defects should be caught during fabrication QC efforts, but sometimes get missed.

Secondary Joints or Lamination Problems and Airborne Carryover of Wastewater Constituents. Discussed previously were the commonplace joining methods for FRP odor control ductwork and equipment. These are the potential weak links most generally found in FRP equipment. Because wet resin adhesion to cured resin relies on mechanical bonding to abraded profile in the cured resin, inadequate surface preparation or a lack of good resin wetting-out of the mating surfaces for the secondary joining lay-up often result in adhesion-related voids or leak paths into the joint. In odor control ductwork, this leads to significant performance problems. The leakage of biogenic-formed sulfuric acid finds its way through the corrosion barrier and into the structural laminate, which can cause resin and reinforcement separation. Also, this contamination means successful, future repair of the leaking joint will be nearly impossible. Ultimately, the acid makes its way through the laminate, often traveling along the spaces between the less-resin-rich structural wall layers or between the HLU layer and the parent FRP.

In addition, carryover of wastewater liquid and particulate, especially from grit classifiers, can often find its way into compromised secondary joints in ductwork and equipment. This particulate carryover can be the result of inadequate baffle design for ductwork intake; excessive airflow velocity at ductwork intake; or problems with the grit removal equipment itself. This carryover of grit can cause abrasive damage to the corrosion liner, later resulting in contamination of the structural wall laminate elements. This contamination, in turn, can result in leakage problems and, as previously noted, it precludes the possibility of proper laminate repairs. This is most commonly problematic in ductwork leading to scrubbers, biofilters, and later absorbers.

Specification and Applicable Standard Compliance

When evaluating the condition of FRP equipment, it is essential to assure that design and fabrication were performed in compliance with both the equipment specifications and all pertinent standards. For the specifications, the critical compliance checks are as follows:

- Proper resin use.
- Resin-to-reinforcement ratios.
- Laminate sequence.
- Minimum laminate thickness requirements for the corrosion liner and the structural wall.
- Surface hardness: the laminate should, when fabricated and cured, have a Barcol hardness of at least 90 percent of the manufacturer's minimum specified hardness for the cured resin, when tested in accordance with ASTM D2583, *Standard Test Method for Indentation Hardness of Rigid Plastics by Means of a Barcol Impressor*.
- Minimum laminate tensile and flexural strength test values (for the new FRP equipment).

For in-service FRP odor control ductwork and equipment, it is crucial to recognize that FRP materials intrinsically lose strength properties over time. Hence, coupon removal and subsequent strength testing is extremely important in the condition assessment of in-service FRP. This test data should be compared to the originally specified strength values. As this loss in strength occurs, the follow-up testing should affirm that the FRP strength continues to meet the minimum specified values from design.

FRPI (Fiberglass Reinforced Plastics Institute) Standard Practice SP 9000, *Laminating Process Certification* should be used (if written into the project specifications) for certifying the FRP laminate process for the equipment. In addition, FRPI has developed guidelines and specifications that are extremely useful in assuring FRP quality assurance during fabrication and for certifying fabricators.

Other standards that should be reviewed for compliance prior to and during FRP equipment design and in-service inspections, when pertinent, are:

- NBS Voluntary Product Standard PS 15-69, *Custom Contact-Molded Reinforced-Polyester Chemical-Resistant Process Equipment*
- ASME RTP-1, *Reinforced Thermoset Plastic Corrosion-Resistant Equipment*

Recommended Condition Assessment Methods

Based on the damage mechanisms and defects (described above) most commonly found in FRP odor control equipment, the following condition assessment methods are recommended:

General Condition Assessment. For general condition assessment purposes, samples of the FRP can be removed by core drilling or cutting through the FRP laminate. This should be done only where evidence of defects or attack has been identified. Sampling can be effectively performed with odor control equipment on-line or off-line depending upon specific plant operating conditions, using the methods depicted in **Photos 2 through 9**. Scrubbers and absorbers need to be off-line for this sampling. Ductwork can often be sampled at lower fan airflow velocities without full shutdown. The removed core samples can be sized according to the test needs. For example, a 6-inch core can be used for laminate glass content and sequence, while an 18-inch core is needed for physical strength testing of the laminate.

Coupon sampling also is a key way to measure and assure that FRP laminate thickness requirements for the equipment are met.

Material Selection. To confirm the proper resin material was used in the FRP, removed samples of the FRP can be analyzed by FT-IR to compare the chemical composition spectrum with the known FT-IR spectrum for the specified resin. **Figure 3** above shows a typical FT-IR comparison.

Again, visual evidence of FRP degradation may lead one to perform this sampling and testing at failed and non-failed locations for comparison purposes.

Proper Cure or Resin Degradation. To check that the resin properly cured or has not degraded, visual inspection and physical probing with a knife or scraper are first recommended. If obvious softening is not indicated in this way, Barcol hardness testing, as described above, can be performed on removed coupons to discern if the FRP has lost hardness and did not initially cure properly.

Mechanical Damage. This is primarily detected by visual examination of both the inside and outside surfaces of the ductwork or equipment. Look for the tell-tales described earlier in this paper—such as cracking or impact damage—and document those indications with digital photographs and written descriptions (Refer to **Photo 10**).

Environmental Defects. These defects, which mostly involve UV degradation of the resin and ultimate fiber reinforcement exposure, are detected visually. **Photo 11** shows a typical case of UV degradation.

Photo 2: Area for coupon removal is taped off, abraded and made ready for coring (or cut-out) removal of the coupon.

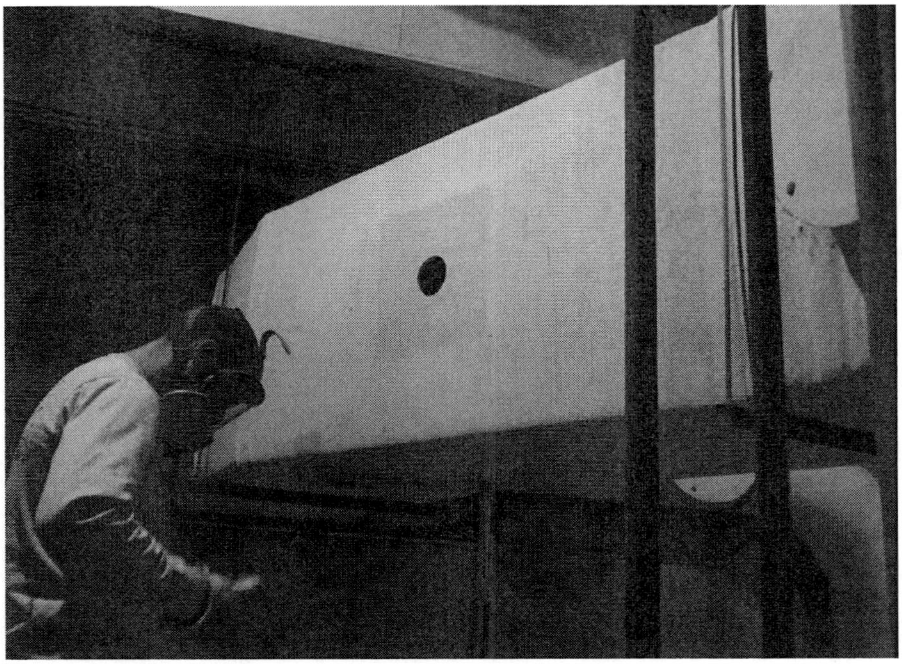

Photo 3: Hole is cored and coupon removed.

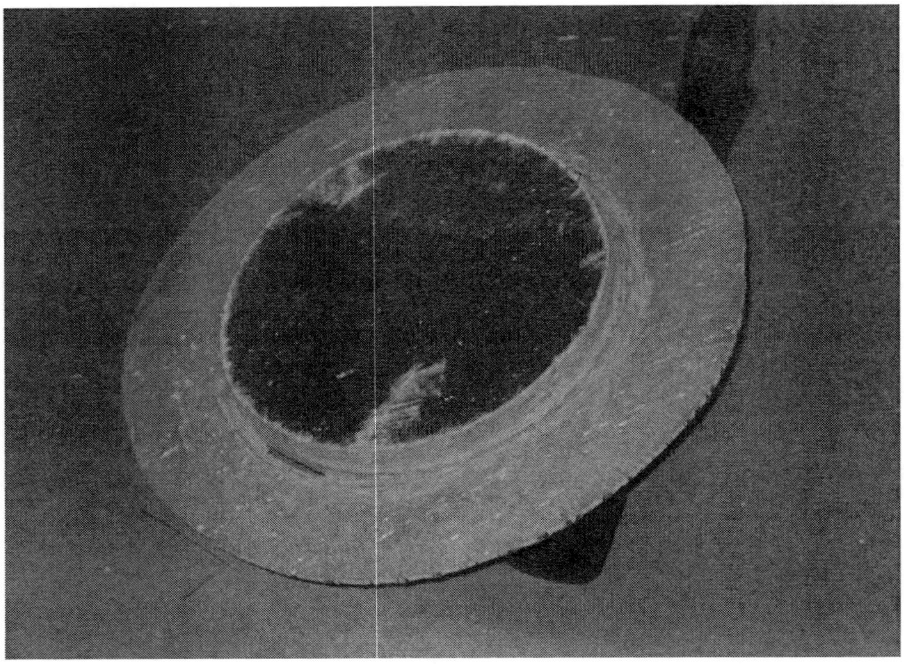

Photo 4: A pre-made flanged plug, slightly undersized to the core hole and having full wall thickness, is made ready for the repair. It includes an integral thinner flange for placement and subsequent HLU mating to the parent ductwork surfaces.

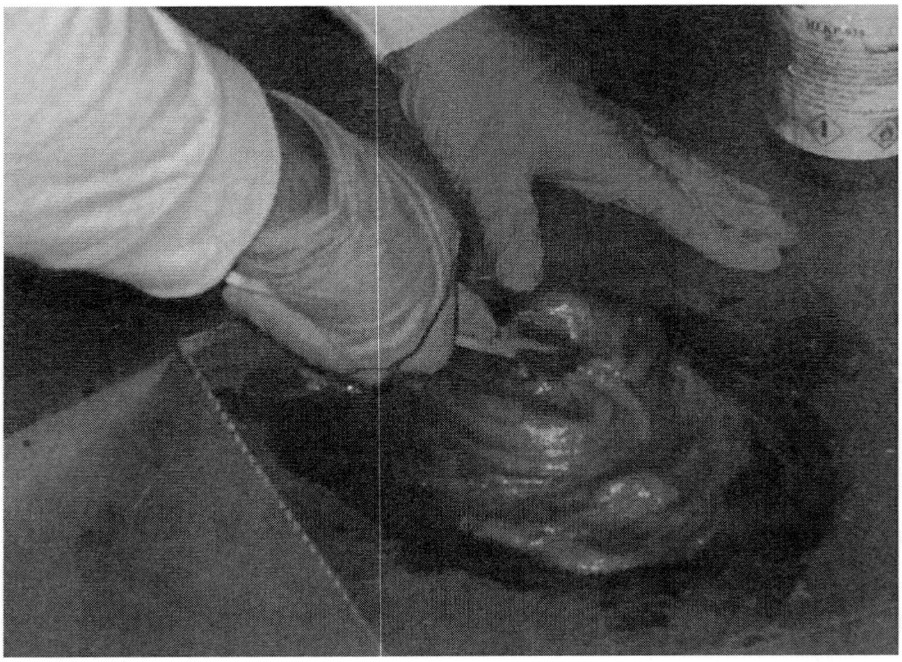

Photo 5: A thixotropic, rapid-setting vinyl ester resin-based paste is mixed and made ready for repair plug installation.

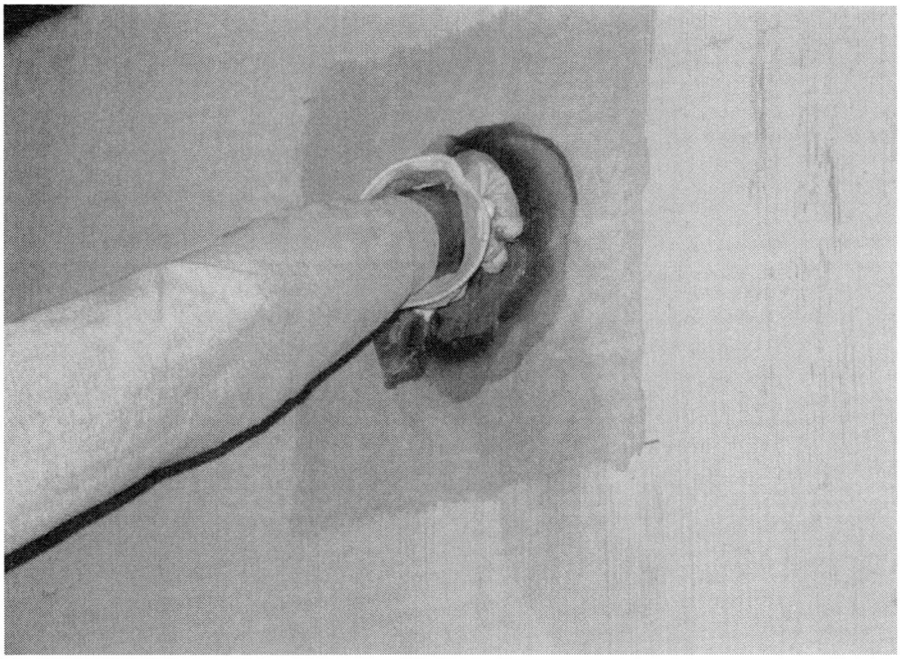

Photo 6: Plug is installed and held in place for initial set.

Photo 7: Excess paste is removed and surface smoothed out.

Photo 8: Hand lay-up of CSM and resin is installed—three layers to seal in the plug repair area.

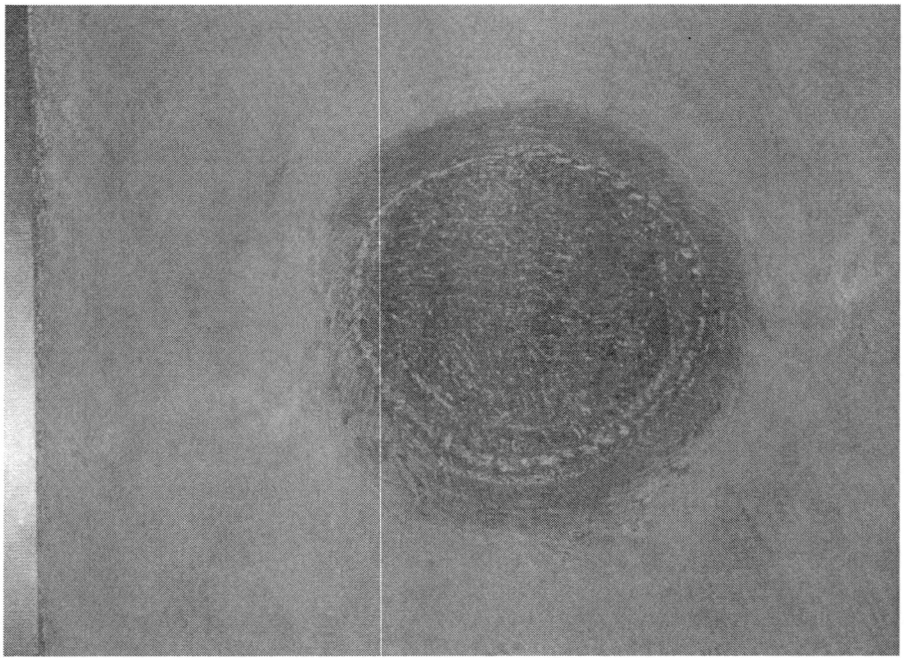

Photo 9: Completed repair of coupon removal area.

Photo 10: Typical gusset cracking which can be very problematic for nozzle connections.

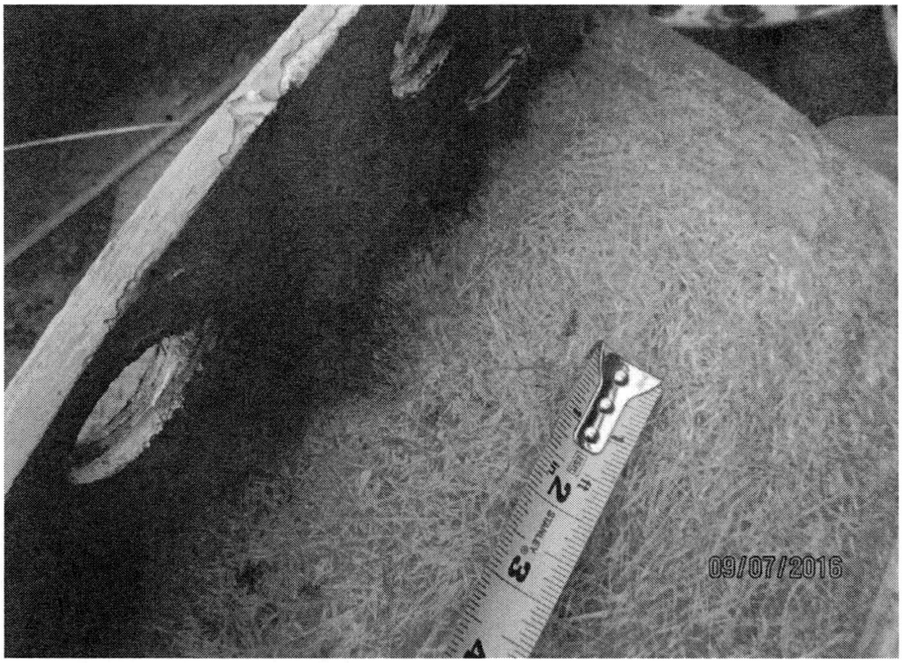

Photo 11: Typical case of UV degradation.

Fabrication Defects. To check for fabrication defects—assuming visual inspection has shown evidence of cracking, crazing, deflection, or laminate delamination—coupons should be removed and tested for resin-to-reinforcement ratio (glass content) and for laminate sequence. These examinations are performed in accordance with ASTM D2584, *Standard Test Method for Ignition Loss of Cured Reinforced Resins*. The resin is burned off and what is left is the glass reinforcement. From this test, the resin content can be checked on a percentage basis to determine compliance with the specifications and pertinent standards. In addition, the laminate sequence can be laid out to assess if all laminate layers and reinforcement types (i.e., WR, CSM, etc.) were actually used in the fabrication as specified and designed.

Refer to **Figures 4 and 5**, which show laminate sequence evaluation for two coupon samples from odor control ductwork at two different odor control facilities.

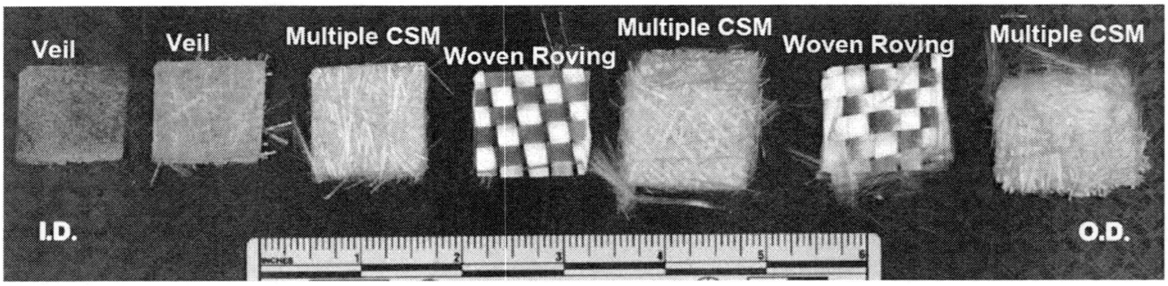

Figure 4. Laminate sequence from FRP core.

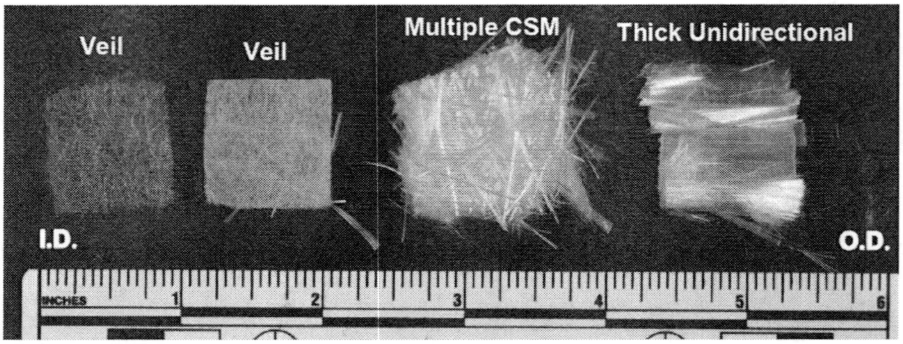

Figure 5. Laminate sequence from FRP core.

The laminate sequence analyses shown in **Figure 4** revealed double resin-rich veils, which exceeded the 20-mil minimum thickness required by the project specifications. The laminate sequence analyses shown in **Figure 5** also revealed double resin-rich veils, which exceeded the 20-mil minimum thickness requested by the project specifications; however, one CSM layer was missing, per the design, which was also reflected in the laminate thickness measurement.

From the Loss-on-Ignition tests, the resin content percentages for the laminates can also be checked and compared to specified values.

Many other fabrication defects can occur in FRP equipment. Experience is essential in knowing what else to look for.

Secondary Laminations or Joint Problems. As discussed above, joint preparation, cleaning, and hand lay-up practices are of critical importance to assuring leak-free joints that provide a continuous process barrier and prevent process contamination of the structural laminate wall.

Leakage of process fluids are dead giveaways that either the corrosion barrier has been compromised or joining quality was poor. Notice the leakage of carry-over of wastewater condensation and biogenic-formed sulfuric acid through the ductwork structures in **Photos 12 and 13**. All of the leakage in **Photo 13** emanated from a poorly made butt joint.

Photo 12: Leakage of carry-over wastewater condensation and biogenic-formed sulfuric acid through the ductwork structure

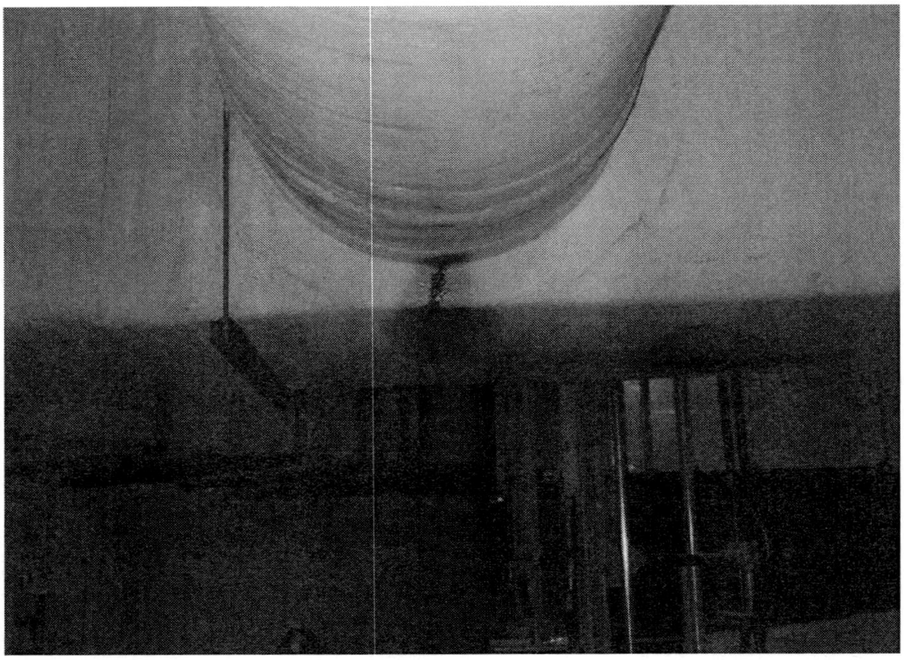

Photo 13: Typical leakage of acidic condensate at a poorly made butt joint where a round-to-square ductwork transition occurred.

Summary

Understanding how FRP odor control ductwork and equipment is fabricated and how it can degrade or otherwise fail due to built-in defects is a prerequisite for knowing where and how to look and assess its in-service condition. This paper has hopefully provided some insight to wastewater engineers and operators as to how best to have condition assessments performed on these important assets.

Converting an Odor Control Plan to Corrosion Control

Elsa Wright (Evoqua Water Technologies), Vaughan Harshman, P.E. (Evoqua Water Technologies)

ABSTRACT

Manatee County, Florida has maintained an odor and corrosion control program for over 20 years. While the program was designed with the capability to control both odors and corrosion, the primary objective of the program has been control of odors, specifically odor complaints. Recently, the County investigated changing the objectives of the program from odor to corrosion control. The County's goal for corrosion control is to protect and prolong the life of master lift stations, larger forcemains, and gravity interceptors. The objective is to reduce H_2S levels to <10 ppm in the headspace of wet wells and gravity lines. A team consisting of personnel from the County and from the County's odor and corrosion control supplier was assembled to evaluate the effectiveness and cost of program expansion to include corrosion control objectives.

The first and most expedient step in the evaluation was to increase treatment levels at existing chemical feed sites. This effort did not bring H_2S levels into the corrosion control range of ≤ 10 ppm. The team then began to investigate the County's master lift station's service areas to pinpoint sulfide generation hotspots. Key elements in identifying sulfide generation hotspots included Manatee County's available data on their collection system. A thorough investigation of the collection system was conducted resulting in recommendations for additional chemical feed locations. Due to challenges in the collection system, including availability of wastewater sampling locations, trials at expert chosen locations proved to be the best way to predict the success of a permanent chemical feed site for corrosion control. An effective corrosion control plan is made possible through a partnership with the utility, an odor and corrosion control supplier, and through the professional services of consulting engineers.

Conclusions drawn are as follows:

1. Liquid samples downstream of a lift station may not always be possible, which would be necessary to predict a chemical dose rate.
2. Some collection systems may not be a good fit for corrosion control due to layout.
3. Trial and error may be the only possible way to set a chemical dose rate for corrosion control.
4. An Odor Control Plan cannot be directly converted into a Corrosion Control Plan through turning up dose rates at existing feed sites.

KEYWORDS

Hydrogen Sulfide, Corrosion Control, Odor Control, Liquid Phase Corrosion Control, Liquid Phase Odor Control, Iterative Process

INTRODUCTION

Manatee County, Florida has had a primary focus on odor control for many years. The County and their odor control partner treat over 80 lift station, manhole, and treatment plant locations with a variety of liquid phase and vapor phase solutions. In early 2016, Manatee County requested their current odor control program be evaluated for conversion to a corrosion control program. In particular, the County wanted to explore the potential for cost savings over infrastructure replacement. The primary objective was to protect the County's master lift stations by reducing H_2S levels to <10 ppm in the headspace of wet wells and gravity lines. A level of <10 ppm H_2S vapors was chosen as this is the accepted level for corrosion control (Morton). The project would be multiphase, beginning with a report on the current state of the collection system corrosion, and ending with implementation of a corrosion control program that included optimized liquid phase feed rates of both existing and, if necessary, new dosing sites.

A project such as this may seem straight forward at first; gather sulfide data from control points, evaluate that data to see if additional corrosion treatment is necessary, adjust dose rates and/ or add feed sites to meet objectives. However, the complexity of the Manatee County collection system presented a challenge. Manatee County's assistance and wide availability of data on their collection system was a key element in making the implementation of an effective corrosion control plan possible.

METHODOLOGY

In the Master Lift Station 13-A collection system, the odor control supplier was currently treating 8 upstream lift stations. Table 1 outlines the stations treated along with the product and feed rates. A diagram of this collection system can be seen in Figure 5 on the last page. When initially evaluated, it was found that MLS 13-A had an average atmospheric sulfide level of 144 ppm and a dissolved sulfide level of 5.2 mg/L. The atmospheric sulfide levels can be seen in Figure 1.

Treated Lift Stations in MLS 13 A System		
Lift Station	Chemical	Feed Rate (GPD)
S4	Bioxide	19
Samoset 1	BP 71	72
42-A	BP71	6
9-A	Bioxide	11
Manatee Woods	BP71	35

Table 1: Odor Control treatment locations, product and feed rate for the Manatee MLS 13-A area.

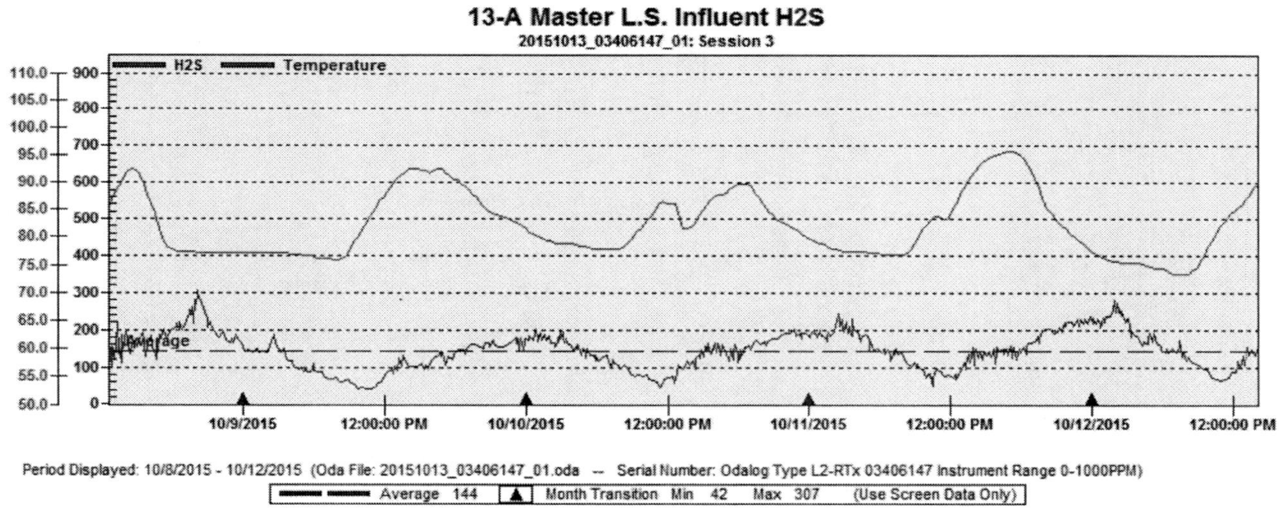

Figure 1: Atmospheric sulfide data from MLS 13-A with odor control on upstream lift stations. Taken from Biofilter inlet.

It was clear from this data that sulfide levels at MLS 13-A were not in corrosion control range of < 10 ppm. The first round of optimization was laid out; current upstream liquid phase systems were reevaluated to not only provide odor control at their downstream control point, but to now provide corrosion control at MLS 13-A. The chemical feed rates during the time data in Figure one was collected can be seen in Table 2. Feed rates were increased to the optimized level seen in Table 1 over June 16th to June 23rd, and the effect of the optimization was measured at MLS 13-A. Results can be seen in Figure 2.

Optimization Feed Rate Changes					
Chemical	Master Lift Station	Lift Station	Current GPD	Optimized GPD	Feed Rate Multiplier
BP 71	13-A	S4	19	30	1.56
	13-A	42-A	6	16	2.67
	13-A	9-A	11	33	3

Table 2: Feed rates prior to and during optimization round 1 of MLS 13-A.

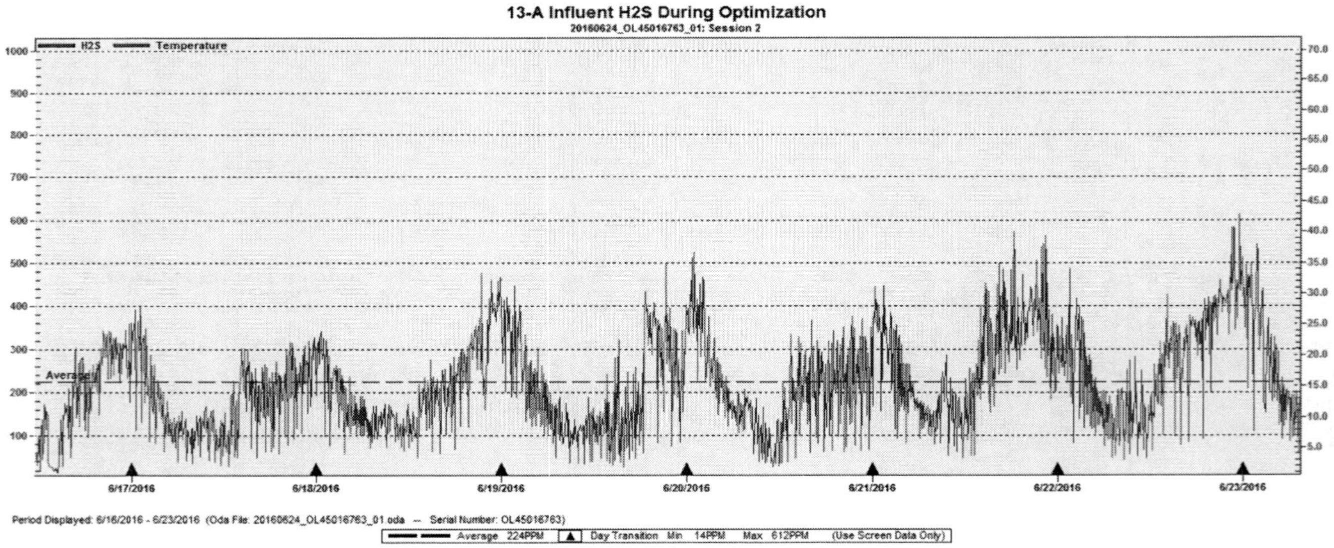

Figure 2: Atmospheric sulfide data from MLS 13-A Biofilter inlet during optimization trial 1.

An average atmospheric sulfide level of 224 ppm was observed at the increased dose rates. This is nearly 100 ppm higher than the level recorded prior to optimization. Note that the initial H_2S levels were recorded in October, when the average temperature in the area is 67° F to 85° F, and the optimization period H_2S levels were recorded in June, when the average temperature is 73° F to 90° F. Biological Oxygen Demand (BOD), and by extension H_2S generation, has a linear relationship with temperature. Therefore, it would be natural to see an increase in H_2S levels when comparing these two times of year, and perhaps more if the dose rates had not been elevated.

With these findings, it was clear the currently in place odor control program was not sufficient to protect MLS 13-A from corrosion. This was due to one main issue with the upstream chemical feed sites; they were located to control odor "hot spots", but not necessarily all major sulfide contributors. This is a key characteristic of an odor control program. While it may treat odors in populated areas and keep odor complaints under control, it leaves sites in less public areas untreated, sending dissolved and atmospheric sulfide down the line to larger lift stations. Additionally, a station where sulfide remains in the liquid phase is less likely to receive odor complaints, and therefore is left untreated in an odor control program. The dissolved sulfide is sent down the line to manifest as corrosive H_2S in the receiving infrastructure.

The next step was to pinpoint key points of sulfide generation within the MLS 13-A collection system for treatment. The most helpful tool in this process proved to be Manatee County's detailed GIS system. Using the GIS, each lift station and forcemain upstream of MLS 13-A was evaluated for its potential to generate sulfide. Forcemain length, diameter and average daily flow were key information taken from the GIS to estimate this. Additionally, liquid samples were taken from potential new feed sites to determine the level of existing dissolved sulfide, which would be sent to the MLS if left untreated.

As forcemains tied into larger trunk lines, it became more challenging to pinpoint which station was the source of the dissolved sulfide, as there were no accessible points to grab a liquid sample isolating one station's flow. Additional challenges were presented when a sample point was available, however located in a heavy traffic area, making it difficult to sample.

Despite the challenges with isolating data from individual lift stations' forcemains, a plan was devised to choose a corrosion control feed point with the greatest level of confidence. Data would have to be collected in the field by turning on each individual stations pumps and collect a liquid sample at the discharge location. Isolating flows would be possible through assistance from the County. If it was observed that a high level of dissolved sulfide was present at a discharge location, computer modeling would be used to find a chemical feed rate necessary to reduce sulfide levels. Forcemain lengths, diameters and flow rates from each station would be used to calculate detention times and model the theoretical sulfide generation. If considerable sulfide contribution was modeled from a station in addition to having sufficient detention time for a chemical reaction, it would be determined to be a site for additional chemical feed.

In Figure 3, the intersection of 53rd and 9th in the MLS 13-A collection system area is shown. At this intersection, a forcemain which captures 14 lift stations' flow coming from the North breaks into a gravity line which captures 7 additional lift stations' flow. Lift stations on both of these lines were suspected of being major sulfide contributors. Prior to this intersection, there are no accessible points for a grab sample upstream on the forcemain. A second forcemain from the South also breaks to gravity at this intersection. Additionally, a second gravity line from the East joins here. The heavy traffic at this area made sampling the gravity line impossible.

Figure 3: Intersection of 53rd & 9th. Forcemain (dotted line) dumps to gravity (solid line) at this heavy traffic intersection. (Manatee GIS)

Though there was no possible way to grab an isolated sample from the North forcemain, it was still possible to identify the sulfide contributing potential of lift stations breaking to the North gravity line before it tied into this intersection. Samples were taken from this line at the manhole pictured in the top right of Figure 3 (circled in yellow) to determine if dissolved sulfide was present in this line. Once determined that dissolved sulfide was present, samples were taken from each forcemain break to gravity from upstream points. This process made it clear which lift stations on this line were contributing the most sulfide, and therefore identified where corrosion control feed systems would be most effective. These findings can be observed in Table 3. This investigative process would not have been possible absent Manatee's GIS system and assistance in the field.

Site	ADF (MGD)	Retention Time (Hrs)	Dissolved Sulfide (mg/L)	Lbs. Sulfide Generated/ day	Effective Feed site and reasoning
S2	0.13	2.4	1.5	1.56	Yes; Sufficient Sulfide generated. Chemical fed here will capture sulfides from other stations with sufficient retention time.
S7	0.02	2.0	6.0	0.85	No. Flow too low. Sulfides generated here can be captured by a chemical feed at 30EE.
Palms Center	0.003	4.7	2.0	0.05	No. Flow too low. Sulfides generated here can be captured by a chemical feed at 30EE.
30EE	0.05	0.4	2.5	1.04	Yes; A station on this line must have a feed to capture sulfides from tie in lines. Maximum flow at this station of all possibilities on line.
Manatee Woods	0.15	1.3	0.0	0.00	Continue with current feed rate as dissolved nitrates are observed at outfall manhole.
Gravity of Manatee Woods, 30EE & S7	0.21	0.2	5.0	8.45	Feed not possible here, must occur at 30EE.

Table 3: Findings of investigation on MLS 13-A upstream lift stations. Key areas of sulfide generation identified with logic as to why these sites would make effective feed locations.

With the identification of lift stations on this gravity line as effective feed sites, there was more confidence in in the plan for permanent systems that would bring sulfide levels at MLS 13-A down to corrosion control range. However, due to the problems previously described on sampling lift station flows in the North Forcemain, the best solution was to propose a weeklong trial period of chemical feed and monitor how this effected sulfide levels at MLS 13-A.

Two additional feed sites, lift stations 30EE and S2, were selected upstream of MLS 13-A due to their larger flow rate and ability to treat incoming flows from other lift stations. The County and the odor control supplier agreed to a demonstration of a nitrate product on these lines to remove existing sulfide as well as prevent the formation of additional sulfide.

RESULTS

Results of the demonstration on additional feed sites on dissolved and atmospheric sulfide levels at MLS 13-A can be seen in Table 4 and Figure 4 below.

Dissolved Sulfide Data from 13-A influent flow collected 1/23/2018 – 1/24/2018			
Time	Dissolved Sulfide	Time	Dissolved Sulfide
11:30 AM	6 mg/L	11:30 PM	Dry
12:30 PM	6 mg/L	12:30 AM	1.5 mg/L
1:30 PM	6 mg/L	1:30 AM	1.5 mg/L
2:30 PM	5 mg/L	2:30 AM	Dry
3:30 PM	6 mg/L	3:30 AM	Dry
4:30 PM	6 mg/L	4:30 AM	Dry
5:30 PM	6 mg/L	5:30 AM	Dry
6:30 PM	6 mg/L	6:30 AM	1 mg/L
7:30 PM	6 mg/L	7:30 AM	1 mg/L
8:30 PM	Dry	8:30 AM	1 mg/L
9:30 PM	6 mg/L	9:30 AM	1 mg/L
10:30 PM	0 mg/L	10:30 AM	1 mg/L

Table 4: Dissolved Sulfide Data collected over 24 hours at MLS 13-A during demonstration period.

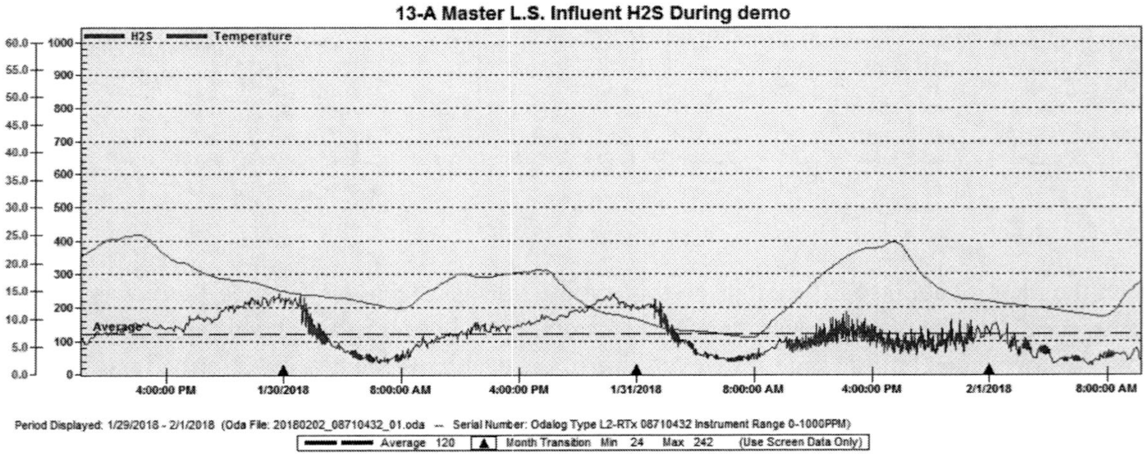

Figure 4: Atmospheric sulfide data from MLS 13-A Biofilter inlet during optimization trial 2, or the demonstration period 1/29-2/1.

MLS 13-A H_2S Data Comparison		
	Average H_2S (ppm)	Max H_2S (ppm)
Prior to Optimization (10/8-10/12)	144	307
During Optimization Phase 1 (6/16-6/23)	224	612
During Optimization Phase 2 (1/29-2/1)	120	242

Table 5: Atmospheric H2S Data at MLS 13-A compared at different phases of the optimization.

DISCUSSION

From the H_2S data in Table 5, it is observed that the additional feed sites from optimization phase 2 provide lower atmospheric sulfide levels than observed prior to the optimization or during phase 1. While these results show progress being made in bringing atmospheric sulfide levels into corrosion control range, the requirement of $\leq$ 10 ppm H_2S is not met. Dose rate increases in combination with pump timer setting adjustments are required to further reduce atmospheric levels. Work continues in this area.

Liquid sulfide data seen in Table 4 shows that the accepted corrosion control range of $\leq$ 1 mg/L dissolved sulfide is not being met at all 24 hours of the day. Further dose rate increases are necessary to bring liquid levels into corrosion control range.

CONCLUSIONS

While corrosion control programs have gained the attention of many municipalities across the United States as cost saving alternatives to infrastructure replacement, it is important to keep in mind the complexity of such a program when compared to an odor control program on the same collection system. Corrosion control is generally more complex, and the proper solution may only be presented after a number of trial and error periods.

A Utility who wishes to treat their system for corrosion control will have the most success when they have accurate and up to date data on their collection system. Flow rates, system layouts, & pump runtimes are just few examples of information Manatee County had available to assist the engineer in putting together a successful corrosion control program. This data combined with the analytical data and expertise provided by a corrosion control expert can lead to an effective and cost-efficient corrosion control program.

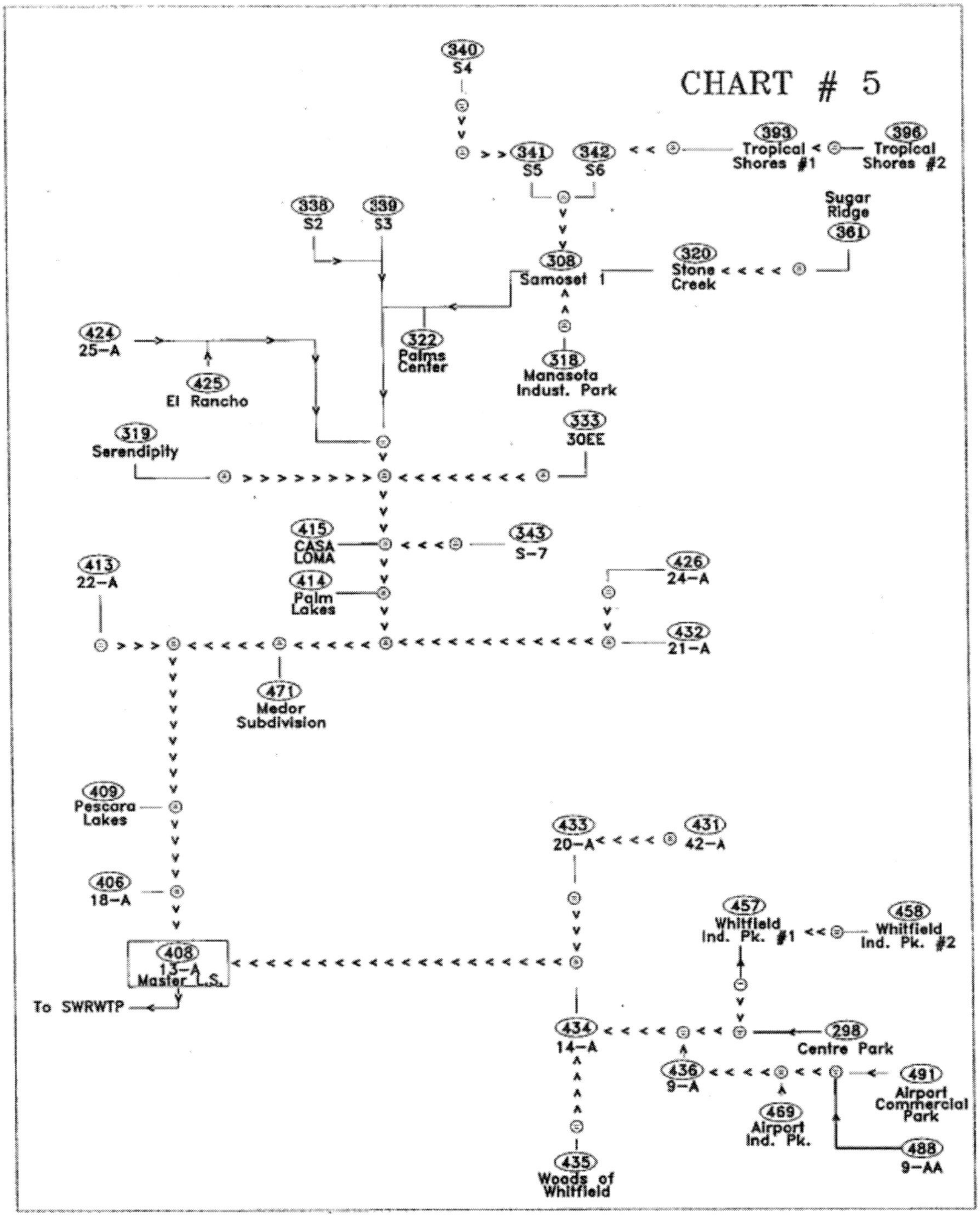

Figure 5: Manatee County Master Lift Station 13-A collection system.

REFERENCES

Manatee County GIS Mobile Server. (2015, October). Retrieved from mymanatee.org: http://www.mymanatee.org/gisapps/mobile/index.html?type=utilities-public

Robert Morton, R. C.-L. (1989). *Study Of Sulfide Generation And Concrete Corrosion Of Sanitary Sewers*. Carson, California: Sanitation Districts of Los Angeles County

Pumping a splash? It's a gas(-gas-gas)! Is your force main vulnerable to interior corrosion?

Dan Porter*[1], Mark Feltner[2]

[1]Brown and Caldwell, Virginia Beach, VA
[2]Hampton Roads Sanitation District, Virginia Beach, VA
dporter@brwncald.com

ABSTRACT

The United States Environmental Protection Agency (EPA) and the Hampton Roads Sanitation District (HRSD) agreed to a Consent Decree in 2010 with the goal of reducing sanitary sewer overflows (SSOs) in the region. HRSD maintains approximately 500 miles of interceptor sewer, 90% of which is pressurized force main (FM). Due to this heavy reliance on force mains, a specific requirement of the Consent Decree and resulting HRSD Management, Operations, and Maintenance (MOM) Program was the need for monitoring of hydrogen sulfide (H_2S) within the wastewater interceptor system. Exactly how to monitor H_2S was not specified in the Consent Decree or by the EPA, nor are there well-documented H_2S monitoring programs in response to the SSO enforcement actions for FM-reliant systems.

In the absence of H_2S monitoring program precedents, HRSD developed a baseline H_2S monitoring program with an intermediate goal of characterizing the relationship between measurable parameters and observed asset corrosion. This baseline was completed in 2013 and showed that H_2S gas alone was a poor indicator of force main risk. Therefore, the operational program, which began in 2015, must measure other corrosion precursors in addition to H_2S gas to assess risk. As the program has been implemented, HRSD has assessed the relative risk on several force mains within their condition assessment program. This paper will present the evolution of the H_2S Monitoring Program and results from follow-up condition assessment activities.

KEYWORDS: Corrosion, force main, hydrogen sulfide, Consent Decree, , MOM, monitoring program, regulatory compliance, asset management

INTRODUCTION

Hampton Roads Sanitation District (HRSD) developed a hydrogen sulfide (H_2S) baseline monitoring program to satisfy a requirement of the Consent Decree between HRSD and the United States Environmental Protection Agency (EPA). The goal of the Consent Decree is to reduce sanitary sewer overflows (SSOs) in the region. One identified cause of SSOs is corrosion in the interceptor system caused by H_2S. H_2S has been identified as one of the most prevalent and important compounds associated with internal corrosion and odor assessments in wastewater systems.

While HRSD operates some gravity sewers, the vast majority (90%) of HRSD's pipelines are pressurized. These pressure pipelines act as manifolds and receive flow from approximately 1,200 pump stations (PS) across the region. While protocols for gravity sewer H_2S monitoring programs are established, force main (FM) programs, if they exist, are largely undocumented in industry literature.

HRSD's H_2S program steering team implemented a baseline monitoring program over 2 years from 2013 to 2015. This baseline program was described by Porter, Serge, Culbertson, and Feltner in the paper *Using a Hydrogen Sulfide Monitoring Program to Assess Corrosion Risk in Interceptor Sewer System Assets. (*Presented at Water and Environment Odors and Air Pollutants Conference, Miami, FL, 2014). HRSD has since implemented an operational program that informs force main condition assessment activities.

BACKGROUND

Most corrosion- and odor-producing compounds found in domestic wastewater have a low molecular weight and are relatively volatile substances. The compounds mainly form as a result of the anaerobic decomposition of organic material containing sulfur (S) and nitrogen (N_2), which produce compounds such as H_2S, ammonia (NH_3), carbon dioxide (CO_2), and methane (CH_4). The level of odor emissions and corrosion potential in a wastewater interceptor system is primarily a function of the production and release of sulfur-containing compounds. The sulfur-containing compounds found in domestic wastewater are food for the bacteria that are found in a living matrix of filamentous microbes and gelatinous material coating the submerged walls of the interceptor pipes, commonly referred to as the slime layer. In gravity sewers, wastewater generally creates anaerobic conditions near the submerged pipe walls that allow bacteria to generate reduced sulfur compounds. Sulfide production from bacteria is partially offset by oxidation caused by air interchange with the wastewater surface. Air interchange is largely the result of turbulent flow and surface mixing. Following generation in the liquid phase, sulfides are volatilized to some degree as H_2S gas within the sewer headspace. The H_2S gas can then be deposited onto the exposed sewer surface not submerged in water, where it is converted to sulfuric acid (H_2SO_4) by aerobic bacteria. This process can lead to corrosion.

Sulfides have a far greater propensity for production in force mains than in gravity sewers because there is no air interchange with the wastewater. Once the dissolved oxygen (DO) in the wastewater is consumed, anaerobic activity accelerates and sulfide production increases. Turbulence in the flow, often due to diurnal fluctuations, releases the sulfides contained in the slime layer into the wastewater. It is quite common for gravity sewer or treatment plant odor problems to ultimately be caused by high sulfide production in upstream force mains and other points of "full pipe" flow, such as inverted siphons. As sulfide production increases, the sulfide combines with hydrogen ions to form H_2S gas, which is released from the wastewater into the headspace. Once the H_2S reaches a minimum of 2–5 parts per million (ppm) in the headspace, a potential for odor impacts and corrosion develops (WERF, 2007). The length of the trunk sewer or force main and corresponding residence time are also significant factors that affect overall sulfide production. When headspace oxygen (O_2) is present, the H_2S can be converted to H_2SO_4, which can attach to the un-submerged pipe wall, causing the potential for corrosion. The corrosion from H_2S can be localized to the crown of the pipe or the springline, or the corrosion may be spread uniformly above the waterline.

Because headspace oxygen is required for acid attack, not all assets are vulnerable. Pump stations and gravity mains operate continuously with headspace oxygen present. Only some areas of force mains have the potential to contain oxygen. Common oxygen entrainment sources in force mains are pump stations and air release vents. Pump stations have the potential for oxygen entrainment due to certain operating conditions, such as when any of the following conditions exist:

- Water is allowed to cascade into a wet well by influent lines, sump pumps, bleed lines, etc.
- The water level becomes so close to the pump suction that vortexes result
- The station is operating with a negative or low suction head so that air is pulled through the pump seals (Aguiar et al., 1995)

Air release valves (ARVs) located where the hydraulic grade line drops below the crown of the pipe have the potential to pull air into the pipeline (creating a condition where the internal pressure is lower than atmospheric which will allow air to enter the pipeline) instead of exhausting air. This can be observed in a small percentage of HRSD force main manual air vents.

METHODOLOGY

HRSD Management, Operations, and Maintenance Program

HRSD has established an EPA-approved Management, Operations, and Maintenance (MOM) Program. This MOM Program tracks and provides for continuous-improvement initiatives. Due to HRSD's heavy reliance on force mains, establishing an H_2S monitoring program was one such initiative.

Internal force main corrosion from H_2S acid attack occurs in areas which have an unwetted, unprotected pipe wall, H_2S gas, oxygen gas and sulfate reducing bacteria. Figure 1 below illustrates this type of phenomena along with other common gas and liquid components. Zones 1 through 3 show the presence of normal wastewater components and full force main flow, the entrainment of atmospheric air from an outside source and the potential corrosion location, respectively.

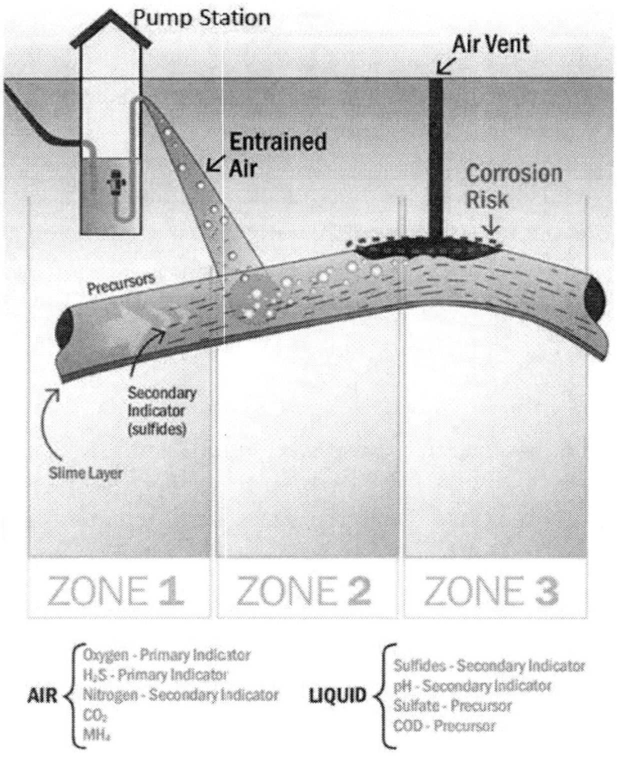

Figure1. Internal corrosion risk in force mains

HRSD used a baseline survey, completed in 2013, to characterize the relationship between corrosion and hydrogen sulfide levels in various asset classes. The results of the baseline program were documented by Porter, et.al. (2014) and did not show an observable correlation between H_2S gas concentrations and force main corrosion. However, the baseline data showed a potential relationship between sites of known corrosion and high proportions of nitrogen gas.

As stated above, oxygen is not a generated by processes within a force main; rather, it is injected or entrained from sources drawing from the ambient atmosphere. Ambient atmosphere consists of close to 78% nitrogen, 20% oxygen and 2% argon and trace gases. This nitrogen component is significant because, unlike oxygen, nitrogen is conserved in a force main. Additionally, in the HRSD system, the denitrification process is assumed to be minimal in most areas. Therefore, the presence of high proportions of nitrogen gas, even in the absence of oxygen gas, within a force main may be an indicator of previous air entrainment.

Corrosion in force mains can be highly localized as shown in Figure 2 below. Additionally, the most common methods for HRSD force main inspection are also highly localized; measuring wall thickness in points with an area of approximately 1 square centimeter. This means that it is difficult to verify the presence or absence of internal corrosion. Therefore, only areas of known internal corrosion can provide a suitable comparison to the measurable gas parameters. Other areas are suitable for providing a population of gas samples but cannot be used to assess the level at which specific gas components may be problematic.

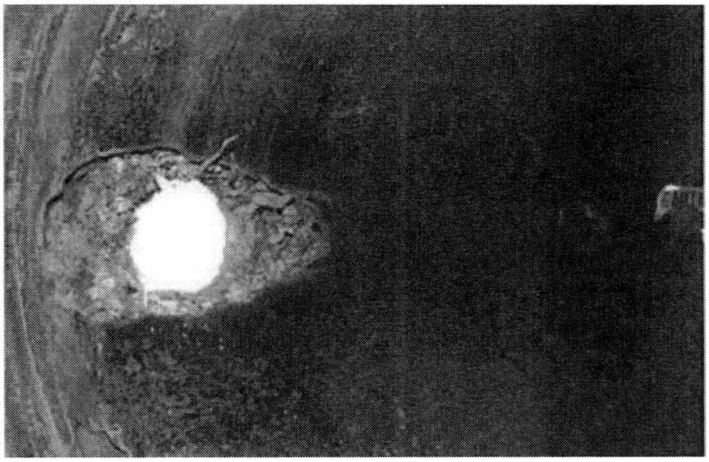

Figure 2. Localized corrosion example (courtesy of Lewis Engineering)

Given the potential relationship between nitrogen gas and force main corrosion as well as the limitations of the available condition data, HRSD developed an intermediate sampling plan in known corrosion areas to test if those areas tended to have higher nitrogen gas proportions. Three areas of known internal corrosion were selected for H_2S gas sampling for the following parameters, consistent with the baseline sampling program:

- H_2S - ppmv
- CO2 - %
- CH4 - %
- O2 - %
- Balance - % (Nitrogen)

RESULTS AND DISCUSSION

HRSD performed 2 sampling events at the three sites for the intermediate sampling plan. Sampling events were conducted in warm and cool wastewater months; similar to the baseline sampling program. The data collected is presented in Table 1. The minimum observed nitrogen gas proportion was 67% at the Long Hill Rd site. Cooler wastewater temperatures corresponded with higher nitrogen in all cases when compared to the higher temperature samples (most likely due to less methane generation in cooler conditions). Five of six measured samples had H_2S concentrations associated with highly corrosive conditions in unpressurized assets (i.e. gravity mains).

Table 1. Intermediate Sampling Results							
	JR1002B-1 Old Menchville Rd		W5016-3 Long Hill Rd		AT1148M-6 Kempsville Rd		
	October 2014	February 2015	October 2014	February 2015	July 2013 (from baseline survey	October 2014	February 2015
H_2S (ppmv)	1258	0	826	439	331	No gas	97
CH_4 (%)	21.5	7.9	31.3	12.8	22.2	No gas	10
CO_2 (%)	2.7	1.8	1.6	1.1	1.7	No gas	0.8
O_2 (%)	0.3	0.6	0.2	1.0	0.5	No gas	0.2
Balance (N_2) (%)	75.6	89.8	67	85.2	75.6	No gas	89

From the baseline sampling program and other initiatives, HRSD had a population of 76 gas samples for comparison with the intermediate sampling data. The balance gas proportion data had a range of 0.1% to 89.8% and a median of 54%. Plotting the intermediate sampling results in a cumulative frequency plot of the population of balance gas proportions (Figure 3) shows all of the intermediate sampling data points are above the 67th percentile and most are above the 75% percentile.

Given that sites with known corrosion showed relatively high proportions of balance gas, HRSD elected to incorporate H_2S gas monitoring into the force main condition assessment program. H_2S gas sampling is conducted on high risk force mains to prioritize sites for direct wall thickness testing and condition assessment. The gas data is considered along with other factors including unvented high points and potential for external corrosion. Nitrogen gas proportion in excess of 50% are considered to be at greater risk of H_2S acid attack corrosion. A process flow is shown in figure 4.

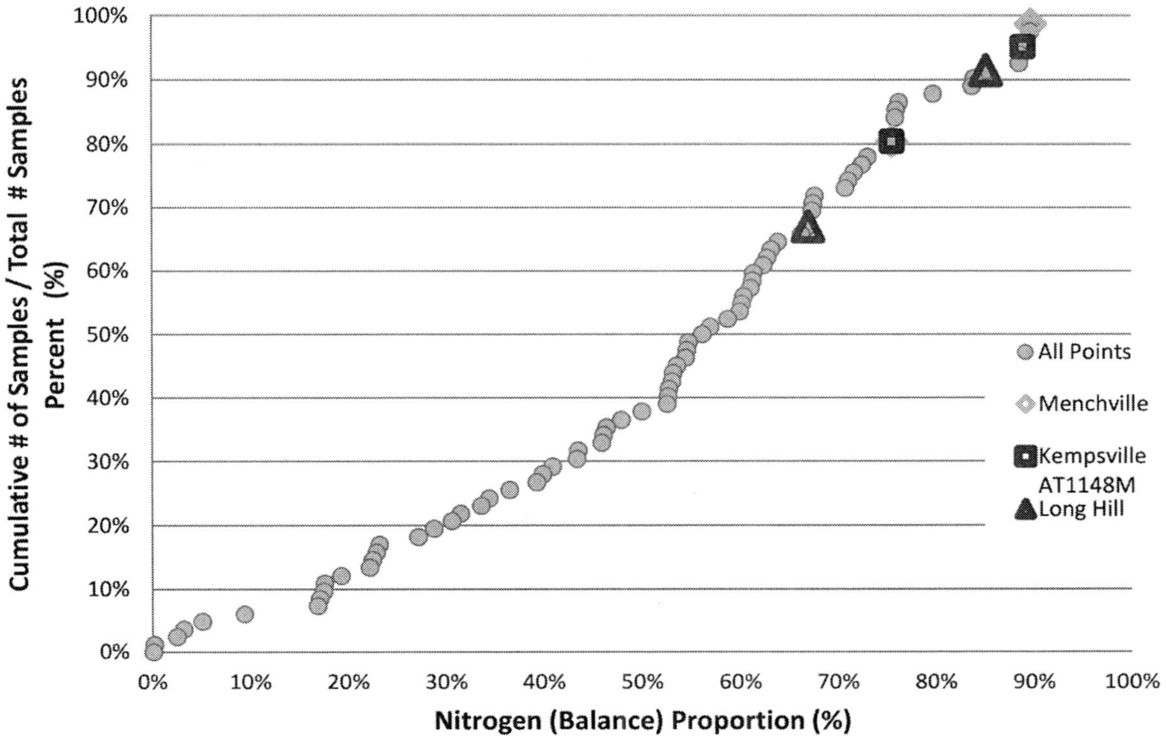

Figure 3. Nitrogen gas proportions in the HRSD interceptor system

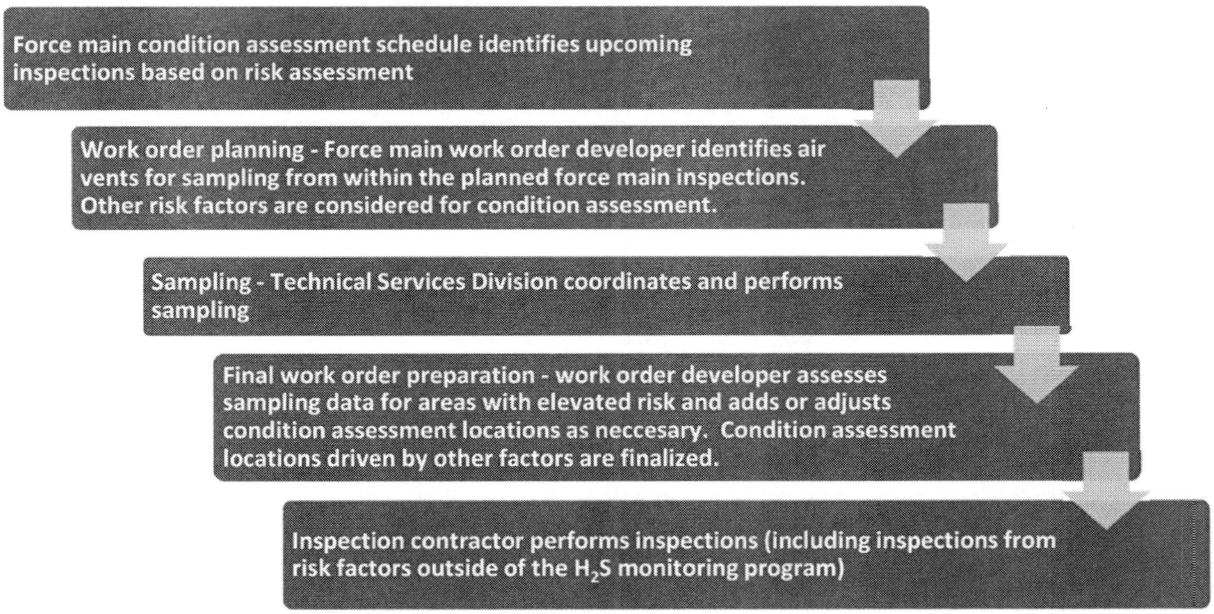

Figure 4. Integration of the H_2S Program with the condition assessment program

The operational program has sampled eight force mains through 2017. This sampling included 51 air vents. Of those 51 air vents:

- 75% had balance gas less than 50%
- Follow-on wall thickness testing at sites with greater than 50% balance gas has found no actionable wall degradation
- Other sites pending condition assessment

The H_2S program has provided increased confidence that H_2S acid attack is not occurring on the eight force mains by focusing the condition assessment near air vents where the conditions for H_2S acid attack are most conducive. Prior to the H_2S program, review of drawings and internal acoustic inspection were the only tools available to determine potentially vulnerable high points.

SUMMARY AND CONCLUSIONS

HRSD was required to implement a long term hydrogen sulfide monitoring program as part of the MOM Program. Without specific guidance for force main systems, HRSD developed a baseline plan to assess H_2S (and other measureable parameters) across the system. From the baseline plan, HRSD determined that gravity asset sampling did not enhance their ability to avoid asset failures. HRSD also determined that follow-up (intermediate) sampling was necessary to determine if measurable force main parameters could be used to help avoid asset failures. This intermediate sampling showed a potential relationship between nitrogen gas (as a surrogate for oxygen) and observed H_2S acid attack corrosion. By applying this potential relationship to high risk force mains already scheduled for inspection, HRSD has created a long term H_2S monitoring program that is not overly resource intensive and provides valuable condition assessment screening information for the highest priority force mains.

REFERENCES

Aguiar, Luis, LeBeau, Armand J., Williams, David B., Schafer, Perry L., Fuller, C. Zachary., Baylis, Jack A. (1995) Stopping Force Main Collapses and Establishing Corrosion Control – A World-Class Utility's Approach. *Presented at Water and Environment Federation Annual Conference*. Alexandria, VA.

American Water Works Association (AWWA) (1996); *Internal Corrosion of Water Distribution Systems*, Denver. Chapters 1 and 2.

Bowker, Robert P.G., Audilbert, Gerald A., Shah, Hemang J., Webster, Neil A. (1992). *Detection, Control, and Correction of Hydrogen Sulfide in Existing Wastewater Systems*; United States Environmental Protection Agency, Washington, D.C.

Lewis, Richard. (2013) *Report of Findings: Forensic Investigation of a PCCP Failure on the HRSD 24-Inch PCCP North Trunk- Part B Interceptor Force Main*; Lewis Engineering and Consulting, Inc., Gainesville, FL.

Porter, Daniel, Serge, Gina, Culbertson, Erin, Feltner, Mark (2014) Using a Hydrogen Sulfide Monitoring Program to Assess Corrosion Risk in Interceptor Sewer System Assets. *Presented at Water and Environment Odors and Air Pollutants Conference*, Miami, FL.

Water Environment Federation (WEF) (2004). *Control of Odors and Emissions from Wastewater Treatment Plants, Manual of Practice No. 25*; Alexandria, VA.

Water Environment Research Foundation (WERF) (2007); Minimization of Odors and Corrosion in Collection Systems; Alexandria, VA. Chapter 9.

Portable Odor Control – A Solution to Wastewater Construction Projects Odor Problems in Albuquerque

Authors: Neil A. Webster, P.E. (Webster Environmental Assoc., Inc.), Jon Erstgaard, P.E. (ABCWUA), Phil Parkins, P.E., (Carollo Engineers), Bruce Koetter, P.E. (WEA)

ABSTRACT

The Albuquerque Bernalillo County Water Utility Authority (Water Authority) needed portable odor control units that would provide temporary odor control during construction of various interceptor, lift station and treatment plant projects. The mobile units would be moved from site to site and would be operated by Water Authority personnel or by on-call contractors. Could they be built to include a fan, air treatment units with redundancy, control panel, ducts, dampers and other components and be kept safe from theft? Power sources needed to be from a generator or a motor control center such as a lift station.

Carollo Engineers and Webster Environmental Associates, Inc. investigated the alternatives, contacting and visiting other agencies like King County, WA and Pima County, AZ that had similar systems and consulted experienced vendors. There are roll-off containers, skid mounted systems and units built on trailers (enclosed or not enclosed) that could be purchased or rented. The consultants and Water Authority team established requirements for the project; during several workshops.

Using a decision matrix with base factors and by weighting the importance of each factor to the Water Authority, including; performance, flexibility, theft resistant, accessibility and cost, trailer mounted non-enclosed units were selected. Some preliminary design layouts were prepared and two systems were selected:

- A 5,000 cubic feet per minute (cfm) unit with three (3) rectangular carbon vessels (two in service – one back-up)
- A 1,000 cfm unit with two 5 ft diameter vertical carbon vessels (one redundant)

Contract documents were prepared for the portable odor control system and were bid by ABCWUA. The project was awarded to PureAir Filtration at a contract price of $182,035 for the 5,000 cfm unit, $62,215 for the 1,000 cfm unit and $28,374 for the generators. The trailers can be moved about with a heavy duty pickup truck using a fifth wheel hook-up. The portable generators have their own trailers.

KEYWORDS – Portable odor control units, Albuquerque odor control, trailer mounted odor control, carbon adsorbers for odor control

INTRODUCTION

Background

In early 2017, the Water Authority (WA) recognized the need for portable Odor Control Units and requested an investigation and recommendations for this concept. The Water Authority will own the units, but an On-Call contractor may also deliver and operate the systems according to standard operating procedures and then return the systems to ABCUA.

The system sizes were identified as 1,000 cubic feet per minute (cfm) capacity and 5,000 cfm capacity to match existing needs. Each system will use carbon adsorbers as the treatment technology. A portable system shall consist of a complete unit that can be moved from site-to-site, and shall come complete with fan, carbon units, electrical power, ducts, dampers and other components. The fan shall run from either a portable generator, or direct power such as from a Motor Control Center at a Lift Station. These systems will be used to provide odor control for various interceptor, lift station and other construction projects, in the City of Albuquerque and Bernalillo County.

Project Objectives

As a new approach to odor control the Water Authority suggested that the consultants first check with other municipalities that have similar problems and needs and that have set up portable odor control systems for temporary control of odors during construction projects or as interim solutions while long term odor control decisions are being made. The Owner asked the consultants to:

- Decide on an approach to portable odor control systems
- Develop bid documents
- Coordinate the purchase of the systems from a qualified vendor.

The Owner had the following requirements for the systems:

- Need good performance and flexibility, for variable applications
- Fan and generator to be quiet
- Meet DOT requirements
- Provide high level of theft protection and tight security
- Use **carbon adsorption** for simplicity and flexibility with redundant vessels
- Provide separate trailer mounted quiet generator

Project Status

A project Technical Memorandum with evaluation and recommendations was finalized in March, 2017, and Contract Documents were prepared and bid in the fall of 2017. A contract was subsequently awarded and fabrication began with a delivery date of late January, 2018. The next steps will be to deliver the units to Albuquerque, commission the systems and conduct performance testing in the spring of 2018.

METHODOLOGY

Evaluation

Calls and site visits were made with Pima County, AZ and with King County, WA to review their portable systems, contacts were made with vendors and meetings were held with the Water Authority to receive input on priorities and requirements. Details of their systems are summarized below:

Pima County, AZ

- They have several portable units, including a small solar powered system
- 1- 500 – 2,000 cfm unit on dual axle trailer – fan is skid mounted off the trailer
- 1 - 5,000 cfm roll-off container built by Tetrasolv
- Suggests using steel as more durable
- Mainly bring power from a station or pole, but has a generator
- Trouble with generator is it has to be refueled often
- Waste Management can move roll-offs

King County, WA

- Has several portable units up to 8,000 cfm (since 2001) portable units on enclosed trailers
- Horizontal flow 24" deep carbon vessel
- Enclosure is part of plenum with air discharged out the back of the vehicle
- Built by Coolbreeze Air
- Uses flexible hoses

- Sometimes used for extended periods of time
- In the process of designing three (3) new ones
- King County also has a trailer mounted 1,000 cfm unit with 2 vertical carbon vessels and sound enclosure

Several manufacturers were contacted to find out about their experiences with portable odor control units. Evoqua has 5, 000 cfm roll-off container types with carbon media that are for sale or lease. The fan package with motor is supplied skid mounted.

Daniel Mechanical has skid-mounted dual-bed carbon adsorbers which can be vertical or horizontal and can be mounted on a trailer or moved around on the skid from site to site. Daniel can fabricate whatever else is specified and submitted a proposal and drawings that were useful in preparing a scope of work in the bid document. ECS can also fabricate any configuration necessary and suggested looking at round horizontal vessels on a trailer, as well. After this project was completed, ECS came out with a portable trailer mounted unit that is 12,000 cfm and is available for lease.

Tetrasolv built the roll-off carbon adsorbers for Pima County and provides carbon replacement services for ABCWUA too. They also built the 1,000 cfm solar powered unit (1.5 HP) with converter and battery – all enclosed ("Solar Taco Truck" – see pic) for Pima County. They are steel fabricators and can fabricate systems to specification

PureAir Filtration has significant experience with carbon adsorber vessels and portable systems and can fabricate according to a set of specifications.

Alternatives Evaluation

Types of Portable Units

The types of portable odor control devices that were considered included the following:

- Skid Mounted – Put units on a skid and transport to each site and unload with fork lift or crane
- Trailer Mounted – Flat bed dual-axle trailers will contain all the equipment and can be moved by a heavy-duty truck with towing capacity of about 25,000 lbs. Trailer is moved from site to site and is left and secured. The trailer can be open or enclosed (like King County)
- Roll-off Container – Carbon is contained in a roll-off container which is moved from site to site with special trucks like Waste Management uses. A separate trailer is needed for the fan and motor control center.
- Generator – Portable generators can be purchased from manufacturers that are quiet and can be hauled from site to site separately. If the generator is not needed (i.e. power available at the temporary location) then there is no need to bring it on site.

Requirements and Design Matrix

During this investigation and in discussion with the Water Authority some decisions were made concerning the direction of the project, and some important factors to be considered were identified. These decisions included:

- Provide mist eliminator on discharge of fan to protect carbon from moisture
- Solar power units were not to be considered further due to lack of flexibility, size limitations and theft risk
- No remote monitoring is anticipated into SCADA
- Use carbon with a high capacity for H_2S (recommend pelletized version for lower pressure drop and less movement during transport)
- Be able to monitor bed conditions with sample ports
- Provide Variable Frequency Drive (VFD) for 5,000 cfm unit due to varying site conditions
- Provide tight security to prevent theft of trailer or skid, including locks and GPS tracking with
- Use FRP vessels instead of lined steel due to corrosion impacts and weight limitations
- Provide lights on the trailer for security
- Trailer and odor control equipment must meet all federal highway safety requirements

Selection Process

A decision matrix analysis was run to select the best option based on important factors that best fit the requirements of the project. There are two elements of the analysis, one is the base score for each factor and one is the weight of each factor. The score is rated on a scale of zero to five (0 = poor, 5 = very good) and the weight is based on the importance, i.e. 0 = Not important at all and 5 = very important. The score multiplied by the weight give a total for each factor and option, and the highest score for each option would be selected, using this approach.

The Options are:

- Skid Mounted
- Open Trailer Mounted
- Roll-off container
- Enclosed Trailer Mounted
- Open Trailer with Redundant Vessels

The selected factors and weights are as follows:

1. Performance (5) – Ability to replace carbon, have some redundancy for continuous operation and treatment efficiency.
2. Flexibility (5) – Be able to be moved by ABCWUA or contractor equipment, not too many separate components, compact.
3. Theft Resistance (4) – Easy to protect, secure, theft prohibitive.
4. Accessibility (3) – Can monitor carbon bed, access the equipment and components for O&M.
5. Cost (2) – Relative capital cost (the Water Authority has not expressed concern over the budget but this may have more importance at some point in time).

The analysis is shown in **Table 1** (scores) and **Table 2**, which provided base scores and weighted total scores, respectively. The two highest scores are the open trailer mounted units. The highest score was the open trailer with spare vessels due primarily to performance standards and the importance of keeping the system running without waiting for carbon replacement.

Table 1

Decision Matrix Base Score[1] for Portable Odor Control Units for ABCWUA

Options	Base Scores for each Factor				
	Performance	Flexibility	Theft Resistant	Accessibility	Cost
Skid Mounted	3	3	4	4	4
Open Trailer Mounted	3	5	5	4	4
Roll-Off	3	3	5	3	3
Enclosed Trailer Mounted	3	3	5	2	3
Open Trailer with Spare Vessels	5	5	5	4	3

[1]Scores – 0 – Poor, 5 – Very Good

Table 2

Decision Matrix Weights for Portable Odor Control Units for ABCWUA

Options	Weighted[1] Scores for each Factor					
	Performance	Flexibility	Theft Resistant	Accessibility	Cost	Totals
Weights	**5**	**5**	**4**	**3**	**2**	
Skid Mounted	15	15	12	15	8	65
Open Trailer Mounted	15	25	20	15	8	83
Roll-Off	15	15	20	9	6	65
Enclosed Trailer Mounted	15	15	20	6	6	62
Open Trailer with Spare Vessels	25	25	20	12	6	88

[1]Scores – 0 – Poor, 5 – Very Good

RESULTS

The Systems

Two trailer mounted systems were bid by three qualified suppliers; Daniel Mechanical, ECS and PureAir Filtration. Specifications were prepared for bidding to include the following:

The preliminary design criteria for each type of system shall include:

- Trailer width is to be 8.5 ft (standard) with maximum weight of 25,000 lbs
- 1,000 cfm fan is 5 HP and 5,000 cfm is 20 HP
- Select 50 feet per minute (fpm) design velocity through carbon
- Use 3 ft bed depth
- 39 KW portable generator for each unit
- Provide flexible hose for duct from source to fan (24 in. for 5,000 cfm unit and 12 in. for 1,000 cfm unit
- Use 460 VAC 3 phase power with a main circuit breaker. Provide 120 V transformer and two (2) 20 amp circuits and two (2) 15 amp circuits. Provide motor starter with circuit breaker and overloads, 100 ft extension cord, ability to plug into generator or connect to utility power, main switch with circuit breaker, start-stop function inside panel, indicator lights, VFD (for 5,000 cfm unit only), hour meter, panel air conditioner, and panel lights
- Measure differential pressure across carbon bed
- Provide drains for carbon beds and mist eliminator
- Install sample ports at 1', 2' depth and outlet of carbon
- Safety and theft protection including motion sensor flood lights, heavy duty locks on the trailer hitch and wheels, GPS tracking and locks on the control panel

The contract was awarded to the qualified low bidder, PureAir Filtration at a contract price of $182,035 for the 5,000 cfm unit, $62,215 for the 1,000 cfm unit and $28,374 for the generators. The trailers can be moved about with a heavy duty pickup truck using a fifth wheel hook-up. The portable generators have their own trailers.

Supply

PureAir provided shop drawings and design criteria for the systems and Figure 1 and Figure 2 are drawings of the 5000 cfm and 1,000 cfm systems, respectively. The carbon units will be performance tested in February, 2018 and shall remove 99% of Hydrogen Sulfide (H_2S) at 10 ppm inlet concentration average. The carbon is high capacity Sulphabsorb XL (0.2g/cc capacity) and PureAir estimates the life of the carbon in each vessel to be 2.5 years at 10 ppm inlet H_2S.

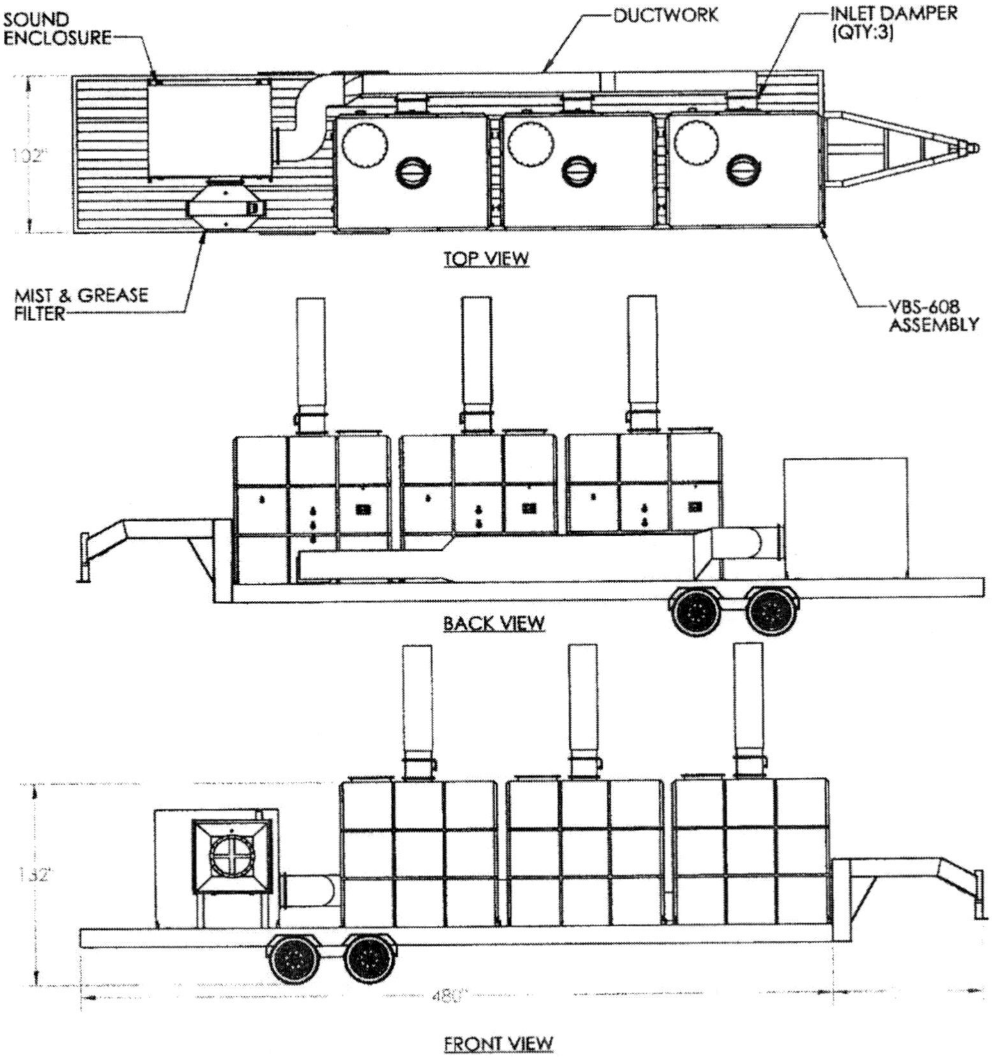

Figure 1

5,000 cfm Portable Odor Control System Built by PureAir for ABCWUA

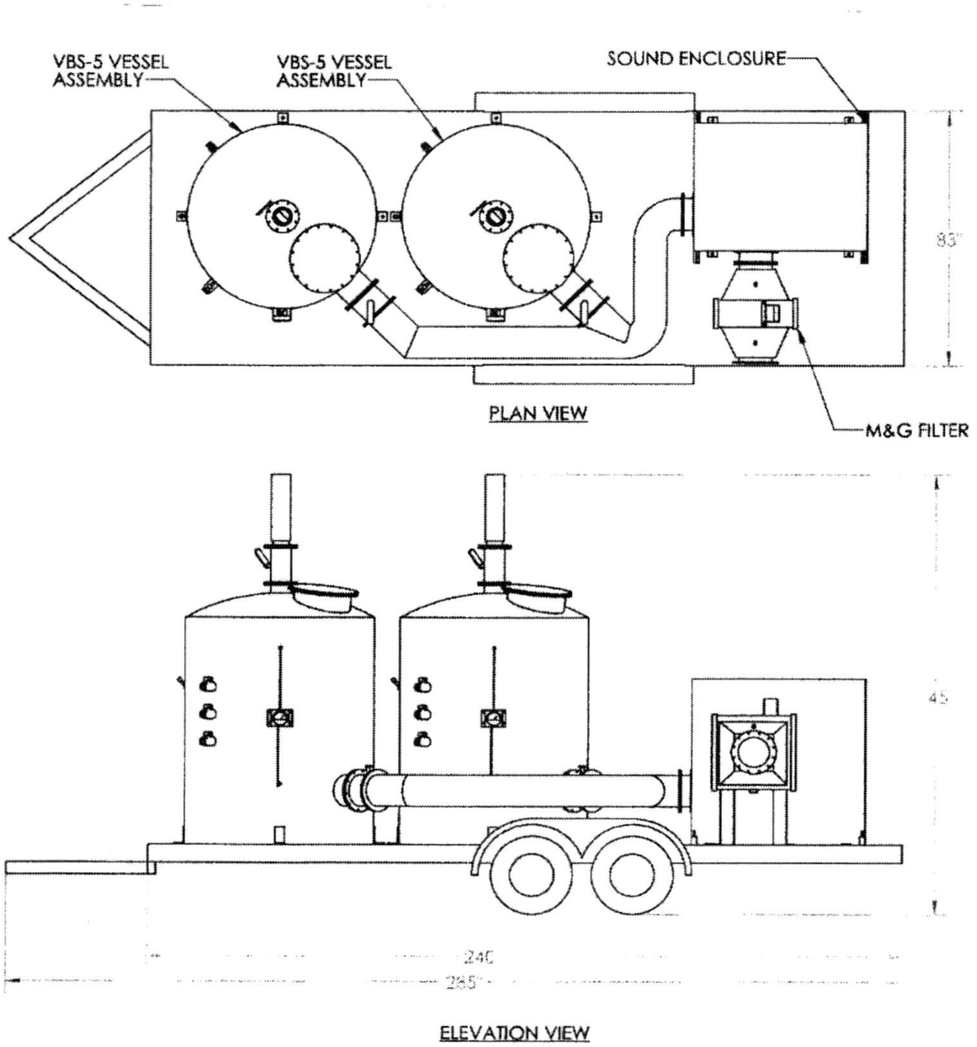

Figure 2

1,000 cfm Portable Odor Control System Built by PureAir for ABCWUA

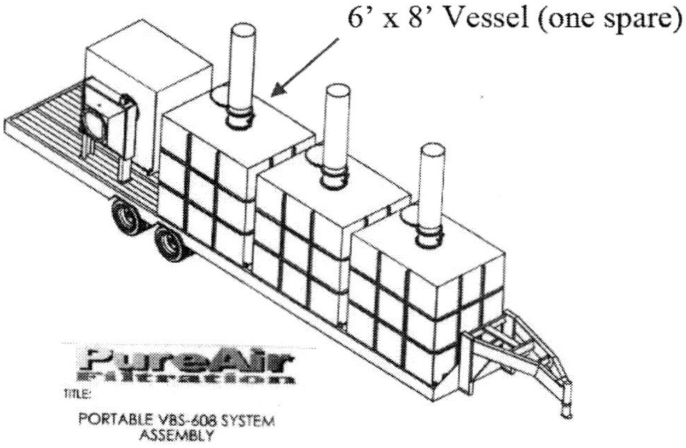

5,000 cfm Unit

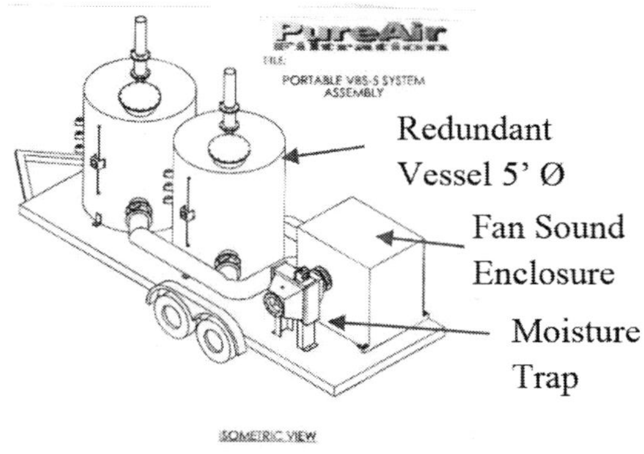

1,000 cfm Unit

Along with the flexible duct, a manhole adaptor lid is provided to pull air from a collection system, for each system, as pictured below.

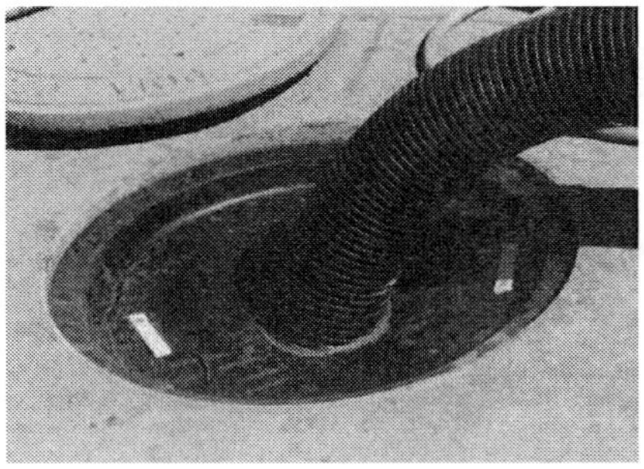

Manhole Adaptor Lid

Three (3) rectangular carbon vessels are provided for the 5,000 cfm system, each 6 ft by 8 ft in size with 3 ft of media depth. Two vessels will be on line at a time with one redundant vessel so that carbon replacement can occur without downtime. A 40 foot long, 8.5 feet wide BigTex tandem axle trailer with Gooseneck is provided with a G.V.W.R. of 30,000 lbs. The estimated weight of the fully loaded trailer is 25,000 lbs. The 1,000 cfm unit will have two (2) 5 ft diameter vertical vessels, which will provide full redundancy with one vessel in service at a time. The trailer for this unit is 7 ft wide and 20 ft long. Each unit will have a separate trailer mounted 39 KW generator.

CONCLUSIONS

ABCWUA will have access to portable control systems for years to come for a variety of temporary odor control projects, mainly in their collection system but certainly that could be used at their wastewater treatment plants. This selection process for the systems was conducted with open meetings and discussions to select the best possible system for the Water Authority. The benefits of this approach allowed the project objectives to be met and provided portable odor control on trailers that can be hauled by heavy duty pick up trucks like the Ram 3500 HD and the Ford F-450 Super Duty. The systems will be used directly by the Water Authority or by contractors. Each trailer mounted system has redundancy, multiple sources of power, noise enclosures for fans, quiet generators, moisture protection for carbon and long life carbon media.

Odor Control Using Pure Oxygen, Raymore, MO

David Clidence*

[1] ECO Oxygen Technologies, Indianapolis, Indiana
*Email: dclidence@eco2tech.com

ABSTRACT

Working in wastewater we've come to accept the fact that flowing wastewater goes septic over time and we're left with managing the consequences of high hydrogen sulfide (H_2S) and sulfide loads and the associated odors and corrosion. Widely accepted management practices are ventilating odorous spaces and scrubbing the air in chemical or biological scrubbers. Alternatively, a variety of chemicals are available for liquid phase sulfide control, creating a dependency on chemical suppliers and causing high O&M costs.

Using pure oxygen for odor and corrosion prevention has gained significant momentum in recent years due to its low operating cost and advancements in technologies to efficiently dissolve oxygen into wastewater. Foul odors emanating from wastewater collection and treatment plants are largely caused by the presence of H_2S, which is formed under anaerobic conditions. A logical solution to H_2S control is therefore the addition of pure oxygen to induce aerobic conditions.

The presentation will give background information on the root cause of odor and corrosion and how it can be prevented through the targeted addition of dissolved oxygen. A case study of an installation in Raymore, MO will be presented.

The City of Raymore, MO has an extensive sewer system including 17,000 feet of force main that discharges in the main downtown business district, causing significant odor and corrosion issues for the City. H_2S concentrations in the downstream gravity interceptor were measured as high as 1,000ppm, causing not only health but also safety risks.

HDR was hired to do a comparison of chemicals and Superoxygenation Systems to combat the problem. As the SuperOxygenation System prevents H_2S formation, rather than treating the extremely high concentrations, the results were significantly better than existing chemical use and they provide a more consistent and lower life cycle cost for budgeting.

Results of the odor control project and pictures of the installation will be presented.

KEYWORDS: Odor control, odor control evaluation, pump stations, odor control chemicals, cost analysis, ferrous chloride, magnesium hydroxide, superoxygenation, hydrogen peroxide

INTRODUCTION

Raymore, MO Force Main Odor Problem

The City of Raymore, MO has an extensive sewer system. Like many cities, Raymore has a long force main that was sized years ago for a projected level of industrial and residential development that has not yet been realized. Consequently, this 17,000 feet force main is oversized with wastewater velocities running less than optimal (less than 2 ft/sec). This allows for significant anaerobic conditions to develop in the force main resulting in the generation of malodorous hydrogen sulfide gas. The force main discharges in the main downtown business district, causing significant odor and corrosion issues for the City. H_2S concentrations in the downstream gravity interceptor were measured as high as 1,000 ppm, causing not only health but also safety risks. The generation of hydrogen sulfide gas in this force main has plagued the City for many years. The City has tried many methods and spent a great deal of money to reduce the odor production. The City sought additional new methods for preventative, non-chemical in-pipe treatment to further reduce the odors to non-detectable levels.

An ECO_2 SuperOxygenation system was designed and installed at the Owen Good pump station. Performance data shows the system successfully reduced hydrogen sulfide gas at the force main discharge. This is being done at a capital and annual operating cost resulting in a 3-year payback versus alternative chemical treatment.

METHODOLOGY

Root Cause of Odor in Force Mains

The root cause of most odor problems associated with wastewater collection and treatment is the presence of naturally created malodorous gases. The gases are formed by Sulfate Reducing Bacteria (SRB's), under anaerobic conditions resulting from the imbalance of high BOD loading and the limited solubility of oxygen into wastewater. This problem is exacerbated in long force mains and slow-cycled residential lift stations due to high microbial oxygen uptake rates, long detention times, and low dissolved oxygen (D.O.) levels. These anaerobic conditions produce significant amounts of hydrogen sulfide (H_2S) which is both odorous and highly corrosive. Since anaerobic conditions are precursors to H_2S formation, a logical solution to H_2S control is to induce aerobic or oxic conditions. SuperOxygenation is a process to sustain aerobic conditions by using an ECO_2 "Speece" Cone (Downflow Bubble Contactor) to dissolve pure oxygen into wastewater to prevent the formation of H_2S and the associated corrosion.

Root Cause of Corrosion in Force Mains

The possibility of extending the life of concrete wastewater infrastructure is critically important economically, environmentally and aesthetically to sanitation districts and management personnel. Concrete pipes which corrode long before the estimated life expectancy is reached, cause complicated engineering problems as well as public inconvenience and increased overall expense. Thus, the elimination of corrosion processes through new technological advances brings welcome opportunity to exploit the full life time of wastewater concrete and steel infrastructure even with heavy use. Non-detectable levels of H_2S conclusively arrest and eliminate corrosion because two principles of bacterial metabolism are axiomatic:

- Microbes in absence of D.O. SO_4 + BOD = H_2S (simplified equation)
- Microbes in presence of D.O. H_2S + D.O. = H_2SO_4 (simplified equation)

The microbial generation of H_2S and related reduced sulfide species in wastewater is due to the following important factors: the very limited solubility of oxygen in water, high BOD concentrations found in wastewater severely restricted oxygen transfer rates in collection systems due to low reaeration rates inside of pipes, low ORP caused by anaerobic conditions increases bacterial activity, fermentation of BOD at warmer temperatures, and prolonged retention times involved in most collection systems. Infrastructure corrosion proceeds when H_2S formed in the wastewater volatilizes into the headspace and is absorbed into the oxic moisture condensed on the crown of the pipe of a gravity sewer. Under these aerobic conditions the ubiquitous Thiobacillus thiooxidans oxidizes H_2S to H_2SO_4, which tenaciously corrodes concrete and steel. The implications of extending the life of a sewer system are noteworthy. Sewer replacement is costly and inconvenient in the extreme, and should not have to be undertaken at all until the estimated life span of the sewer system is reached.

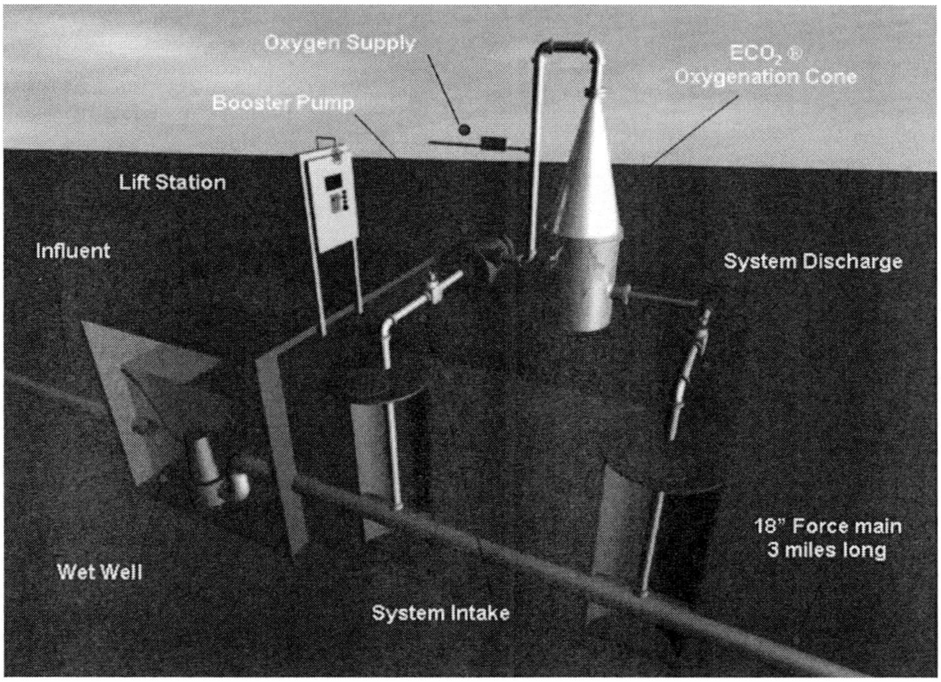

Figure 1: Schematic of an ECO_2 SuperOxygenation System Installed on a Force Main

ECO_2 SuperOxygenation Technology

ECO_2 SuperOxygenation system is technology designed to dissolve pure oxygen into raw wastewater to sustain aerobic conditions to prevent the formation of H_2S. The high oxidation-reduction potential (ORP) created under aerobic conditions prevents the conversion of H_2S to H_2SO_4, thereby preventing the potential for corrosion.

The system operates by pumping a side stream of raw, unscreened wastewater through a conical shaped oxygen transfer reactor. Gaseous oxygen is feed into the cone which has an exceptionally large gas/water interface generated by an intense bubble swarm to achieve rapid SuperOxygenation. The inlet of the Cone sized to provide enough inlet water velocity to shear the oxygen bubbles entering the Cone. The Cone widens to a diameter which decreases the water velocity to a point below the buoyant velocity of the bubbles keeping them suspended inside the Cone until they are dissolved. ECO_2 SuperOxygenation system completely dissolves the gaseous oxygen into the water before it is blended back into the force main. There are no internal baffles or static mixer that could collect rags or stringy material.

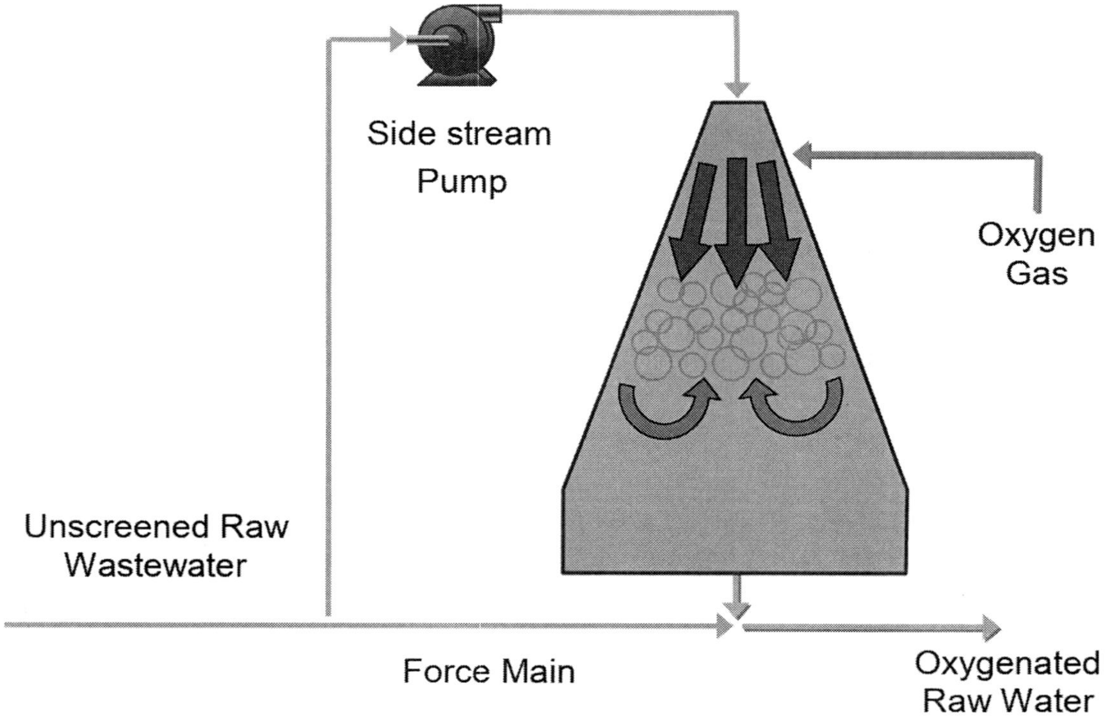

Figure 2: ECO_2 SuperOxygenation Flow Diagram

The ECO_2 SuperOxygenation system is based on Henry's law and has resulted in extremely high rates of oxygen transfer as well as the actual prevention of both odor and corrosion in current applications. Dependent upon the amount of oxygen needed to be dissolved, the ECO2 SuperOxygenation system may involve pressurization by hydrostatic pressure or pumping against a throttling valve at the discharge. The rule of thumb is for every one (1) foot of head pressure inside the force main, one (1) mg/L of dissolved oxygen can be added.

The ECO_2 SuperOxygenation system is designed to meet the dissolved oxygen requirements of the system that it is treating. To prevent the formation of H_2S, a positive D.O. level must be carried through the entire length of the force main or primary clarifier. The amount of oxygen needed to be dissolved is determined by multiplying the hydraulic retention time (HRT) by a theoretical 10 mg/L/hr microbial oxygen uptake rate. Once the amount of oxygen need to be dissolved is determined, the ECO2 SuperOxygenation system and side stream pump is sized to deliver the needed amount of oxygen.

The D.O. demand in the force main must be matched by the Oxygen delivery rate to the SuperOxygenation system. Thus, the mass flow rate of Oxygen fed to the system is controlled by feedback from water flow through the force main. If water flow through the force main is zero, the feedback reduces the Oxygen feed to the system to zero.

The oxygen being dissolved for odor prevention can be supplied to the ECO_2 SuperOxygenation system in two ways; bulk liquid oxygen (LOX) or onsite oxygen generation (Pulse Swing Absorption (PSA) or Vacuum Swing Absorption (VSA).

Chemical Alternatives for Odor Control

Chemical addition for odor control have proven effective on major trunk lines, but does little to enhance biological processes, relying instead on direct reaction with sulfides after they have formed, rather than preventing sulfide formation. SuperOxygenation has the benefit of beginning the treatment process in the collection system, while preventing odors from forming.

SuperOxygenation is a green technology that does not increase TDS or the amount of sludge disposal at the treatment plant. Not only is H_2S generation prevented by oxic conditions, but any H_2S already present in the wastewater is preferentially oxidized within 5 to 20 minutes after oxic conditions are established because S= is a very high energy substrate under aerobic conditions (more so tha carbonaceous BOD). Each pound of D.O. injected into the collection system for odor/corrosion prevention also provides a corresponding savings of an extra pound of D.O. in the subsequent secondary treatment system.

Owen Good Pump Station Installation

Figure 3: ECO_2 SuperOxygenation System Installed on the Owen Good Pump Station

The components of the system are comprised of a 6-foot diameter ECO_2 SuperOxygenation Cone and oxygen flow control system, 15 HP wastewater pump and liquid oxygen tank. The system operates by redirecting a side-stream flow of 2,800 gpm of raw, unscreened wastewater from the force main and pumping it through an ECO2 Cone. Liquid oxygen is delivered to site and stored in a liquid oxygen storage tank. Oxygen gas is fed at a rate of up to 955 lbs per day into the ECO_2 Cone which has an exceptionally large oxygen/wastewater interface generated by an intense bubble swarm to achieve rapid oxygen dissolution. The ECO_2 SuperOxygenation system completely dissolves the gaseous oxygen into the wastewater side-stream before it is blended back into the force main. Enough oxygen is dissolved into the force main to sustain aerobic conditions through the remainder of the force main reducing the amount of H_2S generated.

Figure 4: ECO_2 SuperOxygenation System Equipment Site Layout

RESULTS

Untreated, the force main generated H_2S in excess of 800 – 1,000 ppm at the force main discharge. Since the SuperOxygenation system came online in Oct of 2015, there has been a 10-

fold reduction in average H_2S levels. Figure 5, shows an average H_2S level of 0 ppm at the force main discharge.

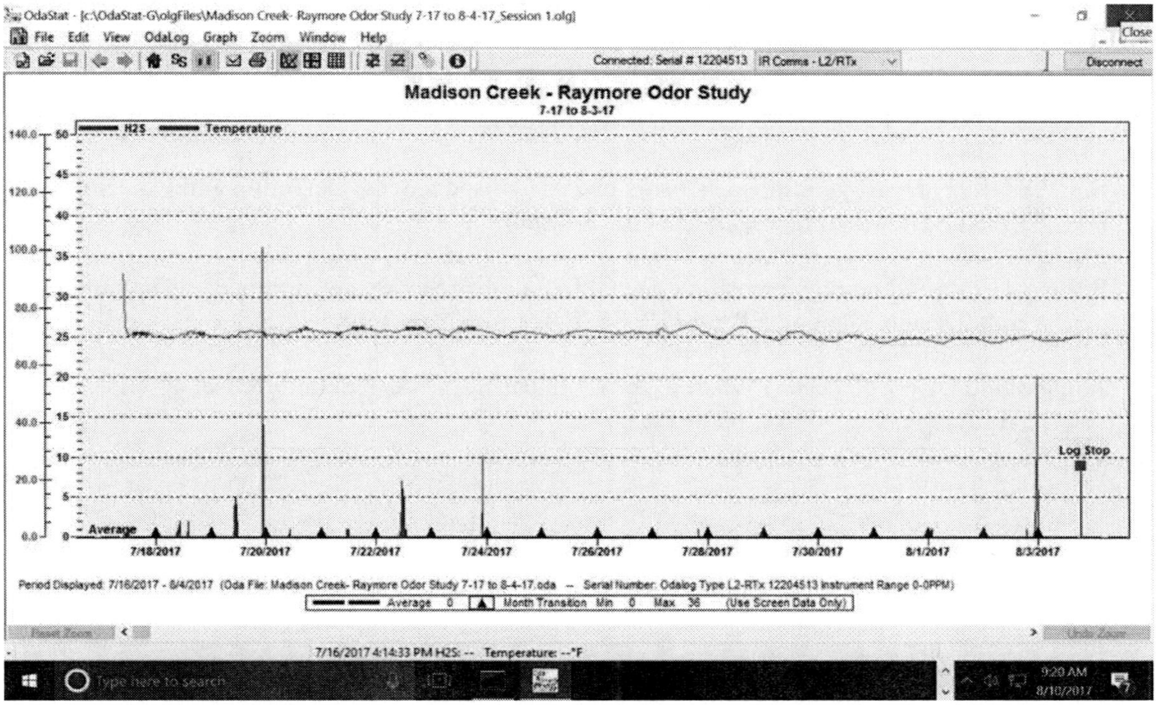

Figure 5: H2S Results from ECO2 SuperOxygenation

CONCLUSIONS

Generation of malodorous hydrogen sulfide gas in the Owen Good force main has plagued the City of Raymore, MO for many years. In 2015, the City sought additional new methods for preventative, non-chemical, in-pipe treatment and chose ECO_2 SuperOxygenation technology to sustain aerobic conditions in the force main to further reduce H_2S to acceptable levels. This is being done at a capital and annual operating cost resulting in a 3-year payback versus previously used chemical treatment.

Recent SuperOxygenation technological advances have refuted commonly held misconceptions about retention of high D.O. levels as well as ushered in a new scenario for successful cost effective and environmentally friendly H_2S abatement applications.

Preventing the formation of H_2S, SuperOxygenation also reduces the potential for corrosion. Infrastructure damage and collapse, often incurred early in the usage history of concrete and steel pipes, can now be minimized by the incorporating SuperOxygenation and extend the usefulness of collection systems for many years.

REFERENCES

U.S. EPA (2000) Collection Systems Technology Fact Sheet for Sewers and Lift Stations; EPA 832-F- 010-073; Washington, District of Columbia.

U.S. EPA (1974) Process Design Manual for Sulfide Control in Sanitary Sewerage Systems; 625174005; Washington, District of Columbia.

U.S. EPA (1991) Hydrogen Sulfide Corrosion: Its Consequences, Detection and Control; 832S91100; Washington, District of Columbia.

U.S. EPA (1985) Odor and Corrosion Control in Sanitary Sewerage Systems and Treatment Plants Design Manual; EPA/625/1-85/018; Cincinnati, Ohio.

Tchobanoglous, George (2003) Wastewater Engineering Treatment and Reuse, 4th ed. Metcalf & Eddy, Inc. New York, New York.

TECHNICAL MEMORANDUM

CASE STUDIES FOR THE IMPLEMENTATION OF VENTILATION STRATEGIES USING NEW VENTILATION MODELLING TOOLS

Josef Cesca[1], Graeme Hamer[1]

[1]CH2M, Level 7, 9 Help St, Chatswood, NSW, 2067, Australia

ABSTRACT

This paper describes the practical application of a Ventilation Tool using the conservation of momentum algorithm (Ward et al, 2010; Hamer et al, 2012) that was developed in association with Water Environment Research Foundation (WERF) and the Sewer Corrosion & Odour Research (SCORe) Project. This paper explores a case study where the Ventilation Tool was used in the planning of an odour control system for sewers that was implemented. The case study focuses on the practical application of this tool, implementation of the recommended strategy and how the actual installations compared to the theoretical recommendations.

KEYWORDS: *Modelling, Collection Systems, Ventilation, Odour Control*

INTRODUCTION

The use of sewer corrosion and modelling in planning odour control strategies is becoming more commonplace as modelling methods evolve and the benefits that modelling brings to odour control strategies are more accurately realised. However, most models and tools employed have limitations in relation to our understanding of several key in-sewer processes. Particularly challenging in sewer corrosion and sulphide modelling is a good understanding of the ventilation rates in sewers which are critical for predicting / estimating hydrogen sulphide (H_2S) concentrations and subsequently corrosion.

There have been several empirical algorithms proposed for predicting natural ventilation air movements, the most popular being the Pescod & Price Equation (Pescod and Price, 1982), but these have not been widely used for the ventilation of sewers as the results have been unreliable. Design of natural ventilation systems has generally been by rules of thumb. However, these systems have sometimes failed to protect nearby residents from odours and in many cases the vents have been closed which resulted in further problems.

The natural forces influencing ventilation are a fine balance between:

- Relative density between sewer air and outside air
- Wastewater flow induced drag
- Friction on pipe walls
- Changes in barometric pressures along a sewer
- Wind velocities over ventilation stacks

In association with WERF, the SCORe Project has developed a new algorithm based on the conservation of momentum (Ward et al, 2010; Hamer et al, 2012) which accounts for the forces outlined above. The model has been proven in field testing during development to be much more reliable than previous models (Hamer et al, 2012). A tool has been developed based on this conservation of momentum algorithm which can be used to predict both natural and forced ventilation air movements in sewers.

This Ventilation Tool is now being used to develop a better understanding of ventilation control strategies. More importantly, however, the new ventilation algorithm is being used to provide improved prediction of air movements that are used for estimating gas phase H_2S concentrations and subsequently corrosion rates within a sewer network. Air movement in collection systems is one of the most critical parameters for estimating both H_2S concentrations and corrosion rates so the new tool provides another critical step to improving corrosion rate predictions.

The Ventilation Tool also provides an improvement on previous methods by calculating pressure throughout a sewer under analysis. This serves to extend the Tool's capabilities, allowing it to predict where in-gassing and out-gassing to atmosphere will occur, and the gas flow rate between sewer and atmosphere. The Ventilation Tool provides a pressure distribution of a sewer, giving the static and differential pressure at each node. The Tool also allows a user to evaluate fan sizes and positions within the sewer, leading to the development of a fan curve to determine a fan's operating point from a system curve derived from the Tool.

The Tool's applications are numerous, ranging from determining the condition of a sewer through to ventilation and odour control equipment design. Specific applications include:

a) Designing new sewer systems

b) Assessing existing systems and proposed modifications

c) Assessing odour impacts and complaint investigations

d) Estimating out-gas flow rates

e) Evaluating effects of projected sewer condition changes

f) Using sulphide fate models to assess odour and corrosion "hot-spots"

g) Assessing ventilation rates within a sewer

h) Estimating the size and location of forced ventilation equipment

i) Sizing vapour-phase odour control systems

j) Sizing air jumpers and other sewer ventilation structures.

METHODOLOGY

Overview

CH2M was engaged to develop a solution to alleviate odour complaints near a sewer that receives a significant volume of trade waste. The three major trade waste producers were investigated and found to be meeting their discharge permit limits and were undertaking a reasonable level of pre-treatment of waste prior to discharge to the sewer.

A range of control options were reviewed to determine the most appropriate solution for the odour issues. The options considered included reconfiguration of the sewer, implementation of trade waste controls, chemical dosing using ferric chloride, forced ventilation and combinations of options. Control options were modelled using SeweX, and the Ventilation Tool in conjunction with INTERCEPTOR sulphide model to predict their impacts on liquid and vapour phase sulphide concentrations along the sewer section. This aspect of the project is not discussed in this paper apart from noting that the understanding of sewer material and using the Ventilation Tool was critical for accurate estimation of gas phase H_2S concentrations. The INTERCEPTOR model's gas phase H_2S concentrations incorporating the Ventilation Tool airflow results and different pipe materials are shown in Figure 1 below, along with actual measured H_2S concentrations.

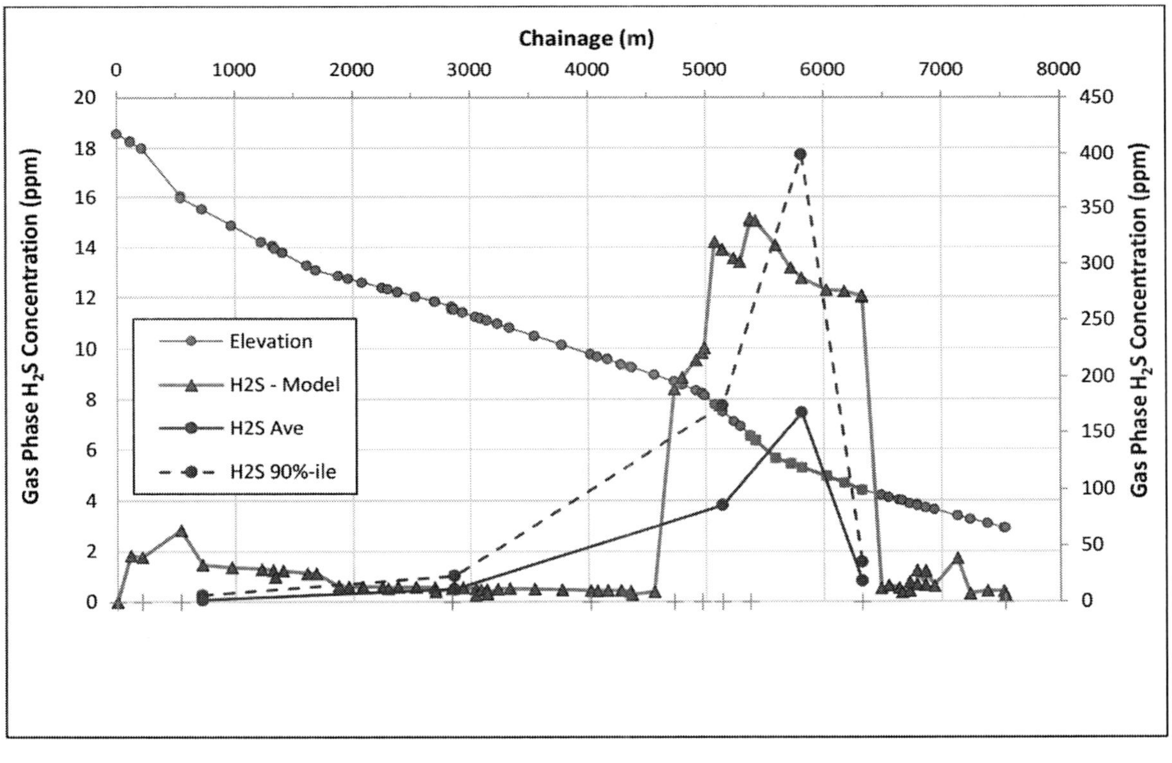

Figure 1 : INTERCEPTOR Model-Predicted H_2S Rates Compared to Field Data During Optioneering Phase.

The preferred option from the optioneering phase of the project was the option involving ventilation at two locations. This paper primarily focuses on the further detailed ventilation modelling using the Ventilation Tool for estimating the optimal ventilation strategy that was subsequently implemented. The results of the modelling were then compared with sampling data for pressure to confirm that the model reflected reality in the base case and to support its use in evaluating control options for odour issues. H_2S data in the sewer before and after the control strategy was implemented is also discussed.

Network Information

A layout of the section of sewer considered in this investigation is shown in Figure 2.

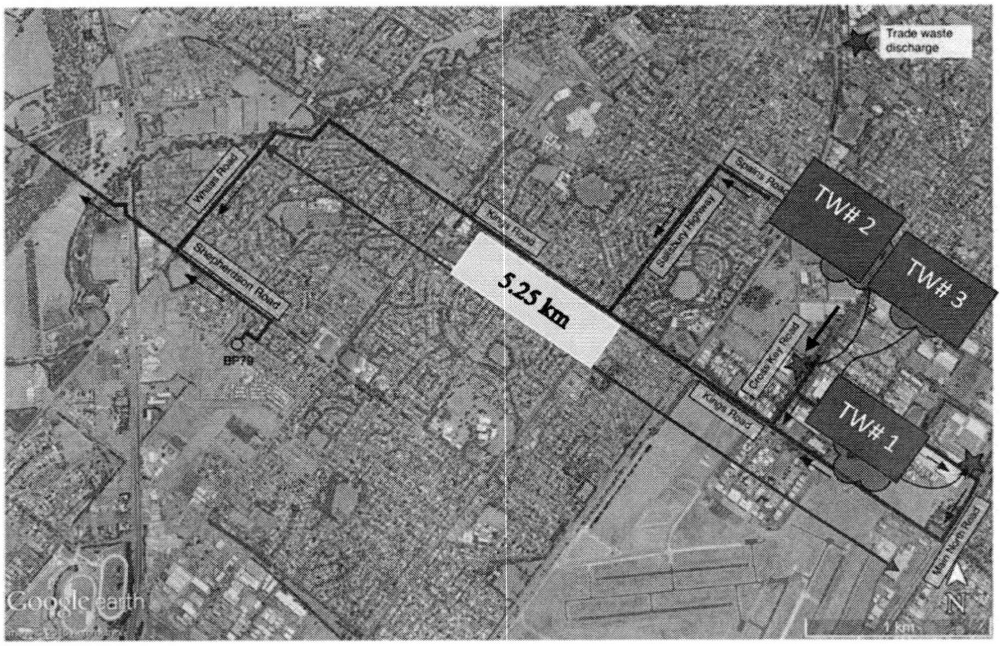

Figure 2 Layout of the of sewer network section investigated

There are 13.1 km of sewer pipes in the network examined and the main trunk sewer is about 6km long. The sewer system includes a single rising main which has negligible impact on the network. There are 122 gravity sewer pipes with diameters ranging from 225 mm to 675 mm. The sewer pipes are made of vitrified clay (VC), polyvinyl chloride (PVC), lined reinforced concrete (RCPL) or unlined reinforced concrete (RC). The sewage in the catchment is a mixture of both domestic and discharges from several industries. The average total flow is 4.4 ML/day.

Sewer Sampling and Monitoring

Sewer wastewater sampling (LS) and headspace H_2S (HS) monitoring was undertaken at several sites to provide concurrent data along the sewer system as well as for use as input and calibration data for modelling.

The following wastewater parameters were sampled as eight grab samples collected over a 48-hour period on 16-17 June 2016, along with some sampling from this time until 8 July 2016:

- Sulphide – total and soluble

- Biochemical oxygen demand (BOD)
- Chemical oxygen demand (COD)
- pH
- Temperature
- Conductivity
- Dissolved oxygen (DO)
- Oxidation-reduction potential (ORP)
- Sulphate

BOD, sulphate, and total and dissolved sulphides were analysed in a laboratory and the remainder of parameters were measured on site using an appropriate probe. Dissolved sulphide was also measured using colorimetric tubes on site.

Gas phase testing (GS) of air pressure and air velocity in the sewer headspace was undertaken at five locations to derive headspace velocities for input into modelling. The air pressure was measured as the differential pressure with respect to atmosphere. The velocity was measured in the sewer pipe, approximately 10-15 cm above the level of the wastewater flow.

Monitoring of airflow and pressure in the sewer headspace was undertaken using a VelociCalc ventilation meter (model 9555-P).

In projects that followed this, differential pressure was monitored using a similar rig to that shown in Figure 3 using purpose-built loggers that incorporate a P350-D-1-inch pressure sensor by Pace Scientific. A pressure sensor (Deltabar M PMD55) from Endress+Hauser connected to a ULTx Remote Telemetry Logger from Thermofisher has also being used on other projects.

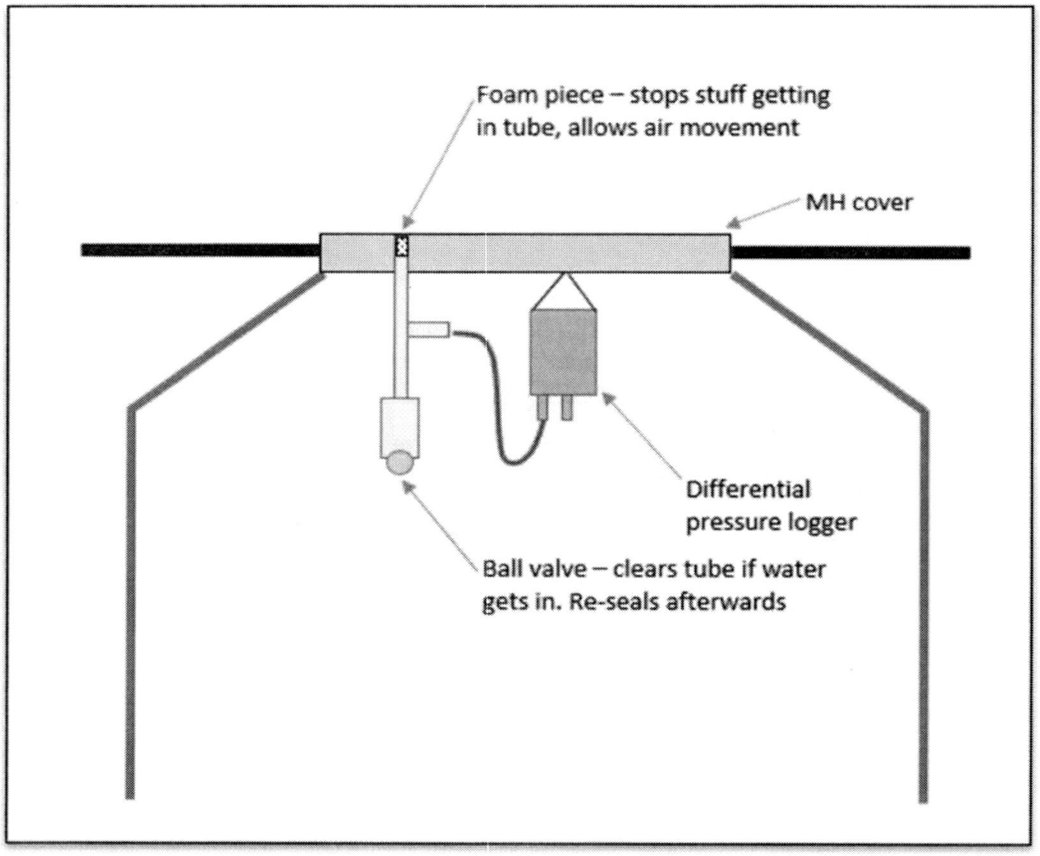

Figure 3: Typical differential pressure monitoring rig

H_2S monitoring in the sewer headspace (HS) was undertaken using high range Odalogs with a nominal range of 0-1000 ppm.

The sampling locations relative to the start of the main trunk sewer are illustrated below in *Figure 4*.

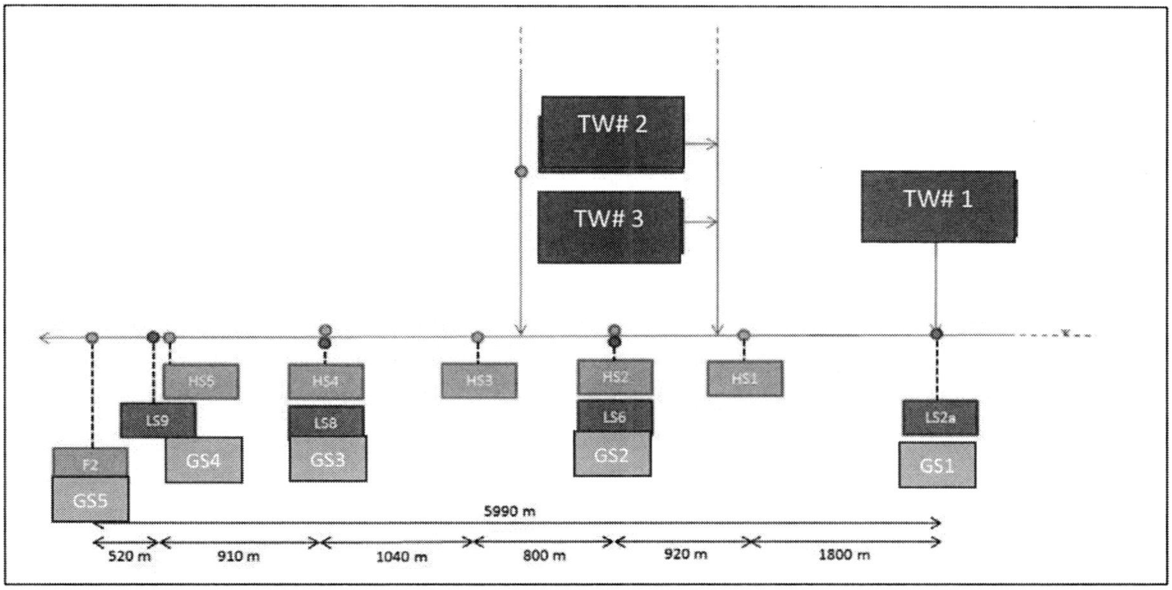

Figure 4: Layout of wastewater and H_2S sampling location.

Ventilation Model Configuration and Control Scenarios

A schematic diagram of the base case scenario is provided in Figure 5. The modelled sewer trunk is shown in black, along with the locations of all inducts (inlet vents), educts (outlet vents), and bends modelled.

Schematic diagrams of Control Scenarios A and B are provided in Figure 6 and Figure 7. The locations of the two (2) odour control units (OCUs) and important air inlets are indicated.

The model configuration for Control Scenarios A and B are very similar, with the main difference between the two models being the ventilation airflow rates and the number of closed vents. Vents blocked off are shown with a cross on them, and additional vents blocked off in Control Scenario B over that in Control Scenario A are highlighted in Figure 7 following.

Table 1 below presents the ventilation airflows adopted for the modelled control scenarios. The airflows presented for Control Scenario A correspond to those that achieve a minimum negative pressure of -20 Pa in the modelling results.

Table 1: Ventilation Airflows for Modelled Control Scenarios

Location	Air Extraction Rate[1]	
	(m^3/s)	(m^3/hr)
Control Scenario A – Minimum Airflows		
OCU at Teasdale Cr	-0.259	-932
Inlet Vent	0.097	348

OCU at end of Whites Rd	-0.260	-936
TOTAL OCU EXTRACTION RATES	**-0.519**	**-1,868**
Control Scenario B – Original Design Airflows		
OCU at Teasdale Cr	-0.564	-2,030
Middle Inlet Vent	0.459	1,652
OCU at end of Whites Rd	-0.553	-1,990
TOTAL OCU EXTRACTION RATES	**-1.117**	**-4,020**

Notes:

1. Air extracted from the sewer is presented as negative values, and air going into the sewer is presented as positive values.

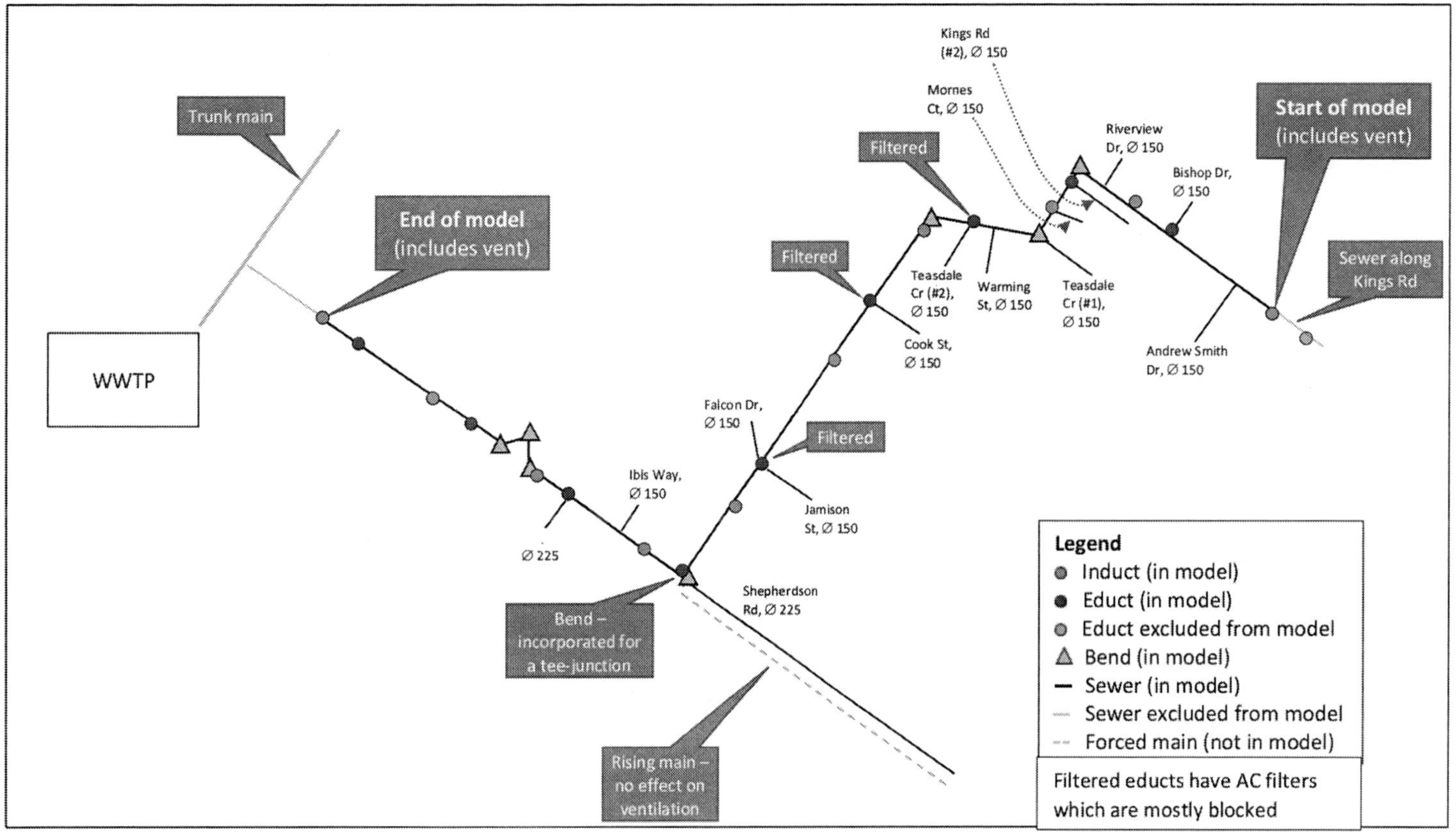

Figure 5: Base Case - Schematic Configuration

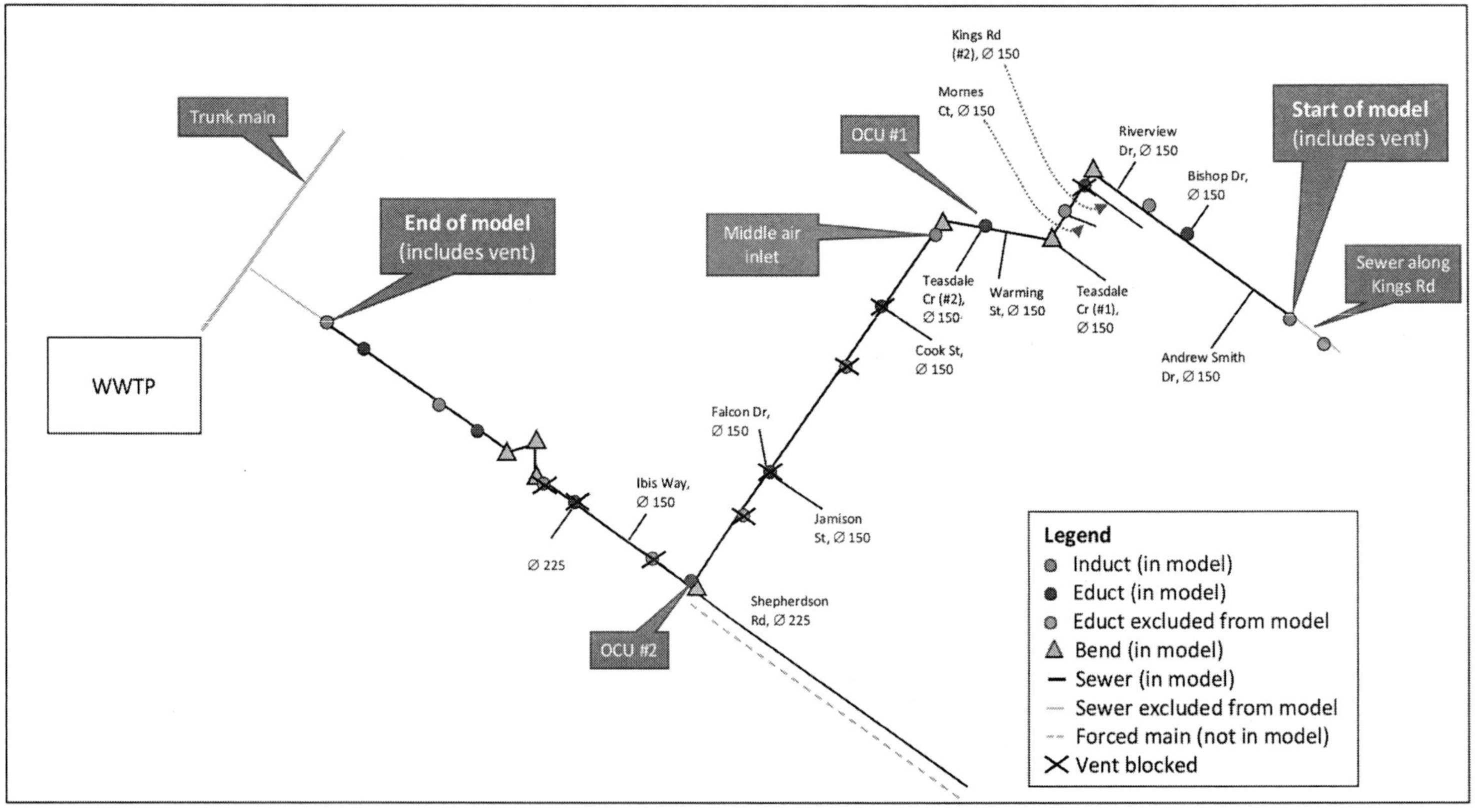

Figure 6: Control Scenario A - Schematic Configuration

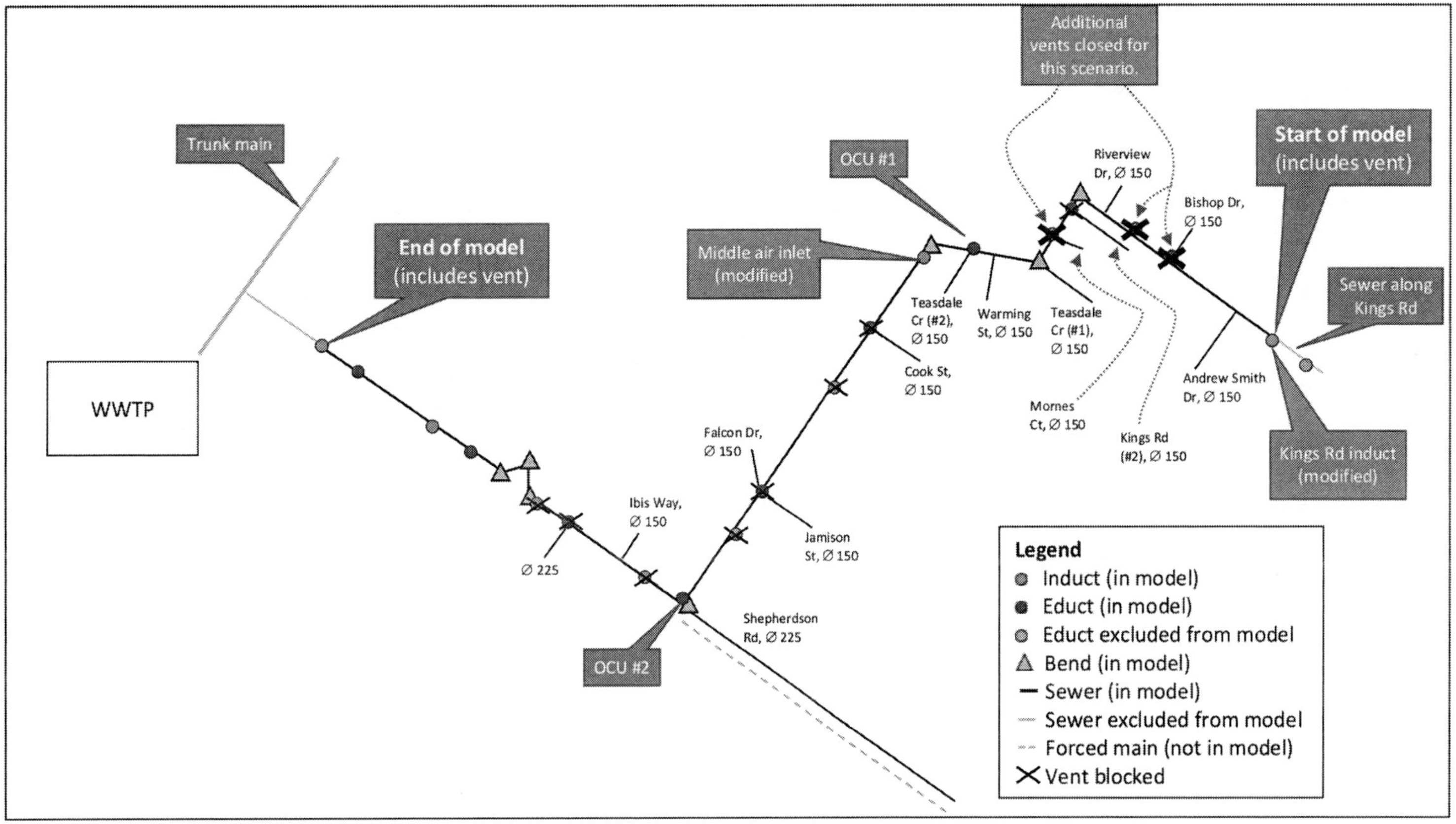

Figure 7: Control Scenario B - Schematic Configuration

RESULTS

Sampling results

A summary of the most relevant liquid phase sampling results are shown in Table 2. We observed that the dissolved sulphide ranged from about 0 mg/l to about 37 mg/l along the main trunk sewer and the BOD ranged from about 1600 mg/l to about 1190 mg/l. The average and 90th percentile pH levels all slightly decreased along the length.

Table 2: Liquid Phase Sampling Results

Parameter	Unit	Statistical Value	LS2a	LS6 -	LS8	LS9
Dissolved sulphide – colorimetric tubes	ppm	90%ile	0	15.1	27.15	23.5
		Average	0.00	10.13	21.75	14.10
Dissolved sulphide - lab	mg/L	90%ile	1.06	19	36.3	37.3
		Average	0.44	9.73	27.31	22.75
Total sulphides - lab	mg/L	90%ile	1.33	22.5	47.4	57.3
		Average	0.59	14.90	38.00	40.58
BOD	mg/L	90%ile	1559	1572	1322	1190
		Average	1281	1216	1013	936
pH	pH units	90%ile	7.50	7.42	7.24	7.27
		Average	7.43	7.27	7.08	7.12
Temperature	°C	90%ile	31.0	28.5	27.7	25.8
		Average	30.3	27.7	26.7	25.2
Sulphate	mg/L	90%ile	2185	1761	1194	1074
		Average	1936	1459	1032	950
DO	mg/L	90%ile	0	0	0	0

Statistical analysis of the H_2S data collected in June-July 2016 is shown in Table 3. We observed that the H_2S concentrations ranged from virtually 0 ppm at the start of the main trunk sewer to over 1600 ppm at the monitoring point HS5 close to the end of the main where most of the odour complaints have been received. This data is presented graphically in Figure 8 to illustrate the diurnal variations in H_2S along the length of the sewer.

Table 3: H_2S Monitoring Statistical Analysis

Location	Monitoring Period		H_2S Concentration (ppm)				
	Start Date	End Date	10th Percentile	25th Percentile	Average	90th Percentile	Maximum
HS1	6-06-2016	29-06-2016	0	1	4.5	11	30

Location	Monitoring Period		H_2S Concentration (ppm)				
	Start Date	End Date	10th Percentile	25th Percentile	Average	90th Percentile	Maximum
HS2	6-06-2016	19-07-2016	3	11	68.5	178	450
HS3	6-06-2016	19-07-2016	0	0	0.0	0	36
HS4	6-06-2016	29-06-2016	28	64	219.0	521	1107
HS5	6-06-2016	17-07-2016	74	170	542.0	1371	1680*

**The loggers used are only accurate to 1000 ppm and this was the maximum reading possible with the logger.*

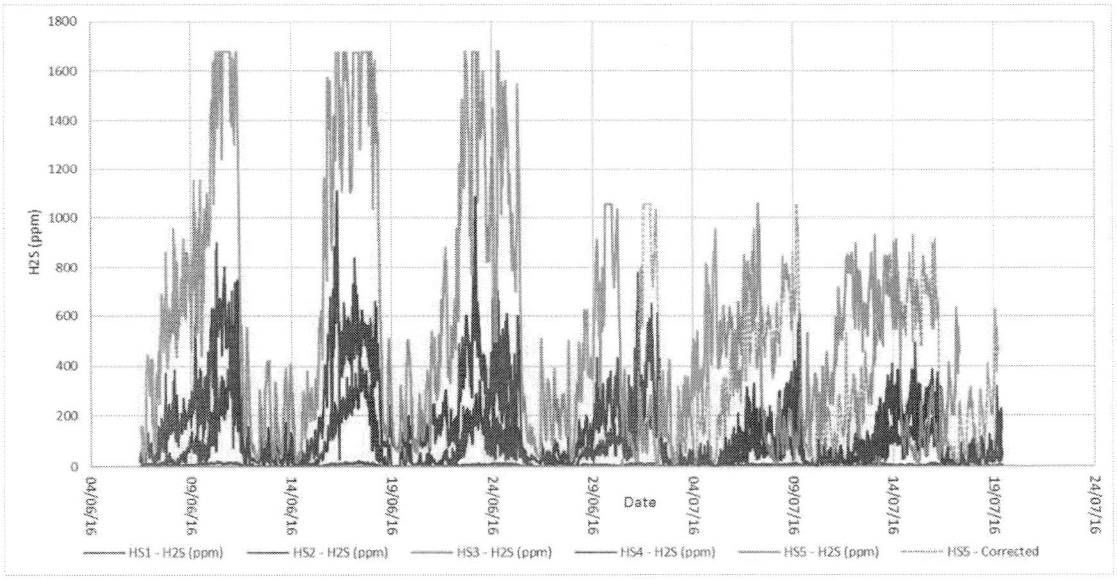

Figure 8: H_2S patterns at the 5 monitoring points along the main trunk sewer

The differential pressure was generally neutral or positive in the order of 0-2 Pa suggesting outgassing of high H_2S concentrations to the local community. Location GS4 had pressures up to +5 Pa and the highest H_2S concentrations. The results between GS3 and GS4 are consistent with a hydraulic jump due to the greater slope of the sewer in this location, compared to the more level slope of the rest of the network. The end of the trunk at GS5 was always under negative pressure which is likely influenced by the downstream ventilation system at the treatment plant.

The air velocities were generally low, with most averages near or less than 0.2 m/s showing that there is very little air movement in the sewer. The greatest average air velocity measured was 0.5 m/s at the end of the trunk sewer before the treatment plant. The variation from the average for most measurements was less than +/- 0.1 m/s. There was no obvious factor that determined the different air flows measured at each location except towards the treatment plant which was influenced by the ventilation system.

Data Analysis Modelling Results

Figure 9 shows the dissolved sulphide concentrations steadily increasing along the sewer. At the steeper sewer grade these concentrations decrease as sulphides are driven from the liquid to the vapour phase. The H_2S concentrations are also shown and are generally low and rise rapidly as dissolved sulphide concentrations decrease in the highlighted steep section of the sewer.

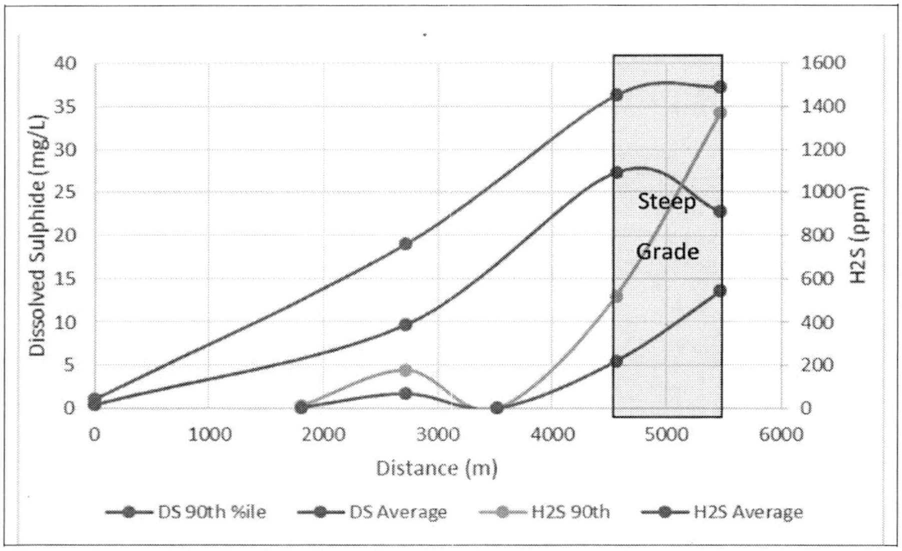

Figure 9: Dissolved sulphide and H_2S vs Distance

The predictions from the Ventilation Tool for the two control scenarios compared against the results of the base case modelling are shown in Figure 10 and Figure 11 below.

Figure 11 suggests that adequate ventilation and extraction will be achieved within the sewer network to achieve the target minimum negative pressure of -20 Pa with implementation of the control options.

This is further supported by Figure 10, which illustrates the OCUs will draw air from upstream and downstream of the OCUs for Control Option B:

- Installation of the first OCU is predicted to result in air flowing in the opposite direction of sewage flow for the section between the OCU and the middle air inlet. An air velocity 0.5 m/s would occur in this section.
- Downstream of the air inlet and down to the second OCU, the air will flow in the direction of sewage flow. An air velocity of 1.0 m/s will occur.
- Downstream of the second OCU, the air within the sewer flows in the opposite direction of sewage flow for most of the sewer. An air velocity of 1.4 m/s will occur between OCU #2 and the next vent downstream.

- The second OCU will provide extraction and treatment of foul air for up to 700 m downstream, after which the air in the sewer will start to change direction to flow in the direction of sewage.

The modelling results of the Control Scenarios suggest that installation of the two OCUs and retainment of a central air inlet between the two will provide the required odour containment to reduce out-gassing and resulting odour complaints.

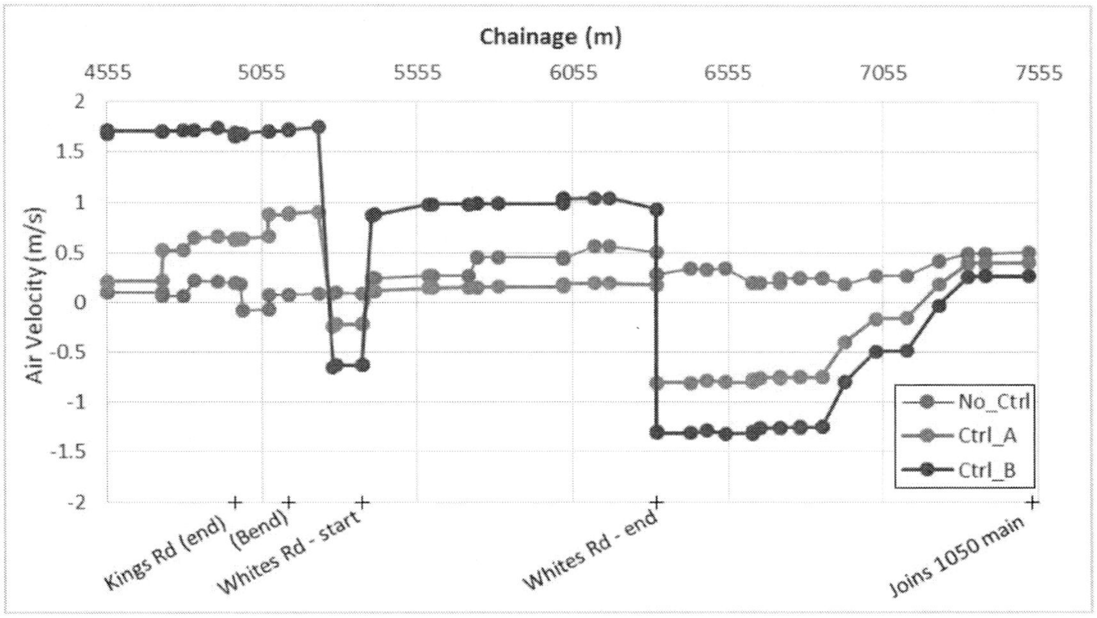

Figure 10: Predicted Air Velocity in Sewer – Control Scenarios vs Base Case

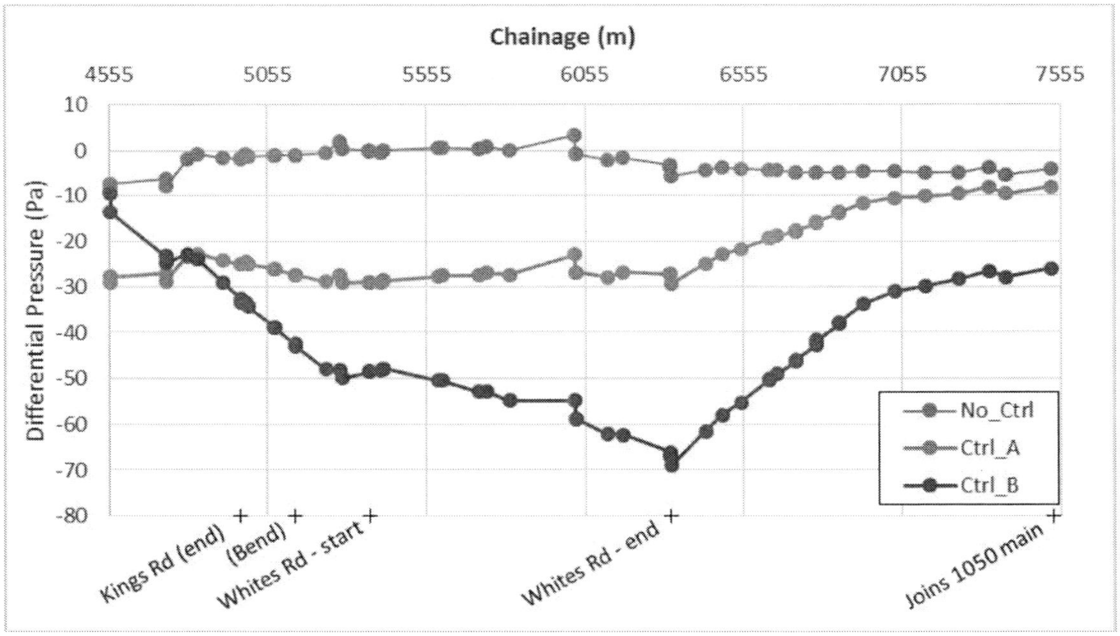

Figure 11: Predicted Differential Pressure in Sewer –Base Case vs Control Scenarios

A negative pressure of between -20 to -70 Pa was predicted to be achievable within the sewer trunk because of the extraction flowrates. No outgassing of odours is expected from the sewer trunk at these levels, and consequently no fugitive odour emissions are expected following installation of the two odour control systems and closure of selected vents.

The maximum extraction airflow rates shown for Scenario B allow for a factor of safety to overcome potential uncertainty associated with unknown aspects of the network. These extraction rates allow for air velocities of 1-1.5 m/s in the sewer, which are suitable for this application. The fan was sized for the maximum extraction rate and includes a variable speed drive. Optimal extraction rates at each ventilation point will then be determined by adjusting the fan to achieve the required differential pressure at the air inducts.

Data Analysis: Comparing Modelling Results with Commissioning Data

The average results of pressure readings in the network after commissioning are shown in Table 4. Figure 12 illustrates the model predicted negative pressure compared to the actual measured negative pressure after commissioning of the new system. We observe that the actual negative pressure is similar to the predicted negative pressure everywhere except of one location just downstream of OCU #2. The stronger negative pressure at this location may be attributed to partial blockage in the sewer.

Table 4: Pressure Readings after Commissioning

Location	2	3	4	MH1*	6	8	10	MH2*	14
Average dP (Pa)	-24	-38	-43	-51	-51	-46	-59	-96.75	-31.5

* MH1 and MH2 are the offtake manholes to OCU #1 and OCU #2 respectively.

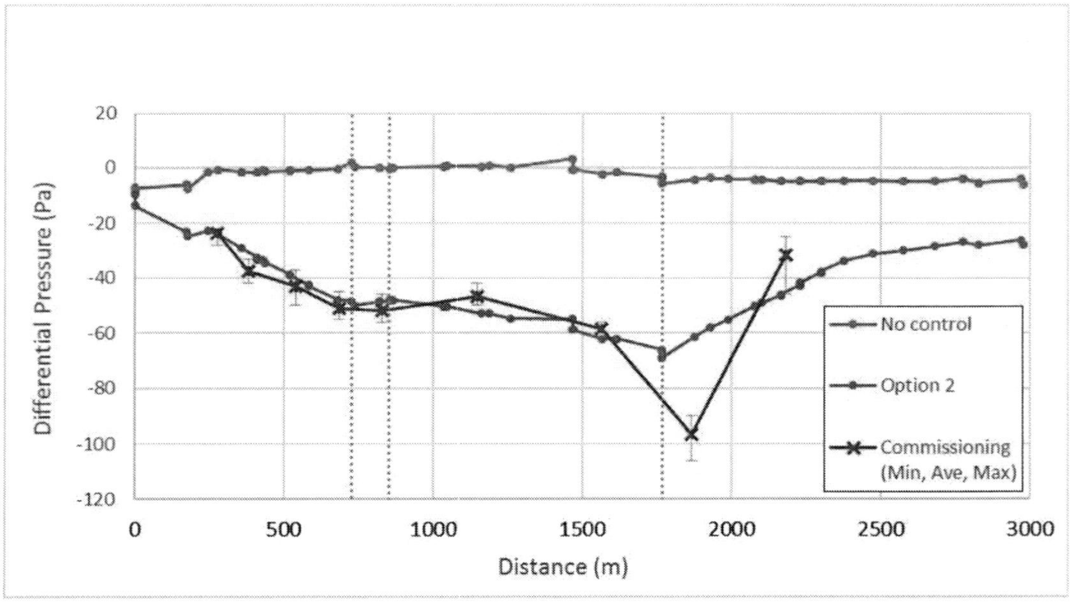

Figure 12: Predicted Differential Pressure in Sewer Compared to Commissioning Data

Data Analysis: Comparing H_2S Results before and after Commissioning

H_2S concentrations were monitored using Odalogs for a two week period after commissioning from the 17th July 2017 up to the 20th July 2017. This was approximately the same time of year that the previous data was collected. The H_2S concentrations before and after commissioning are illustrated in Figure 13.

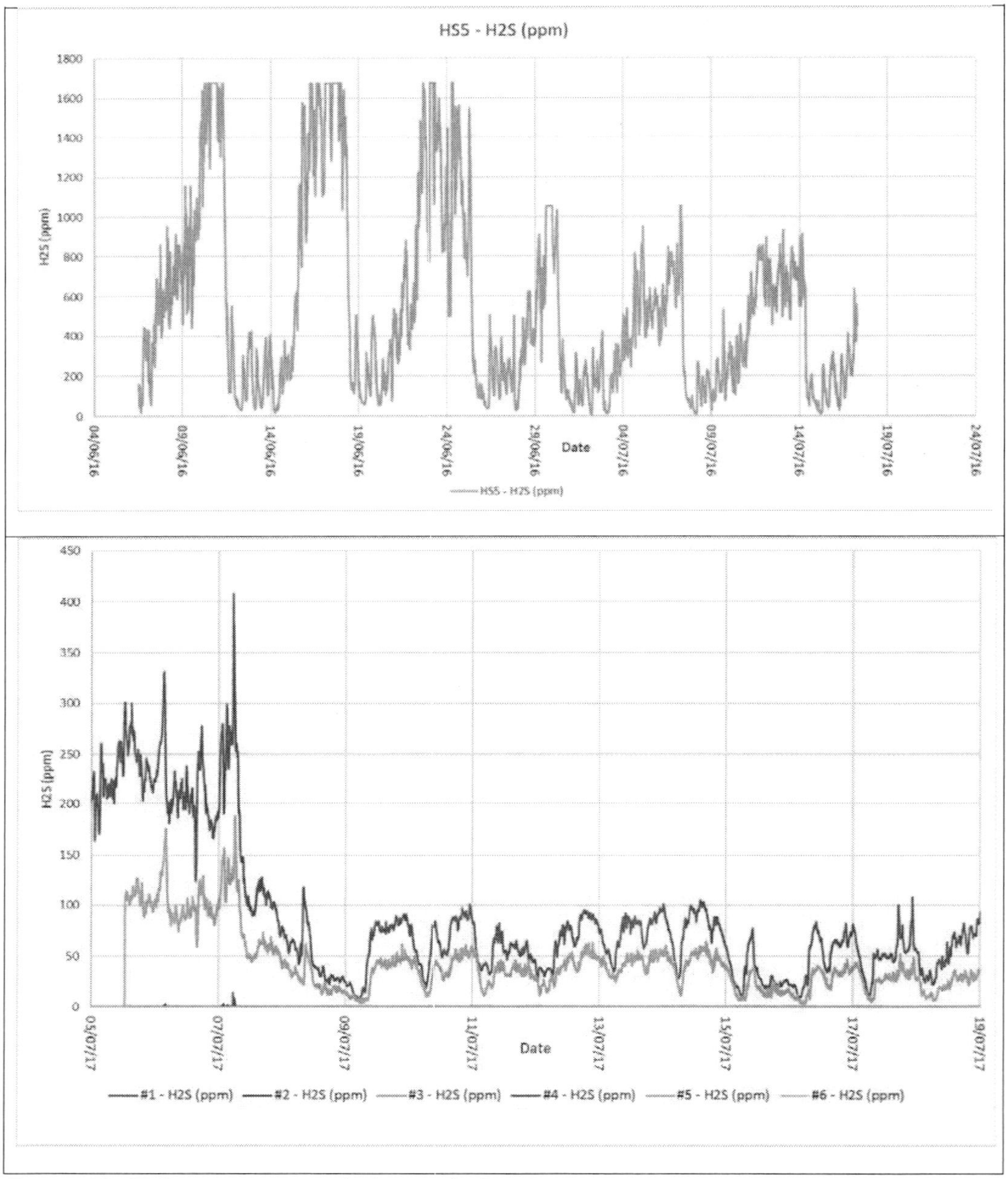

Figure 13: H_2S concentrations before and after Commissioning

We observe from the above that the sewer ventilation approach adopted reduced H_2S concentrations much more than anticipated from over 1000 ppm to less than 100 ppm. This illustrated how very high H_2S concentrations can be substantially alleviated by overcoming stagnate air movement in the sewer and the introduction of fresh air through a well-designed ventilation strategy. This also reduced corrosion, safety and odour risks.

CONCLUSION

By developing a greater fundamental understanding of the processes involved in various aspects of odour and corrosion, the water industry can improve the accuracy of control of odour and corrosion in sewers. Furthermore, the practical tools and approach developed for implementation of these strategies in live sewers provides useful insights to assist in the success of future installations

This paper illustrated that the use of a new Ventilation Tool based on the conservation of momentum model successfully estimated the optimal locations and ventilation rates required to achieve the required negative pressure in a problematic trunk sewer with very high H_2S concentrations and numerous odour complaints. In-sewer H_2S concentrations generally dropped by more than one order of magnitude after commissioning of the new ventilation system and the H_2S concentrations from the odour control units have been below detection limits since commissioning.

REFERENCES

Pescod, M.B.; Price, A.C. Major Factors in Sewer Ventilation, J. Water Pollution. Control Federation, 1982, 54(4), 385–397.

Ward, M., Hamer, G., (2011). Development of Operation Strategies and User Guidelines for Ventilation, Sewer Corrosion and Odour Research Linkage Project, Sub-project 4, Australian Research Council.

Hamer G., Cesca J., Bustamante H., Vitanage D., Ward M., Parker W., Witherspoon J. (2012). Water Industry Collaboration to Improve Sewer Ventilation Knowledge and Planning, Ozwater 2012.

Town of Cary, NC. Preventing a collection system odor issue—from concept through final performance test and beyond

Jarrod Buchanan, Town of Cary, NC. Philip Wolstenholme Brown and Caldwell, Atlanta, GA.
Ryan Sprowl, Brown and Caldwell, Raleigh, NC.

ABSTRACT

The Western Cary Collection System serves a rapidly developing service area for the Town of Cary, North Carolina (Town). This paper describes the various stages of an odor control project from evaluation through startup. A primary project driver was to be proactive in controlling collection system odors when a change in flow direction took place following completion of the new Western Wake Water Reclamation Facility.

The wastewater is conveyed from two pump stations via individual force mains into a combined force main and, finally, into a new 48-inch corrosion-resistant gravity interceptor. A spreadsheet model predicted wastewater sulfide concentrations for years 2015 and 2030. Liquid- and gas-phase data were collected to calibrate the model. A dynamic air analysis was performed to project air movement and odor emissions. Final recommendations for the collection system included:

- Preventive engineering modifications: eliminate turbulent drops, remove some manholes, decrease flow deflection angles inside manholes, design transitions to be "invert-to-invert" rather than "crown-to-crown," sealing of appropriate manholes, and incorporation of air check valves.
- Chemical dosing to limit the generation of hydrogen sulfide.
- Exhaust and treat using single-stage biofiltration with engineered media at two locations along the gravity pipeline.

Despite the high hydrogen sulfide (H_2S) inlet concentration, biofilter performance on H_2S is excellent, reducing an inlet air stream concentration from spikes of more than 200 ppm to less than 1 ppm at outlet. However, there has been an organic sulfide compound breakthrough, which is not surprising because the high-inlet H_2S will prevent the biofilter from acclimating to organic sulfides.

This paper describes the chronological events from conceptual design through commissioning.

KEY WORDS

Odor control, biofilter, force main.

BACKGROUND AND OBJECTIVES

The Western Cary Odor Control project provided the Town of Cary (Town) with a proactive opportunity to address its potential odor situation by reducing odors in advance of Western Wake Water Reclamation Facility (WWWRF) startup, and putting facilities and strategies in place before changing wastewater flows and accompanying odor potential occurred in the Western Cary collection system. After a qualitative and economic analysis, Brown and Caldwell recommended a combination of liquid-stream chemical addition and exhaust air treatment using biofilters at two exhaust air locations.

The scope of the project was limited to a network of force mains and gravity sewer mains between the Kit Creek Pump Station (KCPS), Morris Branch Regional Pump Station (MBRPS), and West Cary Regional Pump Station (WCRPS) that comprise the Western Cary collection system; including existing force mains; proposed new force mains; and the proposed Green Level Interceptor. The modified collection system was proposed to flow in the opposite direction to the existing flow. At strategic manholes 8-inch influent lines were purposefully placed along the interceptor during design so that the developers would never have to tie directly into the interceptor. Figure 1 shows the original Western Cary collection system layout. Other upstream modifications were initiated. The Thomas Brooks PS was abandoned and the gravity sewer was extended along the same route at the force main to the Bachelor Br. Interceptor (BBI). Also, Fieldstone Pump Station was modified to operate as a shedding pump station, and a gravity sewer was extended to the BBI along the same route at the old Thomas Brooks FM. Similarly with White Oak Pump Station, it was modified, and now operates as shedding station, with gravity sewer extending under Highway 55 to the White Oak Interceptor.

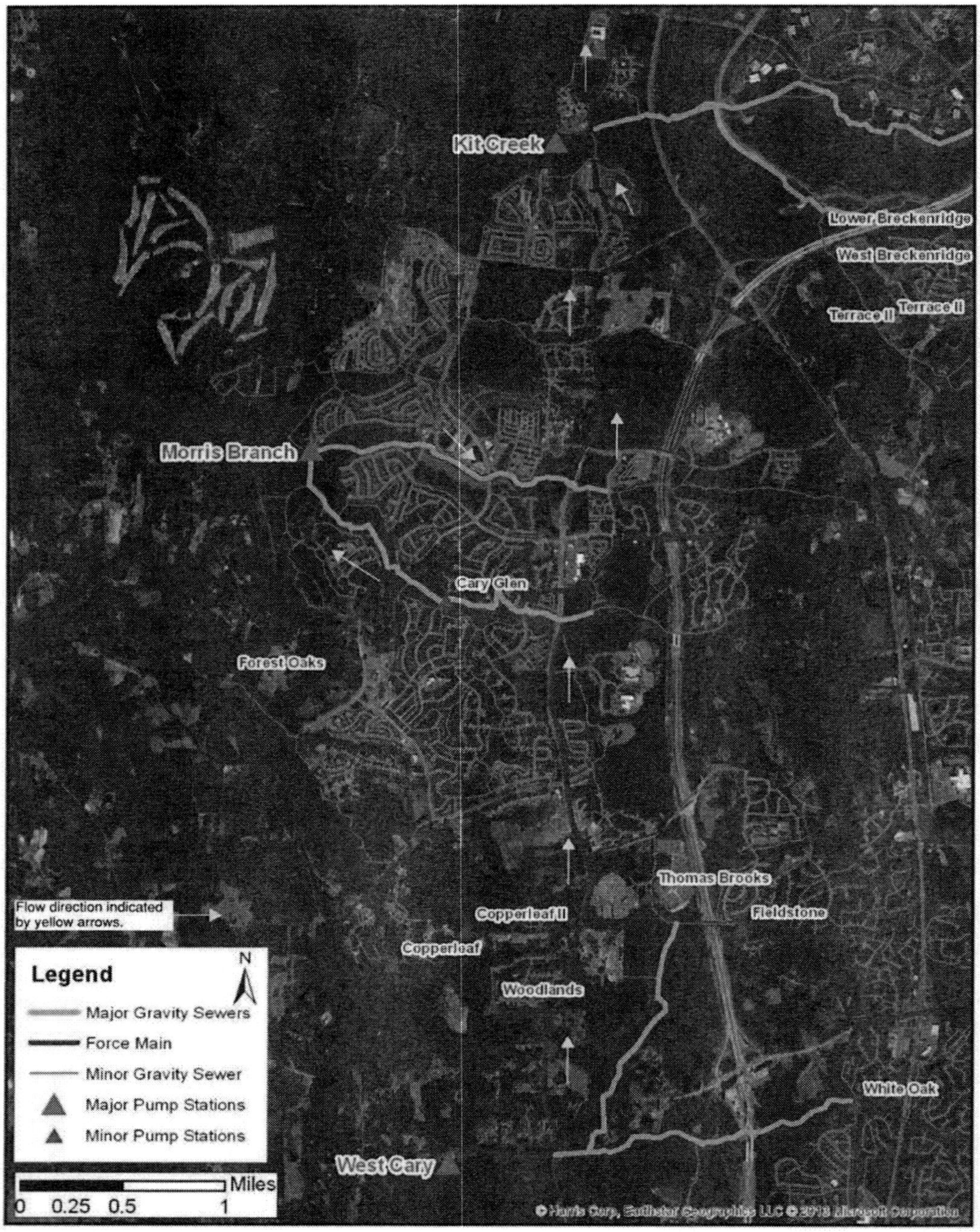

Figure 1. Original Western Cary collection system schematic

The existing and planned infrastructure are within sight of existing and new high-end residential development and in a number of locations, adjacent to existing and proposed greenways. Although the primary objective of the project was to reduce odor impacts, other important ancillary goals were to maintain or improve aesthetics, avoid local fugitive emissions, reduce corrosion, and provide an environmentally friendly solution. The facility needed to be low

maintenance and as "hands off" as possible. Design features supporting this approach include media with a long life to minimize changeout requirements.
To determine the potential for odorous air within the modified Western Cary collection system—and what could be done to effectively control gas creation (and the resulting odors) during the initial low-flow conditions expected following WWWRF startup, as the flow increases, and at buildout—the Town developed alternatives capable of treating potential odors through the life of the project.

Following commissioning of WWWRF, flow will be shifted south rather than north; this will result in some pump stations (e.g., WCRPS) conveying larger flows while reducing flow at others. The wastewater collected at KCPS and MBRPS will be conveyed south to WCRPS via parallel 2.0-mile- and 1.5-mile-long force mains feeding a combined 2-mile-long force main and1-mile-long, 48-inch-diameter gravity interceptor. The new force mains will have two significant high points, each equipped with an air relief valve. WCRPS will then pump the wastewater to the proposed Beaver Creek Pump Station, which will convey wastewater to the WWWRF. Figure 2 shows a schematic of the collection system for the future flow scenario.

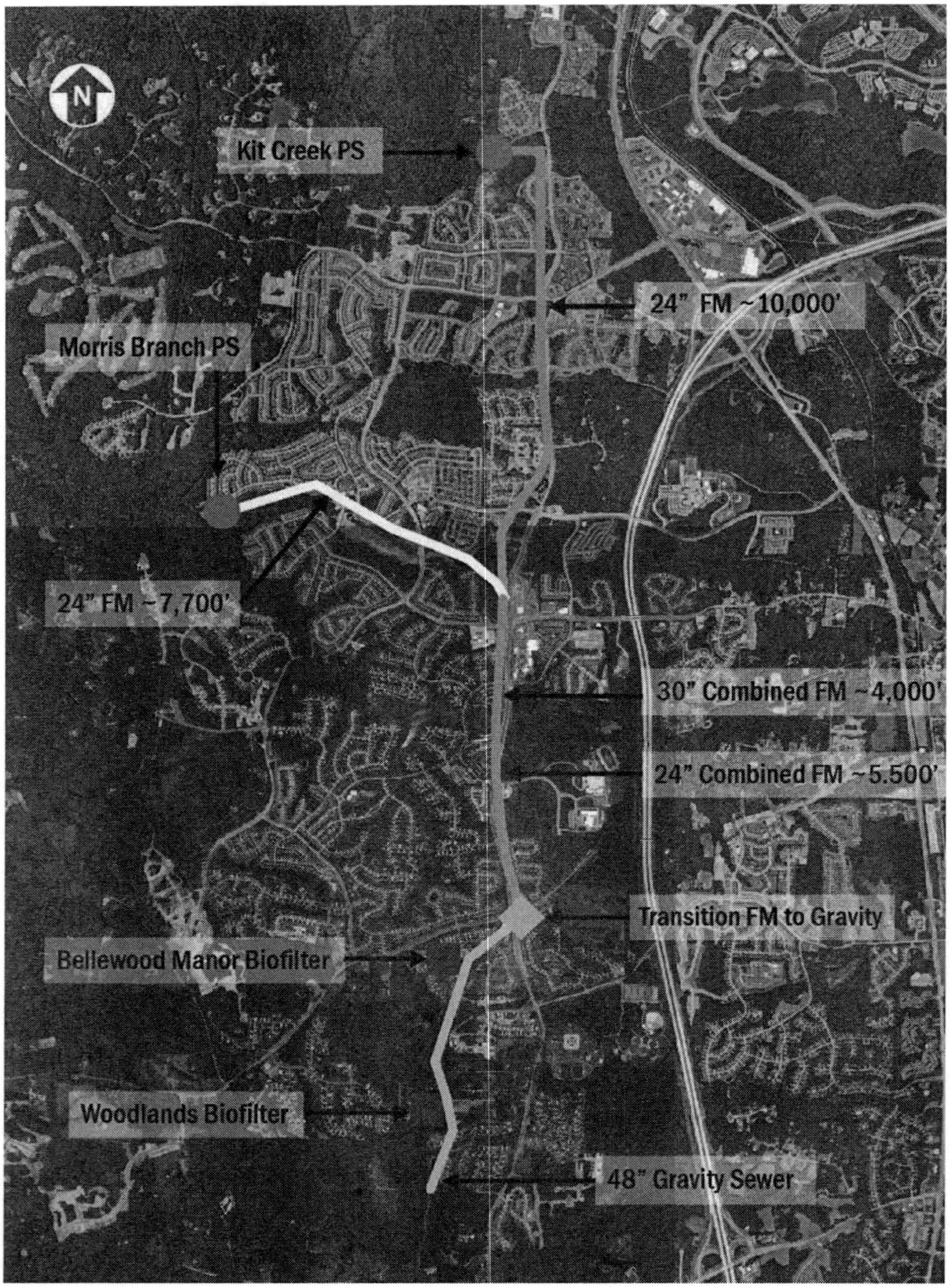

Figure 2. Reversed Western Cary collection system schematic

To ensure a multipronged approach and redundancy, the Town wanted to look at the Western Cary collection system holistically, which required multiple facilities and strategies to mitigate issues in the event of any individual failure, with the goal of controlling odors 100 percent of the time. The Town looked to develop alternatives that could treat potential odors through chemical, vapor-phase, or other treatment technologies with consideration for construction, operation, and annualized costs. The Town wanted to focus on doing the right thing by implementing appropriate technical solution(s) for potential odors in this area, and Operations and Engineering divisions agreed to share budgeting if necessary.

For its current flow conveyance plan, the Town has invested in gas-phase treatment and chemical injection odor control systems at pump stations and other sensitive sites throughout the primary wastewater conveyance corridor. Both KCPS and MBRPS include gas phase odor control using biofilter plus carbon, and chemical addition. Table 1 lists the treatment systems for each pump station within the collection system.

Wastewater pH in this area system was measured from 6.5 to 8.0 with and without chemical addition. In 2013, chemical addition using a 5 percent ferrous sulfate solution was employed with 300 gallons per day (gpd) dosed at KCPS, and 350 gpd dosed at MBRPS (for a total of 230,000 gallons per year).

Table 1. Western Cary Collection System Odor Control

Location	Air Flow Capacity (cfm)	Odor Control System
KCPS	3,000	Biofilter with activated carbon as backup
Force main discharge manhole	1,000	Biofilter
MBRPS	2,000	Biofilter with temporary activated carbon; chemical addition to discharge force main
WCRPS	6,700	2-stage: biofilter and activated carbon; chemical addition to discharge force main

SAMPLING AND ANALYSIS

The Town was diligent in data collection, and made previous data available to Brown and Caldwell on this area of the collection system from 2011; additional sampling was performed in 2012 and 2013. A key parameter for this project was sulfide generation in the force mains, and an easily measured data point was hydrogen sulfide (H_2S) at the force main discharges. Examples of these values with and without force main chemical dosing are shown in the following table and figures. Table 2 lists average and peak H_2S concentrations at the Kit Creek force main discharge.

Table 2. Kit Creek Force Main Discharge H_2S Values

Parameter	Average (ppm)	Peak (ppm)
Without chemical: 11/2012	191	515
With chemical: 6/2011	14	95

The daily H_2S trends on Figures 3 and 4, show the high H_2S concentrations as well as the impact of chemical dosing. H_2S concentrations were at times very high and at levels that could cause odor impacts to the surrounding areas, and corrosion in unprotected infrastructure. The diurnal variation of H_2S concentration corresponds with pumping cycles. KCPS is the most downstream station of the four in series, so it is expected to exhibit high accumulated liquid-phase sulfide concentrations, which often corresponds to high gas-phase H_2S concentrations.

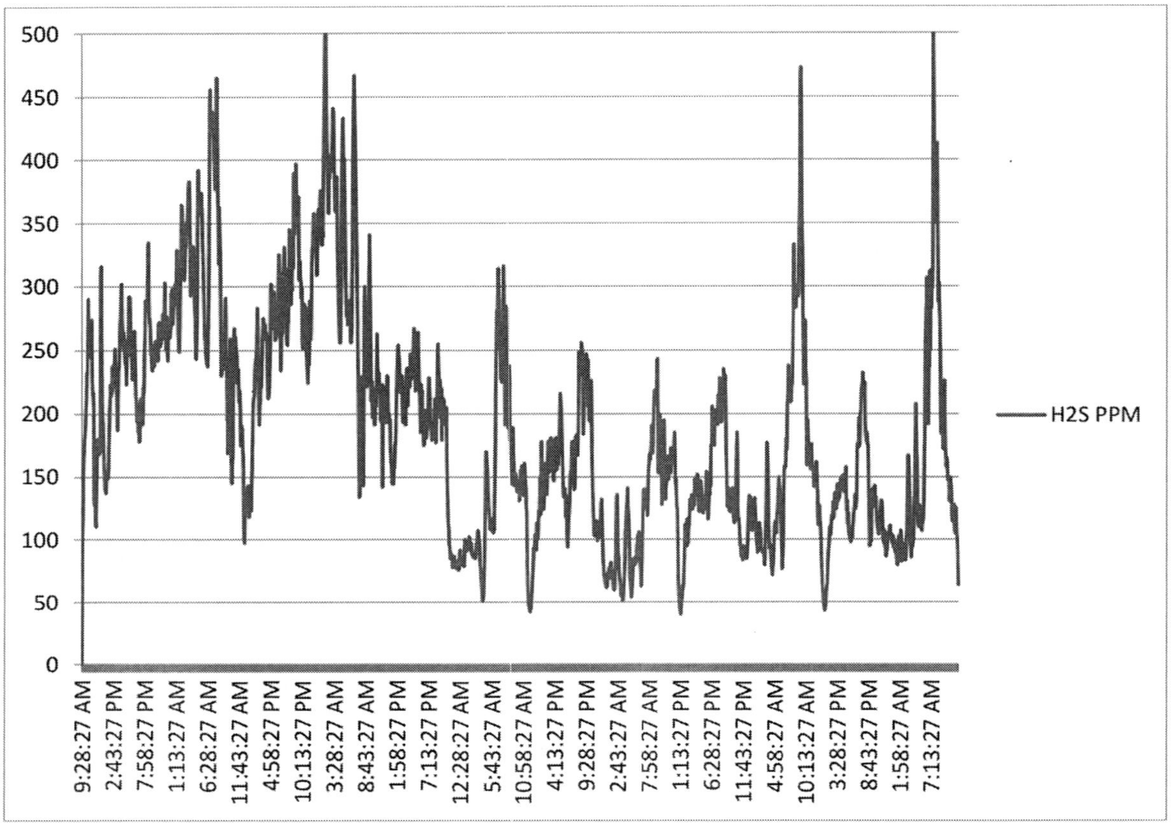

Figure 3. H_2S concentration KCPS force main discharge: 11/27/2011–12/3/2011
(without chemical dosing)

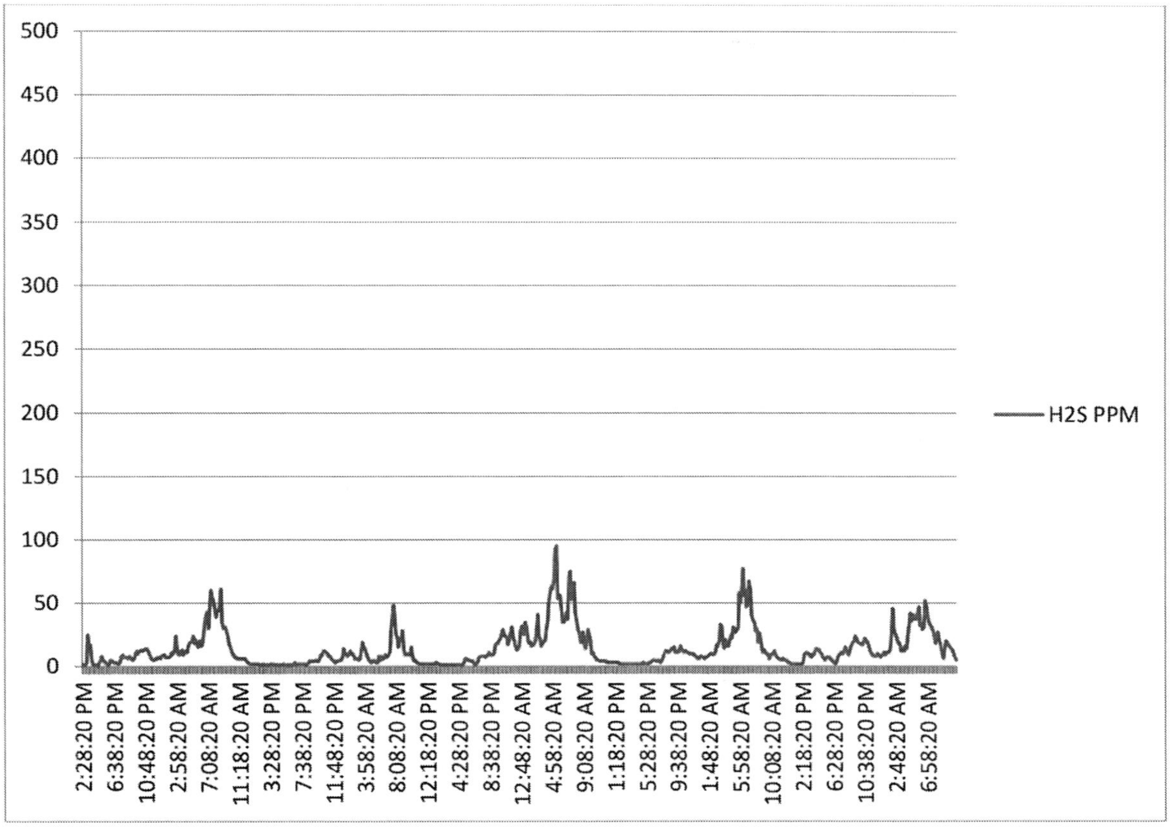

Figure 4. H_2S concentration KSPS force main discharge: 6/23–6/28/2011 (with chemical dosing)

Table 3 shows the liquid-phase data collected in 2009 and again in 2012. Peak total sulfide concentrations were associated with high equivalent dissolved sulfide concentrations. On average, dissolved sulfide concentrations were approximately 28 to 50 percent of total sulfide concentrations, the latter value being at the high end of the typical range for municipal wastewater. As shown in Table 3, dissolved oxygen concentrations were good for a force main discharge. Chemical addition appeared to reduce gas-phase H_2S concentrations during the 2 sampling days.

Table 3. Liquid Sampling Results: Force Main Discharge Locations

Parameter		Winter 2009	September 2009		November 2012		
		Kit Creek	Morris Branch	West Cary	Morris Branch	West Cary	Kit Creek
Average total sulfide concentration (mg/L)		1.6	5.5	2.3	5.1	4.2	6.0
Peak total sulfide concentration (mg/L)		2.25	7.75	3.25	7.00	5.75	8.00
Dissolved sulfide concentration percent of total		50	40	25	40	28	50
Dissolved oxygen concentration (mg/L)					2.4–3.2	3.7–3.9	2.1–1.9
H_2S concentration (ppm) [a]	with chemical				1.5	36.0	1.7
	no chemical(s)				29	> 50	11

a. Kit Creek force main discharge maximum H_2S: summer 2011 = 82 ppm; summer 2012 = 316 ppm.

LIQUID AND AIR PHASE MODELING

Sulfide-generation modeling predicted similar generation in both force mains with total sulfide at the combined force main discharge at 5.4 milligrams per liter (mg/L) for a 2015 average daily flow of 3.4 mgd (5.5 hours detention time), and 3.5 mg/L for a 2030 average daily flow of 6.22 mgd (2.9 hours detention time). Traditionally, data show that approximately 50 percent of the total sulfide was dissolved sulfide. Once the wastewater discharged, the sulfide concentration was predicted to reduce with H_2S gas stripping and finish at the end of the gravity segment with total sulfide values of 2.9 and 2.0 mg/L for 2015 and 2030, respectively. Current 2018 flow rates are 1.14 mgd at Kit Creek and 1.55 mgd at Morris Branch, for a combined flow rate approximately 23 percent lower than anticipated. In main part this reduction is due to water conservation measures implemented by the Town.

The Town provided wastewater flow rate projection data, and a dynamic air analysis of the Green Level Interceptor was performed for year 2015 and 2030 wastewater flow rates and buildout. Local gravity sewers enter the interceptor at five manholes. Table 4 presents the results of the air flow analysis.

According to the model, the total air flow rate to fully depressurize the Green Level Interceptor at 2015 peak hour flow rates is 7,020 cubic feet per minute (cfm). For 2030 peak hour flows the required air flow rate is 10,150 cfm.

Table 4. Maximum Green Level Interceptor 2015 Air Emission Rates from Dynamic Model

Location	Water Depth (inches)	Air Flow Rate (cfm)
Manhole 35	6.65	2,149
Manhole 21	11.00	1,326
Manhole 15	6.50	2,218
Manhole D	11.00	1,328

Note: Water depths and air flow rates in this table were calculated at 2015 peak hour wastewater flow rates in the 48-inch diameter interceptor.

OPTIONS AND RECOMMENDED SOLUTION

Throughout this project, Brown and Caldwell worked with Town staff to make design suggestions for the Town's design of the Green Level Interceptor. With odor mitigation in mind, Brown and Caldwell recommended the following features:

- Reduce sharp transitions and level out any significant slope changes to reduce H_2S gas solution volatilization potential and prevent high-pressure locations. Vortex drops can be useful in these situations but, because there are no deep drop structures in the Green Level Interceptor vortex, inserts are not appropriate.
- When 8- or 12-inch side sewers enter the main 48-inch-diameter interceptor, make those connections with vertical pipe sections to avoid a drop from the entering pipe to the main flow.
- Remove the manhole between manholes 30 and 31 and raise the invert of manholes 31 and 32 to level out slopes.
- Update the force main to gravity-transition to an invert-to-invert configuration—the Town has traditionally used crown-to-crown transitions from force main to gravity. For odor mitigation, it is typically recommended that invert-to-invert transitions be made. This reduces turbulence at the transition and thus the potential release of H_2S gas.
- Seal and vent appropriate manholes at the start of the gravity pipeline to minimize air entry. Sealed manholes prevent air entering the system reducing available oxygen for H_2S conversion to sulfuric acid, as well as reducing total air in the system and subsequent potential for more air escaping downstream.
- Equip manhole 37 at the force main discharge with an air check valve. Seal the four manholes between manhole 37 and the air exhaust at manhole 32 to prevent air emissions.
- Install passive carbon units (either canisters as shown in Figure 5, or saucers under the manhole) at locations where small air flow is expected. Vents are required every 1,000 ft.

Figure 5. Typical carbon canister for passive air vent

After a qualitative and economic analysis, Brown and Caldwell recommended a combination of liquid-stream chemical addition and exhaust air treatment using biofilters at the two exhaust air locations.

Several alternative chemicals were evaluated, and the Town decided to continue with ferrous sulfate for now. The Town prefers ferrous sulfate as opposed to iron chlorides to avoid adding excessive chloride to the wastewater (because high chloride concentrations can affect wastewater effluent quality and treatment processes).

Based on the dynamic air analysis, Brown and Caldwell determined that two exhaust air locations were required, one at manhole 28 and one at manhole 15. Both would require exhaust air treatment.

DESIGN

The design was performed in 2014, during which time pipeline construction began. Several of the recommended pipeline features were included during pipeline construction, including: reducing sharp transitions, adding invert-to-invert connections, removing certain manholes, sealing manholes, venting non-sealed manholes to carbon canisters, and adding air flaps at small diameter sewer connections.

Two biofilters were located, one at manhole 28: the Bellewood biofilter, and one at manhole 15: the Woodlands biofilter (both named after local landmarks). These biofilters were designed with initial air flows of 2,200 and 3,600 cfm, respectively, at 45 seconds biofilter empty-bed retention time, and with the capability to increase air flow to 3,700 and 5,200 cfm by adding extra media and increasing the fan speed. These biofilters were designed for pipe-in-gravel air distribution for reasons described in the next paragraph.

Biofilter system design features included: drip irrigation within the media at the Bellewood biofilter, cleanouts to access each lateral, space for future media, an air bypass connection to enable installation of a temporary odor control system during maintenance or media change, an underground fan, and electrical vaults. Adding drip irrigators within the media was designed to rinse the lower level (which is expected to operate at a low pH for optimum H_2S removal), preventing the acidic conditions from migrating up through the media. The upper media layer can then operate at a higher pH conducive to organic sulfur compound removal.

CONSTRUCTION

Construction took place during 2015–16, immediately following completion of the gravity pipeline, and was performed by the pipeline contractor as a change order. Figure 6 shows a view of the start of construction in the wooded area for the Bellewood biofilter.

Figure 6. Bellewood biofilter location

During construction, Brown and Caldwell and the Town considered several contractor-suggested cost-saving modifications. One suggestion was cleanout removal, but it was decided that if the media lasted the predicted 15 years without changeout, then it is possible that the air distribution laterals can collect deposits, thus cleanouts are important. An analysis of construction costs confirmed that the in-ground construction—especially at confined sites—is expensive. Figure 7 shows the below-grade media structure during drip irrigation system installation.

Figure 7. Installation of drip irrigators within media

STARTUP AND TESTING 2016–17

Startup, commissioning, and performance-testing the biofilters was on schedule and went well. This included instrument calibration, air flow balancing, smoke tests to show air distribution in the media, and performance testing.

As predicted, the biofilter experienced high H_2S inlet concentrations, and although the organic reduced sulfur concentrations were low, their low odor threshold resulted in noticeable odors in the area surrounding the biofilter. Figure 8 shows the plot of H_2S at a manhole upstream of the Bellewood biofilter in October 2016. H_2S concentrations in the pipeline were occasionally very high, reaching 900 parts per million (ppm).

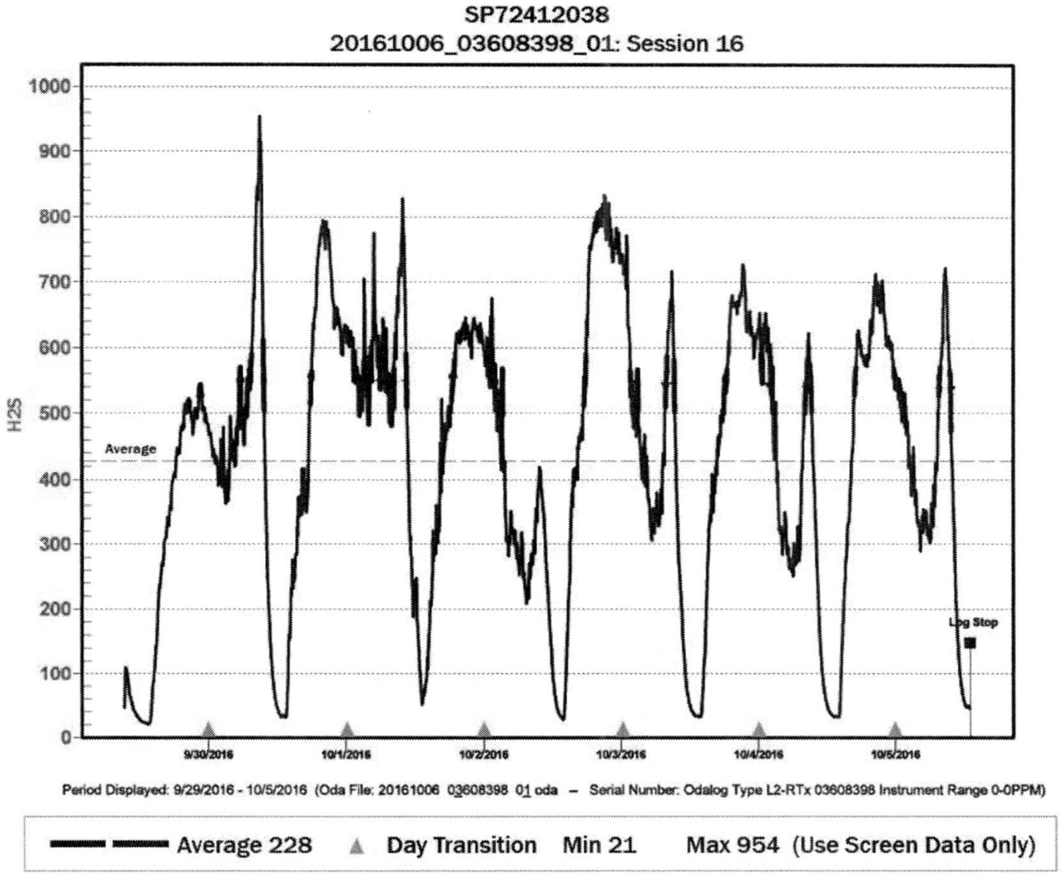

Figure 8. H_2S concentrations in gravity pipeline upstream of biofilter

Since startup, the system has been continually optimized to reduce odors at a reasonable cost. By September 2017, the chemical feed adjustment made a significant impact on the H_2S concentration within the interceptor, with peak values reduced by approximately 50 percent. September 2017 data showed that the biofilter is removing H_2S excellently—in fact better than specified. It reduced the average H_2S of 26 ppm and spikes of up to 270 ppm to less than 1 ppm (see Figure 9). The biofilter H_2S outlet was nearly always non-detect as shown by the blue line.

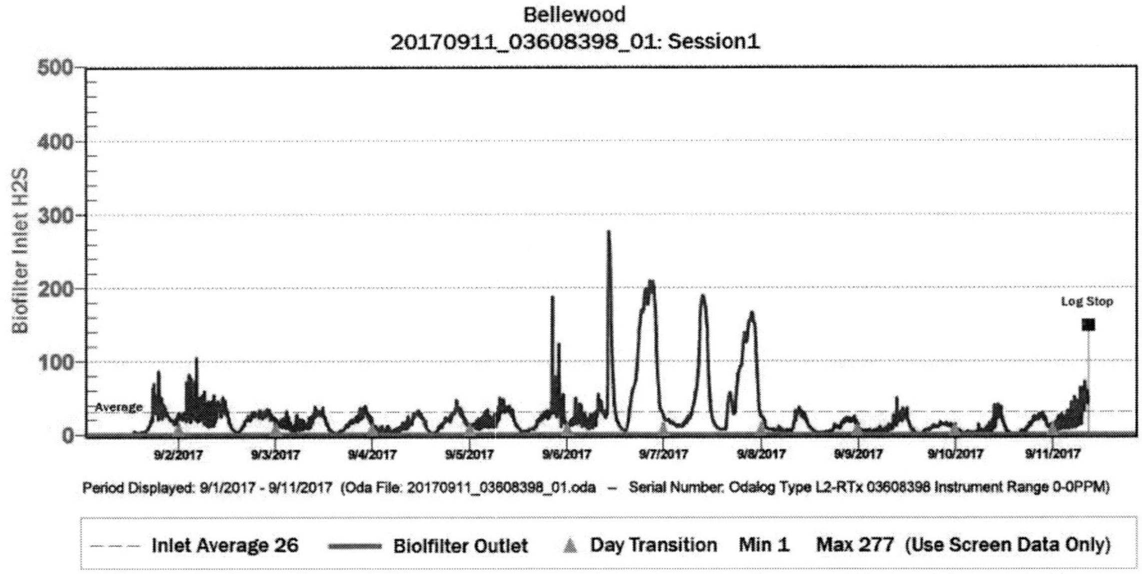

Figure 9. Biofilter inlet and outlet H_2S concentrations

FURTHER PERFORMANCE ENHANCEMENTS

Optimization was hampered by a Bellewood biofilter prolonged shutdown due to a loss of electrical power, and upon restart the biofilter was acclimating when odor complaints were received. This increased neighborhood awareness and prompted more attention on the biofilter. Neighbors noticed organic sulfide emissions and odors, so measures were taken to improve Bellewood biofilter performance on these compounds. The air residence time was increased from 45 seconds to nearly 60 seconds, while still ensuring a negative pressure in the pipeline.

Sampling the air stream and liquid stream is ongoing in March 2018 to more precisely identify and quantify the organic odors. Following completion of the sampling, additional investigations will be conducted to determine whether the organic sulfide contributions are originating from an industrial wastewater source.

Further work was performed to reduce H_2S spikes by modifying the chemical addition dosing profile. This is a challenge because intermittent pump station operation and subsequent long detention times in the force mains, particularly overnight, are the main factors in high sulfide generation.

The Town is considering a temporary second stage of carbon and potassium permanganate to remove organic odors, but this will require covering the biofilter and will violate the concept of all equipment being located underground.

CONCLUSIONS

The Western Cary collection system was reconfigured to accommodate flow in a reverse direction to a new reclamation facility. The collection system is experiencing lower-than-anticipated flows, and hence higher-than-anticipated H_2S. As wastewater flows increase and detention time decreases, the sulfide concentration can be expected to decrease.

Despite the high H_2S concentration, biofilter performance on H_2S is excellent, reducing an inlet air stream concentration from spikes of more than 200 ppm to less than 1 ppm at outlet. However, there has been an organic sulfide compound breakthrough, which is not surprising because the high-inlet H_2S will prevent the biofilter from acclimating to organic sulfides and growing other species of bacteria to remove the organics.

Biofilter restarts followed a trend of higher outlet odors associated with the organic sulfide compounds, which increased neighborhood awareness and prompted more attention on the biofilter.

The Town is optimizing alternate chemicals; however, there are limited chemicals appropriate for high sulfide concentrations that also meet the Town's water quality criteria. Another option under investigation a second-stage treatment addition using activated carbon.

REFERENCES

Water Environment Federation. Odor Conference 2014. *Optimizing Odor Control Solutions for a New Interceptor with Air Flow and Odor Generation Modeling*, contributing authors: Erin Culbertson, Philip Wolstenholme, Brown and Caldwell, Jamie Revels, Town of Cary NC.

UNDERSTANDING MECHANISMS AND SOURCES OF ODORS IN RESOURCE RECOVERY FACILITIES: IMPACT OF COLLECTION SYSTEM, PRIMARY AND SECONDARY TREATMENT

Ulrich Bazemo[1,4], Adrian Romero[2,4], Amber Spriggs[4], Helene Hauduc [3], Ahmed Al-Omari[4], Imre Takacs [3], Sudhir Murthy[4], Alba Torrents[1], Matt Higgins[5] and Haydée De Clippeleir[4]

[1]Department of Civil & Environmental Engineering, University of Maryland, College Park, MD 20742, USA.
[2] CH2M, Charlotte, NC, 28227, USA.
[3] Dynamita, 7 Lieu – dit Eoupe, La Redoute 26110 Nyons, FRANCE.
[4]DC Water and Sewer Authority, 5000 Overlook Ave., SW., Washington, DC 20032, USA.
[5] Department of Civil & Environmental Engineering, Bucknell University, Lewisburg, PA 18837, USA.

KEYWORDS
Methyl mercaptan, hydrogen sulfide, methionine

INTRODUCTION

Controlling odor production and emissions remains as one of the most challenging task for resource recovery facilities (RRFs). In addition, stricter environmental regulations are pressuring RRFs to reach higher levels of treatment without increasing their footprint (Corcoran et al., 2010) nor their indirect impact to the surroundings (i.e. odor nuisance). Odors emitted at the beginning of the RRFs are mostly produced in sewer networks and treatment of odor emitted in screening buildings and grit removal facilities is common practice. Primary and secondary treatment systems are potential sources of significant odor emissions due to the large surface area of the systems. Odors in the latter processes are mainly caused by volatile sulfur compounds (VSCs) emission such as methyl mercaptan (MM), dimethyl sulfide (DMS) and hydrogen sulfide (H_2S) (Estrada et al., 2011; Giri et al., 2014) . VSCs are important sources of unpleasant odors in wastewater systems due of their malodorous characteristics and low odor thresholds (0.07–5.9 ppbv) (Abalos et al., 2002); even a small amount of VSCs can contribute to significant odor pollution.

A previous study by this group focused on the sources and mechanisms of VSCs production in secondary systems to better understand potential alternatives for odor prevention instead of covering and treating the large surface area and high air flows that these systems present (Romero et al., 2016) [5]. From this study, it was found that MM was the main VSC emitted in full scale tests, and that MM was produced from the degradation of the amino acid methionine after hydrolysis of proteins and not from the methylation of H_2S present in the bioreactor influent. Degradation of methionine leading to MM production by the methionine gamma lyase was the preferred pathway under low oxidation reduction potential (ORP)conditions (-20mV to -200mV), such as those found in anaerobic zones (selector zones) and in the sludge blankets in the clarifiers that is then recycled to the aeration basins as return activated sludge (RAS). It was also found that the main source of methionine was the biomass itself and not proteins from the influent.

Methionine degradation leading to MM production was inhibited under aerobic conditions (measured DO > 0 mg/L), and when nitrate was present above 1 mg N/L in the solution, thus inhibiting MM production and emission. These results, however, may not directly translate to primary systems given that it is not considered an activated sludge process yet active biomass can be found that has been transported through the collection systems. Therefore, there is a need to increase the understanding on odor production compared to odor emission in primary systems and differentiate the impact of odors produced in the collection system and those produced on-site. The implications of this increased understanding will potentially lead to operational improvements as a strategy to minimize nuisance odor emissions.

In this study, the source (either being the sewer or the treatment system itself) and the mechanism of odor formation was studied for District of Columbia Water and Sewer Authority (DC water), Blue Plains Advanced Wastewater Treatment Plant (AWWTP). The study focused on the primary systems which have significant odor emission, especially during the summer but don't have odor treatment systems in place. The impact of VSCs formed in sewer systems on odor emissions in the plant was quantified and differentiated from the odor formed in the plant. This understanding would ultimately support the development of tools for better decision making in odor control strategies.

MATERIALS AND METHODS

Sampling campaigns

All samples were collected at Blue Plains AWWTP. A sampling campaign for odor monitoring was conducted on primary settling tanks using a flux chamber (0.13 m^2). Air samples were collected into 1.4 L canisters coated with fused silica (Entech Instruments Inc., Simi Valley, CA, USA) for analysis of organic sulfur compounds. Odalogs were used for quantification of H_2S concentrations. Along with gas sampling, to investigate possible correlation with odor production, wastewater parameters were analyzed for total elemental analysis (TSS/VSS, COD, TKN, TP, TS, total protein) as well their fractionation in soluble and colloidal fraction. Sampling campaigns on the primary treatment systems were performed on 06/12/17, 07/05/17, 10/16/17, 11/13/17, and 12/04/17.

Analytical methods

Analysis of MM, DMS and DMDS were performed by using an Entech Instruments Inc. (Simi Valley, Ca., USA) system consisting of a Model 7500 autosampler followed by a Model 7200 Preconcentrator coupled to an Agilent 5790N Gas Chromatographer-Mass Spectrometer (Agilent Technologies Inc., Wilmington, De., USA). TSS and VSS were determined using standard methods (EPA 160.2, EPA 160.4). For TKN analysis, samples were heated in the presence of sulfuric acid for two and one half hours. The residues were cooled, diluted to 25 mL and analyzed using semi-automated colorimetry method (EPA 351.2). Phosphorous, ammonium, nitrite, nitrate, sulfate, COD were determined using HACH kits. Total sulfur concentrations were determined using an Inductive Coupled Plasma – Atomic Emission Spectroscopy.

Batch experiments

To estimate concentration and loading rate of dissolved organic VSC concentrations, liquid sample from raw influent, grit effluent, primary effluent and primary sludge were collected in 5 gallon buckets. Wastewater samples (10 L) were added to bucket midpoint and gas samples were collected in evacuated canisters at equilibrium using static headspace method (Iliuta et al., 2007) . Dissolved concentrations were calculated based on the head space concentrations and their Henry constants. In addition, to gain understanding of MM formation and the impact of raw wastewater composition on its formation, batch studies were conducted with primary sludge and grit effluent (high proteins content) and primary sludge and final effluent (low proteins content). These tests were carried out in 2 L Plexiglas jars, in which 1.6 L of sample (sludge + wastewater) was added, leaving 0.4 L headspace. Tests were performed at constant stripping rate of 1.5 L/min of nitrogen gas and gas samples were collected at regular intervals (every 15 minutes) over a period of 60 minutes. This method allowed for MM production rate calculations based on gas phase measurements for the given conditions and sludge/food mixture. An aerated and mixed batch method was used to estimate the mass of active heterotrophic organisms in influent wastewater (Wentzel et al., 1997). In this test, the Oxygen Uptake Rate was measured semi – continuously for 24 hours and based on default monod kinetics, an initial biomass concentration was estimated.

RESULTS AND DISCUSSION

The foci of this project were (i) to determine the significance of organic VSCs emission in primary systems and (ii) to elucidate the mechanisms of VSCs formation in primary systems.

Chemically Enhanced Primary Treatment (CEPT) process was applied using Ferric chloride and anionic polymer to enhance coagulation and flocculation for more effective organic and phosphorous removal. The overall CEPT performance for both the east side (60% of the flow) and west side (40% of the flow) is presented in Table 1. This leads to primary effluent quality of 70 mg TSS/L, 160 mg TCOD/L, 2 mg TP/L and 35 mg TN/L. For the first time, a sulfur balance was performed over the primary system, indicating a removal of total sulfur of about 10% and an influent fraction of sulfur of 0.04 g S/g COD.

Table 1: Removal efficiencies for total suspended solids (TSS), total COD (TCOD), total nitrogen (TN), total phosphorous (TP) and total sulfur (TS) over the EAST and WEST CEPT system at Blue Plains AWWTP over a period from 05/31/2017 to 11/20/2017

	TSS	*TCOD*	*TN*	*TP*	*TS*
East CEPT	*75.4 ± 13.3*	*58.8 ± 21.2*	*33 ± 10.7*	*54.5 ± 8.8*	*9.1 ± 3.1*
West CEPT	*66.7 ± 12.4*	*57.5 ± 35.1*	*39.9 ± 30.5*	*46.9 ± 16.3*	*10.4 ± 16*

1. Significance of organic sulfur compounds emission in primary treatment

As expected, high H_2S concentrations were detected in both CEPT systems. However, lower emissions were detected during winter time (starting November) compared to the higher temperature conditions (Figure 1A,B). Of the organic VSC analyzed, only traces (<0.9 pbb) of dimethyl sulfide (DMS) and dimethyl disulfide (DMDS) were detected and their impact could therefore be neglected. However, significant methyl mercaptan (MM) concentrations were detected in both CEPT systems during both winter and summer conditions. Overall, emission concentrations ranges from 100-250 ppb in summer with a decreasing trend towards winter time reaching 25-80 ppb (Figure 1A,B).

When translating the observed emission concentration into their impact on the odor units based on MM and H_2S odor threshold of 2 ppb and 100 ppb (Deshmuckh et al., 2014) , respectively, the significance of MM emission can be clarified. Figure 1C and Figure 1D show that MM emission contributed 94% in odor while only 25% in mass basis (Figure 1A,B). Overall a 50% decrease in odor was observed in winter time versus summer time. The impact of temperature on the H_2S and MM emission concentration is presented in Figure 2. Preliminary Arrhenius fitting of the data showed a theta of 1.17 and 1.21 for MM and H_2S production, respectively. More data needs to be collected to improve the reliability of the model variables obtained.

In addition, it has to be noted that slight differences existed between the EAST and WEST CEPT systems but also most probably between individual clarifiers depending on the operational conditions applied. Therefore, it is important to take into account the variability and uncertainty when translating the obtained knowledge into plant odor impacts.

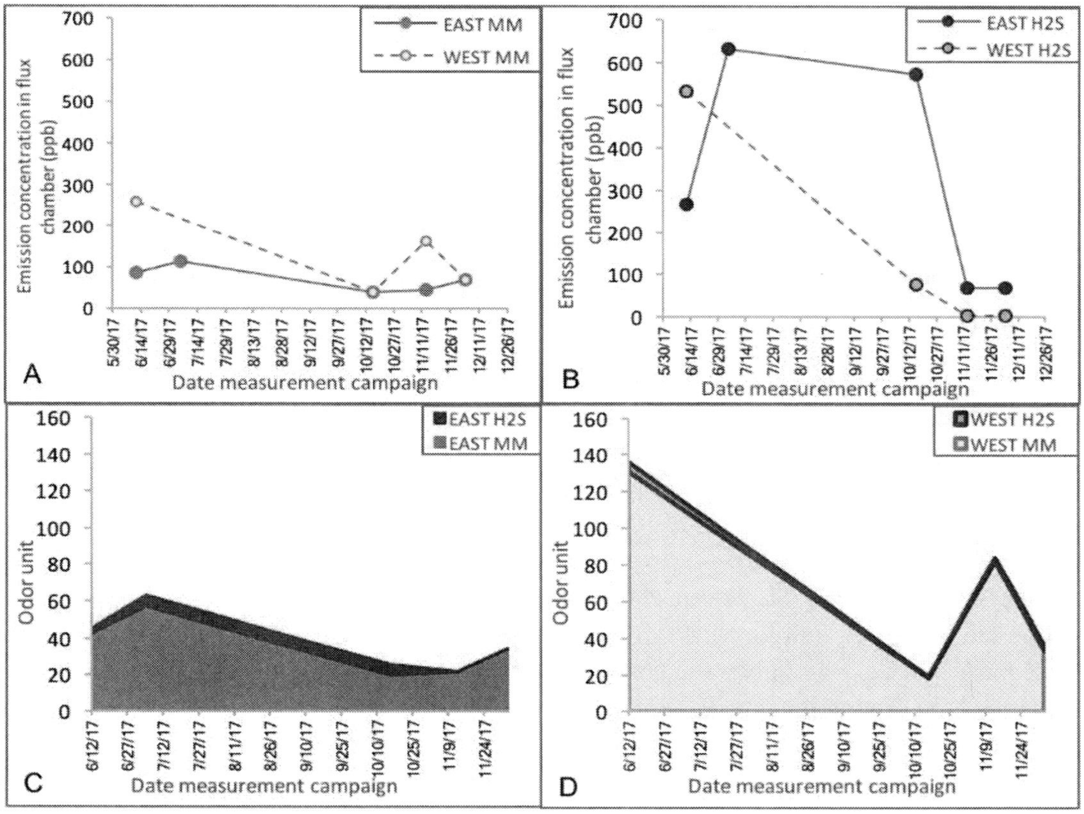

Figure 1: MM (A) and H2S (B) emissions as measured by the applied flux chamber on the EAST and WEST CEPT systems at Blue Plains AWWTP, and their contribution to the odor based on their difference in odor threshold concentrations (C: EAST; D: WEST).

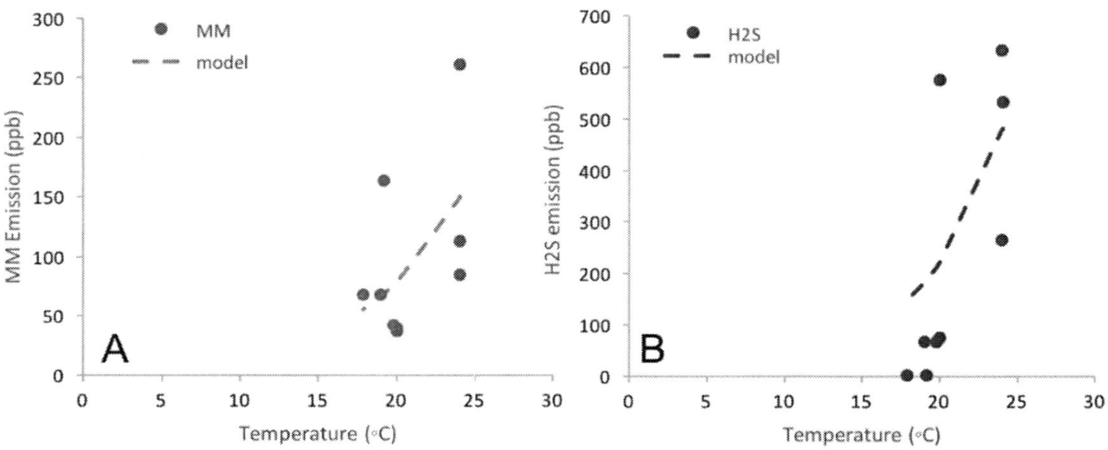

Figure 2: Impact of temperature on the formation and emission of MM and H_2S in primary system at Blue Plains, including Arrhenius model fit resulting in theta of 1.17 for MM and theta of 1.21 for H2S

2. Mechanisms for odor formation in primary treatment

2.a Stripping of collection system odors

The first cause of the emission could be just stripping of dissolved odor compounds formed in the collection system. Based on the total sulfur measurements in the wastewater before and after the grit chamber, a potential emission of sulfur of 275 kg-S/d occurred in the screening and grit chamber building. Of this, a negligible amount of 8.7 ug-MM-S/d can be attributed to stripping of dissolved MM present during summer period (05/31/2017 to 07/30/2017) in the raw wastewater. An additional emission of 0.4 ug MM-S/d over the primary system can be attributed to the stripping of soluble MM present in grit effluent. As the overall emission measured over the primary system during this time was 288 ug-MM-S/d only 0.13% of the MM emission was caused by MM formed in the sewer system, thus the impact of MM formed in collection system on MM emissions in the primary system can be neglected. Overall, MM emission caused by MM formed in the collection system is 9.2 ug-S/d.

2.b Seeding of active biomass and methionine in primary system as key for odor formation

Based on the kinetic expression for MM formation (equation 1) it is clear that both active biomass content (X_bio) as well as substrate concentration (S_methionine) are critical parameters determining the MM production rate. In case of primary treatment both factors would be dependent on the influent characteristics.

$$Rate\ of\ MM\ formation = \mu\max * \frac{X\_bio}{Y} * \frac{S_{methionine}}{S_{methionine}+Ks} * \frac{Ki+ORP}{ORP}\ (equation\ 1)$$

with

μ max: maximum growth rate of the organism (d-1)
X_bio: active biomass
Y: biomass yield
S_methionine: methionine substrate concentration
Ks: half saturation constant for methionine
ORP: oxidation reduction potential
Ki: inhibition constant for ORP

It is known that the sludge retention time within the primary treatment is too short (2 to 4 hours) for significant growth of heterotrophic organisms (OHOs). Therefore, to allow for microbial MM formation, seeding of active biomass to the primary system is essential to allow for significant rates. Measurement of active biomass in the raw wastewater indicated the presence of 5-50 mg active biomass-COD/L or 2-30% of influent COD, with the highest concentrations during the warmer temperatures. The relationship between the odor emission and the active fraction present in the wastewater is presented in Figure 3A. This initial data indicates a strong dependency and linear trend between active biomass and odor emission in primary systems. In addition, as the slope for H_2S is higher compared to MM, the measured active biomass might contain more sulfate reducing bacteria than MM producing organism. Alternatively, sulfate concentration were most probably less limited (sulfate in primary effluent is around 10 mg S/L) than methionine concentrations allowing for higher production rates.

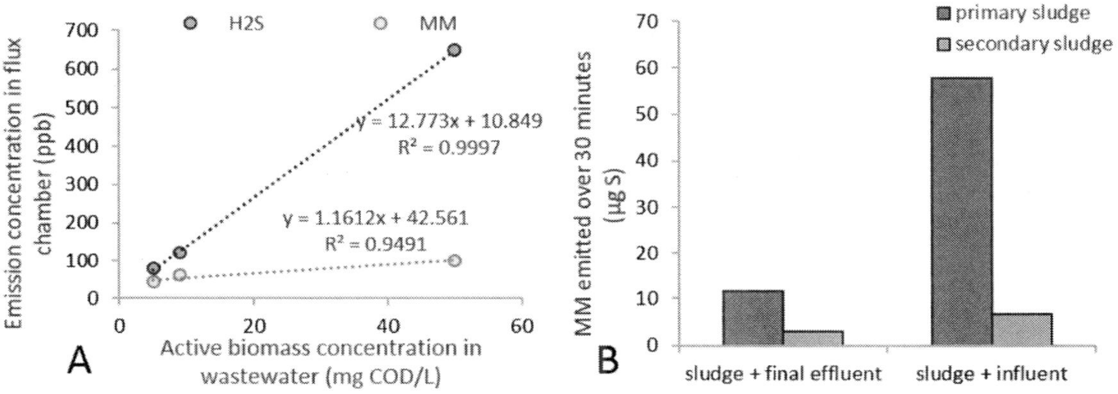

Figure 3: A: Relation between influent active biomass concentration and odor emissions presented as concentration in the flux chamber; B: increased MM production of primary sludge when fed with grit effluent compared to feeding with final effluent indicating the presence of methionine in grit effluent. B: MM production of the sludge with and without addition of the influent wastewater for primary sludge (grit effluent) and secondary sludge (primary effluent).

Methionine is formed in wastewater based on hydrolysis of proteins and is a direct substrate for organisms to produce MM. It is anticipated that the hydrolysis rate might not be fast enough to provide enough protein degradation during primary treatment and it is thus believed that to produce MM in primary treatment methionine need to be formed in the sewer system. A batch experiment was performed to determine the impact of substrate availability in the wastewater versus production of proteins in the sludge (Figure 3B). Test showed that indeed MM production by primary sludge was fully dependent on the influent characteristics, only leading to MM production in the presence of grit effluent. This confirmed that methionine seemed to be present in the wastewater and no significant protein hydrolysis took place in the sludge. Future analysis will be needed to get a better sense of methionine concentration profiles over the course of the year.

2.c Sludge blanket height as way to control ORP conditions needed for odor formation

As presented by equation 1, even with significant active biomass and methionine being present, MM production rate can be controlled by ORP conditions. Low ORP conditions are needed for the organisms to convert methionine into MM and thus elevated ORP inhibits this reaction. Also for sulfate reduction to H_2S, similar ORP conditions are needed. Within primary systems, low ORP conditions are present in the sludge blanket and it was thus expected that by managing the sludge blanket, VSC emissions could be controlled.

At normal operating condition, primary sludge was removed from clarifier every 45 minutes thus allowing for increased sludge blanket heights to be formed over this period, typically up to 0.15 m. To increase the sludge blanket height and study its impact on MM and H_2S emission, the wasting cycle was increased to 90 minutes resulting in final blanket heights up to 0.45 m. The blanket height was measured using a sludge judge as H_2S and MM data were collected. The MM and H_2S emission as a function of sludge blanket height is shown in Figure 4. MM emission was minimal when sludge blanket heights were smaller than 0.1 m, while during this conditions substantial H_2S emission was observed (Figure 4). This indicated the potential significant source of H_2S from the collection system emitted from the primary system through stripping. When operating at increased sludge blanket height, increased MM and H_2S emission were recorded. This indicated that, under lower ORP (measurement range detected was -50 mV to -300 mV), formation of MM and H_2S was achieved. It was concluded that significant reduction of sulfur compounds by anaerobic bacteria occur at the bottom of the primary settling tanks. Previous work showed that sulfur reducing bacteria activity in secondary sludge blanket were significantly present (Sekyiamah et al., 2008), the latter was also observed for primary sludge with H_2S emission rate being much higher than MM emission rate.

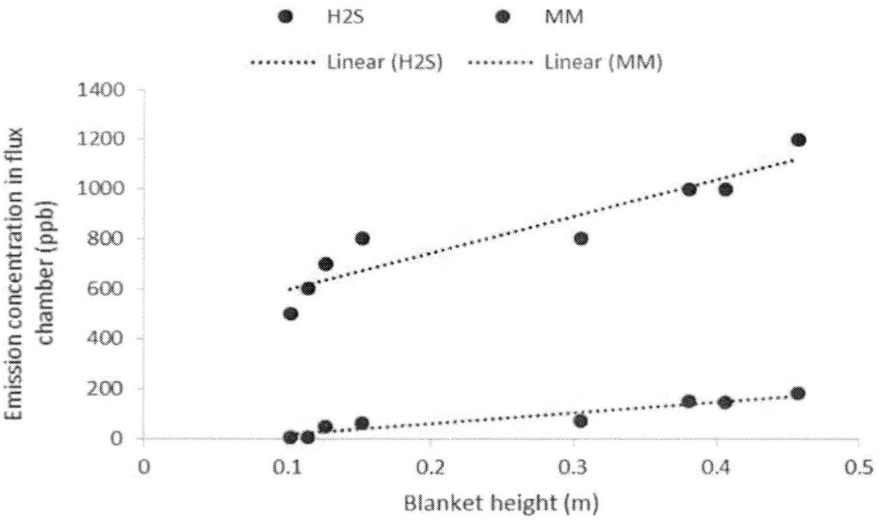

Figure 4: MM and H2S gas phase concentration in primary settling tank in relation to sludge blanket height

CONCLUSION

Total sulfur mass balance over the grit chambers have shown that MM represents a negligible fraction of total sulfur removed in this system. However, this study showed that MM and H_2S emissions are relatively important in primary treatment process. MM emission in primary systems seems to be directly linked to (i) the biomass formation in the collection system impacting the active fraction to be fed to the primary system, (ii) methionine concentration present in the wastewater as a product of protein hydrolysis in the collection system and (iii) ORP conditions applied through management of the sludge blankets. Due to these factors, odor measurements in summer were about 50% higher than in winter. Ongoing year-round measurements will hopefully clarify the different aspects and more clearly link influent characterization and behavior in the collection system to the MM production and emission during primary treatment.

REFERENCES

1. Corcoran, E., C. Nellemann, E. Baker, R. Bos, D. Osborn, H. Savelli (eds). 2010. Sick Water? The central ole of wastewater management in sustainable development. A Rapid Response Assessment United Nations Environment Programme, UN – HABITAT, GRID- Arendal.www.grida.no ISBN: 978- 82- 7tra701- 075-5
2. Estrada, J.M., Kraakman, N.J.R.B., Munoz , R., Lebrero, R., 2011. A comparative analysis of odour treatment technologies in wastwater treatment plants. Environ. Sci. Technol. 45 (3), 1100-1106.
3. Giri BS, Kim KH, Pandey A, Cho J, Song H, Kim YS. 2014. Review of biotreatment techniques for volatile sulfur compounds with emphasis on dimethyl sulfide. Process Biochemistry 49 (2014) 1543- 1554
4. M. Abalos, X. Prieto, J.M. Bayona Detrmination of volatile sulfides in wastewater by headspace solids-phase micoextraction followed by gas chromatography – mass spectrometry J. Chromatogr. A, 963 (1-2)(2002), pp. 249- 257
5. Romero A, De Clippeleir H, Garrido M, Debarbadillo C, Torrents A, Murthy S. 2016. Controlling ORP as the key to reduce odor emission in secondary systems. Proceedings of the Water Environmental Federation WEFTEC Conference 2016, New Orleans LA.
6. M.C. Iliuta, F. Larachi, Gas-liquid partition coefficoents and Henry's law constants of methyl mercatan in aqueous solutions of FE(II)- CDTA chelate complex Fluid Phase Equilib. 253 (2007) 124–129.
7. Wentzel, M.C.; Mbewe, A.; Ekama, G.A. (1995) Batch Test for Measurement of Readily Biodegradable COD and Active Organism Concentrations in Municipal Wastewaters.Water SA, 21 (2), 117-124.
8. Deshmuckh, S.N., Jana A., Bhattacharyya, N., Bandyopadhyay, R., Panday., R.A Quantitative determination of pulp and paper industry emissions and associated odor intensity in methyl mercaptan equivalent using electronic nose. Atmospheric Environment 82 (2014) 401- 409
9. Sekyiamah, K. Kim, H., McConnell, L., Torrents, A., Ramirez, M. (2008). Identification of seasonal variations in volatile sulfur compound formation and release from the secondary treatment system at a large advanced wastewater treatment plant. Water Environmental Research. 80 (12): 2261-7

Assessment of Odor Emissions at a Municipal WWTP in the Middle East

Timur Dunaev, Louis le Roux
BioAir Solutions, LLC
110 Kresson-Gibbsboro Rd., Ste. 303
Voorhees, NJ 08043

Haidar Ezal-deen
BioAir Soluions, LLC
P.O. Box 9354
Sharjah, UAE

Corresponding Author: tdunaev@bioairsolutions.com

Abstract—
Dispersion models are often used to estimate the impact of odor emissions from unit processes at municipal wastewater treatment plants (WWTP) at the fence line of the property. The typical approach is to assess the emission rates from the various odor sources, and then perform air dispersion modeling to determine the expected odor concentration at the fence line. If the odor concentration at the fence line meets the WWTP's treatment objective, then no further treatment is required. However, if it does not meet the fence line odor objective, then the various odor sources are isolated and treated until the desired fence line odor concentrations are being met. This methodology works well for existing WWTPs, but it becomes challenging when a greenfield WWTP is designed, since no actual odor emission data is available for use in the dispersion model.

This paper reports emission data estimations on a greenfield WWTP plant in the Middle East (Plant) to assess the degree of treatment that is required to achieve the Plant's desired fence line odor objectives. Because odor emission data was not available for WWTPs in the Middle East, it created a substantial challenge to perform the dispersion modeling, and emission rates were assumed. After the Plant was constructed, actual emission rates from the entire Plant were measured and compared with the theoretical assumed values, to determine whether the odor capture and treatment methods used at the Plant were sufficient to meet its fence line odor requirements.

A comprehensive odor emissions study was performed at the Plant with the following objectives:
1) **Measure the emission rate of every unit process identified as an emission source during the design phase of the Plant**
2) **Compare the measured emission rates to the previously assumed emission rates**

The data collected during this study forms a valuable database for use with dispersion models for similar future greenfield WWTPs in the Middle Eastern region to ensure that acceptable fence line odor concentrations are achieved. This paper further highlights some of the challenges experienced during the performance of the odor study, and the practical methods developed to perform odor measurements.

***Keywords** — odor measurement; H_2S measurement, dispersion modeling; WWTP odor measurements; odor emissions estimation*

I. Introduction

With the rapid increase in global population, housing and business developments are encroaching on existing wastewater treatment plants (WWTPs). This often results in odor complaints, and neighboring residents are no longer willing to tolerate the adverse effects of odors on their quality of life. Moreover, some odorous compounds could potentially pose a health and safety concern, and negatively impact property values. Therefore, odor control from industrial and municipal facilities has gained increased attention in recent years. Increased air quality in industrialized nations is also thought to have a major impact on perception of odors and emissions from these facilities.

In the past, odor control was not considered part of the master planning for municipal WWTPs; however, newly constructed WWTPs are often required to include odor control as part of their design. Including odor control during the initial design phase assures a comprehensive approach for odor control from WWTPs and avoids any potential nuisance complains once the WWTPs are put in operation.

A prominent municipal WWTP in the Middle East (Plant) has recently expanded its sewage treatment capacity by 50%. Odor control was an important consideration during the design phase to ensure that the Plant is a "good neighbor", and does not adversely impact the local community near the Plant. The Municipality was determined to use globally-recognized standards for its fence line odor concentrations, and their goals are summarized in Table 1 below.

Table 1: Hydrogen Sulfide and Odour Management Standard of the Plant

Parameter	Value	Time Reference	Actions
Hydrogen sulfide concentration at the Site fence limit	< 5 ppbv	4 Hour Average	100% Compliance
Odour at the Site fence limit	< 5 O.U./m^3	4 Hour Average	100% Compliance

During the design stage of the WWTP, BioAir has conducted air dispersion modeling. The modeling included all potential odor sources, which encompassed all new uncovered sources as well as the stack of the new odor control system. Dispersion modeling was performed to demonstrate that odors at the Plant boundary will not exceed 5 ppbv of H_2S and/or 5 Odour Units (D/T), excluding background H_2S and odors from the existing WWTP and surroundings areas.

Since BioAir was not able to source available odor emission data from WWTPs in the Middle East, emission data were used based on the following sources:

- Emission data available from an odor study previously performed at a WWTP in the UAE was used for sources common to the Plant, and
- for sources for which data was not available, the emissions data was calculated using method described in "Measurement of Odor Emissions from Municipal Sewage Sources" [1].

BioAir's process engineers ran two air dispersion models: one for hydrogen sulfide and the other foul odors. The models were identical in setup but differed in the emission rates resulting from the various emission sources. Tables 2 summarizes the calculated emission rates from the various sources as used in the model.

Table 2: Stack, Area, and Fugitive Sources at Ajman WWTP

Unit	Source	Odour emission rate 10^3, OU/m^2/sec	H_2S emission rate, µg/m^2/sec
1. AERATION TANK – 1 2. AERATION TANK – 2	Aeration Tanks	1.3	0.86
3. CLARIFIER TANK – 1 4. CLARIFIER TANK – 2 5. CLARIFIER TANK – 3 6. CLARIFIER TANK – 4	Clarifiers	3.9	1.1
7. RAS/SAS	RAS Collection Tank	7.7	2.1
8. WEIR	Weir at Discfilters	3.9	1.1
9. CHLORINE CONTACT TANK	Chlorine Contact Chamber	1.9	0.5
10. SUMP INLET CHAMBER	Sump Inlet Chamber	1.9	0.5
11. OCU FOR SLUDGE TREATMENT	Biotrickling Filter	0.47	274
12. SCUMSTATION – 1 13. SCUMSTATION – 2	Scum Stations	NA	NA
14. SLUDGE AREA 15. SCUM STATION	Sludge Thickener Tank	NA	NA
16. LIQUOR PS 17. SCUM STATION	Return Liquor Pumping Station	NA	NA
18. DEWATERERING	Dewatering Building	NA	NA

* In order to calculate the emission rates, the flow rate is corrected to an average temperature of 301 K. This temperature is the average temperature at the site as calculated from the 1-hr averaged 5 years available meteorological data

Potential fugitive sources, i.e. Item Nos. 12 – 18 in Table 2, were also identified and included in the model as zero emissions sources. Item 11 describes the proposed odor control system, which would treat air from the sludge treatment area. Other sources were left untreated. Based on the modeling results, which predicted that the Plant's fence line odor targets as outlined in Table 1 would be met, it was decided to proceed with the proposed odor mitigation measures as outline in Table 2.

After the Plant was fully constructed and started up, the actual emission rates from the Plant were measured to validate the odor emission assumptions that were used during the initial modeling.

II. Materials and Methods

BioAir designed an odor study to measure the actual odor emissions from the various sources at the Plant. The study consisted of the following –

A. Categorizing Sources

All the odor sources were divided into three (3) categories, as summarized in Table 3 below:

Source A: Point Sources
Source B: Quiescent and Turbulent Sources, and
Source C: Fugitive Sources

Table 3: Designated verification procedure for area and point sources

Source	Source names	Suggested verification procedure
Odor Control System	ODOR CONTROL UNIT	A
Scum Station #1	SCUM STATION 1	C
Scum Station #2	SCUM STATION 2	C
Aeration Tanks	AERATION TANK-1 AERATION TANK-2	B
Clarifiers	CLARIFIER TANK – 1 CLARIFIER TANK – 2 CLARIFIER TANK – 3 CLARIFIER TANK – 4	B
RAS Collection Tank	RAS/SAS	C
Weir at Disc filters	WEIR	C
Chlorine Contact Chamber	CHLORINE CONTACT TANK	C
Sump Inlet Chamber	SOP SUMP INLET CHAMBER	B
Sludge Thickener Tank	SLUDGE THINKENER TANK	C
Dewatering Building	DE-WATERING BUILDING	C
Return Liquor Pumping Station	RETURN LIQUOR PUMPING STATION	C

1: Point (Stack) Sources (Type A Source)

H_2S: The outlet stack from the odor control system stack (considered a point source) was connected to a continuous H_2S measuring device, i.e. OdaLogs (Detection Instruments, Scottsdale, AZ), to measure the H_2S at 2 min sampling intervals. The OdaLog measurements were checked periodically with a Jerome meter (Model J605, Arizona Instruments, Chandler, AZ) to confirm their accuracy. Since both the airflow and H_2S concentration were known, they were converted to a load (g H_2S/s) for use in the dispersion model.

Odors: Standard accepted methods for odor measurements as defined in ASTM E679 [2] or EN 13725 [3] require that odors be collected in Tedlar bags and analyzed by a certified olfactometry laboratory.

Since there were no olfactometry laboratories available in the Middle East, it was decided to use Nasal Rangers (St. Croix, Minneapolis, MN) for odor measurement. The Nasal Ranger was brought into proximity of the exhaust of the odor control system stack, and the odors were measured directly in the exhaust stream as shown on Figure 1 below. The odor measurement procedure was performed following the Operational Manual for Nasal Ranger [4], and a trained group of four people constituted the "odor panel", and each of the panelist was pre-screened with the n-butanol Odor Sensitivity Test Kit [5].

Since both the airflow and odor concentration were known, the OU/s were calculated and used as an input parameter in the dispersion model. For both the H_2S and odor measurements, the air flow was measured using a portable anemometer (TSI, Velocicalc 9535).

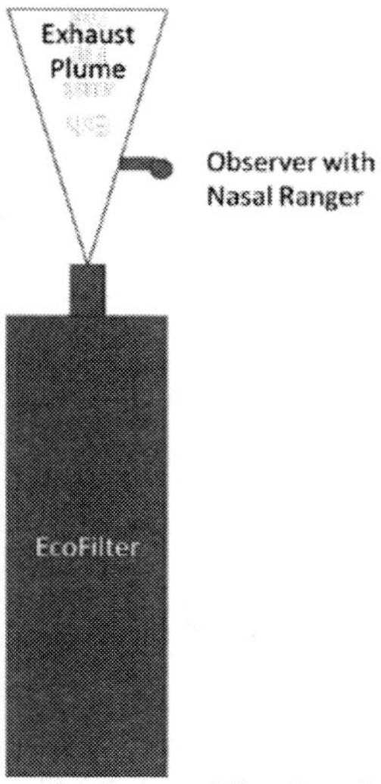

Figure 1: Suggested odor verification for Type A sources

2: Quiescent and Turbulent Sources (Type B Source)

For quiescent and turbulent sources the flux chambers were located at locations that closely represent the odors from these sources, using simplified EPA Flux Chamber Method [6].

Airflow Measurement: Each flux chamber was connected to an air pump, and the airflow rate was measured with a rotameter.

H_2S Measurement: The H_2S was measured at 2 min intervals using OdaLogs and Multiple OdaLog Sampling System (Detection Instruments, MOSS), and the measurements were verified periodically with a Jerome meter.

Odors: Odors were measured on site using Nasal Rangers, by measuring the odors from the exhaust stream of the Flux Chambers as shown on Figure 2 below.

The H_2S and odor flux were calculated using the air flow and respective H_2S and odor measurements.

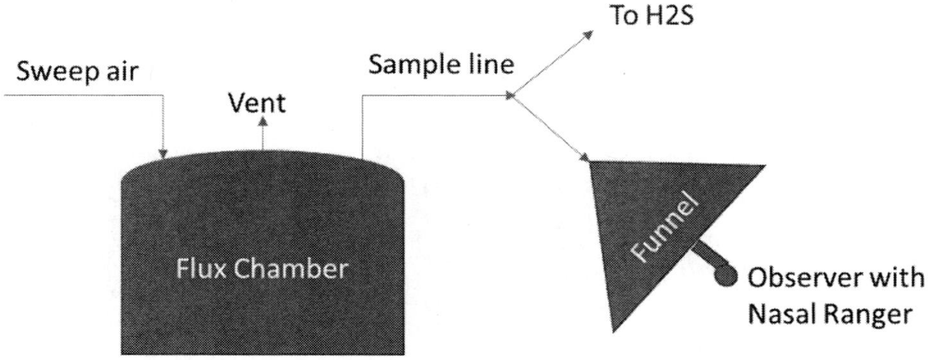

Figure 2: Suggested Odor Verification for Type B Sources

3: Fugitive Sources (Type C Source)

Periodic, scheduled measurements were conducted using the Jerome meter and Nasal Ranger at the fugitive sources as outlined in Table 3. It was decided to consider 5 ppmv and 5 OU for H_2S and odor, respectively, as the criteria to differentiate between fugitive and non-fugitive sources. For example, if the actual measured H_2S concentration at a Type C Source is > 5 ppmv, or the odor concentration is > 5 OU, then the source was considered fugitive.

B. Study Duration

The study was designed to last two (2) workdays (Data Collection Period), where:

- One (1) day was used for planning, preparation of equipment, and location of equipment, and
- One (1) day for actual data collection
- A Portable weather station was installed
- For Type A sources:
    - Airflow from each source was confirmed using a portable anemometer
    - Each source was equipped with a proper range OdaLog H_2S logger
    - Two times per day for one day H_2S measurements from OdaLogs were confirmed by Jerome meter
    - Two times per day for one day odors were measured with a Nasal Ranger
- For Type B sources:
    - One flux chamber was installed at each representative Type B source
    - Each flux chamber was equipped with a proper range OdaLog H_2S logger
    - Two times per day for one day H_2S measurements from OdaLogs were confirmed by Jerome meter
    - Two times per day for one days odors were measured with a Nasal Ranger
- Type C sources:
    - Two times per day for one day H_2S measurements were taken in immediate proximity of these sources using Jerome meters
    - Two times per day for one day odor measurements were taken in immediate proximity of these sources using Nasal Rangers
    - For all sources, random checking of ammonia and total mercaptans were performed using Gastec tubes at least once a day for one day

C. Equipment Installation

The Category B sources (see Table 3) required installation of a Flux Chambers for odor and H_2S measurement. Figure 3 shows the installation of the carbon in the carbon filter used for the sweep air into the Flux Chambers. This is important to prevent introduction of any ambient odor sources other than the emissions captured by the Flux Chamber.

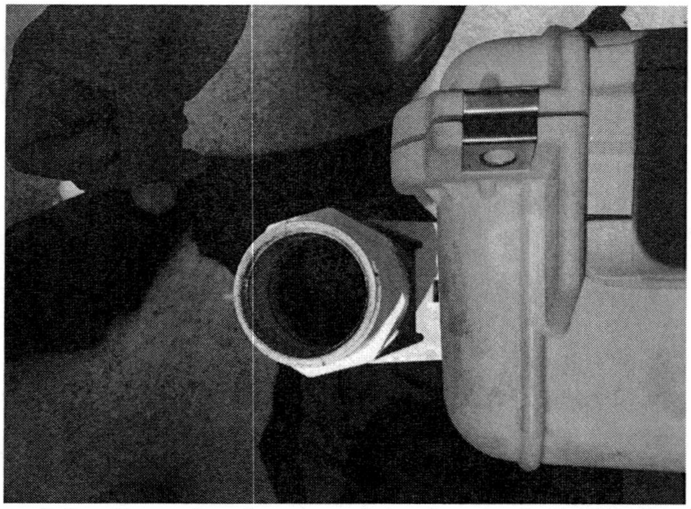

Figure 3: Loading carbon for sweep air treatment prior to flux chamber

For the Aeration Tank, the initial location for the Flux Chamber was decided to be next to inlet chamber wall of Aeration Tank (see Figure 4). The Flux Chamber was later re-located for reasons listed hereafter.

Figure 4: Aeration Tank Flux Chamber

The surface of the Secondary Clarifier was not suitable for installation of the Flux Chamber, since it would interfere with the movement of the scum scraper of the Clarifier travelling bridge. Furthermore, the highest turbulence area for the clarifiers would be the overflow weirs, thus it was decided to locate the flux chambers at the overflow weirs of the Clarifier.

The Flux Chamber for the Outlet Works was located at the Inlet Chamber of the (future) Outlet Works pumps.

A weather station was installed at an elevated location to collect weather data most representative of the entire Plant. The roof of a building at the Plant was identified as the most ideal location. The North direction for the wind direction sensor calibration was determined by using a GPS calibrated smartphone. The wind speed, wind direction, temperature and relative humidity sensors were installed.

D. Data Collection

On the morning of the start of the Data Collection Period, an odor team was assembled as follows:

1) Three (3) BioAir representatives
    a. One (1) person coordinating the study
    b. Two (2) people collecting ammonia, amines, and total mercaptan samples and measurements
2) One (1) contractor representative helping with the study coordination
3) Two (2) operator representatives
4) One (1) outside contractor collecting H_2S samples with a Jerome meter

For the odor measurements, an Odor Panel was selected that consisted of four people, i.e. one person from BioAir, one person from the contractor, and two people from the WWTP operator, all pre-screened with an Odor Sensitivity Kit [5].

Two (2) Nasal Rangers were used during the Study. The four people of the Odor Panel were divided into two pairs, each pair taking odor measurements consecutively of each other. Each Odor Panel member used their fitted face mask for the Nasal Ranger.

All the Category C samples were collected facing South, regardless of the location and direction of the wind, to be consistent with the technique by which the data was collected. The south direction was determined using the same GPS calibrated smart phone as was used for the weather station. The time of every measurement was recorded to later cross reference with the weather data.

Measurements were taken using the following methodology:

i. Odor Team arrives at sample point
ii. Time and date are recorded
iii. Odor Panel, Jerome H_2S and ammonia/amines/total mercaptan sample collection begins simultaneously
iv. Once all the measurements were completed, the Odor Team moves on to the next sampling source

Table 4 below summarizes the sampling order sequence.

Table 4: Sampling Points

SOURCE	TYPE	DATA COLLECTION DETAILS
Dewatering building	C	Team stands in-between Solids Bins facing South
Scum Station #1	C	North side of Scum Station facing South
Scum Station #2	C	North side of Scum Station facing South
Sludge Thickener	C	Top of Sludge Thickener facing South
Odor Control System	A	Per Methods Listed Herein
RLPS	C	Top of RLPS facing South
Clarifiers	B	Per Methods Statement
RAS	C	Top of RAS facing South
Aeration	B	Per Methods Statement
Outlet Works	B	Per Methods Statement
CCTs	C	South side of CCTs facing South
Disc Filter Weirs	C	South side of Disc Filter Weirs facing South

III. Results and Discussion

A. Weather Measurements

The Weather Station operated for two days to gather weather data. Wind speed and direction was very useful when analyzing Category C Sources, since Category A and B Sources were not impacted by the wind due to the sampling method used to collect the data.

Figures 5 and 6 summarize the Wind Rose, Wind Speed and Direction, respectively, and Figure 7 shows the Temperature and Relative Humidity. During the evaluation of Category C Sources, each Source was paired with the wind direction and speed for that specific time of the day.

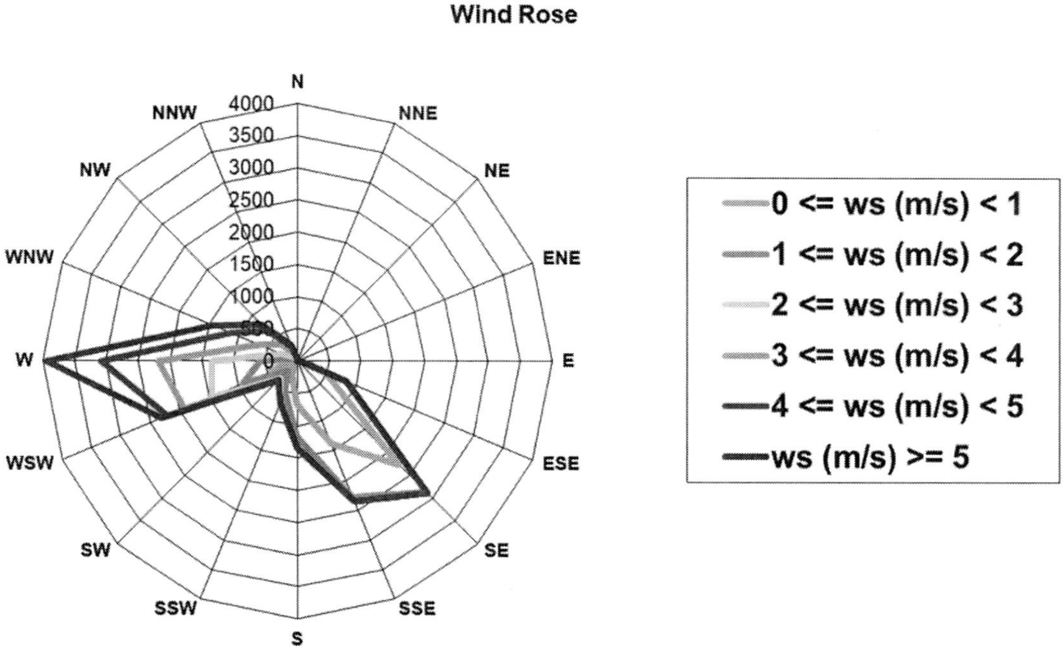

Figure 5: Wind Rose

It is clear from Figure 5 that the predominant wind direction at the Plant during the Data Collection Period was W and SE directions, depending on the time of the day.

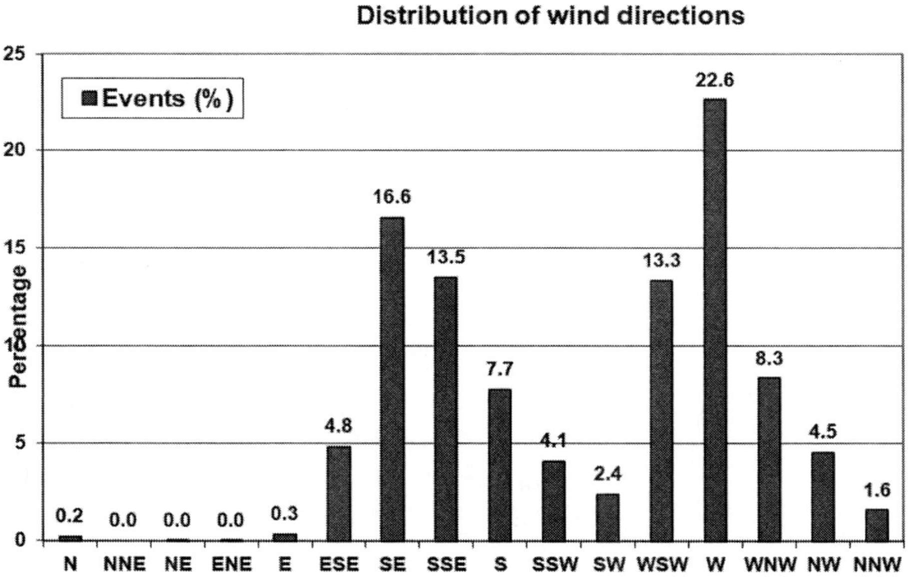

Figure 6: Wind Direction

The Temperature and Relative Humidity ("RH") were observed only to make sure there were no weather abnormalities during the study. Both Temperature and RH profiles appear to be as expected during the Data Collection Period.

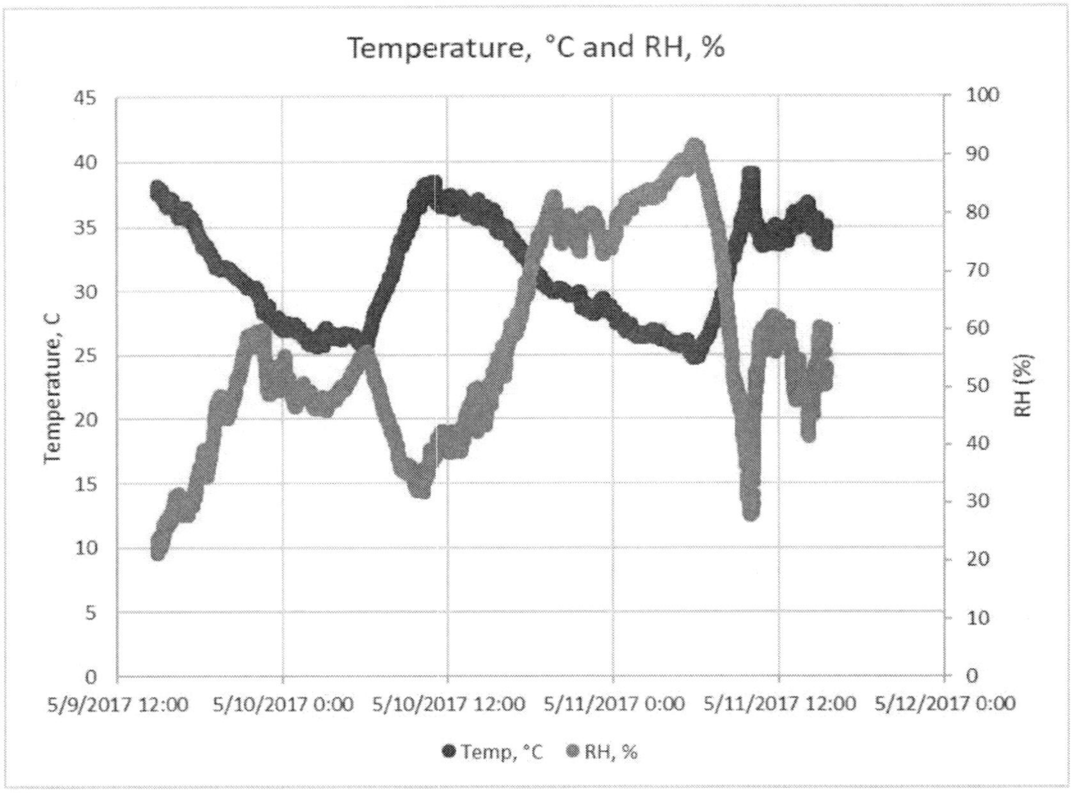

Figure 7: Temperature and Relative Humidity

B. Category A Sources

There was one Category A Source, namely the outlet stack of the odour control system at the Sludge Treatment Area.

Sludge Treatment Odor Control System

The airflow from the ventilated sources were balanced, and then the air was treated in the biological odor control system that serves the sludge treatment area. The airflow through the odor control system was also measured prior to the data collection.

The measured emissions from the EcoFilter® odor control system (BioAir Solutions, Voorhees, NJ) (OCS) during the Data Collection Period were as follows -

Table 5: Odor Measurements at OCS

		MORNING								AFTERNOON							
SOURCE	TYPE	Time	H2S (Jerome) ppb	Odor D/T		NH3 ppmv	R-NH2 ppmv	MM ppmv	Time	H2S (Jerome) ppb	Odor D/T		NH3 ppmv	R-NH2 ppmv	MM ppmv
EF82				1	> 60						1	30-60			
				2	30-60						2	> 60			
				3	30-60						3	30-60			
	A	11:32AM	320 ppb	4	> 60	0	0	0.5	3:30 PM	160 ppb	4	30-60	0	0	0.25

The averaged values from the collected data were used to compare the measured emission rates to the emission rates used during the initial Dispersion Modeling. The results are summarized in Table 6.

Table 6: Actual Emission Rates at EcoFilter odor control system

Odor Control System	Assumed Emission Rate	Actual Emission Rate
Odor Emission Rate 10^3, D/T/sec	0.47	0.044
H_2S Emission Rate, µg/sec	274	293

The measured odor emission rate for the OCS as listed in Table 6 appears to be an order of magnitude lower than the values assumed during the initial odor modeling. On the other hand, the measured H_2S value is higher, i.e. 293 µg/sec vs. 274 µg/sec, but only because the actual airflow rate from the OCS was 11.6% higher than the design airflow. If the actual airflow was set at the design airflow, the actual emission rate in Table 6 would be have been less than the 274 µg/sec that was used in the Odor Modeling. The assumed value for the H_2S concentration in the Odor Modeling was 250 ppb_v, and actual measured H_2S concentration was 240 ppb_v.

C. Category B Sources

The Category B sources were equipped with a Flux Chamber and a continuous H_2S data logger. Spot H_2S measurements were taken with a Jerome meter in accordance with the methods as described herein.

Outlet Works

Due to construction work upstream of the Outlet Works, some of data collection was impaired during the first part of the Data Collection Period. It was decided that this source was not a major contributor to the odors from the Plant and that the emission data collected in the afternoon would be considered representative for this source. Table 7 provides a summary of the data collected for the Outlet Works.

Table 7: Odor Measurements at Outlet Works

		MORNING							AFTERNOON						
SOURCE	TYPE	Time	H2S (Jerome) ppb	Odor	D/T	NH3 ppmv	R-NH2 ppmv	MM ppmv	Time	H2S (Jerome) ppb	Odor	D/T	NH3 ppmv	R-NH2 ppmv	MM ppmv
SOP				1	NA						1	30-60			
				2	NA						2	30-60			
				3	NA						3	30-60			
	B	10:18 AM	NA	4	NA	NA	NA	NA	4:30 PM	0.00 ppb	4	30-60	0	0	0

From the data collected at the Outlet Works, it was clear that no obnoxious odors were present, and all the members of the Odor Panel identified it as "chlorine" smell. Table 8 provides a summary of the actual emissions versus the assumed emissions for the Outlet Works.

Table 8: Actual Emission Rates at Outlet Works

SOP	Assumed Emission Rate	Actual Emission Rate
Odour Emission Rate 10^3, D/T/sec	42	0.028
H_2S Emission Rate, µg/sec	12	0

The measured emission rates as listed in Table 15 for the Outlet Works appear to be an order of magnitude lower than the values assumed during the initial Odor Modeling.

Clarifiers

Data collected from the Clarifier was from the most turbulent area of Clarifiers, i.e. clarifier weirs, which represents the worst possible scenario from an odor release perspective. Data from all other Clarifier sources is expected to be substantially less than what was observed from the collected data.

Table 9: Odor Measurements at Clarifiers

SOURCE	TYPE	MORNING Time	H2S (Jerome) ppb	Odor D/T	NH3 ppmv	R-NH2 ppmv	MM ppmv	AFTERNOON Time	H2S (Jerome) ppb	Odor D/T	NH3 ppmv	R-NH2 ppmv	MM ppmv
Clarifiers	B	9:45 AM	0 ppb	1 5-7 2 7-15 3 4-7 4 4-7	0	0	0.2	1:41 PM	3.99 ppb	1 30-60 2 30-60 3 15-30 4 30-60	1	0	0

Table 10: Actual Emission Rates at Clarifier

Clarifier	Assumed Emission Rate	Actual Emission Rate
Odor Emission Rate 10^3, D/T/sec	266	0.06
H_2S Emission Rate, µg/sec	740	0.00724

Measured emission rates as listed in Table 10 for the Clarifiers are orders of magnitude lower than the values assumed during the Odor Modeling.

Aeration Tanks

Evaluation of the Aeration Tank data revealed an unusually high H_2S and odor concentration. A well operated Aeration System is expected to release very low H_2S emissions, and all observed odors should be non-offensive and classified as musty / earthy smell. The Flux Chamber was originally located close to where the sewage enters to the Aeration Chamber. The measured H_2S concentration and odorous were relatively high and noxious in nature as indicated in Table 11 below.

Table 11: Odor Measurements at Aeration Tanks

SOURCE	TYPE	MORNING Time	H2S (Jerome) ppb	Odor D/T	NH3 ppmv	R-NH2 ppmv	MM ppmv	AFTERNOON Time	H2S (Jerome) ppb	Odor D/T	NH3 ppmv	R-NH2 ppmv	MM ppmv
Aeration	B	10:03 AM	181 ppb	1 30-60 2 30-60 3 30-60 4 30-60	0	0	0.5	1:55 PM	1,126 ppb	1 > 60 2 > 60 3 30-60 4 30-60	0	0	1.5

After the data collection was completed during the Data Collection Period, it was decided to continue collecting data but to move the Flux Chamber to another location in the Aeration Tank. This was done to identify a location in the Aeration Tank that is more representative of the H_2S and odor emissions from the entire Aeration Tank surface, rather than measuring high odors directly after the inlet chambers.

Figure 8: Aeration Tanks Flux Chamber Relocated (Location 1 is on the Left and Location 2 is on the Right)

Based on the BioAir's experience with inlet conditions from similar WWTPs, it was decided to move the Flux Chamber 20 m away from its initial location next to the Inlet Section to the Aeration Tank (Location 1). Since the triangular Inlet Section to the

Aeration Tank was already going to be covered, the new location 20 m away (Location 2, see Figure 8) would provide a more representative assessment of the H_2S and odor that would be released from the Aeration Tank.

It is BioAir's experience that even though the Aeration tanks are designed as completely Stirred Tank Reactors, the mechanical configuration of the tanks' inlet creates a localized Plug-Flow regime, which results in H_2S emissions close to the sewage entrance to the Aeration Tank before the dissolved sulfide is oxidized by sulfur oxidizing bacteria in the system. Therefore, it was concluded that the assumption that Location 2 is a more representative location for the odor coming from the Aeration Tank is reasonable.

Figure 9 shows the H_2S concentration for Location 1 and Location 2 in the Aeration Tank.

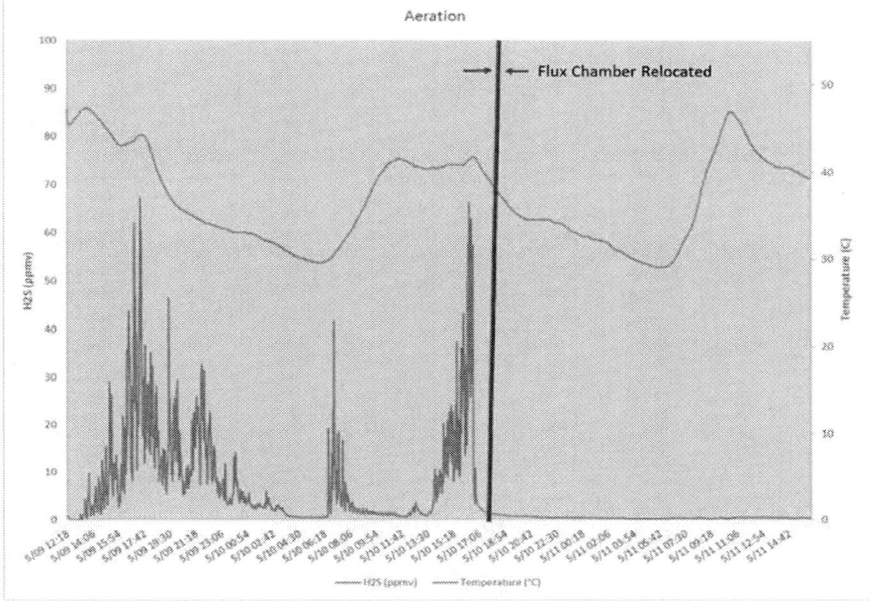

Figure 9: Aeration Tanks H_2S Data

Even though Location 1 was outside of the Inlet Structure, it still continually measured H_2S concentrations of up to 70 ppmv (70,000 ppbv), which is substantially higher than the 80 ppb_v used during the initial Dispersion Modeling. However, as soon as the Flux Chamber was moved 10 m further away from the inlet (Location 2), the H_2S concentration dropped below detectable levels (< 100 ppb_v), which is what should be expected .

Table 12: Actual Emission Rates at Aeration Tanks (Location 1)

Aeration	Assumed Emission Rate	Actual Emission Rate
Odor Emission Rate 10^3, D/T/sec	117	114
H_2S Emission Rate, µg/sec	765	6,254

The measured emission rates for odor as presented in Table 11 for Location 1 of Aeration Tanks appears to be below the value assumed during the Odor Modeling. On the other hand, the H_2S values are substantially higher. No Odor data was collected at Location 2, but it would be safe to assume that the data would have been even lower than what is presented at Table 12. H_2S was below the instrument's detection limit of 0.1 ppmv, which would have put Actual Emission Rate close to or below the assumed values.

D. Category C Sources

Category C sources were identified as potential fugitive odor sources ("Fugitive Sources"). All the Fugitive Sources were modeled as non-odorous during the initial Dispersion Modeling, and thus any values measured during the Study would prove that the assumption was incorrect unless the Fugitive Source could be contained and treated, or the actual odor levels would be below fence line requirements. It should be noted that the Fugitive Odor measurements were affected substantially by the odor sources upstream of the wind direction ("Background Odors")

Based on the site observations there are multiple potential Background Odors, since there were several sources at the existing WWTP that were uncovered.

Table 13: Odor Measurements at Category C sources

		MORNING							AFTERNOON						
SOURCE	TYPE	Time	H2S (Jerome) ppb		Odor D/T	NH3 ppmv	R-NH2 ppmv	MM ppmv	Time	H2S (Jerome) ppb		Odor D/T	NH3 ppmv	R-NH2 ppmv	MM ppmv
Dewatering Building				1	> 60						1	> 60			
				2	>60						2	> 60			
				3	30 - 60						3	> 60			
	C	8:18 AM	146 ppb	4	NA	1	2	0	1:20 PM	0	4	30 - 60	0	0	0.2
Scum Station #1				1	15-30						1	30-60			
				2	2-4						2	30-60			
				3	4-7						3	15-30			
	C	8:28 AM	20.76 ppb	4	7-15	0	0	0	1:25 PM	20.73 ppb	4	15-30	0	0	0
Scum Station #2				1	4-7						1	2-4			
				2	4-7						2	4-7			
				3	2-4						3	7-15			
	C	8:35 AM	16.33 ppb	4	2-4	0	0	0	1:30 PM	25.00 ppb	4	>60	0	0	0
Sludge Thickener				1	15-30						1	15-30			
				2	15-30						2	15-30			
				3	4-7						3	7-15			
	C	8:42AM	46.25 ppb	4	30-60	0	0	0	1:32 PM	3.86 ppb	4	15-30	0	0	0
RLPS				1	15-30						1	15-30			
				2	4-7						2	30-60			
				3	4-7						3	15-30			
	C	8:49 AM	5.88 ppb	4	7-15	0	0	0	1:37 PM	122.4 ppb	4	7-15	0	0	0
RAS				1	30-60						1	30-60			
				2	> 60						2	> 60			
				3	30-60						3	> 60			
	C	9:58 AM	89.55 ppb	4	15-30	0	0	0	1:51 PM	47.13 ppb	4	30-60	0	0	0
Chlorination				1	15-30						1	30-60			
				2	15-30						2	30-60			
				3	7-15						3	15-30			
	C	10:13 AM	6 ppb	4	15-30	0	0	0	4:25 PM	72.57 ppb	4	15-30	0	0	0
Disk Filter Weirs				1	30-60						1	30-60			
				2	30-60						2	30-60			
				3	30-60						3	30-60			
	C	10:20 AM	44.41 ppb	4	7-15	0	0	0	2:09 PM	79.79 ppb	4	30-60	0	0	0

IV. CONCLUSIONS

Measurement of odors at WWTPs is complex, although several general conclusions could be made from this Study:

<u>Category A Sources – Point Sources</u>: Category A sources, i.e. well-ventilated sources that are treated in the odor control system to concentrations < 0.1 ppmv H_2S in the exhaust air that is discharged through a stack, posed no risk to offsite odor release. The measured odor and H_2S concentrations from OCU stack were well below the concentrations used in the initial Dispersion Modeling.

<u>Category B Sources - Quiescent and Turbulent Sources</u>: All Category B Sources, except for the areas immediately after the inlet chambers of the Aeration Tanks, were well below the concentrations used in the initial Dispersion Modeling

<u>Category C Sources – Fugitive Sources</u>: Some of the Category C Sources proved challenging to classify, since they emitted odors that were close to the threshold levels of 5 OU and 5 ppbv H_2S. The main source included the sludge collection bins. This observation reinforces the importance of the operational management of the sludge bins to prevent the sludge from aging and to reduce the risk of odor release. Standard Operating Procedures should include good cleaning practices, containment of the sludge bins, and frequent disposal of the sludge.

It is clear from the results of this study that the initial assumed emission rates used in the dispersion model were mostly more conservative than the actual emission rates measured after the Plant was put in operation.

REFERENCES

[1] "Measurement of Odor Emissions from Municipal Sewage Sources", given at the AWMA Odor Specialty Conference in Detroit, Michigan, USA on 24-26 October, 1989.

[2] "Standard Practice for the Determination of Odor and Taste Thresholds by a Forced-Choice Ascending Concentration Series Method of Limits.", ASTM International E679

[3] "Air Quality – Determination of Odour Thresholds by Dynamic Dilution Olfactometry", EN 13725

[4] "The Nasal Ranger Field Olfactometer Operation Manual", Version 6.2., U.S. Patent No.: 6,595,037. www.nasalranger.com

[5] "Odor Sensitivity Test Kit" TO 2000 06086V2.2, www.fivesenses.com

[6] "Measurement of Gaseous Emission Rates from Land Surfaces Using an Emission Isolation Flux Chamber. User's Guide', US EPA, 1986.

New GHG Methodology to Estimate/Quantify Sewer Methane

John Willis[1,2]; Bill Brower[3]; Walter Graf[4]; Sudhir Murthy[3]; Chris Peot[3]; Pusker Regmi[5]; Keshab Sharma[2]; and Zhiguo Yuan[2]

[1] Brown and Caldwell – 990 Hammond Dr., Suite 400, Atlanta, GA 30328
[2] The University of Queensland
[3] DC Water, Washington, DC - USA
[4] Water Research Foundation (WRF)
[5] Brown and Caldwell
[*] JWillis@BrwnCald.com

Introduction. the Intergovernmental Panel on Climate Change (IPCC) has identified that wastewater in closed, underground sewers in the developed world are not a significant methane (CH_4) source (IPCC 2006). Other greenhouse gas (GHG) protocols have followed IPCC's lead and have ignored sewer CH_4. This simplification has been contradicted by research (Guisasola, 2008; Foley, 2009; Guisasola, 2009; Gutierrez, 2009; van Voorthuizen, 2011; Willis, 2012; and Liu, 2014) where significant CH_4 emissions have been repeatedly detected from both pressure mains and gravity sewers.

The Water Research Foundation's (WRF, formerly WE&RF and WERF) Conveyance Asset Prediction System (CAPS) project (WERF Project #U5R12) developed a methodology for estimating collection-system methane (CH_4) on a sewershed-wide basis. The method uses hydraulic-model output at average flows, actual sewage temperatures and actual flow rates as input.

A Knowledge Development Forum (a workshop that is not for CEU credit) entitled "Sewer-Methane GHG-Methodology Standardization" will be held on Saturday at WEFTEC-2018 in New Orleans. The forum objective is to drive toward WEF-industry consensus on a new greenhouse gas (GHG) accounting methodology for adoption by ICLEI, the California Air Resource Board (CARB), IPCC, and other protocols. This paper summarizes the method and provides specific guidance on its use for estimating sewer-CH_4 generation.

KEYWORDS

Greenhouse Gas (GHG); Collection System; Methane; and GHG Accounting.

INTRODUCTION

DC Water developed a GHG tracking template that they have used to document their Scope-1, -2, and -3 emissions on an annual basis, every year since 2007. Their emphasis has been on developing the most accurate emissions estimate possible. Their position has been that GHG emissions will eventually be regulated and that the most important preparation is to understand what your emissions actually are followed closely by an understanding of what can be done to affect those emissions.

To this end, they have regularly measured actual emissions and used more utility-specific emissions methodologies. As an example, they participated in some of the initial flux-hood based, process-N_2O research (Ahn, 2010) and they still use that site-specific emissions rate for their mainstream nitrification/ dentification processes. DC Water also recognized that carbon dioxide (CO_2) released from respiration of methanol-based carbon (added to enable enhanced denitrification) was anthropogenic CO_2 – and not biogenic like the balance of activated-sludge respired CO_2. To this end, DC Water advocated the development of Method WW.9 for "process carbon dioxide emissions from the use of fossil fuel derived methanol for biological N removal" in ICLEI's most recent US Protocol (ICLEI, 2012).

And in a similar fashion, the common, protocol-endorsed assumption that CH_4 emissions from sewers in the developed world can be assumed to be zero was identified as the single largest vulnerability in their Scope-1 emissions inventory. Of course, this was only after resolving the process N2O emissions (which proved to be minor in their case) and accounting for methanol CO_2 (which proved to be their most significant Scope-1 emission). So that the CAPS project, which was led by principal investigators from Brown and Caldwell (BC) and The University of Queensland (UQ) was created to investigate that vulnerability.

The CAPS project developed a new, simplified, gravity-sewer methodology and verified/calibrated it using full-scale data. The gravity-sewer method was then applied in combination with an already-verified forcemain method to model CH_4 emissions from DC Water's entire collection system over calendar 2014 flows and sewage temperatures. Model output at average flow conditions and determined that in 2014, sewer-CH_4 represented annual emissions of almost 12,800 MT of CO_2e, or 34.4% of DC Water's Scope-1 emissions (these data are presented in Table 1). That percentage is largely understated however, as DC Water's methanol- CO_2 represents an otherwise unheard of 45.6% of their emissions. If the methanol emissions are eliminated, thereby better simulating most other water resource recovery facilities (WRRFs), sewer CH_4 then represents 63.2%. The CAPS project goes on to estimate that sewer CH_4 represents at least 50% of the Scope-1 GHG produced by the centralized sewage-treatment industry and therefore represent an enormous GHG-accounting vulnerability for such utilities.

Description	2014 GHG Emissions Inventory, MT CO_2e/yr	Percentage of Scope-1 Emissions by Source	Percentage of Scope-1 Emissions by Source with Methanol Excluded
Scope 1			
Natural Gas	2,369	6.4%	11.7%
Vehicle Fuel	1,581	4.3%	7.8%
Refrigerants	0	0.0%	0.0%
CO_2 from Addition of Methanol	16,953	45.6%	0.0%
Process N_2O	798	2.1%	3.9%
Effluent Discharge N_2O	2,690	7.2%	13.3%
Sewer CH_4	12,793	34.4%	63.2%
Total Scope 1:	37,183	100.0%	100.0%

Table 1: 2014 Sewer CH_4 compared to Balance of DC Water GHG Emissions

METHOD DESCRIPTION

The CAPS collection-system methodology uses a gravity-sewer equation and two equations for surcharged sewer and/or forcemains. As back-ground, CH_4 generation is affected by the interaction of sewage (as a source of carbon) and biofilms that grow on the surface of pipes in collection systems. Those biofilms are the repository of the hydrolyzing, methanogenic, sulfur-reducing, and other biological populations that affect the liquid-phase carbon- and sulfur-cycle conversions. This biofilm-driven basis runs somewhat counter to popular assumption that sewer- CH_4 generation is driven by deposition of sediments in slow-moving sewers. Recent research by Liu et al. (2015) found that when sediments do occur, biofilms like those on pipe surfaces grow on the surface of the sediment. An additional finding was that while CH_4 production does occur in that biofilm and to a lesser extent in the sediments that the overall combined production is very similar to that of biofilms attached to the floor in sewers without sediments. Even more importantly to the CAPS method, this finding suggests that whether clean or with prolonged, extensive sedimentation, a sewer would produce roughly the same amount of CH_4.

The methods provided below derive CH_4 production through biological activity. The method further assumes that all produced CH_4 is emitted to the atmosphere as Scope-1 GHG and that no CH_4 "sinks" or conversion processes exist. 100% of the CH_4 generated would be released from a combination any of the following "sources":

- Manhole lids, pumping station wet wells, and sewage vacuum-/air-release valves.
- Odor control exhaust. This is believed to be true in all chemical-scrubbing and activated-carbon absorption applications which do not use chemicals that would oxidize or otherwise adsorb CH_4. It is possible, although we are unaware of any publications documenting this effect, that biofilters or bio-trickling filters might temporarily adsorb and then biologically convert a (likely limited) portion of CH_4 to CO_2, eliminating its global warming influence.
- Processes at the receiving Water Resource Recovery Facility (WRRF) like aerated (or other) grit removal, cascades over weirs, turbulent locations like flumes or barscreens, or being stripped through aeration processes.

This assumption may be overly simplistic; and yet no basis for an alternative assumption exists now. One hypothesis is that, due to the higher concentrations of gaseous CH_4 in the sewer headspace and increased dissolved CH_4 in the bulk liquid and slimes, that these environments might be conducive the existence of methanotrophs (organisms that consume CH_4). These organisms would certainly serve as an un-accounted for CH_4 sink and could reduce overall CH_4 emissions.

Gravity-Sewer Method/Equation. The gravity sewer equation was developed using UQ's SeweX collection system process model. SeweX uses a system of at least 41 processes (run on a Matlab® platform) that has been, and continues to be developed by UQ's Advanced Water Management Centre (UQ-AWMC). SeweX-modelled processes include: A) Anaerobic, anoxic, and aerobic transformations of carbon and sulfur in the bulk liquid, biofilms, and exposed pipe surfaces; B) Chemical precipitation of sulfide and other anions by metal ions; C) pH changes from biological and chemical reactions; D) Gaseous transfer of CO_2, CH_4, H_2S and other compounds between liquid and gas phases in gravity sewers and adsorption of H_2S on exposed pipe surface; and E) Convective transport of these same materials in the bulk liquid and headspace gases. The simplified gravity-sewer CH_4-production method is defined by the following equation:

$$r_{CH4\text{-}GS} = 0.419 \times 1.06^{(T-20)} \times Q^{0.26} \times D^{0.28} \times S^{-0.138} \qquad \text{Eq. 1}$$

Where:

$r_{CH4\text{-}GS}$ = CH_4 emission rate in kg CH_4/(km*day) as a function of temperature
T = Temperature in ^{o}C
Q = Flow in m^3/s
D = Pipe diameter in m

S = Pipe slope in m/m

The simplified method was derived from a large set of SeweX model runs. The following findings were also developed during the development and review process:

- The simple algorithm uses slope, diameter, and flow – and an assumption that the pipe is free discharging – to calculate the water depth and wetted area. Because many gravity sewers are not free-discharging, and their hydraulic grade can be increased by downstream hydraulic levels. This is an inherent source of CH_4-generation under-reporting in the method.
- The process was shown to be entirely independent of carbon limitations for soluble COD levels above 100 mg/L with only a 9% decrease in CH_4 production when soluble COD concentrations reached 50mg/L. As such, this method appropriate for all be the very weakest domestic sewage.
- CH_4 production is also independent of sulfate (SO_4^{2-}) concentrations from 2.5 to 50 mg-S/L; SO_4^{2-} was hypothesized to inhibit CH_4 production due to the associated addition of oxygen and the competition for carbon in the biological sulphur reducing process.
- When three 24-hour, continuous simulations were run with the same total flow but three different levels of diurnal variation (shown in Figure 1) the results showed identical daily CH_4 production. As such daily CH_4 production is dependent on the average flowrate, and independent on the degree of variation.
- As discussed previously, CH_4 generation is also thought to be independent of sediment accumulation.

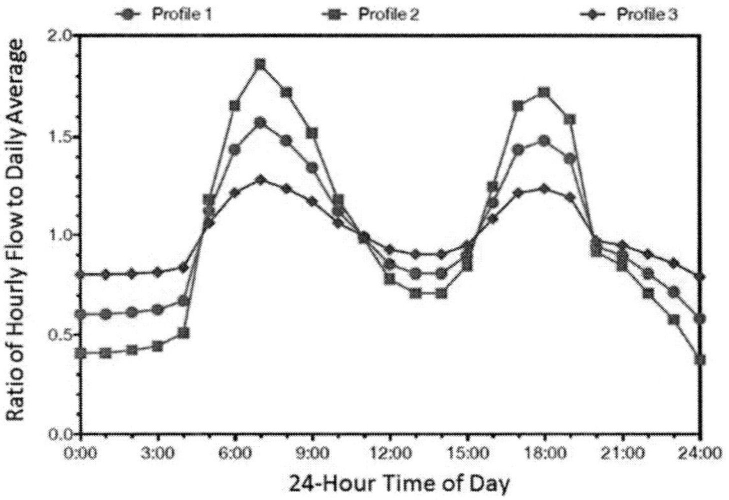

Figure 1. Simulated Diurnal Variation Relative to Average Flow

As a final note, the gravity-sewer simplified method was verified using full-scale data collected at a long-term odor-control facility that treats air exhausted from DC Water's Potomac interceptor (PI). Using the 13,000 to 13,600 actual cubic feet per minute (acfm) flux of exhausted air at and gas-phase measurements for CH_4 and CO_2, allowed the exhausted GHG mass to be determined. These data were collected during three-day summer and winter sampling campaigns. The accuracy of the gravity-sewer method to estimate the daily total emissions was determined to be within -1% to +5% in summer and -6% to +19% in winter (when production is half that in summer). Data for that calibration effort are presented in Table 1.

Monitoring Day Designation:	Summer Day 1	Summer Day 2	Summer Day 3	Winter Day 1	Winter Day 2	Winter Day 3
Date	9/16/2014	9/17/2014	9/18/2014	4/7/2015	4/8/2015	4/9/2015
Measured Data for each Day of Sampling						
Average PI Sewage Temperature, °C	22.1	21.8	21.5	12.7	12.1	12.6
Average Measured Sewage Flow, mgd	29.9	30.8	28.7	33.3	36.9	36.2
Measured CH_4 Emissions, kg CH_4/D	*131*	*131*	*132*	*83*	*69*	*67*
Estimated CH_4 to be Emitted at LTOAF-17						
Methodology-Estimated Emissions	*137*	*137*	*131*	*79*	*77*	*79*
Total Predicted CH_4 as % of Measured	105.0%	104.6%	99.3%	94.2%	111.5%	118.6%
Seasonal Predicted CH_4 as % of Measured		103.0%			107.1%	

Table 1. Method Verification on the Potomac Interceptor

Forcemain/Surcharged-Sewer Method/Equations. Two equations have been developed for forcemains. The original, full equation is based on full-scale operation of smaller, intermittently-operated pumping stations/forcemains. The equation (Eq. 2, below) highlights the impact/need for energy that is required to diffuse sewage carbon into the biofilm. During period of pumping, carbon is continuously driven into the slime by the flow and CH_4 production occurs at a consistently high rate. During periods when pumps are off, the infusion of carbon is stopped, slime biology rather quickly deplete the locally-available carbon, and CH_4 production is curtailed. As such, for smaller systems with intermittent flows, the following equation can be used:

$$r_{CH4\text{-}FM} = 3.452 \times N^{(0.202)} \times D \times 0.396^{(1-N\times PT/1440)} \times 1.06^{(T-20)} \qquad \text{Eq. 2}$$

Where: $r_{CH4\text{-}FM}$ = CH_4 emission rate in kg CH_4/(km*day)
N = Number of pump cycles per day
= 1 for continuously flowing systems
PT = Pump time; or the duration of each pump cycle in minutes
= 1440 for continuous systems

Our conceptualization of surcharged sewers (those gravity-fed segments that are fully submerged, with 180°, fully-wetted perimeters with full-circumference slime areas) is that they are continuously flowing. As such, even at potentially low velocities, adequate energy exists (more than zero) at turbulence-inducing pipe walls to force sewage carbon into the slime and "refeed" carbon-consuming, CH_4- and H_2S-forming organisms.

For continuously operated forcemains or surcharged sewers, Eq. 2 can be simplified by replacing $N^{(0.202)}$ and $0.396^{(1-N\times PT/1440)}$ terms with "1.0" for a once-per-day (N=1), 1,440-minute (PT=1440), continuous operation. Eq. 3 is the further simplified equation for non-intermittent operation:

$$r_{CH4\text{-}FM} = 3.452 \times D \times 1.06^{(T-20)} \qquad \text{Eq. 3}$$

In modeling DC Water's collection system, there are no intermittently-pumped forcemains. As such, only Eq.3 was used exclusively, in lieu of Eq. 2, in our system-wide model. This is likely the case for many systems with the exception being those with multiple, small-capacity stations that cycle on and off based on high- and low-level float-switch actuation.

Combined Collection System Method. A hydraulic-model-generated output "shape file", run at average flow rates is used to estimate collection-system-wide CH_4 production. Average flow rates are used because slimes need regular submergence to become fully established and average conditions representative regularly-wetted conditions. There are likely poorly- to very-poorly-established slimes above average sewage levels and less-than-fully-developed slimes immediately below those same levels; yet full-scale data suggest that these effects "balance".

Each pipe segment is classified into one of the following three categories (with schematic images of each depicted in Figure 2):

1. **Gravity Sewer.** Uses Eq. 1 for CH_4-production.
2. **Surcharged Sewer.** Uses Eq. 3 as the default for continuously flowing systems. Eq. 2 can be used for intermittent pumping situations but it's use will require derivation of a basis for the modeled number and average length of pumping cycles per day.
3. **Hybrid.** This designation represents situations where the hydraulic grade is both below and above the pipe crown in parts of the segment. Under these conditions the change in hydraulic grade from upstream to downstream end can be compared to the change in pipe crown over that same length. In the DC Water model, we assumed that both varied linearly between the segment end elevations so that length upstream and downstream of the transition could be approximated. Eq. 1 and Eq. 3 can then be used for the gravity and surcharged portions of the segment, respectively.

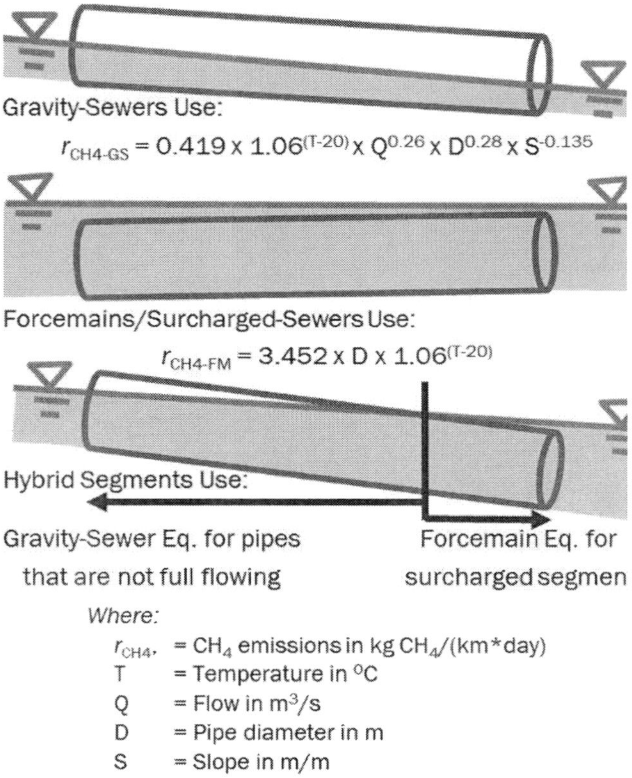

Figure 2: CH_4 -Production Methodology Equations and Pipe Segment Characterization

As the equations are based in metric units, conversion from standard-US to metric units may be required. Alternatively, it is recommended that hydraulic model output be set to metric units for generation of the shape file. To be effectively used, the shape file must have the following data:

- Segment lengths, meters or kilometers – $r_{CH4\text{-}GS}$ and $r_{CH4\text{-}FM}$ are expressed in kg CH_4/(km*day) and production in each segment must be normalized for its length in km.
- Segment diameter in meters.
- Flow rate in each segment in cubic meters per second.
- Upstream- and downstream-end invert and hydraulic grade elevations – these data allow classification of each segment into the above categories and determination of proportional length of gravity and submerged lengths in hybrid segments.
- Segment slope in meters per meter (not percentages) – which can also be calculated from upstream and downstream invert and segment length.

The final data required for a system-wide model, collection-system-sewage temperature, will likely involve location-specific research (as very few utilities measure plant influent wastewater temperature (however, if they do, then satisfying this requirement is straightforward). In DC Water's case, sewage temperatures were collected intermittently during the summer and winter sampling effort. DC Water also continuously monitors effluent temperature at the Blue Plains Advanced Water Reclamation Facility (Blue Plains) on 10-minute intervals. These data were used to derive a correlation between Blue Plains effluent and raw-sewage temperatures. The equation is as follows (the basis for its derivation is shown in Figure 3):

$$T_{CS} = 1.19 \times T_{BP\text{-}Eff} - 6.70 \qquad \text{Eq. 4}$$

Where:

T_{CS} = Collection-system sewage temperature in ^{O}C

$T_{BP\text{-}Eff}$ = Blue Plains average (based on the average of two instruments) effluent temperature, in ^{O}C

Blue Plains effluent ranges from 1 to nearly 6 degrees C greater than the corresponding raw sewage temperature. This is thought to be due to heat addition in the activated-sludge process (as aeration air is heated by approximately $4^{O}C$ per psi of compression prior to its release) and absorption of solar-thermal energy during daylight hours. The temperature differences were greater and the spread of data was also more disparate (and the Eq. 4 fit poorer) in the winter. That said, the correlation for average daily temperatures is very good.

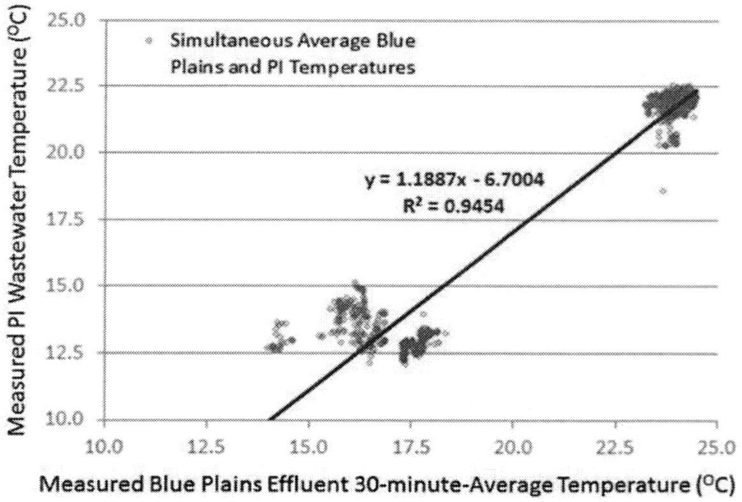

Figure 3. Derivation of DC Water Raw-Sewage to Blue-Plains-Effluent Temperature Correlation

CONCLUSIONS AND NEXT STEPS

The researchers have organized a Knowledge Development Forum at WEFTEC-2018 in New Orleans. This forum will include sewer-process experts from around the world who have conducted related research; interested, progressive-minded utility and consulting staff; and a representative from ICLEI. The goal is to discuss the basis for this method, its proposed use, and determine if it is technically sound and warrants adoption by our industry. If not, vulnerabilities within the method and/or suggestions as how it might be improved will be entertained.

We are also looking to compile a database of 10 or more collection-system-wide CH_4 emissions inventories from different climates, system configurations, and sizes. This effort would be used to create a further-simplified method that utilities or local governments without the extensive, data-intensive demands that are required for the method defined herein. That lower-tier (less accurate) method that could be more readily used to more accurately estimate sewer CH_4 emissions than simply assuming their non-existence, as currently advocated in the protocols. We are looking for interested parties to use the method described herein to populate that database and are willing to anonymize any information provided, as requested by each utility. Alternatively, Brown and Caldwell could implement this method for a $25k to $35k fee using a utility-provided shapefile and temperature data as the basis; the primary product would be a single equation that would estimate collection-system daily CH_4 emissions as a function of flow and sewage temperature. That equation could then be used on a daily, monthly average, or annual average basis to estimate emissions for GHG recording. The product could also include analyzing a specific calendar-years' worth of emissions, if so desired. Finally, we are also looking for progressive utilities who would be willing to use their collection-system shapefile for use as a case study during the WEFTEC Forum. If you are interested in any aspect of this research, please call the primary author, John Willis, at 770-361-6431.

Comparison of Characterization Techniques for Wastewater Treatment Plant Odor Evaluations

Phyllis G. Diosey
Hazen and Sawyer, 498 Seventh Avenue, New York City, NY 10018

Abstract
Wastewater treatment facilities are often the focus of local odor regulations. Since hydrogen sulfide is a typical constituent of the odors from the overall wastewater treatment, these local odor regulations are often based on the control of hydrogen sulfide emissions. However, the treatment of wastewater involves a variety of processes (and associated structures), such as screening for any large material that is mixed in with the influent water and removing grit (sand, gravel, and other heavy particles), primary settling or sedimentation to allow organic solids and material to settle, aeration to encourage bacteria to break down remaining sludge particles after the primary treatment, and final settling and filtration before the effluent leaves the plant. While hydrogen sulfide is a major contributor to offsite odors from wastewater treatment, the odors from these individual processes are often very different both in the character, strength, and intensity from H_2S. In recognition of these differences, some jurisdictions look to control odors from a non-pollutant-specific approach using dilution to threshold (D/T) as an odor identifier. However, even with the use of H_2S or D/T, there are variations both in the allowable odor levels, the applicable receptor location, and the averaging time across jurisdictions. An earlier investigation into the use of these two different odor approaches (Diosey et al., 2004) evaluated the offsite impact due to odorous emissions from a 60 MGD municipal wastewater treatment facility in an urban area. The present analysis examines the results from recent investigations of municipal wastewater treatment plants; one plant in the southeast United States and another plant located southwest United States. Sources at both plants were sampled for H_2S and D/T, and then modeled with refined dispersion modeling using five years of representative hourly meteorological data. The maximum modeled offsite impacts of H_2S and D/T from individual process sources at each of these facilities are presented in the paper and are compared in order to evaluate the effects of using these different odor parameters, and to identify the potential ramifications each approach poses when trying to meet regulatory criteria and minimize impacts on the surrounding communities.

Key Words: Odor impact, odor modeling, H_2S, dilution-to-threshold,

Introduction

Historically, odor assessments for wastewater treatment plants tended to focus on hydrogen sulfide (H_2S), because a number of wastewater processes give off this distinctive unpleasant odor. However, wastewater treatment plants also have a number of processes where the type of odor varies from process to process. As noted in earlier papers, processes at the influent end of the wastewater treatment plant, such as the headworks, and primary tanks tend to have a strong rotten egg odor, characteristic of hydrogen sulfide (H_2S) and a "sewer like" odor. The odors from aeration tanks are generally described as earthy or musty. Dewatering facilities can emit an ammonia-like odor, while disinfection tanks can smell of chlorine. Odors can also be the result of combinations of compounds, where the components can sometimes be identified, but are not necessarily quantified. In order to more accurately quantify complex odors, five parameters are often used:

- **Character** Odor *character* or odor quality is generally reported in terms of standardized descriptors, such as "fruity," "earthy," "musty," etc. Odor panelists and odor observers are trained to use this terminology to identify an odor.
- **Hedonic Tone** *Hedonic tone* measures the pleasantness or unpleasantness of the odor independent of the character. Different scales may be used, such as +10 would be pleasant, 0 would be neutral, and -10 would be unpleasant.
- **Intensity** Odor *intensity* is the relative strength of the odor above the threshold. Using the standard butanol method, an olfactometer delivers varying concentrations of butanol in air to eight sniffing ports. The odor *intensity* is then expressed in terms of parts per million of butanol.
- **Persistence** The *persistence* is defined as the rate of change of intensity with odor concentration.
- **Concentration** The *detection threshold* (DT) is the lowest concentration of a substance that can be detected above a blank sample by an odor panel. The *recognition threshold* (RT) is the lowest concentration of a substance that can be recognized based upon the character of the odor. The *dilution ratio* (D/T, or *dilutions to threshold* value) is the estimated number of dilutions with equal volumes of clean air needed to make the odor non-detect.

Given the variety and complexity of the emissions, the use of a single compound, such as H_2S, with a low odor threshold and high emission rate, can potentially lead to errors in determining odor impacts offsite However, this fact, along with the potential for other types of odor, make it difficult to set appropriate criteria for regulating odor, as **Table 1** indicates. The level of nuisance from odor sources, such as wastewater processes, varies widely among jurisdictions across the United States, and this in turn will affect the level of control proposed.

This paper presents the results of odor evaluations using refined dispersion modeling for specific processes at two different wastewater treatment facilities using both H_2S and D/T to evaluate offsite odor impacts. Extensive onsite monitoring was conducted at both facilities to determine emissions of H_2S and odor (in terms of odor units) from the individual process sources. The processes at the plants were then simulated using a refined atmospheric dispersion model (ISCST3/AERMOD), and both H_2S and odor emissions based on the sampling data were modeled. The two approaches to evaluating odors were then compared for specific process sources, and compared to standard odor criteria in terms of H_2S and D/T.

Table 1. State and Local Odor Regulations and Criteria

Location	Standard / Criteria	Averaging Time	Source/Receptor
Bay Area, CA	5 D/T	5 or more confirmed complaints per year averaged over 3 years	Fenceline
Colorado	7 D/T	Scentometer	Urban
	15 D/T		Rural
	7 D/T; 2 D/T		CAFO
Connecticut	0.0045 ppm H_2S	15 minutes	Ambient air
	7 D/T	3 samples	Ambient air
Illinois	8 D/T	Scentometer	Residential, recreational, retail sales, institutional, educational premises
	24 D/T		Industrial premises
	16 D/T		All other premises
	Pollutant-specific	3-5 seconds (1)	Ambient air
Maricopa County, AZ	0.03 ppm H_2S	30 minutes	Nearest occupied place
Missouri	7 D/T	Nasal Ranger	
New Jersey (2)	5 D/T	5-minutes (or less)	Sensitive receptors
New York State	10 ppb H_2S	1-hour	Ambient air
New York City, NY	1 ppb H_2S	1-hour	Sensitive receptors (3)
North Dakota	7 D/T	Scentometer	Urban/rural
Texas	0.08 ppm H_2S	30 minutes	Residential, business, or commercial
	0.12 ppm	30 minutes	Industrial property and vacant tracts and range lands not normally occupied by people.

Notes: (1) Modeling for state permitting purposes (based on Wang and Skipka, 1993). (2) D/T criterion for POTWs. (3) For wastewater processes.

Methodology

In order to investigate the implications of these two different approaches, the impacts of pollutant-specific (H_2S) emissions were compared to impacts from complex odors (in the form of D/T) for individual sources at two municipal wastewater treatment facilities. Field monitoring was conducted at each facility to identify and quantify the emissions of odors from the various process steps at the plant. Specific sources at each facility were sampled for both H_2S and D/T. The field sampling included measurements of H_2S and odor characterization, including odor concentration (in terms of D/T), odor intensity using the n-butanol scale, and hedonic tone (using a 20-point scale). Direct emissions measurements from open water surfaces were performed using an isolation flux chamber. Weir emissions were measured directly with the use of a specially designed weir flux chamber. Sampling of stacks and vents was performed by filling Tedlar® bags and then sampling the contents for odor parameters (such as D/T and odor intensity) or H_2S. Air samples were analyzed for H_2S using a Jerome Meter® or OdaLog®. Sensory samples were analyzed at an odor laboratory by an odor panel using an olfactometer.

Using the data from the field sampling to estimate the emissions of H_2S or odor units (ou), the identified odor sources at each facility were then modeled for both H_2S and D/T with five years of local hourly meteorological data over a receptor grid that included the boundary of each facility and the surrounding communities. The following facilities were included in the analysis:

1. 60 MGD municipal wastewater treatment facility in an urban area of the northeast U.S. **(Figure 1)**

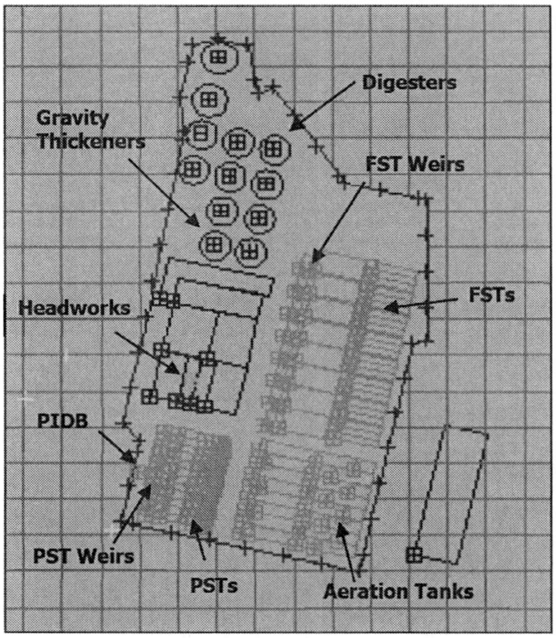

Figure 1. Eastern Urban WWTP

2. 30 MGD municipal wastewater treatment facility in a suburban area of the southwest U.S. (**Figure 2**).

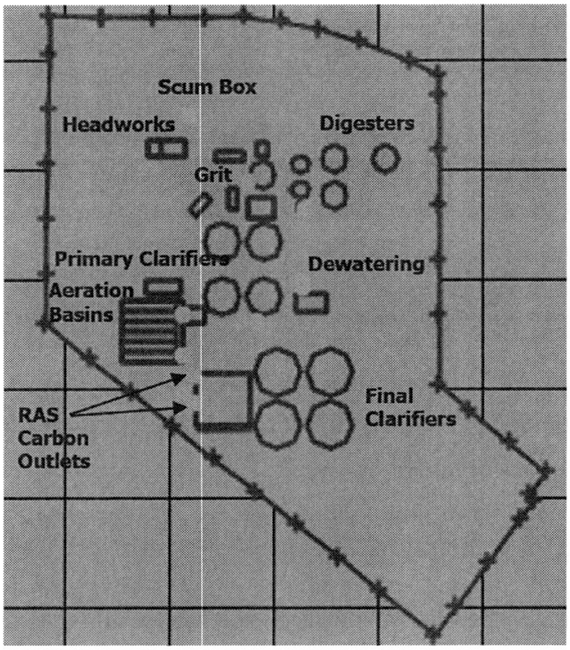

Figure 2. Southwestern Wastewater Facility

Methodology

Specific sources at each facility were sampled for both H_2S and D/T. Field monitoring studies was conducted at each of the facilities in order to identify and quantify the emissions of odors from the various process steps at the plant. The field sampling included measurements of H_2S and odor characterization, including odor concentration (in terms of D/T), odor intensity using the n-butanol scale, and hedonic tone (using a 20-point scale). Direct emissions measurements from open water surfaces were performed using an isolation flux chamber. Weir emissions were measured directly with the use of a specially designed weir flux chamber. Sampling of stacks and vents was performed by filling Tedlar® bags and then sampling the contents for odor parameters (such as D/T and odor intensity) or H_2S. Air samples were analyzed for H_2S using a Jerome Meter® or OdaLog®. Sensory samples were analyzed at an odor laboratory by an odor panel using an olfactometer.

H_2S, emission rates in units of grams per second (g/s) were input into the model for the point sources (stacks) and in units of grams per second per square meter for the open area sources (aeration tanks, weirs, etc.). The model output for H_2S is reported in units of micrograms per cubic meter ($\mu g/m^3$) or parts per billion (ppb).

Using dilution to threshold (D/T) along with the source parameters (e.g., volumetric flowrate), a "pseudo-mass emission rate" in terms of *odor units per second.* The odor unit or D/T emission rate is treated in the model just like the mass emission rate, and the downwind 'concentration' is in units of D/T or odor units (ou/m^3), rather than $\mu g/m^3$. A limitation of the D/T approach is that unlike units of mass concentration of a particular compound where the impacts are additive, the D/T impacts from multiple sources at a receptor cannot be added unless the odors from those different sources are similar in character. Hence, D/T is often used for representing the impact due to an individual odor source. Using the data from the field sampling to estimate the emissions of H_2S or odor units (ou), the identified odor sources at each facility were then modeled for both H_2S and D/T using refined dispersion models (ISCST3/AERMOD) with five years of local hourly meteorological data over a receptor grid that included the boundary of each facility and extended into the surrounding communities.

Results and Discussion

The model output was in the form of maximum hourly odor or H_2S concentration at each receptor over the course of each year of meteorological data. From this output, isopleths (lines of constant odor concentration, either as H_2S concentration in $\mu g/m^3$ or ppb, or as D/T) using data from the year with the maximum offsite impacts for the specified source were developed for each facility. The level and the spread of each type of isopleth was then compared for each of the facilities.

1. 60 MGD Urban WWTP

The odor emissions from the primary settling tanks (PSTs) were selected for the 60 MGD urban wastewater treatment facility. The twelve primary settling tanks were divided into three individual area sources for each tank to represent influent area, mid-tank, and effluent areas before the weir. The PSTs were modeled as ground level area sources. Discrete receptors were located along the facility fenceline at approximately 25-meter spacing. In addition, ground-level receptors were located along a 1-kilometer by 1-kilometer Cartesian grid centered on the plant at intervals of 25-meter spacing, and an additional set of Cartesian receptors at 100-meter spacing extended for 1 kilometer for a total grid size of 2 km by 2 km around the center of the plant. Terrain elevations of the receptor locations were obtained from 7.5' USGS Digital Elevation Model (DEM) files.

The H_2S isopleths for this source group began at 1 part per billion (ppb) and increased in increments of 5 ppb. The 1 ppb minimum was selected to correspond to the local odor criterion. The D/T isopleths started at 3 D/T, and increased in increments of 2 D/T. A value of 3 D/T was selected for this source to be the lower limit of odor detection. **Figures 3a** and **3b** compared impacts of H_2S and D/T for the PSTs in terms of D/T and H_2S from this earlier investigation. For this process source group, the 1 ppb H_2S isopleth extends to more than 1 kilometer offsite, and 10 ppb impacts extends about 400 meters beyond the southwest fenceline of the plant. The D/T isopleth, however, shows significantly higher offsite impacts. The 3 D/T extends less than the 1 ppb H_2S isopleth, and 10 D/T impacts were found out to 800 meters from the fenceline. The comparison of extent of the two different odor parameters, i.e., H_2S concentration versus D/T indicates that the use of the 1 ppb H_2S criterion that is used by the local jurisdiction to evaluate/control odors from wastewater processes impact a significantly larger area than the more generic odor parameter 3 D/T, potentially requiring greater levels of control to meet the more stringent level.

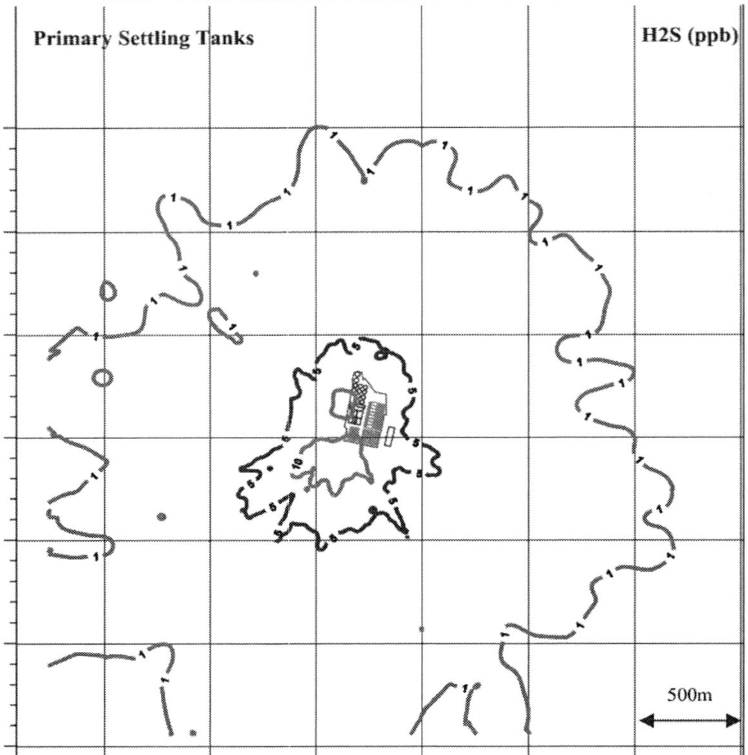

Figure 3a – Primary Settling Tank: H_2S Isopleths (ppb)

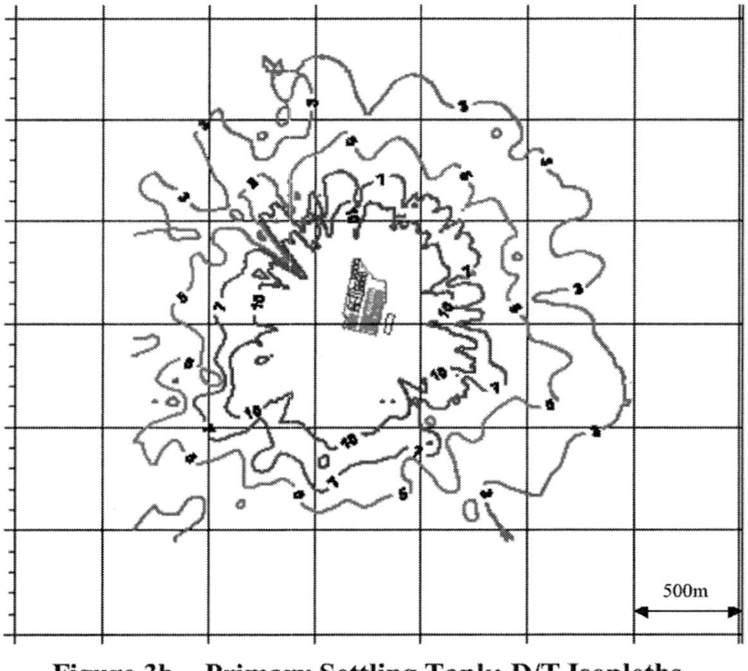

Figure 3b – Primary Settling Tank: D/T Isopleths

2. 30 MGD Southwestern Suburban WWTP

The emissions of H_2S and odor units from the two RAS (return activated sludge) carbon odor control outlets were evaluated for this facility. The outlets were modeled as elevated point sources, approximately 2 meters above grade with exhaust diameters of 1 meter. **Figures 4a** and **4b** present the H_2S and D/T isopleths from the RAS carbon outlets.

Figure 4a. RAS Carbon Outlets: H_2S Isopleths (ppb)

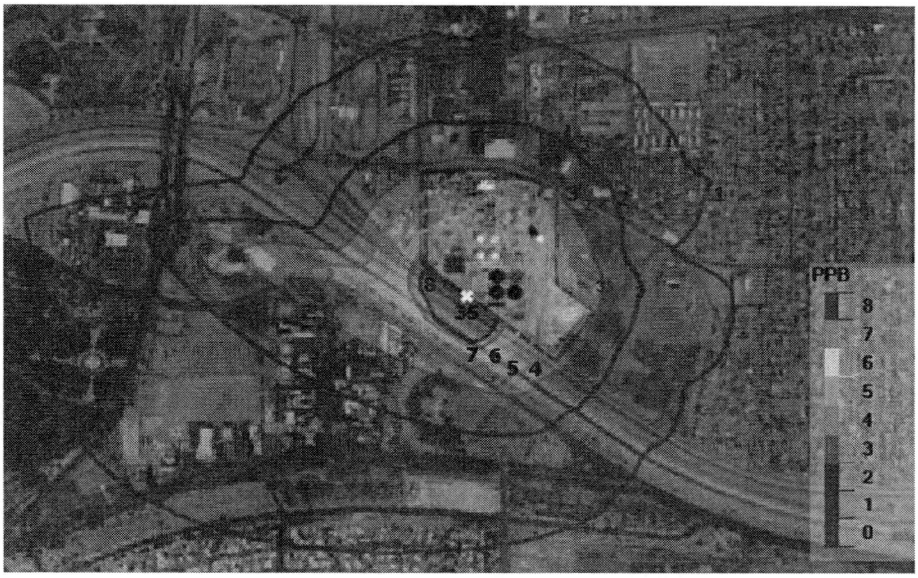

Figure 4b. RAS Carbon Outlets: D/T Isopleths

In this scenario, the 1 ppb H_2S isopleth is almost identical to the 3 D/T isopleth in shape and extent, hence, using either form of emissions for this particular source would result in approximately the same conclusion with respect to the level of odor control needed.

Figures 5a and **5b** present the isopleths for the two dewatering odor control stacks, 6.1 meters above grade, and 0.762 meters in diameter. For this source, the H_2S offsite impact is just under 1 ppb (~1.4 $\mu g/m^3$), but the 3 D/T isopleth extends several hundred meters beyond the facility boundary. Since, as noted earlier, dewatering process odors are ammonia-like, hence, it makes sense for the D/T odors to spread further than H_2S.

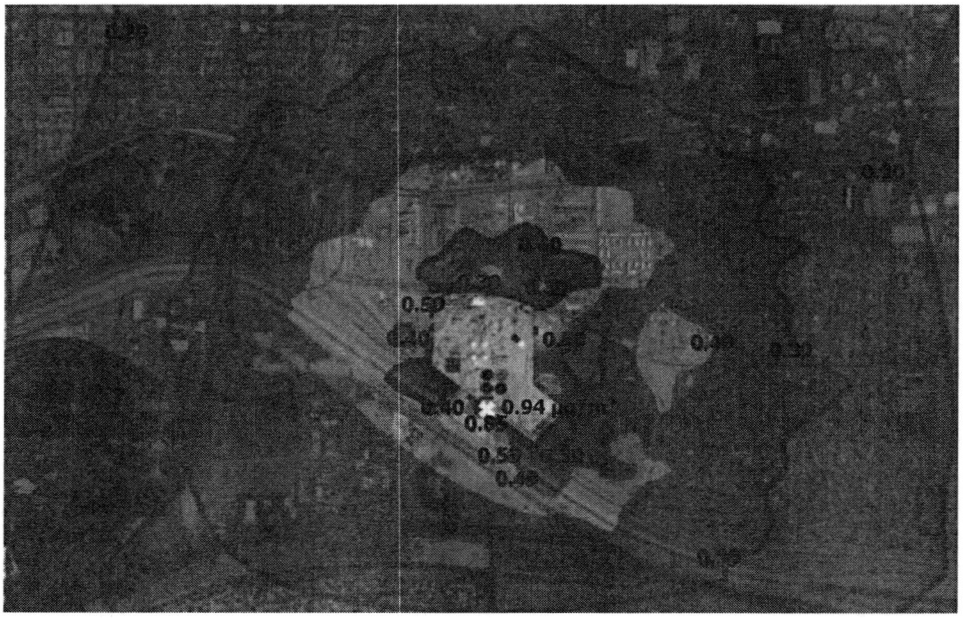

Figure 5a. Dewatering Odor Control Stacks: H_2S Isopleths ($\mu g/m^3$)

Figure 5b. Dewatering Odor Control Stacks: D/T Isopleths

Results

Process sources at two different wastewater treatment facilities were modeled for H_2S and odor in terms of D/T. Analysis of the different processes and the different odor indicators demonstrates how the selection of odor indicator for sampling and modeling can complicate the determination of what sources may require odor control to reduce offsite odors, and also how much control is required. In addition, the variety of odor criteria and regulations from various jurisdictions can potentially result in either excess control, or control of only some of the processes that may require control, pointing to the need to come to a consensus on both the odor indicator, or indicators to consider, and applicable criteria to apply in order to minimize the impact of offsite odors in the surrounding community.

References

Diosey, P.G., Buono, M. E., and Farrell, L. (2004). Modeling of Odors from a Wastewater Treatment Plant: A Comparative Study. Presented at the WEF & AWMA Specialty Conference: Odors and Air Emissions 2004, Bellevue, WA.

Wang, J. and Skipka, K. J. (1993). Dispersion Modeling of Odorous Emissions. Presented at the 86th Annual Meeting & Exhibition, Air & Waste Management Association.

Proactively Monitoring and Managing Odors using Real-Time Odor Plume Prediction

Charlotte G. Smith, P.E., Perkins Engineering Consultants, Inc.
Steven Daniels, Trinity River Authority of Texas

ABSTRACT

The Trinity River Authority Central Regional Wastewater System (CRWS) retained Perkins Engineering Consultants, Inc. to research, specify and assist with procurement of a real-time odor plume prediction system. The real-time model is used to monitor the CRWS odor production and as a tool for assessing strategies for reducing the CRWS odor footprint.

The real-time odor plume prediction system was designed to achieve the following objectives:

- Use field-measured odor and atmospheric values, along with programmed fixed emission values for selected processes, to project, on a real-time basis, and to continuously record, in a graphic format, the projected intensity and locations of impact of odors from the CRWS facility.
- Notify staff when odors approach or exceed predetermined levels at selected locations on or off the plant site.
- Provide information to be used in evaluating the potential effectiveness of alternative strategies for preventing or reducing odors perceived offsite.
- Provide information to project the estimated intensity and frequency of odors experienced by the traveling public on Interstate 30 south of and other selected receptor locations.
- Provide a model to help with optimization of odor control chemical feed in reducing the probable off-site odor impact.

An odor dispersion model predicts the intensity, frequency, and spatial extent of nuisance odors generated by the treatment facilities. Continuous monitoring and modeling systems display odor isopleths in real-time based on live atmospheric data. Real-time odor plume prediction systems can use onsite continuous measurement sensors correlated to odor units (OU) and an onsite weather station as input to calculate the resulting odor plume. The number of onsite sensors depends on the size of the facility and available budget.

The features, capabilities, and limitations of real-time odor plume prediction systems as well as lessons learned through the procurement of a real-time model for the CRWS facility are discussed later in this paper.

KEYWORDS

Real-Time Odor Plume Prediction, Odor Modeling, Odor Control Optimization

INTRODUCTION

Odor dispersion modeling illustrates how odors generated by wastewater conveyance and treatment are transported offsite. Odor dispersion models can be used to depict emissions of specific odorants, such as hydrogen sulfide, or more commonly in recent years, the overall odor impact using odor units. Odor dispersion model output, or odor isopleths, show the spatial extent at which specified levels are predicted to be perceived.

Modeling with odor units allows for emissions to be evaluated for the total impact since numerous compounds can contribute to the odorous character of the ambient air. Caution is required as the odor intensity or strength may not vary proportionally with the concentration of odorants and odors from different treatment processes are not additive. For example, the intensity of an odor at a concentration of 100 may only be twice as great as its intensity at a concentration of 10 and odor emissions from two equal sources typically do not result in double the odor impact on surrounding areas.

Factors impacting odor dispersion include:

- Odor intensity of sources
- Flux and discharge air flow rates
- Meteorological conditions (especially wind speed, wind direction and atmospheric stability)
- Surrounding topography and building obstructions
- Water surface elevations and stack heights
- Stack discharge velocities

ODOR DISPERSION MODELS

The two models used for regulatory dispersion modeling of contaminants, as well as for odor dispersion modeling, are the steady state model AERMOD and the non-steady state model CALPUFF. Details on the two most widely used air dispersion models include:

- AERMOD – The American Meteorological Society (**A**MS)/Environmental Protection Agency (**E**PA) **R**egulatory **Mod**el (AERMOD) was specially designed to support the EPA's regulatory modeling programs. AERMOD was designed for short-range dispersion, up to 50 kilometers, of air pollutant emissions from stationary industrial sources. AERMOD incorporates air dispersion based on planetary boundary layer turbulence structure and scaling concepts.

- CALPUFF – **Cal**ifornia **Puff** model (CALPUFF) is a Lagrangian puff model originally developed by the Sigma Research Corporation (SRC) in the late 1980s. CALPUFF can be applied on scales of tens to hundreds of kilometers. CALPUFF is approved as a long-range model for regulatory purposes, which is typically defined as transport over distances beyond 50 kilometers, as well as on a case-by-case basis for near-field applications involving complex terrain or wind conditions and for coastal applications.

AERMOD has historically been the regulatory dispersion model most commonly used to model wastewater odors as it was designed for short-range dispersion, but CALPUFF is gaining popularity. Historical use of the non-steady state CALPUFF model for regulatory applications for distances within 50 kilometers has been limited because of the lack of appropriate model validation studies.

Under unstable conditions, the results of the two models are generally in agreement. Under calm wind stable conditions, CALPUFF may be more accurate than AERMOD. AERMOD generates a long plume concentrated at the source, whereas CALPUFF generates puffs that stagnate over the source.

If susceptible, odor complaints often occur during inversion conditions. On damp, cold days, odors may collect in low lying areas adjacent to conveyance and treatment facilities and impact offsite receptors as a pocket of odorous air. CALPUFF appears to more accurately model these conditions.

REAL-TIME ODOR PLUME PREDICTION MODEL INPUTS

The inputs to a real-time odor plume prediction system are similar to the inputs for a conventional odor dispersion model, except meteorological and possibly one or more odor source inputs are real-time.

Meteorological Data

Meteorological conditions can be highly localized. An onsite weather station is recommended for real-time applications. The onsite weather station should measure and transmit the following at a minimum:

- Wind speed and direction
- Atmospheric pressure
- Solar radiation
- Relative humidity
- Temperature

The weather station must be installed far from buildings and other structures to prevent interference with collecting accurate atmospheric data. Weather stations typically require 120 V power.

Cellular data communication technology is recommended for transmitting data from the weather station to a cloud server. It is estimated a weather station would consume about 100 to 200 MB of cellular data monthly, if data is collected in nominal four-minute increments.

Meteorological data collected at nearby station(s) could be used during an onsite weather station outage to continue with odor plume prediction. The National Oceanic and Atmospheric Administration (NOAA) National Climatic Data Center (NCDC) and Weather Underground are potential sources of meteorological data.

Terrain and Other Static Data

Terrain, land use and other static data are typically incorporated into a real-time odor plume prediction system during model development. Both dispersion models include data preprocessors to incorporate complex terrain using USGS digital elevation data. The National Elevation Dataset (NED) is the primary elevation data product of the USGS.

Structures, such as buildings, can affect plume rise and odor dispersion. Turbulent wake zones can be created around buildings, forcing emissions to the ground instead of rising freely before dispersing. Stacks which are not tall enough to avoid downwind turbulent wake zones should be modeled considering building downwash.

EPA's Building Profile Input Program (BPIP) program can be used to process building information to account for the influence of building dimensions on plume dispersion. Current building downwash algorithms only model the effects of nearby point sources; area and volume sources are not included.

Building downwash algorithms are currently one of the biggest sources of confusion and unexpected results with AERMOD. An effort to update building downwash algorithms by the Air and Waste Management Association (A&WMA) Atmospheric Modeling and Meteorology subcommittee is underway.

Odor Sources

The wastewater treatment processes that have been found to cause the greatest off-site odor typically include preliminary and primary treatment and solids handling. Activated sludge processes, especially trickling filters, may impact off-site if sensitive receptors are close to the fence line. The anaerobic and anoxic zones associated with biological nutrient removal have the potential to make activated sludge processes more odorous than historically perceived by our industry. Secondary clarifiers, effluent filters and disinfection typically do not impact off-site receptors.

Dispersion models can incorporate multiple point (scrubber stacks or vents), area (uncovered process units) or volume (open doors, etc.) emission sources. Typically, a combination of area and point sources are used to model odors from wastewater treatment facilities.

Real-time odor plume prediction systems can incorporate from one to all of the potential odor sources at a treatment facility. If no online sensors are provided, odor emission rates can be modeled as fixed values or with programmed fluctuations. Potential fluctuations typical to odor emissions include diurnal or seasonal ups and downs.

Fixed or pre-programmed variable odor emission rates could be based on one or more samples. Sampling for OU includes collecting odorous air from selected treatment units in inert Tedlar® bags and testing the collected foul air in a laboratory with an olfactometer and a panel of trained individuals. Probable variabilities in odorous emissions are not accounted for with grab OU sampling.

Using fixed or pre-programmed emission rates significantly reduces the instrumentation and thus the initial cost of a real-time monitoring system, but this approach may result in a loss in resolution in respect to diurnal fluctuations in emission rates. Fixed or pre-programmed odor source assumptions could either overestimate or underestimate emissions. The user must also use care to see that the fixed rates accurately reflect the basins that are out of service.

Online sensors specific to individual treatment units can be used to provide real-time odor emissions to include in the real-time model. Sensors should ideally be surrounded by the odor source to allow for monitoring of selected source regardless of wind direction. Surrounding sensors by each odor source can prove impractical in some situations.

Options for online sensors include electronic odor monitors and gas sensors. Electronic odor monitors include an array of metal oxide and other sensors. System suppliers calibrate the metal oxide and other sensors in the electronic odor monitors to "smell" odor units, as based on OU sampling conducted during equipment calibration. Electronic odor monitors may not accurately account for odorous compounds for which the monitor was not calibrated.

Individual gas sensors could be selected for OU monitoring based on the composition of the emissions from an odor source. The concentration of the most odorous compound from a specific treatment unit could be calibrated to OU. The most odorous compound may not be the compound with the highest concentration, as some odorous compounds have higher detection and recognition thresholds than others.

Actual treatment unit locations, footprints, elevations, stack heights, stack diameters, etc. would be used for modeling selected odor sources. Multiple units carrying out the same treatment process could be modeled using the same odor emission rates but as separate treatment units.

The real-time odor plume prediction system should provide users with the ability to turn off or on individual odor sources to match current operations.

As with the weather station, cellular data communication technology is the currently preferred method for transmitting data from the sensor(s) to a cloud server. It is estimated a single gas sensor would consume about 100 to 200 MB of cellular data monthly, if data is collected in nominal four-minute increments. Multi-gas sensors and electronic nose monitors would use more cellular data than single gas sensors.

REAL-TIME ODOR PLUME PREDICTION MODEL OUTPUTS

The receptor grid associated with a real-time odor plume prediction system is similar to the receptor grid for a conventional odor dispersion model. Rather than a static odor plume, a real-time odor plume prediction system will provide a frequently updated animated odor plume, taking into account online weather station and odor sensor inputs.

Real-time odor monitoring systems can be programmable to alarm when a situation could result in odor complaints. Systems allowing the user to define geographic points with defined odor exceedance thresholds enables the user to make process adjustments to potentially reduce offsite odor impacts for sensitive receptors.

The accuracy of the model results should not be considered absolute. Model results are only as accurate as the aggregate of the inputs and the model itself.

CRWS REAL-TIME ODOR PLUME PREDICTION SYSTEM

CRWS serves approximately 1.2 million people in the Dallas/Fort Worth area. CRWS has a comprehensive and wide-ranging odor-control program including liquid-phase chemical odor control in the collection system and at the treatment plant as well as foul air collection and treatment with bioscrubbers and biofilters.

CRWS procured and installed an OdoWatch™ real-time odor plume prediction system in 2015 consisting of the following:

- Two electronic odor monitors or eNoses with emission rates applicable to three and four treatment units each for a total of seven treatment units monitored by eNoses
- Six fixed odor emission rates applicable to between one to twelve treatment units for a total of 25 fixed sources
- One weather station on a freestanding tower
- One cloud based central computer with the OdoWatch™ platform based on the dispersion model AERMOD

The OdoWatch™ system at CRWS was supplied by a partnership of Kruger, a Veolia Water Solutions company, and Odotech. The OdoWatch™ system was developed by Odotech in Montreal. Odotech has over 70 OdoWatch™ installations worldwide. In December 2017, Odotech was acquired by the Brisbane and San Francisco based Envirosuite.

CRWS installed the electronic odor monitors and weather station with in house personnel. Odotech calibrated the electronic odor monitors and developed the cloud based AERMOD real-time dispersion model. A third eNose to monitor a second set of primary clarifiers was installed by CRWS and calibrated by Odotech in early 2017.

The odor emission rates used for calibrating the electronic odor monitors and designating the fixed sources were based on OU sampling conducted by Odotech in November 2015 as well as OU testing results from odor abatement master planning work in July 2010. The OU measurements from the two sampling events as well as the OUs assumed for the real-time odor plume prediction system are listed in Table 1.

Table 1 – Odor Source OU Inputs and Supporting Data

	Testing by Odotech November 2015	Testing for Master Planning July 2010	OU Selected for OdoWatch™ System
Primary Clarifier Nos. 1 to 4	Approx. 11,000	> 60,000, 14,000, 9,500	eNose 1
Pump Station 6 (Influent Hatches)	Approx. 100,000	> 60,000	eNose 2
Onsite Storage Basin (OSSB)	548	Not Measured	548
Headworks Grit and Screening Dumpster Building	421	1,700	1,061
Grease Concentrator Building	14,426	Not Measured	14,426
Equalization Basin Nos. 1 to 6	20,154	14,000	17,077
Aeration Basin Nos. 1 to 12	4,270	2,500, 1,000	2,590
Primary Clarifier Nos. 5 to 8	21,039	12,000, 4,400, 3,800	eNose 3 (formerly 10,310)

Solids handling was not included in the real-time odor plume prediction system due to ongoing major changes to the process, including but not limited to the addition of thermal hydrolysis. Real-time model output indicates the odor impact of Aeration Basin Nos. 1 through 12 may be currently overestimated. Lowering the odor emission rate from the Aeration Basins in the real-time odor plume prediction system is currently under evaluation.

Figure 1 is photos of the outside and inside of the electronic odor monitor or eNose at an influent pump station.

Figure 1 – Photos of Inside and Outside of eNose

eNoses use an array of gas sensors and mathematical algorithms to provide detailed information on the real-time odor fluctuations of a source. eNoses do not measure individual odorous compounds, but rather correlate electrical and other responses of the gas sensors to odor strength through multiple calibration curves. The eNoses include:

- Metal-oxide semiconductors
- Metal-oxide-silicon field effect transistors
- Conducting polymers
- Surface-acoustic-wave devices
- Quartz resonators
- Fiber-optic chemical sensors

eNoses and other odor sensors should be surrounded by odor source to allow for monitoring of selected source regardless of wind direction.

A screen shot of the real-time odor plume prediction system online interface is provided as Figure 2.

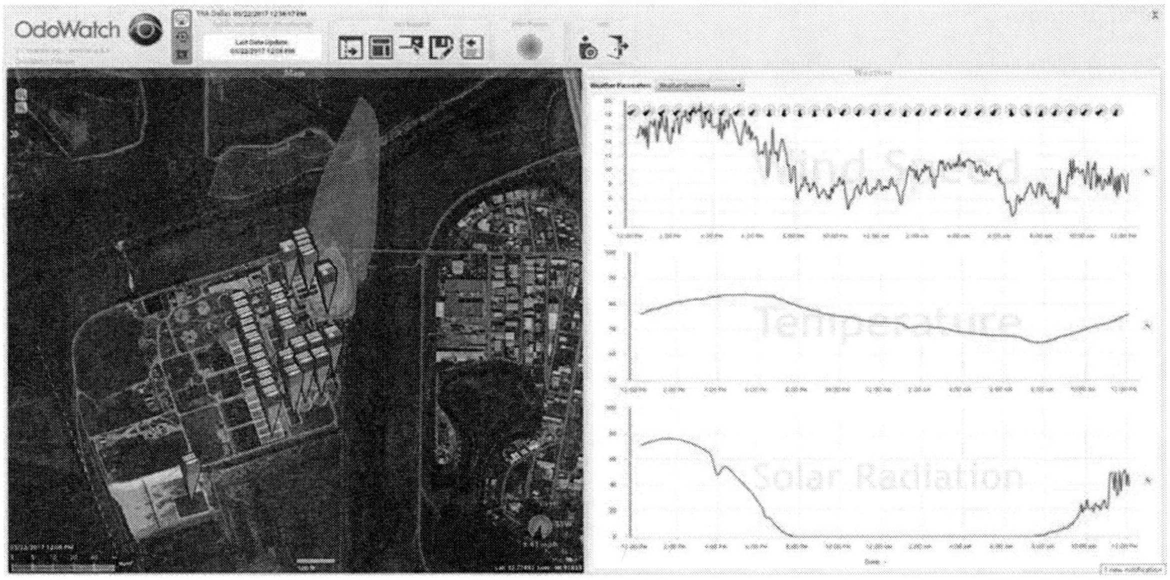

Figure 2 – Real-Time Odor Plume Prediction System Interface

Windows displayed can include odor plume and odor source contributor maps, weather data graphs, historical and current alert tables, alert point odor prediction and source contributor graphs, odor source emission rate graphs, etc. Screens are highly customizable and can be modified depending the preferences of each user type.

Odor plumes are predicted with real-time weather and odor source data every four minutes. Historical predicted odor plumes can be viewed through the platform by selecting the time frame of interest. Predicted odor plumes are displayed sequentially to allow a full day of results to be viewed in about one minute.

Sensor, odor contributor, weather, alert, emission rate and other data can be exported to .csv, .pdf, images and video. Videos can be especially helpful as a public relations tool.

Receptor alert points allow close and continuous monitoring of specific geographic locations. Alert points can be easily added through the OdoWatch™ interface by the user. The system is able to incorporate up to 250 separate alert points. The user can set low, medium and high odor exceedance thresholds at each alert point.

The system can be tailored to send email and/or text alerts to different selected personnel when the model predicts low, medium and high odor exceedances at the specified alert points. The CRWS system sends email alerts to a progressively larger group of users when one or more of the several alert points predict odors exceeding a low of 10 OU, medium of 20 OU and a high of 40 OU.

The alerts allow staff to be proactive rather than reactive, giving them actionable data in real-time to improve operations.

Figure 3 illustrates how the model presents contributing treatment units when odor exceedances are predicted for an alert point via a map display. Other methods of presenting odor exceedance predictions, such as tabular and graphical formats, are available via the user interface.

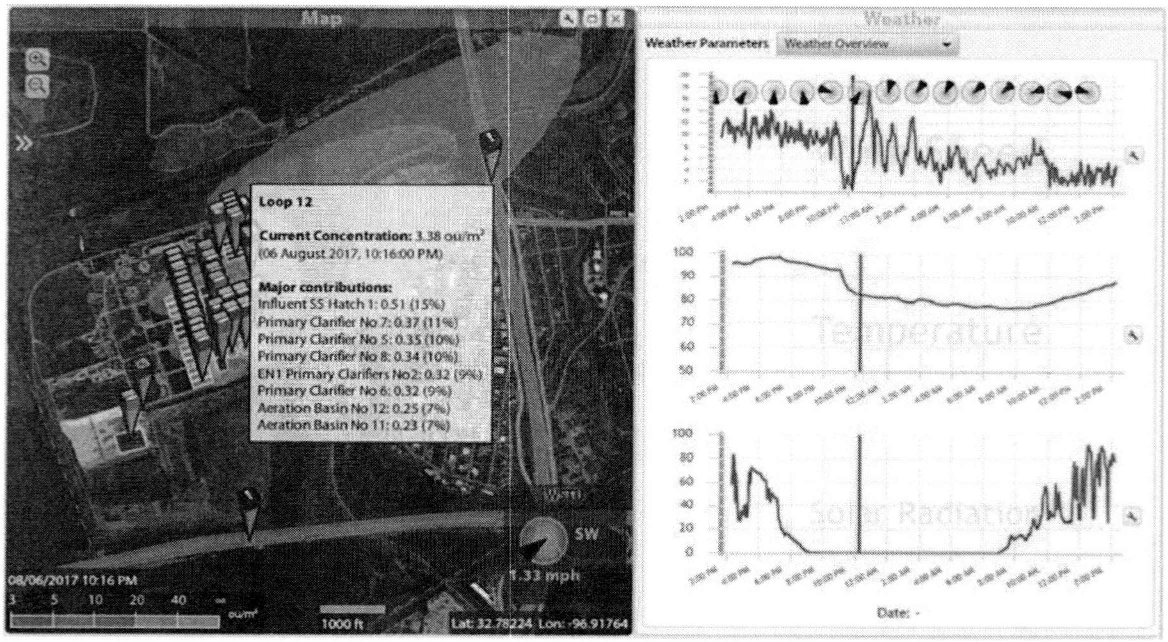

Figure 3 – Identification of Odor Contributors for Alerts

CRWS has several potentially odorous neighbors, including but not limited to landfills, special waste processors, food processors, prepared food distributors and its own collection system. Confusion by the public between odors from the treatment facility and odors generated by its industrial and other neighbors has been an ongoing issue. Real-time odor plume prediction has enabled staff to retroactively evaluate if odor complaints with designated specific times of occurrence are attributable to the treatment facility.

Odor source contributor data will be useful in planning future odor control capital improvements. Model results can be used to identify which of the odorous treatment processes to control to effectively reduce offsite impact.

Costs for the real-time odor plume prediction system are summarized in Table 2.

Table 2 – Cost of Real-Time Odor Plume Prediction System, 2015 Dollars

Item	Budgetary Costs
Initial System including: • Two Electronic Noses • Six Fixed Sources • Weather Station • Cloud Based Control System • Startup and Training	$ 240,000
Additional Electronic Noses	$ 40,000
Additional Fixed Sources	$ 7,000
Annual Electronic Nose and Other System Maintenance Costs	10-percent of Purchase Price
Annual Cloud Hosting Costs	Nominal $3,000 to $ 5,000 depending on Size of System

The real-time odor plume prediction system can be expanded incrementally when budget allows for the addition of more real-time odor sensors and fixed sources. The dispersion model itself and the recommended onsite weather station, if budget allows, are all that is required to start real-time odor plume prediction monitoring.

CURRENT AND FUTURE USES & BENEFITS

As discussed earlier, Envirosuite acquired Odotech in December 2017. The real-time odor plume prediction system available from Envirosuite is subscription based. The subscription covers system access, updates, data storage and support services. Capital expenditures would be needed for an onsite weather station, real-time odor monitoring equipment, and odor emission rate evaluations specific to the treatment facility, if desired by the user, as monitoring and other equipment is not included in the subscription costs.

Several odor control chemical suppliers offer monitoring and control systems which use sulfide concentrations measured at a targeted monitoring location for adjusting odor control chemical dosage rates. Similar control logic could allow for chemical dosage rates to be adjusted up or down to keep one or more sensitive receptors out of the predicted area of odor impact, as based on the results of a real-time odor plume prediction system. Manual adjustment of chemical odor control dosage rates based of odor plume predictions is already attainable.

Software geo-referencing odor complaint data with location and timing information is readily available. At least one supplier offers a timeline feature animating received odor complaints to visually depict how complaints relate with meteorological conditions or process conditions. Odor complaint locations could be evaluated against the predicted odor plume either within or outside of the system.

Odor plume prediction using forecasted meteorological data would allow users to act to control emissions before odors impact sensitive receptor locations. Using forecasted data would provide time to implement process adjustments, such as chemical addition or postponement of intermittent odorous operations, for reducing odor emissions when meteorological conditions could result in offsite impacts.

CONCLUSIONS

The real-time odor plume prediction system has been a useful tool for monitoring the CRWS odor production, alerting management when thresholds may be exceeded, and assessing strategies for reducing the odor footprint. The addition of monitored and fixed odor sources to the existing real-time system is planned as budget allows, particularly after the solids handling improvements project is complete and fully operational as designed.

Cloud based systems allow for data to be accessed from anywhere with an internet connection. System alerts to selected personnel, even the interested public, allow for planning to avoid and if unsuccessful in avoiding, respond to predicted odor impacts.

WATS Sewer Process Model as a tool for Construction Projects Alternative Selection

Nohemy Revilla[1], Guo Ji Chiu[1], Jes Vollertsen[2]
[1]City and County of San Francisco, Public Utilities Commission, Wastewater Enterprise
[2] Aalborg University, Department of Civil Engineering, Thomas Manns Vej 23, 9220 Aalborg, Denmark

Abstract

San Francisco's combined sewer system which transports dry weather (DW) and wet (WW) flows is composed of about 1000 miles of main line sewers and over 2000 miles of side sewers. An important portion of the main sewers are comprised by large transport and storage boxes (T/S), large circular force mains and the respective pump stations that convey sanitary flows and storm runoff to two (2) main treatment facilities (Oceanside Treatment Plant (OSP) and Southeast Plant (SEP)) and one (1) wet weather facility (Northpoint Wet Weather Facility (NPF)) in the City. The City and County of San Francisco, Wastewater Enterprise (Part of San Francisco Public Utilities Commission (SFPUC)) maintains and operates this sewer system. As part of the Sewer System Improvement Program, the City acquired, applied and currently uses the novel sewer processes model, the WATS model, for sewer system project alternative analysis in regards to odor and corrosion control. Recently, since about 2016, the WATS model has been used to analyze odor and corrosion potential for different projects which involved a wide variety of conditions. This report summarizes some of the projects for which the WATS model was used for project scenario simulations. Additionally, this report shows the usefulness of the WATS model to provide a simple way for scenario analysis during planning and design phase of construction projects.

1. WATS modeling for the Mariposa Dry-Weather Pump Station (MPS) and Force Main Improvements Project: The construction plan for MPS improvements involved using a wet weather T/S box to transport dry weather flows (i.e, low flows through a large pipe). Both, existing conditions which involved normal flow pattern, and operation, and potential future conditions during construction (i.e, dry weather flow through wet weather T/S box) were simulated with the WATS model. Results showed an increase on H_2S gas concentrations from existing conditions vs. future conditions. Liquid sulfide and shear stress (tractive force) result analysis showed no difference between existing conditions and future conditions.

2. Marin Street Sewer Improvements: The purpose of this project was to perform WATS model simulations for Marin Street Sewer with current conditions and future conditions, and to look into potential odor problems with the proposed new Sanitary Sewer. Future conditions for this sewer pipe included an increase in sewer flows, a larger sewer pipe and a different pipe material. WATS model simulations for new flow, new pipe diameter and new pipe material (i.e., proposed/future conditions) showed that sulfide and H_2S concentrations in the pipe will still be very low (Sulfide < 0.5 mg/L, and H_2S << 10 ppm). Shear stress simulations in the new pipe are shown to be greater than 2 Pa in the majority of pipe sections (shear stress should be larger than a critical value (here 0.5 Pa for at least 1 hour each day). The only parameter that is increasing with future conditions is pipe corrosion. However, because corrosion is increasing, H_2S concentrations get reduced as compared to present conditions.

3. Salesforce Tower's on-site wastewater treatment sewer discharge impact on collection system odors: Salesforce new building has to follow the new City and County of San Francisco Non-potable Water Ordinance (Onsite Water Reuse for Commercial, Multi-family, and Mixed Use Development Ordinance), and for which the new Salesforce building will include onsite wastewater treatment

and a higher concentration discharge to the sewer system. The purpose of these WATS simulations was to assess the impact on sewer odors and corrosion from the new Salesforce discharge to the sewer. After performing the WATS model analysis, which was based on provided Salesforce lateral pipe location, wastewater flows, and loadings, Salesforce Tower should not negatively impact the existing sewer

Introduction

San Francisco's combined sewer system which transports dry weather (DW) and wet (WW) flows is composed of about 1000 miles of main line sewers and over 2000 miles of side sewers. An important portion of the main sewers are comprised by large transport and storage boxes (T/S), large circular force mains and the respective pump stations that convey sanitary flows and storm runoff to two (2) main treatment facilities (Oceanside Treatment Plant (OSP) and Southeast Plant (SEP)) and one (1) wet weather facility (Northpoint Wet Weather Facility (NPF)) in the City. The City and County of San Francisco, Wastewater Enterprise (Part of San Francisco Public Utilities Commission (SFPUC)) maintains and operates this sewer system. San Francisco is divided into the Westside and Bayside drainages. The Westside basin generates 25 percent of annual combined sewer flows while Bayside generates 75% (Figure 1).

As part of the Sewer System Improvement Program (SSIP), the City acquired, applied and currently uses the novel sewer processes model, the WATS model, for sewer system project alternative analysis in regards to odor and corrosion control. Recently, since about 2016, the WATS model has been used to analyze odor and corrosion potential for different projects which involved a wide variety of conditions. This report summarizes some of the projects for which the WATS model was used for project scenario simulations. Additionally, this report shows the usefulness of the WATS model to provide a simple way for scenario analysis during planning and design phase of construction projects.

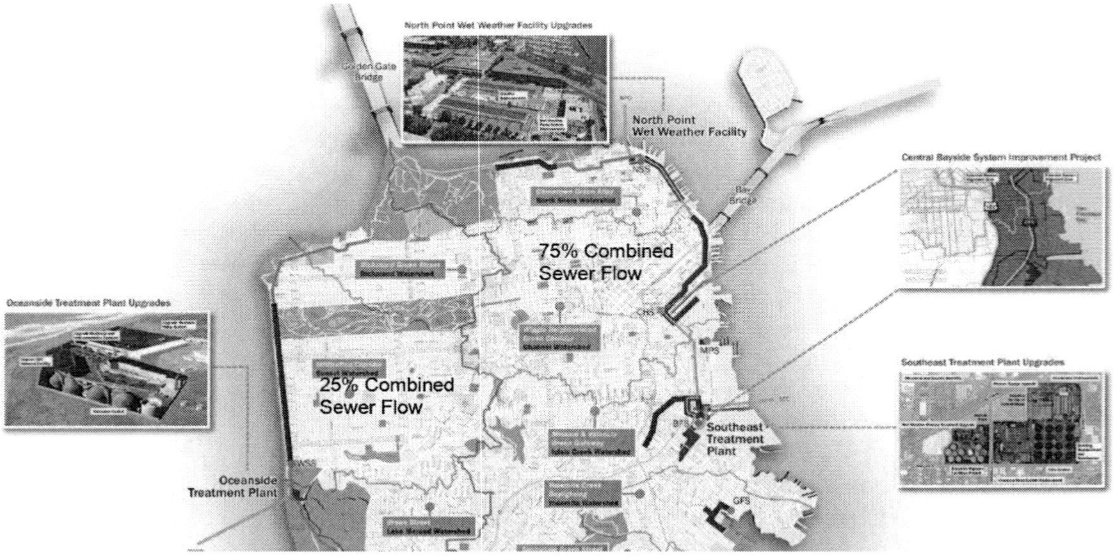

Figure 1. San Francisco's drainage area

Background

The WATS model is a conceptual model that combines sewer model concepts first developed at Alborg University in Denmark in the early 1990s. These concepts are based on suspended growth and biofilm growth systems developed for wastewater treatment models first developed in the 1980s. The numerical model (i.e., the WATS Model) was written simultaneously when the conceptual model was being developed. The relevant sewer process included as part of the WATS Model concept and the numerical model include the following:

- Biological processes - Aerobic, anoxic, anaerobic
- Chemical processes - Oxidation, precipitation
- pH, alkalinity, wastewater buffer system (wastewater composition)
- Hydraulics – sewer network data and flow (routing water and air through the network)
- Gas movement and ventilation to the atmosphere
- Sewer concrete corrosion
- Management recommendations like addition of iron, strong oxidants, nitrate, oxygen
- Stochastic modeling for extreme event statistics

The City had the need to stablish baseline parameters or targets that could allow the Engineering team to decide on areas with the need of improvements in reference to odor control. These baseline parameters are now used on every project requiring a WATS model analysis. The following targets were selected in 2016 for determining the areas of odor concerns and for establish the needs of the City's Collection System:

- Dissolved sulfide or total sulfide concentration (liquid phase) > 0.5 mg S/L

In addition to getting to targets mentioned above, a couple parameters are also considered and overlaid on top of the targets to help with higher taxation of the needs. These parameters include:

- Areas with H_2S (gas) concentration >10 ppm
- Areas highly likely to have permanent sediments: Shear stress (tractive force) < 0.5 Pa

Total sulfide or dissolved sulfide in wastewater is the predominant factor or indicator for odor related problem occurrence in the sewers. As reported in Hvitved-Jacobsen, T., Vollertsen, J., & Nielsen, A. H. (2013), total sulfide concentration greater than 0.5 mg/L in the wastewater will increase the potential of producing odors. Typically, there is a small amount of iron and zinc in the wastewater, precipitating approx. 0.5 mg/L of sulfide. Hence, when total sulfide gets above approximately 0.5 mg S/L, then sulfide begins to become dissolved sulfide in the wastewater, and hence also H_2S-gas in the sewer atmosphere.

According to Hvitved-Jacobsen, T., Vollertsen, J., & Nielsen, A. H. (2013), a range of 0.5 mg/L to 2 mg/L of total sulfide concentration will result in moderate or some odor problems. Therefore, for determining areas of concern in the San Francisco Collection System, it was decided to select a total sulfide concentration greater than 1 mg/L as the primary target, and total sulfide concentrations between 0.5 mg/L and 1 mg/L as the secondary target. Note that dissolved sulfide refers to the sum of H_2S, HS^- and S^{2-} dissolved in the wastewater and it is one of the main parameters represented and calculated by the WATS model (WATS also calculates total sulfide but it is not included in the report analysis). Total sulfide concentration is the parameter and method employed in the laboratory for measuring sulfide concentrations and for calibrating the WATS model. The literature (Water Environment Federation. Control of Odor and Emissions from Wastewater Treatment Plants – WERF Manual of Practice No. 25.) reports a difference between dissolved sulfide and total sulfides of 10% to 30% (i.e. Dissolved sulfide = 0.7 to 0.9 * Total sulfide).

Hydrogen sulfide concentrations in the sewer air above 10 ppm was selected as another parameter for analysis because it is the California Occupational Safety and Health Administration (Cal OSHA) permissible exposure limit (PEL) for H_2S. Note that WATS model assumes all manholes in the collection system are open, which allows the gas to escape or enter freely in and out the system.

Conditions that allow odor generations are similar to those allowing solid deposition in sewers. Normally, for designing self-cleaning sewers for reducing the risk of sewer deposition, a couple of conditions are used:

- Dry weather flow rate should be large enough to minimize solid settling
- Magnitude of shear stress should be large enough to minimize the risk of solids deposition in the sewer. **Note**: "shear stress is the force exerted by the flowing water per unit area of the pipe wall" (t)

Hvitved-Jacobsen T., Vollertsen J., Nielsen A. H. (2010) recommends that the following shear stress t should be larger than t critical (t critical = 0.5 Pa) for at least 1 hour each day:

- t < 0.5 Pa = permanent sediments are highly likely
- 0.5 < t < 1.5 Pa = permanent sediments are possible
- t > 1.5 Pa = permanent sediments are unlikely

Therefore, shear stress is one more parameter that is included as part of the analysis for determining areas of concerns in terms of odors in San Francisco Collection System. In particular, the analysis will include areas falling under the criterion of shear stress "t" < 0.5 Pa = permanent sediments are highly likely.

Following are three case studies which show how WATS model has been used by SFPUC WWE Engineers to analyze collection system construction project odor potential.

1. WATS modeling for the Mariposa Dry-Weather Pump Station (MPS) and Force Main Improvements Project

The Mariposa Pump Station (MPS) pumps both dry- and wet-weather flows from the Mariposa T/S box sewer to the gravity sewer at 21st and Illinois streets, which then flow to Southeast Plant (SEP). This pump station is actually two pump stations: one dry- and one wet-weather pump station. The dry-weather pump station is a separate structure located south of the wet-weather pump station. The dry-weather pump station was originally constructed in 1954; the wet-weather pump station was constructed in 1993 along with the T/S structure. During wet weather, a system of baffles and a weir keep debris in the T/S box and prevent the discharge of solids into the bay. As wastewater in the T/S box rises upstream to a level that exceeds the capacity of the dry-weather pump station, flow spills over a weir into the wet-weather T/S box and into the wet-weather pump station. Refer to Figures 2.a and 2.b.

The construction plan for MPS improvements involved using a wet weather T/S box to transport dry weather flows (i.e, low flows through a large pipe). The purpose of running WATS model simulations for this project was to predict odor generation potential due to construction scope of work. Model simulations were analyzed during the project's early design phase. Both, 1) existing conditions which involved normal flow pattern and operation, and 2) potential future conditions during construction (i.e, dry weather flow through wet weather T/S box) were simulated with the WATS model.

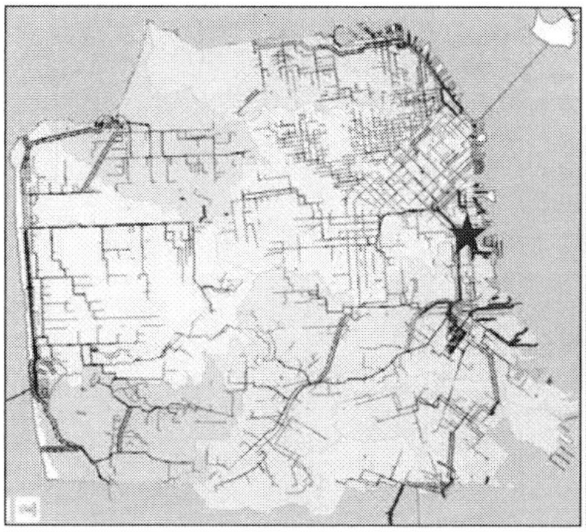

Figure 2a. Mariposa Pump Station Location.

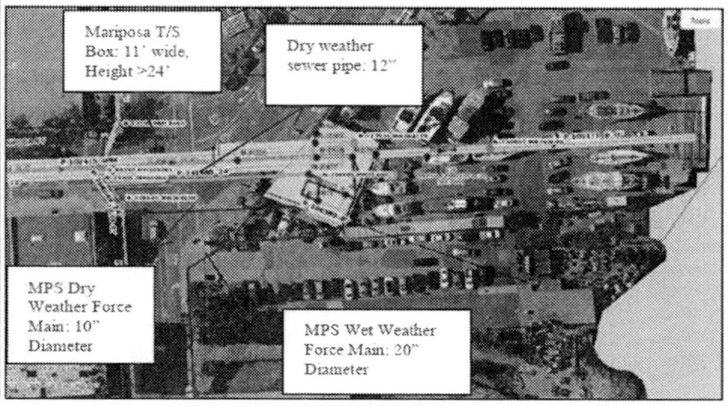

Figure 2b. SFPUC GIS Screen Shot near MPS

1.a. Mariposa Pump Station Original Conditions (Baseline)

The WATS model was used to simulate baseline conditions which only reflected dry weather conditions (Note: WATS model is not designed to simulate wet weather conditions). Simulation results showed that odor levels are low around MPS during normal dry weather flow conditions as shown in Table 1.

Table 1. WATS Model Simulation Results - Baseline

Mariposa Pump Station - Baseline					
Asset ID	Flow in the pipe [L/s]	Bottom shear stress of gravity pipes [Pa] (> 0.5 Pa)	Dissolved sulfide in the wastewater at downstream node [mg-S/L]	H_2S gas in the sewer air at downstream node [ppm]	Average corrosion rate in the pipe [mm/year]
85735-30969	11.57	16.02	0.05	0.73	0.054
30969-30965	11.57	0.13	0.01	0.02	0
350116-85733	45.06	-	0.11	0.35	0.128
85733-209053	45.06	1.46	0.11	0.09	0
209053-30965	45.48	1.48	0.11	0.08	0
18652-209053	0.42	0	0.38	0.4	0
30965-30974	57.05	2.32	0.08	0	0
30974-350124	57.05	23	0.07	0	0
350124-95013	57.05	9.15	0.07	0.02	0
95013-350129	57.05	8.31	0.07	0	0
30980-30982	57.47	18.2	0.03	0.37	0.086
30980-30982	57.47	18.2	0.03	0.37	0.086

Baseline WATS model simulations show that there isn't any pipe around the pump station with sulfide concentrations greater than 0.5mg/L. Additionally, H_2S concentrations are normally less than 1 ppm pipe in pipes located around MPS. There is a pipe upstream MPS has shear stress lower than 0.5 Pa, which indicates high sediment build up. Figure 3 shows a WATS model screen shot representing the area around Mariposa Pump Station.

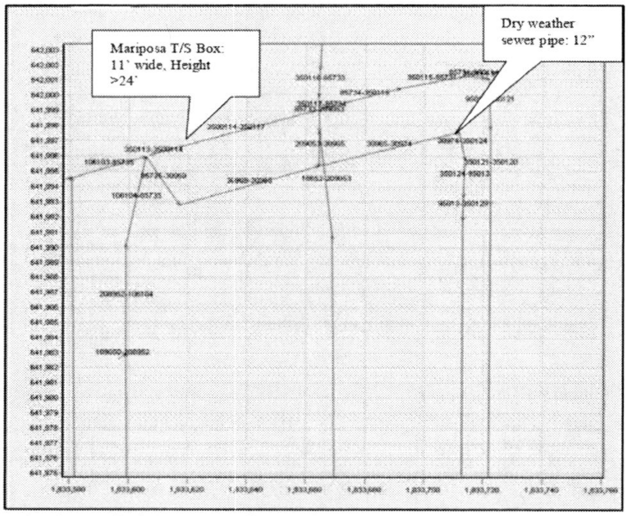

Figure 3. WATS Model Screen shot of area around Mariposa Pump Station

1.b. Mariposa Pump Station Construction Project Conditions

Sewer network changes were incorporated in the WATS model in order to run simulations reflecting construction project scope of work (i.e, routing the dry weather flow through the wet weather box). In this case, a couple of pipes/network links were deleted and a couple links were added or created. Network changes included the following:

- Deleted pipe 30965-30974
- Deleted pipe 209053-30965
- Added pipe 209053-85734
- Added pipe 30965-309053

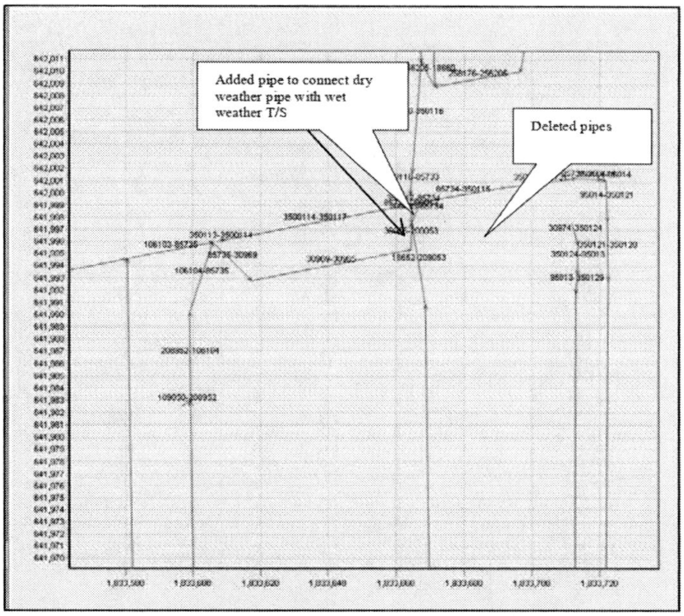

Figure 4. WATS Model Screen shot showing added and deleted pipes (and pipe names). Refer to Figure 3

Figure 4 shows changes made to the collection system network for representing construction project conditions. Additionally, Table 2 includes simulations results under these conditions. Simulation results showed that in general, the construction project sequence could proceed with the plan of temporary routing of the dry weather flow through the wet weather T/S Box. Results showed an increase on H_2S gas concentrations from existing conditions vs. future conditions (from 0.37 ppm to 4.84ppm in one of the links). Liquid sulfide and shear stress result analysis showed some difference between baseline conditions and construction projects conditions, however, the change in values did not represent a negative outcome if compared to target threshold H_2S of 10 ppm, especially because re-routing dry-weather flows through the wet-weather T/S is a temporary setup. In this case study, the WATS model was a very useful tool which provided an analysis on whether a construction project will meet odor threshold targets set for the City's Collection System.

Table 2. WATS Model Simulation Results - Construction Project Conditions

Mariposa Pump Station - Construction Project Conditions					
Asset ID	Flow in the pipe [L/s]	Bottom shear stress of gravity pipes [Pa] (> 0.5 Pa)	Dissolved sulfide in the wastewater at downstream node [mg-S/L]	H_2S gas in the sewer air at downstream node [ppm]	Average corrosion rate in the pipe [mm/year]
85735-30969	11.57	16.02	0.05	0.73	0.054
30969-30965	11.57	2.63	0.01	0.77	0
350116-85733	45.06	7.73	0.11	0.37	0.139
85733-209053	45.06	2.88	0.11	0.42	0
18652-209053	0.42	0.91	0	0.02	0
30980-30982	57.47	18.2	0.11	4.84	0.377
30965-209053 (Added Pipe)	11.57	1.65	0	0.73	0.002
209053-85734 (Added Pipe)	57.05	464.87	0.08	1.25	0
85734-350115 (Wet weather Pipe)	57.05	1.35	0.02	0.1	0.063
350115-85731 (Wet weather Pipe)	57.05	1.43	0	0.06	0.039
85731-350114 (Wet weather Pipe)	57.05	15.89	0	0.01	0.025
350114-95014 (Wet weather Pipe)	57.05	0.26	0	0	0
95014-350121 (Wet weather Pipe)	57.05	1.42	0	0	0
350121-350120 (Wet weather Pipe)	57.05	11.4	0	0	0
350120-30980 (Wet weather Pipe)	57.05	0.26	0.01	0	0
30980-30982	57.47	18.2	0.11	4.84	0.377

2. Marin Street Sanitary Sewer Improvements

Marin Street Sanitary Sewer is located on Marin St. between 3rd Street and the Marin Street Outfall Structure. The sewer is 24 inches in size and approximately 1,980 feet in length. Its primary purpose is to convey sanitary flow from approximately 600 acres of service area which includes Mission Bay South, Potrero Hill, Pier 70, and Pier 80. The sewer conveys combined sanitary and storm runoff during wet-weather. Due to the limited capacity of the 24" sanitary sewer, a new sewer is proposed which will have a minimum capacity of 7.4 MGD (11.6 CFS) having a flow depth that is 50% of the sewer diameter which guarantees open channel flow conditions. This new flow considerations were calculated based on the 2045 population predictions. The recommended

sewer size that will meet the projected flows and maintain open flow conditions was a sewer diameter of 30" in size with a roughness value of 0.01 (applicable for smooth pipe surface).
The purpose of this project was to perform WATS model simulations for Marin Street Sewer with current conditions and future conditions, and to look into potential odor problems with the proposed new sanitary sewer pipe.

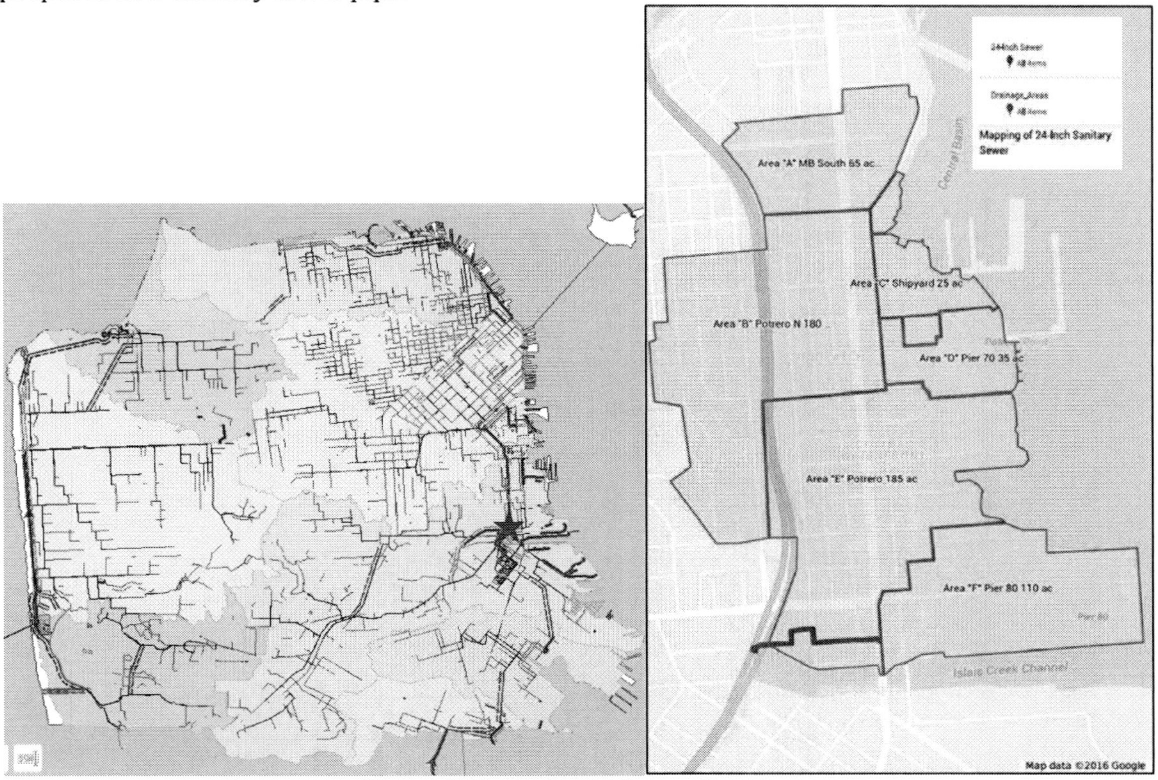

Figure 5. Marin Street Sewer Location

Future conditions for this sewer pipe included an increase in sewer flows, a larger sewer pipe (from a dimeter of 24" to 30") and a different pipe material (from Vitrified Clay (VC) to Reinforced Concrete Pipe (RCP).

2.a. Marin Street Sewer Original Conditions (Baseline)

WATS Model simulation results for the baseline conditions (baseline flows of 1.7 MGD), baseline pipe diameter (24") and material (VC) showed that sulfide concentration in analyzed pipes was no more than 0.11 mg/L. Therefore, it was determined that there is no odor potential in the Marin Street sewer links. Additionally, Marin Street sewer H_2S concentration simulations are between 0 ppm and 2.05 ppm, indicating a low H_2S concertation (i.e, lower than the target threshold of 10 ppm). Shear stress simulations also showed a positive result regarding critical shear stress of 0.5 Pa. In this case, shear stress was greater than 0.5 Pa for all links in the Marin Street network. WATS model simulations show that the existing shear stress is greater than 2 Pa in the majority of this pipe sections. Table 3 shows results of the original of Marin Street Sewer original conditions.

Table 3. WATS Model Simulation Results Marin Street Sewer - Baseline

Marin Streets Sewer Improvements - Baseline Conditions				
Asset ID	Bottom shear stress of gravity pipes [Pa] (> 0.5 Pa)	Dissolved sulfide in the wastewater at downstream node [mg-S/L]	H_2S gas in the sewer air at downstream node [ppm]	Average corrosion rate in the pipe [mm/year]
22593-22594	2.66	0.11	2.22	0
238854-22593	2.65	0.11	2.05	0
22582-238854	2.65	0.1	1.74	0
22548-22582	2.66	0.1	1.41	0.00002
238914-22548	2.66	0.09	1.23	0.000024
22549-238914	2.65	0.09	1.12	0
22550-22549	2.65	0.09	0.98	0
22551-22550	2.61	0.08	0.78	0.000006
249681-22551	2.6	0.08	0.65	0
238913-249681	2.65	0.07	0.77	0
22552-238913	2.65	0.06	0.5	0
22546-22552	2.11	0.06	0.36	0
22534-22546	2.11	0.05	0.22	0
350164-22534	9.99	0.05	0.09	0

2.b. Marin Street Sewer Original Flow Conditions with new Pipe dimensions and material

One other scenario that was analyzed for the new Marin Street pipe was the scenario which considers original flow conditions (present day average flows) with new pipe material (RCP) and diameter (30"). Table 4 shows sulfide concentrations in this scenario simulation remained almost unchanged. On the other hand, H_2S gas concentrations were reduced along the sewer pipe. Pipe corrosion was the only parameter showing an increase with future conditions simulations. Corrosion increased due to the fact that concrete is a sink for H_2S. Concrete helps reducing H_2S concentration by reacting with it, which in turns increases corrosion rates in the pipe.

Table 4. WATS Model Simulation Results – Original Flow, New Pipe diameter = 30" and Pipe material = RCP

Marin Street Sewer Improvements - Construction Project Conditions w/original Flow			
Asset ID	Dissolved sulfide in the wastewater at downstream node [mg-S/L]	H_2S gas in the sewer air at downstream node [ppm]	Average corrosion rate in the pipe [mm/year]
22593-22594	0.12	0.19	0.043
238854-22593	0.11	0.18	0.041
22582-238854	0.11	0.17	0.037
22548-22582	0.1	0.15	0.035
238914-22548	0.09	0.14	0.034
22549-238914	0.09	0.13	0.033
22550-22549	0.09	0.12	0.031
22551-22550	0.08	0.11	0.030
249681-22551	0.08	0.1	0.028
238913-249681	0.07	0.09	0.024
22552-238913	0.06	0.07	0.021
22546-22552	0.06	0.07	0.022
22534-22546	0.05	0.22	0
350164-22534	0.05	0.09	0

2.c. Marin Street Sewer Future Conditions

WATS model simulations for proposed peak flow (7.4 MGD), new pipe diameter of 30" and new pipe material (Reinforce Concrete Pipe - RCP) showed that sulfide and H_2S concentrations in the pipe will still be very low (Sulfide < 0.5 mg/L, and H_2S << 10 ppm). Shear stress simulations in the new pipe are shown to be greater than 3 Pa in the majority of pipe sections. Again, as explain in the previous section, the only parameter that is increasing with future conditions is pipe corrosion. Even when corrosion increased in these simulation results, it was recommended not to provide pipe coating due to the low corrosion rates in the pipes.

Table 5. WATS Model Simulation Results Marin Street Sewer - Future flow = 7.4MGD, New Pipe diameter = 30", Pipe material = RCP

Construction Project Conditions				
Asset ID	Bottom shear stress of gravity pipes [Pa] (> 0.5 Pa)	Dissolved sulfide in the wastewater at downstream node [mg-S/L]	H_2S gas in the sewer air at downstream node [ppm]	Average corrosion rate in the pipe [mm/year]
22593-22594	4.75	0	0	0.0026
238854-22593	4.75	0	0	0.0015
22582-238854	4.75	0	0	0.0008
22548-22582	4.75	0	0	0.0012
238914-22548	4.75	0	0	0.0015
22549-238914	4.75	0	0	0.0017
22550-22549	4.74	0	0	0.0021
22551-22550	4.75	0	0	0.0027
249681-22551	4.75	0	0	0.0032
238913-249681	4.75	0	0	0.0039
22552-238913	4.75	0.01	0.01	0.0057
22546-22552	3.73	0.01	0.02	0.0101
22534-22546	3.74	0.01	0.06	0.0223
350164-22534	18.42	0.01	0.12	0.0591

3. Salesforce Tower's Wastewater Treatment Sewer Discharge to the Collection System

The new Salesforce Tower is located at 415 Mission Street between First and Fremont Streets (Figures 6 and 7). This building is the tallest in the San Francisco skyline and the second-tallest building west of the Mississippi River (the Wilshire Grand Center in Los Angeles is the tallest). Salesforce new building has to follow the new City and County of San Francisco Non-potable Water Ordinance (Onsite Water Reuse for Commercial, Multi-family, and Mixed Use Development Ordinance). The skyscraper includes onsite wastewater treatment and a higher concentration discharge to the sewer system.

New Salesforce building sewer discharge analysis was performed with the WATS model by adding a new pipe to the San Francisco collection system on Mission St. (Figure 7). Water discharge amount was provided as the total building water demand minus the potable water offset. The different scenarios were run with several COD scenarios with greater values than the upstream pipe connection COD concentration.

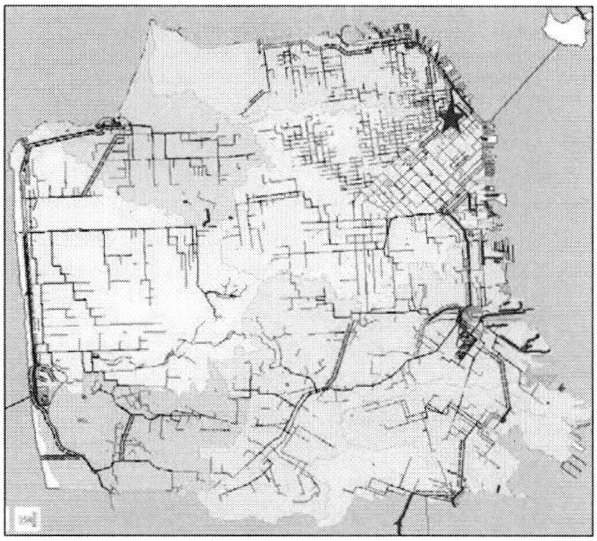

Figure 6. Salesforce Tower Location

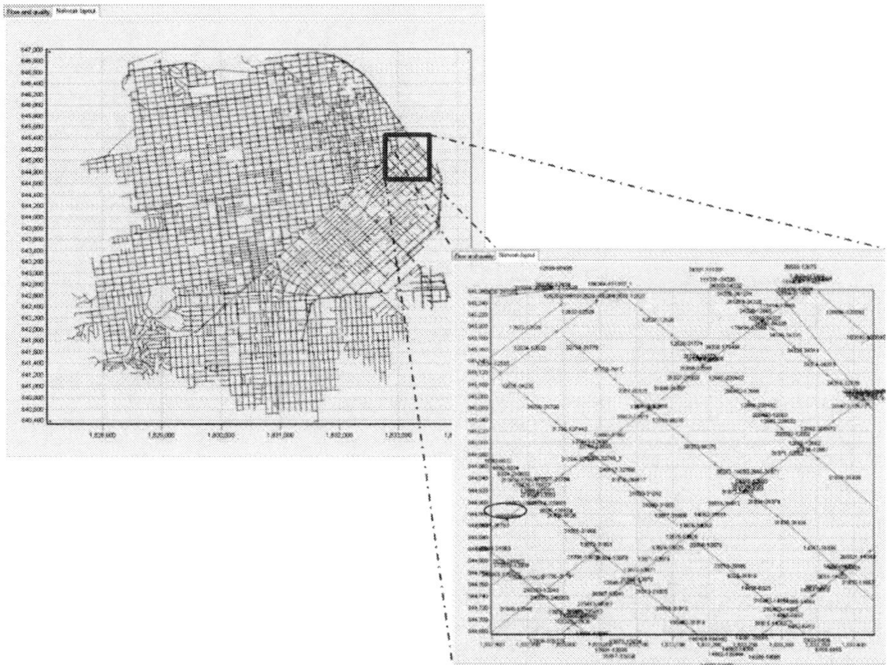

Figure 7. WATS model Screen shots including the new Salesforce discharge pipe.

WATS model simulations results showed that the new Salesforce discharge to the source was not going to negatively impact odors in the collection sewer. All three scenario results (COD of 949 mg/L, 2000mg/L and 3000 mg/L) showed almost no change in liquid sulfide concentration, H_2S gas and corrosion values (Refer to Figures 8 and 9).

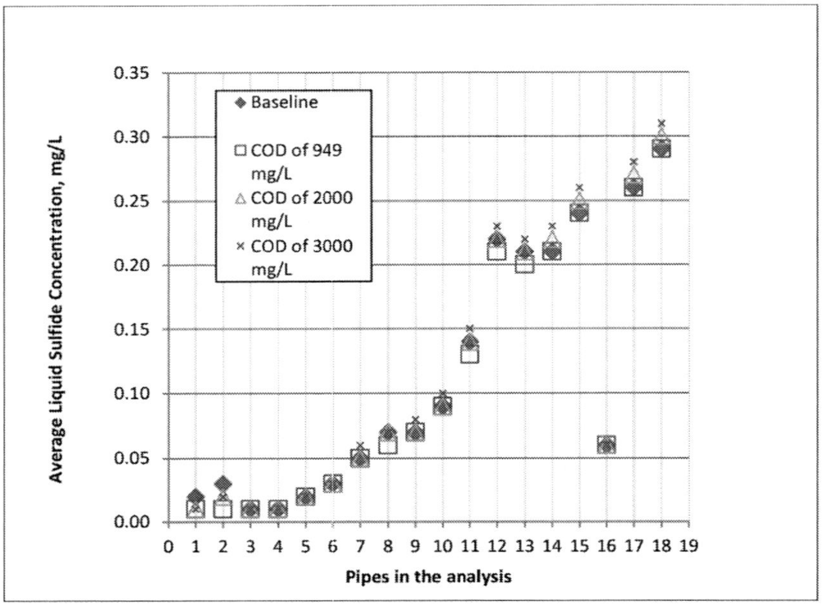

Figure 8. Sulfide WATS model simulation results for Baseline and three different COD loading scenarios

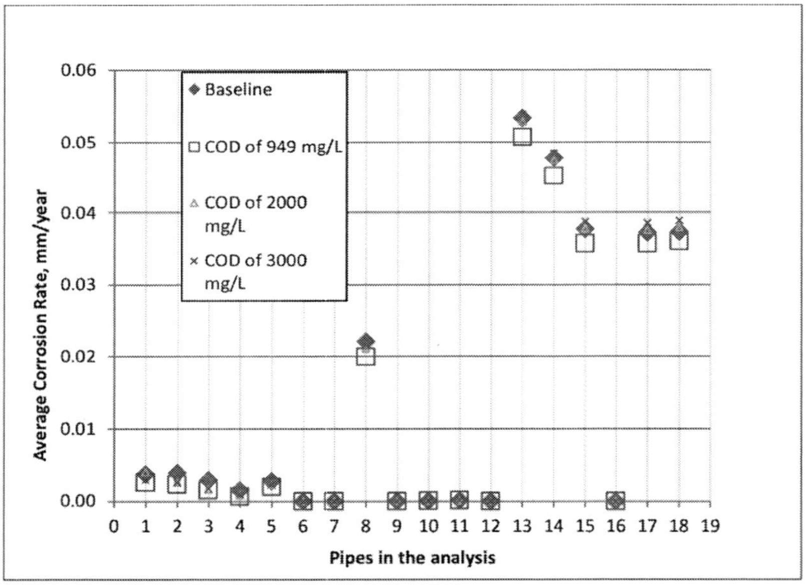

Figure 9. Hydrogen Sulfide WATS model simulation results for Baseline and three different COD loading scenarios

Conclusions

WATS model is a very useful tool which has provided analysis to SFPUC on whether a construction project will meet odor threshold targets set for the City's Collection System. The WATS model can inform a project design team scenario selection or some measure that need to be taken during project's construction phase.

References

Hvitved-Jacobsen T, Vollertsen J, Nielsen A H (2010). *Urban and Highway Stormwater Pollution – Concepts and Engineering*. CRC Press/Taylor & Francis Group, pp 347, (ISBN: 978-1-4398-2685-0.)

Hvitved-Jacobsen T, Vollertsen J, Nielsen AH (2013). Sewer Processes: Microbial and Chemical Process Engineering of Sewer Networks. Second Edition, pp 408, CRC Press, ISBN 978-1-43-988177-4

Water Environment Federation. *Control of Odor and Emissions from Wastewater Treatment Plants – WERF Manual of Practice No. 25.* WEF 2004 (ISBN 1-57278-189-0)

Using Odor Panel Data to Enhance Odor Dispersion Models – Reality Checks Provide Better Results

David McEwen - Brown and Caldwell, Raleigh, NC
Charles McGinley – St Croix Sensory, Stillwater, MN

ABSTRACT

Dispersion modeling is a powerful tool used by engineers to identify odor sources at wastewater treatment facilities (and other industrial and commercial facilities with sources of odor) and to prioritize which sources produce the most significant offsite impacts. Dispersion models like AERMOD and CalPuff are used frequently to assess odor impacts in offsite areas and are also used to project the impacts of apply new or additional odor control to key sources. However, like any model, there are limitations to the desire to apply dispersion model outputs to real-world situations. This occurs because of several reasons, such as fluctuating odor emissions from sources, modeling of only one compound when others may contribute to odor, or modeling odor when some sources are in fact more odorous than others. Furthermore, the output of odor dispersion modeling is inherently conservative, in that it shows worst-case meteorological conditions over what is often a period of 5 years of meteorological data. The result of these issues and conservativism can present problems when sharing the model outputs (in particular odor contour plots) with the general public. This paper provides an overview of how odor dispersion modeling is used, and offers three "reality check" potential enhancements that can provide modelers with tools to either reduce the model run time (which can be long, in particular with large area sources and a large number of offsite receptors) or provide an output plot that is more consistent with actual worst-case impacts offsite, and also more consistent with a history of complaints. In a sense, these suggested enhancements offer a form of calibration of the model to actual conditions, which wastewater treatment facility owners and the general public may find more reasonable and useful.

KEYWORDS: Odor control, dispersion modeling, odor panel, hydrogen sulfide, organic sulfides, detection threshold, hedonic tone, odor characterization.

PURPOSE

The purpose of this paper is to present how dispersion models are used to prioritize odor impacts from sources within wastewater facilities and how certain enhancements and "reality checks" can be used to either decrease the time required for model runs or provide a more realistic output that can better match offsite odor complaints. The strategies developed and presented in this paper have been formulated by Brown and Caldwell following completion of multiple wastewater system odor dispersion models, and from obtaining feedback from peers and wastewater utility representatives on how best to provide optimal results from these powerful tools. The advantage of using these enhancements is that modelers will have some tools to provide a sense of "calibration" to real-world conditions, which often are associated with historical complaints. Given this opportunity to create a more realistic and accepted baseline modeling scenario, successive runs of the model incorporating additional odor control would be more effective.

INTRODUCTION

This section provides the background of how dispersion models are used in odor studies, resulting in prioritization of odor sources with respect to their offsite potential to result in offsite odor impacts.

Dispersion Model Description

Dispersion models are typically run assuming constant emissions from odor sources (though emissions can be changed for different conditions), which are often measured either in the field on in a laboratory from collected bag samples. The dispersion model called AERMOD is the EPA-certified air dispersion model that predicts ground-level concentrations of airborne compounds given input data. The model can be used to predict concentrations of a specific compounds, such as hydrogen sulfide, or of the "total odor" emitted by a source. The total odor value, also referred to as the detection threshold (DT), is either estimated or measured directly by an odor panel. This DT value represents the number of times a volume of air would need to be diluted with odor-free air to render it non-detectable to the average human nose. Odor DT values are unitless, with values presented in dilutions-to-threshold, or D/T.

AERMOD and other dispersion models are straight-line Gaussian plume models that use mathematically and statistically averaged approximations of plume rise and wind-driven dispersion behavior to estimate downwind odor DT or compound concentrations. Results are presented as offsite odor levels calculated by the model, averaged over the course of 1 hour. Generally, dispersion models provide a good estimate of long-term odor impacts (in the range of 1 day to 1 year).

Models provide a two-dimensional representation of ground-level concentrations associated with multiple sources; for example, various sources at a wastewater treatment plant (WWTP). Sources may include point sources, such as stacks or HVAC outputs of buildings that consist of an air stream of a design velocity and airflow rate; area sources that emit odor near ground level and are characterized by two dimensions (such as a primary clarifier); and volume sources, which include buildings that contain odorous process and have a positive pressure that results in fugitive emission from openings.

Dispersion Model Input and Output Parameters

Input parameters common to dispersion models used in odor studies include the following:

- The location of the property line (fence line) of the site inside which odor sources are being modeled. This may actually refer to a boundary beyond a wastewater treatment plant's actual property line, for example, when undeveloped green space is desired not to be considered "offsite" areas where odor impacts would be calculated.
- Meteorological data: up to five consecutive years of data are input into the model (to capture "worst-case" meteorological conditions), including hourly measurements of wind speed, wind direction, and atmospheric stability class. The data can be collected from an onsite weather station, if one exists, or from the nearest weather station to the site, often a local airport.
- Terrain elevations, collected from U.S. Geological Survey Digital Elevation Model quadrangle maps of the site and surrounding area.
- A receptor grid, which includes individual points arranged as a grid both within and outside of the site fence line. The proximity of receptor points within this grid is input by the modeler, but may be a

function of state guidelines or at the discretion of the modeler. Receptor locations are input as UTM coordinates; each receptor also includes a height above ground, which is typically entered as the ground level. An example receptor spacing is 25 meters along the fence line, 100 meters out to 1 kilometer in all directions, and 200 meters out to 2 kilometers in all directions.

- Stack dimensions (for point sources): For stacks that either exit odor control units or are untreated exhausts from buildings that house odor sources, dimensions including stack height above ground elevation and diameter are input into the model.
- Process area dimensions: For area sources, the length and width or diameter or the source is input in addition to the height of the water or solid surface relative to ground level. For volume sources, the height of the building is input along with the size of the opening where odors can escape.
- Foul air parameters, including the airflow rate or velocity either exhausting from a stack (point source), off the surface of an area source, or from an opening in a volume source. Models will also require an ambient temperature and foul air temperature from a point source (exhaust stack).
- Building dimensions within the vicinity of odor sources. These include both onsite and offsite buildings that are determined to potentially impact plume dispersion by causing downwash. The number of buildings included and buildings that can be ignored due to insignificant size or distance from the nearest odor source is at the discretion of the modeler and often corresponds to general "rules-of-thumb" with regard to what buildings to include in the model.

Offsite odor impacts can be plotted on top of a map or an aerial photograph of the area, including the WWTP or other facility with odorous sources. The output typically will include three critical items:

- A plot of odor contours (also referred to as odor isopleths), which represent the results of worst-case meteorological conditions during the time period considered (considering actual hourly meteorological data during that time). The contours are shown in offsite areas and in most cases, will typically decrease in concentration (or D/T) with distance from the odorous sources. The plot usually shows the point of maximum odor impact – a receptor which contains the highest offsite odor values during the modeling period. An example dispersion model output is shown in Figure 1.
- A plot of odor contours or simplified table of the number of exceedances of a given critical threshold occur as predicted during the model over the course of modeling period. For example, a stack source may exceed 10 D/T in offsite areas a maximum of 500 times during the period. The tabular version of this output would just provide the maximum number of exceedances of the threshold; the contour plot would look similar to the odor contour plot, with the values typically decreasing with distance from the sources. An example output is shown in Figure 2.
- A tabular description of the offsite odor impacts of each individual source, describing a situation where that source would be acting alone. This output would include a single concentration or D/T value that identifies the worst-case odor impact for that source.

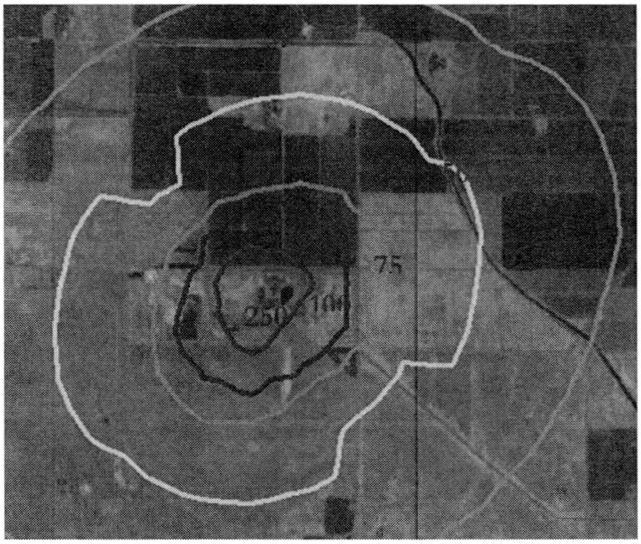

Figure 1. Example odor dispersion model output plot: maximum offsite odor contours (in D/T).

Figure 2. Example odor dispersion model output plot: odor threshold exceedance contours.

ADDRESSING SOURCE PROXIMITY TO FENCE LINES

This section presents the first of three aspects of odor dispersion modeling that this paper will discuss that have presented issues in odor studies in the past, and how these issues can be addressed to produce a more efficient model that more accurately reflects actual offsite impacts.

Problem Sources

In this case, "problem sources" may not actually be a problem at all, at least with respect to odor emissions. Some sources of concern specifically with respect to dispersion modeling that may require some scrutiny are those that are very close to a treatment plant or a pump station fence line. This is of particular interest for a pump station, as almost always a wastewater collection system pump station is on a small footprint. The concern here is both the proximity to the fence line and what is on the other side of that boundary. One should be careful in analyzing offsite odor impacts in areas that are uninhabited and sometimes never occupied, because as long as those areas are outside of the fence line, the dispersion model will count a calculated odor value above an established threshold as an exceedance, when in fact it would never contribute to any complaints simply because of the location of the boundary.

Consider the example in Figure 3. The red point sources represent small drum scrubbers containing activated carbon, each of which treat a small airflow of 400 cubic feet per minute (cfm). Historical data has shown that odor emissions from each scrubber could be as high as 500 D/T, which while significant when standing next to the scrubber, typically the odor value will reduce dramatically when dispersing into offsite areas.

Figure 3. Minor odor sources near a WWTP fence line

However, in this case, with the fence line only 40 feet away, dispersion of, say, an odor value of 300 D/T will not result in as much of a reduced value as it would if the fence line were 100 or 200 feet away. As Figure 3 indicates, the fence line in the vicinity of these sources is bounded by a river and a railroad track, and the nearest reasonable receptor (defined in this case as the nearest location where an individual may live or work that could complain) is 260 feet away. Therefore, if the WWTP's goal is to limit offsite odors to, say 10 D/T, these two sources alone could be responsible for many exceedances of that threshold over the course of a 5-year modeling period, with virtually none of them of significance because the odors are present in areas that are either uninhabited, or not prone to odor complaints (such as a railroad track or a river not commonly used for navigation purposes).

Note that the converse of this problem is also important to note, which is a situation where a property fence line and proximity to an odor source is of extreme importance. This would be applicable to many wastewater pump stations where residential or commercial development is close to the pump station property. Additionally, the "sensitive receptor" in some cases may not be a building at all. Consider the location shown in Figure 4. This is a pump station located very close to a walking path that runs along Monterey Bay in a high-capacity tourist area (Cannery Row) in Monterey, California.

Figure 4. Critical odor source near a WWTP fence line

In this application, it would not be unreasonable to set a very low odor threshold (say 1 to 2 D/T) beyond the pump station property line in a dispersion model because the walking path contains sensitive receptors at all hours of the day and into the evening. For this pump station, it was critical that a robust odor control system be connected to new covers over its wet well to maintain a constant negative pressure, thus providing good odor treatment in addition to eliminating non-treated fugitive emissions.

Resolving the Dispersion Modeling Issue

In odor dispersion models where sources are close to the facility fence line and though even though good odor treatment is provided, numerous offsite impacts remain (in unoccupied areas highly unlikely to contain complainants), one of two strategies are often employed:

- The sources may be simply ignored and removed from the model. While there is potentially a compounding effect that would not be realized if other, larger and/or more highly odorous sources produced plumes that went in the same direction of the smaller source, the addition of smaller sources remains minor. Eliminating these sources may not only make the results more meaningful; they may also moderately reduce the run time for the model.
- The fence line can be essentially artificially extended beyond, sometimes well beyond, the actual facility property boundary. This was done for an odor control master plan for Sacramento Regional County Sanitation District (Figure 5), where the maximum offsite odor would have been approximately 150 D/T if the plant property were used as the fence line, but the maximum offsite odor was approximately 25 D/T when the fence line was extended to include unoccupied green space surrounding the plant.

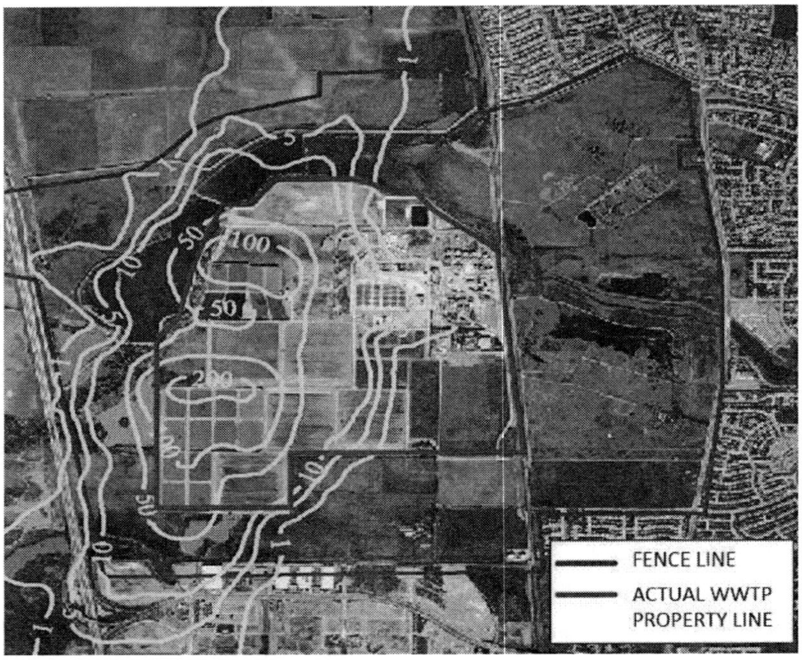

Figure 5. Extension of a fence line beyond uninhabited areas

SCREENING OUT NON-IMPACTFUL SOURCES

The second issue that can over-complicate some dispersion models that consider multiple odor sources at a WWTP is the reality that some sources are actually insignificant with respect to offsite impacts. This section discusses these sources and how to evaluate their significance.

Potential Non-Impactful Sources

Non-impactful sources are defined in this paper as odor sources that have essentially no chance of emitting odors that travel offsite and reaching odor receptors where people live, work, or recreate. A non-impactful source could include the example given in the previous section of small odor control units that emit low airflow rates (on the order of a few hundred cfm) and reasonably low odor (say, less than 500

D/T). These sources could be considered non-impactful particularly if they are isolated from receptors; for example if they are well inside a WWTP boundary. Other examples of likely non-impactful sources, based on case studies completed by Brown and Caldwell and in the literature, including the Water Environment Federation (WEF) Manual of Practice No. 25, include the following:

- Disinfection facilities and treatment plant outfalls contain very low concentrations of odorous compounds that travel offsite. One exception could be a strong bleach odor in a hypochlorite disinfection process, but this is still unlikely to be impactful, as the bleach odor tends to stay more localized near the tanks.
- Secondary clarifiers rarely are incorporated into odor dispersion models because sulfide is typically consumed in upstream liquid-phase processes (influent pumping, headworks, and primary clarifiers, primarily). An exception to this could be if sulfide loading to a WWTP is exceptionally high and there is no chemical injection for sulfide reduction at the head of the plant. In this case, odors may be highest where the water flows over the weirs and produces volatilization or in a center well, where similar volatilization may occur. The significance of secondary clarifiers with respect to odor emissions and potential inclusion in dispersion models can be verified using simple field testing with a hand-held hydrogen sulfide (H_2S) analyzer. Continuous measurements around a secondary clarifier in the part per billion by volume (ppbv) range indicate the process is a non-impactful source.
- Dumpsters storing waste material such as grit may have very high odors at the source, but the source is so small that dispersion models will usually predict no offsite impacts. Additionally, many dumpsters will at times be inside buildings and at other times outside of buildings, and at times loaded with material and at other times empty. This further lessens the potential offsite odor impact.

Note that these are general rules-of-thumb with respect to some odor sources that in the past have not been major contributors to offsite odors from a multi-source treatment plant. Smaller plants and wastewater collection system pump stations with even smaller footprints should consider odor sources on a case-by-case basis, particularly if sensitive receptors live, work, or recreate nearby.

Screening Modeling to Perform Reality Checks

Dispersion models for large multi-source facilities such as WWTPs can take a long time to run, often on the order of hours. This is usually because of the number of sources and the number of offsite receptors that are incorporated into the model (a tighter receptor grid or greater distance away from the source facility both will increase this number). Because of this, any opportunity that a modeler can take to eliminate odorous sources that a projected to have a non-critical offsite impact should be taken.

Some sources can be highly odorous at the emission location, but the emissions do not contribute significantly to offsite impacts because of their location with respect to receptors. This is a similar procedure as was discussed in the previous section of eliminating small sources that are very close to facility fence lines, but in this case a greater amount of analysis is needed to understand whether or not an emission source is significant and should be included in the model.

If a source is suspected of being non-impactful with respect to offsite impacts because of the distance to sensitive receptor locations such as houses, schools, or popular commercial areas, prior to incorporating the source into a full AERMOD model a screening test can be conducted using one of several much simpler models, such as SCREEN3 or AERSCREEN, which do not take into account some of the parameters associated with the larger models, and use worst-case meteorological conditions. Input of source parameters is relatively easy (primarily source emission rate, temperature, and source dimensions associated with point, area, or volume sources), and the model run time is essentially instantaneous.

The result of these screening models is not an intricate contour plot like the ones shown in Figure 5, but instead is a table of compound concentrations at increasing distances radially from the source. Therefore, if plotted, the output would look like a bullseye with concentric circles indicating selected discrete distances away from the source, also input into the model. A good use of the model for screening out non-impactful sources is to use the distance to the nearest sensitive receptor as one of these radial distances from the source. When this is done, the screening model will calculate a concentration of an odorous compound such as H_2S at that sensitive receptor. If that value is low, say on the order of 1 part per billion by volume (ppbv) for H_2S concentration, then it would be safe to eliminate the source from the model (though this could be facility dependent). The reason for this is the generally accepted threshold of human detection for H_2S is approximately 1 ppbv in field conditions; therefore, the model would be projecting that the odor may be barely detected by someone standing at that receptor under the worst-case meteorological conditions.

A theoretical example of this screening method is presented in the remainder of this section. Figure 6 shows a wastewater treatment facility whose primary clarifiers are near its northern fence line. The closest sensitive receptor to the primary clarifier source is identified approximately 0.5 miles away, as shown in the figure. Prior to including the primary clarifiers in a plant-wide AERMOD dispersion model, or even prior to commencing dispersion modeling at all, this screening method can be followed to indicate whether a dispersion model may be an appropriate action.

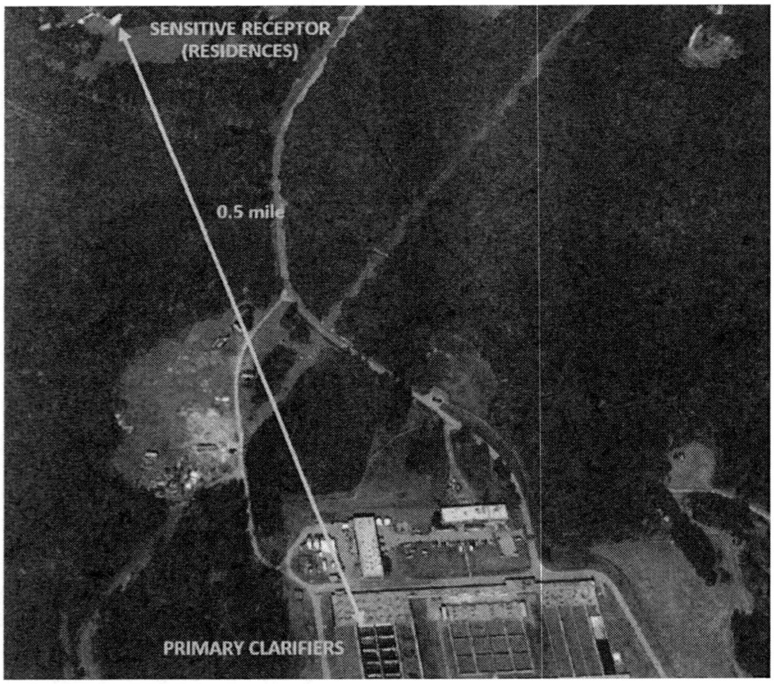

Figure 6. Theoretical example of screening an odor source against a sensitive receptor

In this example, odor emission rates are applied to the primary clarifiers using data collected by St. Croix Sensory (McGinley, 2008), which includes the values shown in Table 1. The composite values shown in the final row of the table are used in the screening model.

Table 1. Odor Values Collected by St. Croix Sensory for Primary Clarifiers

Primary Clarifier Area	Number of Samples	Laboratory-Measured Odor DT of Samples Collected (D/T)		
		1st Quartile	Median	3rd Quartile
Influent Channel	25	1,052	2,741	5,628
Quiescent Surface	69	436	1,277	2,153
Effluent Weir	32	2,262	3,166	5,363
Effluent Channel	50	1,982	2,911	4,884
Composite Value		1,295	2,293	3,955

Given an assumed human detection threshold (the concentration at which the human nose can detect a compound) for H_2S to be 1 part per billion by volume (ppbv), the composite DT values were converted to equivalent H_2S concentrations to be input into the model, and then converted back to DT values (units of D/T) after running the model. Using the dimensions of the primary clarifiers, the distance to the sensitive receptor, worst-case meteorological conditions, and a flat rural terrain, the screening model and associated calculations quickly produces worst-case offsite odor concentrations, as shown in Table 2.

Table 2. Sensitive Receptor Odor Value Calculations for Theoretical Primary Clarifier Source

Assumed Odor Range	Source Odor (D/T)	H_2S Emission Rate (lb/hr)	Receptor Odor (D/T)
1st Quartile	1,295	6.1E-07	0.95
Median	2,293	1.1E-06	1.7
3rd Quartile	3,955	1.9E-06	2.9

The results in Table 2 for Receptor Odor indicate what the screening model predicts will be worst-case odors (assuming worst-case meteorological conditions) at the sensitive receptor given the corresponding source odor. For this example, the decision on whether the primary clarifier source is significant and should be incorporated into a plant-wide dispersion model would be made by the modeler and utility by determining what the threshold for odor tolerance is at that sensitive receptor. If, for example, the goal is to keep odors below 5 D/T at all times off site, the results shown in Table 2 suggest that this would be met if the odor emissions from the primary clarifiers are below the value set as the 3rd Quartile, and likely significantly higher than that.

However, if in this theoretical example the utility is dealing with a complainant at that sensitive source that the utility needs to assure that under no conditions will detectable odors reach their property, additional steps may need to be taken. A first step may be to measure the source odor and compare it to those in Table 2. (If, for example, the measured source odor using an odor panel laboratory is 1,000 D/T, the process may stop at that point as the worst-case predicted offsite odor at the sensitive receptor would be projected to be lower than 1 D/T, the threshold of human detection, per Table 2.)

Absent measurement of a low odor at the source, if deemed appropriate, a full AERMOD model could be run on the single primary clarifier source. The model would use actual meteorological data specific to the site (which may not include the worst-case conditions used in the screening model), terrain data, and building elevations to determine if offsite impacts may be a concern for that sensitive receptor.

SOURCE-SPECIFIC DISPERSION MODELING REALITY CHECKS

Dispersion models that input odor and not specific compounds run a risk of a false equivalence between odor sources based on similar DT values. A reality check on dispersion model input considers sources which should be scrutinized based on personal perceptions of the offensiveness in addition to odor DT.

Problematic Sources when Modeling Odor DT

Odor panels such as St. Croix Sensory will indicate to its clients that the detection threshold (DT) value that they provide as a measurement of essentially the strength of the odor in the sample analyzed in some cases should not be looked at in a vacuum. By this, it is sometimes worth noting that simply because an odor from a source has a relatively high DT value, it may not mean that the odor is extremely unpleasant. Recalling that the DT measurement is simply a measurement of how many times the sample must be diluted with fresh (odor-free) air on a volumetric basis to make it essentially non-detectable to the average nose, the sample analyzed could be highly offensive, or even highly fragrant or even pleasant (such as a perfume or an emission from a food production facility).

Accepting of course that wastewater treatment facility sources are essentially all offensive to some degree, with perhaps the exception of disinfection and outfall facilities, it is fair to also say that some sources are more offensive than others. For example, a headworks facility or influent pump station wet well is likely to emit a more offensive odor than an aeration basin. For most operators, this would be fairly obvious. The odor panel does have a means of quantifying this, and that is by providing a characterization of the odor called “hedonic tone”. Hedonic tone is somewhat subjective, as the odor panelists are asked to rate the odor sample (undiluted) on a scale of -10 to +10, with -10 corresponding to the most offensive odor the panelist has ever smelled, and +10 corresponding to the most pleasant. This hedonic tone measurement can be used in tandem with the odor DT measurement to give a better idea of the actual odor emissions with respect to potential impact on a surrounding community and in particular the complaint potential.

Table 3-5 provides an example of how the odor panel measurements of DT and hedonic tone, and speciation of H_2S organic compounds by a separate laboratory can show that considering only odor DT could be a misleading means of evaluating these potential odor emissions, offsite impacts, and complaints.

Table 3. Odor Panel and Compound Speciation for Processes within the Same WWTP (Case “A”)

Process	Odor Panel DT (D/T)	Hedonic Tone	H_2S Concentration (ppmv)	Organic Sulfide Concentration (ppmv)			Measured Compound Concentration (ppmv)
				Methyl Mercaptan	Dimethyl Sulfide	Dimethyl Disulfide	
Grit Collection Bins	1,400	-0.8	0.192	0.020	0.004	ND	0.216
Primary Clarifier	930	-2.2	0.356	0.024	0.009	ND	0.389
Aeration Basins	660	-2.2	0.036	0.008	ND	ND	0.044

Table 4. Odor Panel and Compound Speciation for Processes within the Same WWTP (Case "B")

Process	Odor Panel DT (D/T)	Hedonic Tone
Headworks	2,900	-3.8
Primary Clarifier	1,300	-3.0
Aeration Basins	5,500	-2.6

Table 5. Odor Panel and Compound Speciation for Processes within the Same WWTP (Case "C")

Process	Odor Panel DT (D/T)	Hedonic Tone	H_2S Concentration (ppmv)	Organic Sulfide Concentration (ppmv)			Measured Compound Concentration (ppmv)
				Methyl Mercaptan	Dimethyl Sulfide	Dimethyl Disulfide	
Primary Clarifier Inlet	7,000	-6.2	0.772	0.046	0.004	ND	0.822
Aeration Basins	3,100	-3.0	0.020	ND	0.002	0.001	0.023

In all cases presented in the three tables, while the aeration basin odor DT values provided by the odor panel are generally similar or at least within an order-of-magnitude of the odor panel DT values for the other upstream (typically more odorous) sources, other measurements of hedonic tone and odor compound concentrations indicate much greater discrepancies between the aeration basins odor emissions and those other sources. This observation is related to how using odor DT alone in dispersion modeling can lead to misleading offsite odor impact results for a process such as an aeration basin.

Reality Check for Less Offensive Odor Sources

The results from Case "C" in Table 5 will be used to compare how dispersion modeling of some odor sources can be misleading when using the odor panel DT value only. Not that the aeration basins DT value (Table 5) is on the same order of magnitude as the primary clarifier inlet, while the primary clarifier inlet hedonic tone measurement of offensiveness was identified as far worse than the aeration basins, and the total measured reduced sulfur compound concentration shown in the table is significantly higher for the primary clarifier inlet as well.

In fairness, there are certainly other compounds not shown in the table that are odorous that also contribute to the 3,100 D/T measured value for the aeration basins, but the reduced sulfur compounds have historically been shown to be particularly contributory for liquid-phase odor emissions due to their low threshold of human detection (all in the range of ppbv). This threshold means that even with dispersion and dilution as the odor plume travels offsite, these compounds will still be present in detectable concentrations, while other compounds with higher thresholds of human detection (perhaps in the ppmv range instead) will not be detectable with similar dilution and dispersion offsite.

Finally, in this and in other examples, the "reality check" of aeration basin odor emissions and their potential offsite impacts needs to be considered from a simple human perception perspective. Experience from studies conducted by Brown and Caldwell has shown that most operators and even some of the general public find aeration basin odors less offensive when standing next to the source as opposed to standing next to the inlet to a primary clarifier, for example. As a result, these same individuals would logically assume that complaints associated with offsite odors are not the result of aeration basins.

Using the data collected by St. Croix Sensory (McGinley 2008), the 3,100 D/T is assumed to be the median value for the sake of this analysis. Given the similar 1st Quartile and 3rd Quartile calculations made in that study corresponding to collected data for aeration basins (113 data points), the corresponding values for the Case C aeration basin odor concentration (at the source) would be as follows:

- 1st Quartile odor DT: 1,143 D/T
- Median odor DT: 3,100 D/T
- 3rd Quartile odor DT: 9,427 D/T

Inputting these data in a screening model results in an odor contour plot as shown in Figure 6, where the odor contour 1,500 meters from the center of the source is shown for each odor DT value.

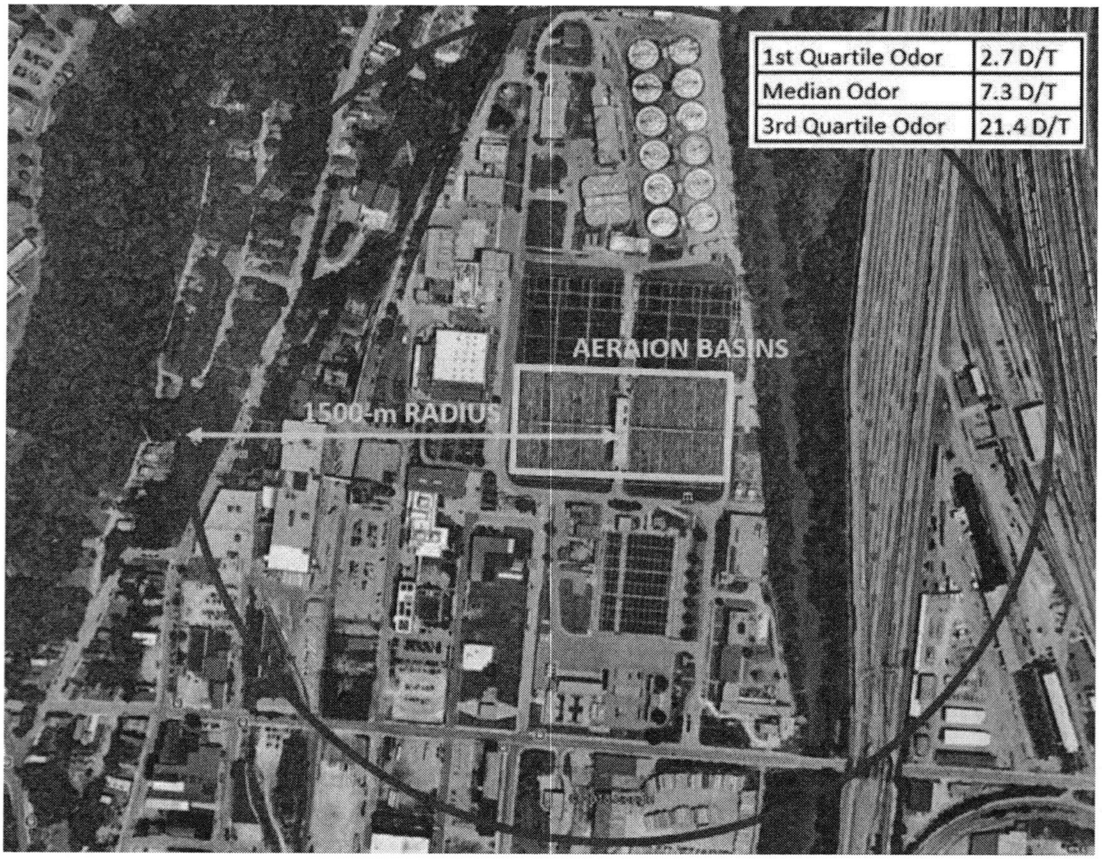

Figure 7. Output contour plot of a theoretical screening dispersion model for an aeration basin

Figure 7 shows that the screening model is predicting offsite impacts of 7.3 D/T at locations 1,500 meters (approximately 1 mile) downwind of the middle of the aeration basin source under worst-case meteorological conditions. Additionally, if the measured 3,100 D/T at the source were lower than the actual odor DT from the source (such as the 9,427 D/T reflective of the 3rd quartile), odors at that offsite location would be more than 20 D/T. These detectable odors would extend into the residential community to the west. Given the nature of the aeration basin odor at the source, which many describe as musty, the dispersion modeler may conclude that the input odor value should be lower than the median. Doing this could be correlated with odor complaints and observations by operation and maintenance staff.

Given the trends noted in Tables 3-5, where the hedonic tone and/or sulfur-containing compounds are considerably lower at the aeration basins than at upstream sources such as primary clarifiers or headworks facilities, it may be useful to adjust the model odor input for the aeration basins. Doing this would account for these discrepancies, rather than only using the DT value, which may not be entirely reflective of the offensiveness of the odor whether the odor emissions can be detected with dispersion and dilution offsite. For example, for the example provided with input data and model output in Table 5 and Figure 7, the odor DT input could be multiplied by the ratio of hedonic tones (-3.0/-6.2), resulting in a new odor input of 1,500 D/T. This would be a value closer to the 1st quartile odor DT, which may be more appropriate and the resulting worst-case 2.7 D/T offsite odors might match better with complaint history and observations by the plant staff.

It is fair to question the use of a screening model for this exercise rather than the full AERMOD model, as AERMOD takes into account more parameters, many of which can influence the results significantly. Some of the differences of using AERMOD model to evaluate and calibrate a single source are as follows:

- AERMOD would include 1 to 5 years of continuously collected meteorological data, and even with the 5 years of data, the worst-case conditions assumed by the screening model may not take place; in this case the screening model would *over-predict* offsite impacts.
- AERMOD takes into account nearby buildings and tall structures that block plume movement and cause downwash; in this case the screening model would *under-predict* offsite impacts.
- AERMOD uses terrain data collected from existing quadrangle maps to assess the effects of topography on plume rise and dispersion; in this case the screening model may *under-predict or over-predict* offsite impacts.
- Setting up an AERMOD model requires much more time than running a screening model; if it is clear than a facility-wide dispersion model is not being produced and a single source, such as an aeration basin, that may be borderline with respect to significant offsite impacts is being evaluated regarding whether or not to install odor control on it, a screening model may be a better tool.

The use of AERMOD for this exercise is encouraged if a dispersion model is being completed for an entire facility, and the odorous process area in question will be a part of the model (thus rendering screening unnecessary). In this case, the impact of, say, an aeration basin offsite impact such as the one in this example would be the only source that is "turned on" in the plant-wide model. This would produce a set of offsite contours as described previously and the modeler and facility owner can suggest whether the offsite impacts for that single source is reflective of actual potential conditions, or if adjustments should be made to the input to account for the non-offensive component of the input odor DT value.

SUMMARY

Dispersion modeling is a powerful tool used by engineers to identify odor sources at wastewater treatment facilities, industrial facilities, and commercial facilities that contain sources of odor. These models can be used to prioritize which sources produce the most significant offsite impacts. However, like any model, there are limitations to the desire to apply dispersion model outputs to real-world situations. This occurs because of several reasons, such as fluctuating odor emissions from sources, modeling of only one compound when others may contribute to odor, or modeling odor when some sources are in fact more odorous than others. The result of these issues and conservativism can present problems when sharing the model outputs (in particular odor contour plots on top of aerial photographs) with the general public.

This paper suggests the following three "reality check" potential enhancements that can provide modelers with tools to either reduce the model run time and/or provide an output plot that is more consistent with actual worst-case offsite odor impacts offsite that have resulted in complaints:

- One of the outputs of odor dispersion modeling as described in this paper is identification of the number of times an established threshold (measured either in odor units or concentration of a given compound) is exceeded by a source's emissions during the modeling period. An issue with some dispersion models is that sometimes minor sources that happen to be located very close to facility fence lines can lead to an overestimate of the number of exceedances of the threshold. This is of particular importance when there are no sensitive receptors near the fence line proximate to the source.
- Some sources within a facility fence line may be non-impactful to offsite odors with respect to projected odor levels at the closest sensitive receptor, such as a residential community or commercial area. These sources can be "screened out" of a dispersion model using relatively simple screening tools and projecting worst-case conditions. In some cases, large area sources or distant receptors may be eliminated, which can decrease model run time (which can be hours for some models).
- When specifically modeling odor dispersion and not a specific compound, the odor detection threshold measured by an odor panel may imply greater significance to some sources with respect potential associated offsite odors than actually may be the case. This is because some sources may have a relatively high odor value, but the odor may not be as offensive as other parts of a facility (for example an aeration basin versus a headworks facility). Analyzing the single offsite impacts of these individual sources (using either a screening tool or a full dispersion model such as AERMOD) is a good means of calibrating and providing a reality check on whether the odor concentration model input is truly appropriate. If it is not, steps can be taken to reduce the odor input value by considering the source's offensiveness or contributing source compounds, thus essentially calibrating the source with complaint history and plant operation and maintenance staff observations.

REFERENCES

McGinley, Michael (2008), *Odors Threshold Emission Factors for Commmon WWTP Processes*, Water Environment Federation Odors and Air Pollutants Specialty Conference, Phoenix, AZ.

Water Environment Federation (2004) Manual of Practice 25 *Control of Odors and Emissions from Wastewater Treatment Plants*, Alexandria, VA.

AUTHOR INDEX

AUTHOR INDEX